Microwave Electronics

Drawing on over 20 years of teaching experience, this comprehensive yet self-contained text provides an in-depth introduction to the field of integrated microwave electronics. Ideal for a first course on the subject, it covers essential topics such as passive components and transistors; linear, low-noise and power amplifiers; and microwave measurements. An entire chapter is devoted to CAD techniques for analysis and design, covering examples of easy-to-medium difficulty for both linear and nonlinear subsystems, and supported online by ADS and AWR project files. More advanced topics are also covered, providing an up-to-date overview of compound semiconductor technologies and treatment of electromagnetic issues and models. Readers can test their knowledge with end-of-chapter questions and numerical problems, and solutions and lecture slides are available online for instructors. This is essential reading for graduate and senior undergraduate students taking courses in microwave, radio-frequency and high-frequency electronics, as well as professional microwave engineers.

Giovanni Ghione is Professor in Electronics at Politecnico di Torino, Italy, and a Fellow of the IEEE. He has authored or co-authored several books, including *Semiconductor Devices for High-Speed Optoelectronics* (Cambridge, 2009), and is the Editor in Chief of *IEEE Transactions on Electron Devices*.

Marco Pirola is Associate Professor at Politecnico di Torino, where he coordinates the Microwave Laboratory.

The Cambridge RF and Microwave Engineering Series

Series Editor
Steve C. Cripps, Distinguished Research Professor, Cardiff University

Editorial Advisory Board
James F. Buckwalter, UCSB
Jenshan Lin, University of Florida
John Wood, Maxim Integrated Products

Peter Aaen, Jaime Plá and John Wood, *Modeling and Characterization of RF and Microwave Power FETs*
Dominique Schreurs, Máirtín O'Droma, Anthony A. Goacher and Michael Gadringer (Eds.), *RF Amplifier Behavioral Modeling*
Fan Yang and Yahya Rahmat-Samii, *Electromagnetic Band Gap Structures in Antenna Engineering*
Enrico Rubiola, *Phase Noise and Frequency Stability in Oscillators*
Earl McCune, *Practical Digital Wireless Signals*
Stepan Lucyszyn, *Advanced RF MEMS*
Patrick Roblin, *Nonlinear RF Circuits and the Large-Signal Network Analyzer*
Matthias Rudolph, Christian Fager and David E. Root (Eds.), *Nonlinear Transistor Model Parameter Extraction Techniques*
John L. B. Walker (Ed.), *Handbook of RF and Microwave Solid-State Power Amplifiers*
Anh-Vu H. Pham, Morgan J. Chen and Kunia Aihara, *LCP for Microwave Packages and Modules*
Sorin Voinigescu, *High-Frequency Integrated Circuits*
Richard Collier, *Transmission Lines*
Valeria Teppati, Andrea Ferrero and Mohamed Sayed (Eds.), *Modern RF and Microwave Measurement Techniques*
Nuno Borges Carvalho and Dominique Schreurs, *Microwave and Wireless Measurement Techniques*
David E. Root, Jason Horn, Jan Verspecht and Mihai Marcu, *X-Parameters*
Earl McCune, *Dynamic Power Supply Transmitters*
Hossein Hashemi and Sanjay Raman (Eds.), *Silicon mm-Wave Power Amplifiers and Transmitters*
Isar Mostafanezhad, Olga Boric-Lubecke and Jenshan Lin (Eds.), *Medical and Biological Microwave Sensors*
Giovanni Ghione and Marco Pirola, *Microwave Electronics*

Forthcoming

Richard Carter, *Microwave and RF Vacuum Electronic Power Sources*
T. Mitch Wallis and Pavel Kabos, *Measurement Techniques for Radio Frequency Nanoelectronics*
Pedro, Root, Xu and Cotimos Nunes, *Nonlinear Circuit Simulation and Modeling*
Michael Schröter and Martin Claus, *Carbon Nanotube Electronics for Analog Radio-Frequency Applications*

Microwave Electronics

GIOVANNI GHIONE
Politecnico di Torino

MARCO PIROLA
Politecnico di Torino

CAMBRIDGE
UNIVERSITY PRESS

University Printing House, Cambridge CB2 8BS, United Kingdom

One Liberty Plaza, 20th Floor, New York, NY 10006, USA

477 Williamstown Road, Port Melbourne, VIC 3207, Australia

314/321, 3rd Floor, Plot 3, Splendor Forum, Jasola District Center, New Delhi – 110025, India

79 Anson Road, #06–04/06, Singapore 079906

Cambridge University Press is part of the University of Cambridge.

It furthers the University's mission by disseminating knowledge in the pursuit of education, learning, and research at the highest international levels of excellence.

www.cambridge.org
Information on this title: www.cambridge.org/9781107170278
DOI: 10.1017/9781316756171

© Cambridge University Press 2018

This publication is in copyright. Subject to statutory exception and to the provisions of relevant collective licensing agreements, no reproduction of any part may take place without the written permission of Cambridge University Press.

First published 2018

Printed in the United Kingdom by TJ International Ltd. Padstow Cornwall

A catalog record for this publication is available from the British Library.

Library of Congress Cataloging-in-Publication Data
Names: Ghione, Giovanni, 1956– author. | Pirola, Marco, author.
Title: Microwave electronics / Giovanni Ghione (Politecnico di Torino), Marco Pirola (Politecnico di Torino).
Other titles: Cambridge RF and microwave engineering series.
Description: Cambridge, United Kingdom ; New York, NY : Cambridge University Press, 2018. | Series: The Cambridge RF and microwave engineering series | Includes bibliographical references and index.
Identifiers: LCCN 2017025023| ISBN 9781107170278 (hardback ; alk. paper) | ISBN 1107170273 (hardback ; alk. paper)
Subjects: LCSH: Microwave devices. | Electronics. | Microwave integrated circuits.
Classification: LCC TK7876 .G48 2018 | DDC 621.381/3–dc23
LC record available at https://lccn.loc.gov/2017025023

ISBN 978-1-107-17027-8 Hardback

Additional resources for this publication at www.cambridge.org/ghione

Cambridge University Press has no responsibility for the persistence or accuracy of URLs for external or third-party internet websites referred to in this publication and does not guarantee that any content on such websites is, or will remain, accurate or appropriate.

To our families: Simonetta and Gianluca Ghione, and Patrizia, Giorgio and Dario Pirola.

Contents

	Preface	*page* xv
	Notation and Symbols	xix
1	**A System Introduction to Microwave Electronics**	1
	1.1 Radio Frequencies, Microwaves, Millimeter Waves	1
	1.2 Transmitting Information Through RF and Microwaves	4
	1.2.1 A Review of Signal Modulation	5
	1.2.2 The RF Interface: Architecture	7
	1.2.3 Linear and Nonlinear Building Blocks of the RF Interface	14
	1.3 Enabling Technologies for RF, Microwaves and mm Waves	18
	1.3.1 Examples of (M)MICs	22
	1.4 Questions and Problems	25
	1.4.1 Questions	25
	1.4.2 Problems	26
	References	26
2	**Passive Elements and Circuit Layout**	28
	2.1 Transmission Lines	28
	2.1.1 Transmission Line Theory	28
	2.1.2 Parameters of Quasi-TEM Lines	38
	2.1.3 The Reflection Coefficient and the Smith Chart	43
	2.2 Planar Transmission Lines in Microwave Integrated Circuits	51
	2.2.1 The Coaxial Cable	53
	2.2.2 The Microstrip	54
	2.2.3 The Coplanar Waveguide	59
	2.2.4 Coupling and Radiation Losses in Planar Lines	62
	2.3 Lumped Parameter Components	65
	2.3.1 Inductors	66
	2.3.2 Capacitors	71
	2.3.3 Resistors	72
	2.3.4 Chip Inductors, Resistors and Capacitors	73
	2.4 Layout of Planar Hybrid and Integrated Circuits	74
	2.4.1 Some Layout-Related Issues	75

Contents

	2.4.2	Hybrid Layout	78
	2.4.3	Integrated Layout	79
2.5	Microwave Circuit Packaging		81
2.6	Questions and Problems		84
	2.6.1	Questions	84
	2.6.2	Problems	84
References			85

3 CAD Techniques — 87

3.1	Modeling of Linear and Nonlinear Blocks	87
3.2	Power Waves and the Scattering Parameters	89
	3.2.1 Representations of Linear Two-Ports	89
	3.2.2 Power Waves	92
	3.2.3 Power Wave *n*-Port Model: The Scattering Matrix	93
	3.2.4 Properties of the S-Matrix	97
	3.2.5 Power Wave Equivalent Circuit	100
	3.2.6 Direct Evaluation of the Scattering Parameters	102
	3.2.7 Reference Plane Shift	103
	3.2.8 Cascade Connection of Two-Ports: The T-Matrix	105
	3.2.9 Solving a Network in Terms of Power Waves	106
3.3	Analysis Techniques for Linear and Nonlinear Circuits	108
	3.3.1 Time-Domain vs. Frequency-Domain Methods	108
	3.3.2 The Harmonic Balance Technique	108
	3.3.3 The HB Technique: Multi-Tone Excitation	112
	3.3.4 The Envelope HB Time-Frequency Technique	119
3.4	Circuit Optimization and Layout Generation	123
3.5	Questions and Problems	124
	3.5.1 Questions	124
	3.5.2 Problems	125
References		126

4 Directional Couplers and Power Dividers — 129

4.1	Coupled Quasi-TEM Lines	129
	4.1.1 Analysis of Symmetrical Coupled Lines	129
	4.1.2 Coupled Planar Lines	134
	4.1.3 Coupled Microstrips	135
4.2	The Directional Coupler	139
	4.2.1 General Properties of Directional Couplers and Power Dividers	141
4.3	The Two-Conductor Coupled Line Coupler	144
	4.3.1 Frequency Behavior of the Synchronous Coupler	149
	4.3.2 Effect of Velocity Mismatch and Compensation Techniques	152
4.4	Multiconductor Line Couplers	155
	4.4.1 The Lange Coupler	160

	4.5	Interference Couplers	162
	4.5.1	Branch-Line Coupler	162
	4.5.2	Lumped-Parameter Branch-Line Couplers	168
	4.5.3	The Hybrid Ring	172
	4.6	Power Combiners and Dividers	173
	4.6.1	The Wilkinson Distributed Divider	174
	4.6.2	Wilkinson Lumped Dividers	179
	4.7	Directional Couplers Summary	181
	4.8	Questions and Problems	182
	4.8.1	Questions	182
	4.8.2	Problems	183
	References		184

5 Active RF and Microwave Semiconductor Devices — 185

5.1	Active Microwave Components	185
	5.1.1	Semiconductor Microwave Components	185
	5.1.2	Device Modeling Approaches	190
5.2	Semiconductors and Semiconductor Alloys for Microwave Transistors	192
	5.2.1	Semiconductor Properties: A Reminder	193
	5.2.2	Transport Properties	197
	5.2.3	Heterostructures and Semiconductor Alloys	199
	5.2.4	The Substrate Issue	202
5.3	Heterojunctions and Reduced Dimensionality Structures	203
5.4	Microwave Schottky-Gate Field-Effect Transistors	205
	5.4.1	FET DC Model	209
	5.4.2	Choice of the DC Working Point	211
	5.4.3	FET Small-Signal Model and Equivalent Circuit	211
	5.4.4	Scaling the Small-Frequency Parameters vs. the Gate Periphery	218
	5.4.5	Frequency Behavior of the Scattering Parameters	219
	5.4.6	High-Speed Compound Semiconductor FETs: The HEMT Family	222
5.5	The RF MOSFET	230
5.6	Microwave FETs: A Comparison	232
5.7	Heterojunction Bipolar Transistors	234
	5.7.1	HBT Equivalent Circuit	237
	5.7.2	HBT Choices and Material Systems	240
5.8	Measurement-Based Microwave FET Small- and Large-Signal Models	242
	5.8.1	Small- and Large-Signal Circuit Models	242
	5.8.2	Extracting the Small-Signal Equivalent Circuit from Measurements	242
	5.8.3	Large-Signal FET Circuit Model Principles	246

		5.8.4 MESFET LS Models	249
		5.8.5 HEMT LS Models	252
	5.9	Measurement-Based Bipolar Small- and Large-Signal Models	254
	5.10	Questions and Problems	255
		5.10.1 Questions	255
		5.10.2 Problems	256
	References		257
6	**Microwave Linear Amplifiers**		**261**
	6.1	Introduction	261
	6.2	Generator-Load Power Transfer	262
		6.2.1 Generator Directly Connected to Load	262
		6.2.2 Power Transfer in Loaded Two-Ports	265
	6.3	Power Gains of a Loaded Two-Port	268
		6.3.1 Maximum Gain and Maximum Power Transfer	268
		6.3.2 Operational Gain	270
		6.3.3 Available Power Gain	274
		6.3.4 Transducer Gain	275
		6.3.5 Power Matching and Unconditional Stability	275
	6.4	Two-Port Stability	276
		6.4.1 Analysis of Stability Conditions	279
	6.5	Two-Port Stability Criteria	283
		6.5.1 Two-Parameter Criteria	283
		6.5.2 Proof of Two-Parameter Stability Criteria	284
		6.5.3 One-Parameter Stability Criterion	289
		6.5.4 Proof of the One-Parameter Criterion	289
	6.6	Two-Port Stability and Power Matching	290
		6.6.1 Unconditional Stability and Simultaneous Power Matching	290
		6.6.2 Managing Conditional Stability	292
		6.6.3 Stability Circles and Constant Gain Contours	292
		6.6.4 Unilateral Two-Port	293
	6.7	Examples of FET Stability Behavior	295
		6.7.1 Stability and Gains at Constant Frequency	295
		6.7.2 Stability and Gains as a Function of Frequency	297
	6.8	Linear Microwave Amplifier Classes	300
		6.8.1 The Linear Amplifier – Design Steps	303
	6.9	The Open-Loop Narrowband Amplifier	304
		6.9.1 Design of Matching Sections	306
	6.10	Active Device Stabilization and Bias	314
		6.10.1 Stabilization	314
		6.10.2 DC Bias	318
	6.11	The Open-Loop Wideband Amplifier	320

		6.11.1 The Balanced Amplifier	321
	6.12	The Wideband Feedback Amplifier	326
	6.13	The Distributed Amplifier	334
		6.13.1 The Continuous Distributed Amplifier	336
		6.13.2 The Discrete Cell Distributed Amplifier	342
	6.14	Questions and Problems	349
		6.14.1 Questions	349
		6.14.2 Problems	349
	References		350
7	**Low-Noise Amplifier Design**		352
	7.1	Introduction	352
	7.2	A Review of Random Processes	353
		7.2.1 First- and Second-Order Statistics	353
		7.2.2 Random Process Through Linear Systems	357
		7.2.3 Symbolic Frequency-Domain (Phasor) Notation	358
	7.3	Analyzing Circuits with Random Generators: The Symbolic Technique	359
	7.4	Equivalent Circuit of Noisy Linear N-Ports	362
		7.4.1 Noisy One-Ports	362
		7.4.2 Noisy N-Ports	363
	7.5	The Physical Origin of Noise	366
		7.5.1 Thermal and Diffusion Noise	367
		7.5.2 RG Noise	367
		7.5.3 1/f or Flicker Noise	368
	7.6	Noise Models of Passive Devices	368
		7.6.1 Noise in Passive Devices: The Nyquist Law	369
		7.6.2 Noise in Junction Diodes	371
	7.7	System-Oriented Device Noise Parameters: The Noise Figure	372
		7.7.1 Evaluating the Noise Figure	374
		7.7.2 The Minimum Noise Figure	381
	7.8	Noise Models of Microwave Transistors	385
		7.8.1 Noise Models of Field-Effect Transistors	385
		7.8.2 Noise Models for Bipolar Transistors	389
	7.9	Noise Figure of Cascaded Two-Ports	390
		7.9.1 Noise Measure	392
	7.10	Low-Noise Amplifiers	392
		7.10.1 Common-Gate (Base) LNA Stage	394
		7.10.2 LNA with Inductive Source/Emitter Series Feedback	398
	7.11	Questions and Problems	407
		7.11.1 Questions	407
		7.11.2 Problems	408
	References		409

8 Power Amplifiers — 411

- 8.1 Introduction — 411
- 8.2 Characteristics of Power Amplifiers — 412
  - 8.2.1 Power, Gain, Distortion — 412
  - 8.2.2 AM–AM and AM–PM Characteristics — 420
  - 8.2.3 Dynamic Range — 422
  - 8.2.4 Efficiency and Power Added Efficiency — 423
  - 8.2.5 Input Matching — 424
- 8.3 Power Amplifier Classes — 424
- 8.4 The Quasi-Linear Power Amplifier (Class A) — 426
  - 8.4.1 Class A Design and the Load-Pull Approach — 432
- 8.5 Analysis of Distortion and Power Saturation in a Class A Amplifier — 437
  - 8.5.1 Power-Series Analysis of Compression and Intermodulation — 437
  - 8.5.2 Power Saturation in Class A — 443
- 8.6 Nonlinear Amplifier Classes: AB, B, C — 444
  - 8.6.1 Single-Ended Class B Amplifier — 445
  - 8.6.2 From Class A to Class C Amplifier — 448
- 8.7 High-Efficiency Amplifiers — 456
  - 8.7.1 Harmonic Loading: The Class F Amplifier — 456
  - 8.7.2 The Switching Class E Power Amplifier — 460
  - 8.7.3 The Doherty Amplifier — 468
- 8.8 Layout and Power Combining Techniques — 481
- 8.9 Linearization Techniques — 481
- 8.10 Questions and Problems — 484
  - 8.10.1 Questions — 484
  - 8.10.2 Problems — 486
- *References* — 487

9 Microwave Measurements — 489

- 9.1 Introduction — 489
- 9.2 Basic Microwave Instrumentation Tools — 489
- 9.3 The Reflectometer — 491
- 9.4 The Vector Network Analyzer — 497
  - 9.4.1 Downconversion Module Solutions — 497
  - 9.4.2 VNA Calibration — 499
  - 9.4.3 Defining the Error Model — 500
  - 9.4.4 Calibration Algorithms — 501
- 9.5 Load and Source Pull Characterization — 507
  - 9.5.1 Scalar Systems — 508
  - 9.5.2 Vectorial Systems — 509
  - 9.5.3 Load Tuning Techniques — 510
- 9.6 System-Level Characterization — 513
- 9.7 Noise Measurements — 514
- 9.8 Questions and Problems — 517

		9.8.1	Questions	517
		9.8.2	Problems	518
	References			519
10	**CAD Projects**			522
	10.1	Introduction		522
	10.2	Microstrip Line and Stub Matching of a Complex Load		522
	10.3	Design of a 3 dB Directional Coupler at 20 GHz		524
	10.4	Fitting and Optimization of a Field-Effect Small-Signal Equivalent Circuit		528
	10.5	Small-Signal FET Stabilization		531
	10.6	Design of a Maximum Gain Amplifier at 15 GHz		532
	10.7	Design of a Two-Stage Balanced Amplifier		538
	10.8	Design of Parallel Resistive Feedback Wideband Amplifiers		543
	10.9	Design of a Uniform DAMP with 40 GHz Bandwidth		546
	10.10	FET Noise, Single-Ended and Balanced LNA Design		550
	10.11	Design of 5 GHz LNA with Source Inductive Series Feedback		554
	10.12	Design of a 3.5 GHz Narrowband Hybrid Class A Power Amplifier		556
	10.13	Design of a 3.5 GHz Narrowband Hybrid Class B and AB Power Amplifier		561
	References			564
	Index			566

Preface

Microwave and millimeter-wave electronics is today far more widespread than it used to be only 20 years ago. Traditional applications based on metal waveguide approaches are still on the market (think about radar systems and some satellite-based systems); however, the introduction of solid-state hybrid and above all monolithic microwave integrated circuits (MMICs) using III-V semiconductors such as gallium arsenide, initially in the low microwave range but now covering frequencies up to millimeter waves, has allowed for a dramatic reduction in the size, weight and cost of many microwave systems in fields ranging from wireless telecommunications to space applications to automotive radars.

Starting from the beginning of this century, a new revolution has taken place in MMICs, with the introduction of RF, microwave and now also mm-wave silicon-based ICs (CMOS but also SiGe). This has finally marked the entrance of microwave systems in the area of low-cost consumer electronics. At the same time, new semiconductor materials for high-power applications (such as gallium nitride) are gradually entering the market of microwave systems, with a promise of size and cost reduction related to the record power densities achievable. And yet, despite the widespread conversion to solid-state electronics, in some areas vacuum tubes are still successfully surviving[1] and expanding their potential, e.g., in the field of THz sources.

The design of microwave circuits is largely based today on Computer-Aided Design (CAD) techniques that have turned the "black magic" associated with the design of distributed (transmission line or waveguide) circuits into a routine that is easily manageable by the designer – of course, provided that he or she has well understood the basics of microwave electronics. Alternative approaches based on lumped parameter components, which make the design of microwave integrated circuits quite similar to that of analog integrated circuits at large, have indeed become increasingly popular in MMICs, at least up to the middle microwave range. On the other hand, high-frequency monolithic and hybrid ICs still have to partly rely on distributed components and the related design styles. In conclusion, today's microwave design is a well-balanced blend of distributed and lumped technological approaches that the designer should be able to master.

---

[1] Think about the high-power sources exploited in long-range radars and also consumer goods like microwave ovens, where the power source is a 1 kW power magnetron – a vintage device invented during the Second World War.

Microwave CAD tools initially emerged in the 1980s, following the early development of circuit simulators such as the open source software, SPICE (Simulation Program with Integrated Circuit Emphasis). Early tools were limited to the analysis, and sometimes optimization, of linear circuits, and used analytical, closed-form models for the components, or measured small-signal parameters in a tabular format. Data input was carried out through quite user-unfriendly ASCII netlists. Since those beginnings, the evolution of CAD tools has been substantial. On the one hand, such tools now benefit from the graphical interfaces available under today's operating systems; netlists have been replaced by the direct entry of graphical schematics, layout generation has been added, and flexible data output and effective visualization tools have been made available. On the other hand, CAD tools have dramatically extended their potential in two directions: the analysis and optimization of nonlinear circuits through techniques such as the harmonic balance method; and the seamless integration of the circuit simulator into fast 3D or 2.5D electromagnetic solvers. As a final step, most microwave circuit CAD tools are now in turn integrated within system-level simulation design suites, thus allowing the loop between circuit-level and system-level design to be effectively closed by the designer.

However powerful and user-friendly microwave CAD tools have become, their user should nevertheless be aware of a number of issues that affect high-frequency analog circuit design. Even a simple problem like a transmission line matching of a load through CAD can lead to unrealistic results if the designer is not aware of all aspects related the behavior of distributed circuits, and also to their physical implementation with a specific technology. Fitting a simple lumped parameter FET model to measured scattering parameters can become a nightmare if the initial guess where to start the optimization is chosen at random.

Taking into account the above remarks, it is not surprising that an introductory textbook on microwave electronics should try to work at several levels, where system-level applications, enabling technologies, component and circuit theory, and CAD techniques seen both in their theoretical basis and in their practical use, should interact in order to provide the student with a realistic vision of microwave design.

To this end, we choose to start (Chapter 1) from a system-level overview, selecting as a target the ubiquitous telecommunication transceiver. A RF or microwave transceiver (besides being in most students' pockets as a part of a smartphone or in most students' backpack as the WLAN transceiver of their notebook) hosts the most important microwave subsystems, such as low-noise, high-gain and power amplifiers, mixers, oscillators and frequency synthesizers, together with passive elements such as switches and filters. In this introductory chapter, a glimpse is also given at the enabling technology choices, with a stress on the planar (hybrid or monolithic) integrated circuit. A broad system level should allow the reader to correctly place the role of the different microwave subsystems, and in particular of the amplifier, that is in fact the main focus of the text. Limiting the treatment to amplifiers within the framework of a first course on microwave electronics is in our opinion a reasonable choice, that allows the material to be kept self-contained and the textbook of a reasonable size.

Linear elements and subsystems can be a convenient starting point for the detailed treatment of microwave circuits. Chapter 2 covers the technology of passive (distributed or lumped) elements and their modeling. Chapter 3 introduces the scattering parameters as a natural and convenient representation of a linear N-port and provides an early introduction to the basic concepts of microwave CAD (linear and nonlinear circuit frequency-domain analysis; optimization techniques; layout). The presentation of harmonic-balance techniques also includes aspects that should not considered any more as "advanced," since they are now available in some widespread CAD tools, like multitone and envelope analysis. Chapter 4 completes the review of passive elements with the presentation of coupled transmission lines and directional couplers and dividers, both in distributed and in lumped form.

Chapter 5 is entirely devoted to microwave active devices and their linear and nonlinear equivalent circuits. Some emphasis is placed on semiconductor theory, material choices and enabling technologies, covering both III-V compound semiconductor FETs (from the vintage MESFET to the HEMT in its more recent varieties) and Si-based RF nanometer-scale MOSFETs; III-N (GaN-based) FETs are also reviewed together with the LDMOS as power device solutions. Although the design examples provided by the text are mostly carried out with FETs, III-V and SiGe bipolars are also introduced in some detail.

Having completed the treatment of passive and active components, Chapter 6 introduces the design of linear, high-gain amplifiers, starting from the theory of loaded two-ports and the issue of two-port stability. The discussion concerns not only general design strategies for the traditional open-loop amplifier, but also a number of specific topics that are popular in microwave amplifier design, such as the balanced amplifier and the distributed amplifier.

Chapter 7 is devoted to a discussion of noise modeling in active devices and of low-noise amplifiers. Taking into account that noise is a difficult subject whose background from communication theory courses has sometimes to be refreshed, the initial part of Chapter 7 includes a review of the theory of random processes. After introducing the device noise models and system-level parameters like the noise figure, the traditional low-noise design approach based on noise figure minimization is introduced, stressing its possible shortcomings in terms of input matching. Then, typical low-noise topologies, like the common-gate and the series inductively degenerated stages, are discussed.

Chapter 8 is entirely devoted to the design of power amplifiers; after a discussion of the characteristic parameters describing power gain, distortion and efficiency, the optimum design of class A amplifiers is introduced with the help of the Cripps analytical load-pull model. The review of traditional high-efficiency classes (AB to B to C) is followed by the introduction of more advanced solutions, such as the harmonic-loaded amplifier (class F), the switching amplifier (class E) and high-efficiency combined solutions (the Doherty amplifier). Chapter 9 covers the topic of microwave measurements (linear, power and noise), with an emphasis on the calibration theory.

A modern introductory course in microwave electronics should include a set of CAD laboratories; however, at the same time, CAD tools exhibit continuous evolution that makes it difficult to create a stable textbook.[2] Although some "paper and pencil" design examples are provided in the text, we devote one last chapter (Chapter 10) to the presentation of a number of "CAD examples" or "CAD projects," to be seen as potential CAD laboratory traces. The level of the traces is intentionally tutorial, even if most features of CAD tools are covered, including linear and nonlinear circuit examples. To avoid early obsolescence, the traces are schematic and only provide a glimpse of the design process, with some significant results. The related projects are made available online as additional material, allowing the reader to run all cases in the original CAD environment. Apart from Chapter 10, each chapter is finally followed by a set of questions and problems that may be helpful to check what has been learned and to apply some of the techniques presented.

This book is the result of work begun almost 20 years ago when the authors started to give lectures in Microwave Electronics at the II Faculty of Engineering of Politecnico di Torino in the (now vanished) Vercelli campus. By and by, lecture notes were collected and an early textbook in Italian, covering some of the material presented here, was assembled. Although this was an entirely new course, part of the material covered was actually adapted from courses given in earlier years by senior (and now retired) colleagues, like Professor Gianpaolo Bava, Professor Carlo Naldi and Professor Claudio Beccari, to whom the authors are indebted.

Later, the course in Microwave Electronics was somewhat transplanted (with a slight downsizing from 10 to 6 credits) to the III Faculty of Engineering of Politecnico di Torino, in the main Turin campus, where it was given for a number of years by Professor Simona Donati Guerrieri; she contributed with helpful suggestions and corrections to the Italian textbook trace, to the course slides and to the development of CAD lab traces. The authors are indebted to her for her contribution and for helpful discussions. Finally, from the beginning of the last decade the lectures on Microwave Electronics became part of a Master degree in Electronics offered entirely in English; this suggested the rewriting in English of the lecture traces, adding and updating material and ultimately leading to the present textbook.

The authors would like to thank a number of colleagues from Politecnico di Torino who have, directly or indirectly, contributed to this book: Professor Vittorio Camarchia, who managed for some years the organization of the Microwave Electronics CAD labs; Professor Andrea Ferrero, for his contribution to Chapter 10 on microwave measurements; Professor Fabrizio Bonani, for helpful discussions, in particular concerning noise. Finally, the authors would like to mention a number of Italian colleagues with whom they have had and continue to have fruitful research collaboration in the design of microwave circuits (in geographical order, north to south): Professor Fabio Filicori and Vito Monaco of Bologna University; Professor Giorgio Vannini of Ferrara University; Professor Franco Giannini, Ernesto Limiti and Paolo Colantonio of University of Roma Tor Vergata; Dr Claudio Lanzieri, now with Leonardo.

---

[2] Also gray literature like CAD lab traces used in the courses have to be updated almost every year.

Notation and Symbols

$x(t)$	scalar variable, time domain
$X(\omega)$	scalar variable, frequency domain
$\underline{x}(t)$	vector variable, time domain
$\underline{X}(\omega)$	vector variable, frequency domain
$\overline{\overline{x}}(t)$	tensor (matrix) variable, time domain
$\overline{\overline{X}}(\omega)$	tensor (matrix) variable, frequency domain
X_0, X_{DC}	scalar variable, DC
$\delta x(t)$	scalar variable fluctuation, time domain
$\delta X(\omega)$	scalar variable fluctuation, frequency domain
$x_n(t)$	random (noise) variable, time domain
$X_n(\omega)$	random (noise) variable, frequency domain
$\langle x(t) \rangle$	time average of deterministic function or random process $x(t)$
$\overline{x(t)}$	ensemble average of random process $x(t)$
$\overline{x}$	ensemble average of random variable x
$S_x(\omega)$	power spectrum of x (also S_{xx})
$\overline{XX^*}$	power spectrum of x in terms of spectral average
$S_{xy}(\omega)$	correlation spectrum between x and y
$R_x(\tau)$	autocorrelation function of x (also R_{xx})
$R_{xy}(\tau)$	correlation function between x and y
P	scalar parameter
$\underline{P}$	vector parameter
a	[eV m^2] derivative of $E_F - E_c(0)$ vs. n_s (modulation doped structure)
a	bipolar transistor common base current gain
a	[m] lattice constant
a, b	[W$^{1/2}$] forward and backward power waves
α	[m^{-1}] attenuation
α_c	[m^{-1}] conductor attenuation
α_d	[m^{-1}] dielectric attenuation
ACPR	Adjacent Channel Power Ratio
A/D	Analog to Digital
AFM	Artificial Frequency Mapping
b	bipolar transistor base transport factor

Notation and Symbols

B	[Hz] bandwidth
BJT	Bipolar Junction Transistor
β	[m^{-1}] propagation constant
β	bipolar transistor common emitter current gain
c_0	[m s^{-1}] speed of light in vacuum, $c_0 = 2.99792458 \times 10^8$ m s^{-1}
C_{ch}	[F m^{-2}] channel capacitance per unit surface (FETs)
C_{eq}	[F m^{-2}] equivalent 2DEG capacitance (HEMTs)
C_{GS}	[F] gate-source capacitance
C_j	[F] junction capacitance
$\mathcal{C}$	[F m^{-1}] capacitance per unit length
$\mathcal{C}_a$	[F m^{-1}] capacitance per unit length in air
CIMR	Carrier to Intermodulation Ratio
δ	[m] skin penetration depth
δ	[rad] loss angle
ΔE_c	[J] [eV] conduction band discontinuity
ΔE_v	[J] [eV] valence band discontinuity
D/A	Digital to Analog
DPA	Doherty Power Amplifier
DUT	Device Under Test
$E(\underline{k})$	[J] [eV] dispersion relation
$\underline{E}(\omega)$	[V m^{-1}] electric field, frequency domain
E_c	[J] [eV] conduction band edge
E_D	[J] [eV] donor energy level
E_F	[J] [eV] Fermi level
E_{Fh}	[J] [eV] quasi-Fermi level, holes
E_{Fi}	[J] [eV] intrinsic Fermi level
E_{Fn}	[J] [eV] quasi-Fermi level, electrons
E_g	[J] [eV] energy gap
E_g	[V] generator open-circuit voltage
E_t	[J] [eV] trap energy level
E_v	[J] [eV] valence band edge
ENR	Excess Noise Ratio
EVM	Error Vector Magnitude
$\underline{\mathcal{E}}$	[V m^{-1}] electric field
$\mathcal{E}_{br}$	[V m^{-1}] breakdown electric field
ϵ	[F m^{-1}] dielectric permittivity
$\epsilon(\omega)$	[F m^{-1}] complex dielectric permittivity, frequency domain; $\epsilon = \epsilon'(\omega) - j\epsilon''(\omega)$
ϵ_{eff}	effective permittivity
ϵ_r	relative dielectric permittivity
ϵ_0	[F m^{-1}] vacuum dielectric permittivity, $\epsilon_0 = 8.854187817 \times 10^{-12}$ F m^{-1}
η	drain or collector efficiency
f	[Hz] frequency
$f_{\max}$	[Hz] maximum oscillation frequency

QAM	Quadrature Amplitude Modulation
QSOLT	Quick Short Open Load Thru calibration
R	power reflectivity
R_g	[Ω] generator internal resistance
R_G	[Ω] generator internal resistance
R_G	[Ω] gate resistance
R_S	[Ω] generator (source) internal resistance
R_{in}	[Ω] input resistance
R_n	[Ω] parallel noise resistance
r_n	[Ω] series noise resistance
$\mathcal{R}$	[Ωm^{-1}] resistance per unit length
$\mathbf{S}$	scattering matrix
SNR	Signal over Noise Ratio
$S_i(\omega)$	[A^{2}Hz^{-1}] current (i) power spectrum
$S_{i_1 i_2}(\omega)$	[A^{2}Hz^{-1}] correlation spectrum between i_1 and i_2
$S_v(\omega)$	[V^{2}Hz^{-1}] voltage (v) power spectrum
$S_{v_1 v_2}(\omega)$	[A^{2}Hz^{-1}] correlation spectrum between v_1 and v_2
σ	[Sm^{-1}] conductivity
SOLR	Short Open Load Reciprocal calibration
T	[s] period
T	[K] absolute temperature
τ_t	[s] transit time
TRL	Thru Reflect Line calibration
U_0	[J] [eV] vacuum level
v_f	[m s^{-1}] phase velocity
v_g	[m s^{-1}] group velocity
$v_{h,\text{sat}}$	[m s^{-1}] hole saturation velocity
v_n	[m s^{-1}] electron drift velocity
$v_{n,\text{sat}}$	[m s^{-1}] electron saturation velocity
V_{br}	[V] breakdown voltage
V_{DS}	[V] drain-source voltage
$V_{DS,br}$	[V] drain-source breakdown voltage
$V_{DS,k}$	[V] drain-source knee voltage (corresponding to drain current saturation)
V_{GS}	[V] gate-source voltage
V_T	[V] thermal voltage
V_{TH}	[V] threshold voltage
VNA	Vector Network Analyzer
W	[m] FET gate width or periphery
$\mathbf{Y}$	[S] admittance matrix
Y	Y-factor in noise measurements
$\mathbf{Z}$	[Ω] impedance matrix
Z_g	[Ω] generator internal impedance
Z_G	[Ω] generator internal impedance

Z_{in}	[Ω]	input impedance
Z_L	[Ω]	load internal impedance
Z_{out}	[Ω]	output impedance
$Z_s(\omega)$	[Ω]	$Z_s = R_s + jX_s$ surface impedance (resistance, reactance)
Z_S	[Ω]	generator (source) internal impedance
Z_0	[Ω]	characteristic impedance
Z_∞	[Ω]	characteristic impedance
z		normalized impedance

1 A System Introduction to Microwave Electronics

1.1 Radio Frequencies, Microwaves, Millimeter Waves

The meaning of high-speed and high-frequency electronics has changed in a century of technological evolution; see [1]. Radio-frequencies (RF) were at first introduced within the framework of radio broadcasting and point-to-point transmission at the beginning of the twentieth century; however, new applications (such as TV broadcasting and above all the radar around 1940), fostered in turn by the availability of RF signal generators at increasing frequency, caused a gradual explosion of new application fields in the RF, microwave and millimeter wave frequency bands. Before 1970, high-frequency systems were based on vacuum tube generators; in the following decades, solid-state semiconductor devices able to operate above 1 GHz, based both on Silicon and on compound semiconductor technologies, were gradually introduced, leading to a new paradigm, the hybrid and then monolithic Microwave Integrated Circuit, (M)MIC, see, e.g., [1, Ch. 16]. Integration and the resulting downsizing of RF and microwave systems ultimately made microwave electronics, at least in high-volume applications, a low-cost commodity well suited to the consumer market.

For the sake of clarity, let us focus on the meaning of the terms RF, microwaves and millimeter waves and on some of the significant applications in each frequency band:

- The *RF band* includes frequencies between a few MHz and 1 GHz, with free space wavelengths of the order of 1 m (30 cm at 1 GHz); the applications include analog radio broadcasting through frequency modulation (FM), and also many other systems, like TV broadcasting, point-to-point communications (for instance, the lowest band in European GSM systems), long-range radar systems, industrial applications like RF heating and RF drying.
- The *microwave* frequency band includes frequencies between 1 GHz and 30 GHz, with free space wavelengths between 30 cm and 1 cm.[1] In the microwave range, several applications exist: below 10 GHz we have all mobile phone systems up to the LTE (Long-Term Evolution, commonly known as 4G LTE) standard, many Wireless Local Area Network (WLAN) applications, terrestrial and satellite radio links, radar systems, positioning systems. Also the electronic part of long-haul optical communication systems operates in this band, although in this case many of the exploited

---

[1] Notice that the name *microwaves* is somewhat misleading, since the related wavelength is much larger than 1 μm!

circuit architectures are digital rather than analog. It should be taken into account that the speed requirement is similar both in fast digital and analog circuits, but digital circuits typically require lower output voltages and power, while analog circuits often have more demanding requirements in terms of output power (think, for instance, of the typical requirement of mobile handset power amplifiers to transmit 1 W of radiated power at the output antenna).

- Beyond microwaves we find the *millimeter wave* band, with frequencies between 30 GHz and 100 (300) GHz, free space wavelengths below 1 cm. Typical applications range from radio links to radar systems (e.g., automotive radar). Conventional electronic circuits (e.g., in integrated form) become increasingly difficult to manufacture above 100 GHz, even if working subsystems (e.g., receivers) have been demonstrated above 500 GHz.

Compound semiconductor technologies like GaAs, SiGe and InP help to push almost conventional integrated circuit concepts beyond the microwave range (GaAs roughly up to 50 GHz, SiGe similar but with reduced output power, InP being the highest performance but also highest cost semiconductor), but with decreasing output power, so that above a few hundred GHz (and certainly above 1 THz) no conventional (i.e., transistor-based) electronics is feasible [3, Ch. 14]. The region above 300 GHz, also referred to as the Teraherz (THz) range, is of great interest today for many applications, but the enabling technologies above 1 THz are fairly different from the ones exploited in conventional electronics, and sometimes similar to optical technologies. A sketch of the frequency regions with the related applications (audio, video, radio broadcasting, radar) is shown in Fig. 1.1. On the top row we find the so-called ITU (formerly CCIR; ITU

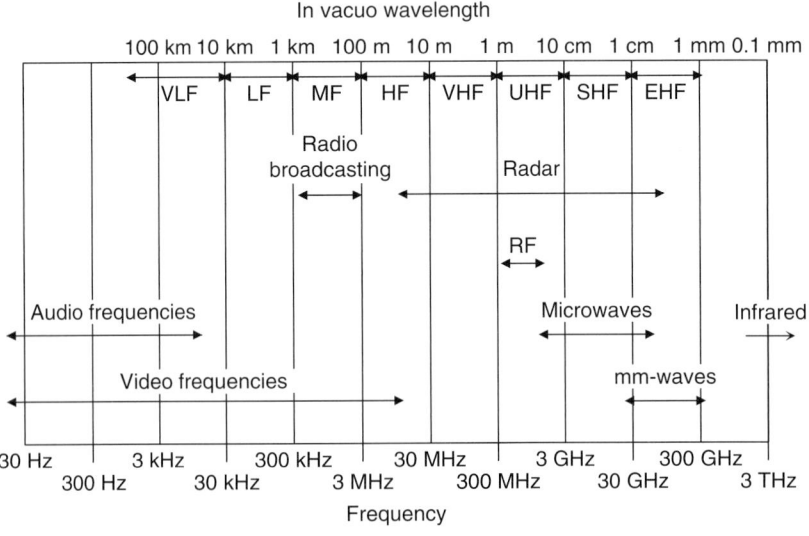

Figure 1.1 Frequency bands and related applications together with the free space wavelengths. The bands in the top row are defined according to the ITU (formerly CCIR) denomination; see [2].

1.1 Radio Frequencies, Microwaves, Millimeter Waves

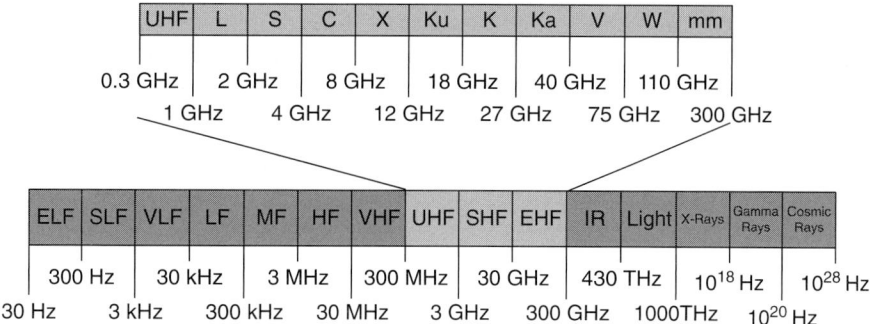

Figure 1.2 Frequency bands in the RF, microwave and millimeter wave range and their IEEE denominations (above). In the row below a more extended range has been considered according to the ITU standard. For the IEEE band names [2] see also Table 1.1.

stands for International Telecommunication Union) radio frequency denominations (the letter F in the band names stands for Frequency, L, M and H stand for low, medium and high, respectively, while the first letter qualifies the second, e.g., ELF is Extremely Low Frequency, UHF Ultra High Frequency etc.).

Taking into account that the ITU system does not allow for specific denominations for the frequency bands in the RF, microwave and millimeter wave range, standard denominations have been proposed for these bands according to a number of different systems. The IEEE standard, see Fig. 1.2, is still the most commonly used denomination; notice that, according to the IEEE standard, millimeter waves actually have a lowest frequency of 40 GHz.[2] Some more details on the IEEE band denomination are given in Table 1.1. According to the IEEE standard, mobile phones can operate, for instance, in the L and S bands, while terrestrial microwave point-to-point links are mainly in the X band and satellite links in the K-Ka band; automotive radars are in the W band.

The allocation of the frequency bands to specific systems and services, that is decided at an international level, is related to many considerations. Increasing the operating frequency of a system from RF to millimeter waves may have advantages:

- Reducing the wavelength, antenna systems become more directive and propagation is quasi-optical, thus allowing for space diversity, i.e., the capability to exploit in the same environment different radiation beams without causing interference.
- With increasing frequency, the capacity of a channel having the same fractional bandwidth increases.
- Finally, the use of higher frequencies is fostered by the gradual occupation of lower frequency bands.

---

[2] Extending the term "microwaves" to denote frequencies above 30 GHz is not without meaning, since the enabling technology, above all at the integrated circuit level, does not significantly change around 30 GHz.

Table 1.1 IEEE denomination of frequency bands [2] The notes are from [4, 5].

Band	Frequency range	Note
HF	3 to 30 MHz	High Frequency
VHF	30 to 300 MHz	Very High Frequency
UHF	300 to 1000 MHz	Ultra High Frequency
L	1 to 2 GHz	Long wave or long-range surveillance
S	2 to 4 GHz	Short wave
C	4 to 8 GHz	Compromise between S and X (?)
X	8 to 12 GHz	X for fire-control crosshair, or for secrecy?
Ku	12 to 18 GHz	K-under
K	18 to 27 GHz	German *Kurz*, meaning *short*
Ka	27 to 40 GHz	K-above
V	40 to 75 GHz	unknown – Very strongly absorbed in atmosphere?
W	75 to 110 GHz	W follows V in the alphabet
mm	110 to 300 GHz	Millimeter

Of course, increasing the operating frequency also has a number of shortcomings:

- The atmospheric attenuation of the Hertzian channel increases with frequency and exhibits sharp peaks in some frequency bands starting from the millimeter waves; see, e.g. [6].[3]
- The development of integrated circuits becomes more and more difficult when increasing the frequency due to the active element performances (such as the available power, which decreases roughly as the inverse of the square of the frequency) and to the losses introduced by passive elements.
- Costs increase, since low-cost silicon circuits have to be replaced by more expensive III-V-based circuits.

1.2 Transmitting Information Through RF and Microwaves

To understand the basic reason for using RF or microwaves within the framework of a communication system exploiting transmission through the Hertzian channel, we have to recall that the basic information content, both in digital and analog form, is mainly a baseband or lowpass signal, i.e., a signal whose spectrum extends from DC (approximately) to some maximum frequency. For instance:

- A digital bit sequence at a given rate has a spectral occupation roughly extending from DC to a frequency of the order of the bit rate.
- An analog audio signal with a bandwidth sufficient for High Fidelity standards has a spectrum extending from almost DC up to about 20 kHz. If the same signal is sampled

---

[3] The atmospheric attenuation at sea level ($T = 20\,°C$, water vapor density 7.5 g/m^3) has a first peak due to H$_2$O absorption around 22 GHz (0.2 dB/km) to increase up to around 20 dB/km near 60 GHz, due to O$_2$ absorption. Two transmission windows exist in the range 60–118 GHz and 118–183 GHz, with minimum attenuation around 0.2 dB/km. At 183 GHz the attenuation reaches 30 dB/km. Due to the large attenuation, the 60 GHz band is not allocated (i.e., free).

1.2 Transmitting Information Through RF and Microwaves

at the Nyquist frequency (40 kHz) and represented with N bits the spectral occupation of the $40 \times N$ kbps digital signal will be of the order of $40 \times N$ kHz.

A baseband signal cannot be typically transmitted as it is through electro-magnetic propagation. While a few transmission media (like a coaxial cable or a two-wire cable) are intrinsically wideband, others (like a rectangular metal waveguide) have a lower cutoff frequency and globally behave like narrowband systems. If, as in many cases in the RF and microwave range, the Hertzian channel is to be exploited through propagation in free space, proper radiator systems (or *antennas*) have to be used to couple a guided signal to free space propagation. Unfortunately, however, most antennas are resonant or narrowband, and the antenna size itself should be of the order of magnitude of the wavelength of the signal (e.g., $\lambda_0/4$). Signals including a significant low-frequency portion would therefore require a huge antenna system. Moreover, low-frequency signals typically propagate in free space all over the Earth (since for them the ionosphere behaves like a resonant cavity), making it impossible for different users to use the same frequency band.

In order to make use of *suitably small antenna systems* and to operate with frequency bands where *more than one user* can be supported, the baseband signal should be upconverted to a suitably high frequency through *modulation* of a carrier. The modulation principle (under the form of analog modulation, amplitude or frequency) dates back to the beginning of the twentieth century [1]; as is well known, in analog *amplitude modulation* the resulting spectrum of the modulated signal has the width of the Nyquist frequency and is located around the carrier. The same approximate result is obtained for low-index frequency modulation.

Thanks to modulation (analog or digital), the low-frequency baseband signal is converted into a (usually) *narrowband* signal around the RF carrier; see Sec. 1.2.1 for a short review.

In conclusion, communication systems making use of propagation through a Hertzian channel transmit and receive some *baseband information* (today typically in digital format) representing data, voice, video signals etc. However, in order to do so they must modulate (according to an analog or digital modulation format) the baseband content to RF before transmitting it through an antenna, and demodulate it to baseband once received.

1.2.1 A Review of Signal Modulation

We may generally assume that a *carrier* at angular frequency ω_c is both phase- and amplitude-modulated (in analog or digital form) by two lowpass or baseband modulating signals $\phi(t)$ and $A(t)$ so that the modulated signal is:

$$x(t) = A(t) \cos(\omega_c t + \phi(t)).$$

By applying trigonometric equalities we can also write:

$$\begin{aligned} x(t) &= A(t) \cos \phi(t) \cos \omega_c t + A(t) \sin \phi(t) \sin \omega_c t \\ &= I(t) \cos \omega_c t + Q(t) \sin \omega_c t, \end{aligned} \quad (1.1)$$

where $I(t)$ and $Q(t)$ are referred to as the *in-phase* and *quadrature* components of the modulating signals. They can also be defined as:

$$I(t) = \text{Re}\left[A(t)\,e^{j\phi(t)}\right], \quad Q(t) = \text{Im}\left[A(t)\,e^{j\phi(t)}\right].$$

Given the spectra $I(\omega)$ and $Q(\omega)$ of the baseband signals $I(t)$ and $Q(t)$, the spectrum of the modulated signal is readily obtained through Fourier transformation:

$$\begin{aligned}
X(\omega) &= \int_{-\infty}^{\infty} I(t)\cos(\omega_c t)\exp(-j\omega t)\,dt + \int_{-\infty}^{\infty} Q(t)\sin(\omega_c t)\exp(-j\omega t)\,dt \\
&= \frac{1}{2}\int_{-\infty}^{\infty} I(t)e^{-j(\omega-\omega_c)t}dt + \frac{1}{2}\int_{-\infty}^{\infty} I(t)e^{-j(\omega+\omega_c)t}dt \\
&\quad + \frac{1}{2j}\int_{-\infty}^{\infty} Q(t)e^{-j(\omega-\omega_c)t}dt - \frac{1}{2j}\int_{-\infty}^{\infty} Q(t)e^{-j(\omega+\omega_c)t}dt \\
&= \frac{1}{2}I(\omega-\omega_c) + \frac{1}{2}I(\omega+\omega_c) - j\frac{1}{2}Q(\omega-\omega_c) + j\frac{1}{2}Q(\omega+\omega_c).
\end{aligned}$$

The resulting modulated signal has the shape of a superposition of two amplitude-modulated signals, as shown in Fig. 1.3.

Examples of modulations are:

- the analog amplitude modulation, AM: $A(t)$ is the modulating signal and $\phi(t)$ is constant;
- the analog phase modulation: $\phi(t)$ is the phase modulating signal and $A(t)$ is constant;
- if the total phase is $\psi = \omega_c t + \phi(t)$, we can define the instantaneous angular frequency as $d\psi/dt = \omega_c + d\phi(t)/dt = \omega_c + \Phi(t)$. In this case, with $A(t)$ constant, we have the analog frequency modulation, FM;
- the digital modulations: in the most general case, the phase ϕ and the amplitude A (giving the signal envelope) assume in the complex plane $I + jQ$ a discrete set of values referred to as the *signal constellation*. As an example, the constellation of a 16QAM modulation is shown in Fig. 1.4.

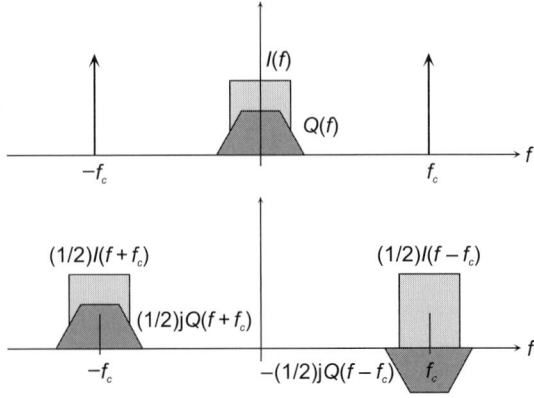

Figure 1.3 Spectrum of I and Q modulated signal.

1.2 Transmitting Information Through RF and Microwaves

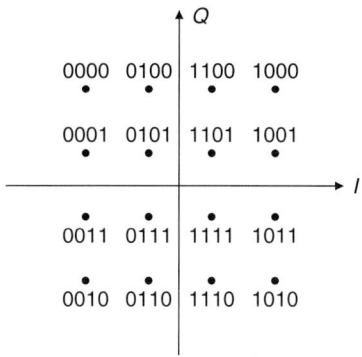

Figure 1.4 Constellation diagram of a 16QAM digital modulation.

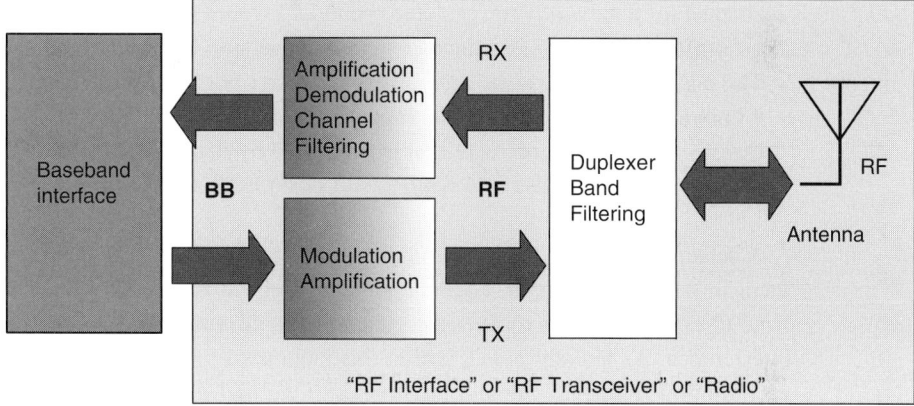

Figure 1.5 The RF interface of a RX/TX system and its main blocks.

1.2.2 The RF Interface: Architecture

The *transmitter* (TX) and the *receiver* (RX) stages are the two basic blocks in the *RF interface* of a communication system, also called the *transceiver*. The main building blocks of the RF interface, see Fig. 1.5, are, for the TX branch, a section providing modulation and amplification, in which the signal is modulated from baseband to RF and boosted to a suitable power level. The last part of the TX chain includes a passband filter, whose main purpose is to avoid the radiation of power outside the system frequency band, thus complying with electromagnetic compatibility standards. The RX branch again shows at the input some passband filter aiming at restricting the signal bandwidth to the system band (mainly to reduce input noise and suppress out-of-band interference), some low-noise amplification, and finally the demodulation needed to bring the signal from RF to baseband. The duplexer is a circuit enabling the antenna to work both as a transmitter and as a receiver. Since, typically, the TX and RX frequencies are different, in many systems two separate antennas (RX and TX) may be used.[4] If the two frequencies are close enough, a single antenna can be used.

Let us have a closer look at the transceiver architecture starting from the *receiver*. In downconverting the signal from RF to baseband, several choices are possible, the main ones being *direct conversion* (from RF to baseband in a single step) and multi-step conversion, called the *heterodyne* technique. In the heterodyne method, the RF signal is downconverted to a lower frequency (typically in the MHz range), called the IF (Intermediate Frequency), and this is in turn converted to baseband. Heterodyne systems can also make use of multiple intermediate frequencies, typically two. The main reason for introducing the heterodyne technique is the impossibility to filter, within the frequency bandwidth allocated to the system, the single channel we intend to receive, for two reasons:

- The quality factor $Q \approx f_0/\Delta f$ of the passband filter required to select at RF the desired channel (of bandwidth Δf, centered around f_0) would be prohibitively large. Filtering becomes possible if f_0 is reduced by downconversion to IF.
- The desired channel to be received is not fixed but is often dynamically allocated within the system bandwidth; an input channel filter should also be variable, another extremely severe requirement that is simplified by demodulating the RF signal into a fixed IF through the use of a syntonized Local Oscillator (LO).

A schematic picture of the heterodyne receiver is shown in Fig. 1.6 (a). The first element in the receiver chain, after the antenna and the antenna duplexer, is a bandpass band filter whose aim is to filter the system band, eliminating all out-of-band noise and interference. As already noted, within this band (assuming that channels are allocated according to frequency domain multiplexing) we find the desired channel. As we will discuss in Chapter 7, in a receiver chain the signal noise is the main concern and, in theory, the first element of the chain should be an amplifier exhibiting low noise and high gain. In fact, a filter must be placed in front of such a low-noise amplifier (LNA) for the reasons already mentioned, with a passband loss as low as possible. After the LNA we have a second filter, called the Image Rejection Filter, whose purpose will be clarified further on (see page 10). Finally, we find the element performing signal downconversion or demodulation: this is a mixer, a two-input device having as a secondary input the Local Oscillator (LO) and as the output the system band downconverted to the Intermediate Frequency (IF). The IF section typically includes *high-gain amplifiers* (HGA) that also appear in the RF section, for instance in cascade to the LNA or as stages of a multi-stage LNA. In some cases, system optimization suggests the use of two IFs rather than one.

Transmitters, see Fig. 1.6 (b), typically exploit direct conversion from the baseband to the RF through an upconversion mixer driven again by the LO. The key element of the transmission chain is the power amplifier (PA), an amplifier whose main purpose is to bring the transmitted signal to an output power level suitable for the system considered.

---

[4] There may be other reasons; for instance, in a radio repeater the received signal and the transmitted signal (uplink and downlink) refer to two different directions in space.

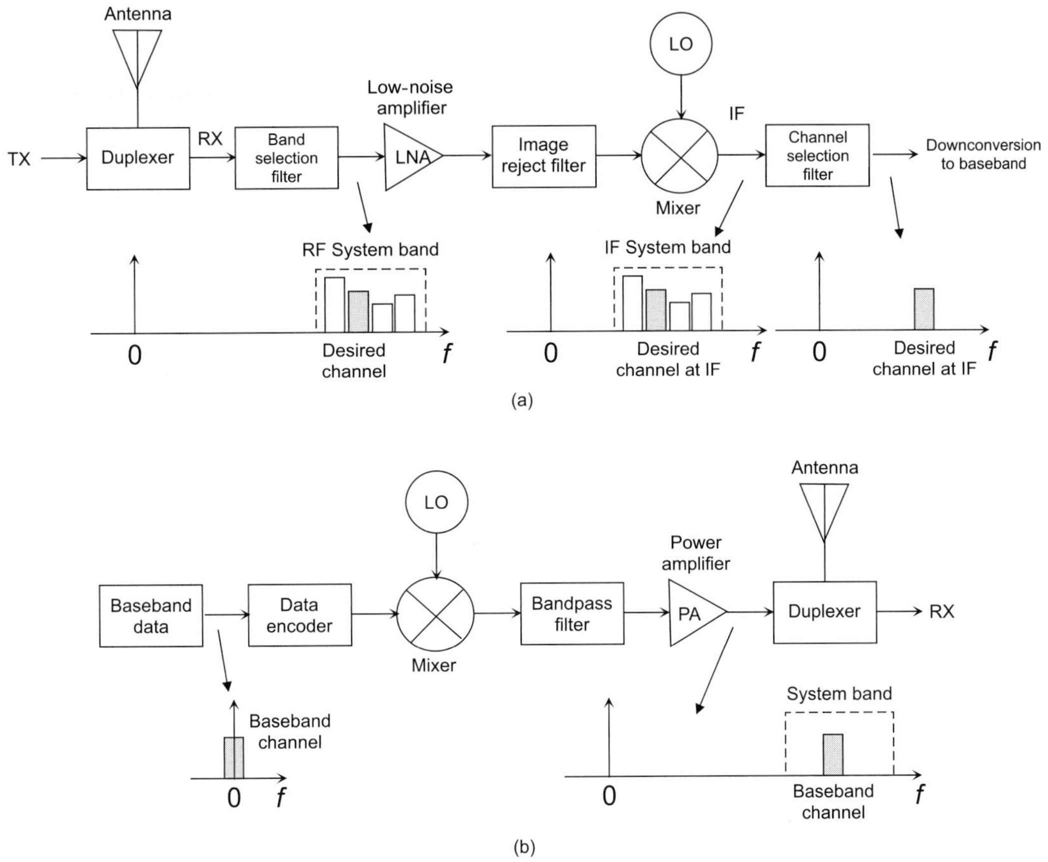

1.2 Transmitting Information Through RF and Microwaves

Figure 1.6 Scheme of a heterodyne RX (a) and of a direct conversion TX (b).

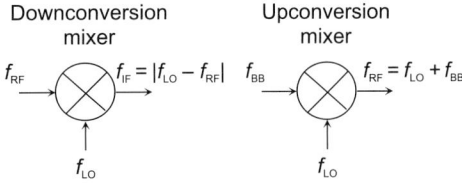

Figure 1.7 Downconversion and upconversion mixers.

While amplifiers (low-noise, high-gain, power) are, at least ideally, linear subsystems, the mixer is (always ideally) a bilinear component whose output is the filtered product of two input signals. In other words, the mixer is an analog multiplier followed by a filter.

Suppose that a phase and amplitude RF modulated signal is multiplied by an (ideally) single-frequency LO; assuming that the mixer amplification A_M and the mixer phase ϕ_M be constant we obtain, as the multiplier output:

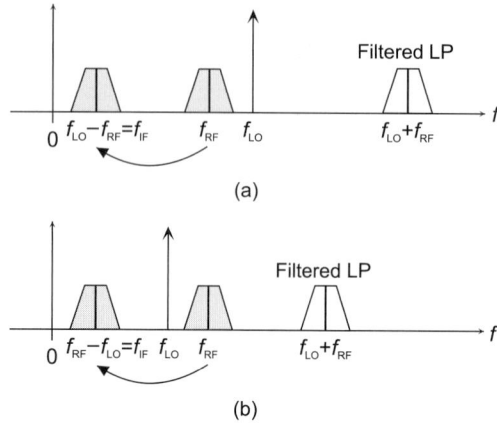

Figure 1.8 Frequency spectrum of a downconversion mixer: high-side injection (a), low-side injection (b).

$$y = A_{\rm RF}(t)\cos(\omega_{\rm RF}t + \phi_{\rm RF}(t)) \times A_{\rm LO}\cos(\omega_{\rm LO}t + \phi_{\rm LO})$$
$$= \frac{A_{\rm RF}(t)A_{\rm LO}}{2}\cos\left[(\omega_{\rm RF} + \omega_{\rm LO})t + \phi_{\rm RF}(t) + \phi_{\rm LO}\right]$$
$$+ \frac{A_{\rm RF}(t)A_{\rm LO}}{2}\cos\left[(\omega_{\rm RF} - \omega_{\rm LO})t + \phi_{\rm RF}(t) + \phi_{\rm LO}\right],$$

where $|\omega_{\rm RF} - \omega_{\rm LO}| = \omega_{IF}$ is the intermediate frequency IF. Notice that two alternatives are possible:

- $\omega_{IF} = \omega_{\rm RF} - \omega_{\rm LO} \rightarrow \omega_{\rm LO} < \omega_{\rm RF}$ (low-side injection or infradyne);
- $\omega_{IF} = \omega_{\rm LO} - \omega_{\rm RF} \rightarrow \omega_{\rm LO} > \omega_{\rm RF}$ (high-side injection or supradyne).

A schematic plot of the frequency distribution in a high-side (a) and low-side (b) injection mixer is shown in Fig. 1.8. In a downconversion mixer the $\omega_{\rm RF} + \omega_{\rm LO}$ product is eliminated through low-pass filtering and the final mixer output is:

$$y_M = \frac{A_M A(t) A_{\rm LO}}{2}\cos\left[\omega_{IF}t + \phi(t) + \phi_{\rm LO} + \phi_M\right], \qquad (1.2)$$

i.e., the original signal is downconverted in frequency but retains its information content. Notice that the multiplying (pumping) effect of the LO can globally lead to overall signal amplification, at least in certain mixer classes. In the upconversion mixing the operation is similar, but the low-frequency mixing product is eliminated by high-pass filtering, while the high-frequency (RF) mixing product is the output.

A significant problem affecting heterodyne receivers stages is the so-called *image frequency* issue. Consider a baseband signal with positive and negative spectral content; following upconversion around the RF, the modulated signal has a symmetric sideband around the *image frequency* $f_{\rm IM}$, see Fig. 1.9 (a), and so has the negative spectrum part. This symmetric (lower in this case) sideband can be eliminated by high-pass filtering, leading to a single sideband RF signal, and its presence is of no consequence.

1.2 Transmitting Information Through RF and Microwaves

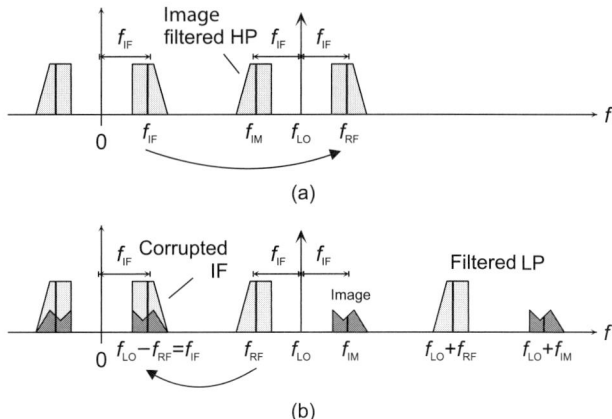

Figure 1.9 The image issue: in upconversion there is no aliasing (a) while aliasing of the desired band occurs with the image in downconversion (b).

A completely different situation arises in downconversion; see Fig. 1.9 (b). In fact, the unwanted image of the RF signal is downconverted to IF together with the desired signal, so that aliasing occurs between the desired signal and the image. We have in fact (for simplicity we use unmodulated signals):

$$y = [A_{RF} \cos \omega_{RF} t + A_{IM} \cos \omega_{IM} t] \times A_{LO} \cos \omega_{LO} t$$
$$= \frac{A_{RF} A_{LO}}{2} \cos(\omega_{RF} + \omega_{LO}) t + \frac{A_{RF} A_{LO}}{2} \cos(\omega_{RF} - \omega_{LO}) t$$
$$+ \frac{A_{IM} A_{LO}}{2} \cos(\omega_{IM} + \omega_{LO}) t + \frac{A_{IM} A_{LO}}{2} \cos(\omega_{IM} - \omega_{LO}) t,$$

where the image frequency is symmetrical to the RF with respect to the LO. In this case $\omega_{RF} > \omega_{LO}$ and therefore:

$$\omega_{IF} = \omega_{RF} - \omega_{LO} \rightarrow \omega_{RF} = \omega_{LO} + \omega_{IF}$$
$$\omega_{IM} = \omega_{LO} - \omega_{IF} \rightarrow \omega_{IM} - \omega_{LO} = -\omega_{IF};$$

after low-pass filtering it follows that:

$$y = \frac{A_{RF} A_{LO}}{2} \cos \omega_{IF} t + \frac{A_{IM} A_{LO}}{2} \cos(\omega_{IM} - \omega_{LO}) t$$
$$= \frac{A_{RF} A_{LO}}{2} \cos \omega_{IF} t + \frac{A_{IM} A_{LO}}{2} \cos(-\omega_{IF}) t$$
$$= \left(\frac{A_{RF} A_{LO}}{2} + \frac{A_{IM} A_{LO}}{2} \right) \cos \omega_{IF} t,$$

i.e., aliasing occurs of the RF and IM spectral contents. Notice that the image problem exists, as such, only in heterodyne receivers. In direct conversion receivers, having zero intermediate frequency, the LO frequency coincides with the RF frequency. If the RF signal results from a simple amplitude modulation (e.g., from the modulation of the in-phase or quadrature components only) the right and left sidebands of the RF signal (i.e.,

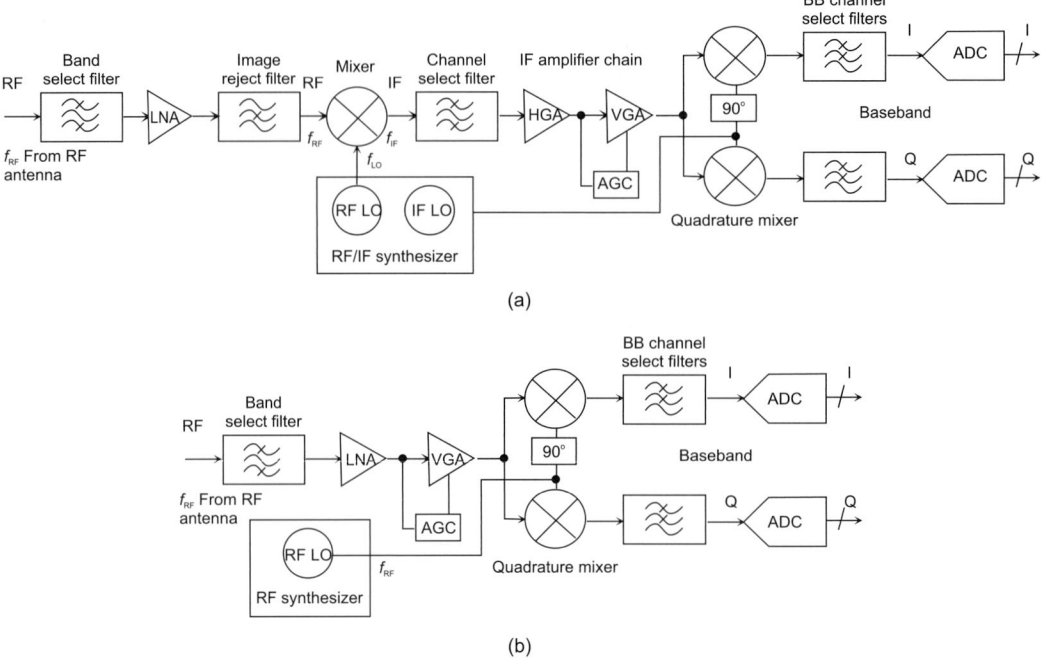

Figure 1.10 Receiver architectures: (a) heterodyne; (b) direct conversion or homodyne.

the spectral content lower or higher than the RF frequency, respectively) are symmetric, and the image coincides with either sideband; after downconversion, no aliasing arises because the spectral content of the image is the same as that of the desired signal. The situation is different if the RF signal originates from the modulation of the in-phase and quadrature component; in such case a simple direct conversion to DC would not allow both components to be recovered, and *quadrature demodulation* technique is needed, as discussed in the next paragraphs.

In heterodyne receivers, the image frequency spectral content can be canceled through filtering (the *image rejection filters*) or through proper receiver architectures called *image rejection receivers*; see, e.g., [7]. Notice that in some microwave receivers operating at high frequency the image rejection filter is avoided due to the large distance between the RF and the IF. Suppose for instance a RF at 60 GHz and the IF at 20 GHz with a LO (low side) at 40 GHz; in this case the image frequency is at 20 GHz: far enough away from the RF to allow for rejection from the band filter.

A more detailed structure of a heterodyne receiver is shown in Fig. 1.10 (a). After the band select filter and the LNA we find as expected the image reject filter and the mixer. Notice that, more correctly, the LO is labeled "RF/IF synthesizer" since not only does it provide the two local oscillators at RF and IF, but it also allows the RF frequency to be tuned in order to syntonize the receiver to a prescribed channel, which is then downconverted always to the same IF. After the dowconversion mixer we find the channel selection filter (as opposed to the band selection filter at the system input) together with a high-gain amplifier and a variable-gain amplifier (VGA) that is intended to provide the demodulation stage from IF to baseband with a constant peak voltage signal. The last demodulation block from IF to baseband includes a *quadrature demodulator* implemented through two mixers fed by a sine and a cosine Local Oscillator (LO).

The need for quadrature demodulation arises from the fact that a general amplitude and phase modulated signal can be expressed as ω_c (the carrier frequency is here the IF), as in (1.1). Clearly, mixing $x(t)$ with a pure sine or cosine would yield only the Q or I baseband component, respectively, while mixing with a combination would lead to aliasing of the two components. Separately mixing with a sine and a cosine having the same frequency instead yields (for simplicity we assume that the LO has unit amplitude and that the mixer gain is one):

$$[I\cos\omega_c t + Q\sin\omega_c t]\cos\omega_c t$$
$$= \frac{1}{2}I + \left[\frac{1}{2}Q\sin 2\omega_c t + \frac{1}{2}I\cos 2\omega_c t\right]_{\text{LP filtered}}$$
$$[I\cos\omega_c t + Q\sin\omega_c t]\sin\omega_c t$$
$$= \frac{1}{2}Q + \left[-\frac{1}{2}Q\cos 2\omega_c t + \frac{1}{2}I\sin 2\omega_c t\right]_{\text{LP filtered}}.$$

The scheme is shown in Fig. 1.11. Notice that quadrature mixing is only needed in downconverting to baseband, not to IF. Going back to Fig. 1.10, after quadrature mixing and low-pass filtering the I and Q signal components are typically sampled and converted to digital format.

In direct-conversion (or homodyne) systems, see Fig. 1.10 (b), the RF is directly converted to baseband; this architecture has the advantage of requiring only lowpass filters that, contrarily to bandpass filters that require inductors or, even worse, surface acoustic wave (SAW) components, can be monolithically integrated; although the design of the quadrature demodulator is more critical due to the higher input frequency, this architecture (or its variations) has become increasingly popular in recent years since it allows for the complete monolithic integration of the receiver (and, sometimes, of the transmitter as well). Notice that in conventional transceivers the conversion from analog to digital takes place at baseband; with the availability of fast Analog to Digital Converters

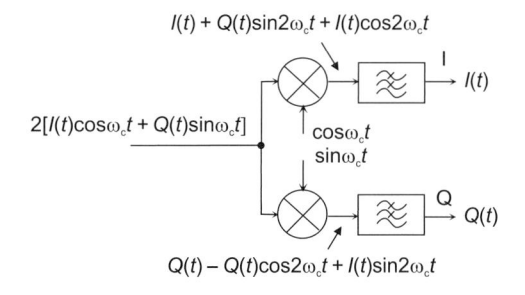

Figure 1.11 Quadrature mixer. The angular frequency ω_c corresponds to the IF frequency in heterodyne systems and to the RF frequency in direct conversion systems.

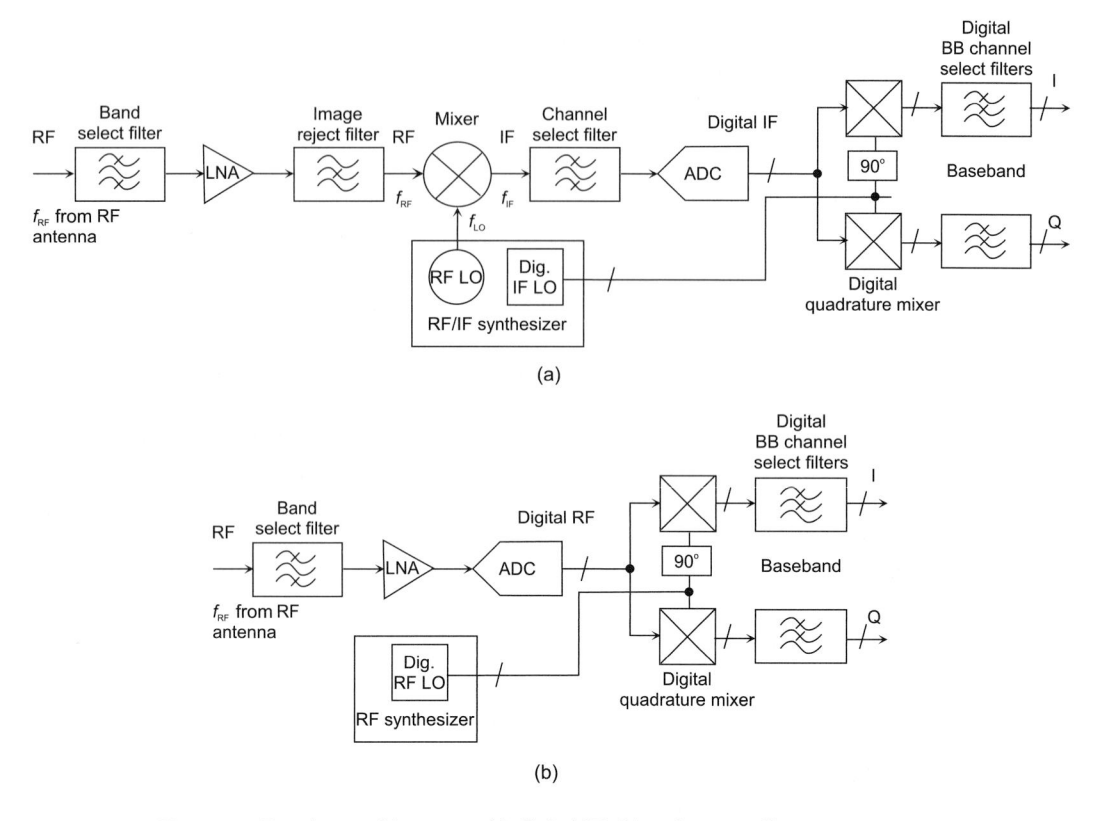

14 **A System Introduction to Microwave Electronics**

Figure 1.12 Receiver architectures: (a) digital IF; (b) software radio.

(ADCs) the possibility arises of performing this conversion at IF or even at RF (provided that the RF is low enough). The first approach is termed Digital IF, see Fig. 1.12 (a), and is comparatively popular, e.g., in basestation receivers. The last approach, the so-called software radio, see Fig. 1.12 (b), allows for the greatest flexibility in terms of multi-standard receivers, since the signal processing is entirely digital; however, the need for high-speed ADCs makes the approach infeasible when the RF is large (e.g., in the microwave range), although problems can be alleviated by decreasing the generality of the receiver (e.g., if the signal to be received is, on the whole, narrowband, subsampling can be performed in the conversion process).

1.2.3 Linear and Nonlinear Building Blocks of the RF Interface

In all architectures a number of building blocks systematically appear:

- amplifiers:
  - low noise (LNA);
  - high gain (HGA);
  - power (PA);
- mixers (modulators, demodulators);

1.2 Transmitting Information Through RF and Microwaves

- local Oscillators (as the engine of Frequency Synthesizers);
- filters (bandpass, lowpass);
- couplers;
- duplexers, RF switches, circulators.

Some blocks (like couplers and circulators) have not been mentioned yet and are listed for the sake of completeness. From a system standpoint, some analog subsystems are approximately *linear* (like the low-noise amplifier, the high-gain amplifier, the filters, the duplexers, the coupler); others are quasi-linear (e.g., the power amplifiers), in the sense that nonlinear effects are a serious concern in design, although the main operation of the device is linear (in fact, nonlinear effects are unwanted). Oscillators and frequency multipliers[5] are intrinsically based on nonlinear effects. Finally, mixers exploit nonlinear effects to implement a *time-varying linear component* essential to their operation.

Let us review, from a system and signal standpoint, the main features of linear vs. nonlinear blocks. We suppose all blocks to be *non-autonomous*, i.e., the output is zero for zero input.

In linear blocks the output $y(t)$ is proportional to the input $x(t)$. If the linear block is *memoryless* (non-dispersive) we simply have $y(t) = Ax(t)$; if the block is *with memory* (dispersive) a linear relation exists between the Fourier transforms of $y(t)$ and $x(t)$, $Y(\omega)$ and $X(\omega)$, respectively:

$$Y(\omega) = H(\omega)X(\omega),$$

where $H(\omega)$ is the system transfer function, the Fourier transform of the pulse response $h(t)$. Notice that in linear blocks *frequencies do not mix*, i.e., sinusoidal input at f leads to a sinusoidal output at the same frequency: no other frequencies are generated (harmonics, intermodulation products). This behavior is shown in Fig. 1.13 (a).

Nonlinear blocks can in turn be *memoryless* (non-dispersive) or *with memory* (dispersive). Let us discuss first the memoryless case, since it yields all the main features of nonlinear blocks. In such a case we simply have:

$$y(t) = f(x(t)),$$

where f does not explicitly depend on time. Assuming that f can be expanded in power series (with $f(0) = 0$):

$$f(x) = f_1 x + f_2 x^2 + f_3 x^3 + \ldots$$

and supposing a sinusoidal input $x = \sin \omega t$, we immediately have:

$$y(t) = f_1 \sin \omega t + f_2 (\sin \omega t)^2 + f_3 (\sin \omega t)^3 + \ldots$$
$$= \frac{1}{2}f_2 + \left(f_1 + \frac{3}{4}f_3\right) \sin \omega t - \frac{1}{2}f_2 \cos 2\omega t - \frac{1}{4}f_3 \sin 3\omega t \ldots .$$

We point out a number of effects of the nonlinearity:

- generation of harmonics (second and third if we truncate the series to the third power);

---

[5] The frequency multiplier is a subsystem exploited to increase the output frequency of an oscillator.

- generation of a DC term or rectification (depending on all even-order terms);
- generation of additional contributions to the first harmonic (depending on all odd-order terms), i.e., distortion at the fundamental.

The relationship between the input and output spectrum for a single-tone input is shown in Fig. 1.13 (c). If we make the input signal slightly more complex, introducing, for example, a simple modulated input made of two sinusoidal tones having the same amplitude, instead of a single one:

$$x = \sin \omega_1 t + \sin \omega_2 t,$$

we obtain:

$$\begin{aligned}
y(t) =& f_1 (\sin \omega_1 t + \sin \omega_2 t) + f_2 (\sin \omega_1 t + \sin \omega_2 t)^2 \\
&+ f_3 (\sin \omega_1 t + \sin \omega_2 t)^3 + \ldots \\
=& f_2 + \left(f_1 + \frac{9}{4}f_3\right) \sin \omega_1 t + \left(f_1 + \frac{9}{4}f_3\right) \sin \omega_2 t \\
&- \frac{1}{2}f_2 \cos 2\omega_1 t - \frac{1}{2}f_2 \cos 2\omega_2 t - \frac{1}{4}f_3 \sin 3\omega_1 t - \frac{1}{4}f_3 \sin 3\omega_2 t \\
&- f_2 \cos (\omega_1 + \omega_2) t + f_2 \cos (\omega_1 - \omega_2) t \\
&+ \frac{3}{4}f_3 \sin (2\omega_2 - \omega_1) t - \frac{3}{4}f_3 \sin (\omega_1 + 2\omega_2) t \\
&- \frac{3}{4}f_3 \sin (2\omega_1 + \omega_2) t + \frac{3}{4}f_3 \sin (2\omega_1 - \omega_2) t \ldots.
\end{aligned}$$

Apart from the harmonics, the rectifying terms and the distortion at the fundamental, a new class of terms appears, with frequency $\pm n f_1 \pm m f_2$, called the intermodulation products (IMP); see Fig. 1.13 (d). If the two frequencies are close and mimic a transmission channel with bandwidth from ω_1 to $\omega_2 = \omega_1 + \Delta\omega$ we see that some of the IMPs have frequencies much lower or much higher than the channel, but two are close to the channel:

$$2\omega_2 - \omega_1 = 2\omega_1 + 2\Delta\omega - \omega_1 = \omega_2 + \Delta\omega$$
$$2\omega_1 - \omega_2 = 2\omega_1 - \omega_1 - \Delta\omega = \omega_1 - \Delta\omega.$$

If the channel is continuous between ω_1 and ω_2, see Fig. 1.13 (e), it can be shown that a continuous distribution of IMPs originating from the third-order term of the power series expansion as frequency differences ("third-order IMPs") arises between $\omega_1 - \Delta\omega$ and $\omega_2 + \Delta\omega$, i.e., such IMPs cause in-band *intermodulation distortion*. The same behavior is found for all-odd power terms from the power series. While the generation of harmonics is exploited in frequency multipliers, in-band distortion is the main factor limiting the dynamics of a power amplifier for large values of the input.

While in linear blocks an exact system model always exists, in nonlinear blocks this is only true if the block is memoryless (non-dispersive). For nonlinear dispersive systems the situation is complex: the exact model (the so-called *Volterra series*, see Chapter 3) is impractical, and simplified models have to be devised by combining linear dispersive and nonlinear memoryless blocks. In any case, harmonic generation

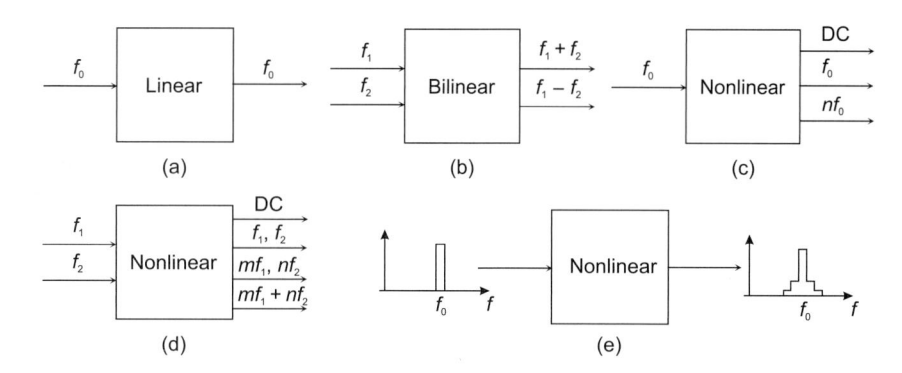

Figure 1.13 Linear (a) and bilinear (b) subsystems; nonlinear subsystems with single-tone (c), two-tone (d) and narrowband (e) input. In case (e) only the neighborhood of the centerband frequency is shown, neglecting DC and harmonics.

and intermodulation distortion are distinctive features also of nonlinear blocks with memory.

A last remark concerns components like the mixer. Consider as a starting point an amplifier exhibiting a variable gain; the gain A depends on a control variable $z(t)$, so that the system input–output relation reads $y(t) = A[z(t)]x(t)$. Assuming that the dependence of the gain on $z(t)$ can be approximated as linear and homogeneous, i.e., $A[z(t)] \approx A'z(t)$, we obtain:

$$y(t) = A[z(t)]x(t) \approx A'z(t)x(t). \tag{1.3}$$

The relationship (1.3) clearly is *bilinear* and can be exploited as a paradigm of a mixer having input x, local oscillator z and output y (to complete the mixer of course further elements, like filters, are needed). Notice that the above relation can be written more compactly as:

$$y(t) = A(t)x(t),$$

where we stress that mixing occurs in a linear, time-varying system. Mixing of course also takes place in a nonlinear system, as the generation of IMPs demonstrates. As we have already shown, if a bilinear system has two single-tone inputs at f_1 and f_2, $f_1 > f_2$, the output will contain the frequencies $f_1 \pm f_2$; see Fig. 1.13 (b).

Finally, many analog circuits (e.g., low-noise and high-gain amplifiers) operate in small-signal conditions, i.e., with an instantaneous working point in the neighborhood of the DC bias point. In such conditions, even if the component is nonlinear, a linear model is enough to characterize operation. From a signal standpoint, and using for simplicity a memoryless nonlinear system, we have that, separating the DC component from the small-signal component:

$$x(t) = X_0 + \hat{x}(t), \quad y(t) = Y_0 + \hat{y}(t),$$

so that:

$$y(t) = F(x(t)) \rightarrow Y_0 + \hat{y}(t) = F(X_0 + \hat{x}(t))$$

$$\approx F(X_0) + \frac{dF}{dx}\bigg|_{X_0} \hat{x}(t) = Y_0 + \hat{y}(t),$$

i.e., separating the DC and time-varying components:

$$\hat{y}(t) = \frac{dF}{dx}\bigg|_{X_0} \hat{x}(t),$$

that yields the small-signal model.

In general, models able to completely characterize a nonlinear device (e.g., a transistor) in **any operating condition** are denoted as large-signal models, often assuming the form of a large-signal equivalent circuit; from large-signal models, linear models, typically with memory, can be extracted through small-signal linearization, sometimes again as a linear equivalent circuit.

1.3 Enabling Technologies for RF, Microwaves and mm Waves

RF, microwave and millimeter wave subsystems can be implemented today in a number of technologies, that can be divided into two classes:

- the planar circuits, characterized by comparatively low cost and small size; can be hybrid (integrating all circuit elements apart from the active ones, i.e., the transistors) or monolithic (integrating on a semiconductor substrate all circuit elements, passive and active), according to an approach very similar to the one pursued in low-frequency monolithic or hybrid Integrated Circuits (ICs). Planar circuits exploit as signal transducers planar transmission lines such as the *microstrip* or the *coplanar* line;
- the waveguide circuits, where the signal transducers and passive components are made with hollow metal waveguides (e.g., rectangular or circular). Due to the weight, bulk and high cost of processing (and the inherently narrowband characteristics, due to the waveguide dispersion features), closed metal waveguide circuits are used in applications where the waveguide has superior properties with respect to the transmission lines used in hybrid circuits, namely: ability to handle extremely high powers without damage and without compatibility problems (high isolation); low attenuation, and, consequently, possibility of building circuits with high quality factor (for example, very frequency-selective resonators and filters); direct and efficient interfacing to antenna systems.

The present book focuses on planar circuits, either hybrid or monolithic. The main difference between the monolithic and hybrid approaches consists in the level of integration of the elements:

- in hybrid microwave circuits, integration concerns all distributed elements (transmission lines and components made with these), the lumped elements (such as resistors and capacitors) are sometimes integrated but, often, introduced in discrete form; the active elements (the transistors) are always discrete. The hybrid circuits often make

use of a design strategy based on distributed elements, and therefore have dimensions proportional to the operating wavelength (large at low frequency, smaller at high frequency);
- in monolithic microwave circuits (MMICs) all elements are integrated with the semiconductor substrate, although some elements may be added externally in discrete form.[6] The design is typically based on lumped elements as far as possible, since such elements are more compact, and only above 20–30 GHz are distributed elements exploited (also due to the increasing losses of lumped elements).

The choice between a hybrid and a monolithic implementation is then influenced by a number of considerations, which can be summarized as follows:

- **cost**: monolithic and hybrid circuits have very different manufacturing and processing costs. The hybrid realization has higher cost per unit, but the process setup costs are low (and so is the turnaround time); conversely, the monolithic approach has a low-cost per unit of output, but high costs and long times for the project setup (typically, months). The monolithic realization is therefore only suitable for large-scale production, while the hybrid approach is well suited to low-volume production;
- **repeatability**: integrated circuits suffer from variability due to the technological process, but on the whole have a more repeatable behavior, also depending on process maturity;
- **performance**: certain performance benefits can only be obtained through tuning of the realized circuit. This is possible only in hybrid circuits. There are therefore circuit categories (typically narrowband) that are less critical to design in hybrid than in monolithic form. On the other hand, parasitics (e.g., related to interconnects) are less important in MMICs than in hybrid circuits.

Both hybrid and monolithic microwave integrated circuits are *planar*, i.e., realized on a planar insulating substrate. In *hybrid circuits* the substrate is a thin dielectric layer; dielectrics commonly used are Teflon and its derivatives (with relative dielectric constants ranging from 2 to about 4) and ceramics (such as alumina), mechanically rigid, and characterized by higher dielectric constants (around 10). The substrate is usually bottom coated by a metal layer which serves as a ground plane. In *monolithic circuits* the substrate is a semi-insulating semiconductor, for example, intrinsic GaAs (dielectric constant 13). Silicon integrated circuits make use of low-doping substrates coated with insulating dielectric layers (e.g., silicon oxide) to reduce substrate losses. In monolithic integrated circuits the substrate is typically thin, for example, less than 300 μm thick.

Due to the comparatively small wavelength in the field of microwaves and millimeter waves, hybrid and monolithic microwave integrated circuits follow, as already noted, two different strategies:

---

[6] Consider, for example, the DC blocking capacitors or RF blocking inductors whose large value makes them inconvenient to realize in monolithic form.

- the lumped approach, that exploits concentrated elements (like resistors, capacitors, inductors or subsystems made thereof, like filters, couplers, power dividers, matching sections) that must be small compared to the operating wavelength (typically $< \lambda_g/8$). The lumped design is preferred in monolithic integrated circuits due to its compactness.
- the distributed approach, making use of elements (couplers, filters, power dividers, matching sections) based on planar transmission lines (e.g., the microstrip); the size of the distributed elements is indicatively $\lambda_g/4$ at centerband, and therefore such elements are by no means extremely small. For this reason, the distributed approach is preferred in hybrid circuits, and exploited in monolithic integrated circuits only above a certain frequency (e.g., 30 GHz), or when indispensable.

The microstrip (consisting of a metal strip placed on a dielectric layer, with a ground plane below) is by far the most common transmission line solution. Coplanar waveguides (in which the strip and the ground planes are on the top surface of the substrate) are sometimes used, e.g., at millimeter waves; see Chapter 2 for details.

Active microwave devices are either bipolar (conventional or heterojunction) or field-effect (MESFETs, HEMTs, MOSFETs) transistors; see Chapter 5. The enabling active device and monolithic IC semiconductor technologies can be summarized as follows:

- Silicon-based ICs cover the lower microwave range (up to a few GHz) in the conventional bipolar version; however, bipolar Si MMICs are now obsolete and MOSFET ICs cover all applications (apart from the power ones) up to 5–10 GHz and beyond, reaching in some cases the millimeter wave range [8]. Power MOSFETs (LDMOS) are also used up to 4–5 GHz.
- ICs based on Silicon–Germanium (SiGe) Heterojunction Bipolar Transistors (HBT) provide low-noise and low-power solutions up to millimeter waves. SiGe is, however, a quite complex and expensive technology when compared to CMOS ICs.
- GaAs (Gallium Arsenide) ICs (HEMTs and HBTs) are the traditional solution for microwave monolithic integrated circuits up to 50–70 GHz, covering both low-noise and power applications. The technology is mature but more expensive than the Si-based equivalent.
- InP (Indium Phosphide) ICs (HEMTs and HBTs) permit electronics to be pushed well into the mm-wave range; however, they are expensive and confined to niche applications.
- GaN (Gallium Nitride) devices and ICs are a rising, but not yet completely mature, high-power microwave technology.
- GaSb (Gallium Antimonide) provides exceptionally low-noise circuits up to mm waves; the technology is confined so far to niche applications; see [9] for a review.

Finally, vacuum tubes still successfully cover high-power applications (signal generation and amplification), sometimes coupled with extremely wide bandwidth (e.g., in the Traveling Wave Tube (TWT) amplifiers).

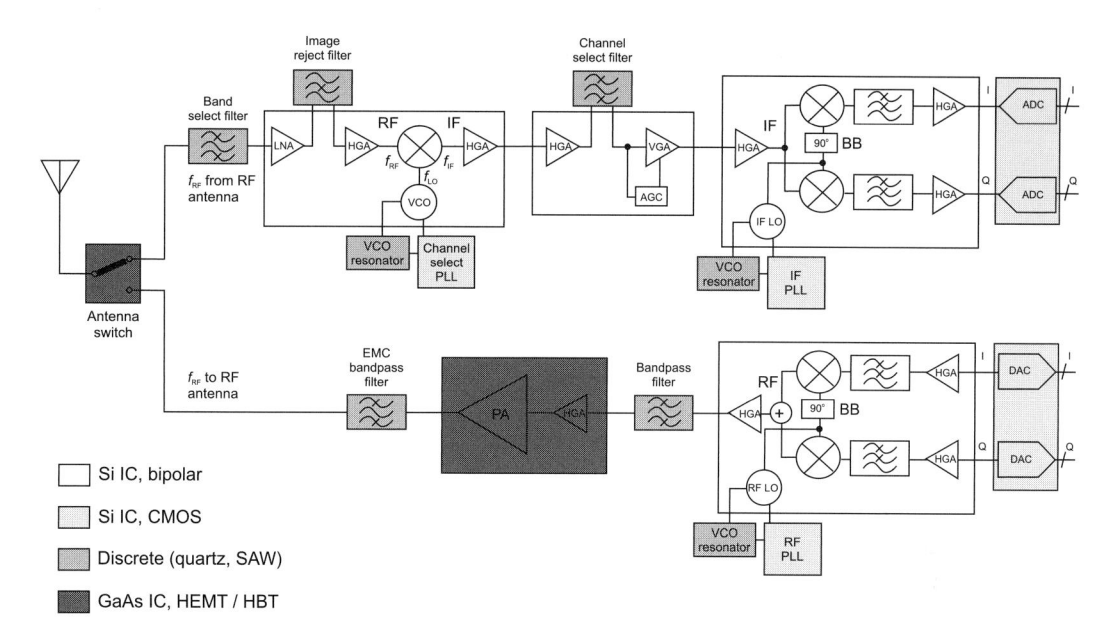

1.3 Enabling Technologies for RF, Microwaves and mm Waves

Figure 1.14 The RF interface of a GSM single-band mobile phone handset from around 2000: notice the variety of integrated circuits having different technologies.

An example of the progress in integration for very similar transceivers is shown in Fig. 1.14 and Fig. 1.15. In the first figure, the RF transceiver is shown for a GSM handset manufactured around 2000; the receiver stage with a heterodyne architecture is in the above branch, below is the transmitter stage. In the RX chain the RF, IF and IF to baseband blocks are realized with conventional bipolar circuits (the operating frequency is low, assuming the phone will work in the initial European GSM band below 1 GHz), most filters are external and realized in the SAW (Surface Acoustic Wave) technology, while the digital parts and the ADC–DAC block, together with the Phase Locked Loop (PLL, one of the main components of the frequency synthesizer), are in MOS Si technology. A few components however are in GaAs: the power amplifier (Si bipolars are not able to achieve the $\approx$1 W maximum output power required by 2–4G standards) and the switches (electronic switches with enough isolation cannot be realized in Si). *Tanks* are the resonators of the RF and IF oscillators, usually implemented through SAW or quartz discrete technologies.

Let us consider a more recent example of transceiver, to highlight the progress made in technology and system integration. The simplified block scheme of the MAX2830 transceiver [10] is shown in Fig. 1.15. This is a direct conversion RF interface for Wireless Local Area Network (WLAN) applications at 2.4 to 2.5 GHz. It may be noticed that the system includes a CMOS integrated circuit implementing all functions, including low-noise and power amplification, exploiting external components only for the voltage-controlled oscillator (VCO) resonator, for the input and output filters, and for the antenna switch, that is realized in GaAs technology to allow for better isolation. Thanks to the much lower power prescribed by WLAN when compared to mobile phone standards, the

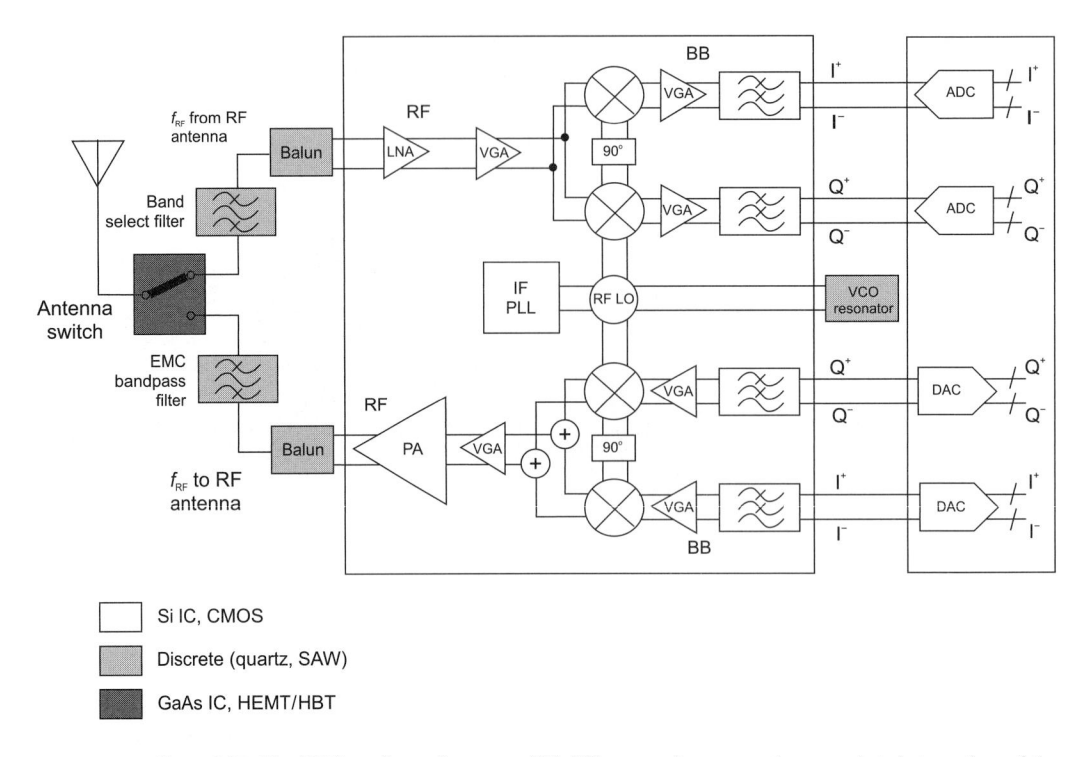

Figure 1.15 The RF interface of a recent WLAN transceiver: note the complete integration of the transceiver in Si CMOS technology.

circuit adopts a MOS power stage. Finally, notice that the architecture is balanced, i.e., in differential form, to enhance robustness to electromagnetic (EM) disturbances. Balancing Units (Baluns) are used to transform the balanced signals into single-ended ones.

1.3.1 Examples of (M)MICs

In the present section we discuss a few examples of hybrid and monolithic realizations of microwave circuits. Details will be provided mainly to highlight the features of the technological solutions.

We start in Fig. 1.16 with a laboratory prototype of hybrid *balanced amplifier*, a tandem wideband architecture that will be discussed in detail in Sec. 6.11.1 [11]. The balanced amplifier is made of two equal one-stage amplifiers in tandem; see the center of Fig. 1.16; the active device (a GaAs FET) is mounted on a metal ridge providing the electrical ground and heat sinking. On the left of the transistor, we have a distributed input matching section made of microstrip lines and stubs. A similar output matching section is located at the right of the active element. The bias circuit ("Bias T"), separating the RF from the DC part and *vice versa*, is made of RF chokes, i.e., large inductors blocking the RF signal realized with a short wire coil in air, and bypass capacitors shorting the RF signal to ground. The upper and lower external wires are the DC bias connectors. The two tandem amplifiers are RF fed at the input by a separate packaged hybrid

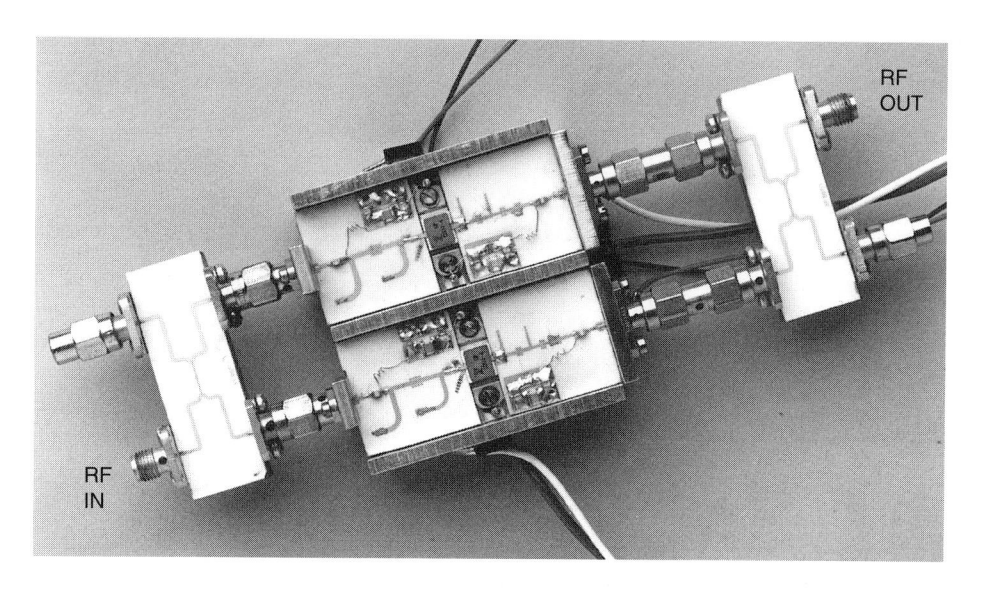

1.3 Enabling Technologies for RF, Microwaves and mm Waves

Figure 1.16 A hybrid microwave balanced amplifier. Courtesy of prof. C. Naldi [11].

implementing a Lange 3 dB directional coupler; see Sec. 4.4.1. The two input and output couplers are connected to the amplifier by means of four coaxial connectors, while the global input is lower left; the global output is in the upper right corner of the picture. The bolts screwed on the auxiliary input and output coaxial connectors are matched loads. The amplifier operation is roughly as follows: the coupler splits the input power in two equal parts, that are amplified by the two tandem amplifiers and recombine at the output coupler. However, thanks to the properties of the coupler, the two outputs are in phase quadrature; this allows the power reflected back by the two amplifier stages to be deviated on the matched loads rather than directly reflected into the amplifier input or output. The balanced amplifier therefore allows good input and output matching while combining two amplifiers that may be considerably mismatched at the input or output – a condition that often arises in the design of wideband amplifiers. In the example shown, all transmission lines are microstrips on an alumina substrate (the white material visible on top).

GaAs-based MMICs are not confined to the implementation of simple analog circuits. A typical example of mixed digital–analog circuit is the so-called *core chip*, that performs, on a transmitter or receiver stage, the function of a digitally controlled phase shift and amplitude modulation. Such a function is typically required in the control circuitry of steered-beam antenna arrays. The digital part consists of a serial-to-parallel (S2P) converter, while commutation from the transmitter to the receiver stage is performed by a switch. Although the design of III-V FET-based digital circuits is challenging [12], the integration of the S2P converter allows the core chip architecture to be considerably simplified, avoiding the introduction of a large number of digital parallel inputs.

An example of X-band core chip is shown in Fig. 1.17; the left inset in Fig. 1.17 identifies the separate functional blocks of the MMIC [13, 14]. The TX amplifier coplanar

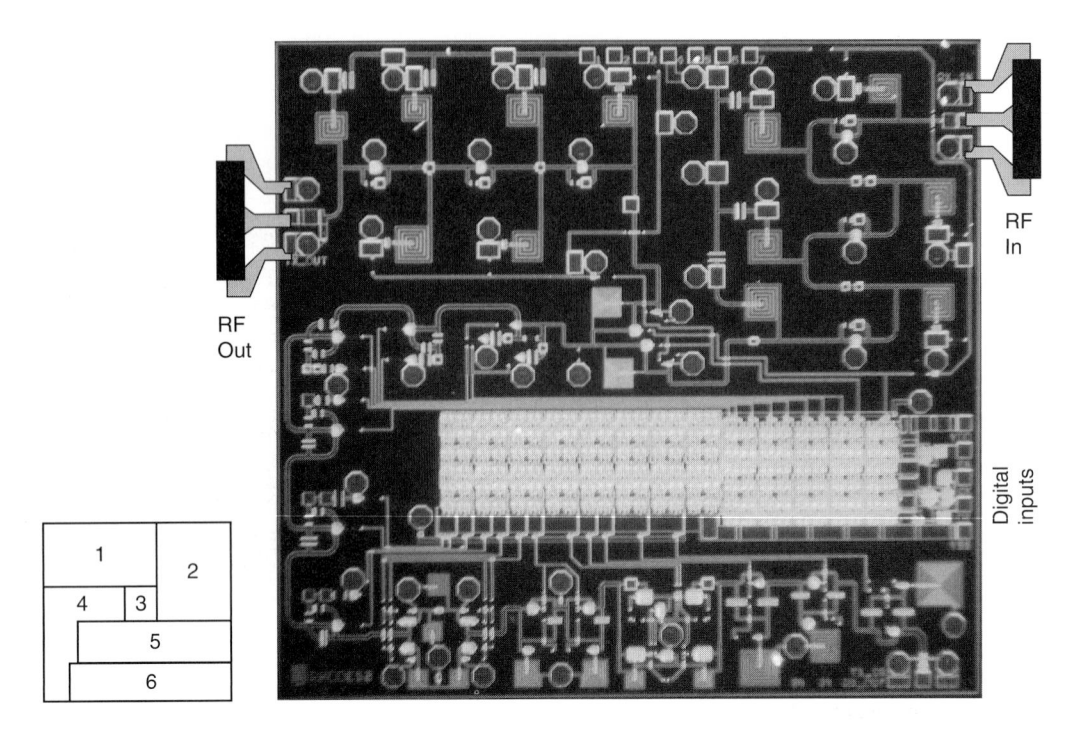

Figure 1.17 A GaAs integrated core chip; in the left inset a scheme of the internal functional blocks is shown: (1) TX stage amplifier; (2) RX stage amplifier; (3) switch between the TX and RX stages; (4) 6-bit variable attenuator; (5) serial to parallel converter; (6) 6-bit phase shifter. Copyright © 2012 IEEE. All rights reserved. Adapted, with permission, from [14, Fig. 2].

output is on the upper left of the circuit, while on the upper right we find the RX amplifier coplanar input. The digital input of the serial to parallel converter is on the right of block (5) in the inset of Fig. 1.17. Both the RX and the TX amplifiers are multistage, each stage using parallel feedback; spiral inductors and MIM capacitors are used in the circuit to implement bias circuits and matching networks. The total chip size is around 16 mm^2 and the technology is a 180 nm PHEMT process from the OMMIC foundry [14] allowing for depletion and enhancement devices.

Fig. 1.18 shows a three-stage low-noise GaN HEMT MMIC [15]. The layout reveals the presence of spiral inductors (the large coils) both for the DC bias circuits and for interstage matching. GaN low-noise amplifiers have recently been introduced; they exhibit robustness to input signal spikes without requiring, as in other technologies, the introduction of protection circuitry. The low-noise nature of the circuit, that again makes use of a microstrip technology but with coplanar input and output, is revealed by the first stage (left) where the source of the field-effect GaN transistor is connected to the ground through an inductor (in the dashed white rectangle) and an additional inductor is connected to the gate (in the dashed white rectangle at the left of the first stage). The amplifier with a source inductive series feedback is in fact one of the most common microwave low-noise topologies; see Sec. 7.10.2.

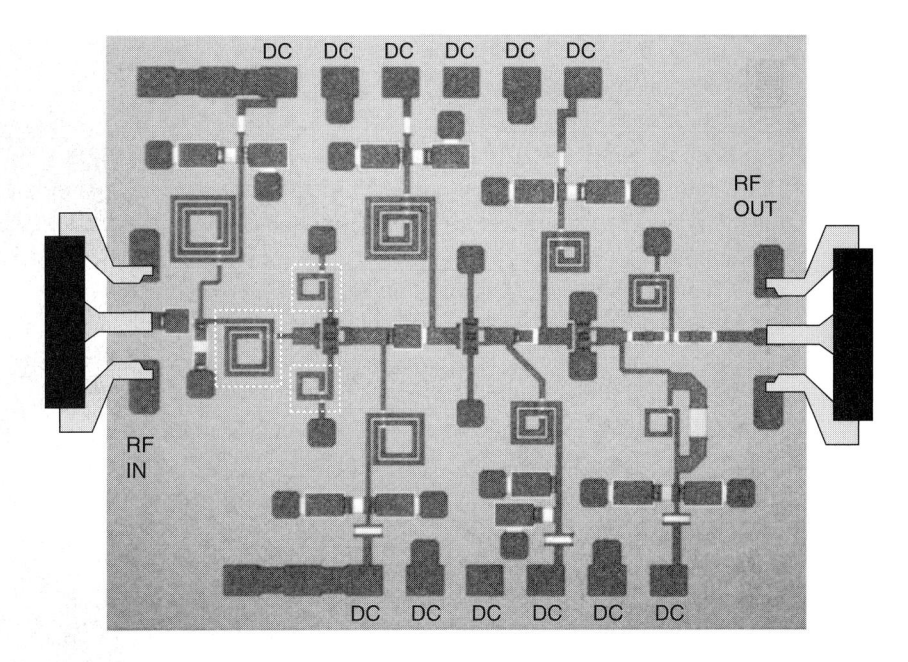

Figure 1.18 A three-stage monolithic low-noise GaN amplifier. On the coplanar input (left) and output (right) the coplanar measurement probes are schematically shown. Above and below we show the bonding wires for DC bias. Adapted from [15, Fig. 16]. Copyright © 2013 IEEE. All rights reserved. Reprinted with permission.

1.4 Questions and Problems

1.4.1 Questions

1. Define RF, microwaves and millimeter waves from the standpoint of frequency allocation.
2. Suppose a radar has to be designed to detect objects of an average size of 1 cm: is an RF operating frequency adequate for this? Explain why/why not. Suggest (if the case) a more suitable frequency range.
3. Identify the L and the K bands (frequency limits). In what frequency band do mobile phones operate?
4. Explain why signals cannot conveniently be transmitted in baseband through a Hertzian channel, but rather they have to be upconverted through analog or digital modulations. Assume as an example a hi-fi signal with frequency between (approximately) DC and 20 kHz.
5. Explain why transmitting the human voice in baseband through a portable phone would for many reasons be impractical.
6. In a cellular system the same frequency channels (e.g., around 2 GHz) are used in several cells. Explain why there is no interference between different users sharing, in nearby cells, the same channel.
7. Describe a basic RF RX/TX (receiver/transmitter) scheme.

8. List the basic RX section building blocks, starting from the antenna down to the downconversion mixer.
9. List the basic TX section building blocks, starting from the upconversion mixer up to the antenna.
10. What is the difference between a homodyne and a heterodyne receiver?
11. What is the difference between a low-noise, high-gain and maximum power amplifier?
12. What are the typical features of planar vs. waveguide microwave circuits?
13. Explain the differences between a hybrid and a monolithic microwave circuit.
14. List in order of increasing frequency range the following semiconductors: indium phosphide, silicon, gallium arsenide.
15. Explain the difference between a lumped and a distributed-parameter circuit. Why can distributed elements not be integrated in an RF circuit?
16. Quote a few microwave field-effect or bipolar transistors with the related semiconductor material.
17. Explain the difference between analog large-signal and small-signal models. Clarify which models are linear and which models are nonlinear.
18. Explain the difference between a memoryless model and a model with memory.
19. Assume that the balanced amplifier in Fig. 1.16 has a centerband frequency of 4 GHz. Estimate the size of the amplifier. (Hint: the length of the multiconductor part of the Lange amplifier is a quarter wavelength at centerband and the average guided wavelength is the free space wavelength divided by the effective refractive index; see Example 4.2 for an estimate of the coupler length. From the coupler length the total size of the amplifier can be approximately estimated.)

1.4.2 Problems

1. An antenna is working at a frequency of 100 kHz. Assuming the antenna length is equal to $L = \lambda_0/100$, evaluate L.
2. A dielectric medium has $\epsilon_r = 9$. Evaluate the free-space wavelength at 10 GHz and the wavelength in the dielectric medium.
3. Estimate the typical size of a distributed element operating at 100 MHz and at 50 GHz. Assume, as the centerband dimension, a quarter of the wavelength, and as the wavelength $\lambda = \lambda_0/n$ where λ_0 is the free space wavelength and $n = 2.5$ is the effective refractive index.

References

[1] T. K. Sarkar, R. Mailloux, A. A. Oliner, M. Salazar-Palma, and D. L. Sengupta, *History of wireless*. Hoboken, NJ: Wiley-IEEE Press, 2006.
[2] "IEEE standard letter designations for radar-frequency bands," *IEEE Std 521-2002 (Revision of IEEE Std 521-1984)*, pp. 1–3, 2003.
[3] J. N. Burghartz, *Guide to state-of-the-art electron devices*, ser. Wiley-IEEE. Wiley, 2013.
[4] "Radio spectrum," https://en.wikipedia.org/wiki/Radio_spectrum#ITU.

References

[5] T. K. Sarkar, M. S. Palma, and E. L. Mokole, "Echoing across the years: a history of early radar evolution," *IEEE Microwave Magazine*, vol. 17, no. 10, pp. 46–60, Oct. 2016.

[6] N. C. Currie and C. E. Brown, *Principles and applications of millimeter-wave radar*, ser. Artech House radar library. Artech House, 1987.

[7] B. Razavi, *RF microelectronics (2nd Edition)*, Upper Saddle River, NJ, USA: Prentice Hall Press, 2011.

[8] A. Siligaris, O. Richard, B. Martineau, C. Mounet, F. Chaix, R. Ferragut, C. Dehos, J. Lanteri, L. Dussopt, S. Yamamoto, R. Pilard, P. Busson, A. Cathelin, D. Belot, and P. Vincent, "A 65 nm CMOS fully integrated transceiver module for 60 GHz wireless HD applications," in *Solid-State Circuits Conference Digest of Technical Papers (ISSCC), 2011 IEEE International*, Feb. 2011, pp. 162–164.

[9] B. R. Bennett, R. Magno, J. B. Boos, W. Kruppa, and M. G. Ancona, "Antimonide-based compound semiconductors for electronic devices: a review," *Solid-State Electronics*, vol. 49, no. 12, pp. 1875–1895, 2005.

[10] "MAX2830," www.maximintegrated.com/en/products/comms/wireless-rf/MAX2830.html.

[11] C. Naldi, private communication.

[12] S. I. Long and S. E. Butner, *Gallium arsenide digital integrated circuit design*. McGraw-Hill, 1990.

[13] W. Ciccognani, M. Ferrari, G. Ghione, E. Limiti, P. E. Longhi, M. Pirola, and R. Quaglia, "A compact high performance X-band core-chip with on board serial-to-parallel conversion," in *The 40th European Microwave Conference*, Sep. 2010, pp. 902–905.

[14] A. Bentini, W. Ciccognani, M. Palomba, D. Palombini, and E. Limiti, "High-density mixed signal RF front-end electronics for T-R modules," in *2012 IEEE First AESS European Conference on Satellite Telecommunications (ESTEL)*, Oct. 2012, pp. 1–6.

[15] S. Colangeli, A. Bentini, W. Ciccognani, E. Limiti, and A. Nanni, "GaN-based robust low-noise amplifiers," *IEEE Transactions on Electron Devices*, vol. 60, no. 10, pp. 3238–3248, Oct. 2013.

2 Passive Elements and Circuit Layout

2.1 Transmission Lines

Transmission lines (TXLs), simple or multiconductor, are a key distributed element in microwave circuits. They operate as signal transducers between components but are also the building block for passive distributed elements such as couplers, filters, matching sections, power dividers. In hybrid and monolithic microwave circuits the preferred guiding structures are the so-called TEM or quasi-TEM lines, characterized by broadband behavior and by the absence of a cutoff frequency, that is found in metal waveguides.

From a theoretical standpoint, N metal conductors with a ground plane support N TEM or quasi-TEM propagation modes. In TEM (Transverse ElectroMagnetic) modes, the electric and magnetic fields are transverse with respect to the propagation direction, i.e., orthogonal to the line axis. A purely TEM mode propagates in a line without conductor losses and with a homogeneous cross section, while lossy metal lines and lines with a non-homogeneous cross section support quasi-TEM modes with small longitudinal field components. An example of quasi-TEM line is the microstrip, where the cross section is partly filled by a dielectric, and partly by air. TEM and quasi-TEM lines may also support upper propagation modes with a cutoff frequency; however, the excitation of those modes is to be avoided because they contribute to radiation losses and coupling.

2.1.1 Transmission Line Theory

Transmission lines are a simple but convenient model for one-dimensional wave propagation, serving as a bridge between circuit theory and electromagnetics. Suppose first that the line is made of two lossless parallel conductors, one acting as the signal conductor, the other as the return or ground conductor, surrounded by a homogeneous, lossless medium. Such a structure supports a TEM propagation mode in which the electric and the magnetic fields lie in the line cross section and are orthogonal to the line axis and wave propagation direction; see Fig. 2.1 (a). In a TEM mode the transverse electric field can be derived from a potential function satisfying, in the line cross section, the Laplace equation. This is uniquely determined by the potential of the signal line $v(z,t)$ with respect to the ground line, where the coordinate z is parallel to the line axis and propagation direction. The transverse magnetic field is in turn related to the total current $i(z,t)$ flowing in the signal conductor; $-i(z,t)$ flows instead in the return conductor. The

2.1 Transmission Lines

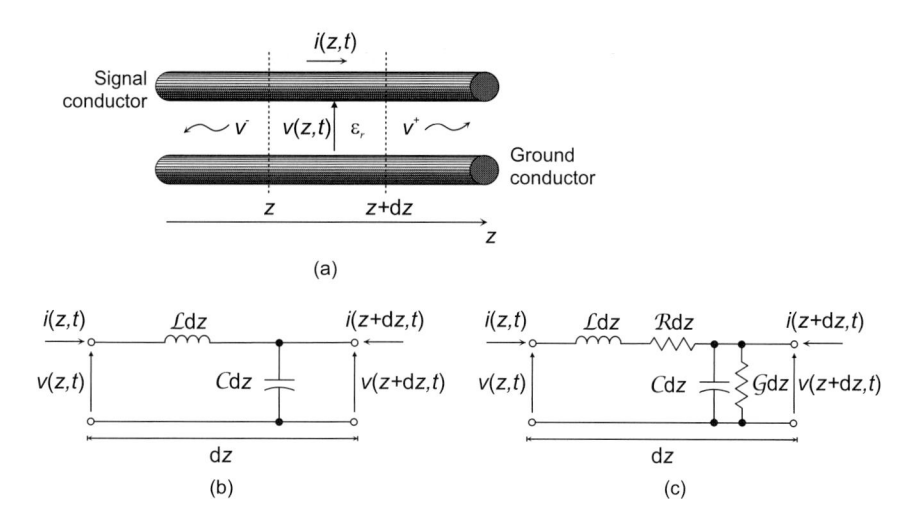

Figure 2.1 TEM transmission line (a); equivalent circuit of a line cell of length dz in the lossless (b) and lossy (c) cases.

voltage v and current i satisfy the following partial differential equation system, called the *telegraphers' equations*:

$$\frac{\partial}{\partial z}i(z,t) = -\mathcal{C}\frac{\partial}{\partial t}v(z,t) \qquad (2.1)$$

$$\frac{\partial}{\partial z}v(z,t) = -\mathcal{L}\frac{\partial}{\partial t}i(z,t), \qquad (2.2)$$

where $\mathcal{L}$ is the *per-unit-length* (p.u.l.) line inductance, $\mathcal{C}$ the p.u.l. line capacitance. Eqs. (2.1) and (2.2) can be derived from the Kirchhoff equations applied to the lumped equivalent circuit of a lossless line cell of infinitesimal length, shown in Fig. 2.1 (b). The p.u.l. parameters are the total conductor series inductance and the total parallel capacitance between conductors of a unit length cell. In real lines, conductors exhibit series conduction losses, while parallel losses are associated with the dissipation mechanisms in the dielectric substrate; such series and parallel losses can be included in the model through a series p.u.l. resistance and parallel p.u.l. conductance; see Fig. 2.1 (c).

In the lossless case, system (2.1), (2.2) admits a general solution as a superposition of forward propagating (V^+, I^+) and backward propagating (V^-, I^-) waves:

$$v(z,t) = V^{\pm}(z \mp v_f t)$$
$$i(z,t) = I^{\pm}(z \mp v_f t),$$

where v_f is the *phase velocity* describing the propagation velocity of the wave and $V^{\pm}(\cdot)$ is an arbitrary function of the argument, related to $I^{\pm}(\cdot)$ through (2.3).

Substituting into system (2.1), (2.2) and evaluating the derivatives vs. t and z we obtain:

$$\frac{\partial}{\partial z}I^{\pm}(z \mp v_f t) = -\mathcal{C}\frac{\partial}{\partial t}V^{\pm}(z \mp v_f t) \rightarrow I^{\pm'} = \pm \mathcal{C} v_f V^{\pm'}$$

$$\frac{\partial}{\partial z}V^{\pm}(z \mp v_f t) = -\mathcal{L}\frac{\partial}{\partial t}I^{\pm}(z \mp v_f t) \rightarrow V^{\pm'} = \pm \mathcal{L} v_f I^{\pm'},$$

where $V^{\pm'}$ ($I^{\pm'}$) is the derivative of $V^{\pm}$ ($I^{\pm}$) with respect to the argument. Eliminating the voltage unknown we have:

$$I^{\pm'} = \pm \mathcal{C} v_f V^{\pm'} = \mathcal{L}\mathcal{C} v_f^2 I^{\pm'},$$

from which the propagation velocity results as:

$$v_f = \frac{1}{\sqrt{\mathcal{L}\mathcal{C}}},$$

while the voltage and current waves are related by the *characteristic impedance* Z_0, also denoted as Z_∞ or Z_c, as:

$$V^{\pm} = \pm\sqrt{\frac{\mathcal{L}}{\mathcal{C}}}I^{\pm} \rightarrow V^{\pm} = \pm Z_0 I^{\pm}, \quad Z_0 = \sqrt{\frac{\mathcal{L}}{\mathcal{C}}}. \quad (2.3)$$

Notice that a lossless TXL supports undistorted wave propagation. If the voltages and currents are time-harmonic at a frequency f (angular frequency $\omega = 2\pi f$) the propagating waves will be both time- and space-harmonic:

$$v(z,t) = V^{\pm}(z \mp v_f t) = \sqrt{2}\mathrm{Re}\left[V_0^{\pm}\exp\left(j\omega t \mp j\frac{\omega}{v_f}z\right)\right]$$

$$= \sqrt{2}\mathrm{Re}\left[V_0^{\pm}\exp(j\omega t \mp j\beta z)\right]$$

$$i(z,t) = I^{\pm}(z \mp v_f t) = \sqrt{2}\mathrm{Re}\left[I_0^{\pm}\exp(j\omega t \mp j\beta z)\right]$$

$$= Z_0^{-1}\sqrt{2}\mathrm{Re}\left[V_0^{\pm}\exp(j\omega t \mp j\beta z)\right].$$

The voltage $V_0^{\pm}$ is a complex, constant phasor (normalized with respect to the voltage effective, rather than peak, value) to be determined through the initial and boundary conditions. The parameter:

$$\beta = \frac{\omega}{v_f} = \omega\sqrt{\mathcal{L}\mathcal{C}}$$

is the *propagation constant* of the line. The time periodic waveform with period $T = 1/f$ is also periodic in space with periodicity given by the *guided wavelength* λ_g such as:

$$\beta = \frac{2\pi}{\lambda_g} \rightarrow \lambda_g = \frac{v_f}{f} = \frac{\lambda_0}{n_\mathrm{eff}},$$

where $n_\mathrm{eff} = \sqrt{\epsilon_\mathrm{eff}}$ is the line effective refractive index and ϵ_eff is the line effective (relative) permittivity.[1]

Losses can be taken into account by introducing a series p.u.l. resistance $\mathcal{R}$, associated with ohmic losses in the conductors, and a parallel p.u.l. conductance $\mathcal{G}$, associated with

---

[1] The effective permittivity will be, in the present book, always intended in a relative sense.

2.1 Transmission Lines

the dielectric losses in the surrounding medium, see Fig. 2.1 (c). Series losses, but also an inhomogeneous cross section, induce small longitudinal field components; in such cases the TXL TEM model can be heuristically extended as the so-called quasi-TEM mode, where transverse components are dominant, but also small, frequency-dependent longitudinal components exist.

For lossy lines the telegraphers' equations read:

$$\frac{\partial}{\partial z}i(z,t) = -\mathcal{C}\frac{\partial}{\partial t}v(z,t) - \mathcal{G}v(z,t) \qquad (2.4)$$

$$\frac{\partial}{\partial z}v(z,t) = -\mathcal{L}\frac{\partial}{\partial t}i(z,t) - \mathcal{R}i(z,t). \qquad (2.5)$$

Lossy lines do not generally support undistorted propagation. However, harmonic waveforms in a lossy line propagate as such, albeit with decreasing amplitude. For time-harmonic v and i we assume:

$$v^\pm(z,t) = \sqrt{2}\mathrm{Re}\left[V^\pm(z,\omega)\exp(j\omega t)\right]$$
$$i^\pm(z,t) = \sqrt{2}\mathrm{Re}\left[I^\pm(z,\omega)\exp(j\omega t)\right],$$

where $V^\pm(z,\omega)$ and $I^\pm(z,\omega)$ are now *space-dependent phasors* associated with $v^\pm$ and $i^\pm$. Since the system is linear, the solution $v(z,t)$ or $i(z,t)$, with associated phasors $V(z,\omega)$ and $I(z,\omega)$, can be obtained by superposition of the forward and backward propagating waves, as:

$$v(z,t) = v^+(z,t) + v^-(z,t) \rightarrow V(z,\omega) = V^+(z,\omega) + V^-(z,\omega)$$
$$i(z,t) = i^+(z,t) + i^-(z,t) \rightarrow I(z,\omega) = I^+(z,\omega) + I^-(z,\omega).$$

For time-harmonic signals, system (2.4), (2.5) reduces to first-order differential equations in space:

$$\frac{\partial}{\partial z}V(z,\omega) = -(j\omega\mathcal{L} + \mathcal{R})I(z,\omega) \qquad (2.6)$$

$$\frac{\partial}{\partial z}I(z,\omega) = -(j\omega\mathcal{C} + \mathcal{G})V(z,\omega), \qquad (2.7)$$

whose solution in terms of forward and backward wave phasors reads:

$$V^\pm(z,\omega) = V_0^\pm \exp(\mp\alpha z \mp j\beta z) = V_0^\pm \exp(\mp\gamma z) \qquad (2.8)$$

$$I^\pm(z,\omega) = I_0^\pm \exp(\mp\alpha z \mp j\beta z) = \pm\frac{V_0^\pm(z,\omega)}{Z_0} = \pm\frac{V_0^\pm}{Z_0}\exp(\mp\gamma z), \qquad (2.9)$$

where α is the line attenuation, $\gamma = \alpha + j\beta$ is the *complex propagation constant*, Z_0 is the (now possibly complex) characteristic impedance, and $V_0^\pm$ a complex constant to be determined from initial and boundary conditions. Substituting (2.8) and (2.9) in system (2.6), (2.7), we obtain for the complex propagation constant γ and for the complex characteristic impedance Z_0 the expressions:

$$\alpha + j\beta = \gamma = \sqrt{(j\omega\mathcal{L} + \mathcal{R})(j\omega\mathcal{C} + \mathcal{G})}$$

$$\frac{V^{\pm}(z,\omega)}{I^{\pm}(z,\omega)} = \pm\sqrt{\frac{j\omega\mathcal{L} + \mathcal{R}}{j\omega\mathcal{C} + \mathcal{G}}} \equiv \pm Z_0.$$

The dispersive behavior of the line clearly appears from the frequency dependence of the propagation parameters. The real part of the complex propagation constant $\gamma = \alpha + j\beta$, α, can be further split (in the so-called high-frequency regime, see further on) into a conductor or metal attenuation α_c and a dielectric attenuation α_d; the imaginary part is the real propagation constant, $\beta = \omega/v_f$, where v_f is the phase velocity. The propagation constant β is measured in rad/m, while the attenuation is expressed in Np/m (neper per meter), or, more commonly, in dB/m or dB/cm (decibel per meter or per centimeter); the logarithmic units are related as:

$$\alpha|_{\mathrm{dB/m}} = 8.6859\alpha \qquad (2.10)$$

$$\alpha|_{\mathrm{dB/cm}} = 0.086859\alpha, \qquad (2.11)$$

where α is in natural units, i.e., in Np/m. A forward propagating voltage $V^+(z)$ is attenuated while propagating according to the law:

$$|V^+(z+L)| = |V^+(z)|\exp(-\alpha L),$$

where α is in Np/m and L in m. In dB we obtain:

$$\left|\frac{V^+(z)}{V^+(z+L)}\right|_{\mathrm{dB}} = 20\log_{10}\left|\frac{V^+(z)}{V^+(z+L)}\right| = 20\log_{10}\exp(\alpha L)$$
$$= 20\log_{10} e \times \alpha L = 8.6859\alpha L = \alpha|_{\mathrm{dB/m}} L.$$

The electromagnetic theory shows that the high-frequency current density penetrates the line conductors to an average thickness called the *skin penetration depth* δ [1]:

$$\delta = \sqrt{\frac{2}{\mu\sigma\omega}} = \sqrt{\frac{1}{\pi\mu\sigma f}}, \qquad (2.12)$$

where $\mu \approx \mu_0 = 4\pi \times 10^{-7}$ H/m is the metal permittivity (we assume that conductors are non-magnetic). If δ is much smaller than the conductor thickness the current flow is limited to a surface sheet having sheet impedance (see Example 2.1):

$$Z_s(\omega) = R_s + jX_s = \frac{1+j}{\sigma\delta} = (1+j)\sqrt{\frac{\omega\mu}{2\sigma}}. \qquad (2.13)$$

The high-frequency p.u.l. resistance, see (2.17), therefore increases as a function of frequency as:

$$\mathcal{R}(f) \approx \mathcal{R}(f_0)\sqrt{\frac{f}{f_0}},$$

while the high-frequency p.u.l. inductance has two contributions: the external inductance $\mathcal{L}_{ex}$, related to the magnetic energy stored in the dielectric surrounding the line, and the frequency-dependent internal inductance $\mathcal{L}_{in}$, related to the magnetic

2.1 Transcription Lines

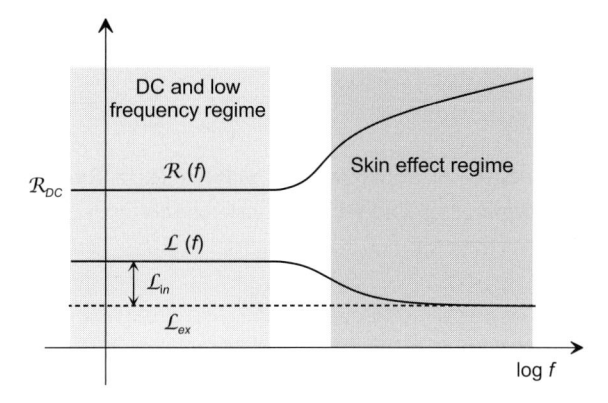

Figure 2.2 Qualitative frequency behavior of the line resistance and inductance from the DC to the high-frequency (skin effect) regime.

energy stored within the conductors. Since the corresponding reactance is given by $X_{in}(f) \approx X_{in}(f_0)\sqrt{f/f_0}$, one has:

$$\mathcal{L}(f) = \mathcal{L}_{ex} + \mathcal{L}_{in}(f) \approx \mathcal{L}_{ex} + \mathcal{L}_{in}(f_0)\sqrt{\frac{f_0}{f}} \underset{f \to \infty}{\approx} \mathcal{L}_{ex}.$$

At very high frequency the total inductance can be therefore approximated by the external inductance. The behavior of the p.u.l. resistance and inductance of a lossy line as a function of frequency are shown in Fig. 2.2; microwave circuits typically operate in the high-frequency, skin effect regime.

For a good conductor δ is of the order of a few μm at frequencies of the order of a few GHz. The frequency behavior of R_s and δ is shown in Fig. 2.3 for several values of conductivity; the maximum value $\sigma = 1 \times 10^8$ S/m is slightly larger than the typical value of good conductors (for copper $\sigma \approx 6 \times 10^7$ S/m).

In the planar microwave technology, composite metal layers are used. They are obtained first through a thin sputtered or evaporated adhesion layer (0.1–0.2 μm) followed by sputtered gold or Al (up to 500 nm). Larger thicknesses (from a few μm to 15–20 μm) can be obtained by gold electroplating. For such thicknesses, the transition between the low- and high-frequency regimes (with fully developed skin effect) takes place in the microwave range.

Example 2.1 Find the relationship between the surface impedance Z_s and the per-unit-length parameters of a conductor of periphery p.

Solution
From the definition, the surface impedance Z_s is the impedance of a metal patch of width w and of length $l = w$ (often expressed in $\Omega/\square$, ohm per square). For a conductor of periphery p and length $l = p$ the total input impedance will be:

$$Z = \mathcal{Z}l = \mathcal{Z}p = \mathcal{R}l + j\mathcal{X}l \equiv Z_s; \tag{2.14}$$

it follows that the p.u.l. impedance of the conductor $\mathcal{Z}$ is:

$$\mathcal{Z} = \frac{Z_s}{p}. \tag{2.15}$$

For example, for a circular wire of radius r ($p = 2\pi r$) and for a strip of width w and thickness t ($p = 2t + 2w$) we have, respectively:

$$\mathcal{Z}_{\text{wire}} = \frac{Z_s}{2\pi r}, \quad \mathcal{Z}_{\text{strip}} = \frac{Z_s}{2(w+t)}. \tag{2.16}$$

The same law holds for the p.u.l. resistance:

$$\mathcal{R} = \frac{R_s}{p}. \tag{2.17}$$

For different reasons, also the p.u.l. conductance will be frequency dependent; in fact, this is associated with the complex permittivity of the surrounding dielectrics:

$$\epsilon_{rc} = \epsilon_r - j\epsilon_2 = \epsilon_r \left[1 - j \tan \delta (f)\right],$$

where δ, generally a weak function of frequency, is the dielectric *loss angle* (nothing to do with the skin penetration depth!). The loss angle of good dielectrics exploited

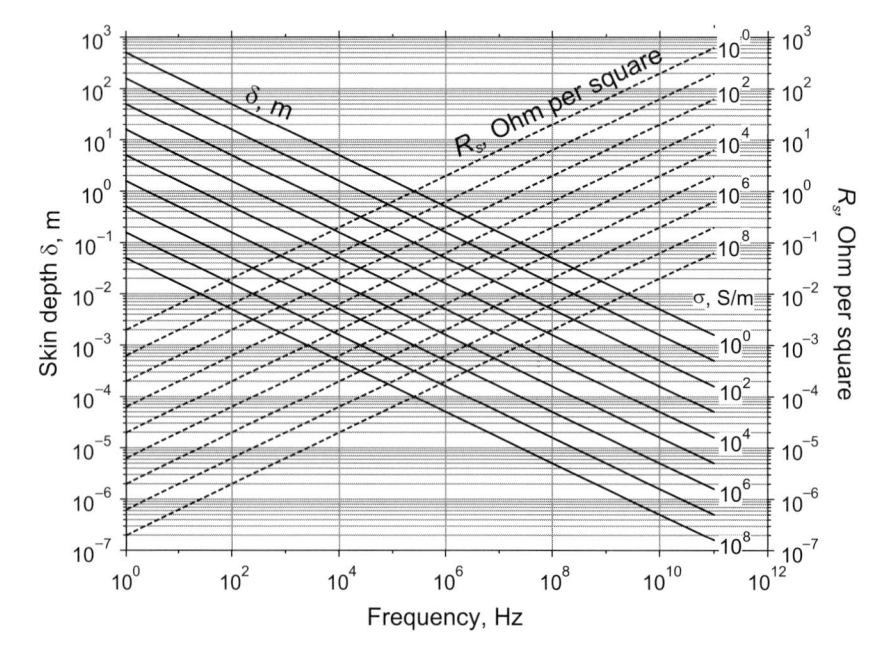

Figure 2.3 Frequency behavior of surface resistance and skin-effect depth for different values of conductivity.

as substrates is typically small, $10^{-2} - 10^{-4}$. The conductivity of the dielectric will therefore be:

$$\sigma = \omega\epsilon_2\epsilon_0 = \omega\epsilon_r\epsilon_0 \tan\delta.$$

The dielectric response of a material is caused by the interaction between the EM wave and microscopic mechanisms in the material. The main interaction with the EM field are with dipolar molecules (e.g., water), atoms and electrons. Each interaction is characterized by a low-pass behavior: at low frequency, the interaction is active and provides a contribution to the dielectric response; at high frequency, the interacting agents (dipoles, atoms, electrons) are unable to follow the time variations of the field, and the related contribution to the dielectric response vanishes. However, the interaction is also affected by losses that are maximized around the transition frequency. The transition frequencies of many molecular interactions fall in the microwave range, where atomic and electronic contributions are anyway active (such contributions disappear in the UV range mainly). In the microwave range, most dielectrics used as substrates exhibit losses roughly proportional to the frequency, since each cycle leads to a loss of energy and the dissipated power increases with the number of cycles per unit time.

To investigate the effect of a frequency-dependent complex permittivity on the p.u.l. parameters, consider a parallel-plate capacitor of area A and electrode spacing h; the capacitor impedance will be:

$$Y = j\omega\epsilon\frac{A}{h} = j\omega\epsilon'\frac{A}{h} + \omega\epsilon''\frac{A}{h} = j\omega C + G(\omega).$$

The result can be generalized to a transmission line with transversally homogeneous (or also inhomogeneous) lossy dielectrics, where in general:

$$\mathcal{G}(f) \approx \frac{f}{f_0}\mathcal{G}(f_0),$$

i.e., the line conductance linearly increases with frequency.

In a TEM (or transversally homogeneous) line, the p.u.l. parallel admittance can be evaluated simply as:

$$\mathcal{Y} = j\omega\mathcal{C} = j\omega\epsilon_{rc}\mathcal{C}_a = j\omega\epsilon_r\mathcal{C}_a + \omega\epsilon_r \tan\delta\mathcal{C}_a;$$

where $\mathcal{C}_a$ is the capacitance p.u.l. in air (i.e., with $\epsilon_{rc} = 1$), the second (real) term is a conductance, yielding:

$$\mathcal{G} = \omega\epsilon_r \tan\delta\mathcal{C}_a = \sigma\mathcal{C}_a/\epsilon_0. \tag{2.18}$$

Materials characterized by significant conductor losses (like doped semiconductors or intrinsic Si) have, on the other hand, a frequency-independent conductivity, leading to a frequency-independent line conductance. If the line cross section is inhomogeneous, like in a quasi-TEM line, evaluating $\mathcal{G}$ is slightly more involved, see the discussion in Sec. 2.1.2.

The remarks made so far on the relative importance of dielectric and metal losses hold for most of the low-loss dielectric of semiconductor substrates, see Table 2.1; in Si,

Table 2.1 Characteristics of some dielectric substrates for hybrid and integrated circuits.

Material	Alumina	Quartz	Teflon	GaAs	InP	Si
ϵ_r	$9.6 - 10$	3.78	2	12.9	12.4	11.9
$\tan \delta$	$10^{-3} - 10^{-4}$	$10^{-4} - 10^{-5}$	10^{-4}	10^{-3}	10^{-3}	10^{-2}

conduction losses can be significant (also depending on the doping level), and for this reason Si circuits exploit oxide layers to screen the top conductors from the substrate.

Taking into account the frequency dependence of the propagation constant in a lossy line and of the frequency dependence of the p.u.l. parameters associated with losses, and assuming narrowband signals, we can approximately model the transmission line in three frequency ranges, the *low frequency*, the *intermediate frequency* and the *high-frequency* range, where most microwave circuits operate.

In the *low-frequency* range the propagation parameters γ and Z_0 are approximately real:

$$\alpha + j\beta \approx \sqrt{\mathcal{R}\mathcal{G}}$$

$$Z_0 \approx \sqrt{\frac{\mathcal{R}}{\mathcal{G}}},$$

and the line works as a resistive distributed attenuator. In the *intermediate frequency* range, $j\omega\mathcal{C} + \mathcal{G} \approx j\omega\mathcal{C}$ while $j\omega\mathcal{L} + \mathcal{R} \approx \mathcal{R}$ in most lines, since typically series losses prevail over parallel losses. The line performances are therefore dominated by the p.u.l. resistance and capacitance (*RC regime*), with parameters:

$$\alpha + j\beta \approx \frac{1+j}{\sqrt{2}} \sqrt{\omega \mathcal{C} \mathcal{R}}$$

$$Z_0 \approx \frac{1-j}{\sqrt{2}} \sqrt{\frac{\mathcal{R}}{\omega \mathcal{C}}}.$$

In the *RC* regime the line is strongly dispersive and the characteristic impedance complex. The *RC* model is adequate, e.g., for low-speed digital interconnects in Si integrated circuits.

Finally, in the *high-frequency range* we have $j\omega\mathcal{C} \gg \mathcal{G}$ and $j\omega\mathcal{L} \gg \mathcal{R}$; the imaginary part of Z_0 can be neglected and the complex propagation constant can be approximated as:

$$Z_0 \approx Z_{0l} = \sqrt{\frac{\mathcal{L}}{\mathcal{C}}}$$

$$\gamma = \alpha + j\beta \approx \frac{\mathcal{R}(f)}{2Z_0} + \frac{\mathcal{G}(f) Z_0}{2} + j\omega\sqrt{\mathcal{L}\mathcal{C}} = \alpha_c(f) + \alpha_d(f) + j\beta_l,$$

where Z_{0l} is the characteristic impedance of the lossless line, $\alpha_c \propto \sqrt{f}$ and $\alpha_d \propto f$ are the conductor and dielectric attenuation, respectively (usually $\alpha_c \gg \alpha_d$ in the RF and microwave range, while dielectric losses may become dominant at millimeter waves), and β_l is the propagation constant of the lossless line. Therefore, in the high-frequency regime a wideband signal (e.g., a passband pulse) propagates almost undistorted, apart from the signal attenuation. The onset frequency for the high-frequency regime (also called the *LC* lossy regime) depends on the line parameter values; integrated structures with micron-scale dimensions can operate in the *RC* range for frequencies as high as a few GHz. Moreover, the impact of losses is related to the length of the TXL; in short structures signal distortion can be modest even though the line operates under very broadband excitation. While the *low-frequency* or *RG* range is of little interest for microwaves, the transition between the *RC* and the lossy *LC* behavior often occurs within the microwave frequency range, particularly in monolithic microwave circuits with small feature size. An example of behavior is discussed in Example 2.2.

Example 2.2 A transmission line has 50 Ω high-frequency impedance, effective permittivity equal to $\epsilon_{\text{eff}} = 6$, conductor attenuation of $\alpha_c = 0.5$ dB/cm, dielectric attenuation of $\alpha_d = 0.01$ dB/cm at $f_0 = 10$ GHz. Suppose that at $f_0 = 10$ GHz the line is already in the high-frequency regime and that the DC p.u.l. line resistance is 1/2 of the resistance at f_0. Evaluate the line parameters (β, α, Z_0) in the *RG*, *RC* and *LC* regime, specifying the frequency ranges of validity.

Solution

Assuming that at f_0 the line is already in the high-frequency regime, in the *LC* approximation we have:

$$Z_0 \approx \sqrt{\mathcal{L}/\mathcal{C}}$$
$$v_f = 1/\sqrt{\mathcal{LC}} = 3 \times 10^8/\sqrt{\epsilon_{\text{eff}}},$$

i.e.:

$$1/\mathcal{C} = 50 \times 3 \times 10^8/\sqrt{\epsilon_{\text{eff}}} \longrightarrow \mathcal{C} = \sqrt{6}/\left(150 \times 10^8\right) = 1.633 \times 10^{-10} \text{ F/m}$$

and thus:

$$\mathcal{L} = \mathcal{C}Z_0^2 = 4.0825 \times 10^{-7} \text{ H/m}.$$

The attenuations in the high-frequency approximation yield:

$$\alpha_c(f_0) \approx \frac{\mathcal{R}(f_0)}{2Z_0} \quad \alpha_d(f_0) \approx \frac{\mathcal{G}(f_0)Z_0}{2},$$

i.e., since $\alpha_c = 0.5$ dB/cm $= 1/0.086859 = 5.75$ Np/m; $\alpha_d = 0.01$ dB/cm $= 0.115$ Np/m:

$$\mathcal{R}(f_0) = 2Z_0\alpha_c = 100 \times 5.75 = 575 \text{ Ω/m}$$
$$\mathcal{G}(f_0) = 2\alpha_d/Z_0 = 2 \times 0.115/50 = 0.0046 \text{ S/m}.$$

Let us verify that the line actually is in the high-frequency regime at 10 GHz; for this we require:

$$2\pi f_0 \mathcal{L} \gg \mathcal{R} \rightarrow 6.28 \times 10 \times 10^9 \times 4.0825 \times 10^{-7} = 25638 \gg 575$$
$$2\pi f_0 \mathcal{C} \gg \mathcal{G} \rightarrow 6.28 \times 10 \times 10^9 \times 1.633 \times 10^{-10} = 10.255 \gg 0.0046$$

and both conditions are verified. We can approximately estimate the external inductance since the total reactance p.u.l. at f_0 is:

$$2\pi f_0 \mathcal{L} = 2\pi f_0 \mathcal{L}_{ex} + 2\pi f_0 \mathcal{L}_{in} = 2\pi f_0 \mathcal{L}_{ex} + \mathcal{R}(f_0) \rightarrow \mathcal{L}_{ex} = \mathcal{L} - \frac{\mathcal{R}(f_0)}{2\pi f_0}$$

$$= 4.0825 \times 10^{-7} - \frac{575}{6.28 \times 10 \times 10^9} = 3.9909 \times 10^{-7} \text{ H/m},$$

while:

$$\mathcal{L}_{in}(f_0) = \frac{\mathcal{R}(f_0)}{2\pi f_0} = 9.1561 \times 10^{-9} \text{ H/m}.$$

Since from the data given we cannot estimate the low-frequency behavior of the inductance we approximate the total inductance with a constant, frequency independent value; the line parameters are therefore:

$$\mathcal{L} = 4.0825 \times 10^{-7} \text{ H/m}$$
$$\mathcal{C} = 1.633 \times 10^{-10} \text{ F/m}$$
$$\mathcal{R}(f_0) = 575 \text{ }\Omega/\text{m}$$
$$\mathcal{G}(f_0) = 0.0046 \text{ S/m}.$$

The frequency behavior of the p.u.l. resistance and conductance can be approximated as follows:

$$\mathcal{G}(f) = \mathcal{G}(f_0)\frac{f}{f_0}$$

$$\mathcal{R}(f) \approx \frac{\mathcal{R}(f_0)}{2}\left[1 + \left(\frac{f}{f_0}\right)^{1/2}\right].$$

The frequency behavior of the propagation constant and attenuation are shown in Fig. 2.4. Notice that, due to the vanishing DC p.u.l. conductance only the intermediate and high-frequency regimes are actually present. In logarithmic scale the high-frequency slope of β is one, the slope of α is asymptotically 1/2 due to the prevailing skin-effect metal attenuation; the intermediate frequency slopes of both α and β are 1/2. The characteristic impedance is shown in Fig. 2.5; in the intermediate frequency range the real and imaginary parts of the impedance are approximately the same in magnitude, and the high-frequency impedance is real.

2.1.2 Parameters of Quasi-TEM Lines

In planar microwave circuits transmission lines made by one signal conductor deposited on a dielectric substrate backed by a ground plane are the most common technological solution. Such lines have an inhomogeneous cross section and therefore support a quasi-TEM propagation mode. Examples of non-TEM (rectangular waveguide), TEM (striplines) and quasi-TEM (microstrips) guiding structures are shown in Fig. 2.6. Non-TEM waveguides have a simply connected metal cross section and therefore do not

2.1 Transmission Lines

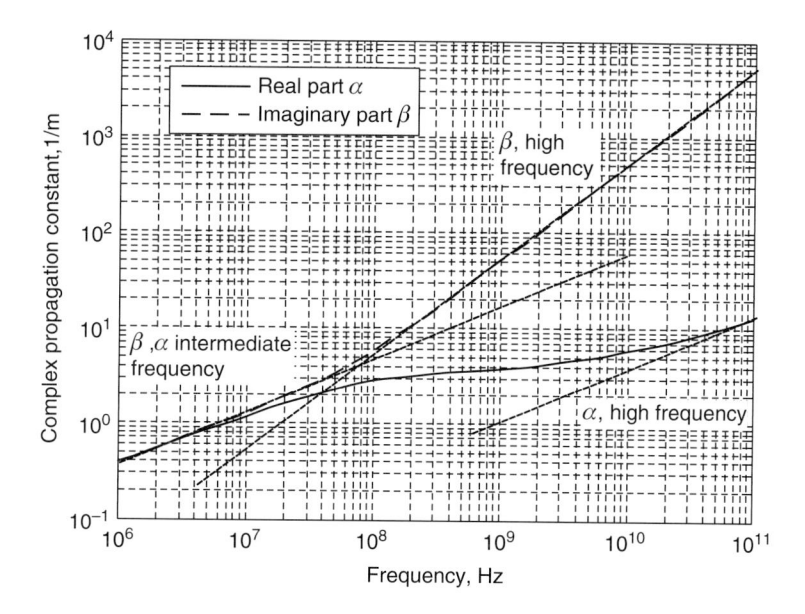

Figure 2.4 Frequency behavior of attenuation and propagation constant from Example 2.2.

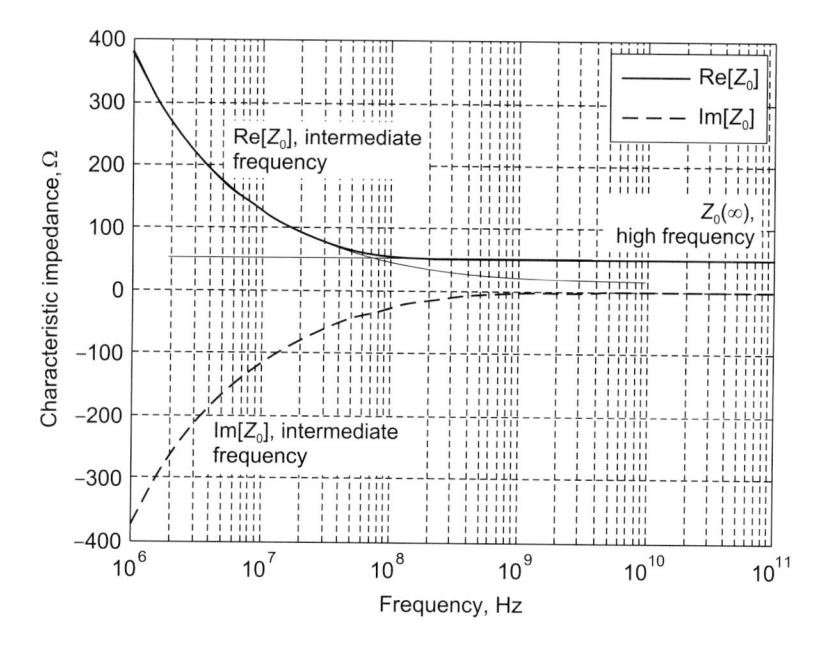

Figure 2.5 Frequency behavior of characteristic impedance from Example 2.2.

support DC conduction, while both TEM and quasi-TEM lines allow for DC conduction through two separated metal conductors. The main difference between the TEM and quasi-TEM case is the fact that, in the latter, the phase velocity and characteristic impedance are weak functions of frequency. The frequency dispersion of the quasi-TEM

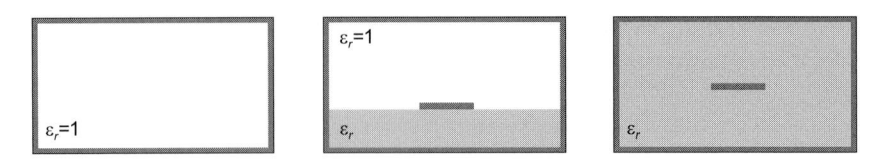

Figure 2.6 Examples of non-TEM (rectangular waveguide), TEM (shielded stripline), quasi-TEM (shielded microstrip) guiding structures.

parameters can be significant for the phase velocity (or effective permittivity) and can be properly modeled in the operating frequency range; above a certain frequency (that increases with decreasing line dimensions and substrate thickness) higher-order modes can appear, leading to radiation losses.

Neglecting in the first approximation losses and confining the treatment to the high-frequency regime, we have that the quasi-TEM line characterization amounts to evaluating $\mathcal{L}$ and $\mathcal{C}$. First of all, we show that in a TEM line with homogeneous cross section and relative medium permittivity ϵ_r, $\mathcal{L}$ does not depend on the dielectric permittivity. In fact, let us call $\mathcal{L}$ the inductance with dielectrics and $\mathcal{L}_a$ the inductance in air; the phase velocity with dielectrics coincides with the phase velocity in the medium, i.e.:

$$v_f = \frac{1}{\sqrt{\mathcal{LC}}} = \frac{c_0}{\sqrt{\epsilon_r}},$$

where c_0 is the velocity of light in air; on the other hand, the phase velocity of the line in which $\epsilon_r = 1$ (line in air or in vacuo) will be:

$$c_0 = \frac{1}{\sqrt{\mathcal{L}_a \mathcal{C}_a}}; \tag{2.19}$$

however, $\mathcal{C} = \mathcal{C}_a \epsilon_r$, and therefore $\mathcal{L} = \mathcal{L}_a$, i.e., the inductance with dielectrics is the inductance in air. The result is quite obvious since the inductance of a set of conductors is influenced by the magnetic permeability of the medium only. The same result holds for quasi-TEM lines.

In a quasi-TEM line the p.u.l. inductance is a function of the p.u.l. capacitance in air; in fact, from (2.19) we obtain:

$$\mathcal{L} = \mathcal{L}_a = \frac{1}{c_0^2 \mathcal{C}_a}. \tag{2.20}$$

Therefore we can express the characteristic impedance and phase velocity as a function of the capacitances in air and with dielectric as follows:

$$Z_0 = \sqrt{\frac{\mathcal{L}}{\mathcal{C}}} = \frac{1}{c_0\sqrt{\mathcal{CC}_a}} = \frac{Z_{0a}}{\sqrt{\epsilon_r}}$$

$$v_f = \frac{1}{\sqrt{\mathcal{LC}}} = c_0\sqrt{\frac{\mathcal{C}_a}{\mathcal{C}}} = \frac{c_0}{\sqrt{\epsilon_r}},$$

where Z_{0a} is the impedance in air. In the quasi-TEM case we can introduce an effective permittivity ϵ_{eff} such as:

$$\mathcal{C} = \epsilon_{\text{eff}} \mathcal{C}_a;$$

we also define the effective refractive index $n_{\text{eff}} = \sqrt{\epsilon_{\text{eff}}}$. Thus, in a quasi-TEM line:

$$Z_0 = \frac{1}{c_0 \mathcal{C}_a \sqrt{\epsilon_{\text{eff}}}} = \frac{Z_{0a}}{\sqrt{\epsilon_{\text{eff}}}}$$

$$v_f = \frac{c_0}{\sqrt{\epsilon_{\text{eff}}}}$$

and, furthermore:

$$\lambda_g = \frac{\lambda_0}{\sqrt{\epsilon_{\text{eff}}}} \tag{2.21}$$

$$\beta = \beta_0 \sqrt{\epsilon_{\text{eff}}}. \tag{2.22}$$

A simple example of evaluation of the effective permittivity of a quasi-TEM line is discussed in Example 2.3.

Example 2.3 Consider a quasi-TEM line made by two parallel metal conductors; the dielectric in not homogeneous, as shown in Fig. 2.7. Evaluate the effective permittivity and impedance of the line supposing that the field lines are orthogonal to the conductors.

Solution
An elementary line section with length $\mathrm{d}z$ is made of two parallel capacitors of size h, W, $\mathrm{d}z$, one in air and the other with a dielectric constant ϵ_r. The total capacitance p.u.l. will be:

$$\mathcal{C} = \frac{C}{\mathrm{d}z} = \frac{1}{\mathrm{d}z}\left(\frac{W\mathrm{d}z}{h}\epsilon_0 + \frac{W\mathrm{d}z}{h}\epsilon_0\epsilon_r\right) = \frac{W}{h}(1+\epsilon_r)\epsilon_0,$$

while:

$$\mathcal{C}_a = \frac{W}{h} 2\epsilon_0.$$

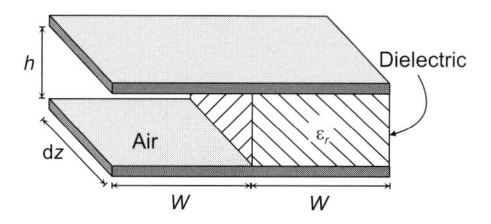

Figure 2.7 Example of parallel plate quasi-TEM line with inhomogeneous dielectric (see Example 2.3).

The effective permittivity is the ratio between the capacitance with dielectrics and the capacitance in air; we obtain:

$$\epsilon_{\text{eff}} = \frac{C}{C_a} = \frac{1 + \epsilon_r}{2}.$$

For the impedance:

$$Z_0 = \frac{1}{c_0 C_a \sqrt{\epsilon_{\text{eff}}}} = \sqrt{\epsilon_0 \mu_0} \times \frac{h}{2W\epsilon_0} \times \sqrt{\frac{2}{1 + \epsilon_r}} = 120\pi \sqrt{\frac{2}{1 + \epsilon_r}} \frac{h}{2W},$$

where $\sqrt{\mu_0/\epsilon_0} = 120\pi$ is the characteristic impedance of vacuum.

In general, the effective dielectric constant of a planar line on a dielectric substrate has values between 1 (the air constant) and the dielectric constant of the substrate. A non-quasi-static analysis permits to find that in a quasi-TEM line the effective permittivity is frequency dependent according to the typical behavior shown in Fig. 2.8, see [2]. The effective permittivity grows slowly with frequency from the quasi-static value $\epsilon_{\text{eff}}(0)$; the increase becomes fast after the inflection frequency f_{infl} which also corresponds (approximately) to the cutoff frequency of the first higher-order mode of the structure. Higher-order modes are mainly guided by the dielectric substrate and therefore lead to power leakage and radiation; for this reason, the inflection frequency limits the useful operation range of the structure. For very high frequency, the effective permittivity tends to the substrate permittivity ϵ_r.

The useful frequency range of the line is limited by the inflection frequency. The behavior can be approximated through empirical expressions, e.g.:

$$\epsilon_{\text{eff}}(f) \approx \left[\epsilon_{\text{eff}}^{\alpha}(0) + \frac{\epsilon_r^{\alpha} - \epsilon_{\text{eff}}^{\alpha}(0)}{1 + (f_{\text{infl}}/f)^{\beta}} \right]^{1/\alpha}, \tag{2.23}$$

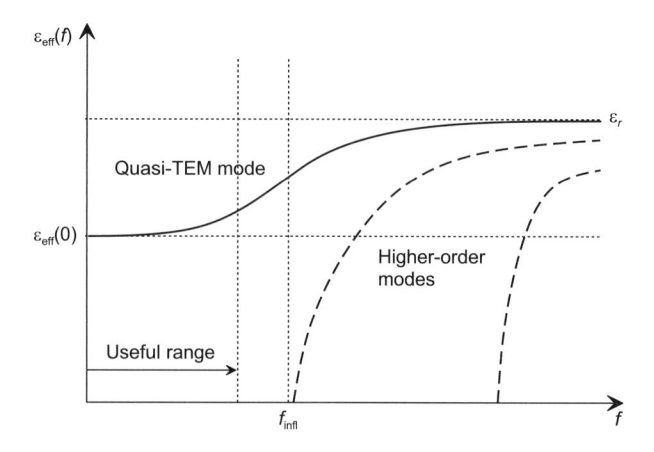

Figure 2.8 Behavior of the effective permittivity of a quasi-TEM line as a function of frequency.

where α and β are fitting parameters. The characteristic impedance does not exactly follow the behavior of the effective permittivity; however, its variation with respect to the quasi-static condition is less important.

The evaluation of dielectric losses in quasi-TEM lines is made more complex by the fact that the electric field energy in the substrate is subject to losses while the energy stored in air is not. As an example, consider a microstrip and imagine associating a substrate p.u.l. capacitance with the substrate electric field energy and an air capacitance to the energy stored in air, so that $C = C_{air} + C_{sub} = C_{air} + \epsilon_{rc}C_{sub,a}$ where $C_{sub,a}$ is the substrate capacitance with a dielectric having unit permittivity. In certain lines, as in coplanar waveguides, the splitting is straightforward. The p.u.l. admittance is therefore:

$$\mathcal{Y} = j\omega C = j\omega C_{air} + j\omega \epsilon_{rc} C_{sub,a}$$
$$= j\omega C_{air} + j\omega \epsilon_r C_{sub,a} + \omega \epsilon_r \tan \delta C_{sub,a},$$

leading to:

$$\mathcal{G} = \omega \epsilon_r \tan \delta C_{sub,a} = \sigma C_{sub,a}/\epsilon_0,$$

with the same frequency behavior as for a TEM line.

2.1.3 The Reflection Coefficient and the Smith Chart

Transmission lines can be exploited as circuit elements using the line solution in terms of forward and backward waves. A circuit including transmission lines can be easily shown to be amenable to a well posed mathematical problem, since each line is completely characterized by two unknowns (the forward and backward voltage phasors V^+ and V^-) and two linear relationships are imposed at the line input and output by the generator and load conditions.

As a simple example, let us evaluate the input impedance of a line of length l closed on a load Z_L (Fig. 2.9). The voltage and current phasors can be written as a superposition of forward and backward waves as:

$$V(z) = V_0^+ \exp(-\gamma z) + V_0^- \exp(\gamma z)$$
$$I(z) = \frac{V_0^+}{Z_0} \exp(-\gamma z) - \frac{V_0^-}{Z_0} \exp(\gamma z),$$

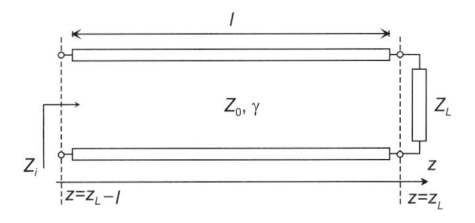

Figure 2.9 Input impedance of a loaded transmission line.

with boundary condition (the line current is directed towards increasing z):

$$V(l) = Z_L I(l) \rightarrow V_0^+ \exp(-\gamma l) + V_0^- \exp(\gamma l)$$
$$= \frac{Z_L}{Z_0}\left[V_0^+ \exp(-\gamma l) - V_0^- \exp(\gamma l)\right],$$

i.e.:

$$\frac{V_0^-}{V_0^+} = \Gamma_L \exp(-2\gamma l),$$

where:

$$\Gamma_L = \frac{Z_L - Z_0}{Z_L + Z_0} = \frac{z_L - 1}{z_L + 1}$$

is the *load reflection coefficient* with respect to Z_0 and $z = Z/Z_0$ is the *normalized impedance*. The line input impedance will be:

$$Z_i = Z(0) = \frac{V(0)}{I(0)} = Z_0 \frac{V_0^+ + V_0^-}{V_0^+ - V_0^-} = Z_0 \frac{1 + \Gamma_L \exp(-2\gamma l)}{1 - \Gamma_L \exp(-2\gamma l)}.$$

Expanding the load reflection coefficient and expressing the exponential in terms of hyperbolic functions we obtain:

$$Z_i = Z_0 \frac{Z_L \cosh(\gamma l) + Z_0 \sinh(\gamma l)}{Z_L \sinh(\gamma l) + Z_0 \cosh(\gamma l)}.$$

Notice that for $l \rightarrow \infty$, $Z_i \rightarrow Z_0$ independent of the load.[2] For a lossless line, however, the input impedance is periodic vs. the line length, with periodicity $\lambda_g/2$ (due to the tan function):

$$Z_i = Z_0 \frac{Z_L + jZ_0 \tan(\beta l)}{Z_0 + jZ_L \tan(\beta l)}. \tag{2.24}$$

The input impedance of a lossless line of infinite length does not converge therefore to the characteristic impedance. This is of course purely theoretical, since no lossless line exists and an infinitely long line would never reach a steady state condition in a finite time: thus a reflected signal would appear only after an infinitely long time. We will consider now some particular but important cases: if the load is a short ($Z_L = 0$) or an open ($Y_L = 0$) we have:

$$Z_i(Z_L = 0) = jZ_0 \tan(\beta l)$$
$$Z_i(Y_L = 0) = -jZ_0 \cot(\beta l);$$

a reactive load is therefore obtained, alternatively inductive and capacitive, according to the value of the *line electrical angle* $\phi = \beta l = 2\pi l/\lambda_g$. For a lossy line we have:

$$Z_i(Z_L = 0) = Z_0 \tanh(\gamma l)$$
$$Z_i(Y_L = 0) = -Z_0 \coth(\gamma l).$$

---

[2] This justifies the symbol Z_∞ for the characteristic impedance.

It can be readily shown by inspection that the input impedance of a shorted lossy line for $l \to 0$ is $Z_i \approx j\omega \mathcal{L}l + \mathcal{R}l$, while the input impedance of a short line in open circuit is $Z_i \approx (j\omega \mathcal{C}l + \mathcal{G}l)^{-1}$. For a more accurate computation of the input impedance of a short line in short or open, see Example 2.4.

Example 2.4 Evaluate the input admittance of a short line in short and the input impedance of a short line in open. Let the p.u.l. impedance and admittance be $\mathcal{Z}$ and $\mathcal{Y}$, respectively, and the line length l.

Solution
The line complex propagation constant and characteristic impedance are:

$$\gamma = \sqrt{\mathcal{Z}\mathcal{Y}}, \quad Z_0 = \sqrt{\frac{\mathcal{Z}}{\mathcal{Y}}},$$

while the input impedance of a line in open, for $l \to 0$, can be expanded as:

$$Z_{\text{in,open}} = Z_0 \coth(\gamma l) \approx Z_0 \left(\frac{1}{\gamma l} + \frac{1}{3}\gamma l\right)$$

$$= \sqrt{\frac{\mathcal{Z}}{\mathcal{Y}}} \frac{1}{\sqrt{\mathcal{Z}\mathcal{Y}}l} + \frac{1}{3}\sqrt{\frac{\mathcal{Z}}{\mathcal{Y}}}\sqrt{\mathcal{Z}\mathcal{Y}}l = \frac{1}{\mathcal{Y}l} + \frac{1}{3}\mathcal{Z}l,$$

i.e., the series of an impedance equal to $1/(\mathcal{Y}l)$ and of an impedance equal to $\mathcal{Z}l/3$. Notice that the first term is dominant unless the p.u.l. admittances and impedances are one purely real and the other purely imaginary. Consider, e.g., an RC line in open where:

$$\mathcal{Z} = \mathcal{R}, \quad \mathcal{Y} = j\omega\mathcal{C},$$

then:

$$Z_{\text{in,open}} \approx \frac{1}{j\omega\mathcal{C}l} + \frac{1}{3}\mathcal{R}l$$

is the series of a resistor and a capacitor. Similarly, for the admittance of a short line in short we have:

$$Y_{\text{in,short}} = Y_0 \coth(\gamma l) \approx Y_0 \left(\frac{1}{\gamma l} + \frac{1}{3}\gamma l\right)$$

$$= \sqrt{\frac{\mathcal{Y}}{\mathcal{Z}}} \frac{1}{\sqrt{\mathcal{Z}\mathcal{Y}}l} + \frac{1}{3}\sqrt{\frac{\mathcal{Y}}{\mathcal{Z}}}\sqrt{\mathcal{Z}\mathcal{Y}}l = \frac{1}{\mathcal{Z}l} + \frac{1}{3}\mathcal{Y}l, \quad (2.25)$$

i.e., the parallel of an admittance equal to $1/(\mathcal{Z}l)$ and of an admittance equal to $\mathcal{Y}l/3$.

The input impedance of a quarter-wave and half-wave line are particularly important. For a quarter-wave line ($l = \lambda_g/4$) we have:

$$Z_i = Z_0 \left[\frac{Z_L \sinh(\alpha l) + Z_0 \cosh(\alpha l)}{Z_0 \sinh(\alpha l) + Z_L \cosh(\alpha l)}\right],$$

while for a half-wavelength line ($l = \lambda_g/2$):

$$Z_i = Z_0 \left[\frac{Z_L \cosh(\alpha l) + Z_0 \sinh(\alpha l)}{Z_0 \cosh(\alpha l) + Z_L \sinh(\alpha l)} \right].$$

For a lossless quarter-wave line the input impedance is:

$$Z_i = Z_0 \left[\frac{Z_L \cdot 0 + Z_0 \cdot 1}{Z_0 \cdot 0 + Z_L \cdot 1} \right] = \frac{Z_0^2}{Z_L},$$

yielding the so-called quarter-wave impedance transformer; while for a lossless half-wave line the input impedance is equal to the load impedance:

$$Z_i = Z_0 \left[\frac{Z_L \cdot 1 + Z_0 \cdot 0}{Z_0 \cdot 1 + Z_L \cdot 0} \right] = Z_L.$$

In the above treatment we made use of the reflection coefficient simply as a notational shortcut. However, this is able to yield a simpler and possibly more significant picture of the line. As already remarked, the forward and backward wave amplitudes uniquely determine the voltages and currents on the whole line. We often prefer to identify the two amplitudes by assigning, e.g., the forward wave phasor V^+ in a certain line section z and the ratio $\Gamma(z) = V^-(z)/V^+(z)$, the reflection coefficient at section z. The variation of the reflection coefficient with position is immediately found; we have:

$$\Gamma(z) = \frac{V^-(z)}{V^+(z)} = \frac{V^-(0)\exp(j\beta z)}{V^+(0)\exp(-j\beta z)} = \Gamma(0)\exp(2j\beta z); \quad (2.26)$$

in other words $\Gamma(z)$ is periodic along the line with a periodicity of $\lambda_g/2$ (voltages and currents have a periodicity λ_g). The reflection coefficient is known everywhere if it is known in one section of the line. The evolution of $\Gamma(z)$ as a complex number is simple: the corresponding phasor rotates with constant magnitude in the complex plane with periodicity $\lambda_g/2$. In the presence of losses the magnitude changes as well, in fact:

$$\Gamma(z) = \frac{V^-(z)}{V^+(z)} = \frac{V^-(0)\exp(\alpha z + j\beta z)}{V^+(0)\exp(-\alpha z - j\beta z)} = \Gamma(0)\exp(2\alpha z + 2j\beta z).$$

The impedance or admittance seen from a section of the line can be immediately identified as follows:

$$Z(z) = \frac{V(z)}{I(z)} = \frac{V^+(z) + V^-(z)}{I^+(z) + I^-(z)} = Z_0 \frac{V^+(z) + V^-(z)}{V^+(z) - V^-(z)} = Z_0 \frac{1 + \Gamma(z)}{1 - \Gamma(z)} \quad (2.27)$$

$$Y(z) = \frac{I(z)}{Z(z)} = \frac{I^+(z) + I^-(z)}{V^+(z) + V^-(z)} = \frac{1}{Z_0} \frac{V^+(z) - V^-(z)}{V^+(z) + V^-(z)} = \frac{1}{Z_\infty} \frac{1 - \Gamma(z)}{1 + \Gamma(z)},$$

with inverse formulae:

$$\Gamma(z) = \frac{Z(z) - Z_0}{Z(z) + Z_0} = \frac{z(z) - 1}{z(z) + 1} \quad (2.28)$$

$$\Gamma(z) = \frac{Y_0 - Y(z)}{Y_0 + Y(z)} = \frac{1 - y(z)}{1 + y(z)}, \quad (2.29)$$

where $Y_0 = Z_0^{-1}$ and $z(z) = Z(z)/Z_0$ and $y(z) = Y(z)/Y_0$ are the normalized impedances (admittances).

Let us review the input impedance problem in terms of the reflection coefficient. A line with length l is closed on Z_L; we want to evaluate the input impedance in $z = 0$. Assume for simplicity that the line is lossless. We have:

$$\Gamma(l) = \frac{Z_L - Z_0}{Z_L + Z_0}, \qquad (2.30)$$

from which:

$$\Gamma(0) = \frac{Z_L - Z_0}{Z_L + Z_0} \exp(-2\mathrm{j}\beta l).$$

Thus:

$$Z(0) = Z_0 \frac{1 + \dfrac{Z_L - Z_0}{Z_L + Z_0} \exp(-2\mathrm{j}\beta l)}{1 - \dfrac{Z_L - Z_0}{Z_L + Z_0} \exp(-2\mathrm{j}\beta l)} = Z_0 \frac{Z_L + \mathrm{j} Z_0 \tan\beta l}{Z_0 + \mathrm{j} Z_L \tan\beta l},$$

as in (2.24).

The relation (2.28) between the normalized impedance $z = r + \mathrm{j}x$ and the reflection coefficient Γ, repeated here:

$$\Gamma = \frac{z - 1}{z + 1}$$

is an analytical mapping from the complex plane z to the complex plane Γ, with the following properties:

- the angle between two intersecting curves in z plane is preserved in the Γ plane, i.e., the mapping is *conformal*;
- circles or straight lines in plane z are transformed in circles or straight lines in plane Γ;
- constant resistance lines $r = $ const. transform into circles in Γ plane with center on the real axis;
- constant reactance lines $x = $ const. transform into circles in Γ plane going through the origin;
- the half plane $\mathrm{Re}(z) > 0$ is transformed into the circle $|\Gamma| \leq 1$, while purely reactive impedances are transformed into the unit circle $|\Gamma| = 1$.

Some important points of the Γ plane are the following ones. For $z = 1$, the normalized reference impedance, we have $\Gamma = 0$, the center of the reflection coefficient complex plane. Short and open circuits correspond to $\Gamma = \mp 1$, respectively. Reactive impedances yield:

$$\Gamma = \frac{\mathrm{j}x - 1}{\mathrm{j}x + 1} \rightarrow |\Gamma| = 1;$$

in particular, inductive impedances have reflection coefficients in the upper Γ plane, capacitive impedances in the lower Γ plane. The above remarks are summarized in Fig. 2.10. Notice that from its definition $\Gamma \equiv \Gamma_V$, the voltage reflection coefficient; the current reflection coefficient trivially is $\Gamma_I = -\Gamma_V$.

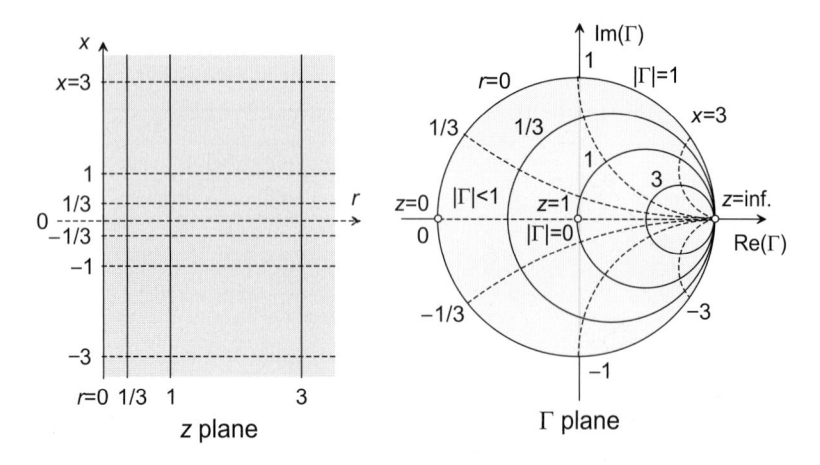

Figure 2.10 Properties of the $z \to \Gamma$ transformation.

A well-known graphical representation of the mathematical transformation between the impedance and the reflectance planes is provided by the so-called *Smith chart* [4], see Fig. 2.11. The Smith chart reproduces, in the Γ plane, a number of circles corresponding to impedances with constant real or imaginary parts, and can be used both to identify the reflection coefficient corresponding to a certain impedance, and to carry out graphical computations, exploiting the fact that along a (lossless) transmission line the reflection coefficient rotates with constant magnitude. The Smith chart as a design graphical aid has been somewhat made obsolete by the advent of CAD tools, but its use in the representation of parameters amenable to reflection coefficients (including the scattering parameters corresponding to reflectances) is widespread both in the instrumentation and in CAD tools.

Example 2.5 Design a line and short-circuit stub matching section to match $Z_L = 25 + j10$ Ω to a $Z_{in} = 50$ Ω impedance. The transmission line impedance is $Z_0 = 50$ Ω and the guided wavelength is $\lambda_g = 1$ cm.

Solution
The matching section adopted is shown in Fig. 2.12. The goal of the design is to evaluate L_1 (the line length) and L_2 (the stub length). The purpose of the line is to transform the load impedance into an input admittance Y'_{in} with real part equal to $G'_{in} = 1/50$ S by the line of length L_1. Then, we compensate the imaginary part B'_{in} by adding in parallel a stub with input susceptance $B_s = -B'_{in}$. The length L_2 is designed in such a way to obtain the correct B_s exploiting a stub with minimum length.

We start from identifying on the Smith chart the point corresponding to the normalized load impedance, see Fig. 2.13:

$$z_L = \frac{Z_L}{Z_0} = 0.5 + j0.2;$$

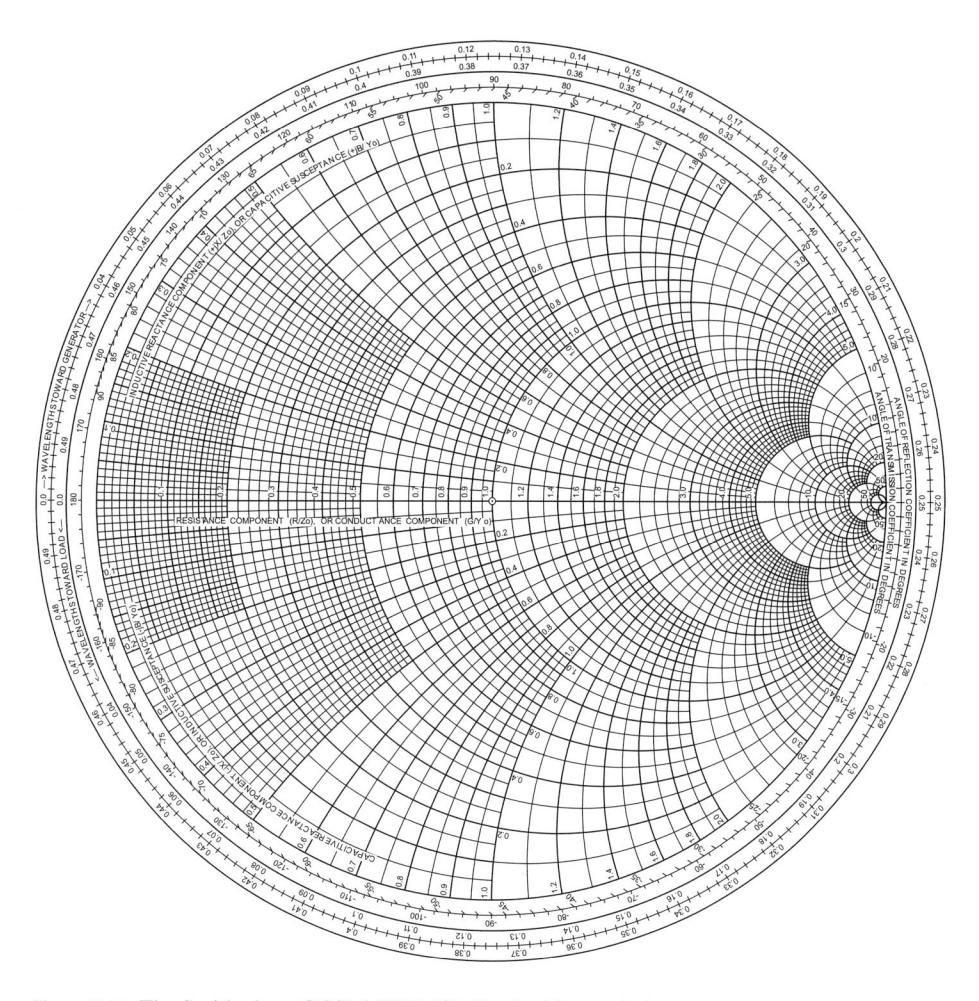

Figure 2.11 The Smith chart ©2015 IEEE [3]. Used with permission.

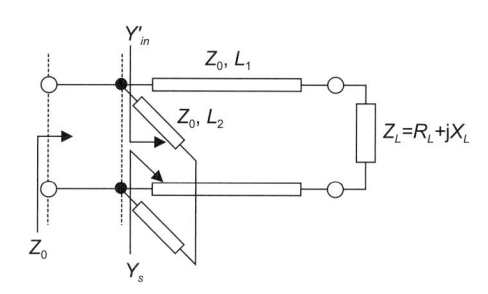

Figure 2.12 Line and stub matching section.

Passive Elements and Circuit Layout

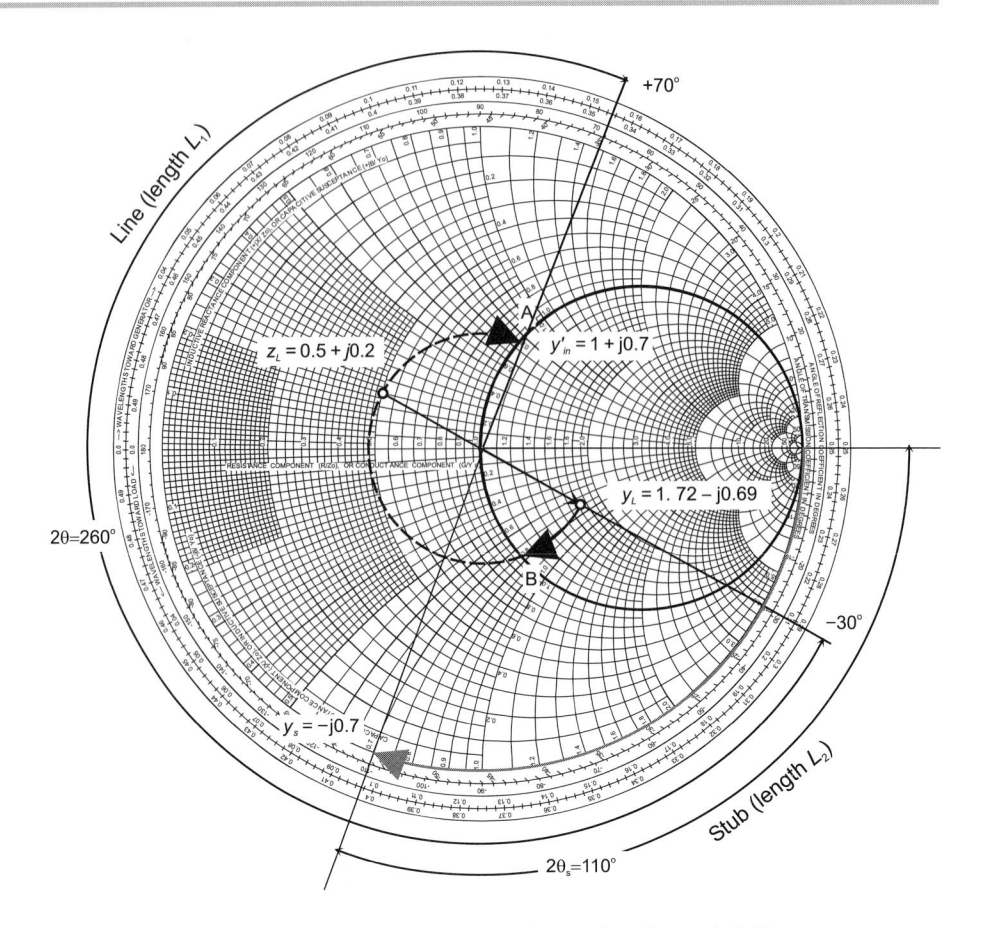

Figure 2.13 Smith chart design of line and stub matching section (Example 2.5).

however, since the stub is in parallel we have to work on admittance values and therefore we have to obtain from Γ_V the corresponding $\Gamma_I = -\Gamma_V$, see Fig. 2.13. Notice that we have:

$$\Gamma_V = \frac{z_L - 1}{z_L + 1} = \frac{0.5 + j0.2 - 1}{0.5 + j0.2 + 1} = -0.310 + j0.175$$
$$= 0.356 \angle 150.60°$$
$$y_L = \frac{1}{z_L} = \frac{1}{0.5 + j0.2} = 1.724 - j0.670$$
$$\Gamma_I = \frac{y_L - 1}{y_L + 1} = \frac{1.724 - j0.690 - 1}{1.724 - j0.690 + 1} = 0.310 - j0.175 = -\Gamma_V$$
$$= 0.356 \angle -29.40°.$$

On the Smith chart we rotate from Γ_I clockwise ("Towards generator") till we intersect the circle of unit normalized admittance $y'_{in} = 1 + jb'_{in}$, see Fig. 2.13, dashed circular arcs. Notice that intersection can occur in two points, point B corresponding to a negative b'_{in} (inductive), point A corresponding to a positive b'_{in} (capacitive). Since we want to exploit a shorted stub whose input impedance is inductive (if the stub is short), we prefer to use point A leading to a longer line but a shorter stub. The electrical length of the line $2\pi L_1/\lambda_g = \theta$ is a solution of the equation:

$$g'_{in} = \text{Re}\left[\frac{1 + \Gamma_I \exp(-j2\theta)}{1 - \Gamma_I \exp(-j2\theta)}\right] = \text{Re}\left[\frac{1 + (0.310 - j0.175) \exp(-j2\theta)}{1 - (0.310 - j0.175) \exp(-j2\theta)}\right] = 1.$$

Numerically we find the two solutions:

A: $\theta = 130.7° \rightarrow \dfrac{L_1}{\lambda_g} = \dfrac{\theta}{2\pi} = \dfrac{130.7}{360} = 0.36 \rightarrow L_1 = 0.36$ cm

B: $\theta = 19.9° \rightarrow \dfrac{L_1}{\lambda_g} = \dfrac{\theta}{2\pi} = \dfrac{19.9}{360} = 5.53 \times 10^{-2} \rightarrow L_1 = 5.53 \times 10^{-2}$ cm.

Graphically, see Fig. 2.13, we go for Solution A from $\approx -30°$ to $\approx 70°$ with a total phase shift $2\theta \approx 260°$ corresponding to $\theta \approx 130°$. The normalized admittance value obtained is:

$$y'_{in} = 1 + j0.76 \approx 1 + j0.7.$$

We have now to design a shorted stub with normalized admittance $y_s = -j0.7$. On the Smith chart the design can be carried out by starting from the short-circuit admittance corresponding to $\Gamma_I = 1$, see Fig. 2.13, gray arrow, and rotating on the unit circle to the point on the constant susceptance curve labeled -0.7; this corresponds to a stub length $2\theta_s = 110°$, i.e.:

$$\frac{L_2}{\lambda_g} = \frac{\theta_s}{2\pi} = \frac{1}{2}\frac{110}{360} = 0.152 \rightarrow L_1 = 0.152 \text{ cm}.$$

We can also analytically derive the stub electrical angle from the expression of the input impedance of a shorted line:

$$b_s = -\cot 2\theta_s = -0.7 \rightarrow \theta_s \approx 55°.$$

2.2 Planar Transmission Lines in Microwave Integrated Circuits

Fig. 2.14 shows some TEM, quasi-TEM and non-TEM microwave waveguides. Apart from the *slot line*, that may be exploited in antenna transitions, hybrid and monolithic microwave integrated circuits are based on quasi-TEM or TEM lines. Among the advantages of quasi-TEM transmission lines are their intrinsic wideband behavior, as opposed to the more dispersive nature of non-TEM media, and the fact that they introduce a well-defined ground plane also available for the active device biasing. A disadvantage of the quasi-TEM lines, which limits their use in the millimeter wave range, are their heavy ohmic losses due to the use of strip conductors. Closed TEM structures like the *stripline* are exploited in particular applications where radiation losses are important, or to obtain high-directivity directional couplers; they are not compatible however with MMICs.

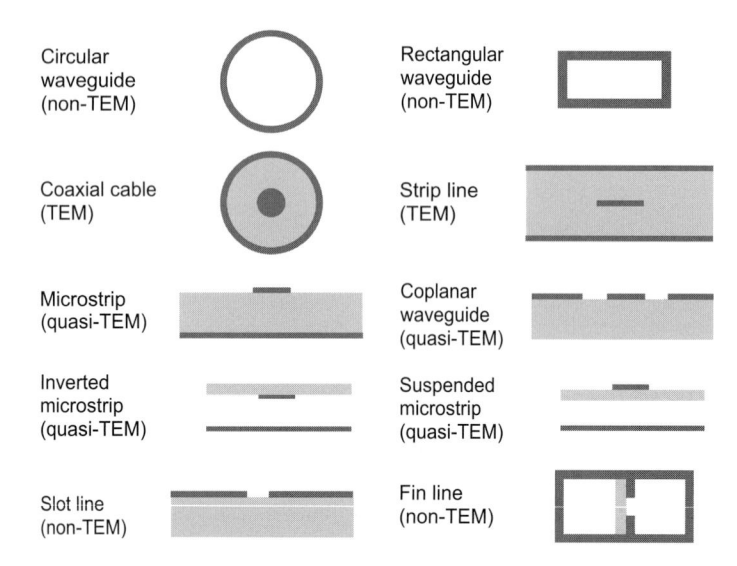

Figure 2.14 Waveguides and transmission lines in microwave circuits.

Other shielded lines like the *finline* are important at millimeter waves where low conductor losses are difficult to obtain. However, most of the integrated microwave circuits are based on the *microstrip* or the *coplanar waveguide* (CPW). CPWs have propagation characteristics almost independent from the substrate thickness and are popular at millimeter wave frequencies, although their layout is less compact than for the microstrip. *Suspended microstrips* are sometimes exploited in hybrid circuits at millimeter waves.

The *microstrip* approach is by far the most popular in MMICs: microstrip lines on compound semiconductor substrates of thickness ranging from 50 μm to 400 μm exhibit acceptable loss and dispersion up to the lower millimeter wave range. The realizable characteristic impedances are able to cover most applications; series element connection is very easy, whereas parallel connection is more troublesome, owing to the absence of an upper ground plane. Suitable techniques exist to circumvent this problem, such as the use of wrap-around ground planes or via holes making the bottom plane accessible from the top of the substrate; however, the first solution imposes constraints on circuit layout and introduces parasitics, while the second one increases the integrated circuit process complexity.

The main advantages offered by the *coplanar* approach can be listed as follows:

- the CPW allows easy series and shunt element connection of passive and active elements;
- the CPW impedance is almost insensitive to substrate thickness;
- on-wafer measurements through coplanar probes are easier and more direct than in microstrip circuits, where a local coplanar input or output can be implemented, but with the help of via holes.

Moreover, the coplanar approach allows for greater flexibility in the use of mixed structures and transitions to slot lines, coupled slot lines etc. which can be profitably exploited in some applications (e.g., mixers, balancing units). However, in coplanar circuits heat dissipation is less efficient than in microstrip circuits, due to the absence of a lower heat sink and to the need to keep the substrate thick.[3] Finally, the lateral ground planes have to be connected periodically by means of airbridges (at intervals less than $\lambda_g/4$) to suppress spurious slot propagation modes, thereby making the process more complex.

2.2.1 The Coaxial Cable

The coaxial cable has a particular role in microwave systems and in instrumentation, although it is not amenable to integration. It can be realized either in rigid or flexible form. It has, a comparatively low attenuation and high immunity to electromagnetic disturbances, being a shielded structure. Let us call a and b the inner and outer conductor radii; the p.u.l. capacitance can be expressed as [5]:

$$\mathcal{C} = \frac{55.556 \epsilon_r}{\log(b/a)} \text{ pF/m},$$

and the p.u.l. inductance:

$$\mathcal{L} = 200 \log \frac{b}{a} \text{ nH/m}.$$

The characteristic impedance is:

$$Z_0 = \frac{60}{\sqrt{\epsilon_r}} \log\left(\frac{b}{a}\right) \ \Omega, \tag{2.31}$$

while the effective permittivity is the dielectric permittivity. The dielectric and conductor attenuations are, respectively:[4]

$$\alpha_d = 27.3 \sqrt{\epsilon_r} \frac{\tan \delta}{\lambda_0} \text{ dB/m} \tag{2.32}$$

$$\alpha_c = \frac{9.5 \times 10^{-5} (a+b) \sqrt{\epsilon_r}}{ab \log(b/a)} \sqrt{f_{\text{GHz}}} \text{ dB/m}. \tag{2.33}$$

The useful frequency range of a coaxial cable is limited by the cutoff frequency of the first higher-order propagation mode, corresponding to the cutoff wavelength:

$$\lambda_c = \pi \sqrt{\epsilon_r} (a+b), \tag{2.34}$$

where the cutoff frequency is $f_c = c_0/\lambda_c$.

---

[3] Substrate thinning would introduce a parasitic lower ground plane, leading to a mixed coplanar–microstrip propagation.
[4] We assume a copper conductor, for a different one the attenuation scales according to the square root of resistivity; in many cases, however, the inner and outer conductors are different, e.g., an inner copper wire and an outer aluminum jacket.

Example 2.6 Consider a copper coaxial cable with a Teflon ($\epsilon_r = 2$) dielectric. Find the ratio b/a corresponding to the minimum conductor losses and evaluate the resulting impedance. Dimension the cable so that the maximum operating frequency is 50 GHz.

Solution

The attenuation is proportional to a function $f(b/a)$ that can be minimized with respect to the line shape ratio by looking for a zero of its first derivative:

$$\alpha_c \propto \frac{1+x}{\log x} = f(x), \quad x = \frac{b}{a} > 1$$

$$\frac{df(x)}{dx} = \frac{1}{\log x} - \frac{1+x}{(\log x)^2}\frac{1}{x} = \frac{x \log x - (1+x)}{x(\log x)^2} = 0.$$

The zero corresponds to $x = 3.5911$ and it can be easily shown by inspection that this is a minimum of f and therefore of the attenuation. The corresponding impedance is, from (2.31), $Z_0 = 54.24\ \Omega$. We then have from (2.34):

$$f_{\max} = \frac{c_0}{\lambda_c} = \frac{c_0}{\pi\sqrt{\epsilon_r}\,(a+b)} = \frac{c_0}{\pi a\sqrt{\epsilon_r}\,(1+b/a)} \rightarrow$$

$$a = \frac{c_0}{\pi f_{\max}\sqrt{\epsilon_r}\,(1+b/a)} = \frac{3 \times 10^8}{\pi \cdot 50 \times 10^9 \sqrt{2}\,(1+3.5911)} = 0.294 \text{ mm}$$

$$b = 3.5911 \cdot a = 3.5911 \cdot 0.294 = 1.06 \text{ mm}.$$

Thus the coaxial cable outer diameter is 2.12 mm while the inner diameter is ≈ 0.6 mm. The conductor attenuation at 50 GHz is from (2.33) $\alpha_c = 3.223$ dB/m.

2.2.2 The Microstrip

The microstrip (see Fig. 2.15 for the cross section) is a quasi-TEM transmission line, due to the inhomogeneous cross section. In the present section, analysis and design formulae are presented for the line parameters; the CAD tools for microwave circuit design have built-in microstrip line calculators or anyway seamlessly embed analytical or numerical microstrip line models.

Approximate expressions are given below for the microstrip parameters; the line geometry is defined in Fig. 2.15. Notice that quite a large number of approximation can be found in the literature; achieving an extremely high accuracy through analytical expression is not today an important goal, since accurate computer-design tools based on electromagnetic models typically are available to the designer.

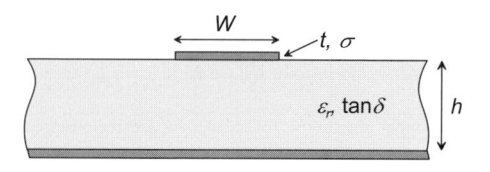

Figure 2.15 Microstrip cross section.

2.2 Planar Transmission Lines in Microwave Integrated Circuits

- Characteristic impedance [6, 7]:

$$Z_0 = \begin{cases} \dfrac{60}{\sqrt{\epsilon_{\text{eff}}}} \log\left[\dfrac{8h}{W'} + \dfrac{W'}{4h}\right], & \dfrac{W'}{h} \leq 1 \\ \dfrac{120\pi}{\sqrt{\epsilon_{\text{eff}}}} \left[\dfrac{W'}{h} + 1.393 + 0.667 \log\left(\dfrac{W'}{h} + 1.444\right)\right]^{-1} & \dfrac{W'}{h} > 1, \end{cases} \quad (2.35)$$

where the effective strip width W' accounts for the strip thickness t according to the correction [6]:

$$\dfrac{W'}{h} = \begin{cases} \dfrac{W}{h} + \dfrac{1.25t}{\pi h}\left[1 + \log\dfrac{4\pi W}{t}\right], & \dfrac{W}{h} \leq \dfrac{1}{2\pi} \\ \dfrac{W}{h} + \dfrac{1.25t}{\pi h}\left[1 + \log\dfrac{2h}{t}\right], & \dfrac{W}{h} > \dfrac{1}{2\pi}. \end{cases} \quad (2.36)$$

- Effective permittivity [8]:

$$\epsilon_{\text{eff}} = \dfrac{1 + \epsilon_r}{2} + \dfrac{\epsilon_r - 1}{2} F\left(\dfrac{W}{h}\right) - \dfrac{\epsilon_r - 1}{4.6} \dfrac{t}{h}\sqrt{\dfrac{h}{W}}, \quad (2.37)$$

where:

$$F\left(\dfrac{W}{h}\right) = \begin{cases} \left[1 + \dfrac{12h}{W}\right]^{-1/2} + 0.04\left[1 - \dfrac{W}{h}\right]^2, & \dfrac{W}{h} \leq 1 \\ \left[1 + \dfrac{12h}{W}\right]^{-1/2} & \dfrac{W}{h} > 1. \end{cases} \quad (2.38)$$

- Dielectric attenuation [9, 8]:

$$\alpha_d = 27.3 \dfrac{\epsilon_r}{\sqrt{\epsilon_{\text{eff}}}} \left(\dfrac{\epsilon_{\text{eff}} - 1}{\epsilon_r - 1}\right) \dfrac{\tan\delta}{\lambda_0} \text{ dB/m}. \quad (2.39)$$

- Conductor attenuation in dB/m [8]:

$$\alpha_c = \begin{cases} 1.38 \dfrac{R_s}{hZ_0} \dfrac{32 - (W')^2}{32 + (W')^2} \Lambda, & \dfrac{W'}{h} \leq 1 \\ 6.1 \times 10^{-5} \dfrac{R_s Z_0 \epsilon_{\text{eff}}}{h}\left(\dfrac{W'}{h} + \dfrac{0.667 W'/h}{W'/h + 1.444}\right) \Lambda, & \dfrac{W'}{h} \geq 1, \end{cases} \quad (2.40)$$

where:

$$\Lambda = \begin{cases} 1 + \dfrac{h}{W'}\left(1 + \dfrac{1.25t}{\pi W} + \dfrac{1.25}{\pi}\log\dfrac{4\pi W}{t}\right), & \dfrac{W'}{h} \leq \dfrac{1}{2\pi} \\ 1 + \dfrac{h}{W'}\left(1 - \dfrac{1.25t}{\pi h} + \dfrac{1.25}{\pi}\log\dfrac{2h}{t}\right), & \dfrac{W'}{h} \geq \dfrac{1}{2\pi}. \end{cases} \quad (2.41)$$

$R_s = \sqrt{\omega\mu/2\sigma} = (\sigma\delta)^{-1}$ is the surface resistance.

- Frequency dispersion of effective permittivity [10]:

$$\epsilon_{\text{eff}}(f) \approx \left(\dfrac{\sqrt{\epsilon_r} - \sqrt{\epsilon_{\text{eff}}}}{1 + 4F^{-1.5}} + \sqrt{\epsilon_{\text{eff}}}\right)^2, \quad (2.42)$$

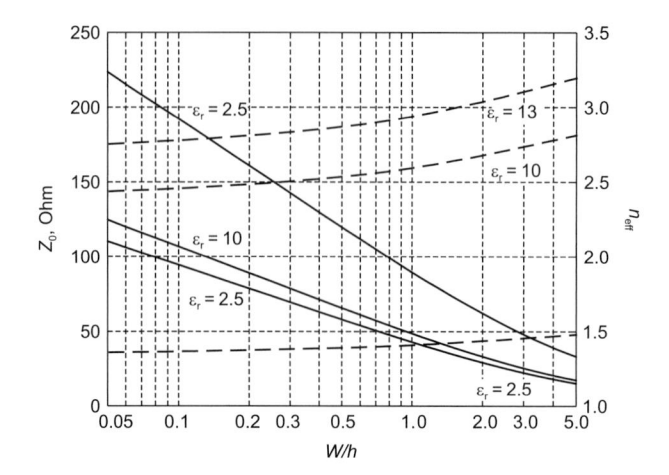

Figure 2.16 The microstrip characteristic impedance and effective refractive index vs. W/h for different substrate permittivities.

where:

$$F = \frac{4h\sqrt{\epsilon_r - 1}f}{c_0} \left\{ 0.5 + \left[1 + 2\log\left(1 + \frac{W}{h}\right) \right]^2 \right\} \equiv K_F f. \qquad (2.43)$$

The parameter F defines the inflection frequency f_{infl}, corresponding to $4F^{-1.5} = 4K_F^{-1.5} f^{-1.5} = 1$, i.e., $f_{\text{infl}} = 2^{4/3}/K_F$.

Fig. 2.16 show examples of the microstrip parameters (impedance and refractive index) as a function of the W/h ratio for different substrate permittivities (GaAs, $\epsilon_r = 13$; alumina, 10; Teflon, 2.5). The minimum W/h is suggested by technological constraints (strips cannot be narrower than $10 - 20$ μm) while the maximum is related to the onset of transversal resonances. Fig. 2.17 presents an example of the metal and substrate losses for a microstrip on alumina, substrate thickness 0.5 mm. At 1 GHz, conductor losses prevail, but, due to the different frequency behavior, dielectric losses can be important at millimeter waves. Notice that the conductors losses decrease with the strip width, i.e., are important for high impedance (narrow) lines.

Eq. (2.35) can be approximately inverted to yield the W/h ratio needed to obtain, with a substrate having given permittivity, a certain characteristic impedance Z_0. A simple set of approximate design formulae is [7]:

- For $Z_0 \geq 44 - 2\epsilon_r$ Ω:

$$\frac{W}{h} = \left(\frac{\exp(B)}{8} - \frac{1}{4\exp(B)} \right)^{-1} \qquad (2.44)$$

$$B = \frac{Z_0}{60}\sqrt{\frac{\epsilon_r + 1}{2}} + \frac{1}{2}\frac{\epsilon_r - 1}{\epsilon_r + 1}\left(0.4516 + \frac{0.2416}{\epsilon_r} \right). \qquad (2.45)$$

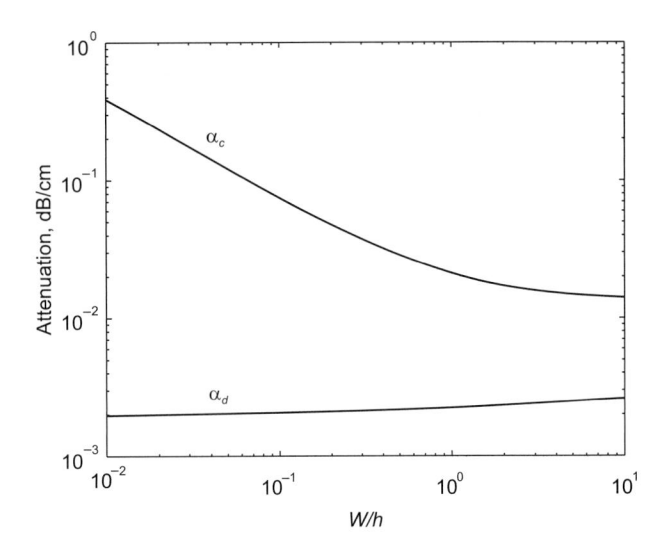

Figure 2.17 Behavior of the dielectric and conductor attenuation for a microstrip on alumina ($\epsilon_r = 10$) as a function of W/h. The substrate thickness is 500 μm and the loss angle is 10^{-3}. The strip thickness is $t = 5$ μm, with metal conductivity $\sigma = 4.1 \cdot 10^7$ S/m. The frequency is 1 GHz.

- For $Z_0 < 44 - 2\epsilon_r$ Ω:

$$\frac{W}{h} = \frac{2}{\pi}(d-1) - \frac{2}{\pi}\log(2d-1) + \frac{\epsilon_r - 1}{\pi \epsilon_r}\left[\log(d-1) + 0.293 - \frac{0.517}{\epsilon_r}\right] \quad (2.46)$$

$$d = \frac{60\pi^2}{Z_0 \sqrt{\epsilon_r}}. \quad (2.47)$$

Example 2.7 Design a 50 Ω microstrip using the substrates (a) CER-10-0250 and (b) TLY-5-0620, data in Table 2.2. Assume gold metallizations, conductivity $\sigma = 4.1 \times 10^7$ S/m and thickness $t = 7$ μm. Plot the frequency behavior of the effective permittivity and of the conductor and dielectric attenuation in the two cases.

Solution
With reference to the design formulae (2.45) and (2.46) we always have $Z_0 = 50 > 44 - 2\epsilon_r$; thus the expression to be used is (2.45). In case (a) $\epsilon_r = 9.5$ and we obtain $B = 2.1025$ from which, see (2.44), $W/h = 1.0073$, yielding $W = 1.0073 \times 0.63 = 0.63$ mm. In case (b) $\epsilon_r = 2.2$ we obtain $B = 1.16$ from which $W = 3.12h = 3.12 \times 1.57 = 4.9$ mm. The low-frequency effective permittivities result, respectively, (a) $\epsilon_{\text{eff}} = 6.41$, (b) $\epsilon_{\text{eff}} = 1.87$. Let us evaluate now the dispersion; in case (a) the k coefficient appearing in $F = kf$, see (2.43) is:

$$k = \frac{4h\sqrt{\epsilon_r - 1}}{c_0}\left(0.5 + \left[1 + 2\log\left(1 + \frac{W}{h}\right)\right]^2\right) = 1.517 \times 10^{-10},$$

Table 2.2 Characteristics of a few Taconic commercial substrates. In the table, h is the substrate thickness, t the thickness of the lower ground plane, ρ the metal resistivity, tan δ the substrate loss angle. CER and RF are ceramic-filled polytetrafluoroethylene (PTFE) laminates, TLα (α = C, T, Y) woven glass reinforced PTFE laminates. See [11] for further details and examples.

ϵ_r	h, mm	t, μm	ρ/ρ_{Au}	tan δ	Name
9.5	0.63	35	0.7118	.0035	CER-10-0250
10.0	1.57	35	0.7118	.0035	CER-10-0620
9.8	1.27	35	0.7118	.0035	CER-10-0500
2.20	1.57	35	0.7118	.0009	TLY-5-0620
2.20	0.78	35	0.7118	.0009	TLY-5-0310
2.20	0.51	35	0.7118	.0009	TLY-5-0200
2.33	1.57	35	0.7118	.0009	TLY-3-0620
2.33	0.51	35	0.7118	.0009	TLY-3-0200
2.55	1.52	35	0.7118	.0006	TLT-8-0600
2.55	0.76	35	0.7118	.0006	TLT-8-0300
3.00	1.57	35	0.7118	.0030	TLC-30-0620
3.00	0.78	35	0.7118	.0030	TLC-30-0310
3.00	0.51	35	0.7118	.0030	TLC-30-0200
3.20	1.57	35	0.7118	.0030	TLC-32-0620
3.20	0.78	35	0.7118	.0030	TLC-32-0310
3.50	1.52	35	0.7118	.0025	RF-35-0600
3.50	0.76	35	0.7118	.0025	RF-35-0300
3.50	0.51	35	0.7118	.0025	RF-35-0200
3.50	0.25	35	0.7118	.0025	RF-35-0100

while in case (b) $k = 3.4823 \times 10^{-10}$. The inflection frequencies $f_{\text{inf}} = 2^{4/3}/k$ are in case (a) $f_{\text{infl}} = 2^{4/3}/(1.517 \times 10^{-10}) = 16.6$ GHz, in case (b) $f_{\text{inf}} = 2^{4/3}/(3.4823 \times 10^{-10}) = 7.23$ GHz. The second substrate, being thicker, is more dispersive. In fact the frequency behavior given by (2.42) shown in Fig. 2.18, confirms that case (b) has a lower inflection frequency, but case (a) exhibits a larger absolute variation of the effective permittivity.

Concerning attenuation, we have (a) tan $\delta = 0.0035$ and (b) tan $\delta = 0.0009$; thus at 1 GHz the attenuations are (a) $\alpha_d = 0.0076$ dB/cm, $\alpha_c = 0.0166$ dB/cm; for case (b) $\alpha_d = 9.57 \times 10^{-4}$ dB/cm, $\alpha_c = 0.0027$ dB/cm. Line (b) has lower conductor losses because of the wider strip. The behavior of losses vs. frequency is reported in Fig. 2.19; again, conductor losses prevail at low frequency but dielectric losses become important at high frequency.

In conclusion, a 50 Ω microstrip line on a heavy substrate (alumina) with $\epsilon_r = 10$ requires $W = h$ while on GaAs the same impedance level can be achieved for $W/h \approx 0.7$. The microstrip ohmic losses decrease for increasing strip width, and are therefore large for high-impedance, narrow-strip lines. The impedances that can be realized on alumina substrates approximately range from 20 to 120 Ω.

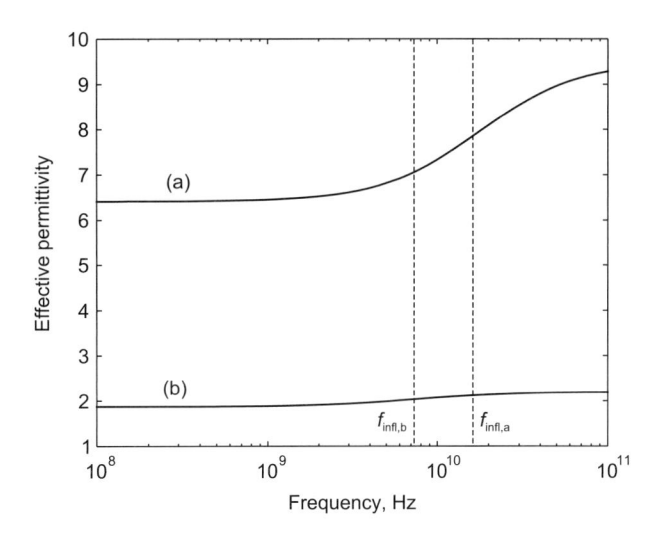

2.2 Planar Transmission Lines in Microwave Integrated Circuits

Figure 2.18 Frequency behavior of effective permittivity, microstrip lines in Example 2.7.

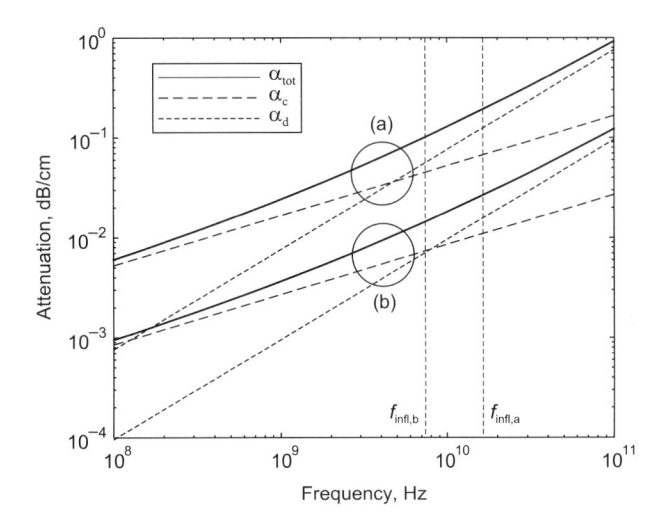

Figure 2.19 Frequency behavior of attenuation, microstrip lines in Example 2.7.

2.2.3 The Coplanar Waveguide

Ideal *coplanar waveguides* (Fig. 2.20) on an infinitely thick substrate are characterized by $\epsilon_{\text{eff}} = (\epsilon_r + 1)/2$, independent of geometry and line impedance; moreover, the line impedance only depends on the ratio between the slot width and the strip width, or, equivalently, on the ratio a/b, where $a = w/2$, $b = s + a$. Notice that $2b$ is the overall lateral extent of the line. The property $\epsilon_{\text{eff}} = (\epsilon_r + 1)/2$ is partly lost in practical lines on non-ideal (finite-thickness, Fig. 2.20) substrates; moreover, in this case the line impedance also becomes sensitive to the substrate thickness. As a rule of thumb, the substrate should be *at least* as thick as the overall lateral extent of the line, i.e., $h > 2b$, to

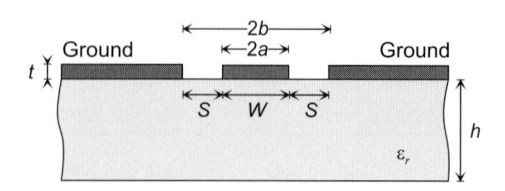

Figure 2.20 Coplanar waveguide (CPW).

make the influence of h on Z_0 almost negligible. For a given h, this imposes a limitation on the maximum line dimensions: e.g., for $h = 300$ μm one must have $b < 150$ μm. Although no limitation occurs to the impedance range, which only depends on the shape ratio a/b, thin substrates require small lines which in turn are affected by heavy ohmic losses. A further cause of non-ideal behavior is the finite extent of the lateral ground planes, leading to an increase of the line impedance. In practical circuits, the overall lateral line extent should be kept as small as possible, provided that no spurious coupling arises between neighboring lines and that the impedance level of the line is not seriously affected. A reasonable compromise is to have $c \approx 3b$ at least. A last variety of coplanar waveguide is the so-called conductor-backed CPW, where an additional lower ground plane is present. Conductor backing improves the mechanical strength and thermal dissipation of the substrate, but affects the line impedance, that becomes dependent on the substrate thickness h; to avoid this, one should have a suitably thick substrate, which in turn vanifies the advantages on thermal dissipation (that requires very low h, e.g., $h \approx 30 - 60$ μm). For additional details on the coplanar waveguide approach see [12, 13].

Quasi-static expressions for the line parameters (**analysis formulae**) have been derived through approximate conformal mapping techniques, and are accurate if the substrate thickness is not small with respect to the line width $2b$. Although the exact asymptotic limit is obtained for $h \to 0$, it is advisable to confine the use of these expressions to $h > b/2$. One has for the characteristic impedance (Z_0) and effective permittivity (ϵ_{eff}) [14]:

- coplanar waveguide (CPW) on infinitely thick substrate:

$$Z_0 = \frac{30\pi}{\sqrt{\epsilon_{\text{eff}}}} \frac{K(k')}{K(k)} \qquad (2.48)$$

$$\epsilon_{\text{eff}} = \frac{\epsilon_r + 1}{2}$$

$$k = a/b.$$

- coplanar waveguide with finite-thickness substrate:

$$Z_0 = \frac{30\pi}{\sqrt{\epsilon_{\text{eff}}}} \frac{K(k')}{K(k)}$$

$$\epsilon_{\text{eff}} = 1 + \frac{\epsilon_r - 1}{2} \frac{K(k')}{K(k)} \frac{K(k_1)}{K(k'_1)},$$

where:
$$k = a/b$$
$$k_1 = \frac{\sinh(\pi a/2h)}{\sinh(\pi b/2h)}.$$

In the above formulae, $K(k)$ is the complete elliptic integral of the first kind, of argument k, while $k' = \sqrt{1-k^2}$. The ratio $K(k)/K(k')$ can be accurately approximated as follows:

$$\frac{K(k)}{K(k')} \approx \frac{1}{\pi} \log\left(2\frac{1+\sqrt{k}}{1-\sqrt{k}}\right), \quad 0.5 \leq k^2 < 1$$

$$\frac{K(k')}{K(k)} \approx \frac{1}{\pi} \log\left(2\frac{1+\sqrt{k'}}{1-\sqrt{k'}}\right), \quad 0 < k^2 \leq 0.5.$$

Analysis formulae for other coplanar waveguides (conductor backed, with finite-extent ground planes) are reported in [15, 16].

Concerning the frequency dispersion of the effective permittivity, an analytical expression of the frequency-dependent behavior of the effective permittivity for a coplanar waveguide with finite substrate is [17]:

$$\sqrt{\epsilon_{\text{eff}}(f)} = \sqrt{\epsilon_{\text{eff}}(0)} + \frac{\sqrt{\epsilon_r} - \sqrt{\epsilon_{\text{eff}}(0)}}{1 + A\left(\frac{f}{f_{TE}}\right)^{-1.8}},$$

where:

$$A = \exp\left\{\left[0.54 - 0.64 \log\frac{2a}{h} + 0.015\left(\log\frac{2a}{h}\right)^2\right]\log\frac{2a}{b-a}\right.$$
$$\left. + 0.43 - 0.86 \log\frac{2a}{h} + 0.540\left(\log\frac{2a}{h}\right)^2\right\}$$

$$f_{TE} = \frac{c_0}{4h\sqrt{\epsilon_r - 1}}.$$

The conductor and dielectric attenuation can be expressed in dB per unit length as [18, 19]:

$$\alpha_c = \frac{8.68 R_s(f)\sqrt{\epsilon_{\text{eff}}}}{480\pi K(k)K(k')\left(1-k^2\right)}$$
$$\times \left\{\frac{1}{a}\left[\pi + \log\left(\frac{8\pi a(1-k)}{t(1+k)}\right)\right] + \frac{1}{b}\left[\pi + \log\left(\frac{8\pi b(1-k)}{t(1+k)}\right)\right]\right\} \quad (2.49)$$

$$\alpha_d = 27.83 \frac{\tan\delta}{\lambda_0} \frac{\epsilon_r}{2\sqrt{\epsilon_{\text{eff}}}} \frac{K(k_1)}{K(k_1')} \frac{K(k')}{K(k)}. \quad (2.50)$$

Fig. 2.21 shows the behavior of the characteristic impedance and attenuation of a CPW on a GaAs substrate as a function of the shape ratio $0 < a/b < 1$. Increasing a the capacitance increases; therefore, the impedance decreases. On the other hand, losses are maximum for $a \to 0$ (large strip resistance due to the narrow strip) and for $a \to b$ (the

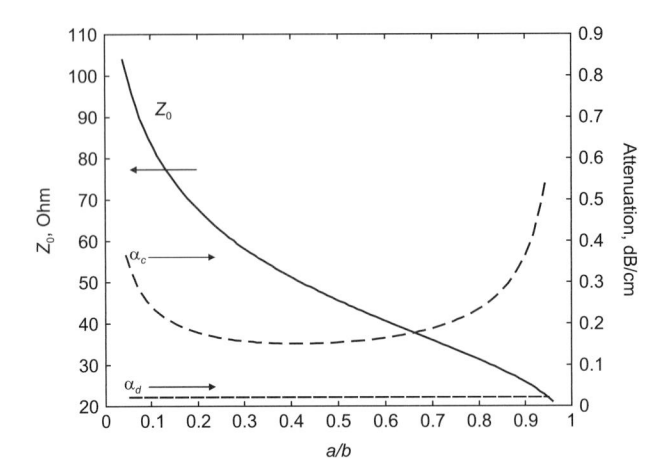

Figure 2.21 Characteristic impedance, conductor and dielectric attenuation of a CPW on GaAs vs. the aspect ratio a/b. We have $2b = 600$ μm and the substrate is infinitely thick. The metal strips are gold with thickness $t = 5$ μm; the dielectric loss angle is $\tan \delta = 0.001$; the frequency is 10 GHz.

impedance vanishes in the limit and therefore $\alpha_c \propto 1/Z_0$ diverges). Minimum conductor losses occur around $a/b \approx 0.4$, that also corresponds to a 50 Ω impedance on GaAs. Notice that dielectric losses at 10 GHz are almost negligible.

2.2.4 Coupling and Radiation Losses in Planar Lines

Parasitic coupling on uniform lines can occur either because the quasi-TEM field of the line couples with other quasi-TEM fields (line-to-line coupling) or because coupling occurs with surface waves or free-space radiation. Coupling between quasi-TEM modes and other guided or radiated waves is significant only in the presence of phase velocity synchronism; if this condition occurs, circuit operation is severely deteriorated owing to power conversion to spurious modes or radiation.

A different sort of coupling takes place in the presence of line discontinuities. In such cases, higher-order line modes are excited and the related power can easily be converted into surface waves or free-space radiation. Such effects modify the circuit behavior of line discontinuities and can be approximately modeled by means of concentrated radiation conductances, see Fig. 2.34. In what follows, some examples of spurious coupling mechanisms are described [20].

Both the microstrip and the coplanar waveguide show the possibility of spurious couplings with surface waves supported by the (grounded) dielectric substrate. As shown in Fig. 2.22, a possible parasitic waveguide can be associated with every guiding structure. The spurious coupling can assume two forms:

- synchronous coupling, when the spurious and the original mode travel with the same phase velocity;
- asynchronous coupling, that may arise only in the presence of discontinuities.

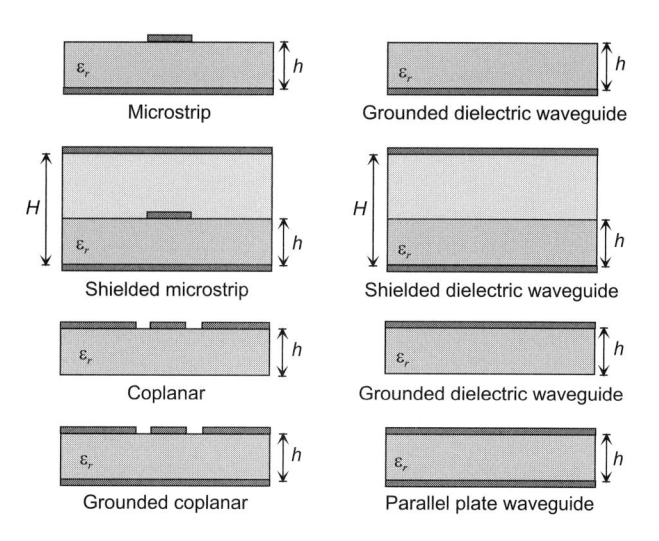

2.2 Planar Transmission Lines in Microwave Integrated Circuits

Figure 2.22 Quasi-TEM lines and associated surface wave dielectric waveguides.

A grounded dielectric slab carries TE or TM waves; in the TE case, the electric field is parallel to the ground plane, in the TM case orthogonal. This second field topology is more compatible with the microstrip field topology, while the CPW has both horizontal and vertical field components. There is an infinite set of surface waves TE_n and TM_n, $n = 0, 1, 2 \ldots$, with cutoff frequencies:

$$f_{cTM_n} = \frac{c_0 n}{2h\sqrt{\epsilon_r - 1}} = 2n \cdot f_{cTE_0}$$

$$f_{cTE_n} = \frac{c_0(2n+1)}{4h\sqrt{\epsilon_r - 1}} = (2n+1) \cdot f_{cTE_0},$$

i.e., the fundamental TM_0 mode has zero cutoff frequency. The same remark holds for the fundamental mode in a metallized dielectric layer shielded, at a distance H, by a second metal plane (Fig. 2.22). The dispersion relationship of the TM_0 mode for this structure can be shown to be:

$$\frac{1}{\epsilon_r}\sqrt{\frac{\epsilon_r - \epsilon_{eff}}{\epsilon_{eff} - 1}} = \frac{\tanh\left[2c_0 f (H - h)\sqrt{\epsilon_{eff} - 1}\right]}{\tan\left(2c_0 f h \sqrt{\epsilon_r - \epsilon_{eff}}\right)},$$

where c_0 is the velocity of light in vacuo. Finally, a dielectric slab metallized on both sides carries a TEM mode (requiring however a potential difference between the two planes) and TE and TM modes similar to the single-sided metallized slab.

Fig. 2.23 shows the dispersion curves for several waveguiding structures on a GaAs substrate ($\epsilon_r = 13$) with $h = 300$ μm. We report a 50 Ω microstrip effective permittivity, a microstrip permittivity in the limit of large impedance, and of two coplanar waveguides with 50 Ω and different transverse dimensions. For surface waves, we show the dispersion curve of the fundamental mode of a metallized dielectric slab with a cover $H = 1.5$ mm above the latter; we also show the TM_1 mode of a dielectric slab with lower and upper metal planes. Concerning coupling with microstrip or coplanar lines, the following remarks hold:

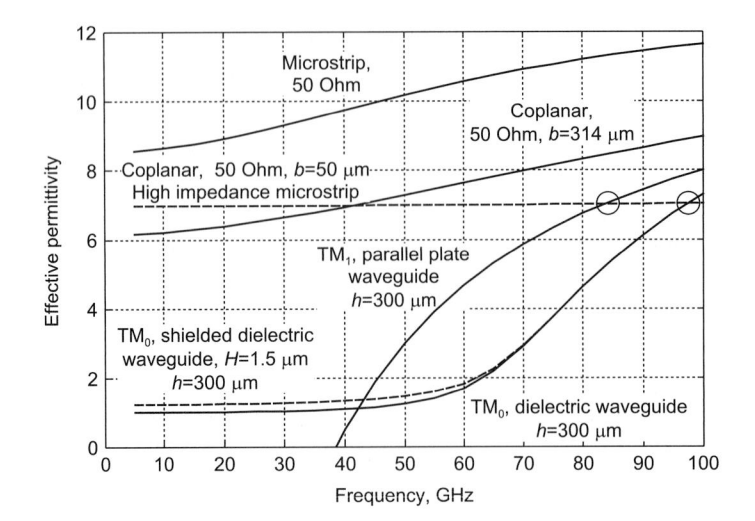

Figure 2.23 Dispersion curves for quasi-TEM modes and surface waves.

- Synchronous coupling with the TM$_0$ mode in a microstrip is possible only for very high impedance lines; on the other hand, synchronous coupling is possible in coplanar lines, with a synchronous frequency f_s given by:

$$f_s = \frac{c_0 \tan^{-1}(\epsilon_r)}{\pi h \sqrt{2(\epsilon_r - 1)}},$$

that, for heavy substrates like alumina, simplifies to:

$$f_s \approx \frac{106}{h_{[mm]} \sqrt{\epsilon_r - 1}} \text{ GHz}.$$

- Synchronous coupling with the TE$_0$ mode is possible but takes place at a higher frequency $f'_s > f_s$:

$$f'_s = \frac{3c_0}{4h \sqrt{2(\epsilon_r - 1)}}.$$

- Finally, coplanar lines with a lower ground plane have a synchronous coupling with the TM$_1$ mode of the dielectric waveguide with upper and lower ground planes; such a coupling occurs at the frequency:

$$f_{ss} = \frac{c_0}{h(\epsilon_r - 1)}.$$

Spurious surface waves can be controlled by inserting dissipative materials in critical points of the circuit, e.g., in the package.

2.3 Lumped Parameter Components

Lumped parameter components (resistors, capacitors and also, at RF and beyond, inductors) are exploited in planar circuits to implement several functions: bias and stabilization resistances, DC blocks (capacitors) or RF blocks (inductors), feedback resistances, lumped parameter couplers, power dividers and matching sections. Although lumped components can be monolithically integrated, components with large values often have to be inserted in hybrid form as discrete elements. Such elements are often called chip capacitors, resistors and inductors, since they are realized on a dielectric chip. Integrated resistors and capacitors have a compact layout, while integrated inductors are large and with poor quality factor.

In principle, lumped elements can be obtained from short transmission lines. In fact, the input impedance of a short line closed on Z_L can be written, taking into account that $\tanh x \approx x$ for small argument, in the form:

$$Z_{in} = Z_0 \frac{Z_L + Z_0 \gamma\, l}{Z_0 + Z_L \gamma\, l}.$$

Taking into account the expression of the characteristic impedance and complex propagation constant:

$$Z_0 = \sqrt{\frac{\mathcal{R} + j\omega\mathcal{L}}{\mathcal{G} + j\omega\mathcal{C}}}, \quad \gamma = \sqrt{(\mathcal{R} + j\omega\mathcal{L})(\mathcal{G} + j\omega\mathcal{C})},$$

we have:

$$Z_{in} = \frac{Z_L + (\mathcal{R} + j\omega\mathcal{L})l}{1 + Z_L(\mathcal{G} + j\omega\mathcal{C})l};$$

therefore, for a line closed on a short circuit ($Z_L = 0$):

$$Z_{in} = \mathcal{R}l + j\omega\mathcal{L}l = R + j\omega L,$$

which synthesizes, for a low-loss line, an inductor. For a short line in open we have:

$$Y_{in} = \mathcal{G}l + j\omega\mathcal{C}l = G + j\omega C$$

corresponding to a capacitor. The values obtained this way are, however, small.

An overview of the parameters of several classes of lumped elements can be found in [20, Table 4.4.2]. Spiral inductors cover the approximate range 0.5–50 nH with quality factors in the range 20–40; high-impedance line inductors are limited to 0.5 nH but may have slightly higher quality factors.

The quality factor Q of a reactive element is defined as the ratio between the resistive and reactive element component as follows:

$$Q = \frac{R}{\omega L} = \frac{\omega C}{G}.$$

Taking into account that RF inductors are in air (no magnetic losses) and that the high-frequency resistance is proportional to $\sqrt{f}$ we have that the inductor Q decreases at high frequency like $f^{-1/2}$. On the other hand, since for a capacitor the main loss mechanism are dielectric losses, for which $G = \omega C \tan \delta$, we have that for a capacitor $Q \approx 1/\delta$. MIM capacitors can reach values of 100 pF with quality factors around 50, while interdigitated capacitors, while allowing for a more precise control of the capacitance value, require much larger areas and cannot reach values above 0.5 pF due to the onset of distributed effects, see [20].

Very high Q resonators and filters cannot typically be obtained with lumped elements, but require external components such as quartz resonators and surface acoustic wave (SAW) components.

2.3.1 Inductors

Microwave inductors are in air, since all magnetic materials, including ferroelectric materials as the ferrites, are restricted to operation below 1 GHz. The operating frequency range of microwave integrated inductors is limited by ohmic losses and by the resonance with the element parasitic capacitance. In spiral inductors, the upper range of values is typically limited to 10-50 nH, since large inductors have a large parasitic capacitance and therefore low resonance frequency.

Small values of inductance, up to 2 nH approximately, can be obtained through high impedance lines (strip inductors, Fig. 2.24 (a)). For the strip inductor several models have been proposed [21, 22, 8], providing slightly different results; we use here the model in [8] with the ground correction in [23] for the inductance and the model in [8] for the resistance:

$$L_{\text{strip}} = 2 \cdot 10^{-1} l \left[\log \frac{l}{W+t} + 1.193 + \frac{W}{3l} \right] K_g \quad \text{nH} \tag{2.51}$$

$$K_g = \begin{cases} 0.57 - 0.145 \log (W/h), & W/h > 0.05 \\ 1, & W/h \leq 0.05 \end{cases} \tag{2.52}$$

$$R_{\text{strip}} = \left[1.4 + 0.217 \log \left(\frac{W}{5t} \right) \right] \frac{R_s l}{2(W+t)} \quad \Omega, \quad 5 < W/t < 100, \tag{2.53}$$

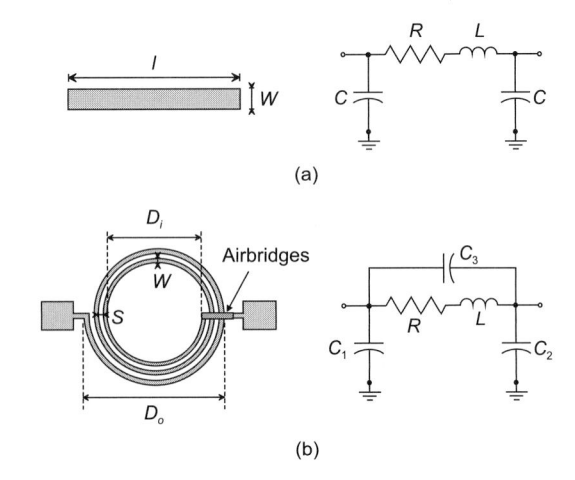

Figure 2.24 Strip inductor (a), spiral inductor (b) and related equivalent circuits.

2.3 Lumped Parameter Components

where lengths are in mm and R_s is the surface resistance of the metallization, t the metal thickness, h the distance of the strip with respect to the substrate. The factor K_g accounts for the reduction of the inductance due to the effect of a lower ground plane; (2.52) is derived in [8] from numerically fitted experimental data reported in [23].

Example 2.8 Evaluate the inductance of a strip inductor of length $l = 0.5$ mm, $W = 50$ μm and $t = 5$ μm, on a 300 μm GaAs substrate ($\epsilon_r = 13$). Assume a metal conductivity of $\sigma = 4 \cdot 10^7$ S/m, and $\tan \delta = 0.001$ for the GaAs substrate. Compare the input impedance of the strip inductor with the input impedance of a microstrip line having the same dimensions as a function of frequency, and find the resonant frequency of the inductor from the transmission line model. Evaluate from the lumped and distributed model the frequency behavior of inductor quality factor.

Solution
For the strip inductor we have $L_{\text{strip}} = L_0 K_g$, where from (2.51) and (2.52) we obtain:

$$L_0 = 2 \cdot 10^{-1} \times 0.5 \times \left[\log \frac{0.5}{5 \cdot 10^{-3} + 50 \cdot 10^{-3}} + 1.193 + 0.2235 \cdot \frac{50 \cdot 10^{-3}}{0.5} \right]$$
$$= 0.3422\,6 \text{ nH},$$

$K_g = 0.829\,81$, $L_{\text{strip}} = L_0 K_g = 0.284\,01$ nH. The skin-effect strip resistance can be evaluated from the surface resistance $R_s = 3.141\,6 \cdot 10^{-7} \sqrt{f}$ using (2.53):

$$R_{\text{strip}} = \left[1.4 + 0.217 \log \left(\frac{50 \cdot 10^{-3}}{5 \times 5 \cdot 10^{-3}} \right) \right] \frac{0.5 \times 3.141\,6 \cdot 10^{-7} \sqrt{f}}{2(50 \cdot 10^{-3} + 5 \cdot 10^{-3})}$$
$$= 2.214\,0 \times 10^{-6} \sqrt{f} \quad \Omega.$$

Notice that the DC resistance of the strip is:

$$R_{\text{strip,DC}} = \frac{l}{Wt\sigma} = \frac{0.5 \cdot 10^{-3}}{50 \cdot 10^{-6} \times 5 \cdot 10^{-6} \times 4 \cdot 10^7} = 5 \cdot 10^{-2} \quad \Omega,$$

i.e., the skin-effect resistance at $f = 0.51$ GHz. The input impedance finally is $Z_{\text{strip}} = R_{\text{strip}} + j\omega L_{\text{strip}}$. We can model the same inductor with a microstrip model, using the high-frequency expression of the characteristic parameters. Since $h/W = 300/50 = 6 < 2\pi$, we exploit the width correction (see (2.36), above) obtaining $W'/h = 0.20505$ and therefore $W' = 0.20505h = 61.515$ μm; then we use the narrow strip formula (2.37) for the effective permittivity with $F = 0.144$ (2.38) and $\epsilon_{\text{eff}} = 7.76$. The impedance Z_0 and impedance in air Z_{a0} are, from (2.35), $Z_0 = 78.9$ Ω and $Z_{a0} = Z_0 \sqrt{\epsilon_{\text{eff}}} = 220$ Ω, respectively. This yields the total inductance $lZ_{a0}/c_0 = 0.36$ nH, in fair agreement with the value obtained through the concentrated model. The propagation constant is $\beta = 2\pi f \sqrt{\epsilon_{\text{eff}}} c_0 = 5.83 \times 10^{-8} f$ m^{-1} with the frequency in Hz. We have from (2.39) the dielectric attenuation

$$\alpha_d = \frac{27.3}{8.6859} \frac{13}{\sqrt{7.76}} \left(\frac{7.76 - 1}{13 - 1} \right) \frac{0.001}{3 \cdot 10^8} f = 2.7543 \times 10^{-11} f \text{ Np/m},$$

while for the conductor attenuation we use (2.34) yielding $\Lambda = 15.134$ and from (2.33):

$$\begin{aligned}\alpha_c &= \frac{1.38}{8.6859} \frac{R_s}{hZ_0} \frac{32 - (W'/h)^2}{32 + (W'/h)^2} \Lambda \\ &= \frac{1.38}{8.6859} \frac{3.1416 \cdot 10^{-7} \sqrt{f}}{300 \cdot 10^{-6} \cdot 78.9} \frac{32 - 0.20505^2}{32 + 0.20505^2} \times 15.134 \\ &= 3.1829 \times 10^{-5} \sqrt{f} \text{ Np/m}.\end{aligned}$$

The input impedance of the shorted line therefore is:

$$Z_{\text{line}} = Z_0 \tanh\left[(\alpha_c + \alpha_d + j\beta)\, l\right].$$

The magnitudes of the reactance and resistance evaluated from the lumped and the microstrip model are shown in Fig. 2.25; the microstrip model yields a resonance around 55 GHz. The limit $l < \lambda_g/10$ would confine the frequency range of the inductor to frequencies below 22 GHz. Concerning the quality factor, we have:

$$Q_L = \frac{\omega L_{\text{strip}}}{R_{\text{strip}}} = \frac{2\pi f \times 0.28401 \cdot 10^{-9}}{2.2140 \times 10^{-6}\sqrt{f}} = 8.06 \times 10^{-4}\sqrt{f},$$

while from the microstrip model $Q_L = \text{Im}\left[Z_{\text{line}}\right] / \text{Re}\left[Z_{\text{line}}\right]$. The behavior of the quality factor evaluated from the lumped and the microstrip model is shown in Fig. 2.26; while according to the lumped model the quality factor increases monotonically, in the microstrip model this has a maximum and then drops owing to the element resonance.

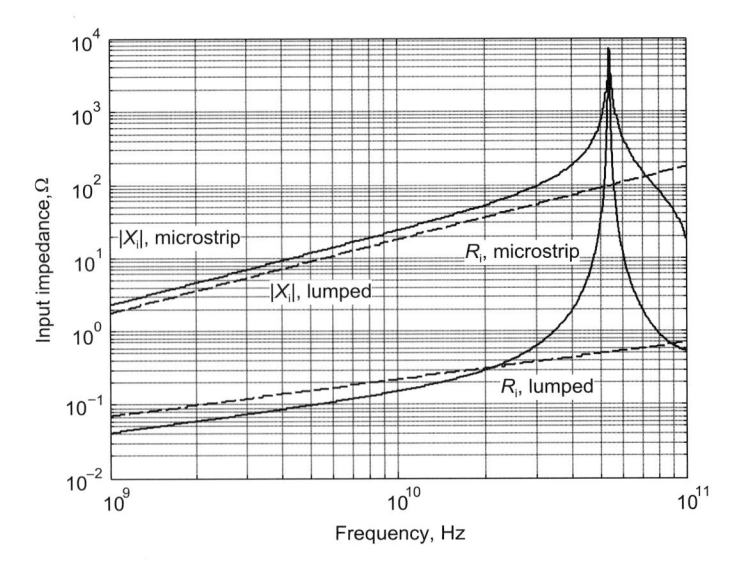

Figure 2.25 Behavior vs. frequency of the reactance and resistance of a strip inductor according to the lumped and microstrip model (Example 2.8).

2.3 Lumped Parameter Components

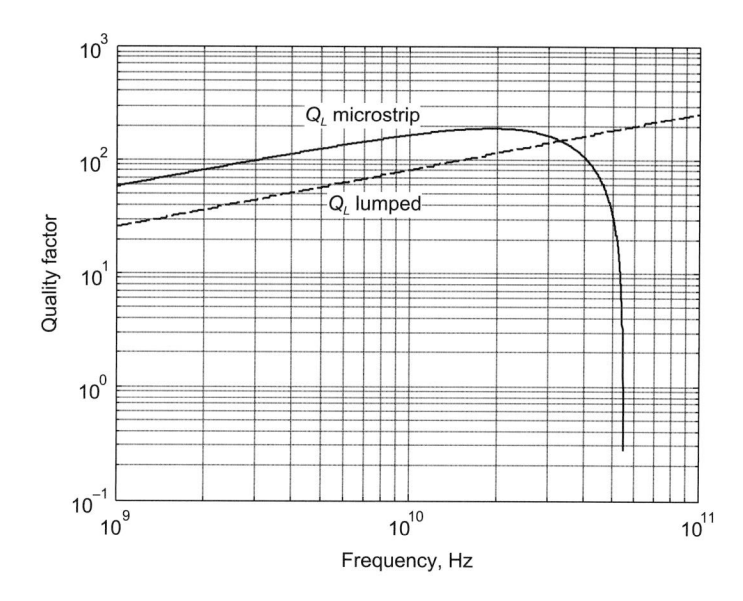

Figure 2.26 Behavior vs. frequency of the quality factor of a strip inductor according to the lumped and microstrip model (Example 2.8).

Larger inductances are typically obtained through spiral inductors; see Fig. 2.24 (b), where the spiral shape can be square, circular, octagonal. Integrated spiral inductors require airbridges to bring the signal from the center of the spiral to the output while minimizing the overpass parasitic capacitance. Several approximate formulae exist for the parameters of spiral inductors; see the discussion in [24]. For a circular inductor of inner radius D_i and outer radius D_o the inductance [24], parasitic resistance and parasitic capacitance [8] are given by:

$$L_{sp} = 0.62832 \times n^2 \overline{D} \left[\log \frac{2.46}{\rho} + 0.2\rho^2 \right] K_g \text{ nH},$$

$$\overline{D} = \frac{D_o + D_i}{2}, \quad \rho = \frac{D_o - D_i}{D_o + D_i}$$

$$R_{sp} = \left[1 + 0.333 \left(1 + \frac{S}{W} \right) \right]^{-1.7} \frac{\pi \overline{D} n R_s}{2W}$$

$$C_3 \approx 3.5 \cdot 10^{-2} D_o + 0.06 \text{ pF},$$

where all dimensions are in mm. The parameters are defined in Fig. 2.24 (b); n is the number of turns, related to the strip and slot widths as:

$$\frac{D_o - D_i}{2} \approx nW + (n-1)S \rightarrow n \approx \frac{1}{W+S} \left(\frac{D_o - D_i}{2} + S \right).$$

The factor K_g from (2.52) approximately takes into account the effect of ground plane coupling. Another popular approximation for the inductance of a circular spiral inductor is the so-called Wheeler formula [25, 8].

Example 2.9 We want to implement a family of spiral inductors with external diameter $D_o = 1$ mm, $W = 50$ μm, $S = W$, varying the number of turns n, with $t = 5$ μm. Evaluate the inductance that can be obtained on a 300 μm substrate varying the number of turns, with a constant external diameter D_o, the quality factor, and the resonant frequency. Metal conductors are made of gold ($\sigma = 4 \cdot 10^7$ S/m).

Solution
We have, with constant D_o and $W = S$:

$$\frac{D_o - D_i}{2} \approx nW + (n-1)S = (2n-1)W,$$

i.e., for the internal diameter:

$$D_i = D_o - 2(2n-1)W.$$

The maximum turn number corresponds to $D_i \approx 0$, i.e.:

$$n = \frac{1}{2} + \frac{D_o}{4W} = 5.5 \approx 5.$$

We then have:

$$\overline{D} = \frac{D_o + D_i}{2} = D_o - (2n-1)W = 1.05 - 0.1n \text{ mm}$$

$$\rho = \frac{D_o - D_i}{D_o + D_i} = \frac{(2n-1)W}{D_o - (2n-1)W} = \frac{2n-1}{21-2n}.$$

The substrate correction factor is:

$$K_g = 0.57 - 0.145 \log\left(\frac{W}{h}\right) = 0.57 - 0.145 \log\left(\frac{50}{300}\right) = 0.83,$$

while the gold surface resistance is:

$$R_s = \sqrt{\frac{2\pi f \mu}{2\sigma}} = 0.093\sqrt{f_{\text{GHz}}}.$$

We therefore obtain for the inductor parameters, with $W = 0.05$ mm:

$$L_{\text{sp}} = 0.52151 n^2 (1.05 - 0.1n) \left[\log\frac{21-2n}{2n-1} + 0.9 + 0.2\left(\frac{2n-1}{21-2n}\right)^2\right] \text{ nH}$$

$$R_{\text{sp}} = 1.2268 n (1.05 - 0.1n) \sqrt{f_{\text{GHz}}}$$

$$C_3 = 0.095 \text{ pF}.$$

Neglecting dielectric losses the quality factor $Q_L = 2\pi f L_{\text{sp}}/R_{\text{sp}}$ increases like the square root of frequency, while (neglecting the capacitance towards ground) the inductor resonant frequency can be approximated as $f_0 = 1/(2\pi\sqrt{LC_3})$. The inductance, quality factor at $f_0/2$ and resonance frequency for $n = 1\ldots 5$ are reported in Table 2.3. Notice that the maximum inductance is of the order of 10 nH and that the resonance frequency decreases with increasing number of turns. The quality factor evaluated at half the resonance frequency can be assumed to be an approximation of the maximum quality factor; the parameter is almost independent of the number of turns. The expression for C_3 is somewhat approximate, since it does not account for the number of turns.

Table 2.3 Inductance, resistance, resonance frequency and quality factor of spiral inductors varying the number of turns from Example 2.9.

n	1	2	3	4	5
L_{sp} (nH)	1.90	4.68	7.11	8.55	8.85
R_{sp} (Ω)	2.83	4.05	4.82	5.32	5.59
f_0 (GHz)	11.83	7.55	6.12	5.58	5.49
Q_L @ $f_0/2$	24.99	27.40	28.33	28.15	27.31

Bonding wires can be modeled as a series parasitic inductance, sometimes to be used as a component, e.g., as an RF choke in bias circuits. Instead of circular wires, ribbons can be exploited with the advantage of a lower inductance. The resistance and inductance of a wire of radius r and length l (in mm) are [26, 27, 22]:

$$L_{\text{wire}} = 0.2l \left[\log\left(\frac{2l}{r}\right) - x \right] \text{ nH}$$

$$x = \begin{cases} 1, & 4r^2 f \gg 1 \text{ m}^2\text{Hz} \\ 3/4, & 4r^2 f \ll 1 \text{ m}^2\text{Hz} \end{cases}$$

$$R_{\text{wire}} = \frac{R_s}{2\pi r} l.$$

Typical bonding wires with a length of 1 mm and a diameter of 25 μm at a frequency above 10 GHz have $4r^2 f > 4 \times (12.5 \cdot 10^{-6})^2 \times 10^{10} = 6.25 > 1$ so that at high frequency the value $x = 1$ should be used in the above formulae, while at RF and in the lower microwave range the value $x = 3/4$ would be more appropriate. According to the two expressions, the inductance with $2r = 25$ μm and $l = 1$ mm is 726 pH/mm (low frequency) and 676 pH/mm (high frequency), a not at all negligible value.

An expression for the wire inductance taking into account a ground plane correction is [8]:

$$L_{\text{filo}} = 0.2l \left[\log\left(\frac{4h}{d}\right) + \log\left(l + \sqrt{l^2 + d^2/4}\right) - \log\left(l + \sqrt{l^2 + 4h}\right) \right.$$
$$\left. + \sqrt{1 + 4h^2/l^2} - \sqrt{1 + d^2/4l^2} - 2\frac{h}{l} + \frac{d}{2l} \right] \text{ nH,}$$

where h is the distance between the wire and the ground plane.

2.3.2 Capacitors

Capacitors can be realized through passive structures or through junctions (Schottky or pn). Passive capacitors can be microstrip patches, interdigitated capacitors, MIM capacitors. Microstrip patches or gaps have a low capacitance, while interdigitated capacitors have a capacitance of the order of 0.5 pF/mm^2. Larger values can be obtained (up to around 30 pF) with MIM (Metal Insulator Metal) capacitors, see Fig. 2.27 (b), for which

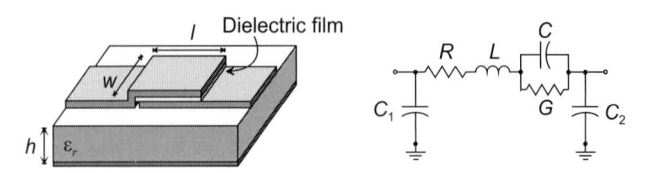

Figure 2.27 MIM capacitor.

the parallel plate formula holds:

$$C = \epsilon \frac{Wl}{d},$$

where Wl is the capacitor area, ϵ the absolute dielectric constant of the dielectric, d the dielectric thickness. For example, a MIM capacitor with a 1 μm silicon oxide layer (relative permittivity 4) has a specific capacitance around 35 pF/mm². The parallel conductance is related to dielectric losses.

Two common figure of merits for dielectrics exploited in realizing capacitors are the product between the capacitance and the breakdown voltage:

$$F_{cv} = C_a V_b = \epsilon_0 \epsilon_r E_b \quad \text{F} \cdot \text{V/m}^2,$$

typically in the range $(8 - 30) \times 10^3$ pF-V/mm², and the product specific capacitance – capacitor Q:

$$F_{cq} = C_a / \tan \delta_d \quad \text{F/m}^2,$$

where C_a is the capacitance per unit surface, V_b the breakdown voltage, E_b the breakdown electric field and δ_d the loss angle. Typical values are $E_b = 1 - 2$ MV/cm, $\epsilon_r = 4 - 20$ for the dielectric constant, $\tan \delta_d = 10^{-1} - 10^{-3}$. A summary of the properties of some relevant dielectrics is reported in [28, Table 4.1]. The tolerance of MIM capacitors is limited by the lack of accurate control of the dielectric thickness, that is of the order of 100 nm for materials such as silicon oxide, silicon nitride or aluminum oxide (maximum specific capacitance of the order of 500 pF/mm²; Chemical Vapor Deposition (CVD) is the process technique), while it can be as large as 1 μm for organic polymide dielectrics (with maximum specific capacitance of the order of 50 pF/mm², deposed by spinning).

2.3.3 Resistors

Planar integrated resistors can be obtained either deposing a thin film on a dielectric substrate (thin film resistors) or doping a semi-insulating substrate (implanted resistors), see Fig. 2.28; a summary of the relevant materials can be found in [28, Table 4.2]. The conductive material can be a high-resistivity metal or compound (Cr, Ta, Ti, TaN, NiCr) with sheet (surface) resistances as low as 5–15 $\Omega/\square$ (Cr) or as high as 60–220 $\Omega/\square$ (TaN). GaAs may exhibit a sheet resistance of the order of 100–1500 $\Omega/\square$. The deposition process is evaporation or sputtering, apart from GaAs that can be grown epitaxially.

2.3 Lumped Parameter Components

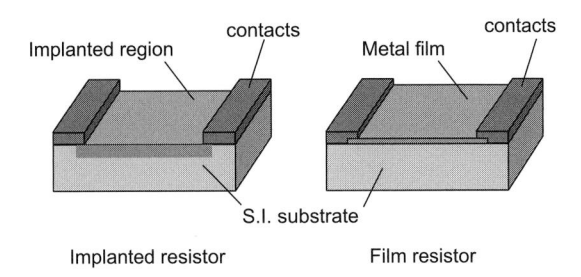

Figure 2.28 Integrated microwave resistors. *S.I. substrate* stands for Semi-Insulating (semiconductor) substrate.

The input impedance is approximated through a short (shorted) *RC* line model, see (2.25), with $\mathcal{Z}l = R$ and $\mathcal{Y}l = j\omega C$; we have:

$$Z_{in} = \left(\frac{1}{\mathcal{Z}l} + \frac{1}{3}\mathcal{Y}l\right)^{-1} = \frac{\mathcal{Z}l}{1 + \frac{1}{3}\mathcal{Y}l\mathcal{Z}l} = \frac{R}{1 + \frac{1}{3}j\omega RC},$$

where $R = \mathcal{R}l$, $C = \mathcal{C}l$ is the parasitic capacitance. The series resistance can be evaluated in the DC or skin effect range according to the operating frequency, resistor thickness and material.

2.3.4 Chip Inductors, Resistors and Capacitors

Discrete lumped components can be externally inserted in *hybrid integrated circuits* (i.e., circuits where the substrate is dielectric, and the active semiconductor devices are not monolithically integrated), usually as surface-mount chip resistors, inductors or capacitors.

Chip resistors are obtained by deposing a resistive thin film over a dielectric (e.g., ceramic) chip. Wrap-around or flip-chip contacts are then added, allowing for surface mounting (SM) on a microstrip or coplanar circuit. An example of such structures (shown bottom up) can be found in Fig. 2.29 (b) and (c); the side size of the component often is well below 1 mm. The resistance of chip resistors typically ranges from a few Ω to 10 kΩ. Chip capacitors can be obtained by depositing a dielectric layer (e.g., SiO_2) on a conductor or semiconductor (e.g., Si); the dielectric layer is then coated with metal so as to define the external contacts, which can be surface mounted through flip-chip (i.e., by connecting the component upside down), see Fig. 2.28. The capacitance of chip capacitors typically ranges from a few pF to 1 μF.

While chip resistors can be properly manufactured so as to achieve spectacular bandwidths (e.g., from DC to millimeter waves), thus making it possible to provide ultrabroadband matched terminations, broadband inductors are difficult to obtain, due to the increase of losses with frequency and to the upper limitation related to the *LC* resonant frequency. The quality factor of RF and microwave inductors typically peaks in a very narrow band, with maximum values well below 10^2. An example of microwave chip inductor is shown in Fig. 2.29 (c); achievable inductance values typically decrease

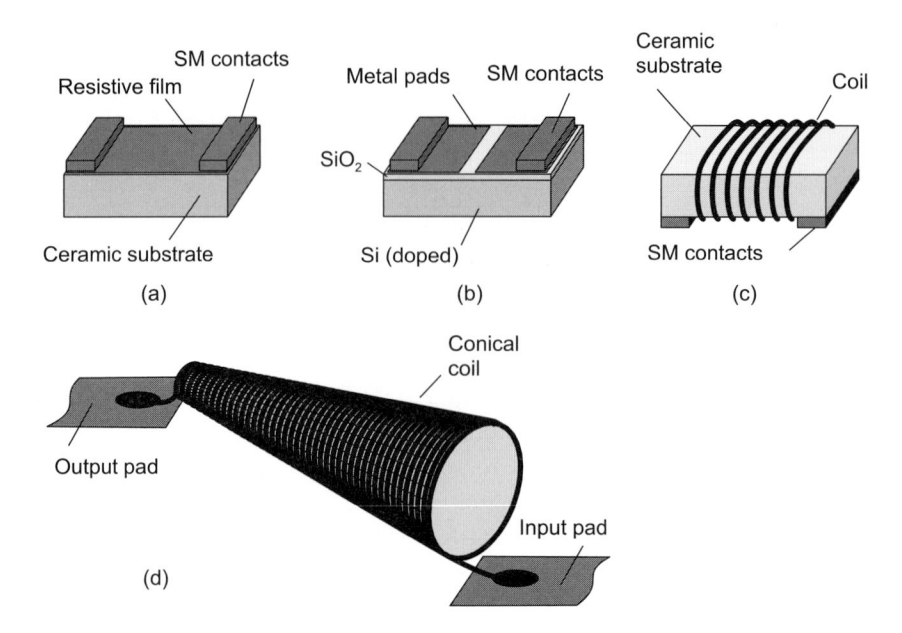

Figure 2.29 Examples of discrete RF lumped components: (a) thin film chip resistor for surface mount; (b) chip capacitor; (c) chip inductor; (d) ultrabroadband conical inductor.

with increasing operating frequency and are limited to 500 nH approximately, with maximum operating frequencies below 10 GHz. However, ultrabroadband bias Ts,[5] typically for instrumentation, require broadband inductors as RF blocks. Conical inductors, see Fig. 2.29 (d), adopt a particular technology allowing for very broadband behavior, due to a strong reduction of the parasitic capacitance and to the scaling invariance of the design [29].

2.4 Layout of Planar Hybrid and Integrated Circuits

High-speed electronic integrated circuits (ICs) can be implemented through two complementary approaches, the hybrid IC and the monolithic IC. Integrated circuits operating in the microwave range are denoted as (Monolithic) Microwave Integrated Circuits, (M)MICs. In the hybrid approach, the circuit is realized on a dielectric substrate, integrating all distributed components and, possibly, some lumped components, which may, however, also be inserted as discrete lumped elements through wire bonding or surface mount techniques. In the hybrid approach, the semiconductor active elements are inserted as lumped components and connected again through wirebonding or surface mount. On the other hand, monolithic circuits integrate, on a semiconductor substrate, all active and passive elements. While hybrid circuits often exploit, at least for narrowband applications, distributed components based on transmission lines, in

---

[5] A bias T, also called *bias Tee*, is a circuit block used to separate the RF circuit from the DC bias; see Sec. 2.4 and Sec. 6.10.2.

monolithic circuits the lumped approach is preferred, owing to the possibility of reducing the circuit size: lumped components are much smaller than the guided wavelength, while distributed elements have, as already recalled, characteristic sizes of the order of $\lambda_g/4$ at centerband. Monolithic integrated circuits can be based on GaAs, InP or Si substrates and may exploit, as active elements, FETs or bipolar transistors, typically Heterojunction Bipolar Transistors, HBTs (see Chapter 5).

According to the transmission medium used, we may have *microstrip* or *coplanar* integrated circuits. Microstrip circuits are more compact in size due to the lower ground plane, but require a precise control of the dielectric thickness, while coplanar circuits are often preferred at very high frequency (mm waves) and do not require control on the substrate thickness. Also in microstrip circuits, the on-wafer high-frequency characterization requires to connect the integrated circuit to the measurement setup through *coplanar probes*; to this purpose, coplanar ground planes must be made available (e.g., through via holes) at the circuit input and output, see Fig. 2.36. The probes are coaxial cables with a coplanar tip transferring by pressure contact the ground planes (lateral) on the lateral ground plane coplanar pads of the circuit connected to the backplane through via holes, while the center conductor is the signal conductor.

A qualitative example of a hybrid or monolithic integrated circuit in the microstrip or coplanar technology can be introduced as a simple, single-stage open-loop amplifier with two lumped bias Ts and input and output matching section. Fig. 2.30 shows a simplified schematic of the single-stage amplifier, with an input matching section, two bias circuits connected to the active element, and an output matching section. The purpose of the matching sections is to transform the load impedance (typically 50 Ω) in the optimum impedance that must be seen at the amplifier input and output port according to a maximum gain, maximum power or minimum noise criterion. The bias circuits, also called bias Ts, are a combination of a DC block (the capacitor) and an RF block (the inductor) whose aim is to separate the paths of the RF and DC currents in such a way that the RF circuit is not loaded by the DC supply and the RF load is isolated from the DC bias. Active devices require in principle two bias sources, although with proper bias schemes these can be reduced to one. Fig. 2.31 presents two possible circuit implementations, with distributed matching sections (a), typically (but not necessarily) hybrid, or with lumped matching sections (b), usually monolithic. Fig. 2.31 (a) also shows the equivalent circuit of two microstrip to coaxial connectors, modeled through a low-pass filter. The external generator and load impedances are connected though coaxial connectors and coaxial to microstrip transitions.

2.4.1 Some Layout-Related Issues

Before discussing details on the hybrid and monolithic layout, let us discuss some layout related problems. Series and parallel elements can be connected to the transmission lines; in microstrip circuits, series connections are easy, while parallel connections require *via holes* (common also in integrated circuits) or *wrap-arounds*, see Fig. 2.32 (a); the use of bonding wires is inconvenient due to the parasitic inductance. In coplanar

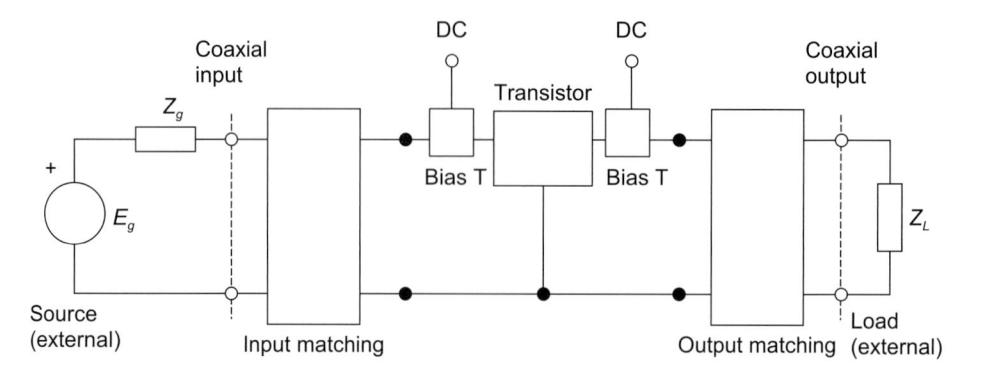

Figure 2.30 Open-loop single-stage amplifier schematic.

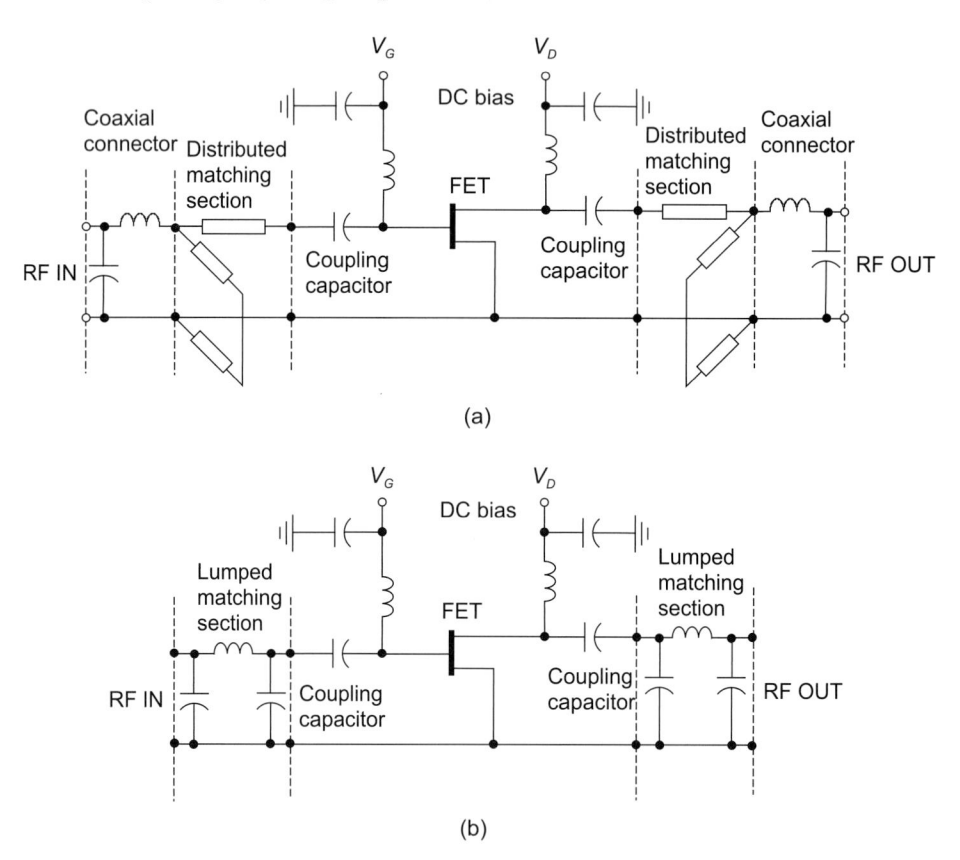

Figure 2.31 Simplified circuit of open-loop amplifier with distributed (a) or lumped (b) matching sections.

circuits, Fig. 2.32 (b) both the series and parallel insertion is possible; symmetric parallel elements are preferred at high frequency.

Stubs are short lines exploited to synthesize reactive elements; open and shorted stubs are easy in coplanar circuits, while in microstrip circuits the implementation of a shorted

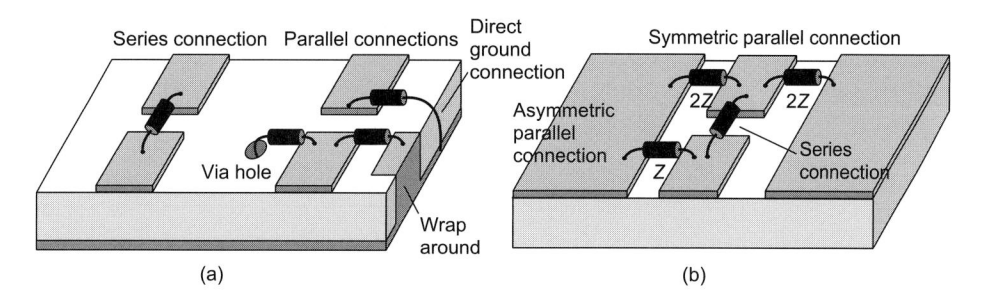

2.4 Layout of Planar Hybrid and Integrated Circuits

Figure 2.32 Series and parallel insertion of lumped elements in a microstrip (a) and coplanar (b) line.

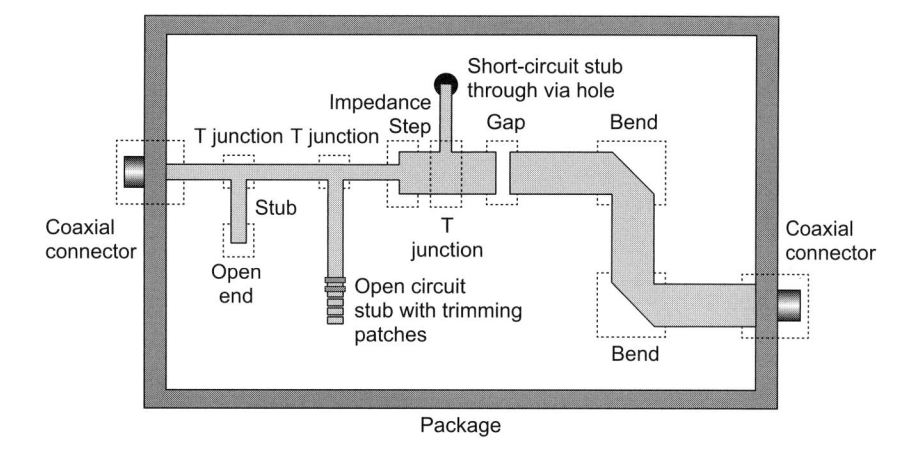

Figure 2.33 Example of microstrip hybrid layout with stubs and discontinuities.

stub is difficult, due to the need to connect the end of the stub to the ground plane (see Fig. 2.33). In hybrid circuits, open circuit stubs can be trimmed through the use of small metal patches that can be connected at the end of the stub.

In hybrid circuits, active elements can be mounted *in chip* or *in package*. In the first case bonding wires or ribbons are needed, in the second case high-frequency packages are often of the flatpack kind, with microstrip connectors; for symmetry, the package has two source or emitter contacts besides the input (gate or base) and the output (drain or collector) terminals.

Planar *line discontinuities* are a general name addressing all those transmission line structures that deviate from the ideal model of a uniform transmission line. In fact, in the design of microstrip and coplanar circuits the layout naturally leads to features that in turn introduce additional parasitics with respect to the ideal optimized schematic made of transmission lines and ideal lumped parameter circuits. For instance, a microstrip stub connected to a line introduces in the layout a T-junction, that is the source of capacitive and inductive parasitics; the same happens when a microstrip bend causes an extra parasitic capacitance to be introduced in correspondence to the bend (minimizing the capacitance is possible by *chamfering* the bend). As a further example, an

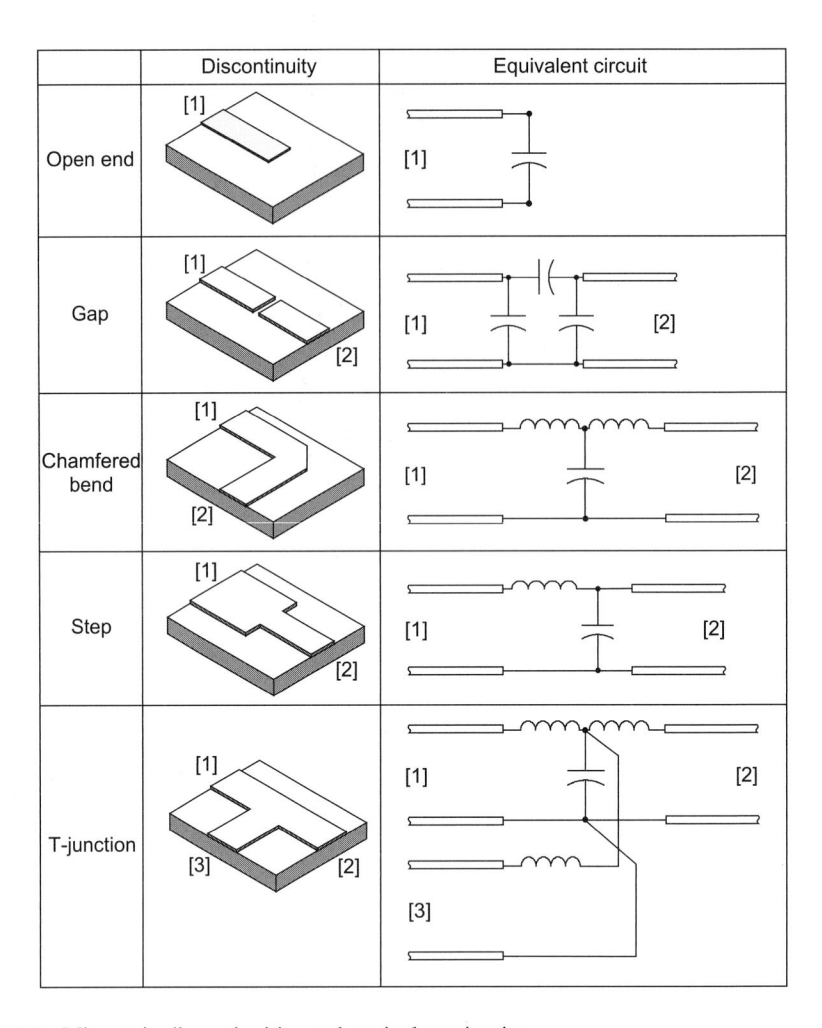

	Discontinuity	Equivalent circuit
Open end	[1]	[1]
Gap	[1] [2]	[1] [2]
Chamfered bend	[1] [2]	[1] [2]
Step	[1] [2]	[1] [2]
T-junction	[1] [3] [2]	[1] [2] [3]

Figure 2.34 Microstrip discontinuities and equivalent circuits.

open circuit stub has an additional fringing capacitance towards ground, that can be compensated for by adjusting the length of the stub. Finally, the microstrip gap cannot be, strictly speaking, considered a discontinuity, since it is exploited to implement a (small) series capacitance. Examples of microstrip discontinuities are shown in Fig. 2.33 while a set of discontinuities, together with the related models, is shown in Fig. 2.34. In general, the circuit optimization is carried out working with ideal elements; the layout is extracted and then an augmented netlist including discontinuity models is generated from it. The augmented netlist is used to verify the design, and, if needed, to tweak and further optimize the circuit.

2.4.2 Hybrid Layout

Fig. 2.35 shows a simplified hybrid microstrip implementation of the single-stage amplifier, in which the input and output matching sections have been separately realized

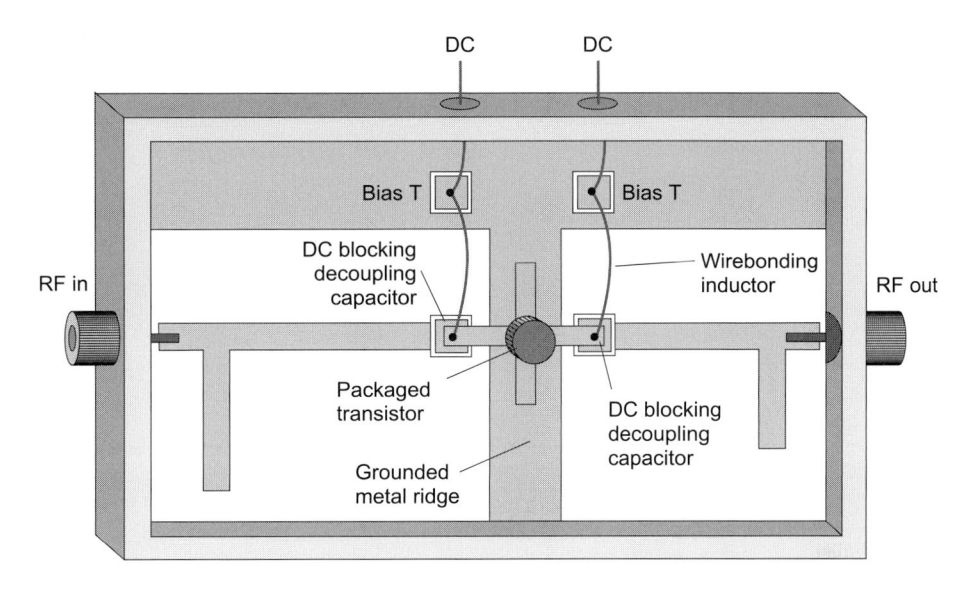

2.4 Layout of Planar Hybrid and Integrated Circuits

Figure 2.35 Qualitative example of microstrip hybrid layout of a single-stage amplifier.

on two different ceramic substrates.[6] The active device is introduced in packaged form and exploits as the ground plane (and also as the heat sink) a ridge in the metal package. Bias Ts are implemented exploiting, as series inductors, the parasitic wire bonding inductance; chip capacitors connected to the grounded package are also part of the bias T. The microstrip lines are connected to the exterior of the circuit through coaxial connectors.

2.4.3 Integrated Layout

In the monolithic layout, no external elements can be integrated within the circuit (although lumped element may be connected externally); as a qualitative example, the monolithic implementation of the single-stage amplifier already described in shown (microstrip form) in Fig. 2.36; lumped input and output matching sections are used. Via holes are introduced to provide local grounding, besides the ground pads needed for the input and output coplanar connectors; the circuit is shown as unpackaged. Finally, Fig. 2.37 shows a coplanar waveguide monolithic implementation exploiting distributed matching sections. Due to the typically small size of MMICs, such a solution is realistic only if the frequency is high enough to make distributed elements compact, e.g., for millimeter wave operation. While coplanar waveguides easily allow for open- and short-circuit line stubs (i.e., short pieces of transmission lines for the implementation of the distributed matching sections), the layout is globally less compact, and ground planes have to be connected together by airbridges (rather than bonding wires, as they would in the hybrid implementation) with a spacing small with respect to the wavelength (e.g., $\lambda_g/8$) to suppress spurious slot-like modes. Both in the microstrip and in the

---

[6] Since coplanar hybrid circuits are uncommon, we only discuss the microstrip hybrid case.

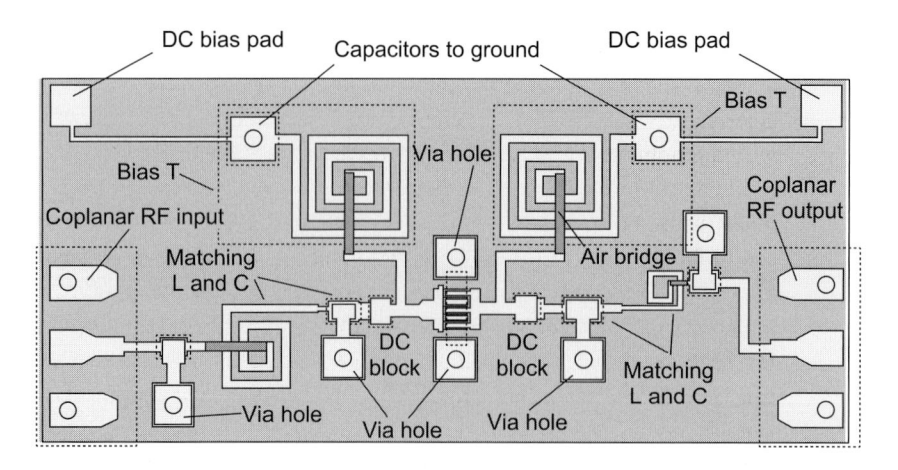

Figure 2.36 Qualitative example of microstrip integrated implementation of single-stage amplifier with DC bias.

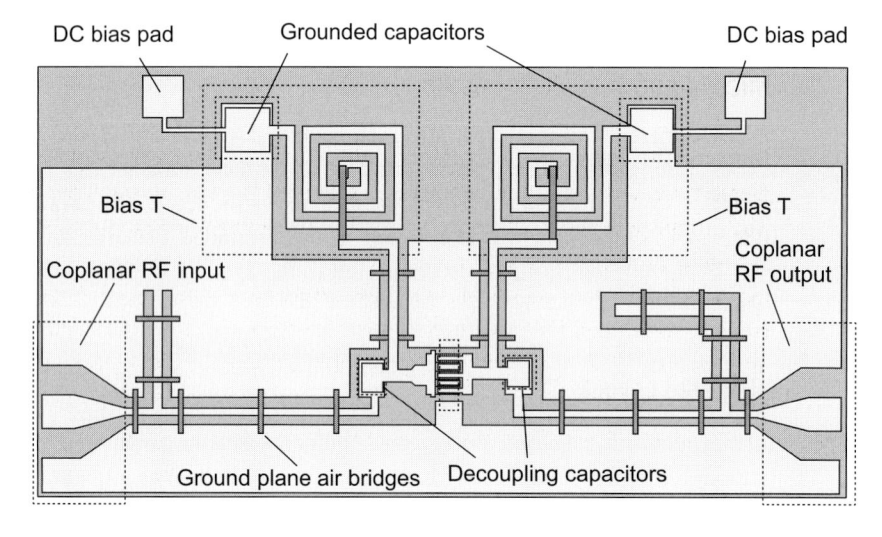

Figure 2.37 Qualitative example of coplanar integrated implementation of single-stage amplifier with DC bias.

coplanar layout, source air bridges are used in the active components. The active device layout has been kept the same in the microstrip and coplanar version, although the difference in operation frequency (microwave vs. millimeter wave) also has an impact on the FET layout (e.g., on the length of the gate fingers, which decreases with increasing frequency).

The Computer-Aided Design (CAD) of MMIC is today a well-developed technology, although analog circuit CAD has a degree of automatic design that is less developed than for digital circuit CAD. Microwave CAD tools make use of a database including libraries of element models (passive and active), a circuit simulator (small signal, large signal steady state, large signal time-domain, noise), optimization tools, often a layout generator and editor. The circuit is internally described by a low-level ASCII format (like a netlist) specifying in the minimal case the network connectivity and the element characteristics (value, library, associated layout files, etc.). The designer assembles, typically with a graphical interface, a circuit where the elements are interconnected, making use of element libraries. Optimization with respect to some prescribed design goal is then made, using however, in the first design phase, ideal elements with a minimal parasitic set. In fact, the use of full models with complex topologies makes optimization critical, not only because the element count is too large, but also because the element parasitics are correlated with each other. Consider the single-stage amplifier with lumped matching sections shown in Fig. 2.31 (b); the schematic only includes ideal elements and can be exploited as a basis for circuit optimization and design. After extraction of a suitable layout, however, an augmented netlist can be obtained, where each element is modeled by a complete equivalent circuit (see Fig. 2.38). The augmented schematic with parasitic elements may be used in order to check the circuit performances and perform further optimization and tuning.

2.5 Microwave Circuit Packaging

Both hybrid and monolithic integrated circuits are typically packaged (in a metal or dielectric enclosure) and connected to other subsystems through electrical connectors. The circuit package is an important part of the microwave circuit, also in terms of cost; it should protect the circuit mechanically, offer electromagnetic (EM) shielding, protect the circuit from chemicals and allow for heat dissipation. Packages can be hermetic (sealed, sometimes filled with inert gases to avoid oxygen contamination through leakage), or open. The two most common approaches are the metal package (often aluminum or brass) and the dielectric package. The metal package is a high-Q solution that often exhibits internal resonances that may be suppressed by locating dissipative media (e.g.,

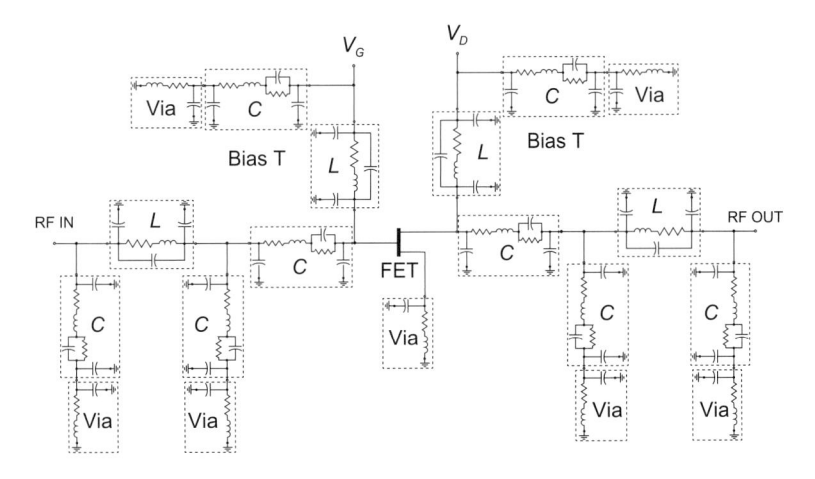

Figure 2.38 Single-stage MMIC amplifier: extended schematic with real elements.

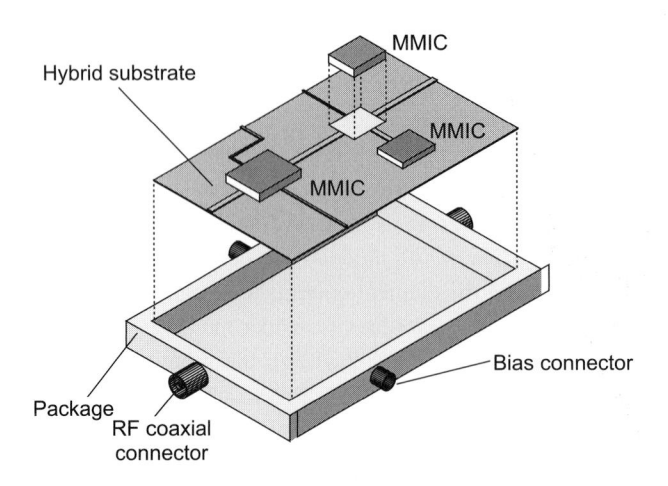

Figure 2.39 Hybrid MMIC mounting in a metal package with coaxial connectors.

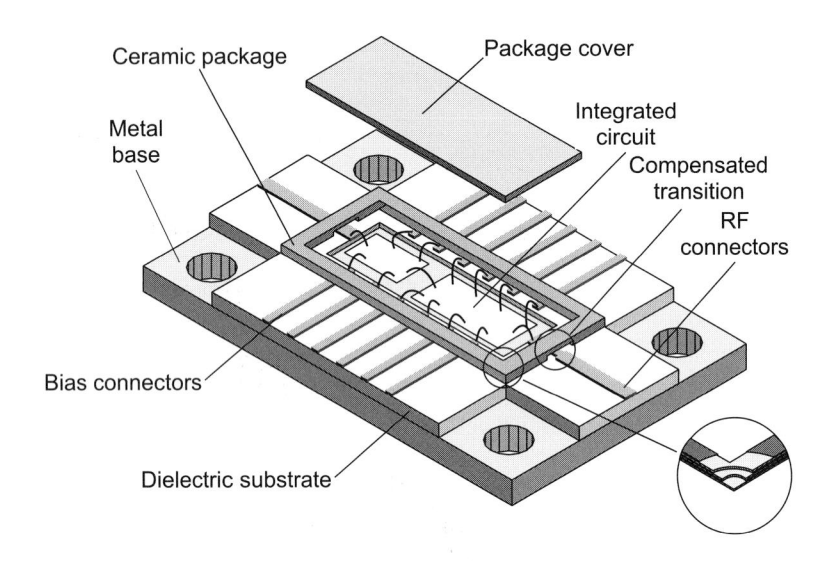

Figure 2.40 Ceramic package with flatpack compensated connectors.

layers of carbon loaded foam) in critical positions. Dielectric packages are low-Q and therefore resonances are less dangerous; often they have a MIM multilayered structure to improve EM shielding. Metal packages can be easily tailored (i.e., fabricated ad hoc) while ceramic packages typically are standard, off-the-shelf products. MMICs can also be mounted package-free within a hybrid circuit: see Fig. 2.39; the package shown is metal, while a ceramic package with a flatpack compensated connector is shown in Fig. 2.40. Notice the layered structure of the package with metal sheets, with the purpose of creating a waveguide system below cutoff able to electromagnetically isolate the package.

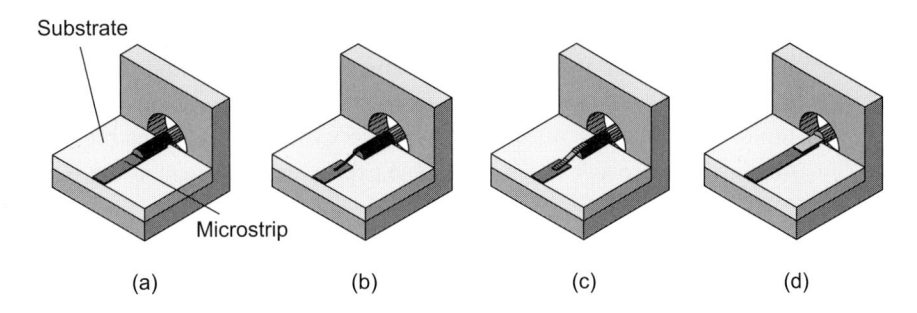

Figure 2.41 Coaxial-microstrip transitions ordered for increasing operating frequency: (a) soldered; (b) wire bonded; (c) ribbon bonded; (d) connected by contact.

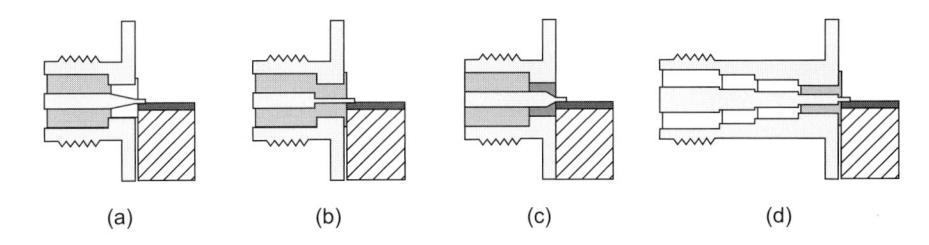

Figure 2.42 Examples of transitions between 3/7 coax or SMA connector and microstrip on alumina substrate: (a) Air transition; (b) Step transition; (c) Hermetic step transition; (d) Multiple step transition.

Package connectors make use of microstrip or flatpack transitions that are compatible with interconnecting planar lines. Metallic packages and system-level modules typically exploit coaxial transitions; see Fig. 2.41 and Fig. 2.42. A few coaxial to microstrip transitions, ordered by increasing frequency operating range, are shown in Fig. 2.41. Soldered connectors can be used up to a few GHz, while at higher frequency wire bonding or ribbon bonding (having lower parasitic p.u.l. inductance) are exploited. High-frequency connectors operating, e.g., at 40 GHz and beyond are not based on wire or ribbon bonding, but rather on contact connectors. High-frequency coaxial connectors are denoted by conventional names, some of them referring to the frequency band they were initially meant to cover. Thus we have the K connectors (up to 40 GHz), the V connectors (up to 60 GHz) and the so-called W1 connectors (Anritsu name, 1 mm radius) up to 110 GHz, which currently is the highest frequency exploited in standard instrumentation. The connector size decreases with increasing frequency.

Finally, thermal and thermo-mechanical issues have a strong impact on the choice of the package. The thermal behavior of a package can be defined by its *thermal resistance*, defined as the ratio between the temperature rise of the circuit and the dissipated power. Complex cooling techniques are often needed in power modules to keep the circuit temperature to an acceptable level. Heating during circuit operation also may cause mechanical problems due to the different expansion coefficients of the materials involved in the package, the circuit and the soldering materials. Synthetic materials made by dispersion of metal (e.g., copper or tungsten) powders in an epoxy matrix can be manufactured with the aim to equalize the expansion coefficient of semiconductors like GaAs. In fact, the expansion coefficients of alumina and GaAs are of the order of $6-7 \times 10^{-6}/°C$ at ambient temperature, the copper expansion coefficient is about three times larger, the one of tungsten lower ($5 \times 10^{-6}/°C$). Copper–tungsten alloys can be obtained that are able to have the same expansion coefficient as the substrates.

2.6 Questions and Problems

2.6.1 Questions

1. Explain the difference between a TEM and a quasi-TEM transmission line.
2. A lossy transmission line has per-unit-length parameters $\mathcal{L}, \mathcal{C}, \mathcal{R}, \mathcal{G}$. Express the characteristic impedance and complex propagation constant of the line in terms of the parameters for the general case and in the high-frequency approximation. Identify, in the high-frequency approximation, the propagation constant and the attenuation.
3. A lossless line is infinitely long. Is the input impedance always equal to the characteristic impedance? Explain.
4. A quasi-TEM line has a per-unit-length capacitance of 5 pF/mm and an in-vacuo capacitance of 2 pF/mm. What is the effective permittivity?
5. Sketch the cross section of a microstrip and of a coplanar waveguide.
6. A microstrip on 0.5 mm thick alumina substrate has a strip width of 0.5 mm. What is (approximately) the characteristic impedance?
7. Sketch the attenuation of a microstrip and of a coplanar waveguide as a function of the strip width.
8. Sketch the behavior of the attenuation of a transmission line as a function of frequency.
9. Sketch a strip, a loop and a spiral inductor. What usually limits the frequency range on which integrated RF and microwave inductors can operate?
10. List some possible uses of inductors in integrated RF circuits.
11. Sketch an interdigitated and a MIM capacitor.
12. What are *chip* inductors, capacitors and resistors? Are they used in hybrid or integrated implementations?
13. What is a coaxial-to-microstrip transition?
14. What are the main differences between a coplanar and a microstrip circuit layout?

2.6.2 Problems

1. A lossless quasi-TEM line has a 50 Ω impedance and an effective permittivity $\epsilon_{\text{eff}} = 2$. Evaluate the per-unit-length parameters $\mathcal{L}, \mathcal{C}$. Compute the guided wavelength at 10 GHz.
2. A lossy quasi-TEM line has a 50 Ω impedance. The dielectric attenuation is 0.1 dB/cm while the conductor attenuation is 1 dB/cm at 1 GHz. Evaluate the per-unit-length parameters $\mathcal{R}, \mathcal{G}$. Estimate their values and the resulting dielectric and

conductor attenuation at 10 GHz. Assuming an effective permittivity $\epsilon_{\text{eff}} = 7$, evaluate the total loss over one guided wavelength at 10 GHz.

3. The conductivity of a 2 μm thick conductor is $\sigma = 1 \times 10^5$ S/m. Evaluate the frequency at which the skin-effect penetration depth is equal to the conductor thickness.
4. A lossless transmission line with 50 Ω characteristic impedance and 5 mm guided wavelength is closed on $Z_L = 50 + j50$ Ω. Compute the input impedance for a 2.5 and 1.25 mm long line.
5. In a MIM capacitor the dielectric is 100 nm thick, with permittivity equal to 2. What is the capacitance per mm^2 area?

References

[1] J. D. Cockcroft, "Skin effect in rectangular conductors at high frequencies," *Proc. R. Soc. Lond. A, Math Phys. Sci*, vol. 122, no. A 790, p. 533–542, Feb. 1929.

[2] R. Garg, I. Bahl, and M. Bozzi, *Microstrip lines and slotlines, Third Edition*, ser. Microwave & RF. Artech House, 2013.

[3] T. Lee, "The Smith chart comes home [president's column]," *IEEE Microwave Magazine*, vol. 16, no. 10, pp. 10–25, Nov. 2015.

[4] P. H. Smith, "Transmission line calculator," *Electronics*, vol. 12, no. 1, pp. 29–31, Jan. 1939.

[5] S. Ramo, J. R. Whinnery, and T. Van Duzer, *Fields and waves in communication electronics*, 3rd ed. New York: John Wiley & Sons, 1994.

[6] M. V. Schneider, "Microstrip lines for microwave integrated circuits," *Bell System Technical Journal*, vol. 48, no. 5, pp. 1421–1444, 1969.

[7] E. Hammerstad, "Equations for microstrip circuit design," in *Microwave Conference, 1975. 5th European*, Sep. 1975, pp. 268–272.

[8] I. Bahl and P. Bhartia, *Microwave solid state circuit design*. Wiley-Interscience, 2003.

[9] R. A. Pucel, D. J. Masse, and C. P. Hartwig, "Losses in microstrip," *IEEE Transactions on Microwave Theory and Techniques*, vol. 16, pp. 342–350, 1968.

[10] E. Yamashita, K. Atsuki, and T. Ueda, "An approximate dispersion formula of microstrip lines for computer-aided design of microwave integrated circuits," *IEEE transactions on Microwave Theory and Techniques*, vol. 27, no. 12, pp. 1036–1038, Dec. 1979.

[11] "Taconic RF & microwave laminates," www.taconic-add.com/en–index.php.

[12] R. Simons, *Coplanar waveguide circuits, components, and systems*. Hoboken, NJ: John Wiley & Sons, 2001.

[13] I. Wolff, *Coplanar microwave integrated circuits*. Hoboken, NJ: John Wiley & Sons, 2006.

[14] G. Ghione and C. Naldi, "Analytical formulas for coplanar lines in hybrid and monolithic MICs," *Electronics Letters*, vol. 20, no. 4, pp. 179–181, Feb. 1984.

[15] G. Ghione and C. Naldi, "Parameters of coplanar waveguides with lower ground plane," *Electronics Letters*, vol. 19, no. 18, pp. 734–735, Sep. 1983.

[16] G. Ghione and C. U. Naldi, "Coplanar waveguides for MMIC applications: effect of upper shielding, conductor backing, finite-extent ground planes, and line-to-line coupling," *IEEE Transactions on Microwave Theory and Techniques*, vol. MTT-35, no. 3, pp. 260–267, Mar. 1987.

[17] S. Gevorgian, T. Martinsson, A. Deleniv, E. Kollberg, and I. Vendik, "Simple and accurate dispersion expression for the effective dielectric constant of coplanar waveguides,"

Microwaves, Antennas and Propagation, IEE Proceedings, vol. 144, no. 2, pp. 145–148, Apr. 1997.

[18] G. Owyang and T. Wu, "The approximate parameters of slot lines and their complement," *IRE Transactions on Antennas and Propagation*, vol. 6, no. 1, pp. 49–55, Jan. 1958.

[19] G. Ghione, "A CAD-oriented analytical model for the losses of general asymmetric coplanar lines in hybrid and monolithic MICs," *IEEE Transactions on Microwave Theory and Techniques*, vol. MTT-41, no. 9, pp. 1499–1510, Sep. 1993.

[20] R. Goyal, *Monolithic microwave integrated circuits: technology & design*, ser. Artech House microwave library. Artech House, 1989.

[21] F. Grover, *Inductance calculations: working formulas and tables*, ser. Dover phoenix editions. Dover Publications, 2004.

[22] F. E. Terman, *Radio engineering handbook*. New York: McGraw-Hill, 1943.

[23] R. Chaddock, "The application of lumped element techniques to high frequency hybrid integrated circuits," *Radio and Electronic Engineer*, vol. 44, pp. 414–420(6), Aug. 1974.

[24] S. Mohan, M. del Mar Hershenson, S. Boyd, and T. Lee, "Simple accurate expressions for planar spiral inductances," *IEEE Journal of Solid-State Circuits*, vol. 34, no. 10, pp. 1419–1424, Oct. 1999.

[25] H. Wheeler, "Simple inductance formulas for radio coils," *Proceedings of the Institute of Radio Engineers*, vol. 16, no. 10, pp. 1398–1400, Oct. 1928.

[26] H. Greenhouse, "Design of planar rectangular microelectronic inductors," *IEEE Transactions on Parts, Hybrids, and Packaging*, vol. 10, no. 2, pp. 101–109, Jun. 1974.

[27] T. Lee, *Planar microwave engineering: a practical guide to theory, measurement, and circuits*. Cambridge University Press, 2004, no. v. 1.

[28] R. Goyal, *High frequency analog integrated circuit design*, ser. Wiley Series in Microwave and Optical Engineering. New York: Wiley, 1995.

[29] T. Winslow, "Conical inductors for broadband applications," *Microwave Magazine, IEEE*, vol. 6, no. 1, pp. 68–72, Mar. 2005.

3 CAD Techniques

3.1 Modeling of Linear and Nonlinear Blocks

In an integrated microwave circuit, passive devices (such as resistors, inductors, capacitors, transformers, simple and coupled transmission lines) are linear blocks, and can therefore be characterized by linear models. Transistors (field effect or bipolar) and diodes are, instead, nonlinear components also in DC or quasi-static conditions, although they can behave in an approximately linear way under small-signal operation. Linear passive reactive components (such as linear inductors and capacitors) are with memory in the (v, i) port variables; nonlinear reactive elements, like the capacitances of *pn* or metal-semiconductor junctions, on the other hand are also associated with active components, such as transistors. Thus, MMICs generally include linear and nonlinear elements, memoryless or with memory.

The small-signal approximation, where the transistor characteristics are linearized in the neighborhood of the DC operating point, is well suited to the design of high-gain or low-noise amplifiers; in small signal conditions the active device is approximated as a linear block with memory. The simulation of distortion and power saturation occurring in power amplifiers (but also in low-noise and high-gain amplifiers in the upper limit of their spurious free dynamic range, see Sec. 8.2.3) require instead a nonlinear transistor model, often including linear and nonlinear memory effects.

In analog circuits, linear n-ports are conveniently modeled in the frequency domain, selecting conventional variables (such as voltages and currents) as the component input and output, or exploiting circuit variables that are more suited to the analysis and characterization of microwave circuits, such as the *power waves*, see Sec. 3.2. A linear n-port can be represented through a linear relationship between input and output variables; if we identify the component status in terms of the Fourier transforms of port voltages and currents, according to the choice of the input and output variable sets we obtain well-known frequency-domain models such as the series (or current-driven, or impedance) representation and the parallel (or voltage-driven, or admittance) representation [1].

Although a linear element is completely described by a proper set of parameters conveniently sampled over the frequency band of interest, in many cases an *equivalent circuit approach* is preferred. The equivalent circuit is an approximate model of the component, whose topology and element parameters have to be fitted on measured (or physically simulated) characteristics. However, the equivalent circuit has a number of advantages: it allows the device to be decomposed into domains suggested by the device physics – think of the parasitic and intrinsic elements in a transistor; it has a broadband response, rather than being limited to the measurement frequency band; it is often scalable vs. the device geometry; finally, it is more computationally efficient when performing a circuit analysis than an interpolated look-up table, and can be easily exploited both in frequency-domain and time-domain linear circuit solvers.

When coming to nonlinear components, two approaches have emerged so far. *Circuit-level* Computer-Aided Design (CAD) tools often exploit nonlinear equivalent circuits made of nonlinear resistors, capacitors, controlled sources; microwave inductors are typically linear, since their core is non-magnetic. Notice that also reactive components are, when described in terms of a proper state variable, memoryless; for instance, nonlinear capacitors are modeled with an instantaneous relation between the capacitor charge and the controlling voltage. Nonlinear resistors and controlled sources are also described by a memoryless relationship between the input and output variables; for instance, the drain current dependence on the gate-source and drain-source voltages in a common-source FET can be modeled as a memoryless function $i_D(t) = f_D(v_{GS}(t), v_{DS}(t))$.

System-level models of nonlinear components are often based on a different approach, since at the system level (e.g., considering a whole amplifier, or even a whole receiver or transmitter) a physics-based analysis deriving from a detailed understanding of the component internal structure is often difficult or too computationally intensive. The problem then arises to model a whole subsystem on the basis of measurements carried out at its terminals. For the sake of definiteness, suppose that the subsystem considered is a power amplifier; the relevant measurements can be, for instance: the small-signal response of the amplifier as a function of frequency; the input–output power sweep at a constant frequency (single tone test, providing information on the amplifier output power compression and saturation); the input–output power sweep with two input tones (two-tone test, providing information on third-order intermodulation distortion); finally, the input–output power sweep in the presence of a modulated input signal whose format is consistent with the specific system application for which the amplifier was designed, see Sec. 8.2.1. Including all of the above information in a single system-level model providing an accurate response to any input signal clearly is a formidable task. A theoretically exact mathematical model for a generic nonlinear system is provided by the so-called *Volterra series* approach [2, 3]; unfortunately, this model has the format of a power series of the input variables with frequency-dependent kernels (the first-order kernel is indeed the linear transfer function) which is difficult to characterize experimentally, unless the system is weakly nonlinear. For a memoryless system the Volterra series model reduces to the power series model (see Sec. 8.5.1), in which the output is an instantaneous function of the inputs, to be expanded in a truncated power series. Although simplified, this model can be effective in obtaining approximate closed-form results in the analysis of compression, harmonic generation and intermodulation distortion in power amplifiers at low enough frequency to make reactive effects negligible, see Sec. 8.5.1. Approximate alternative approaches (the so-called behavioral models [4]) make use of a combination of linear and nonlinear blocks put in parallel, series or cascade. Behavioral models are typically restricted to narrowband systems (e.g., a narrowband power amplifier) where the behavioral model (also called envelope, baseband or descriptive function model) is indeed exact when in-band memory effects are negligible, see Sec. 8.2.2. In state-of-the art CAD tools, circuit-oriented simulators exploiting equivalent circuit models often are seamlessly integrated into system-level simulators.

During the last few years, the increased computational power available also on personal computers has fostered the development of numerical electromagnetic (EM) models integrated within circuit design and analysis CAD tools. A typical example are EM models of passive elements [5], concentrated or distributed (like transmission lines, waveguides, line discontinuities, passive resistive or reactive elements), analyzed by means of numerical techniques such as the Finite Element method [6], the Finite Difference Time Domain technique [7], or spectral-domain Green's function approaches [8]. Such numerical techniques can be used to model critical portions of the circuit, and are anyway able to provide a more accurate and flexible picture of the device response with respect to closed-form analytical approaches.

3.2 Power Waves and the Scattering Parameters

3.2.1 Representations of Linear Two-Ports

Consider an electronic subsystem interacting with the rest of the circuit through two electrical ports whose instantaneous electrical state is given by the port currents and voltages, $i_k(t)$, $v_k(t)$, $k = 1, 2$ (see Fig. 3.1). Such an element is denoted as a *two-port*; a *linear* two-port only includes linear elements. A *non-autonomous* two-port does not include any independent voltage or current sources, while an *autonomous* two-port does. A non-autonomous two-port including controlled sources may be able to provide power gain and it is often called *active*; a typical example is an active device (a transistor) operating in small-signal conditions. Two-ports entirely made of resistors, capacitors, inductors, coupled inductors, single or coupled transmission lines cannot provide amplification and are therefore called *passive*.

In a circuit made by the connection of *linear* elements (that may include active elements operating in small-signal conditions), the superposition principle applies, and therefore the analysis can be carried out considering a steady-state, single frequency sinusoidal excitation. Such a frequency-domain analysis is also practically meaningful, since many RF electronic systems are narrowband.

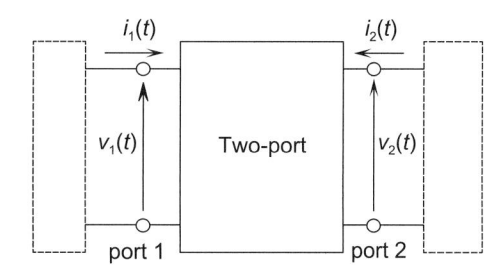

Figure 3.1 Two-port and its input/output voltages and current.

The mathematical model of a linear two-port characterized by the current phasors I_1 and I_2 and the voltage phasors V_1 and V_2 consists of two equations expressing a linear relationship between two independent variables and two dependent variables in the current and voltage set. For generality, we will consider here autonomous two-ports. In the set of four variables (I_1, I_2, V_1, V_2) we have six possible ways to select a pair of dependent and a pair of independent variables, thus obtaining six possible representations of the two-port, namely:

- the **current-driven** or **series** representation, where I_1 and I_2 are the independent variables and V_1 and V_2 the dependent variables:

$$V_1 = Z_{11}I_1 + Z_{12}I_2 + V_{01}$$
$$V_2 = Z_{21}I_1 + Z_{22}I_2 + V_{02},$$

where Z_{ij} are the elements of the impedance matrix (measured in Ω) and V_{0i} is the open-circuit voltage at port i, zero for a non-autonomous two-port;

- the **voltage-driven** or **parallel** representation, where V_1 and V_2 are the independent variables and I_1 and I_2 the dependent variables; the model reads:

$$I_1 = Y_{11}V_1 + Y_{12}V_2 + I_{01}$$
$$I_2 = Y_{21}V_1 + Y_{22}V_2 + I_{02},$$

where Y_{ij} are the elements of the admittance matrix (measured in S) and I_{0i} is the short-circuit current at port i, zero for a non-autonomous two-port. The model is particularly suited to represent field-effect transistors in small-signal common-source operation since it describes, through the transadmittance Y_{21}, the drain current control from the gate voltage;

- the **hybrid-I** representation, where I_1 and V_2 are the independent variables and V_1 and I_2 the dependent variables; the model reads:

$$V_1 = H_{11}I_1 + H_{12}V_2 + V_{01}^{H1}$$
$$I_2 = H_{21}I_1 + H_{22}V_2 + I_{02}^{H1},$$

where H_{ij} are the elements of the hybrid-I matrix (the diagonal elements are impedances and admittances, respectively, the non-diagonal elements pure numbers), V_{01}^{H1} is the open-circuit voltage at port 1 when port 2 is shorted and I_{02}^{H1} is the short-circuit current at port 2 when port 1 is open; again, both are zero for a non-autonomous two-port. This representation is well suited to model bipolar transistors in small-signal common-emitter operation, since the parameter H_{21} is the current gain between base and collector;

- the **hybrid-II** representation, where V_1 and I_2 are the independent variables and I_1 and V_2 the dependent variables, the model is the inverse of the hybrid-I, we omit the equations for brevity;

- finally, the so-called **transmission-I** and **transmission-II** models, that use as independent variables the variables at port 2 (I_2 and V_2) or 1 (I_1 and V_1) and the other set as dependent variables; it is useful in evaluating the representation of two-ports in cascade.[1]

The above representations, whose describing variables are voltages or currents, are not particularly well suited to the modeling and characterization of RF or microwave circuits, as the following remarks point out. Firstly, in the measurement of conventional (impedance, admittance, hybrid) two-port parameters, short and open circuits are nominally required as loads. However, these are difficult to implement at RF over a broad band of frequencies, so that a wideband characterization of a component becomes difficult or impossible. Moreover, most RF transistors cannot be measured in short or open-circuit conditions because they are unstable with such reactive loads. These serious shortcomings can be in fact alleviated, since the evaluation of the impedance (admittance) parameters does not necessarily require that open-circuit (short-circuit) conditions be imposed at the two-ports; any load conditions can be used to derive any set of parameters, see Example 3.1. A far more serious problem is the fact that at RF and microwaves, total voltages and currents are difficult or impossible to measure through conventional instruments, and even their very existence as observable quantities is not to be taken for granted, as in a non-TEM waveguide environment.

Example 3.1 A non-autonomous two-port is loaded with an input and output generator with open circuit voltages E_1 and E_2 and internal impedances Z_1 and Z_2, respectively. Show that the impedance parameters can be derived by measuring I_1 and I_2 in two conditions: (a) first we set $E_2 = 0$ and measure I_{1a} and I_{2a}; then (b) we set $E_1 = 0$ and measure I_{1b} and I_{2b}.

Solution
Taking into account the constitutive relations of the two-port and the equations of the input and output generators, we have in case (a) and (b), respectively:

$$V_{1a} = E_1 - Z_1 I_{1a} = Z_{11} I_{1a} + Z_{12} I_{2a}$$
$$V_{2a} = -Z_2 I_{2a} = Z_{21} I_{1a} + Z_{22} I_{2a}$$
$$V_{1b} = -Z_1 I_{1b} = Z_{11} I_{1b} + Z_{12} I_{2b}$$
$$V_{2b} = E_2 - Z_2 I_{2b} = Z_{21} I_{1b} + Z_{22} I_{2b},$$

therefore we obtain two linear systems:

$$I_{1a} Z_{11} + I_{2a} Z_{12} = E_1 - Z_1 I_{1a}$$
$$I_{1b} Z_{11} + I_{2b} Z_{12} = -Z_1 I_{1b}$$

and:

$$I_{1a} Z_{21} + I_{2a} Z_{22} = -Z_2 I_{2a}$$
$$I_{1b} Z_{21} + I_{2b} Z_{22} = E_2 - Z_2 I_{2b}.$$

Solving the first system we obtain Z_{11} and Z_{12}; from the second Z_{21} and Z_{22}.

---

[1] A similar approach can be implemented within the framework of a power-wave representation of the two-port, see Sec. 3.2.8.

CAD Techniques

To overcome the problems associated with the definition and measurement of conventional two-port parameters, a different representation technique was devised, that exploits, instead of port voltages and currents, a set of progressive and regressive waves (called *power waves* [9, 10], see Sec. 3.1). From a physical standpoint, the approach derives from the theory of transmission lines and waveguides, but equally applies to lumped-parameter circuits. The representative small-signal parameters are denoted as scattering parameters or S parameters; they can be experimentally derived by measuring the two-port terminated by resistive loads, thus overcoming bandwidth and stability problems. Moreover, while the series and parallel representations may be singular, the S parameter representation exists for almost any circuit.

3.2.2 Power Waves

Consider the linear n-port in Fig. 3.2.[2] The state of the n-port is determined by the set of current and voltage phasors at port k, V_k and I_k. Let us associate with port k the so-called *normalization impedance* Z_{0k} that is arbitrary in principle, provided it has positive real part; however, we will in what follows assume Z_{0k} to be real and call it *normalization resistance* R_{0k}. Then, let us introduce for port k the *power waves* a_k and b_k defined as a linear combination of V_k and I_k:

$$\begin{cases} a_k = \dfrac{V_k + R_{0k} I_k}{2\sqrt{R_{0k}}} \\ b_k = \dfrac{V_k - R_{0k} I_k}{2\sqrt{R_{0k}}}. \end{cases} \tag{3.1}$$

Inverting system (3.1) we have:

$$\begin{cases} V_k = \sqrt{R_{0k}}\,(a_k + b_k) \\ I_k = \dfrac{a_k - b_k}{\sqrt{R_{0k}}}. \end{cases} \tag{3.2}$$

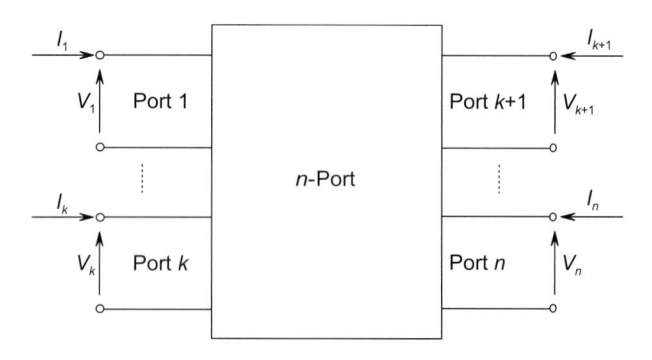

Figure 3.2 Linear n-port.

---

[2] Notice that for $n > 2$ only the impedance and admittance representations can be used by directly extending the two-port case.

3.2 Power Waves and the Scattering Parameters

From a physical standpoint, the power waves a_k and b_k can be easily interpreted with reference to the theory of transmission lines; in fact, in a line with characteristic impedance Z_0 the forward and backward waves propagate with voltages (V^+, V^-) and currents (I^+, I^-), related as:

$$\begin{cases} V^+ = Z_0 I^+ \\ V^- = -Z_0 I^-, \end{cases} \quad (3.3)$$

while the total voltage and current are obtained by superposition as:

$$\begin{cases} V = V^+ + V^- \\ I = I^+ + I^-. \end{cases} \quad (3.4)$$

The power transmitted on the line finally is:

$$P = \operatorname{Re}\left[VI^*\right] = |V^+|^2/Z_0 - |V^-|^2/Z_0.$$

By comparing (3.4) and (3.2), we can readily associate, assuming $R_{0k} \equiv Z_0$, the two power waves with the normalized forward and backward voltages:

$$a_k = V_k^+/\sqrt{R_{0k}}, \quad b_k = V_k^-/\sqrt{R_{0k}};$$

while the power entering port k (or, in the analogy, flowing on the line) is:

$$P_k = \operatorname{Re}\left[VI^*\right] = |V_k^+|^2/Z_0 - |V_k^-|^2/Z_0 = |a_k|^2 - |b_k|^2. \quad (3.5)$$

Thus, a_k is related to the *incident* power, b_k to the *reflected* power, both having dimension $\sqrt{W}$, from which the name "power waves" derives. Notice that the definition of power waves is independent of whether any propagation actually takes place – they can be defined also in lumped-parameter circuits.

From (3.1), if $V_k = -R_{0k}I_k$, then $a_k = 0$. This happens if port k is loaded by the normalization resistance: we denote this as resistance matching, or simply matching, that does not generally imply power matching, i.e., maximum power transfer between the port and the load. In such conditions we also have:

$$b_k = V_k/\sqrt{R_{0k}}. \quad (3.6)$$

3.2.3 Power Wave n-Port Model: The Scattering Matrix

For the sake of generality, let us consider an autonomous n-port, where $V_{0k} \neq 0$ and $I_{0k} \neq 0$. Define the vector of port voltages and currents:

$$\underline{V} = \begin{pmatrix} V_1 \\ \vdots \\ V_k \\ \vdots \\ V_n \end{pmatrix}, \quad \underline{I} = \begin{pmatrix} I_1 \\ \vdots \\ I_k \\ \vdots \\ I_n \end{pmatrix}.$$

For a linear n-port we have the impedance or series representation:

$$\underline{V} = \mathbf{Z}\underline{I} + \underline{V}_0, \quad (3.7)$$

where $\mathbf{Z}$ is the impedance matrix and $\underline{V}_0$ the open-circuit voltage vector; similarly we have the parallel representation:

$$\underline{I} = \mathbf{Y}\underline{V} + \underline{I}_0, \qquad (3.8)$$

where $\mathbf{Y}$ is the admittance matrix and $\underline{I}_0$ is the short-circuit current vector. One of the two matrices may be singular, making the corresponding representation not defined.

Let $\underline{a}$ and $\underline{b}$ be the power wave vectors and let $\mathbf{R}_0$ be the diagonal matrix of normalization resistances:

$$\underline{a} = \begin{pmatrix} a_1 \\ \vdots \\ a_k \\ \vdots \\ a_n \end{pmatrix}, \quad \underline{b} = \begin{pmatrix} b_1 \\ \vdots \\ b_k \\ \vdots \\ b_n \end{pmatrix}, \quad \mathbf{R}_0 = \begin{pmatrix} R_{01} & \cdots & 0 & \cdots & 0 \\ \vdots & \ddots & 0 & & \vdots \\ 0 & 0 & R_{0k} & 0 & 0 \\ \vdots & & 0 & \ddots & \vdots \\ 0 & \cdots & 0 & \cdots & R_{0n} \end{pmatrix}.$$

The power wave vectors $\underline{a}$ and $\underline{b}$ are related by a linear relationship that can be identified as follows. Eq. (3.2) can be written in matrix form as:

$$\begin{cases} \underline{V} = \mathbf{R}_0^{1/2} \left(\underline{a} + \underline{b} \right) \\ \underline{I} = \mathbf{R}_0^{-1/2} \left(\underline{a} - \underline{b} \right). \end{cases} \qquad (3.9)$$

Since $\mathbf{R}_0$ is diagonal, also $\mathbf{R}_0^{1/2}$ and $\mathbf{R}_0^{-1/2}$ are diagonal (the function of a diagonal matrix is the diagonal matrix of the functions of the diagonal elements, see, e.g., [11]). Substituting (3.9) in (3.7) we obtain:

$$\mathbf{R}_0^{1/2} \left(\underline{a} + \underline{b} \right) = \mathbf{Z}\mathbf{R}_0^{-1/2} \left(\underline{a} - \underline{b} \right) + \underline{V}_0, \qquad (3.10)$$

that is:

$$\underline{b} = \left(\mathbf{R}_0^{-1/2} \mathbf{Z} \mathbf{R}_0^{-1/2} + \mathbf{I} \right)^{-1} \left(\mathbf{R}_0^{-1/2} \mathbf{Z} \mathbf{R}_0^{-1/2} - \mathbf{I} \right) \underline{a} \\ + \left(\mathbf{R}_0^{-1/2} \mathbf{Z} \mathbf{R}_0^{-1/2} + \mathbf{I} \right)^{-1} \mathbf{R}_0^{-1/2} \underline{V}_0.$$

The power wave constitutive relationships of the *n*-port therefore is:

$$\underline{b} = \mathbf{S}\underline{a} + \underline{b}_0, \qquad (3.11)$$

where the *scattering matrix* $\mathbf{S}$ is defined as:

$$\begin{aligned} \mathbf{S} &\equiv \left(\mathbf{R}_0^{-1/2} \mathbf{Z} \mathbf{R}_0^{-1/2} + \mathbf{I} \right)^{-1} \left(\mathbf{R}_0^{-1/2} \mathbf{Z} \mathbf{R}_0^{-1/2} - \mathbf{I} \right) \\ &= \left(\mathbf{R}_0^{-1/2} \mathbf{Z} \mathbf{R}_0^{-1/2} - \mathbf{I} \right) \left(\mathbf{R}_0^{-1/2} \mathbf{Z} \mathbf{R}_0^{-1/2} + \mathbf{I} \right)^{-1} \\ &= \mathbf{R}_0^{1/2} \left(\mathbf{Z} + \mathbf{R}_0 \right)^{-1} \left(\mathbf{Z} - \mathbf{R}_0 \right) \mathbf{R}_0^{-1/2} \\ &= \mathbf{R}_0^{-1/2} \left(\mathbf{Z} - \mathbf{R}_0 \right) \left(\mathbf{Z} + \mathbf{R}_0 \right)^{-1} \mathbf{R}_0^{1/2}. \end{aligned} \qquad (3.12)$$

The above equations in (3.12) are all equivalent, since functions of the same matrix commute. The vector of forward wave generators $\underline{b}_0$ is then obtained as:

$$\underline{b}_0 \equiv \left(\mathbf{R}_0^{-1/2} \mathbf{Z} \mathbf{R}_0^{-1/2} + \mathbf{I}\right)^{-1} \mathbf{R}_0^{-1/2} \underline{V}_0 = \mathbf{R}_0^{1/2} (\mathbf{Z} + \mathbf{R}_0)^{-1} \underline{V}_0. \quad (3.13)$$

If $n = 1$, the n-port reduces to a one-port; we thus obtain:

$$S = \frac{Z - R_0}{Z + R_0} \equiv \Gamma, \quad (3.14)$$

where Z is the one-port impedance, and:

$$b_0 = \frac{\sqrt{R_0}}{Z + R_0} V_0. \quad (3.15)$$

In a one-port, S simply is the reflection coefficient Γ of the impedance Z with respect to the normalization resistance.

The normalization resistance matrix $\mathbf{R}_0$ is arbitrary (provided it is not singular); however, in many cases all normalization resistances are chosen as equal ($\mathbf{R}_0 = R_0 \mathbf{I}$) and often $R_0 = 50\ \Omega$ as a default. If the normalization resistance $R_0 = 1/G_0$ is the same for all ports we obtain the following simpler relation:

$$\mathbf{S} = (\mathbf{Z} - R_0 \mathbf{I})(\mathbf{Z} + R_0 \mathbf{I})^{-1} = (G_0 \mathbf{I} - \mathbf{Y})(G_0 \mathbf{I} + \mathbf{Y})^{-1}.$$

Concerning the arbitrariness of the normalization resistance, we show in Example 3.2 that, indeed, the solution of a circuit in terms of voltages and currents is independent from the choice of R_0. Two solutions are compared, the first one where the normalization resistance is chosen as the load resistance and therefore the load appears to be "matched" (but in fact, no power matching to the generator actually takes place), the second one where the normalization resistance is arbitrary; we see that the load voltage is independent from the normalization resistance.

Example 3.2 Consider a real generator of internal impedance Z_g and open circuit voltage E connected to a load resistance R_L. Show that the choice of the normalization resistance has no impact on the solution of a circuit.

Solution
Circuit theory immediately leads to the solution for the load voltage V_L:

$$V_L = \frac{E R_L}{Z_g + R_L}. \quad (3.16)$$

Let us find the solution using power waves. First, assume $R_0 = R_L$. In this case, $\Gamma_L = 0$ and the generator and load constitutive relations are:

$$b_{L1} = \Gamma_L a_{L1} = 0$$
$$b_{g1} = \frac{\sqrt{R_L} E}{Z_g + R_L} + \frac{Z_g - R_L}{Z_g + R_L} a_{g1}.$$

Moreover, since the normalization resistance is the same for the generator and the load, the power waves are continuous at the interface between the two-ports (see Sec. 3.2.9); we then have:

$$a_{L1} = b_{g1}$$
$$a_{g1} = b_{L1} = 0$$
$$b_{g1} = a_{L1} = \frac{\sqrt{R_L}E}{Z_g + R_L}.$$

Thus the load voltage is:

$$V_{L1} = (a_{L1} + b_{L1})\sqrt{R_L} = \frac{\sqrt{R_L}E}{Z_g + R_L}\sqrt{R_L} = \frac{ER_L}{Z_g + R_L}.$$

The result is consistent with (3.16). Imagine now (case 2) that the normalization resistance is $R_0 \neq R_L$; we now have the constitutive relations:

$$b_{L2} = \Gamma_L a_{L2} = \frac{R_L - R_0}{R_L + R_0} a_{L2}$$
$$b_{g2} = \frac{\sqrt{R_0}E}{Z_g + R_0} + \frac{Z_g - R_0}{Z_g + R_0} a_{g2}.$$

Moreover, the power wave continuity implies:

$$a_{L2} = b_{g2} = \frac{\sqrt{R_0}E}{Z_g + R_0} + \frac{Z_g - R_0}{Z_g + R_0} a_{g2}$$
$$a_{g2} = b_{L2} = \frac{R_L - R_0}{R_L + R_0} a_{L2},$$

from which, solving, we obtain:

$$a_{L2} = \frac{\dfrac{\sqrt{R_0}E}{Z_g + R_0}}{1 - \dfrac{Z_g - R_0}{Z_g + R_0}\dfrac{R_L - R_0}{R_L + R_0}}, \quad b_{L2} = a_{L2}\frac{R_L - R_0}{R_L + R_0}.$$

With straightforward manipulations we see that, although the two power waves depend on R_0, the load voltage is independent from the normalization resistance. In fact, in case (2) we have:

$$V_{L2} = (a_{L2} + b_{L2})\sqrt{R_0} = \frac{R_0 E (R_L + R_0 + R_L - R_0)}{(Z_g + R_0)(R_L + R_0) - (Z_g - R_0)(R_L - R_0)}$$
$$= \frac{2R_0 R_L E}{R_0 R_L + R_0 Z_g + R_0 R_L + R_0 Z_g} = \frac{R_L E}{R_L + Z_g} \equiv V_{L1}.$$

The result is the same, and so is the power to the load evaluated according to normalization 1 or 2 using the power wave definition:

$$P_{L1} = |a_{L1}|^2 - |b_{L1}|^2 = \frac{R_L |E|^2}{|Z_g + R_L|^2}$$

$$P_{L2} = |a_{L2}|^2 - |b_{L2}|^2$$

$$= \frac{\dfrac{R_0 |E|^2}{|Z_g + R_0|^2}}{\left|1 - \dfrac{Z_g - R_0}{Z_g + R_0}\dfrac{R_L - R_0}{R_L + R_0}\right|^2} \left(1 - \left|\frac{R_L - R_0}{R_L + R_0}\right|^2\right)$$

$$= \frac{R_0 |E|^2 \left[(R_L + R_0)^2 - (R_L - R_0)^2\right]}{4 R_0^2 |R_L + Z_g|^2} = \frac{4 R_0^2 |E|^2 R_L}{4 R_0^2 |R_L + Z_g|^2} = P_{L1}.$$

Tables 3.1 and 3.2 report the conversion formulae between scattering, admittance and impedance parameters for a two-port having normalization resistance R_0 at both ports. We denote with lowercase symbols the admittance, impedance or hybrid parameters normalized vs. the normalization resistance or conductance.

3.2.4 Properties of the S-Matrix

From (3.5) the net power entering port k can be expressed in terms of power waves; the total power P_{tot} entering the n-port will therefore be:

$$P_{tot} = \sum_{k=1}^{n} P_k = \sum_{k=1}^{n} \left(|a_k|^2 - |b_k|^2\right) = \underline{a}^T \underline{a}^* - \underline{b}^T \underline{b}^*, \qquad (3.17)$$

where ^{T} denotes the transpose. If the circuit is non-autonomous (no internal independent sources) $\underline{b} = \mathbf{S}\underline{a}$; substituting in (3.17) we obtain:

$$P_{tot} = \underline{a}^T \left(\mathbf{I} - \mathbf{S}^T \mathbf{S}^*\right) \underline{a}^*. \qquad (3.18)$$

For a reactive n-port $P_{tot} = 0$ independent of the excitation; this can only be obtained by imposing $\mathbf{S}^T \mathbf{S}^* - \mathbf{I} = 0$; thus, for a reactive n-port:

$$\mathbf{S}^{-1} = \mathbf{S}^{*T} \equiv \mathbf{S}^{\dagger}, \qquad (3.19)$$

i.e., the scattering matrix is *hermitian* (the inverse equals the complex conjugate of the transpose).

Reciprocity characterizes most networks made of passive components (although some microwave passive components including magnetic materials, such as circulators, are non-reciprocal); in terms of the impedance matrix the reciprocity condition reads:

$$\mathbf{Z} = \mathbf{Z}^T, \qquad (3.20)$$

but, from (3.12), we obtain:

$$\mathbf{Z} = \mathbf{R}^{1/2} \left(\mathbf{I} - \mathbf{S}\right)^{-1} \left(\mathbf{I} + \mathbf{S}\right) \mathbf{R}^{1/2}, \qquad (3.21)$$

Table 3.1 Conversion between the Z, Y and H and the scattering parameters for a two-port with normalization resistance R_0 at both ports. We have $z_{ij} = Z_{ij}/R_0$, $y_{ij} = Y_{ij}R_0$, $h_{11} = H_{11}/R_0$, $h_{22} = H_{22}R_0$, $h_{12} = H_{12}$, $h_{21} = H_{21}$.

$$S_{11} = \frac{(z_{11} - 1)(z_{22} + 1) - z_{12}z_{21}}{(z_{11} + 1)(z_{22} + 1) - z_{12}z_{21}} \qquad S_{12} = \frac{2z_{12}}{(z_{11} + 1)(z_{22} + 1) - z_{12}z_{21}}$$

$$S_{21} = \frac{2z_{21}}{(z_{11} + 1)(z_{22} + 1) - z_{12}z_{21}} \qquad S_{22} = \frac{(z_{11} + 1)(z_{22} - 1) - z_{12}z_{21}}{(z_{11} + 1)(z_{22} + 1) - z_{12}z_{21}}$$

$$S_{11} = \frac{(1 - y_{11})(1 + y_{22}) + y_{12}y_{21}}{(1 + y_{11})(1 + y_{22}) - y_{12}y_{21}} \qquad S_{12} = \frac{-2y_{12}}{(1 + y_{11})(1 + y_{22}) - y_{12}y_{21}}$$

$$S_{21} = \frac{-2y_{21}}{(1 + y_{11})(1 + y_{22}) - y_{12}y_{21}} \qquad S_{22} = \frac{(1 + y_{11})(1 - y_{22}) + y_{12}y_{21}}{(1 + y_{11})(1 + y_{22}) - y_{12}y_{21}}$$

$$S_{11} = \frac{(h_{11} - 1)(h_{22} + 1) - h_{12}h_{21}}{(h_{11} + 1)(h_{22} + 1) - h_{12}h_{21}} \qquad S_{12} = \frac{2h_{12}}{(h_{11} + 1)(h_{22} + 1) - h_{12}h_{21}}$$

$$S_{21} = \frac{-2h_{21}}{(h_{11} + 1)(h_{22} + 1) - h_{12}h_{21}} \qquad S_{22} = \frac{(1 + h_{11})(1 - h_{22}) + h_{12}h_{21}}{(h_{11} + 1)(h_{22} + 1) - h_{12}h_{21}}$$

Table 3.2 Conversion between the Z, Y and H and the scattering parameters for a two-port with normalization resistance R_0 at both ports. We have $z_{ij} = Z_{ij}/R_0$, $y_{ij} = Y_{ij}R_0$, $h_{11} = H_{11}/R_0$, $h_{22} = H_{22}R_0$, $h_{12} = H_{12}$, $h_{21} = H_{21}$.

$$z_{11} = \frac{(1 + S_{11})(1 - S_{22}) + S_{12}S_{21}}{(1 - S_{11})(1 - S_{22}) - S_{12}S_{21}} \qquad z_{12} = \frac{2S_{12}}{(1 - S_{11})(1 - S_{22}) - S_{12}S_{21}}$$

$$z_{21} = \frac{2S_{21}}{(1 - S_{11})(1 - S_{22}) - S_{12}S_{21}} \qquad z_{22} = \frac{(1 - S_{11})(1 + S_{22}) + S_{12}S_{21}}{(1 - S_{11})(1 - S_{22}) - S_{12}S_{21}}$$

$$y_{11} = \frac{(1 - S_{11})(1 + S_{22}) + S_{12}S_{21}}{(1 + S_{11})(1 + S_{22}) - S_{12}S_{21}} \qquad y_{12} = \frac{-2S_{12}}{(1 + S_{11})(1 + S_{22}) - S_{12}S_{21}}$$

$$y_{21} = \frac{-2S_{21}}{(1 + S_{11})(1 + S_{22}) - S_{12}S_{21}} \qquad y_{22} = \frac{(1 + S_{11})(1 - S_{22}) + S_{12}S_{21}}{(1 + S_{11})(1 + S_{22}) - S_{12}S_{21}}$$

$$h_{11} = \frac{(1 + S_{11})(1 + S_{22}) - S_{12}S_{21}}{(1 - S_{11})(1 + S_{22}) + S_{12}S_{21}} \qquad h_{12} = \frac{2S_{12}}{(1 - S_{11})(1 + S_{22}) + S_{12}S_{21}}$$

$$h_{21} = \frac{-2S_{21}}{(1 - S_{11})(1 + S_{22}) + S_{12}S_{21}} \qquad h_{22} = \frac{(1 - S_{11})(1 - S_{22}) - S_{12}S_{21}}{(1 - S_{11})(1 + S_{22}) + S_{12}S_{21}}$$

and therefore, taking into account that $\mathbf{R}^{1/2}$ is diagonal and $\mathbf{R}^{1/2} = \left(\mathbf{R}^{1/2}\right)^T$:

$$\mathbf{Z}^T = \mathbf{R}^{1/2}\left(\mathbf{I} + \mathbf{S}^T\right)\left(\mathbf{I} - \mathbf{S}^T\right)^{-1}\mathbf{R}^{1/2}. \tag{3.22}$$

Substituting (3.21) and (3.22) into (3.20) we obtain:

$$(\mathbf{I} - \mathbf{S})^{-1}(\mathbf{I} + \mathbf{S}) = \left(\mathbf{I} + \mathbf{S}^T\right)\left(\mathbf{I} - \mathbf{S}^T\right)^{-1},$$

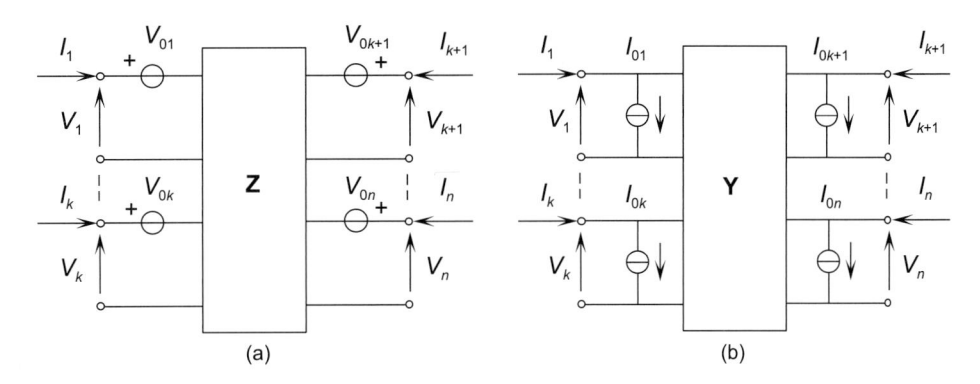

Figure 3.3 Series (a) and parallel (b) equivalent circuit of an *n*-port.

i.e.:
$$\left(\mathbf{I} + \mathbf{S}^T\right)^{-1} (\mathbf{I} - \mathbf{S})^{-1} (\mathbf{I} + \mathbf{S}) \left(\mathbf{I} - \mathbf{S}^T\right) = \mathbf{I},$$

or, equivalently:
$$(\mathbf{I} + \mathbf{S}) \left(\mathbf{I} - \mathbf{S}^T\right) = (\mathbf{I} - \mathbf{S}) \left(\mathbf{I} + \mathbf{S}^T\right),$$

and finally:
$$\mathbf{S} = \mathbf{S}^T. \tag{3.23}$$

Formally, therefore, the reciprocity condition for the scattering matrix coincides with the conditions for **Z** and **Y**, i.e., **S** is symmetric. Additionally, for a reciprocal and reactive *n*-port we finally have:

$$\mathbf{S}^{-1} = \mathbf{S}^*. \tag{3.24}$$

Example 3.3 In a reactive and reciprocal two-port, make the relations between the scattering parameters from (3.24) explicit.

Solution
From (3.24) we obtain:
$$\mathbf{SS}^* = \mathbf{I},$$

i.e., developing the product:
$$|S_{11}|^2 + |S_{12}|^2 = 1$$
$$S_{11}S_{12}^* + S_{12}S_{22}^* = 0$$
$$S_{12}S_{11}^* + S_{22}S_{12}^* = 0$$
$$|S_{22}|^2 + |S_{12}|^2 = 1;$$

thus, S_{11} and S_{22} have the same magnitude (note that the second and third equation are equivalent). This leads to a relationship between phases:

$$\phi_{11} - \phi_{12} = -\phi_{22} + \phi_{12} + n\pi,$$

with n odd, i.e.:

$$\phi_{11} + \phi_{22} = 2\phi_{12} + n\pi.$$

3.2.5 Power Wave Equivalent Circuit

From the series and parallel representations (3.7) and (3.8) we can derive the equivalent circuits in Fig. 3.3. We can proceed in a similar way with the power wave representation (3.11). In order to do this we need to introduce two new components, the forward and the backward wave generators shown in Fig. 3.4 (a) and (b).

The forward and backward wave generators operate by adding a contribution to the forward or backward wave, respectively, where their direction is denoted by the generator arrow. The scattering matrix of the forward and backward wave generators can be easily derived by inspection. For the forward wave generator we have:

$$\begin{cases} b_1 = a_2 \\ b_2 = a_1 + a_0, \end{cases} \quad (3.25)$$

while for the backward wave generator:

$$\begin{cases} b_1 = a_2 + b_0 \\ b_2 = a_1. \end{cases} \quad (3.26)$$

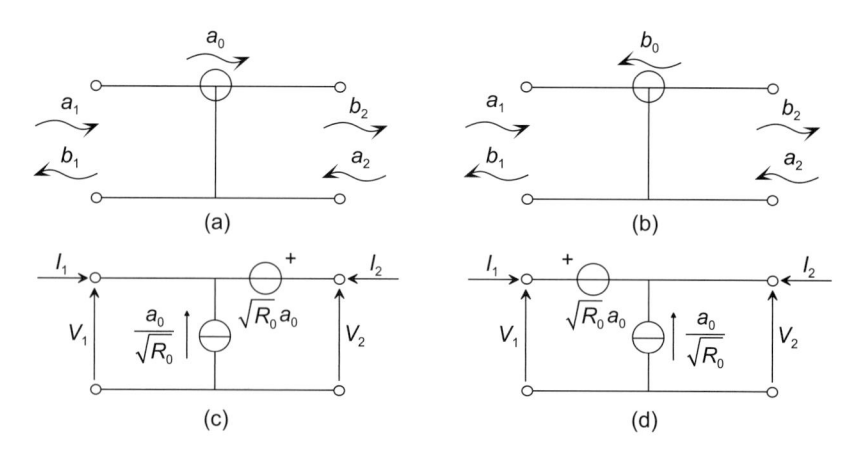

Figure 3.4 Forward (a) and backward (b) generator; the same series voltage and current parallel generators, (c) and (d), respectively.

3.2 Power Waves and the Scattering Parameters

Somewhat arbitrarily, we have associated the term "forward" with an added wave propagating towards the right (and denoted with a_0) while the backward generator is associated with a wave propagating to the left (and denoted with b_0).

The forward and backward wave generators can be easily shown to be a set of correlated series voltage and parallel current generators. Let us consider for example the forward wave generator, whose constitutive relationships are in (3.25). Using the same normalization resistance R_0 at both ports we obtain, expressing the power waves in terms of voltages and currents:

$$\frac{V_1 - R_0 I_1}{2\sqrt{R_0}} = \frac{V_2 + R_0 I_2}{2\sqrt{R_0}}$$

$$\frac{V_2 - R_0 I_2}{2\sqrt{R_0}} = \frac{V_1 + R_0 I_1}{2\sqrt{R_0}} + a_0,$$

i.e.:

$$V_1 - V_2 = R_0 (I_1 + I_2)$$
$$V_1 - V_2 = -R_0 (I_2 + I_1) - 2\sqrt{R_0} a_0.$$

Substituting we obtain:

$$I_2 = -I_1 - \frac{a_0}{\sqrt{R_0}} \qquad (3.27)$$

$$V_2 = V_1 + a_0 \sqrt{R_0}, \qquad (3.28)$$

corresponding to the circuit in Fig. 3.4 (c). Therefore, the forward wave generator corresponds to a set of two correlated voltage (in series) and current (in parallel) generators.

The equivalent circuits of autonomous n-ports in Fig. 3.3 can be suitably rephrased in terms of scattering matrix and forward/backward wave generators, see Fig. 3.5. In fact, if all ports are closed on the normalization resistance and no external forcing terms are present, $a_k = 0 \; \forall k$ and therefore $b_k = b_{0k} \; \forall k$. The result obtained is consistent with the definition of the scattering matrix in (3.11).

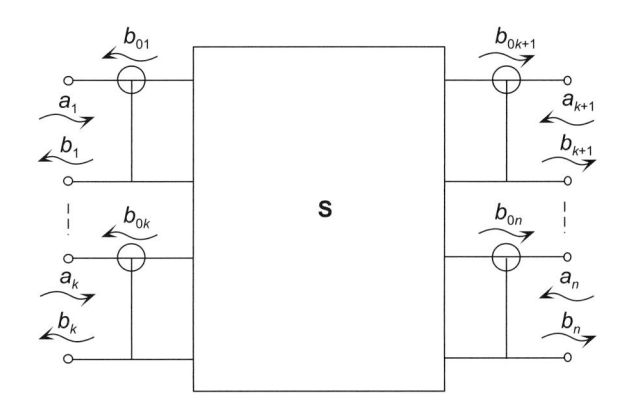

Figure 3.5 Power wave equivalent circuit.

3.2.6 Direct Evaluation of the Scattering Parameters

Eqs. (3.12) and (3.13) enable to evaluate **S** and the forward wave generators $\underline{b}_0$ from **Z** and the open-circuit voltage vector $\underline{V}_0$. Similar relations exist for the admittance matrix and the short-circuit currents. **S** and $\underline{b}_0$ can be, however, measured (or evaluated analytically) by directly exploiting their definition. From (3.11), we obtain $\underline{b} = \underline{b}_0$ when $\underline{a} = 0$, i.e., when all ports are closed on their normalization resistances. The elements of $\underline{b}_0$ derive from the total port voltages; from (3.6) we obtain:

$$b_{0i} = \frac{V_i}{\sqrt{R_{0i}}}. \tag{3.29}$$

Suppose now to set $\underline{b}_0 = 0$ by turning off all independent internal sources (thus making the n-port non-autonomous). From (3.11) we have $\underline{b} = \mathbf{S}\underline{a}$. For the elements of **S** we have:

$$S_{ij} = \left.\frac{b_i}{a_j}\right|_{a_k=0 \forall k \neq j}. \tag{3.30}$$

Condition $a_k = 0 \;\forall k \neq j$ is achieved by closing all ports, apart from the j-th one, on the corresponding normalization resistances, and exciting with a real generator (with an internal impedance that can be conveniently chosen as the port normalization resistance) the j-th port, see Fig. 3.6. The diagonal element S_{ii} is immediately derived from (3.30) as the reflection coefficient at port i when all other ports are closed on their normalization resistance, i.e.:

$$S_{ii} \equiv \Gamma_i = \frac{b_i}{a_i} = \frac{Z_i - R_{0i}}{Z_i + R_{0i}} = \frac{G_{0i} - Y_i}{G_{0i} + Y_i}, \tag{3.31}$$

where $G_{0i} = 1/R_{0i}$ is the normalization conductance of port i and Y_i is the input admittance of the one-port obtained by closing all ports but the i-th one on their normalization resistances.

The out-of-diagonal elements of the scattering matrix are transmission coefficients. To identify them, consider the circuit in Fig. 3.6; the power wave b_G coming out of the generator connected to port j is derived from (3.15) setting $Z = R_0$:

$$b_G = \frac{V_{0j}}{2\sqrt{R_{0j}}};$$

but the same wave enters port j, thus:

$$a_j = b_G = \frac{V_{0j}}{2\sqrt{R_{0j}}}. \tag{3.32}$$

We also have from (3.6):

$$b_i|_{a_k=0 \forall k \neq j} = \frac{V_i}{\sqrt{R_{0i}}}; \tag{3.33}$$

finally, then:

$$S_{ij}|_{i \neq j} = \left.\frac{b_i}{a_j}\right|_{a_k=0 \forall k \neq j} = 2\frac{V_i}{V_{0j}}\sqrt{\frac{R_{0j}}{R_{0i}}}. \tag{3.34}$$

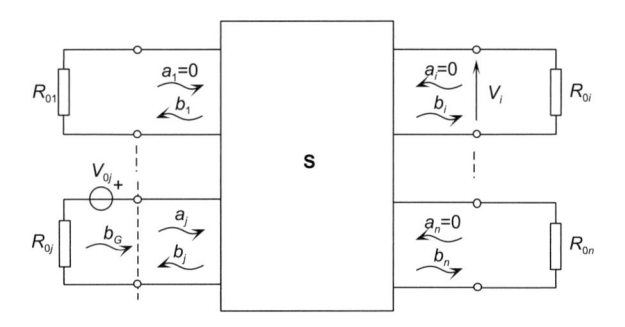

Figure 3.6 Evaluating the out-of-diagonal elements of the scattering matrix.

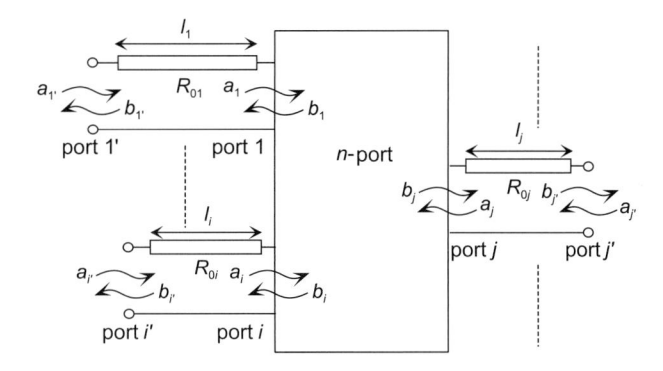

Figure 3.7 Reference plane shift.

Therefore, to evaluate the out-of-diagonal terms of **S** we simply need to compute or measure voltage ratios according to circuit theory.

3.2.7 Reference Plane Shift

Let us connect to port k ($k = 1 \ldots n$) of an n-port a transmission line of electrical length $\beta_k l_k$ and characteristic impedance Z_0 equal to the port normalization resistance R_{0k}, see Fig. 3.7. A new set of ports k' is obtained, such that the power waves at the old (unprimed) and new (primed) ports are related as:

$$a_k = a_{k'} \exp(-j\beta_k l_k)$$
$$b_k = b_{k'} \exp(j\beta_k l_k).$$

Since:

$$S_{ij} = \left.\frac{b_j}{a_i}\right|_{a_j=0},$$

and taking into account that the matching condition at port k is unchanged if a line with $Z_0 = R_{0k}$ is connected to the port, we have for the new set:

$$S'_{ij} = \left.\frac{b_{j'}}{a_{i'}}\right|_{a_{j'}=0} = \left.\frac{b_j \exp(-j\beta_j l_j)}{a_i \exp(j\beta_i l_i)}\right|_{a_j=0} = \exp\left(-j\beta_j l_j - j\beta_i l_i\right) S_{ij}.$$

The relationship between the scattering matrix at the new and at the old set is straightforward. Besides being an additional tool for analysis, see Example 3.4, the reference plane shift technique is exploited in measurements. Consider for instance a coplanar probe system, where the physical location of the ports is not well specified; such a location can be conveniently chosen in a standard position (e.g., in the center of the Device Under Test, DUT) by de-embedding a set of input and output lines.

Example 3.4 Derive the scattering parameters of a line of length l (propagation constant β) with respect to its characteristic impedance R_0 using the reference plane shift.

Solution
We start from evaluating the scattering parameters of a two-port made of two short circuits connecting the upper and lower nodes, respectively, of ports 1 and 2; see Fig. 3.8 (a). Since the structure is symmetrical and reciprocal we have $S_{11} = S_{22} = 0$, as the input reflection coefficient at port 1 is zero (the input impedance is R_0). Concerning $S_{12} = S_{21}$ we have:

$$V_2 = V_{01}/2,$$

and therefore, from (3.34):

$$S_{12} = \frac{2V_2}{V_{01}} = 1.$$

We now shift the reference plane, Fig. 3.8 (b), from port 1 to port 1', with a line of length l, while no shift is made on port 2. We have:

$$S'_{11} = S'_{22} = \exp(-j\beta l)S_{11} = 0$$
$$S'_{12} = S'_{21} = \exp(-j\beta l)S_{12} = \exp(-j\beta l),$$

and thus:

$$S = \begin{pmatrix} 0 & \exp(-j\beta l) \\ \exp(-j\beta l) & 0 \end{pmatrix}.$$

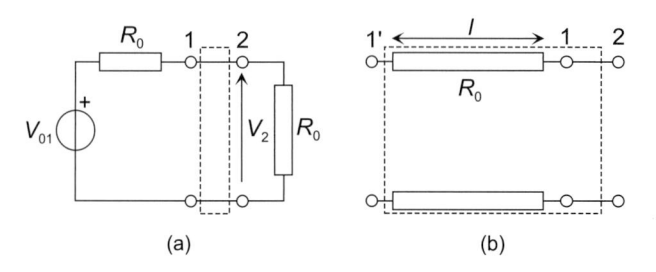

Figure 3.8 Evaluation of the scattering parameters of a line through the plane shift method: starting structure (a), reference plane shift at port 1 (b).

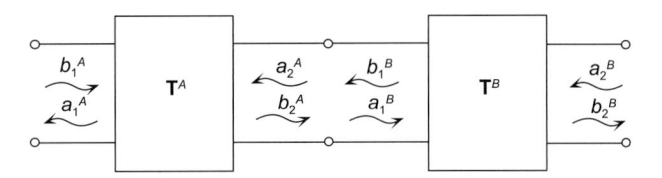

Figure 3.9 Cascade connection of two two-ports characterized by their transmission matrices.

3.2.8 Cascade Connection of Two-Ports: The T-Matrix

The cascade connection of two-ports, see Fig. 3.9, is a very common configuration in microwave systems but also in measurement setups. To describe the cascade of two two-ports it is convenient to define a *transmission matrix* **T** yielding the input power waves in terms of output power waves:

$$\begin{pmatrix} a_1 \\ b_1 \end{pmatrix} = \mathbf{T} \begin{pmatrix} b_2 \\ a_2 \end{pmatrix}. \tag{3.35}$$

We can similarly define:

$$\begin{pmatrix} b_2 \\ a_2 \end{pmatrix} = \mathbf{T}' \begin{pmatrix} a_1 \\ b_1 \end{pmatrix}, \tag{3.36}$$

where $\mathbf{T}' = \mathbf{T}^{-1}$. The two matrices are sometimes denoted as $\overrightarrow{\mathbf{T}} = \mathbf{T}'$ and $\overleftarrow{\mathbf{T}} = \mathbf{T}$, respectively. Suppose, for simplicity, that the normalization resistance is the same for all connected two-ports; in this case $a_1^B = b_2^A$ and $a_2^A = b_1^B$ for the power wave continuity, but we also have, from the definition of the transmission matrices of two-ports A and B:

$$\begin{pmatrix} a_1^A \\ b_1^A \end{pmatrix} = \mathbf{T}^A \begin{pmatrix} b_2^A \\ a_2^A \end{pmatrix} = \mathbf{T}^A \begin{pmatrix} a_1^B \\ b_1^B \end{pmatrix} = \mathbf{T}^A \mathbf{T}^B \begin{pmatrix} b_2^B \\ a_2^B \end{pmatrix}.$$

Therefore, the overall transmission matrix is $\mathbf{T} = \mathbf{T}^A \mathbf{T}^B$. The process can be iterated for the cascade of an arbitrary number of two-ports. The elements of the transmission matrix can be easily obtained from the scattering matrix as:

$$\mathbf{T} = \frac{1}{S_{21}} \begin{pmatrix} 1 & -S_{22} \\ S_{11} & -\Delta_S \end{pmatrix}, \tag{3.37}$$

where Δ_S is the determinant of the S-matrix. Similarly:

$$\mathbf{S} = \frac{1}{T_{11}} \begin{pmatrix} T_{21} & \Delta_T \\ 1 & -T_{12} \end{pmatrix}, \tag{3.38}$$

where Δ_T is the determinant of the transmission matrix.

Example 3.5 Suppose the transmission matrix **T** of a two-port is known. Evaluate the transmission matrix $\mathbf{T}_a$ obtained exchanging the input with the output port. What is the relationship between $\mathbf{T}_a$ and $\mathbf{T}'$?

Solution

From the definition of the transmission matrix, we have:

$$\begin{pmatrix} a_1 \\ b_1 \end{pmatrix} = \begin{pmatrix} T_{11} & T_{12} \\ T_{21} & T_{22} \end{pmatrix} \begin{pmatrix} b_2 \\ a_2 \end{pmatrix}. \quad (3.39)$$

Exchanging the two-port means defining matrix $\mathbf{T}_a$ as follows:

$$\begin{pmatrix} a_2 \\ b_2 \end{pmatrix} = \begin{pmatrix} T_{11a} & T_{12a} \\ T_{21a} & T_{22a} \end{pmatrix} \begin{pmatrix} b_1 \\ a_1 \end{pmatrix}.$$

Notice that $\mathbf{T}_a$ differs from $\mathbf{T}'$ defined in (3.36) because of the different ordering of the power wave vectors. Inverting (3.39) we obtain:

$$b_2 = (T_{22}/\Delta_T)a_1 - (T_{12}/\Delta_T)b_1$$
$$a_2 = -(T_{21}/\Delta_T)a_1 + (T_{11}/\Delta_T)b_1,$$

where Δ_T is the determinant of $\mathbf{T}$. Rearranging we obtain:

$$\begin{pmatrix} a_2 \\ b_2 \end{pmatrix} = \frac{1}{\Delta_T} \begin{pmatrix} T_{11} & -T_{21} \\ -T_{12} & T_{22} \end{pmatrix} \begin{pmatrix} b_1 \\ a_1 \end{pmatrix},$$

that defines the elements of $\mathbf{T}_a$. Notice that $\mathbf{T}_a$ is closely related to $\mathbf{T}'$, where:

$$\mathbf{T}' = \mathbf{T}^{-1} = \frac{1}{\Delta_T} \begin{pmatrix} T_{22} & -T_{12} \\ -T_{21} & T_{11} \end{pmatrix};$$

since the determinant is the same, we have:

$$T_{11a} = T'_{22} \quad T_{22a} = T'_{11}$$
$$T_{12a} = T'_{21} \quad T_{21a} = T'_{12}.$$

3.2.9 Solving a Network in Terms of Power Waves

Consider a network consisting of interconnected m-ports. With each m-port we can associate m sides with $2m$ unknowns, the port voltages and the current entering (and exiting) each port. Suppose the total number of ports is k (interconnected together) with a total number of unknowns $2k$ (port voltages and currents). Consider first what happens when connecting two-ports belonging to two different n-ports (see Fig. 3.10). From the Kirchhoff voltage and current laws we obtain:

$$\begin{cases} V_{ja} = V_{ib} \\ I_{ja} = -I_{ib}. \end{cases} \quad (3.40)$$

For any pair of connected ports, we can write two such equations; therefore, the total number of topological equations is $2k/2 = k$. Since for each port we have one constitutive equation, the problem is well posed, as we have k topological plus k constitutive equations in terms of port voltages and currents, equal to the number of unknowns ($2k$).

3.2 Power Waves and the Scattering Parameters

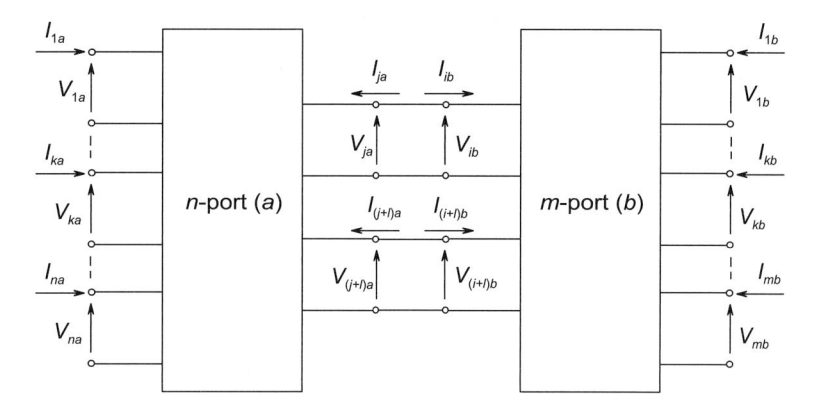

Figure 3.10 Connecting two multiport components.

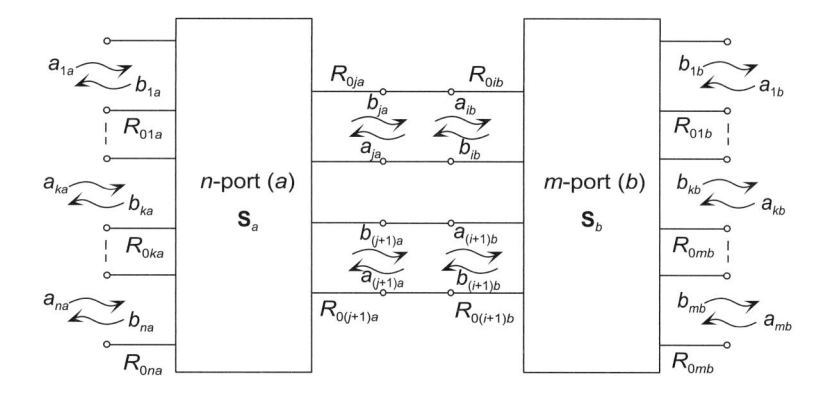

Figure 3.11 Connecting two multiport components in terms of power waves.

We can reformulate the problem in terms of power waves. Since each port hosts a forward and backward power wave, see Fig. 3.11, the total number of unknowns is again $2k$. Again, the constitutive relations based on scattering parameters (3.11) yield k relations, and k topological relationships can be obtained by expressing (3.40) in terms of power waves; from the definition of the power waves at ports i and j connected together we immediately have:

$$\begin{cases} \sqrt{R_{0ja}}(a_{ja} + b_{ja}) = \sqrt{R_{0ib}}(a_{ib} + b_{ib}) \\ (a_{ja} - b_{ja})/\sqrt{R_{0ja}} = -(a_{ib} - b_{ib})/\sqrt{R_{0ib}}, \end{cases} \quad (3.41)$$

that is, solving:

$$\begin{pmatrix} a_{ja} \\ b_{ja} \end{pmatrix} = \frac{1}{2\sqrt{R_{0ja}R_{0ib}}} \begin{pmatrix} R_{0ja} - R_{0ib} & R_{0ja} + R_{0ib} \\ R_{0ja} + R_{0ib} & R_{0ja} - R_{0ib} \end{pmatrix} \begin{pmatrix} a_{ib} \\ b_{ib} \end{pmatrix}. \quad (3.42)$$

If the normalization resistances are the same for ports i and j, (3.42) reduces to:

$$\begin{cases} a_{ja} = b_{ib} \\ b_{ja} = a_{ib}, \end{cases} \quad (3.43)$$

i.e., power waves are continuous across the interconnecting port. Generally speaking, therefore each couple of connected ports provides two topological relationships, yielding in total k relations that, added to the constitutive ones, finally give $2k$ relations, equal to the number of unknowns. Therefore, the problem is well posed also in terms of power waves.

3.3 Analysis Techniques for Linear and Nonlinear Circuits

3.3.1 Time-Domain vs. Frequency-Domain Methods

The CAD analysis of linear and nonlinear circuits typically exploits approaches based on the nodal (or modified nodal [12]) approach, where the unknowns of the problem are the circuit nodal voltages [1]. The direct solution in terms of power waves, though possible, is not commonly found in microwave circuit solvers. With the modified nodal analysis, the resulting equation system is, in time domain, a system of linear or nonlinear ordinary differential equations in time, according to whether the circuit under examination is purely linear or includes nonlinear elements. The solution of the system can be carried out by means of numerical quadrature or time-stepping algorithms, whose convergence and accuracy are ensured by a number of well-known numerical techniques. Such methods are currently implemented in the available circuit simulators, starting from the well-known SPICE [13, 14]. In nonlinear circuits, each time integration step requires in principle the solution of a nonlinear system through iterative techniques, such as the Newton–Raphson method.

The time-domain analysis of RF or microwave narrowband circuits is fraught with a number of difficulties: (1) distributed components can be simulated, but not in a completely straightforward way; (2) in periodic or quasi-periodic operation, the circuit initial turn-on transient is typically very long (due to the slow time constant of the bias circuits) with respect to the period of the excitation. Taking into account the above problems, fast numerical techniques to compute the circuit steady-state operation were developed, both in the time domain (the so-called shooting or extrapolation approaches [15]) and in the frequency domain. For linear circuits, the frequency domain analysis reduces to the well-known phasor approach, where the ordinary differential equation system is Fourier transformed into a complex algebraic system. Nonlinear circuits can be solved in the frequency domain through the so-called Harmonic Balance technique, first proposed in the early seventies [16, 17]; the detailed implementation for single-tone excitation is described in Sec. 3.3.2. Multi-tone excitation or the analysis in the presence of modulated signal required more advanced numerical methods that are described in Sec. 3.3.3 and Sec. 3.3.4, respectively.

3.3.2 The Harmonic Balance Technique

Frequency-domain techniques for the analysis of nonlinear analogue networks aim at directly deriving a periodic or quasi-periodic steady state solution of the circuit, thus

3.3 Analysis Techniques for Linear and Nonlinear Circuits

avoiding the (often long) transient analysis required by time-domain methods. The extension of phasor-like approaches to nonlinear circuits and their widespread diffusion in CAD tools has been, during the last decades, a major breakthrough in the analysis and design of quasi-linear or nonlinear microwave circuits. Here, we will confine the treatment to the *harmonic balance* (HB) technique that is today the preferred simulation method, starting from its initial formulation, in which all frequencies generated by the circuit belong to the set nf_0. This corresponds to a single-tone sinusoidal excitation at f_0 or to a set of commensurate excitations (e.g., two tones at f_0 and $2f_0$). Extensions to the general case of multi-tone excitation will be discussed in Sec. 3.3.3.

Let us consider a single tone excitation at f_0; the output spectrum is suitably truncated at an upper harmonic N. All branch currents and voltages can be written as a truncated Fourier expansion as:

$$i_l(t) = I_{l0} + \sum_{i=1}^{N} [I_{lci} \cos(i\omega_0 t) + I_{lsi} \sin(i\omega_0 t)] \qquad (3.44a)$$

$$v_l(t) = V_{l0} + \sum_{i=1}^{N} [V_{lci} \cos(i\omega_0 t) + V_{lsi} \sin(i\omega_0 t)]. \qquad (3.44b)$$

We can collectively describe such quantities with a $2N+1$ array including the DC component and the in-phase (cosine) and quadrature (sine) components as:

$$i_l(t) \leftrightarrow [I_{l0}, I_{lc1}, I_{ls1}, I_{lc2}, I_{ls2}, \ldots, I_{lcN}, I_{lsN}]^T = \mathbf{I}_l^T$$
$$v_l(t) \leftrightarrow [V_{l0}, V_{lc1}, V_{ls1}, V_{lc2}, V_{ls2}, \ldots, V_{lcN}, V_{lsN}]^T = \mathbf{V}_l^T.$$

We know, however, that the sampling theorem allows a band-limited signal to be recovered, whose spectrum includes $N+1$ frequencies, by sampling it into $2N+1$ equispaced times. Let us call T_0 the fundamental period; we can associate with the branch currents and voltages the time sample arrays:

$$i_l(t) \leftrightarrow \left[i_l(t_1), i_l(t_2), \ldots, i_l(t_{2N+1})\right]^T = \mathbf{i}_l^T$$
$$v_l(t) \leftrightarrow \left[v_l(t_1), v_l(t_2), \ldots, v_l(t_{2N+1})\right]^T = \mathbf{v}_l^T.$$

Taking into account that:

$$i_l(t_n) = I_{l0} + \sum_{i=1}^{N} [I_{lci} \cos(i\omega_0 t_n) + I_{lsi} \sin(i\omega_0 t_n)] = \mathbf{F}_n \cdot \mathbf{I}_l$$

$$v_l(t_n) = V_{l0} + \sum_{i=1}^{N} [V_{lci} \cos(i\omega_0 t_n) + V_{lsi} \sin(i\omega_0 t_n)] = \mathbf{F}_n \cdot \mathbf{V}_l,$$

where:

$$\mathbf{F}_n = [1, \cos(\omega_0 t_n), \sin(\omega_0 t_n), \cos(2\omega_0 t_n), \sin(2\omega_0 t_n), \ldots, \cos(N\omega_0 t_n), \sin(N\omega_0 t_n)].$$

Defining the matrix $\mathbf{F}$ having $\mathbf{F}_n$ as rows (where $\theta_{n,k} = k\omega_0 t_n$):

$$\mathbf{F} = \begin{pmatrix} 1 & \cos\theta_{1,1} & \sin\theta_{1,1} & \cdots & \cos\theta_{1,N} & \sin\theta_{1,N} \\ 1 & \cos\theta_{2,1} & \sin\theta_{2,1} & \cdots & \cos\theta_{2,N} & \sin\theta_{2,N} \\ \vdots & \vdots & \vdots & \ddots & \vdots & \vdots \\ 1 & \cos\theta_{2N+1,1} & \sin\theta_{2N+1,1} & \cdots & \cos\theta_{2N+1,N} & \sin\theta_{2N+1,N} \end{pmatrix}, \quad (3.45)$$

we obtain:

$$\mathbf{i}_l = \mathbf{F} \cdot \mathbf{I}_l, \quad \mathbf{v}_l = \mathbf{F} \cdot \mathbf{V}_l,$$

and, inversely:

$$\mathbf{I}_l = \mathbf{F}^{-1} \cdot \mathbf{i}_l, \quad \mathbf{V}_l = \mathbf{F}^{-1} \cdot \mathbf{v}_l.$$

Consider now a nonlinear network made of linear components (concentrated and distributed) and nonlinear concentrated elements, such as lumped nonlinear inductors and capacitors, or nonlinear dependent generators. A general strategy can be now used to express the constitutive relationships of such elements in terms of the frequency or time domain samples of the branch voltages and currents.

Consider first a *linear resistor* of resistance R; we have:

$$\mathbf{v}_l = R\mathbf{i}_l \rightarrow \mathbf{V}_l = \mathbf{R}\mathbf{I}_l,$$

where $\mathbf{R} = R\mathbf{I}$, $\mathbf{I}$ identity matrix. Linear reactive elements such as *capacitors* and *inductors* require to express in the HB formalism the time derivatives of voltages and currents, respectively. For the sake of definiteness, let us discuss the current time derivative; we have, using the truncated Fourier series:

$$\frac{di_l(t)}{dt} = i'_l(t) = \sum_{i=1}^{N} [-i\omega_0 I_{lci} \sin(i\omega_0 t) + i\omega_0 I_{lsi} \cos(i\omega_0 t)] \,;$$

thus, the frequency domain array associated with the time derivative of the current is:

$$i'_l(t) \leftrightarrow [0, -\omega_0 I_{ls1}, \omega_0 I_{lc1}, -2\omega_0 I_{ls2}, 2\omega_0 I_{lc2}, \ldots, -N\omega_0 I_{lsN}, N\omega_0 I_{lcN}]^T .$$

The frequency-domain sample array corresponding to i'_l, $\mathbf{I}'_l$, can be obtained from $\mathbf{I}_l$ as follows:

$$\mathbf{I}'_l = \begin{pmatrix} 0 \\ -\omega_0 I_{ls1} \\ \omega_0 I_{lc1} \\ \vdots \\ -N\omega_0 I_{lsN} \\ N\omega_0 I_{lcN} \end{pmatrix} = \begin{pmatrix} 0 & 0 & 0 & \cdots & 0 & 0 \\ 0 & 0 & -\omega_0 & \cdots & 0 & 0 \\ 0 & \omega_0 & 0 & \cdots & 0 & 0 \\ \vdots & \vdots & \vdots & \ddots & \vdots & \vdots \\ 0 & 0 & 0 & \cdots & 0 & -N\omega_0 \\ 0 & 0 & 0 & \cdots & N\omega_0 & 0 \end{pmatrix} \begin{pmatrix} I_{l0} \\ I_{lc1} \\ I_{ls1} \\ \vdots \\ I_{lcN} \\ I_{lsN} \end{pmatrix}, \quad (3.46)$$

i.e., in compact form:

$$\mathbf{I}'_l = \Omega \cdot \mathbf{I}_l,$$

where matrix Ω is defined in (3.46). Similarly, we have:

$$\mathbf{V}'_l = \Omega \cdot \mathbf{V}_l.$$

For a *linear capacitor* the charge–voltage relation holds:

$$\mathbf{Q}_l = \mathbf{C} \cdot \mathbf{V}_l,$$

where $\mathbf{C} = C\mathbf{I}$; however $\mathbf{I}_l = \mathbf{Q}'_l = \mathbf{\Omega} \cdot \mathbf{Q}_l$, i.e.:

$$\mathbf{I}_l = \mathbf{\Omega} \cdot \mathbf{Q}_l = \mathbf{\Omega} \cdot \mathbf{C} \cdot \mathbf{V}_l.$$

Similarly, for a *linear inductor*:

$$\mathbf{V}_l = \mathbf{\Omega} \cdot \mathbf{\Phi}_l = \mathbf{\Omega} \cdot \mathbf{L} \cdot \mathbf{I}_l.$$

Using the frequency–time formalism we can also easily describe nonlinear elements. Consider first a *nonlinear resistor* described by the memoriless relationship:

$$\mathbf{v}_l = \mathbf{f}_R(\mathbf{i}_l),$$

where the function $\mathbf{f}_R(\bullet)$ applies to the single elements of the argument vector. The frequency-domain constitutive relationship is easily derived as:

$$\mathbf{F} \cdot \mathbf{V}_l = \mathbf{f}_R(\mathbf{F} \cdot \mathbf{I}_l) \rightarrow \mathbf{V}_l = \mathbf{F}^{-1} \cdot \mathbf{f}_R(\mathbf{F} \cdot \mathbf{I}_l).$$

In a similar way, a *nonlinear capacitor* is described by the charge–voltage nonlinear constitutive relationship, leading to the frequency-domain version:

$$\mathbf{q}_l = \mathbf{f}_C(\mathbf{v}_l) \rightarrow \mathbf{Q}_l = \mathbf{F}^{-1} \cdot \mathbf{f}_C(\mathbf{F} \cdot \mathbf{V}_l),$$

from which the frequency-domain current–voltage relation:

$$\mathbf{I}_l = \mathbf{\Omega} \cdot \mathbf{F}^{-1} \cdot \mathbf{f}_C(\mathbf{F} \cdot \mathbf{V}_l).$$

Similarly, for a *nonlinear inductor* with constitutive relationship $\phi_l = \mathbf{f}_L(\mathbf{i}_l)$ we have the frequency-domain current–voltage relation:

$$\mathbf{V}_l = \mathbf{\Omega} \cdot \mathbf{F}^{-1} \cdot \mathbf{f}_L(\mathbf{F} \cdot \mathbf{I}_l).$$

Generators can be characterized in a similar way; an independent source (current and voltage) yields the relations:

$$\mathbf{I}_l = \mathbf{A}_l, \quad \mathbf{V}_l = \mathbf{E}_l,$$

where $\mathbf{A}_l$ and $\mathbf{E}_l$ are harmonic component vectors. For dependent linear generators the representation is trivial; consider instead, e.g., a current nonlinear voltage-dependent generator defined by:

$$i_l(t) = f_i(v_1, \ldots, v_n) \rightarrow \mathbf{i}_l = \mathbf{f}_i(\mathbf{v}_1, \ldots, \mathbf{v}_2),$$

in the frequency domain one has:

$$\mathbf{F} \cdot \mathbf{I}_l = \mathbf{f}_i(\mathbf{F} \cdot \mathbf{v}_1, \ldots, \mathbf{F} \cdot \mathbf{v}_2) \rightarrow \mathbf{I}_l = \mathbf{F}^{-1} \cdot \mathbf{f}_i(\mathbf{F} \cdot \mathbf{v}_1, \ldots, \mathbf{F} \cdot \mathbf{v}_2).$$

In conclusion, all constitutive relationships can be directly described in the frequency domain. Since the frequency basis adopted to represent voltages and currents is, within the limits of the truncation adopted, a complete orthogonal set, also the topological relationships such as the Kirchhoff Current Law (KCL) and Kirchhoff Voltage Law (KVL) can be imposed on a frequency-by-frequency basis:

$$\sum_k i_k(t) = 0 \rightarrow \sum_k \mathbf{i}_k = 0 \rightarrow \sum_k \mathbf{I}_k = 0$$

$$\sum_l v_l(t) = 0 \rightarrow \sum_l \mathbf{v}_l = 0 \rightarrow \sum_l \mathbf{V}_l = 0,$$

where summations are extended to all currents entering a node (KCL) or to all voltages along a closed loop (KVL), respectively. Notice that both the nodal KCL and the loop KVL have dimension $2N + 1$ in the frequency-domain representation.

Finally, consider the solution of a network where the above frequency-domain approach has been used to define the constitutive relationships of the components; as well known, in a network with n nodes and l sides we can write:

- $n - 1$ linearly independent KCL
- $l - n + 1$ linearly independent KVL
- l constitutive relationships.

In the multi-frequency representation each of the above relations has dimension $2N + 1$; therefore, the total number of scalar equations that can be written in terms of the harmonic components of side currents and voltages is $M = (2N + 1) \times 2l$. The nonlinear system obtained this way is well posed (i.e., the number of equations balances the number of unknowns) and can be solved numerically, e.g., through Newton iteration.

3.3.3 The HB Technique: Multi-Tone Excitation

The harmonic balance technique described so far for single-tone excitation can be, as a matter of principle, applied also for multi-tone excitation, provided that all tones are *commensurate*. We say that the tones belonging to a set f_i, $i = 1 \ldots L$ are commensurate if $D_i f_i = N_i f_0$, where N_i and D_i are mutually prime integers and f_0 is some frequency. It can be shown that a set of commensurate tones yields a strictly periodic waveform with period:

$$T' = T_0 \frac{\text{mcm}(D_1, D_2, \ldots, D_L)}{\text{MCD}(N_1, N_2, \ldots, N_L)},$$

where $T_0 = 1/f_0$, and the symbols mcm and MCD denote the minimum commune multiple and maximum common divisor, respectively. Taking into account that any set of input tones represented in finite precision arithmetic is, strictly speaking, commensurate, we see however that for such "practically non-commensurated" tones T' is extremely large and therefore the corresponding frequency $f' = 1/T'$ extremely small. We therefore see that, in such cases, a very large number of frequencies nf', $n = 1 \ldots N$ has to be considered, whose phasors are, mostly, zero. Indeed, in narrowband systems the excitation is on a narrow frequency band around some carrier frequency f_c, and therefore the output spectrum resulting from a nonlinear subsystem, tends to cluster around DC and the harmonics of the carrier frequency, while most of the remaining spectrum is empty;

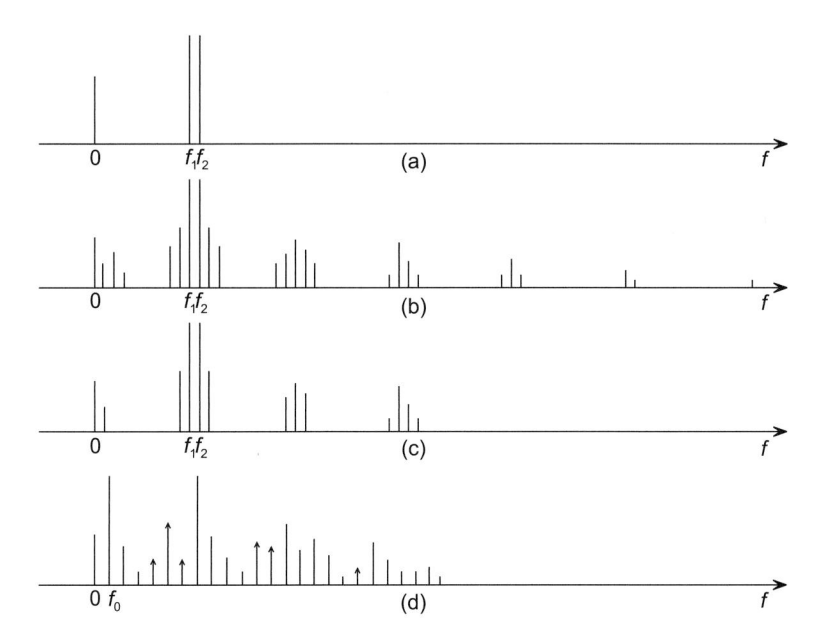

Figure 3.12 Two-tone excitation plus DC (a), and resulting output spectrum with clustering around the input harmonics with box truncation at $N = 3$ (b) or diamond truncation (c); artificial remapping of (b) into a strictly periodic spectrum with fundamental f_0 (d). The normalized input frequencies are 1 and 1.1, respectively, and the spectral amplitudes are chosen so as to qualitatively mimic the behavior of a nonlinear system (e.g., a power active device). The remapping algorithm is the same as in Example 3.6; the spectral lines with an arrow in (d) refer to lines being at negative frequencies in the original spectrum.

a qualitative example relative of a two-tone excitation and the resulting output spectrum is shown in Fig. 3.12 (a) and (b).

A more efficient approach would be to handle directly the multi-tone excitation case (also for non-commensurate tones) through a technique avoiding the unnecessary use of a large number of spectral unknowns that are known *a priori* to be zero. In the presence of an M tone excitation, this corresponds to using the series representation:

$$x(t) = X_0 + \sum_{k=1}^{M} \sum_{i=1}^{N} [X_{cki} \cos(i\omega_k t) + X_{ski} \sin(i\omega_k t)]$$

$$= X_0 + \sum_{l=1}^{L=M \times N} [X_{cl} \cos(\omega_l t) + X_{sl} \sin(\omega_l t)], \quad (3.47)$$

where $x(t)$ denotes a generic voltage or current. The format of the Fourier expansion (3.47) of $x(t)$ is deceivingly similar to (3.44), the basic difference being that (3.47) generally describes a non-periodic signal. It follows that the transformation between frequency and time samples in (3.45) becomes ill-conditioned and therefore approximate only. To overcome the problem, a number of techniques have been suggested; we will discuss here briefly the *oversampling* method [18, 19], the Multi-Dimensional

Fourier Transform (MDFT) approach (see [20] and references therein) and the Artificial Frequency Mapping (AFM) technique [21, 22].

The oversampling technique is based on the fact that any signal $x(t)$, also non-periodic, can be approximated as a superimposition of sinusoidal signals, since the error can be reduced at will including more sinusoidal terms. Therefore:

$$x(t) \approx X_0 + \sum_{i=1}^{K} X_i \cos(\omega_i t + \varphi_i) = X_0 + \sum_{i=1}^{K} X_i^c \cos \omega_i t + \sum_{i=1}^{K} X_i^s \sin \omega_i t, \quad (3.48)$$

where X_0 accounts for the mean value of $x(t)$. The ω_i values are in principle arbitrary, but we will focus on the specific case of circuits excited by multi-tone inputs, where the circuit voltages and currents have frequency components deriving from the input tone intermodulation. This allows the most suitable ω_i values to be derived in a consistent way. Eq. (3.48) can be used to establish a relationship between the $2K + 1$ spectral amplitudes $\left[X_0, \left(X_i^c, X_i^s\right), i = 1 \ldots K\right]$, and the time samples $x(t_i)$. We can in fact write a linear system of H equations in the spectral $2K + 1$ unknowns:

$$x(t_1) = X_0 + X_1^c \cos \omega_1 t_1 + X_1^s \sin \omega_1 t_1 + \cdots + X_K^c \cos \omega_K t_1 + X_1^s \sin \omega_K t_1$$
$$x(t_2) = X_0 + X_1^c \cos \omega_1 t_2 + X_1^s \sin \omega_1 t_2 + \cdots + X_K^c \cos \omega_K t_2 + X_1^s \sin \omega_K t_2$$
$$\vdots$$
$$x(t_H) = X_0 + X_1^c \cos \omega_1 t_H + X_1^s \sin \omega_1 t_H + \cdots + X_K^c \cos \omega_K t_H + X_1^s \sin \omega_K t_H.$$

If we choose $H = 2K + 1$, the system can be in principle inverted and the $2K + 1$ frequency phasors can be computed from the $2K + 1$ time samples, and *vice versa*. In presence of single-tone excitation at ω_0 and for uniformly spaced time samples, the system reduces to the classical Discrete Fourier Transform (DFT) already introduced in single tone harmonic balance analysis, where the $\omega_i = i\omega_0$ and the system matrix is well conditioned. For multi-tone excitation, however, the system can become extremely ill conditioned, unless the time samples are properly chosen according to the ω_i. To overcome this issue, more than $2K + 1$ samples can be used, thus obtaining a system having more equations than unknowns; a typical rule of thumb is to adopt an oversampling factor of 10. After that, a Gram–Schmidt procedure is adopted to select, among all the rows of the system matrix, the $2K + 1$ most orthogonal ones; this allows a well-conditioned DFT matrix to be obtained. Further details on this technique can be found on [18] and [19].

The MDFT is based upon introducing a formal multi-time representation for the signal:

$$x(t_1, \ldots t_M) = X_0 + \sum_{k=1}^{M} \sum_{i=1}^{N} \left[X_{cki} \cos(i\omega_k t_k) + X_{ski} \sin(i\omega_k t_k) \right].$$

For each time variable, a proper uniform sampling can be introduced; in this way, the time-discretized signal is associated with an $M-$ dimensional matrix $\mathbf{x}$, related to a an $M-$ dimensional matrix $\mathbf{X}$ of frequency sine and cosine samples by a transformation matrix $\mathbf{F}$. Time derivatives are expressed through the help of partial derivatives:

$$\frac{dx(t)}{dt} = \frac{dx(t_1,\ldots t_M)}{dt_1} + \ldots + \frac{dx(t_1,\ldots t_M)}{dt_M}\bigg|_{t_1,\ldots t_M=t}$$

and the relevant frequency samples are derived from the frequency samples of x through a proper matrix Ω. Although elegant and reasonably efficient when the number of input tones is low, the MDFT approach becomes heavy for a large number of input tones, needed, e.g., to emulate a realistic band-continuous modulated signal.

Contrarily to the MDFT, the AFM technique remains simple, and compatible with a standard HB implementation, also for a very large number of input tones. In short, the AFM maps the output spectrum (the physical set of frequencies), that is sparse and clustered, into a set of equally spaced frequencies nf_0, $n = 0 \ldots N$ (the artificial set). Notice however that, while f_0 is arbitrary, the mapping between the physical and the artificial set has to be properly defined so as to ensure that if $f_a = Mf_b$ in the physical set, also $f'_a = Mf'_b$ in the artificial set, where $f_a \to f'_a$ and $f_b \to f'_b$, see [18, Sec. 3.3.4.3]; this point is discussed further on when dealing with the truncation issue.

Apparently, the mapping introduced by AFM is unphysical, but it exploits in fact an important property of nonlinear networks made of conventional lumped elements: each component, also nonlinear, is described, with a proper choice of state variables, by a memoriless model. This means that, if the spectral amplitudes in (3.47) are associated with the artificial spectrum (that describes a *strictly periodic* signal) and transformed into an equally spaced time samples vector $\mathbf{x}$ of dimension $2N+1$, satisfying the Nyquist criterion related to Nf_0, as $\mathbf{x} = \mathbf{F} \cdot \mathbf{X}$, where $\mathbf{F}$ is defined in (3.45), this vector can be transformed through a memoriless relationship $\mathbf{f}(\bullet)$ into an output time sample vector $\mathbf{y}$ as:

$$\mathbf{y} = \mathbf{f}(\mathbf{x})$$

and finally the frequency samples vector is obtained as $\mathbf{Y} = \mathbf{F}^{-1} \cdot \mathbf{y}$. This allows the relationship between the input and output frequency samples to be written as:

$$\mathbf{Y} = \mathbf{F}^{-1} \cdot \mathbf{f}(\mathbf{F} \cdot \mathbf{X}).$$

In Example 3.7 a simple demonstration of the correctness of this procedure is discussed for a cubic nonlinearity having as the input a two-tone signal. From the frequency sample vector $\mathbf{Y}$ the corresponding vector $\mathbf{Y}'$ describing the frequency-domain representation of dy/dt can be now recovered through the matrix Ω_t, this time associated with the *true* spectrum of the signal:

$$\Omega_t = \begin{pmatrix} 0 & 0 & 0 & \ldots & 0 & 0 \\ 0 & 0 & -\omega_1 & \ldots & 0 & 0 \\ 0 & \omega_1 & 0 & \ldots & 0 & 0 \\ \vdots & \vdots & \vdots & \ddots & \vdots & \vdots \\ 0 & 0 & 0 & \ldots & 0 & -\omega_L \\ 0 & 0 & 0 & \ldots & \omega_L & 0 \end{pmatrix},$$

where $L = M \times N$ and the angular frequencies are those appearing in (3.47). We have then:
$$\mathbf{Y}' = \mathbf{\Omega}_t \cdot \mathbf{Y}.$$

Thus, the AFM approach allows for multiple tone excitations in an efficient way, exploiting exactly the same framework as for single-tone HB.

We discuss now in some more detail the implementation of the mapping procedure. First of all, some truncation criterion has to be adopted in handling the multi-tone spectrum. While truncation in the single tone case is trivial, in the multiple tone case several choices are possible, the most popular ones being the *box* and the *diamond* truncation. For the sake of simplicity, let us consider the two-tone case, where the possible generated frequencies (positive and negative) resulting from excitation at f_1 and f_2 are:

$$f_{m,n} = mf_1 + nf_2, \quad m, n = -\infty \ldots +\infty.$$

Assume to limit the range of m and n to:

$$-M \leq m \leq M, \quad -N \leq n \leq N,$$

and that (as reasonable) $M = N$; according to *box truncation* we keep all frequencies such as:

$$|m| \leq M, \quad |n| \leq M,$$

while according to *diamond truncation* we keep all frequencies such as:

$$|m| + |n| \leq M.$$

An example of the result from box truncation, in the two-tone case, with $M = 2$ is:

$$\begin{array}{ccccc}
-2f_1 + 2f_2 & -f_1 + 2f_2 & 2f_2 & f_1 + 2f_2 & 2f_1 + 2f_2 \\
-2f_1 + f_2 & -f_1 + f_2 & f_2 & f_1 + f_2 & 2f_1 + f_2 \\
-2f_1 & -f_1 & 0 & f_1 & 2f_1 \\
-2f_1 - f_2 & -f_1 - f_2 & -f_2 & f_1 - f_2 & 2f_1 - f_2 \\
-2f_1 - 2f_2 & -f_1 - 2f_2 & -2f_2 & f_1 - 2f_2 & 2f_1 - 2f_2,
\end{array} \quad (3.49)$$

whereas the corresponding diamond truncation frequency set will be:

$$\begin{array}{ccccc}
 & & 2f_2 & & \\
 & -f_1 + f_2 & f_2 & f_1 + f_2 & \\
-2f_1 & -f_1 & 0 & f_1 & 2f_1. \\
 & -f_1 - f_2 & -f_2 & f_1 - f_2 & \\
 & & -2f_2 & &
\end{array} \quad (3.50)$$

It is quite clear that for a large M the diamond truncation technique leads to a number of frequencies that is approximately one half than in the box truncation case. The extension to the N-tone case is trivial, the frequencies to be considered being:

$$f_{m_1,\ldots m_N} = m_1 f_1 + m_2 f_2 + \ldots + m_N f_N,$$

with the uniform limiting range:

$$-M \leq m_i \leq M;$$

3.3 Analysis Techniques for Linear and Nonlinear Circuits

according to box and diamond truncation we have, respectively:

$$|m_i| \leq M \quad \forall i$$
$$|m_1| + |m_2| + \ldots + |m_N| \leq M.$$

As already recalled, the artificial mapping is not completely arbitrary, since it must preserve the ratios between frequencies in the physical set. Moreover, each frequency in the physical set f_i ($i = 1 \ldots K + 1$) resulting from truncation has to be univoquely mapped into one frequency in the artificial set $(i - 1)f_0$, $i = 1 \ldots K + 1$. Notice that if two frequencies of the physical set coincide, they are mapped in different frequencies of the artificial set; this may occur when in fundamental tones are *commensurate*.[3] A mapping procedure satisfying the above-mentioned requirements is discussed, for a general N tone excitation and for box or diamond truncation in [18, Sec. 3.3.4.3]. For the sake of simplicity we will only provide an example for the two-tone case with box truncation (see Example 3.6).

Focusing for simplicity to the box scheme, a last remark concerns the relationship between the truncation order M and the order of the nonlinearity under consideration, N. In general, the AFM provides an exact result only if $M \geq N$, while if $M < N$ errors are introduced, caused by the aliasing of some of the frequencies in the artificial spectrum. This imposes some care when considering, e.g., exponential nonlinearities, where the approximate order actually depends on the amplitude of the excitation.

Example 3.6 Consider a two-tone excitation (f_1 and f_2) and suppose to box-truncate the physical set to $M = 2$. Derive the artificial set, with fundamental frequency f_0.

Solution

For the two-tone case the procedure to define the artificial set is graphically summarized in Fig. 3.13. The artificial set mapping the physical set in (3.49) is:

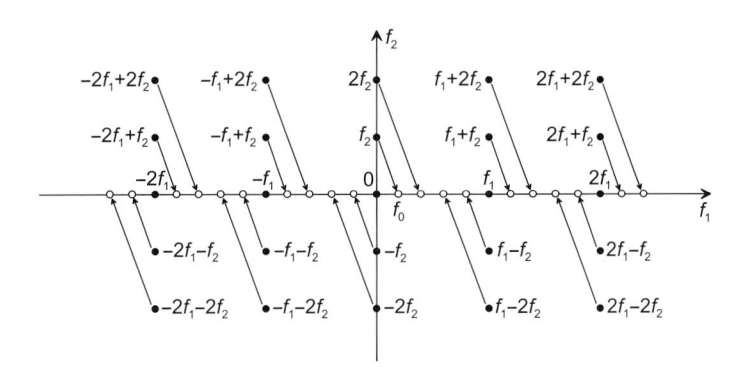

Figure 3.13 Artificial mapping scheme for a box-truncated ($M = 2$) two-tone excitation.

---

[3] As a trivial example, if in the two-tone case one has $f_1 = 1$ and $f_2 = 2$, the first-order IMP $f_2 - f_1 = 1$ coincides with the fundamental f_1.

$$\begin{array}{ccccc} -8f_0 & -3f_0 & 2f_0 & 7f_0 & 12f_0 \\ -9f_0 & -4f_0 & f_0 & 6f_0 & 11f_0 \\ -10f_0 & -5f_0 & 0 & 5f_0 & 10f_0 \\ -11f_0 & -6f_0 & -f_0 & 4f_0 & 9f_0 \\ -12f_0 & -7f_0 & -2f_0 & 3f_0 & 8f_0. \end{array} \qquad (3.51)$$

The artificial set (3.51) preserves the original frequency ratios, e.g., $f_1 - f_2 \to 4f_0$, $2f_1 - 2f_2 \to 8f_0$. Moreover, even if two frequencies in the physical set are the same, they will be mapped into different frequencies of the artificial set: suppose, e.g., again that $f_1 = 1$ and $f_2 = 2$, while $f_2 - f_1 = 1 = f_1$, we have $f_1 = 1 \to 5f_0$, $f_2 - f_1 = 1 \to -6f_0$ (notice that in the artificial set frequencies that are positive in the physical set can be mapped into negative frequencies).

Example 3.7 Consider a two-tone cosine signal $x(t)$ with angular frequencies ω_1 and ω_2 and a memoriless cubic system $y = x^3$. Show that the spectral amplitudes directly derived from $x(t)$ in its original form coincide with those resulting from remapping x and y into strictly periodic signals z and w with fundamental angular frequency ω_0. In evaluating y assume $M = N = 3$.

Solution
Since the input signal x only has cosine components, the same will occur for y; in the natural spectrum we therefore have, with straightforward algebraic and trigonometric manipulations:

$$x(t) = X_1 \cos \omega_1 t + X_2 \cos \omega_2 t$$
$$y(t) = (X_1 \cos \omega_1 t + X_2 \cos \omega_2 t)^3$$
$$= \left(\frac{3}{4}X_1^3 + \frac{3}{2}X_1 X_2^2\right) \cos \omega_1 t + \left(\frac{3}{4}X_2^3 + \frac{3}{2}X_1^2 X_2\right) \cos \omega_2 t$$
$$+ \frac{3}{4}X_1^2 X_2 \cos (2\omega_1 - \omega_2) t + \frac{3}{4}X_1 X_2^2 \cos (2\omega_2 - \omega_1) t$$
$$+ \frac{3}{4}X_1 X_2^2 \cos (2\omega_2 + \omega_1) t + \frac{3}{4}X_1^2 X_2 \cos (2\omega_1 + \omega_2) t$$
$$+ \frac{1}{4}X_1^3 \cos 3\omega_1 t + \frac{1}{4}X_2^3 \cos 3\omega_2 t.$$

Let us now remap the spectrum of x following the artificial mapping (we call z the remapped x function and w the remapping of y). With $M = 3$ we have (to keep the notation compact we synthesize $m\omega_1 + n\omega_2 \to m+n$, with ω_1 horizontal axis; only the underlined frequencies are generated by the third-order nonlinearity):

$$\begin{array}{ccccccc} -3+3 & -2+3 & -1+3 & +0+3 & +1+3 & +2+3 & +3+3 \\ -3+2 & -2+2 & \underline{-1+2} & +0+2 & \underline{+1+2} & +2+2 & +3+2 \\ -3+1 & \underline{-2+1} & -1+1 & +0+1 & +1+1 & \underline{+2+1} & +3+1 \\ -3+0 & -2+0 & -1+0 & +0+0 & +1+0 & +2+0 & +3+0 \\ -3-1 & \underline{-2-1} & -1-1 & +0-1 & +1-1 & \underline{+2-1} & +3-1 \\ -3-2 & -2-2 & \underline{-1-2} & +0-2 & \underline{+1-2} & +2-2 & +3-2 \\ -3-3 & -2-3 & -1-3 & +0-3 & +1-3 & +2-3 & +3-3, \end{array}$$

while the corresponding frequencies in the artificial set are:

$$\begin{array}{ccccccc}
-18\omega_0 & -11\omega_0 & -4\omega_0 & 3\omega_0 & 10\omega_0 & 17\omega_0 & 24\omega_0 \\
-19\omega_0 & -12\omega_0 & -5\omega_0 & 2\omega_0 & 9\omega_0 & 16\omega_0 & 23\omega_0 \\
-20\omega_0 & -13\omega_0 & -6\omega_0 & \omega_0 & 8\omega_0 & 15\omega_0 & 22\omega_0 \\
-21\omega_0 & -14\omega_0 & -7\omega_0 & 0 & 7\omega_0 & 14\omega_0 & 21\omega_0 \\
-22\omega_0 & -15\omega_0 & -8\omega_0 & -\omega_0 & 6\omega_0 & 13\omega_0 & 20\omega_0 \\
-23\omega_0 & -16\omega_0 & -9\omega_0 & -2\omega_0 & 5\omega_0 & 12\omega_0 & 19\omega_0 \\
-24\omega_0 & -17\omega_0 & -10\omega_0 & -3\omega_0 & 4\omega_0 & 11\omega_0 & 18\omega_0.
\end{array}$$

According to the box truncation scheme artificial mapping, and limiting ourselves to the (positive) frequencies generated by the cubic nonlinearity, we have:

$$\begin{aligned}
\omega_1 &\to 7\omega_0 & \omega_2 &\to \omega_0 \\
2\omega_1 - \omega_2 &\to 13\omega_0 & -\omega_1 + 2\omega_2 &\to -5\omega_0 \\
\omega_1 + 2\omega_2 &\to 9\omega_0 & 2\omega_1 + \omega_2 &\to 15\omega_0 \\
3\omega_1 &\to 21\omega_0 & 3\omega_2 &\to 3\omega_0.
\end{aligned} \quad (3.52)$$

Applying the mapping and we obtain:

$$z(t) = X_1 \cos 7\omega_0 t + X_2 \cos \omega_0 t$$
$$w(t) = (X_1 \cos 7\omega_0 t + X_2 \cos \omega_0 t)^3$$
$$= \left(\frac{3}{4}X_1^3 + \frac{3}{2}X_1 X_2^2\right) \cos 7\omega_0 t + \left(\frac{3}{4}X_2^3 + \frac{3}{2}X_1^2 X_2\right) \cos \omega_0 t$$
$$+ \frac{3}{4}X_1^2 X_2 \cos 13\omega_0 t + \frac{3}{4}X_1 X_2^2 \cos 5\omega_0 t$$
$$+ \frac{3}{4}X_1 X_2^2 \cos 9\omega_0 t + \frac{3}{4}X_1^2 X_2 \cos 15\omega_0 t$$
$$+ \frac{1}{4}X_1^3 \cos 21\omega_0 t + \frac{1}{4}X_2^3 \cos 3\omega_0 t.$$

Taking into account the artificial mapping (3.52) we readily see, by inspection, that the amplitudes of the frequency components of $y(t)$ coincide with those of $w(t)$.

3.3.4 The Envelope HB Time-Frequency Technique

Most analog RF and microwave interface circuits operate with narrowband excitation around a carrier at frequency f_0. In a nonlinear circuit, such an excitation will produce a spectrum including the sidebands of all harmonics nf_0, $n = 0 \ldots \infty$.[4] Such sidebands include the harmonics and the intermodulation products originated by the excitation. Although the nonlinearity causes spectral regrowth (i.e., broadening) of the sideband spectra with respect to the input signal, it may be assumed that such spectra, centered around nf_0, do not overlap with each other. In such narrowband conditions (that depend

---

[4] The term sideband (left and right) is used here to refer to the neighborhood of a certain frequency, where a portion of the output spectrum is located. It is usually exploited in the analysis of mixers to denote the upconversion of a baseband signal around all the harmonics of the local oscillator signal.

on the relative bandwidth of the input signal) we can express the time-domain current flowing circuit branch l and the time-domain voltage across each circuit side l by separating a fast time dependence related to the carrier and its harmonics nf_0 and a slow time dependence related to the current and voltage envelopes, as:

$$i_l(t) = I_{l0}(t) + \sum_{i=1}^{N} [I_{lci}(t)\cos(i\omega_0 t) + I_{lsi}(t)\sin(i\omega_0 t_n)] = \mathbf{F}(t) \cdot \mathbf{I}_l(t) \tag{3.53}$$

$$v_l(t) = V_{l0}(t) + \sum_{i=1}^{N} [V_{lci}(t)\cos(i\omega_0 t) + V_{lsi}(t)\sin(i\omega_0 t)] = \mathbf{F}(t) \cdot \mathbf{V}_l(t), \tag{3.54}$$

where $\mathbf{F}(t)$ is a fast varying function of time, defined as:

$$\mathbf{F}(t) = [1, \cos(\omega_0 t), \sin(\omega_0 t), \cos(2\omega_0 t), \sin(2\omega_0 t), \ldots, \cos(N\omega_0 t), \sin(N\omega_0 t)],$$

while $\mathbf{I}_l(t)$ and $\mathbf{V}_l(t)$ are slowly varying envelope functions associated with each harmonic of the fundamental. Notice that the spectrum has been truncated to the N-th harmonic. In this context, *slow* means that the bandwidth of the envelopes is narrower than the spacing between tones, so that the spectra of the envelopes of nearby harmonics will have zero overlap. Similarly, we can express instantaneous charges and fluxes; for simplicity, we will confine the treatment to charges, where:

$$q_l(t) = Q_{l0}(t) + \sum_{i=1}^{N} [Q_{lci}(t)\cos(i\omega_0 t) + Q_{lsi}(t)\sin(i\omega_0 t_n)] = \mathbf{F}(t) \cdot \mathbf{Q}_l(t). \tag{3.55}$$

The goal of the present approach is to deal with the fast time variation through a time sampling technique similar to the Harmonic Balance with single frequency excitation, while the evolution of the slow envelopes is recovered through a time-stepping numerical integration. In other words, the technique yields a slowly time-varying spectrum of the voltages and current for each sideband of harmonic nf_0. The treatment provided here closely follows [23, 20] but, rather than starting from a state variable approach, it directly builds the circuit solution from a set of constitutive equations time-discretized in the slow time scale.

To this aim, suppose to introduce a fast sampling within the period T_0, denoting the fast sampling times $\tau_1 \ldots \tau_{2N+1}$, and define a $2N + 1 \times 2N + 1$ matrix $\mathbf{F}$ whose i-th row is $\mathbf{F}(\tau_i)$. Notice that this matrix coincides with the one defined in (3.45) but with $\theta_{n,k} = k\omega_0 \tau_n$. Since the envelope is almost constant over the period T_0, we have, defining the vectors:

$$\mathbf{i}_l(t) = [i_l(\tau_1, t) \ldots i_l(\tau_{2N+1}, t)]^T$$
$$\mathbf{v}_l(t) = [v_l(\tau_1, t) \ldots v_l(\tau_{2N+1}, t)]^T$$
$$\mathbf{q}_l(t) = [q_l(\tau_1, t) \ldots q_l(\tau_{2N+1}, t)]^T,$$

that, from (3.53), (3.54) and (3.55):

$$\mathbf{i}_l(t) = \mathbf{F} \cdot \mathbf{I}_l(t) \tag{3.56}$$

$$\mathbf{v}_l(t) = \mathbf{F} \cdot \mathbf{V}_l(t) \tag{3.57}$$

$$\mathbf{q}_l(t) = \mathbf{F} \cdot \mathbf{Q}_l(t), \tag{3.58}$$

Let us derive, as a first example, the constitutive relationships of linear memoriless elements (e.g., resistors) and of linear elements with memory (e.g., a capacitor) in terms of the envelope representation. For a resistor, the instantaneous constitutive relationship yields:

$$v_l(t) = R_l i_l(t),$$

i.e., introducing matrix $\mathbf{R}_l = R_l \mathbf{I}$, $\mathbf{I}$ identity matrix, we have, substituting from (3.56) and (3.57) at integration time t_n:

$$\mathbf{F} \cdot \mathbf{V}_l(t_n) = \mathbf{R}_l \cdot \mathbf{F} \cdot \mathbf{I}_l(t_n) \rightarrow \mathbf{V}_l(t_n) = \mathbf{R}_l \cdot \mathbf{I}_l(t_n). \quad (3.59)$$

Notice that the envelope representation is in the slow time variable; since the device is memoriless, the envelope representation is itself instantaneous.

Let us consider now the case of an element with memory, i.e., a linear capacitor. The current, equal to the time derivative of the charge, can be expressed as follows:

$$i_l(t) = \frac{dq_l(t)}{dt} = \sum_{i=1}^{N} [-i\omega_0 Q_{lci} \sin(i\omega_0 t) + i\omega_0 Q_{lsi} \cos(i\omega_0 t)]$$

$$+ \sum_{i=1}^{N} \left[\frac{dQ_{lci}(t)}{dt} \cos(i\omega_0 t) + \frac{dQ_{lsi}(t)}{dt} \sin(i\omega_0 t_n) \right],$$

i.e., introducing the matrix $\mathbf{\Omega}$ defined in (3.46), the time derivative can be expressed as:

$$i_l(t) = \frac{dq_l(t)}{dt} = \mathbf{F}(t) \cdot \mathbf{\Omega} \cdot \mathbf{Q}_l(t) + \mathbf{F}(t) \cdot \frac{d\mathbf{Q}_l(t)}{dt}.$$

The constitutive relationship of a linear capacitor therefore becomes:

$$i_l(t) = C_l \frac{dv_l(t)}{dt} \rightarrow \mathbf{F}(t) \cdot \mathbf{I}_l(t) = \mathbf{C}_l \cdot \mathbf{F}(t) \cdot \mathbf{\Omega} \cdot \mathbf{V}_l(t) + \mathbf{C}_l \cdot \mathbf{F}(t) \cdot \frac{d\mathbf{V}_l(t)}{dt},$$

where $\mathbf{C}_l = C_l \mathbf{I}$, $\mathbf{I}$ identity matrix. Let us now introduce again the fast sampling within the period T_0, and the matrix $\mathbf{F}$ whose i-th row is $\mathbf{F}(\tau_i)$. Furthermore, let us discretize the (slow varying) time t so as to apply an unconditionally stable backward Euler algorithm to the discretization of the time derivative of the envelope. The constitutive relationship of the nonlinear capacitor becomes, after sampling at fast time τ:

$$\mathbf{F} \cdot \mathbf{I}_l(t) = \mathbf{C}_l \cdot \mathbf{F} \cdot \mathbf{\Omega} \cdot \mathbf{V}_l(t) + \mathbf{C}_l \cdot \mathbf{F} \cdot \frac{d\mathbf{V}_l(t)}{dt},$$

i.e., taking into account that $\mathbf{C}_l$ is diagonal:

$$\mathbf{I}_l(t) = \mathbf{C}_l \cdot \mathbf{\Omega} \cdot \mathbf{V}_l(t) + \mathbf{C}_l \cdot \frac{d\mathbf{V}_l(t)}{dt}.$$

Finally, applying a backward Euler algorithm at the sampling time t_n, we obtain:

$$\mathbf{I}_l(t_n) = \mathbf{C}_l \cdot \mathbf{\Omega} \cdot \mathbf{V}_l(t_n) + \mathbf{C}_l \cdot \frac{\mathbf{V}_l(t_n) - \mathbf{V}_l(t_{n-1})}{t_n - t_{n-1}}, \quad (3.60)$$

where $\mathbf{V}_l(t_{n-1})$ is known from the solution at the previous time step. Eq. (3.60) provides the desired constitutive relationship.

Using the envelope formalism we can also describe nonlinear elements. Consider first a nonlinear resistor such as:

$$i_l(t) = f_R(v_l(t)).$$

Exploiting the envelope representation we have:

$$\mathbf{F}(t) \cdot \mathbf{V}_l(t) = f_R(\mathbf{F}(t) \cdot \mathbf{I}_l(t)),$$

i.e., introducing the fast sampling on $\mathbf{F}(t)$ and the slow sampling on the envelopes, and defining $\mathbf{f}_R(\bullet)$ as a function that applies to the single elements of the argument vector, we have at time t_n the nonlinear constitutive relationship:

$$\mathbf{V}_l(t_n) = \mathbf{F}^{-1} \cdot \mathbf{f}_R(\mathbf{F} \cdot \mathbf{I}_l(t_n)). \tag{3.61}$$

In a similar way, a nonlinear capacitor is described by the charge–voltage nonlinear constitutive relationship:

$$q_l(t) = f_C(v_l(t)),$$

that becomes, at time t, introducing the fast sampling on $\mathbf{F}(t)$:

$$\mathbf{Q}_l(t) = \mathbf{F}^{-1} \cdot \mathbf{f}_C(\mathbf{F} \cdot \mathbf{V}_l(t)).$$

Evaluating the time derivative of the charge we obtain, as for the linear case:

$$\mathbf{F} \cdot \mathbf{I}_l(t) = \mathbf{F} \cdot \mathbf{\Omega} \cdot \mathbf{Q}_l(t) + \mathbf{F} \cdot \frac{\mathrm{d}\mathbf{Q}_l(t)}{\mathrm{d}t} =$$
$$= \mathbf{F} \cdot \mathbf{\Omega} \cdot \mathbf{F}^{-1} \cdot \mathbf{f}_C(\mathbf{F} \cdot \mathbf{V}_l(t)) + \mathbf{F} \cdot \frac{\mathrm{d}\left[\mathbf{F}^{-1} \cdot \mathbf{f}_C(\mathbf{F} \cdot \mathbf{V}_l(t))\right]}{\mathrm{d}t},$$

i.e., applying a backward Euler discretization and simplifying:

$$\mathbf{I}_l(t_n) = \mathbf{\Omega} \cdot \mathbf{F}^{-1} \cdot \mathbf{f}_C(\mathbf{F} \cdot \mathbf{V}_l(t_n)) + \frac{\mathbf{F}^{-1} \cdot \mathbf{f}_C(\mathbf{F} \cdot \mathbf{V}_l(t_n)) - \mathbf{F}^{-1} \cdot \mathbf{f}_C(\mathbf{F} \cdot \mathbf{V}_l(t_{n-1}))}{t_n - t_{n-1}},$$
$$\tag{3.62}$$

that provides the algebraic nonlinear relationship (of dimension $2N + 1$) at integration time t_n. The treatment of nonlinear inductor or of nonlinear voltage or current dependent sources is similar, and so is the treatment of a nonlinear voltage-driven conductor. Thus, for each slow time stepping interval $t_{n-1} \to t_n$ we have a system of $2N + 1$ equations for any linear or nonlinear component.

Concerning the topological relationships such as the Kirchhoff Current Laws (KCL) and the Kirchhoff Voltage Laws (KVL), these will read as:

$$\sum_k i_k(t) = 0 \to \sum_k \mathbf{F}(t) \cdot \mathbf{I}_k(t) = 0$$
$$\sum_l v_l(t) = 0 \to \sum_l \mathbf{F}(t) \cdot \mathbf{V}_l(t) = 0,$$

i.e., applying the fast sampling on $\mathbf{F}(t) \to \mathbf{F}$ and the slow sampling on the envelopes:

$$\sum_k \mathbf{F} \cdot \mathbf{I}_k(t_n) = \mathbf{F} \cdot \sum_k \mathbf{I}_k(t_n) = 0 \rightarrow \sum_k \mathbf{I}_k(t_n) = 0$$

$$\sum_l \mathbf{F} \cdot \mathbf{V}_l(t_n) = \mathbf{F} \cdot \sum_l \mathbf{V}_l(t_n) = 0 \rightarrow \sum_l \mathbf{V}_l(t_n) = 0,$$

where summations are extended to all currents entering a node (KCL) or to all voltages along a closed loop (KVL). Notice that both the nodal KCL and the loop KVL have dimension $2N + 1$ in the envelope representation.

Consider now again the solution of the network at time t_n, the solution at time t_{n-1} being known; in a network with n nodes and l sides we can generally write:

- $n - 1$ linearly independent KCL
- $l - n + 1$ linearly independent KVL
- l constitutive relationships.

In the envelope representation each, of the above relations has, at time t_n, dimension $2N + 1$; therefore, the total number of scalar equations that can be written in terms of the slowly varying harmonic components of side currents and voltages is $M = (2N+1) \times 2l$, that exactly corresponds to the number of unknowns (notice that we use for capacitor the voltage as the state variable). The system obtained this way is therefore well posed. In general, the resulting equation system has to be solved numerically, for each time step, e.g., through Newton iteration. The mixed time-frequency domain envelope simulation described here can also be coupled with techniques like the Artificial Frequency Mapping, see, e.g., [24].

3.4 Circuit Optimization and Layout Generation

The design of microwave circuits has been based for a long time on *circuit synthesis* techniques. With this term we generally mean any deterministic procedure leading from a design goal (e.g., the frequency response of a filter, or the power gain of a narrow-band amplifier) to the implementing circuit (structure and element values). Microwave filters [25] are a typical example of a circuit class that can be designed this way, on the basis of classical lumped-parameter network synthesis techniques [26] coupled to lumped-distributed transformation methods such as the Richards' transformation [27], in association when needed to the so-called Kuroda identities (see, e.g., [28] and references therein). While direct synthesis is possible for a number of circuit classes, its result is not always satisfactory altogether, both because the circuit topologies obtained are often non-optimal for a specific implementation (e.g., microstrip), and because synthesis cannot allow for non-ideal behaviors, e.g., in lossy components.

Starting from the early seventies, Computer-Aided Design (CAD) tools for the analysis, design and optimization of microwave circuits began their diffusion; early examples are the SPEEDY and then COMPACT simulators developed by L. Besser [29, 30] (for a history of the early development of microwave CAD see, e.g., [31]). During the eighties, most academic microwave research groups, besides exploiting the first commercially available software packages, began developing their own CAD programs; the

Microwave Electronics group of the Polytechnic University of Torino was no exception, with the development of a FORTRAN package named ACCAD [32], that also included some integration between optimization and synthesis tools. In the long run, however, optimization became the preferred method in the design of microwave analog circuits, often coupled to a first-approximation design that was meant to provide a reasonable initial guess to the numerical tools.[5] Starting from the late sixties, a large number of optimization methods was applied to the design of microwave circuits [33, 34]; for a review see [35] and references therein. Despite the large number of optimization parameters, efficient methods were also devised for the direct geometrical optimization of electromagnetic structures [36, 37, 38, 39]; last but not least, optimization is also a tool of choice for the fitting of small- or large-signal models for active devices on the basis of measured data. State of the art microwave CAD tools today offer the user a battery of optimization techniques that, with some amount of care and experience on the part of the user, typically allow a meaningful solution to be obtained.

Already in the early eighties the problem of automatic layout generation was considered as a part of microwave CAD tools devoted to planar circuits [40]. Also along these lines, the evolution of the state of the art in automatic layout generation was fostered by general developments in analog circuit layout generation; see, e.g., [41, 42]. Today's microwave CAD suites typically offer automatic layout generation capabilities, also providing a seamless link to EM simulators able to check possibly critical parts of the layout.

3.5 Questions and Problems

3.5.1 Questions

1. Does the choice of normalization resistances affect the solution of a network in terms of (1) power waves and (2) voltages and currents?
2. To experimentally characterize the admittance matrix of a two-port do we mandatorily need to connect it with voltage generators and short circuits? Explain.
3. Does the choice of normalization resistances affect the solution of a network in terms of (1) power waves and (2) voltages and currents?
4. Discuss the advantages of a frequency-domain approach to the solution of nonlinear analog networks when compared to time-domain solutions.
5. Explain why the harmonic balance method is difficult to apply in the presence of multi-tone excitation with incommensurate tones.
6. Explain the oversampling method of harmonic balance analysis in the presence of incommensurate input tones.
7. Explain the artificial mapping method of harmonic balance analysis in the presence of incommensurate input tones.

---

[5] Synthesis tools for certain classes of circuits, like filters or matching sections, are however often available in today's microwave CAD suites, sometimes in the form of *wizards*.

3.5 Questions and Problems

8. Discuss the mixed time-frequency domain simulation technique for nonlinear circuits.

3.5.2 Problems

1. A resistive two-port has the following impedance matrix:

$$\mathbf{Z} = R \begin{pmatrix} 2 & 1 \\ 1 & 2 \end{pmatrix}.$$

   Sketch a possible structure (implementing the above impedance matrix and evaluate the scattering matrix (assume the normalization impedance $R_0 = R$).

2. A reactive two-port has the following impedance matrix:

$$\mathbf{Z} = jX \begin{pmatrix} 2 & 1 \\ 1 & 2 \end{pmatrix}.$$

   Evaluate the scattering matrix assuming $R_0 = X$ and check that the properties of the S-matrix of a lossless two-port are verified.

3. A real generator has internal impedance $Z_G = 50 - j50$ Ω and open circuit voltage $V_0 = 10$ V. Assuming $R_0 = 50$ Ω derive the power wave equivalent circuit (Γ_G and b_0).

4. A one-port has the power wave model:

$$b = \Gamma a + b_0.$$

   Exploiting the coupled current-voltage generator model for the power wave generator b_0, show that the power-wave model is equivalent to the series representation:

$$V = ZI + V_0.$$

   Assume R_0 as the normalization resistance; Γ is the reflection coefficient of Z with respect to R_0.

5. A two-port has scattering matrix:

$$\mathbf{S} = \frac{1}{\sqrt{2}} \begin{pmatrix} 1 & j \\ j & 1 \end{pmatrix}.$$

   Discuss whether the two-port is (1) reciprocal; (2) reactive. Derive the Z matrix of the two-port with $R_0 = 50$ Ω.

6. Consider a quadratic nonlinearity $y = x^2$ excited by two tones f_1 and f_2. Discuss the output spectrum when including harmonics up to the order 5 (a) using a box truncation approach; (b) using a diamond truncation approach. For this component only is the inclusion of odd-order harmonics and intermodulation products indispensable? Explain.

7. Consider a quadratic nonlinearity $y = x^2$ excited by two sine tones f_1 and f_2. Confining the spectrum to the second harmonics, DC and second-order intermodulation products, evaluate the output spectrum directly and by remapping the frequencies on

the artificial spectrum nf_0, $n = 0 \ldots N$. Use a box truncation scheme with a proper order.
8. Consider two tones $f_1 = \pi f_0$ and $f_2 = 2f_0$. Are they commensurate? Suppose now to represent them in a finite-precision arithmetic as $f_1 = 3.1415 f_0$ and $f_2 = 2.0000 f_0$. Are they commensurate now? What would be the period of the resulting two-tone excitation?

References

[1] L. Chua, C. A. Desoer, and E. S. Kuh, *Linear and nonlinear circuits*. Macgraw-Hill, New York, 1987.
[2] J. Rugh, *Nonlinear model theory: the Volterra/Wiener approach*. Johns Hopkins Press, 1981.
[3] X. Jing and Z. Lang, *Frequency domain analysis and design of nonlinear systems based on Volterra series expansion: a parametric characteristic approach*. Springer, 2015.
[4] D. Schreurs, M. O'Droma, A. A. Goacher, and M. Gadringer, Eds., *RF Power amplifier behavioral modeling*. Cambridge University Press, 2008.
[5] A. F. Peterson, S. L. Ray, and R. Mittra, *Computational methods for electromagnetics*. IEEE Press, New York, 1998.
[6] J. L. Volakis, A. Chatterjee, and L. C. Kempel, *Finite element method electromagnetics: antennas, microwave circuits, and scattering applications*. John Wiley & Sons, 1998, vol. 6.
[7] A. Taflove and S. C. Hagness, *Computational electrodynamics: the finite-difference time-domain method, 3rd edition*. Artech House Books, 2004.
[8] D. Mirshekar-Syahkal, *Spectral domain method for microwave integrated circuits*. Research Studies Pr Ltd, 1990.
[9] R. B. Marks and D. F. Williams, "A general waveguide circuit theory," *Journal of Research-National Institute of Standards and Technology*, vol. 97, pp. 533–533, 1992.
[10] K. Kurokawa, *An introduction to the theory of microwave circuits*. Academic Press, 1969.
[11] N. J. Higham, *Functions of matrices: theory and computation*. Siam, 2008.
[12] C.-W. Ho, A. Ruehli, and P. Brennan, "The modified nodal approach to network analysis," *IEEE Transactions on Circuits and Systems*, vol. 22, no. 6, pp. 504–509, Jun. 1975.
[13] L. W. Nagel and D. O. Pederson, *SPICE: Simulation Program with Integrated Circuit Emphasis*. Electronics Research Laboratory, College of Engineering, University of California, 1973.
[14] S. M. Sandler, *SPICE circuit handbook*. McGraw-Hill, Inc., 2006.
[15] S. Skelboe, "Computation of the periodic steady-state response of nonlinear networks by extrapolation methods," *IEEE Transactions on Circuits and Systems*, vol. 27, no. 3, pp. 161–175, Mar. 1980.
[16] T. J. Aprille and T. N. Trick, "Steady-state analysis of nonlinear circuits with periodic inputs," *Proceedings of the IEEE*, vol. 60, no. 1, pp. 108–114, Jan. 1972.
[17] M. Nakhla and J. Vlach, "A piecewise harmonic balance technique for determination of periodic response of nonlinear systems," *IEEE Transactions on Circuits and Systems*, vol. 23, no. 2, pp. 85–91, Feb. 1976.
[18] J. C. Pedro and N. B. Carvalho, *Intermodulation distortion in microwave and wireless circuits*. Artech House, 2003.

References

[19] S. A. Maas, *Non linear microwave and RF circuits*. Artech House, 2003.

[20] K. S. Kundert, "Introduction to RF simulation and its application," *IEEE Journal of Solid-State Circuits*, vol. 34, no. 9, pp. 1298–1319, 1999.

[21] P. J. C. Rodrigues, "A general mapping technique for fourier transform computation in nonlinear circuit analysis," *IEEE Microwave and Guided Wave Letters*, vol. 7, no. 11, pp. 374–376, Nov. 1997.

[22] J. C. Pedro and N. B. Carvalho, "Efficient harmonic balance computation of microwave circuits' response to multi-tone spectra," in *Microwave Conference, 1999. 29th European*, vol. 1, Oct. 1999, pp. 103–106.

[23] P. Feldmann and J. Roychowdhury, "Computation of circuit waveform envelopes using an efficient, matrix-decomposed harmonic balance algorithm," in *IEEE/ACM International Conference on Computer-Aided Design, 1996. ICCAD-96. Digest of Technical Papers., 1996*. IEEE, 1996, pp. 295–300.

[24] N. B. Carvalho, J. C. Pedro, W. Jang, and M. B. Steer, "Nonlinear RF circuits and systems simulation when driven by several modulated signals," *IEEE Transactions on Microwave Theory and Techniques*, vol. 54, no. 2, pp. 572–579, Feb. 2006.

[25] A. Matsumoto, *Microwave filters and circuits: advances in microwaves*. Academic Press, 2015, vol. 1.

[26] M. Van Valkenburg, *Introduction to modern network synthesis*. CBLS Publishers, 1960.

[27] P. I. Richards, "Resistor-transmission-line circuits," *Proceedings of the IRE*, vol. 36, no. 2, pp. 217–220, Feb. 1948.

[28] K. Kobayashi, Y. Nemoto, and R. Sato, "Kuroda's identity for mixed lumped and distributed circuits and their application to nonuniform transmission lines," *IEEE Transactions on Microwave Theory and Techniques*, vol. 29, no. 2, pp. 81–86, Feb. 1981.

[29] L. Besser and R. Newcomb, "A scattering matrix program for high frequency circuit analysis," in *IEEE Conference on Systems, Networks, and Computers*. IEEE, Jan. 1971.

[30] L. Besser, F. Ghoul, and C. Hsieh, "Computerized optimization of transistor amplifiers and oscillators Using COMPACT," in *Microwave Conference, 1973. 3rd European*, vol. 1, Sep. 1973, pp. 1–4.

[31] L. Besser, "Compact - microwave circuit optimization through commercial time sharing," in *Microwave Symposium Digest, 2008 IEEE MTT-S International*, Jun. 2008, pp. 711–714.

[32] C. Beccari and C. Naldi, "ACCAD, a new microwave circuit CAD package," in *Proc. Int. Symp. Microwave Technol. Ind. Develop., Campinas, Brazil*, Jul. 1985, pp. 269–273.

[33] J. W. Bandler, "Optimization methods for computer-aided design," *IEEE Transactions on Microwave Theory and Techniques*, vol. 17, no. 8, pp. 533–552, Aug. 1969.

[34] J. W. Bandler and S. H. Chen, "Circuit optimization: the state of the art," *IEEE Transactions on Microwave Theory and Techniques*, vol. 36, no. 2, pp. 424–443, Feb. 1988.

[35] J. W. Bandler and Q. S. Cheng, "Retrospective on microwave cad and optimization technology," in *Microwave Symposium Digest (MTT), 2012 IEEE MTT-S International*, Jun. 2012, pp. 1–3.

[36] V. Rizzoli, A. Costanzo, D. Masotti, A. Lipparini, and F. Mastri, "Computer-aided optimization of nonlinear microwave circuits with the aid of electromagnetic simulation," *IEEE Transactions on Microwave Theory and Techniques*, vol. 52, no. 1, pp. 362–377, Jan. 2004.

[37] S. Koziel, S. Ogurtsov, J. W. Bandler, and Q. S. Cheng, "Reliable space-mapping optimization integrated with EM-based adjoint sensitivities," *IEEE Transactions on Microwave Theory and Techniques*, vol. 61, no. 10, pp. 3493–3502, Oct. 2013.

[38] S. Koziel and J. W. Bandler, "Rapid yield estimation and optimization of microwave structures exploiting feature-based statistical analysis," *IEEE Transactions on Microwave Theory and Techniques*, vol. 63, no. 1, pp. 107–114, Jan. 2015.

[39] S. Koziel, X.-S. Yang, and Q.-J. Zhang, *Simulation-driven design optimization and modeling for microwave engineering*. World Scientific, 2013.

[40] S. L. March and L. Besser, "Artwork generation and mask production for microwave circuits," in *Microwave Conference, 1984. 14th European*, Sep. 1984, pp. 87–96.

[41] R. A. Hastings, *The art of analog layout*. Prentice Hall, 2006.

[42] L. Lavagno, L. Scheffer, and G. Martin, *EDA for IC implementation, circuit design, and process technology*. CRC Press, 2006.

4 Directional Couplers and Power Dividers

4.1 Coupled Quasi-TEM Lines

Multiconductor transmission lines formed by N conductors plus a ground plane are referred to, particularly in analog applications, as *coupled transmission lines*. Multiconductor lines have different applications: as multiconductor buses in high-speed digital circuits, suitable to transmit N parallel digital signals in high-speed digital circuits, and as coupled structures used in many distributed microwave analog components, such as directional couplers and coupled-line filters. While in digital multiconductor buses the electric and magnetic field coupling between lines is unwanted, since it causes *crosstalk*, in analog coupled line components coupling is essential to the operation. In analog circuit components, coupled lines are usually made of two or four conductors; examples are two- or four-conductor coupled microstrips and the coupled coplanar lines shown in Fig. 4.1.

4.1.1 Analysis of Symmetrical Coupled Lines

A multiconductor line with N conductors plus ground supports N TEM or quasi-TEM propagation modes. The analysis will be limited here to two-conductor lines (plus ground), operating in sinusoidal steady state. Therefore, voltages and currents will be denoted by their associated phasors.

Two coupled lines have voltages $V_1(z)$ and $V_2(z)$ and currents $I_1(z)$ and $I_2(z)$, as shown in Fig. 4.2. Voltages and currents can be combined into a vector of voltages $\underline{V}$ and currents $\underline{I}$ such that:

$$\underline{V} = \begin{pmatrix} V_1 \\ V_2 \end{pmatrix}, \quad \underline{I} = \begin{pmatrix} I_1 \\ I_2 \end{pmatrix}.$$

The line is defined, per unit length, by specific parameters that describe not only the inductive and capacitive phenomena typical of a single line (per unit length inductance and capacitance) but also the capacitive coupling (mutual capacitance) and the inductive coupling (mutual inductance) between neighboring lines. From now on, we assume for simplicity, and because this is the most significant case, that the two lines are geometrically (and electrically) symmetric. Moreover, we neglect losses.

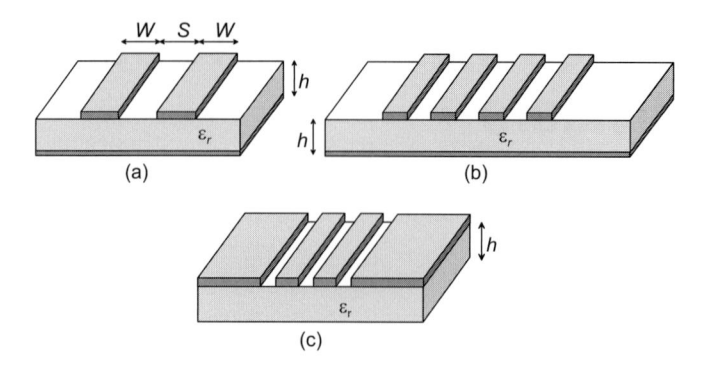

Figure 4.1 Two-conductor coupled microstrip (a), four-conductor coupled microstrip (b), two-conductor coupled Coplanar Waveguide, CPW (c).

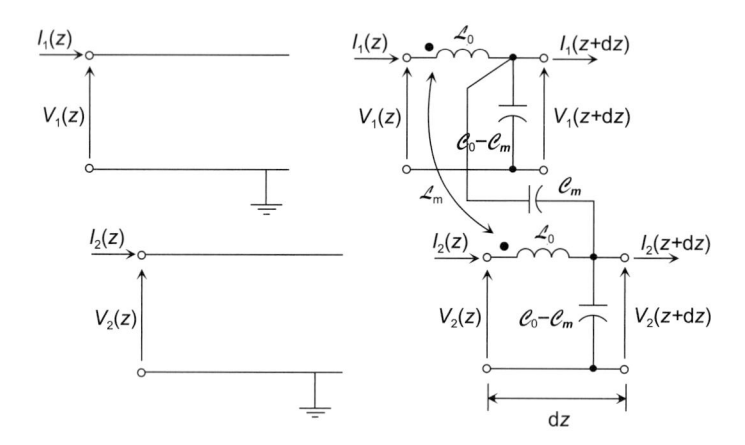

Figure 4.2 Scheme of coupled lines and elementary equivalent circuit.

Starting from the per-unit-length equivalent circuit we can describe the capacitive part (a π capacitor circuit, see Fig. 4.2) with the per-unit-length capacitance matrix:

$$\mathbf{C} = \begin{pmatrix} C_0 & -C_m \\ -C_m & C_0 \end{pmatrix};$$

similarly, the inductive part is described by the per-unit-length inductance matrix:

$$\mathbf{L} = \begin{pmatrix} \mathcal{L}_0 & \mathcal{L}_m \\ \mathcal{L}_m & \mathcal{L}_0 \end{pmatrix}.$$

The p.u.l. capacitance matrix relates the vector of the charges $\underline{Q}$ (per unit length) induced in the two lines to the line voltages $\underline{V}$ as:

$$\underline{Q} = \mathbf{C}\underline{V},$$

4.1 Coupled Quasi-TEM Lines

while the p.u.l. inductance matrix relates the vector of the p.u.l. magnetic fluxes $\underline{\Phi}$ to the vector of the currents flowing in the lines, $\underline{I}$, as:

$$\underline{\Phi} = \mathbf{L}\underline{I}.$$

The voltages and currents satisfy the generalized telegraphers' equation:

$$\frac{d\underline{V}(z)}{dz} = j\omega \mathbf{L}\underline{I}$$

$$\frac{d\underline{I}(z)}{dz} = j\omega \mathbf{C}\underline{V},$$

from which, eliminating, e.g., $\underline{I}$ from the second into the first equation, we obtain:

$$\frac{d^2\underline{V}(z)}{dz^2} = -\omega^2 \mathbf{LC}\underline{V}.$$

We look for exponential solutions for the complex voltage phasors under the form of propagating waves:

$$\underline{V}(z) = \underline{V}_0 \exp(-j\beta z),$$

where $\underline{V}_0 = (V_{01}, V_{02})^T$ is a constant vector. Substituting, we obtain the linear homogeneous system:

$$\left(\beta^2 \mathbf{I} - \omega^2 \mathbf{LC}\right)\underline{V}_0 = 0. \tag{4.1}$$

Nontrivial solutions exist if the system determinant is set to zero; this allows the values of the propagation constant β to be derived from system (4.1). Developing the product **LC** we have:

$$\mathbf{LC} = \begin{pmatrix} \mathcal{L}_0 & \mathcal{L}_m \\ \mathcal{L}_m & \mathcal{L}_0 \end{pmatrix} \begin{pmatrix} \mathcal{C}_0 & -\mathcal{C}_m \\ -\mathcal{C}_m & \mathcal{C}_0 \end{pmatrix}$$

$$= \begin{pmatrix} \mathcal{L}_0\mathcal{C}_0 - \mathcal{L}_m\mathcal{C}_m & -\mathcal{L}_0\mathcal{C}_m + \mathcal{L}_m\mathcal{C}_0 \\ -\mathcal{L}_0\mathcal{C}_m + \mathcal{L}_m\mathcal{C}_0 & \mathcal{L}_0\mathcal{C}_0 - \mathcal{L}_m\mathcal{C}_m \end{pmatrix} \equiv \begin{pmatrix} A & B \\ B & A \end{pmatrix}. \tag{4.2}$$

Substituting in (4.1) and imposing that the system determinant be zero, we obtain:

$$\beta^2 = \omega^2 (A \pm B).$$

Thus, the propagation constant β can assume four values, two with positive sign (β_1 and β_2) and two with negative sign ($-\beta_1$ and $-\beta_2$), implying forward and backward propagating waves, respectively; the absolute value of the propagation constants is however in general different, i.e., two propagation modes can exist with different phase velocities. Developing and substituting A and B from (4.2) we obtain:

$$\beta_1^2 = \omega^2 (\mathcal{L}_0 + \mathcal{L}_m)(\mathcal{C}_0 - \mathcal{C}_m)$$
$$\beta_2^2 = \omega^2 (\mathcal{L}_0 - \mathcal{L}_m)(\mathcal{C}_0 + \mathcal{C}_m).$$

Before giving a physical interpretation of the two propagation modes, let us eliminate the two p.u.l. inductances by introducing, as done for a single line, the in air or in vacuo capacitances. In fact, for an in vacuo line or set of coupled lines, all propagation modes have phase velocity equal to the velocity of light in vacuo, c_0. Thus, for a line in air we have:

$$\beta_{1a}^2 = \frac{\omega^2}{c_0^2} = \omega^2(\mathcal{L}_0 + \mathcal{L}_m)(C_{0a} - C_{ma})$$

$$\beta_{2a}^2 = \frac{\omega^2}{c_0^2} = \omega^2(\mathcal{L}_0 - \mathcal{L}_m)(C_{0a} + C_{ma}).$$

As for a single line, the inductances are not affected by the presence of dielectrics. Deriving the p.u.l. inductances as a function of the in-air capacitances, we finally obtain for the propagation constants of the coupled line the values $\pm\beta_1$ and $\pm\beta_2$, where the double sign refers to the forward and backward modes and:

$$\beta_1 = \frac{\omega}{c_0}\sqrt{\frac{C_0 - C_m}{C_{0a} - C_{ma}}}$$

$$\beta_2 = \frac{\omega}{c_0}\sqrt{\frac{C_0 + C_m}{C_{0a} + C_{ma}}}.$$

We can finally express the propagation constants by introducing two effective permittivities:

$$\beta_1 = \frac{\omega}{c_0}\sqrt{\epsilon_{\text{eff1}}}$$

$$\beta_2 = \frac{\omega}{c_0}\sqrt{\epsilon_{\text{eff2}}},$$

where:

$$\epsilon_{\text{eff1}} = \frac{C_0 - C_m}{C_{0a} - C_{ma}}$$

$$\epsilon_{\text{eff2}} = \frac{C_0 + C_m}{C_{0a} + C_{ma}}.$$

To obtain a physical interpretation of the two propagation modes, we derive from (4.1) the eigenvectors corresponding to the propagation constants. Since the determinant of system (4.1) is zero, the two equations are linearly dependent; substituting in the first equation the value β_1^2, we obtain for the eigenvector $\underline{V}_0$ the relationship:

$$V_{01} = V_{02},$$

while for the value β_2^2 we obtain:

$$V_{01} = -V_{02}.$$

In the first propagation mode the potential of the two lines is equal, section by section; this mode is called the *even* mode and the electric field topology is symmetrical, see Fig. 4.3 (a) for a coupled microstrip line. In the second propagation mode the line potentials are equal in magnitude but opposite, section by section; we call this the *odd* propagation mode, with an antisymmetric electric field pattern as shown in Fig. 4.3 (b).

4.1 Coupled Quasi-TEM Lines

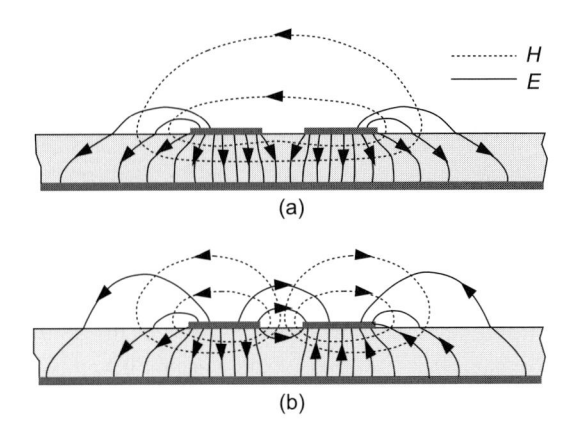

Figure 4.3 Field topologies for the even (a) and odd modes (b) in coupled microstrips.

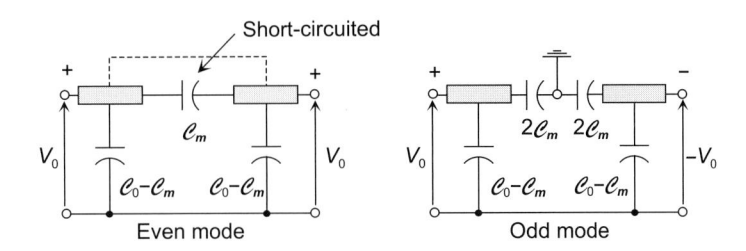

Figure 4.4 Definition of the even and odd mode capacitances in a two-conductor line.

For the even mode we can therefore set $V_{01} = V_{02} = V_0$, while for the odd mode $V_{01} = V_0$, $V_{02} = -V_0$.

To simplify the notation, let us introduce the concept of the even and odd-mode p.u.l. capacitances. Those are defined as the capacitances towards ground of *one* line when the two lines have the same potential or opposite potentials, respectively. With reference to Fig. 4.4, we obtain that the capacitance of a single line towards the ground is, for the two modes:

$$\mathcal{C}_e = \mathcal{C}_0 - \mathcal{C}_m$$
$$\mathcal{C}_o = \mathcal{C}_0 + \mathcal{C}_m.$$

Note that we always have:

$$\mathcal{C}_o \geq \mathcal{C}_e,$$

since $\mathcal{C}_m \geq 0$; the equal sign holds only if the coupling between the strips vanishes, i.e., their distance tends to infinity implying $\mathcal{C}_m \to 0$.

The propagation constants of the two modes can be therefore identified in terms of the even and odd mode capacitances. We obtain:

$$\beta_1 = \beta_e = \frac{\omega}{c_0}\sqrt{\frac{C_e}{C_{ae}}}$$

$$\beta_2 = \beta_o = \frac{\omega}{c_0}\sqrt{\frac{C_o}{C_{ao}}},$$

from which the even and odd mode effective permittivities result as:

$$\epsilon_{\text{eff}e} = \frac{C_e}{C_{ae}}$$

$$\epsilon_{\text{eff}o} = \frac{C_o}{C_{ao}}.$$

In a purely TEM coupled line, the effective permittivities of the two modes coincide and are equal to the medium relative permittivity. This does not happen in general to a quasi-TEM line with a inhomogeneous cross section, since the field patterns of the two modes are different.

Substituting the voltage solutions in the telegraphers' equations we find that for both modes the forward wave current is proportional to the forward wave voltage V_0. The proportionality factor can be interpreted as the even (odd) mode characteristic admittance, i.e., the inverse of the even (odd) mode characteristic impedance Z_{0e} (Z_{0o}). These turn out to be expressed as:

$$Z_{0e} = \frac{1}{c_0\sqrt{C_e C_{ae}}} = \frac{Z_{0ae}}{\sqrt{\epsilon_{\text{eff}e}}}$$

$$Z_{0o} = \frac{1}{c_0\sqrt{C_o C_{ao}}} = \frac{Z_{0ao}}{\sqrt{\epsilon_{\text{eff}o}}},$$

where the index a denotes in-air quantities. The two characteristic impedances follow the inequality:

$$Z_{0e} \geq Z_0 \geq Z_{0o}, \tag{4.3}$$

where Z_0 is the impedance of an isolated line and the equality sign only holds if the distance between the coupled lines tends to infinity. In fact, we have $C_o \geq C_0 \geq C_e$ but also $C_{ao} \geq C_{a0} \geq C_{ae}$ from which $C_o C_{ao} \geq C_0 C_{a0} \geq C_e C_{ae}$; taking into account the definition of impedances, we immediately obtain (4.3). No general relationship exists instead for the even and odd mode effective permittivities. Increasing the spacing between lines, the even and odd mode impedances asymptotically tend to the isolated line impedance as shown in Fig. 4.5.

4.1.2 Coupled Planar Lines

Examples of coupled planar lines are shown in Fig. 4.6. When adopted in the realization of directional couplers, such structures should allow for a large difference between the even and odd mode characteristic impedances, implying a strong coupling C, see the definition in (4.9), while the even and odd mode phase velocities should be (ideally) equal. In practice, two-conductor microstrip or coplanar lines do not allow for the

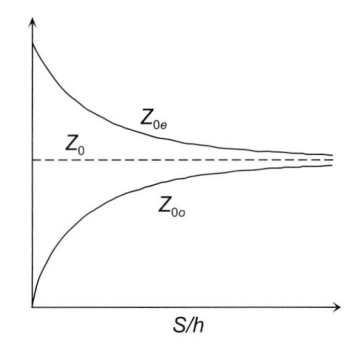

Figure 4.5 Qualitative behavior of the even and odd mode characteristic impedances as a function of the line spacing.

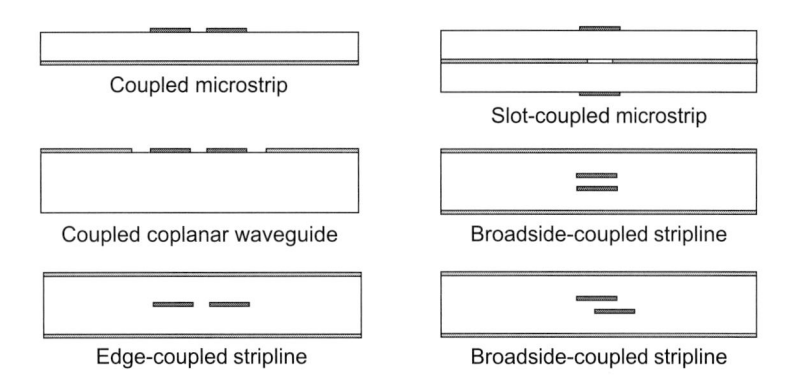

Figure 4.6 Examples of coupled microstrips, coplanar lines and striplines. The signal lines are dark gray, the ground planes pale gray.

first condition, and are therefore well suited to fabricate low-coupling couplers only. Strong coupling can be achieved by broadside-coupled striplines or by multiconductor lines (e.g., the Lange coupler described in Sec. 4.4.1). However, while the coupled stripline is a TEM structure, microstrip-based couplers are not ideal from the standpoint of synchronous coupling, that requires the same even and odd mode phase velocities.

4.1.3 Coupled Microstrips

Two-conductor coupled microstrips are a classical example of coupled quasi-TEM line. However, for technological reasons related to the minimum separation S that can be achieved between strips, such a structure can be only adopted in the realization of low-coupling couplers.

The following quasi-static analysis formulae are from [1, 2], where a frequency dispersion model for the effective permittivities is also presented. Let us call h the substrate thickness of dielectric constant ϵ_r, W the strip width, S the slot width, and further define the normalized strip and slot widths:

$$u = W/h$$
$$g = S/h.$$

The effective permittivity of the even and odd modes can be expressed as follows:

$$\epsilon_{\text{effe}} = \frac{\epsilon_r + 1}{2} - \frac{\epsilon_r - 1}{2}\left(1 + \frac{10}{v}\right)^{-a_e(v)b_e(\epsilon_r)}$$

$$\epsilon_{\text{effo}} = \left[\frac{\epsilon_r + 1}{2} + a_o(u, \epsilon_r) - \epsilon_{\text{eff}}\right]\exp(-c_o g^{d_o}) + \epsilon_{\text{eff}},$$

while we have for the impedances:

$$Z_{0e} = Z_0 \sqrt{\frac{\epsilon_{\text{eff}}}{\epsilon_{\text{effe}}}} \frac{1}{1 - \frac{Z_0\sqrt{\epsilon_{\text{eff}}}Q_4}{377}}$$

$$Z_{0o} = Z_0 \sqrt{\frac{\epsilon_{\text{eff}}}{\epsilon_{\text{effo}}}} \frac{1}{1 - \frac{Z_0\sqrt{\epsilon_{\text{eff}}}Q_{10}}{377}}.$$

The previous formulae use the following parameters of the isolated line of width W:

$$\epsilon_{\text{eff}} = \frac{\epsilon_r + 1}{2} + \frac{\epsilon_r - 1}{2}\left[\left(1 + \frac{12}{u}\right)^{-1/2} + 0.04(1-u)^2\right] \qquad u \le 1$$

$$\epsilon_{\text{eff}} = \frac{\epsilon_r + 1}{2} + \frac{\epsilon_r - 1}{2}\left(1 + \frac{12}{u}\right)^{-1/2} \qquad u \ge 1,$$

moreover:

$$Z_0 = \frac{60}{\sqrt{\epsilon_{\text{eff}}}}\log\left(\frac{8}{u} + \frac{u}{4}\right) \qquad u \le 1$$

$$Z_0 = \frac{120\pi}{\sqrt{\epsilon_{\text{eff}}}}\frac{1}{1.393 + u + 0.667\log(1.444 + u)} \qquad u \ge 1.$$

The following parameters are also used:

$$v = u\frac{20 + g^2}{10 + g^2} + g\exp(-g)$$

$$a_e(v) = 1 + \frac{1}{49}\log\left[\frac{v^4 + \left(\frac{v}{52}\right)^2}{v^4 + 0.432}\right] + \frac{1}{18.7}\log\left[1 + \left(\frac{v}{18.1}\right)^3\right]$$

$$b_e(\epsilon_r) = 0.564\left(\frac{\epsilon_r - 0.9}{\epsilon_r + 3}\right)^{0.053}$$

$$a_o(u, \epsilon_r) = 0.7287\left(\epsilon_{\text{eff}} - \frac{\epsilon_r + 1}{2}\right)[1 - \exp(-0.179u)]$$

$$b_o(\epsilon_r) = \frac{0.747\epsilon_r}{0.15 + \epsilon_r}$$

$$c_d = b_d(\epsilon_r) - [b_d(\epsilon_r) - 0.207]\exp(-0.414u)$$

$$d_o = 0.539 + 0.694\exp(-0.562u),$$

together with the fitting function set:

$$Q_1 = 0.8695 u^{0.194}$$

$$Q_2 = 1 + 0.7519 g + 0.189 g^{2.31}$$

$$Q_3 = 0.1975 + \left[16.6 + \left(\frac{8.4}{g}\right)^6 \right]^{-0.387} + \frac{1}{241} \log \left[\frac{g^{10}}{1 + (g/3.4)^{10}} \right]$$

$$Q_4 = \frac{2 Q_1}{Q_2} \frac{1}{\exp(-g) u^{Q_3} + [2 - \exp(-g)] u^{-Q_3}}$$

$$Q_5 = 1.794 + 1.14 \log \left[1 + \frac{0.638}{g + 0.517 g^{2.43}} \right]$$

$$Q_6 = 0.2305 + \frac{1}{281.3} \log \left[\frac{g^{10}}{1 + (g/5.8)^{10}} \right] + \frac{1}{5.1} \log \left(1 + 0.598 g^{1.154} \right)$$

$$Q_7 = \frac{10 + 190 g^2}{1 + 82.3 g^3}$$

$$Q_8 = \exp \left[-6.5 - 0.95 \log g - \left(\frac{g}{0.15}\right)^5 \right]$$

$$Q_9 = \log Q_7 \left(Q_8 + \frac{1}{16.5} \right)$$

$$Q_{10} = \frac{Q_2 Q_4 - Q_5 \exp \left(Q_6 u^{-Q_9} \log u \right)}{Q_2}.$$

An example of the behavior of the even and odd mode characteristic impedances and of the related permittivities is shown in Fig. 4.7 and Fig. 4.8.

Note that the even and odd mode impedances tend, for large values of the ratio S/h, to the value proper of the isolated microstrip. On the behavior of the even and odd mode impedances, the following remarks hold:

- for increasing W/h, both impedances decrease because the capacitance towards ground of both modes increases;
- for increasing S/h, the odd mode capacitance decreases and therefore the odd mode impedance increases. At the same time, the even mode capacitance increases and therefore the even mode impedance decreases.

The behavior of the effective permittivity is less straightforward. In coupled microstrips, the even mode permittivity is larger than the odd mode permittivity, because, in the even mode, most field lines go through the substrate. The odd mode permittivity is close (for high coupling) to $(\epsilon_r + 1)/2$, because the field lines are approximately distributed in an equal way in the substrate and in the air. Furthermore:

- with increasing W/h, the field is increasingly confined in the substrate and both permittivities grow;

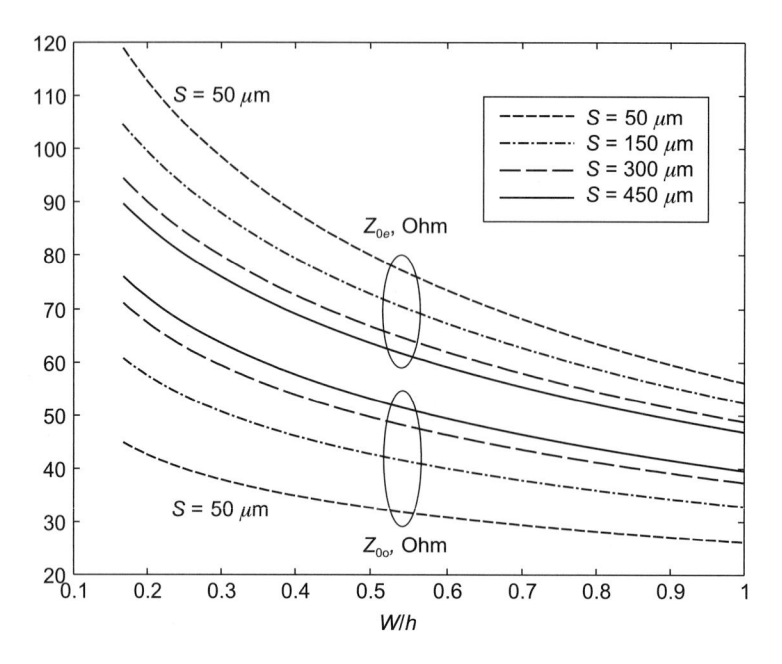

Figure 4.7 Behavior of the even and odd mode characteristic impedances of coupled microstrips on a GaAs substrate ($h = 300$ μm) as a function of the ratio W/h and for different values of S.

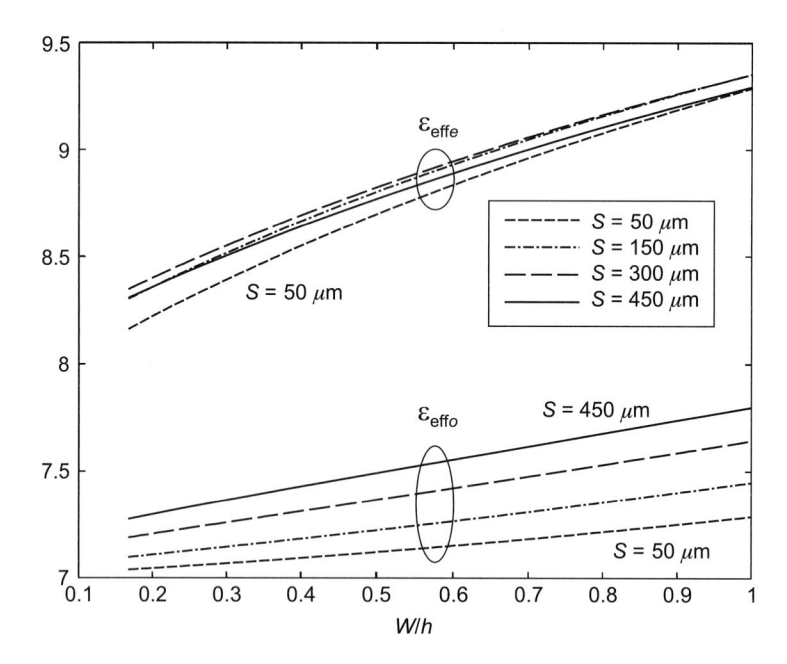

Figure 4.8 Behavior of even and odd mode effective permittivities of coupled microstrips on a GaAs substrate ($h = 300$ μm) as a function of the ratio W/h and for different values of S.

- with decreasing by S/h, the odd mode field is increasingly concentrated in the slot between the two strips, while the odd mode permittivity decreases approaching $(\epsilon_r +1)/2$, and is little affected by S/h.

4.2 The Directional Coupler

The directional coupler is a four-port network, having the purpose of distributing, according to some repartition scheme, the power delivered to a given port (for example port 1, called the input port) between two other ports (for example 2 and 3, called the coupled and transmission or *thru* ports, respectively), keeping port 4 isolated, see Fig. 4.9. From this point of view, the directional coupler would appear similar to the so-called *power divider*: this is an *n*-port where the power entering port 1 is divided, equally or according to some repartition scheme, into the remaining $n - 1$ ports. While couplers can be exploited as power dividers, couplers allow for a more specific function, i.e., they impose a *phase relationship* between the power wave at the coupled and transmitted ports, typically either 90 or 180 degrees.

The directional coupler has several applications in the field of microwave circuits; it is used (in passive circuits) in the realization of delay lines, filters and matching networks; in active circuits, it is a major building block in balanced amplifiers, mixers, attenuators, modulators and phase shifters. It should be remembered that directional couplers have a frequency-dependent behavior, that is ideal at the design frequency, and exhibits a gradual deterioration when moving away from centerband. Usually, directional couplers are moderately narrowband, with maximum typical bandwidths of the order of one octave (i.e., from f to $2f$).

The coupler is characterized by a number of parameters. Consider port 1 as the incident port, port 2 as the coupled port, port 3 as the transmission or thru port and port 4 as the isolated port. We define the power *coupling* coefficient as:

$$K|_{\text{dB}} = -10 \log_{10}\left(\frac{P_2}{P_1}\right),$$

while the *isolation* of port 4 (ideally zero in natural units or ∞ in dB) is:

$$I|_{\text{dB}} = -10 \log_{10}\left(\frac{P_4}{P_1}\right).$$

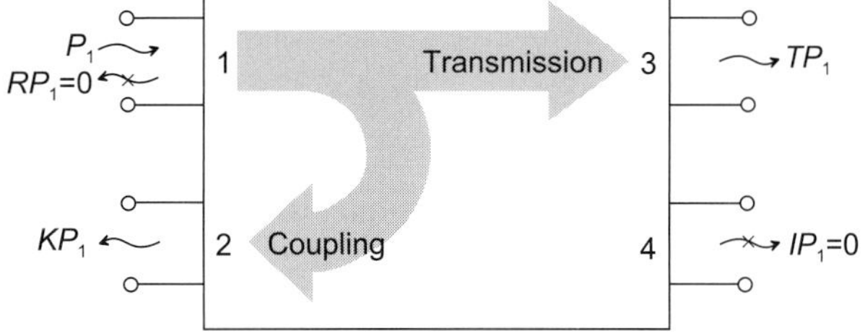

Figure 4.9 Scheme of a directional coupler fed at port 1.

The (power) *transmission* coefficient to port 3 is:

$$T|_{\text{dB}} = -10\log_{10}\left(\frac{P_3}{P_1}\right).$$

Finally, the power *reflection* coefficient R at port 1 (ideally zero, or $-\infty$ in dB) is:

$$R|_{\text{dB}} = 20\log_{10}(\Gamma_1),$$

where Γ_1 is the voltage reflection coefficient. Usually, directional couplers are reactive elements, i.e., show low-power dissipation; this implies $P_1 \approx P_2 + P_3$: the input power is partitioned between ports 2 and 3. A further figure of merit of the coupler is the *directivity* (ideally infinite):

$$D|_{\text{dB}} = 10\log_{10}\left(\frac{P_2}{P_4}\right) = I|_{\text{dB}} - K|_{\text{dB}}.$$

Fig. 4.10 shows some examples of directional couplers. The couplers can be divided into various categories, listed below:

- **Simple coupled-line couplers**; can be uniform or nonuniform. In nonuniform couplers the bandwidth can be increased, at the expense of a larger footprint (in uniform couplers the centerband length is of the order of a quarter wavelength). According to the transmission line choice, high coupling (typically corresponding to the input power being divided in two equal parts, leading to the *3 dB coupler*) can be easily obtained, as in broadside-coupled striplines, or low coupling only (as in coupled microstrips or coplanar lines) can be achieved due to technological constraints. Coupled line couplers have a centerband phase shift of 90 degrees between the coupled port and the transmission port; often they are referred to as *90 degrees hybrids*.
- **Interdigitated couplers**, such as the Lange coupler; they are similar in behavior and operation principle to coupled line couplers, but can achieve higher coupling (in particular 3 dB);
- **Branch-line couplers** are based on an interference principle (the signal is split into two paths that combine at each port in or out of phase) and permit to easily obtain 3 dB couplers, but have a large footprint (typically square with a centerband side of a quarter wavelength) and narrower bandwidth. The branch-line coupler is another example of 90 degrees hybrid.
- The **hybrid ring** (or rat race coupler) is also based on an interference principle, but has a 180 degrees shift between the output ports at centerband. Hybrid rings are often used to generate differential signals balanced with respect to ground; this is particularly useful in the design of mixers. The footprint is large, since the periphery is of the order of one wavelength at centerband.

Other types of directional couplers, that will be not discussed in detail, are the tandem coupler, the meander line coupler and the transformer coupler (based on coupled coils). It should be emphasized that both the branch-line coupler and the hybrid ring can also be implemented in a concentrated form, thus obtaining structures that are much more compact than the distributed versions.

4.2 The Directional Coupler

4.2.1 General Properties of Directional Couplers and Power Dividers

Let us consider a reactive, reciprocal four-port that is, independent of frequency, matched at all ports ($S_{ii} = 0$). We want to investigate the intrinsic properties of the structure in the presence of symmetries. From condition $\mathbf{SS}^\dagger = \mathbf{I}$ where $\mathbf{I}$ is the identity matrix, we obtain the following equations:

$$|S_{12}|^2 + |S_{13}|^2 + |S_{14}|^2 = 1 \tag{4.4a}$$

$$|S_{12}|^2 + |S_{23}|^2 + |S_{24}|^2 = 1 \tag{4.4b}$$

$$|S_{13}|^2 + |S_{23}|^2 + |S_{34}|^2 = 1 \tag{4.4c}$$

$$|S_{14}|^2 + |S_{24}|^2 + |S_{34}|^2 = 1 \tag{4.4d}$$

and:

$$S_{13}S_{23}^* + S_{14}S_{24}^* = 0$$
$$S_{12}S_{23}^* + S_{14}S_{34}^* = 0$$
$$S_{12}S_{24}^* + S_{13}S_{34}^* = 0$$
$$S_{12}S_{13}^* + S_{24}S_{34}^* = 0$$
$$S_{12}S_{14}^* + S_{23}S_{34}^* = 0$$
$$S_{13}S_{14}^* + S_{23}S_{24}^* = 0.$$

The structures exploited as directional couplers typically exhibit a twofold symmetry, with two symmetry planes (vertical and horizontal); examples are the two-conductor or interdigitated coupler and the branch-line coupler (see Fig. 4.10). For the port numbering, refer to the convention in Fig. 4.11. The layout symmetry implies that some of the elements of the scattering matrix are equal:

$$S_{12} = S_{34}$$
$$S_{13} = S_{24}$$
$$S_{14} = S_{23}.$$

From (4.4) we obtain:

$$|S_{12}|^2 + |S_{13}|^2 + |S_{14}|^2 = 1$$
$$S_{13}S_{14}^* + S_{14}S_{13}^* = 0$$
$$S_{12}S_{14}^* + S_{14}S_{12}^* = 0$$
$$S_{12}S_{13}^* + S_{13}S_{12}^* = 0.$$

If we further assume that pairs of uncoupled ports exist, for instance ports 1 and 4 (and therefore, by symmetry, 2 and 3), we have $S_{14} = S_{23} = 0$ and therefore the above system simplifies to:

$$|S_{12}|^2 + |S_{13}|^2 = 1 \tag{4.5}$$
$$S_{12}S_{13}^* + S_{13}S_{12}^* = 0,$$

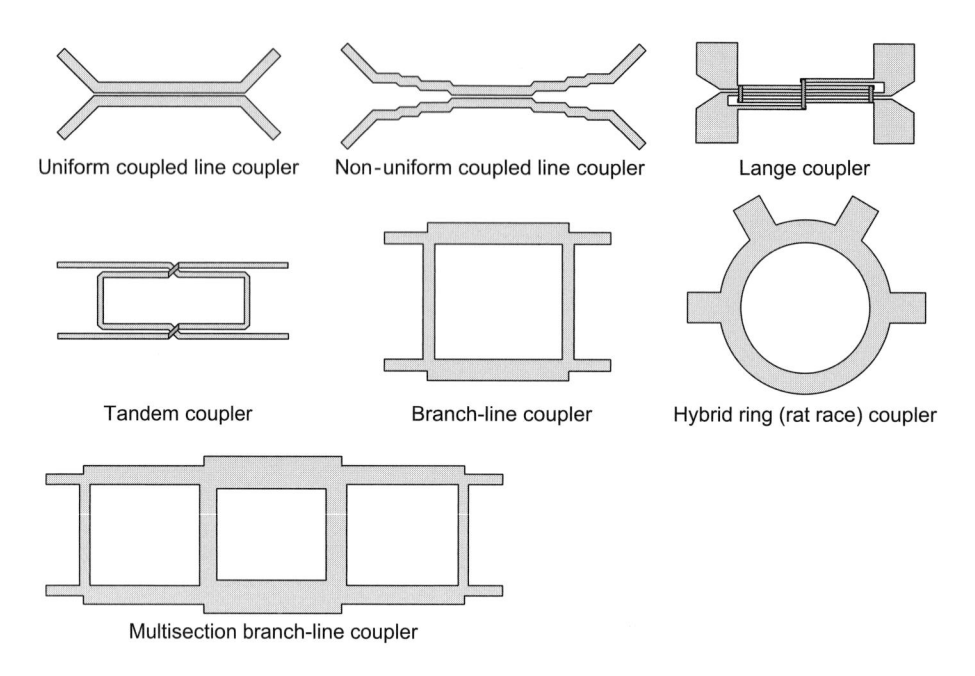

Figure 4.10 Examples of directional couplers.

from which:

$$|S_{13}| = \sqrt{1 - |S_{12}|^2}$$

$$\angle S_{13} = \frac{\pi}{2} + \angle S_{12} + n\pi,$$

where n is an integer number. We conclude that any lossless structure exhibiting twofold symmetry, that is matched at all ports, and allows for port isolation, automatically behaves like a directional coupler imposing 90 degrees phase difference between the output ports (coupled and transmission). Notice that the phase property is frequency independent, provided that the isolation of the uncoupled port is ideal. The *90 degrees hybrid* thus obtained only has one degree of freedom, corresponding to the frequency-dependent coupling $|S_{12}|$. Note that the magnitude of $|S_{12}|$ generally is frequency dependent, but the phase relation between signals at ports 2 and 3 is not. The scattering matrix will read:

$$\mathbf{S} = \alpha \begin{pmatrix} 0 & |S_{12}| & \pm j\sqrt{1-|S_{12}|^2} & 0 \\ |S_{12}| & 0 & 0 & \pm j\sqrt{1-|S_{12}|^2} \\ \pm j\sqrt{1-|S_{12}|^2} & 0 & 0 & |S_{12}| \\ 0 & \pm j\sqrt{1-|S_{12}|^2} & |S_{12}| & 0 \end{pmatrix},$$

where $\alpha = \exp(j\angle S_{12})$ and the double sign depends on the value of the integer n.

Let us consider now the same four-port (reactive and matched at all ports), but with just one symmetry plane; an example is the hybrid ring coupler, discussed in Sec. 4.5.3,

where only a vertical symmetry plane exist. Using the port numbering as in Fig. 4.32, we have:
$$S_{12} = S_{34}$$
$$S_{13} = S_{24}.$$

Substituting the above relations in (4.4) we obtain:
$$|S_{12}|^2 + |S_{13}|^2 + |S_{14}|^2 = 1$$
$$|S_{12}|^2 + |S_{23}|^2 + |S_{13}|^2 = 1$$
$$S_{13}S_{23}^* + S_{14}S_{13}^* = 0$$
$$S_{12}S_{23}^* + S_{14}S_{12}^* = 0$$
$$S_{12}S_{13}^* + S_{13}S_{12}^* = 0$$
$$S_{12}S_{13}^* + S_{13}S_{12}^* = 0.$$

Assuming again that port pairs are decoupled (e.g., ports 1 and 3 are isolated, and so are for symmetry port 2 and 4) we have $S_{13} = S_{24} = 0$ and therefore:
$$|S_{12}|^2 + |S_{14}|^2 = 1$$
$$|S_{12}|^2 + |S_{23}|^2 = 1$$
$$S_{12}S_{23}^* + S_{14}S_{12}^* = 0.$$

From these equations we obtain:
$$|S_{23}| = |S_{14}| = \sqrt{1 - |S_{12}|^2}$$
$$\angle S_{12} = \frac{\angle S_{23} + \angle S_{14}}{2} + \frac{\pi}{2} + n\pi,$$

i.e.,
$$\angle S_{23} = 2\angle S_{12} - \angle S_{14} + m\pi,$$

where m is a positive or negative integer. The coupler scattering matrix now reads:

$$\mathbf{S} = \alpha \begin{pmatrix} 0 & |S_{12}| & 0 & \beta\sqrt{1-|S_{12}|^2} \\ |S_{12}| & 0 & -\beta^{-1}\sqrt{1-|S_{12}|^2} & 0 \\ 0 & -\beta^{-1}\sqrt{1-|S_{12}|^2} & 0 & |S_{12}| \\ \beta\sqrt{1-|S_{12}|^2} & 0 & |S_{12}| & 0 \end{pmatrix},$$

where again $\alpha = \exp(j\angle S_{12})$ and $\beta = \exp(j\angle S_{14} - j\angle S_{12})$. If by design $\alpha = -j$ and $\angle S_{14} - \angle S_{12} = 0$ or $\angle S_{14} - \angle S_{12} = \pi$, i.e., $\beta = \pm 1$, we obtain, respectively:

$$\mathbf{S} = \begin{pmatrix} 0 & -j|S_{12}| & 0 & \mp j\sqrt{1-|S_{12}|^2} \\ -j|S_{12}| & 0 & \pm j\sqrt{1-|S_{12}|^2} & 0 \\ 0 & \pm j\sqrt{1-|S_{12}|^2} & 0 & |S_{12}| \\ \mp j\sqrt{1-|S_{12}|^2} & 0 & |S_{12}| & 0 \end{pmatrix}.$$

The two cases correspond to the upper and lower sign. From a physical standpoint, the two choices imply to exchange the role of the ports where the input signal is transmitted with plus or minus sign. (The hybrid ring analyzed in Sec. 4.5.3 in fact refers to the second case.) Notice that, contrarily to the 90 degrees hybrid, the phase relationship between ports is connected to the condition $\angle S_{14} - \angle S_{12} = 0$ or $\angle S_{14} - \angle S_{12} = \pi$, that is imposed at centerband but may be frequency dependent.

We have demonstrated that a reactive four port matched at all ports yields a directional coupler having a well defined phase relationship (90 or 180 degrees) between outputs. We will now show that a *reciprocal three-port* operating as a power divider (with input at port 1 and output at ports 2 and 3) or combiner cannot be reactive, if we suppose it to be matched at all ports, and we moreover ask for ports 2 and 3 to be isolated. In fact, for a reactive three port the relation $\mathbf{SS}^\dagger = \mathbf{I}$ becomes:

$$|S_{11}|^2 + |S_{12}|^2 + |S_{13}|^2 = 1$$
$$|S_{12}|^2 + |S_{22}|^2 + |S_{23}|^2 = 1$$
$$|S_{13}|^2 + |S_{23}|^2 + |S_{33}|^2 = 1$$
$$S_{11}S_{12}^* + S_{12}S_{22}^* + S_{13}S_{23}^* = 0$$
$$S_{11}S_{13}^* + S_{12}S_{23}^* + S_{13}S_{33}^* = 0$$
$$S_{12}S_{13}^* + S_{22}S_{23}^* + S_{23}S_{33}^* = 0.$$

Imposing matching at all ports, $S_{11} = S_{22} = S_{33} = 0$, and isolation of the output ports with respect to each other, $S_{23} = 0$, we obtain a system of equations that clearly admits no solution:

$$|S_{12}|^2 + |S_{13}|^2 = 1$$
$$S_{12}S_{13}^* = 0$$
$$|S_{12}|^2 = 1$$
$$|S_{13}|^2 = 1.$$

The problem becomes well posed if we relax one of the constraints (thus obtaining a lossy or non-reciprocal or non-matched two-port). The non-reciprocal choice leads to the so-called circulator, see, e.g., [3, 4], while in the Wilkinson divider, see Sec. 4.6.1, a resistance of proper value is connected between ports 2 and 3.

4.3 The Two-Conductor Coupled Line Coupler

We start from considering a symmetric two-conductor coupled line carrying an even and an odd TEM or quasi-TEM propagation mode. In this symmetric case, the coupler can be analyzed through the superposition of even and odd mode excitations.

Let us consider a coupled line section of length l (i.e., a four-port) closed on the reference resistance R_0 on all ports, see Fig. 4.11, top. Port 1 is connected to a signal generator with internal resistance R_0. The excitation can be decomposed into an even and odd excitation as shown in Fig. 4.11; each of them only excites the corresponding even or odd mode.

4.3 The Two-Conductor Coupled Line Coupler

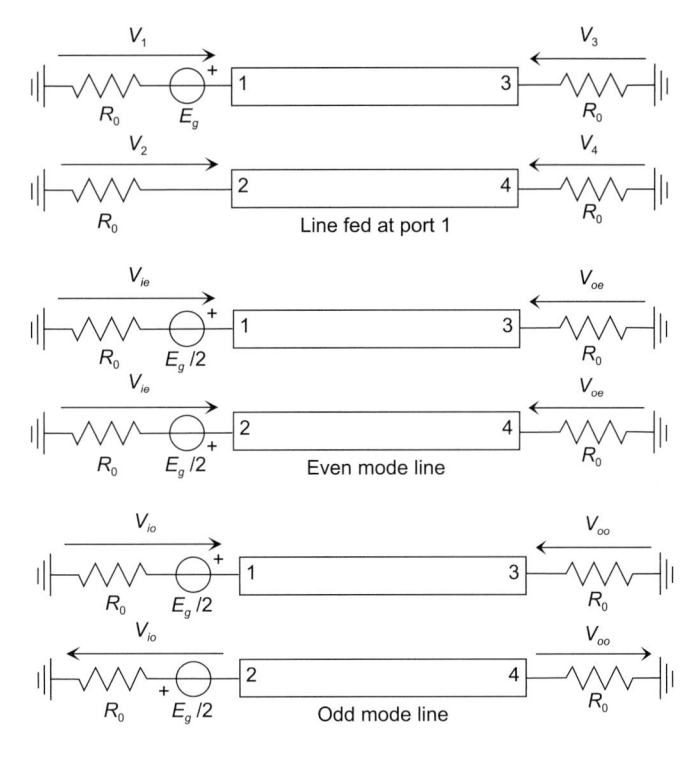

Figure 4.11 Coupled line coupler excited at port 1 (top). The excitation is then decomposed in even (center) and odd (bottom) modes.

Let us call V_{io} and V_{ie} the odd and even mode input voltages and V_{oo} and V_{oe} the output voltages; we have:

$$V_1 = V_{ie} + V_{io}$$
$$V_2 = V_{ie} - V_{io}$$
$$V_3 = V_{oe} + V_{oo}$$
$$V_4 = V_{oe} - V_{oo}.$$

The even and odd mode voltages can be evaluated from the analysis of the loaded two-port in Fig. 4.12. For the sake of definiteness, let us refer to the even mode. The scattering matrix of the even mode line (characteristic impedance Z_{0e} and guided wavelength λ_e) vs. the reference impedance Z_{0e} can be written as:

$$\mathbf{S}_e = \begin{pmatrix} 0 & \exp(-j\theta_e) \\ \exp(-j\theta_e) & 0 \end{pmatrix},$$

where $\theta_e = 2\pi l/\lambda_e$ is the even mode electrical line length.

The reflection coefficients of the load and generator with impedance R_0 are:

$$\Gamma_{0e} = \frac{R_0 - Z_{0e}}{R_0 + Z_{0e}},$$

Directional Couplers and Power Dividers

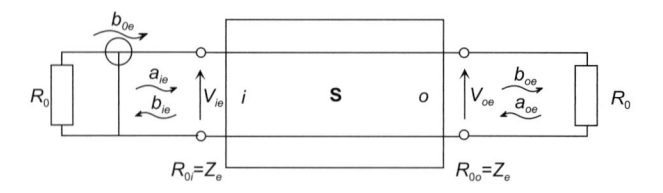

Figure 4.12 Even-mode line as a loaded two-port.

while the even mode forward wave generator at the input has value:

$$b_{0e} = \frac{E_g}{2} \frac{\sqrt{Z_{0e}}}{Z_{0e} + R_0}.$$

From the analysis of the loaded two-port we find that:

$$b_{oe} = b_{0e} \frac{e^{-j\theta_e}}{1 - \Gamma_{0e}^2 e^{-2j\theta_e}}$$

$$a_{oe} = b_{0e} \frac{\Gamma_{0e} e^{-j\theta_e}}{1 - \Gamma_{0e}^2 e^{-2j\theta_e}}$$

$$b_{ie} = b_{0e} \frac{\Gamma_{0e} e^{-2j\theta_e}}{1 - \Gamma_{0e}^2 e^{-2j\theta_e}}$$

$$a_{ie} = b_{0e} \frac{1}{1 - \Gamma_{0e}^2 e^{-2j\theta_e}},$$

while the input and output voltages can be recovered from:

$$V_{ie} = a_{ie}\sqrt{Z_{0e}} + b_{ie}\sqrt{Z_{0e}}$$
$$V_{oe} = b_{oe}\sqrt{Z_{0e}} + a_{oe}\sqrt{Z_{0e}}.$$

A similar results holds for the odd mode line:

$$b_{oo} = b_{0o} \frac{e^{-j\theta_o}}{1 - \Gamma_{0o}^2 e^{-2j\theta_o}}$$

$$a_{oo} = b_{0o} \frac{\Gamma_{0e} e^{-j\theta_o}}{1 - \Gamma_{0e}^2 e^{-2j\theta_o}}$$

$$b_{io} = b_{0o} \frac{\Gamma_{0e} e^{-2j\theta_o}}{1 - \Gamma_{0e}^2 e^{-2j\theta_o}}$$

$$a_{io} = b_{0o} \frac{1}{1 - \Gamma_{0o}^2 e^{-2j\theta_o}},$$

where:

$$\Gamma_{0o} = \frac{R_0 - Z_{0o}}{R_0 + Z_{0o}}$$

and the odd mode forward wave generator is:

$$b_{0o} = \frac{E_g}{2} \frac{\sqrt{Z_{0o}}}{R_0 + Z_{0o}},$$

while $\theta_o = 2\pi l/\lambda_o$. Substituting the forward wave generator value we finally obtain:

$$V_{ie} = \frac{E_g}{2} \frac{1 + \Gamma_{0e} e^{-2j\theta_e}}{1 - \Gamma_{0e}^2 e^{-2j\theta_e}} \frac{Z_{0e}}{R_0 + Z_{0e}}$$

$$V_{io} = \frac{E_g}{2} \frac{1 + \Gamma_{0o} e^{-2j\theta_o}}{1 - \Gamma_{0o}^2 e^{-2j\theta_o}} \frac{Z_{0o}}{R_0 + Z_{0o}}$$

$$V_{oe} = \frac{E_g}{2} \frac{(1 + \Gamma_{0e}) e^{-j\theta_e}}{1 - \Gamma_{0e}^2 e^{-2j\theta_e}} \frac{Z_{0e}}{R_0 + Z_{0e}}$$

$$V_{oo} = \frac{E_g}{2} \frac{(1 + \Gamma_{0o}) e^{-j\theta_o}}{1 - \Gamma_{0o}^2 e^{-2j\theta_o}} \frac{Z_{0o}}{R_0 + Z_{0o}}.$$

As a first step, let us evaluate the resistance R_0 allowing for matching at port 1. To this purpose, suppose that the phase velocities of the even and odd modes are equal:

$$\theta_e = \theta_o = \theta.$$

We anticipate that the centerband frequency of the coupler corresponds to $l = \lambda_g/4$, i.e., $e^{-j\theta} = -j$, $e^{-2j\theta} = -1$. In this condition, we have for the port 1 voltage $V_1 = V_{ie} + V_{io}$:

$$V_1 = \frac{E_g}{2} \frac{Z_{0e}}{R_0 + Z_{0e}} \frac{1 - \Gamma_{0e}}{1 + \Gamma_{0e}^2} + \frac{E_g}{2} \frac{Z_{0o}}{R_0 + Z_{0o}} \frac{1 - \Gamma_{0o}}{1 + \Gamma_{0o}^2}$$

$$= \frac{E_g}{2} \left(\frac{Z_{0o}^2}{R_0^2 + Z_{0o}^2} + \frac{Z_{0e}^2}{R_0^2 + Z_{0e}^2} \right).$$

Port 1 is matched if $V_1 = E_g/2$, i.e., if:

$$\frac{Z_{0e}^2}{R_0^2 + Z_{0e}^2} + \frac{Z_{0o}^2}{R_0^2 + Z_{0o}^2} = 1,$$

from which the matching condition:

$$R_0 = \sqrt{Z_{0e} Z_{0o}}. \tag{4.6}$$

Taking into account this condition, the even and odd mode reflection coefficients result as:

$$\Gamma_{0o} = \frac{R_0 - Z_{0o}}{R_0 + Z_{0o}} = \frac{\sqrt{Z_{0e}} - \sqrt{Z_{0o}}}{\sqrt{Z_{0e}} + \sqrt{Z_{0o}}} = \Gamma \tag{4.7}$$

$$\Gamma_{0e} = \frac{R_0 - Z_{0e}}{R_0 + Z_{0e}} = \frac{\sqrt{Z_{0o}} - \sqrt{Z_{0e}}}{\sqrt{Z_{0o}} + \sqrt{Z_{0e}}} = -\Gamma.$$

Away from the centerband frequency, taking into account that:

$$1 + \Gamma = 1 + \frac{\sqrt{Z_{0e}} - \sqrt{Z_{0o}}}{\sqrt{Z_{0e}} + \sqrt{Z_{0o}}} = \frac{2\sqrt{Z_{0e}}}{\sqrt{Z_{0e}} + \sqrt{Z_{0o}}}$$

$$1 - \Gamma = 1 - \frac{\sqrt{Z_{0e}} - \sqrt{Z_{0o}}}{\sqrt{Z_{0e}} + \sqrt{Z_{0o}}} = \frac{2\sqrt{Z_{0o}}}{\sqrt{Z_{0e}} + \sqrt{Z_{0o}}},$$

we obtain that the even and odd mode voltages can be expressed as:

$$V_{ie} = \frac{E_g}{4}(1+\Gamma)\frac{1-\Gamma e^{-2j\theta}}{1-\Gamma^2 e^{-2j\theta}}$$

$$V_{io} = \frac{E_g}{4}(1-\Gamma)\frac{1+\Gamma e^{-2j\theta}}{1-\Gamma^2 e^{-2j\theta}}$$

$$V_{oe} = \frac{E_g}{4}(1+\Gamma)\frac{(1-\Gamma)e^{-j\theta}}{1-\Gamma^2 e^{-2j\theta}}$$

$$V_{oo} = \frac{E_g}{4}(1-\Gamma)\frac{(1+\Gamma)e^{-j\theta}}{1-\Gamma^2 e^{-2j\theta}}.$$

We can readily evaluate V_1, V_2, V_3, V_4 summing or subtracting the even and odd mode input and output voltages; we obtain:

$$V_1 = V_{ie} + V_{io} = \frac{E_g}{2}$$

$$V_2 = V_{ie} - V_{io} = \frac{E_g}{2}\frac{\Gamma(1-e^{-2j\theta})}{1-\Gamma^2 e^{-2j\theta}}$$

$$V_3 = V_{oe} + V_{oo} = \frac{E_g}{2}\frac{(1-\Gamma^2)e^{-j\theta}}{1-\Gamma^2 e^{-2j\theta}}$$

$$V_4 = V_{oe} - V_{oo} = 0.$$

We conclude that port 1 is always matched, independent of frequency, while port 4 is always isolated from port 1, i.e., $S_{11} = S_{41} = 0$ at all frequencies. Power transmission takes place from port 1 to port 2 (the coupled port) and to port 3 (the transmission or thru port). Taking into account the twofold symmetry of the structure, we have:

$$S_{11} = S_{22} = S_{33} = S_{44} = 0$$

(all ports are matched):

$$S_{41} = S_{32} = S_{23} = S_{14} = 0$$

(defining isolation of ports with respect to the input), and then:

$$S_{21} = S_{12} = S_{34} = S_{43}$$

(defining coupling), and finally:

$$S_{31} = S_{13} = S_{42} = S_{24}$$

(defining transmission). Some of those relationships are anyway imposed by reciprocity. Taking into account that the reference impedance is the same for all ports and that port 1 is matched (the total voltage V_1 coincides with the progressive value) we simply have $S_{ji} = V_j/V_i$. Therefore, in the synchronous case the scattering matrix of the coupler as a function of the electrical length θ results as:

$$\mathbf{S}(\theta) = \begin{pmatrix} 0 & \dfrac{\Gamma(1 - e^{-2j\theta})}{1 - \Gamma^2 e^{-2j\theta}} & \dfrac{(1 - \Gamma^2)e^{-j\theta}}{1 - \Gamma^2 e^{-2j\theta}} & 0 \\ \dfrac{\Gamma(1 - e^{-2j\theta})}{1 - \Gamma^2 e^{-2j\theta}} & 0 & 0 & \dfrac{(1 - \Gamma^2)e^{-j\theta}}{1 - \Gamma^2 e^{-2j\theta}} \\ \dfrac{(1 - \Gamma^2)e^{-j\theta}}{1 - \Gamma^2 e^{-2j\theta}} & 0 & 0 & \dfrac{\Gamma(1 - e^{-2j\theta})}{1 - \Gamma^2 e^{-2j\theta}} \\ 0 & \dfrac{(1 - \Gamma^2)e^{-j\theta}}{1 - \Gamma^2 e^{-2j\theta}} & \dfrac{\Gamma(1 - e^{-2j\theta})}{1 - \Gamma^2 e^{-2j\theta}} & 0 \end{pmatrix}, \quad (4.8)$$

where:

$$\theta = \frac{2\pi l}{\lambda_g} = \frac{\omega \sqrt{\epsilon_{\text{eff}}} l}{c_0}$$

and the reflectance Γ is defined in (4.7).

4.3.1 Frequency Behavior of the Synchronous Coupler

At centerband, i.e., at the frequency where the line length is a quarter wavelength, $\theta = \pi/2$ and the nonzero scattering matrix elements are:

$$S_{21} = \frac{2\Gamma}{1 + \Gamma^2} = \frac{Z_{0e} - Z_{0o}}{Z_{0e} + Z_{0o}} \equiv C \qquad (4.9)$$

$$S_{31} = -j\frac{1 - \Gamma^2}{1 + \Gamma^2} = -j\frac{2\sqrt{Z_{0e} Z_{0o}}}{Z_{0e} + Z_{0o}} = -j\sqrt{1 - C^2}.$$

The parameter C is the *coupling* (or centerband coupling) of the coupler; the power coupling is $K = C^2$. For 3 dB couplers (equal power division between the coupled and the transmission port) $K = 1/2$ and $C = 1/\sqrt{2}$. The centerband scattering matrix can be therefore written as:

$$\mathbf{S}(\pi/2) = \begin{pmatrix} 0 & C & -j\sqrt{1 - C^2} & 0 \\ C & 0 & 0 & -j\sqrt{1 - C^2} \\ -j\sqrt{1 - C^2} & 0 & 0 & C \\ 0 & -j\sqrt{1 - C^2} & C & 0 \end{pmatrix}.$$

Notice that the structure of the scattering matrix corresponds to the general result from Sec. 4.2.1. From (4.8) we can also conveniently derive the following frequency-dependent expression of the scattering matrix elements S_{21} and S_{41}:

$$S_{21} = \frac{jC \sin \theta}{\sqrt{1 - C^2} \cos \theta + j \sin \theta}$$

$$S_{31} = \frac{\sqrt{1 - C^2}}{\sqrt{1 - C^2} \cos \theta + j \sin \theta}.$$

The above expressions also allow the coupler bandwidth to be derived. By inspection, $|S_{21}|$ decreases with respect to the centerband $\theta_0 = \pi/2$ for increasing or decreasing θ in a symmetrical way, while $|S_{31}|$ increases. Let us consider for instance $|S_{21}|$ and identify the arguments $\theta_{1,2}$ for which $|S_{21}|$ decreases by a factor $\alpha < 1$ with respect to the centerband value:

$$\frac{|S_{21}(\theta_{1,2})|^2}{|S_{21}(\theta_0)|^2} = \frac{1}{(1-C^2)\cot^2\theta_1 + 1} = \alpha^2;$$

we obtain:

$$\tan\theta_{1,2} = \pm\alpha\sqrt{\frac{1-C^2}{1-\alpha^2}},$$

where the positive sign corresponds to the lower limit of the bandwidth (θ_1), while the negative sign to the upper limit of the bandwidth ($\theta_2 = \pi - \theta_1$). Considering that $\pi/2 - \arctan(x) = \arctan(1/x)$, the bandwidth is then given by:

$$\theta_2 - \theta_1 = \pi - 2\theta_1 = 2\left[\frac{\pi}{2} - \arctan\left(\alpha\sqrt{\frac{1-C^2}{1-\alpha^2}}\right)\right]$$

$$= 2\arctan\left(\frac{1}{\alpha}\sqrt{\frac{1-\alpha^2}{1-C^2}}\right). \tag{4.10}$$

Since $\theta/\theta_0 = f/f_0$ and $\theta_0 = \pi/2$, we can readily derive from (4.10) the frequency range $B(\alpha, C)$ where $|S_{21}| > \alpha |S_{21}(\theta_0)|$:

$$B(\alpha, C) = f_2 - f_1 = f_0 \cdot \frac{4}{\pi}\arctan\left(\frac{1}{\alpha}\sqrt{\frac{1-\alpha^2}{1-C^2}}\right).$$

$B(\alpha, C)$ can be normalized with respect to f_0 to yield the relative coupler bandwidth $B_r(\alpha, C)$:

$$B_r(\alpha, C) = \frac{B(\alpha, C)}{f_0} = \frac{4}{\pi}\arctan\left(\frac{1}{\alpha}\sqrt{\frac{1-\alpha^2}{1-C^2}}\right). \tag{4.11}$$

Having assigned α, the bandwidth will depend on centerband coupling; since the arctan function increases with increasing argument, the relative bandwidth increases for increasing coupling (C gets closer to 1), or, in other words, large couplings correspond to a wider bandwidth, see, e.g., [5, Fig. 3]. From (4.11) we can estimate the coupling penalty vs. centerband for a bandwidth of an octave in the 3 dB coupler ($C^2 = 1/2$), that will be useful in the discussion of balanced amplifiers. In this case, $f_2 = 2f_1$ and therefore $f_0 = 3f_1/2$ and $f_2 - f_1 = f_1$; it follows that $B_r(\alpha, 1/\sqrt{2}) = 2/3$, and inverting (4.11) we have:

$$\frac{4}{\pi}\arctan\left[\frac{1}{\alpha}\sqrt{2(1-\alpha^2)}\right] = \frac{2}{3} \rightarrow \frac{2(1-\alpha^2)}{\alpha^2} = \tan^2\frac{\pi}{6} = \frac{1}{3},$$

from which we obtain $\alpha = \sqrt{6/7} \simeq 0.93$, corresponding to a coupling deviation vs. centerband of 0.67 dB.

4.3 The Two-Conductor Coupled Line Coupler

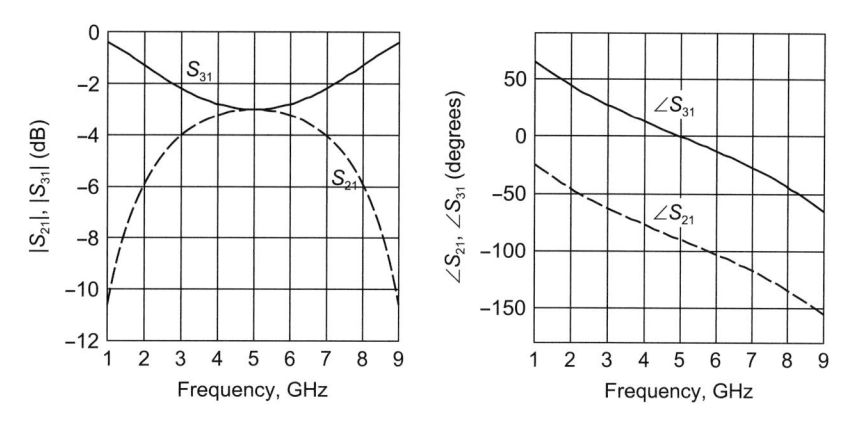

Figure 4.13 Magnitude and phase of S_{21} and S_{31} for an ideal (synchronous) 3 dB coupler.

We summarize here for convenience the formulae for the design and analysis of a coupler, relating the termination resistance R_0 and coupling with the even and odd mode impedances:

$$C = \frac{Z_{0e} - Z_{0o}}{Z_{0e} + Z_{0o}}, \quad R_0 = \sqrt{Z_{0o}Z_{0e}},$$

from which, inverting:

$$Z_{0e} = R_0\sqrt{\frac{1+C}{1-C}}, \quad Z_{0o} = R_0\sqrt{\frac{1-C}{1+C}}.$$

In practice, 3 dB couplers cannot be realized with coupled microstrips. In fact, to obtain a 3 dB coupler on 50 Ω we need $Z_{0e} = 121$ Ω, $Z_{0o} = 21$ Ω; such values would require an extremely close spacing (a few microns) between the strips, that is technologically inconvenient to obtain in hybrid or integrated implementations (besides leading to large losses), cfr. Example 4.1. Integrated 3 dB couplers can be implemented with multiconductor structures, such as the Lange coupler (Sec. 4.4.1). Fig. 4.13 shows the magnitude (in dB) of S_{21} and S_{31} for a 3 dB coupler. The coupler makes use of a coupled line with even and odd effective permittivities equal to 5; the centerband frequency is 5 GHz.

Example 4.1 Design a 3 dB coupler on 50 Ω in a coupled microstrip using a GaAs substrate with thickness $h = 300$ μm. Is the resulting structure technologically realizable?

Solution
As already seen, we have $Z_{0e} = 121$ Ω, $Z_{0o} = 21$ Ω. From Fig. 4.16, the needed width and slot values are $W/h = 0.2$, $S/h = 0.0018$, i.e., $W = 60$ μm, $S = 0.5$ μm. The slot width is far too small to be implemented from a technological standpoint, not to mention the high losses resulting from a very narrow gap.

4.3.2 Effect of Velocity Mismatch and Compensation Techniques

In quasi-TEM couplers, the even and odd mode velocities are in principle different. Supposing $R_0 = \sqrt{Z_{0o}Z_{0e}}$, if $\theta_e \neq \theta_o$ we have for the input voltages:

$$V_{ie} = \frac{E_g}{4}(1+\Gamma)\frac{1-\Gamma e^{-2j\theta_e}}{1-\Gamma^2 e^{-2j\theta_e}}$$

$$V_{io} = \frac{E_g}{4}(1-\Gamma)\frac{1+\Gamma e^{-2j\theta_o}}{1-\Gamma^2 e^{-2j\theta_o}},$$

while for the output voltages:

$$V_{oe} = \frac{E_g}{4}(1+\Gamma)\frac{(1-\Gamma)e^{-j\theta_e}}{1-\Gamma^2 e^{-2j\theta_e}}$$

$$V_{oo} = \frac{E_g}{4}(1-\Gamma)\frac{(1+\Gamma)e^{-j\theta_o}}{1-\Gamma^2 e^{-2j\theta_o}},$$

from which V_1, V_2, V_3, V_4 can be recovered as usual. We then obtain:

$$S_{j1} = 2\frac{V_j}{E_g}, \quad j = 2, 3, 4,$$

taking into account that R_0 is the same for all ports. The reflection coefficient at port 1 can be evaluated considering that:

$$I_1 = \frac{E_g - V_1}{R_0},$$

but the forward and backward voltages at port 1 can be written as:

$$V_1^+ = \frac{V_1 + R_0 I_1}{2} = \frac{E_g}{2}$$

$$V_1^- = \frac{V_1 - R_0 I_1}{2} = V_1 - \frac{E_g}{2},$$

from which:

$$S_{11} = \frac{V_1^-}{V_1^+} = \frac{2V_1}{E_g} - 1.$$

We do not carry out the computations in detail but come to the main consequences. Velocity mismatch leads to an impedance mismatch at port 1 and to a decrease of the isolation at port 4, while coupling and transmission are affected, but not dramatically. Due to the lack of impedance matching, the ideal properties derived in Sec. 4.2.1 do not exactly hold any more. Fig. 4.14 shows the parameters of the already analyzed 3 dB coupler; the coupler length was evaluated from the even and odd mode guided wavelengths as:

$$l = \left(\frac{\lambda_e}{4} + \frac{\lambda_o}{4}\right)\frac{1}{2}.$$

In the example, $\epsilon_{\text{eff}} = 5$ for the even and $\epsilon_{\text{eff}} = 4.5$ for the odd mode. Among the unfavorable consequences of mismatch, perhaps the most serious issue is the decreased isolation, that may cause trouble in systems where the coupler is the interface between a

4.3 The Two-Conductor Coupled Line Coupler

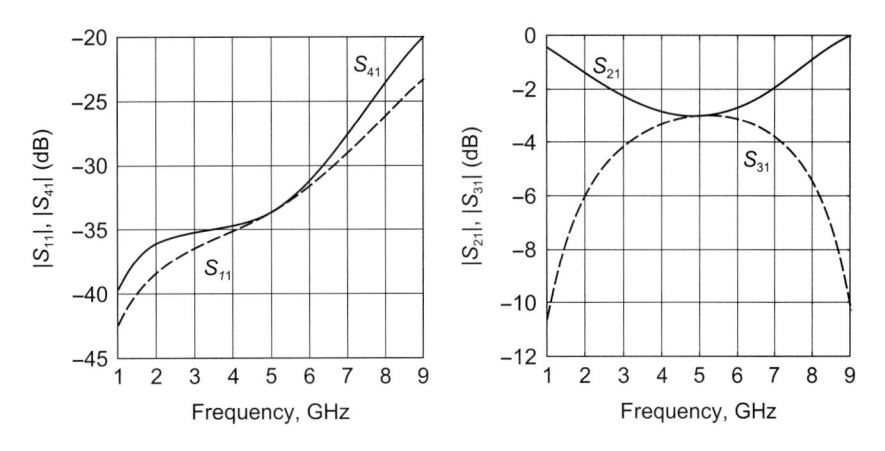

Figure 4.14 Magnitude of S_{11}, S_{41} (left) and S_{21}, S_{31} (right) of a real (non-synchronous) 3 dB coupler. The ratio between the effective permittivities of the two modes is 0.9.

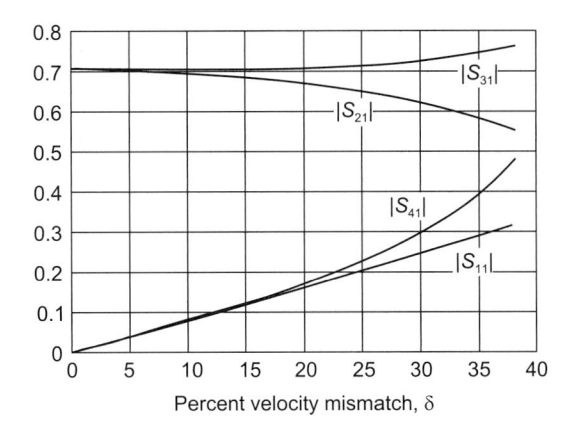

Figure 4.15 Magnitude of the S parameters of a 3 dB coupler in the presence of velocity mismatch.

transmitter (with strong signals) and a receiver (with weak signals). In this case, power leakage from the high-power transmitter to the low-power receiver can impair the system operation.

An approximate estimate of the centerband directivity vs. velocity mismatch is as follows [6]:

$$S_{41} \approx \delta(1 - C^2),$$

where:

$$\delta = \frac{|v_p - v_d|}{v_p + v_d}.$$

An example of the behavior of the scattering parameters magnitude as a function of velocity mismatch is shown in Fig. 4.15.

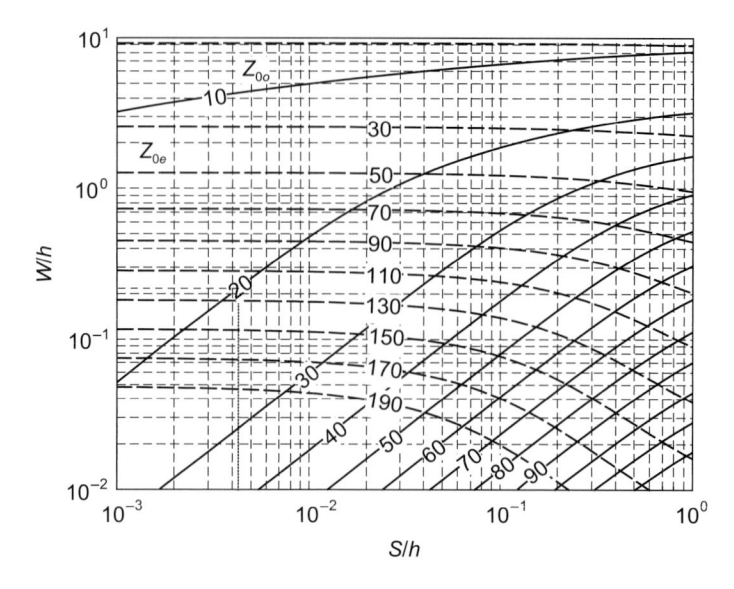

Figure 4.16 Level curves of Z_{0e} and Z_{0o} (Ω) as a function of S/h, W/h (log units) for $\epsilon_r = 13$.

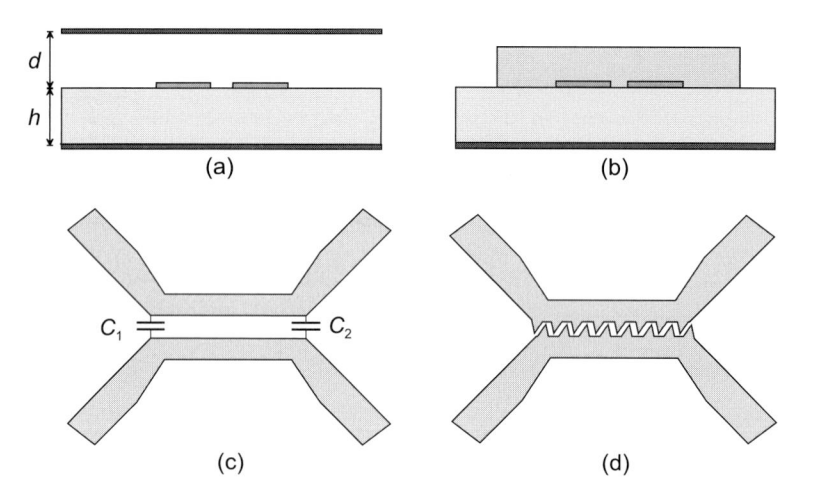

Figure 4.17 Velocity compensation techniques: shielded directional coupler (a) and with dielectric overlay (b); compensation through concentrated capacitances (c) e and line *wiggle* (d).

The directivity of a non-synchronous microstrip directional coupler can be improved by equalizing the phase velocity of the even and odd modes through a number of expedients:

- the use of grounded metal screens. For symmetry, if $d = h$ the effective permittivities of the two modes become equal to $(\epsilon_r + 1)/2$, see Fig. 4.17, (a);
- the use of dielectric overlays. A dielectric layer of suitable thickness and permittivity can compensate for the phase velocity mismatch, as shown in Fig. 4.17 (b), because it independently changes odd and even capacitances.

There are other methods to correct the phase velocity mismatch through distributed or concentrated techniques:

- introducing external loading capacitances, as shown in Fig. 4.17 (c). This modifies the odd mode capacitance only. Take into account that the odd effective permittivity is typically lower than the even one, so that the odd mode electrical length is lower than the even mode electrical length for the same physical line length. Inserting a concentrated additional capacitance $C = C_1 + C_2$ increases the equivalent odd-mode line electrical length; in fact, if $\mathcal{C}_o$ is the p.u.l. odd mode capacitance, we have $\Delta l \mathcal{C}_o = C$, from which an odd mode phase shift results:

$$\Delta \theta_o = \Delta l \omega \sqrt{\mathcal{C}_o \mathcal{L}_o} = \omega C Z_{0o}.$$

This makes it possible to design, at least at centerband, the compensation capacitance.
- the use of serrated of *wiggling* coupled lines, see Fig. 4.17 (d); the wiggling does not greatly affect the even mode capacitance, but has a strong impact on the odd mode capacitance, that depends on the edge coupling between lines.

4.4 Multiconductor Line Couplers

Multiconductor (or interdigitated) line couplers exploit multiconductor lines whose conductors are connected together through wires or air bridges so as to emulate a two-conductor line. Such structures allow a larger difference between the even and odd mode impedances to be obtained, and therefore high coupling with technologically feasible geometries.

Let us start from the analysis of a multiconductor line with an even number of k parallel strips, see Fig. 4.18. As a first approximation, we neglect the mutual capacitance between non-neighboring strips and we suppose that the mutual capacitance between neighboring strips ($\mathcal{C}_{12}$) and the strip capacitance to ground ($\mathcal{C}_{20}$) is the same, independent of the strip position, apart from the two external strips, whose capacitance towards ground will be denoted as $\mathcal{C}_{10}$. The following approximate relationship can be shown to hold [7]:

$$\mathcal{C}_{20} \approx \mathcal{C}_{10} - \frac{\mathcal{C}_{10}\mathcal{C}_{12}}{\mathcal{C}_{10} + \mathcal{C}_{12}}.$$

In fact, the capacitance towards ground of the external strip is similar to the capacitance towards ground of a strip at whose right an infinite number of floating strips (i.e., having zero total charge) is present (see Fig. 4.19). Such capacitance can be estimated as the *iterative capacitance* of an infinite set of metal strips, as shown Fig. 4.20. We thus obtain:

$$\mathcal{C}_{10} = \mathcal{C}_{20} + \frac{\mathcal{C}_{10}\mathcal{C}_{12}}{\mathcal{C}_{10} + \mathcal{C}_{12}},$$

from which we derive $\mathcal{C}_{20}$. Notice that $\mathcal{C}_{12}$ and $\mathcal{C}_{10}$ also are the capacitances between two strips and between each strip and the ground, respectively, of two coupled microstrips.

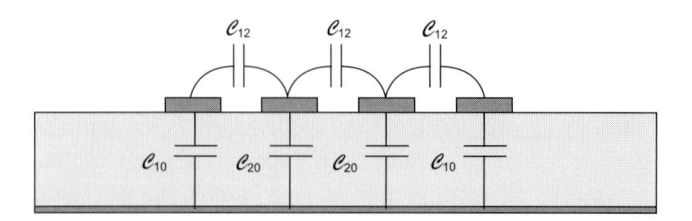

Figure 4.18 Multiconductor line and associate capacitances.

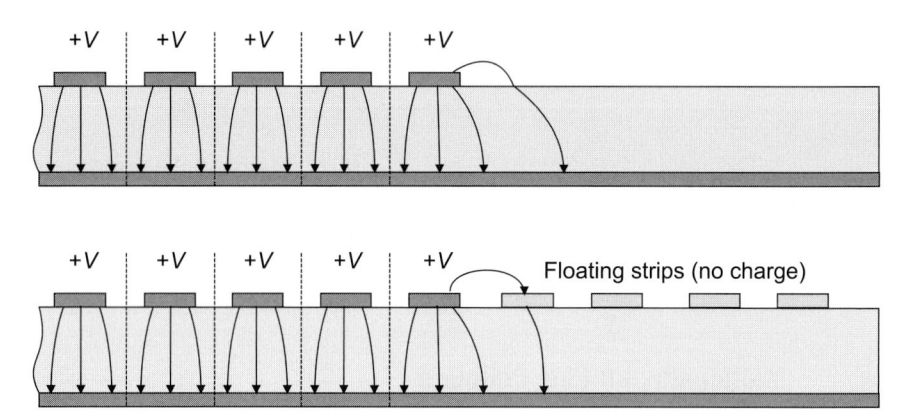

Figure 4.19 Evaluating the capacitance of the two extreme lines in a multiconductor line – I.

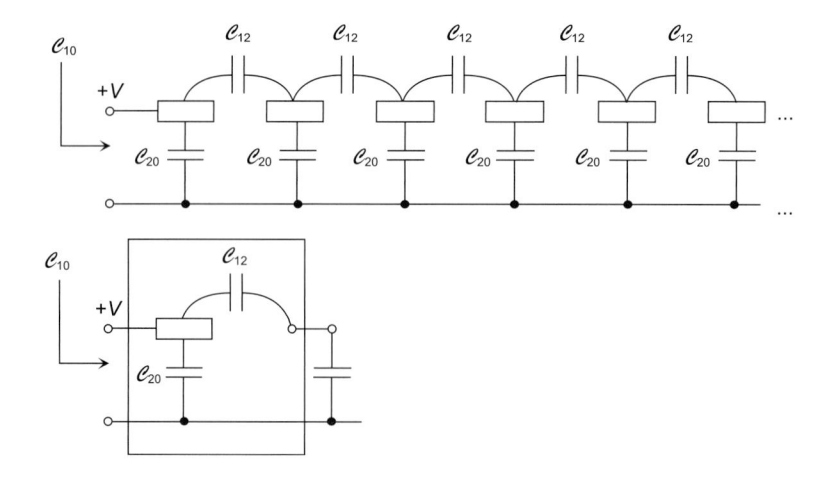

Figure 4.20 Evaluating the capacitance of the two extreme lines in a multiconductor line – II.

These capacitances can be approximated as a function of the line geometry and dielectric substrate parameters, as discussed further on.

Suppose now that the lines of the multiconductor structure are connected in an alternate way, so as to yield two equipotential conductors. The connection is typically not done in a continuous way, but only at intervals, close enough (e.g., less than a quarter wavelength), by means of airbridges or bonding wires (see Fig. 4.23).

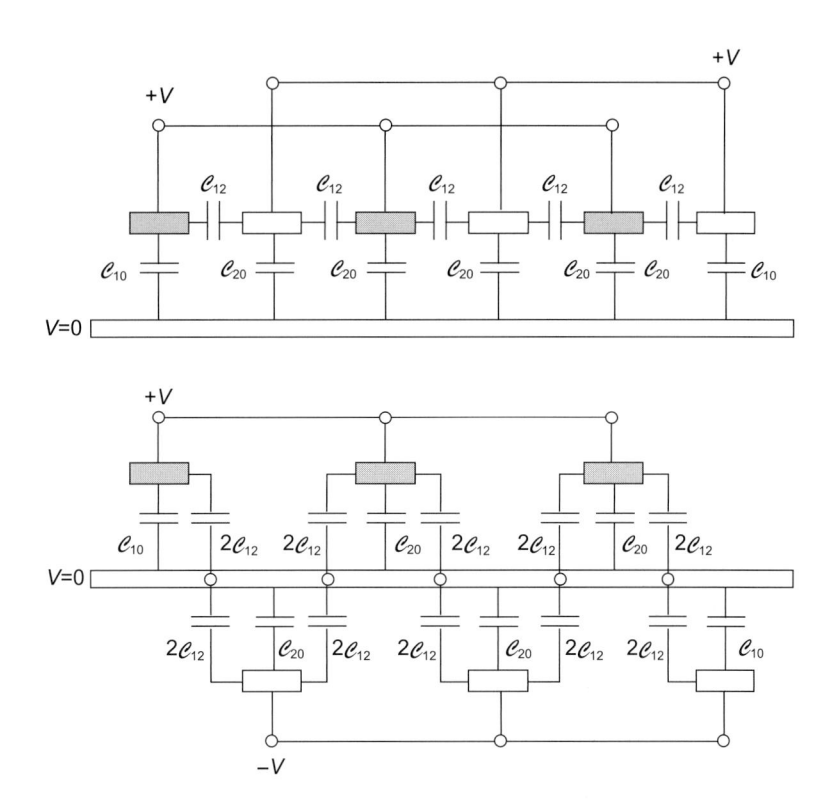

Figure 4.21 Evaluating the even and odd mode capacitances in a multiconductor line with strips connected alternately together; in the case shown $k = 6$.

We can now define an equivalent two-conductor line, able to carry an even and an odd mode. The even and odd mode capacitances are the capacitance towards ground of a set of $k/2$ (with k even) strips when the two sets have the same or opposite potential, respectively. From Fig. 4.21 we obtain:

$$\mathcal{C}_e(k) = (k/2 - 1)\mathcal{C}_{20} + \mathcal{C}_{10}$$
$$\mathcal{C}_o(k) = (k/2 - 1)\mathcal{C}_{20} + \mathcal{C}_{10} + 2(k - 1)\mathcal{C}_{12},$$

where all capacitances are per unit length. Taking into account that for two strips we have:

$$\mathcal{C}_e(2) = \mathcal{C}_{10}$$
$$\mathcal{C}_o(2) = \mathcal{C}_{10} + 2\mathcal{C}_{12},$$

we can obtain $\mathcal{C}_{10}$ and $\mathcal{C}_{12}$ as a function of $\mathcal{C}_e(2)$ and $\mathcal{C}_o(2)$, and therefore express the even and odd mode capacitances of the equivalent two-conductor line derived from the multiconductor one by connecting strips as a function of the even and odd mode capacitances of the two-conductor line:

$$\mathcal{C}_e(k) = \frac{\mathcal{C}_o(2)\mathcal{C}_e(2) + (k-1)\mathcal{C}_e^2(2)}{\mathcal{C}_o(2) + \mathcal{C}_e(2)}$$

$$\mathcal{C}_o(k) = \frac{\mathcal{C}_o(2)\mathcal{C}_e(2) + (k-1)\mathcal{C}_o^2(2)}{\mathcal{C}_o(2) + \mathcal{C}_e(2)}.$$

We can now derive the characteristics (termination impedance and centerband coupling) of the coupler. Taking into account that:

$$C(k) = \frac{Z_{0e}(k) - Z_{0o}(k)}{Z_{0e}(k) + Z_{0o}(k)} = \frac{\mathcal{C}_o(k) - \mathcal{C}_e(k)}{\mathcal{C}_e(k) + \mathcal{C}_o(k)}$$

and that, for velocity matching, $Z_{0e}/Z_{0o} = \mathcal{C}_o/\mathcal{C}_e$, we have [8]:

$$C(k) = \frac{(k-1)(1-R^2)}{(k-1)(1+R^2) + 2R}, \qquad (4.12)$$

where R is defined as:

$$R = \frac{Z_{0o}(2)}{Z_{0e}(2)}. \qquad (4.13)$$

We can similarly show that the matching impedance is:

$$R_0^2 = Z_{0e}(2)Z_{0o}(2)\frac{(1+R)^2}{[(k-1)R+1][(k-1)+R]}. \qquad (4.14)$$

The behavior of the coupling (in dB) as a function of R and for different values of the line number k is shown in Fig. 4.22, left. To obtain large couplings, we need small values of R (and therefore very different even and odd mode impedances) for $k = 2$, while R increases with growing k. The improvement is however marginal for $k > 8$. Similar remarks can be made on the closing impedance normalized vs. the two-strip closing (or matching) impedance, see Fig. 4.22, right. Note that for a termination impedance of 50 Ω we have (for two conductors) $\sqrt{Z_{0e}(2)Z_{0o}(2)} = 50$ Ω, while for four conductors $\sqrt{Z_{0e}(2)Z_{0o}(2)} \approx 100$ Ω, i.e., the needed strip width decreases with increasing k. The effect is negative for large k, since it implies that very narrow strips have to be used to obtain a convenient impedance level.

In conclusion, multiconductor couplers allow a 3 dB coupler compatible with technological constraints to be designed; the benefits are maximum with four or six conductors, while increasing the number of conductors beyond this value is inconvenient due to the complexity of the structure and to the increased losses. The length of the coupler can be determined from the effective permittivity of the even and odd modes. Taking into account that, for a two-line coupler, we have $\mathcal{C}_e(2) = \mathcal{C}_{ae}(2)\epsilon_{\text{eff}e}(2)$ and $\mathcal{C}_o(2) = \mathcal{C}_{ao}(2)\epsilon_{\text{eff}o}(2)$, and defining as $R_a \leq 1$ the ratio between the odd and even mode impedances:

$$R_a = \frac{Z_{0ao}(2)}{Z_{0ae}(2)} = \frac{\mathcal{C}_{ae}(2)}{\mathcal{C}_{ao}(2)},$$

we can write:

4.4 Multiconductor Line Couplers

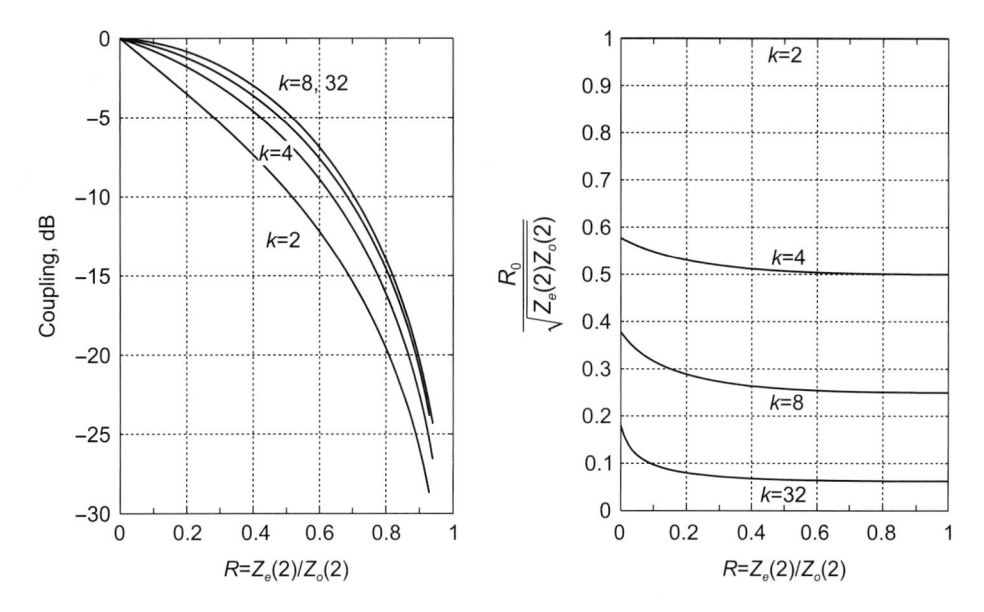

Figure 4.22 Coupling (left) and closing impedances (right) as a function of R for several values of k.

$$C_e(k) = C_{ao}(2)\frac{\epsilon_{\text{effe}}(2)\epsilon_{\text{effo}}(2)R_a + R_a^2(k-1)\epsilon_{\text{effe}}^2(2)}{\epsilon_{\text{effo}}(2) + R_a\epsilon_{\text{effe}}(2)}$$

$$C_{ae}(k) = \frac{C_{ao}(2)C_{ae}(2) + (k-1)C_{ae}^2(2)}{C_{ao}(2) + C_{ae}(2)} = C_{ao}(2)\frac{R_a + R_a^2(k-1)}{1 + R_a}$$

$$C_o(k) = C_{ao}(2)\frac{\epsilon_{\text{effe}}(2)\epsilon_{\text{effo}}(2)R_a + (k-1)\epsilon_{\text{effe}}^2(2)}{\epsilon_{\text{effo}}(2) + R_a\epsilon_{\text{effe}}(2)}$$

$$C_{ao}(k) = \frac{C_{ao}(2)C_{ae}(2) + (k-1)C_{ao}^2(2)}{C_{ao}(2) + C_{ae}(2)} = C_{ao}(2)\frac{R_a + (k-1)}{1 + R_a},$$

from which:

$$\epsilon_{\text{effe}}(k) = \frac{C_e(k)}{C_{ae}(k)}$$

$$= \left(\frac{\epsilon_{\text{effe}}(2)\epsilon_{\text{effo}}(2)R_a + R_a^2(k-1)\epsilon_{\text{effe}}^2(2)}{\epsilon_{\text{effo}}(2) + R_a\epsilon_{\text{effe}}(2)}\right)\left(\frac{1 + R_a}{R_a + R_a^2(k-1)}\right)$$

$$\epsilon_{\text{effo}}(k) = \frac{C_o(k)}{C_{ao}(k)}$$

$$= \left(\frac{\epsilon_{\text{effe}}(2)\epsilon_{\text{effo}}(2)R_a + (k-1)\epsilon_{\text{effo}}^2(2)}{\epsilon_{\text{effo}}(2) + R_a\epsilon_{\text{effe}}(2)}\right)\left(\frac{1 + R_a}{R_a + (k-1)}\right).$$

In multiconductor lines the even and odd permittivities are rather different; for large coupling we have $R_a \to 0$ which implies that the even and odd mode permittivities are similar to the case of the two-conductor line. Also in this case, we can approximate the centerband length through the average of the even and odd mode quarter wavelengths.

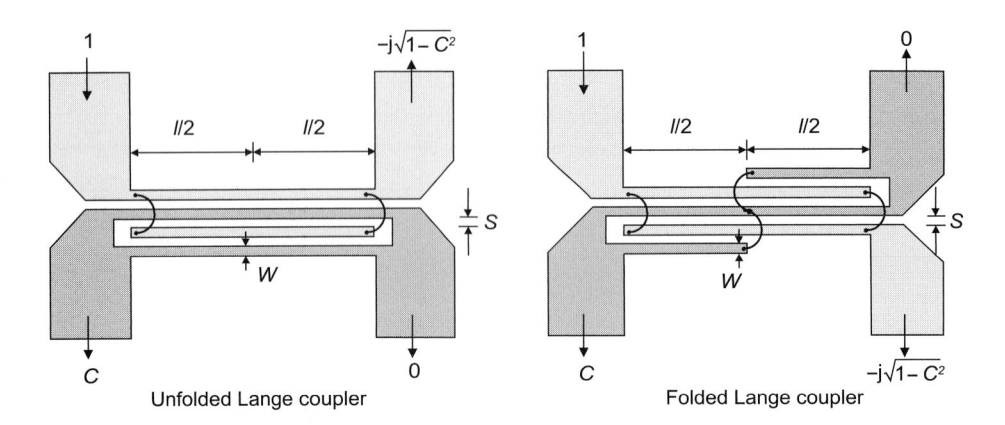

Figure 4.23 Lange coupler, *unfolded* version (left) and *folded* (right) version.

4.4.1 The Lange Coupler

The Lange coupler, named after Julius Lange, who proposed it in 1969 [9], is an interdigitated microstrip coupler consisting of four parallel lines alternately connected in pairs, as shown in Fig. 4.23, left (the *unfolded* version) and in Fig. 4.23, right (the *folded* version, more common in applications). Note that the unfolded version behaves like a two-conductor directional coupler; it is however made by four lines connected two by two at the ends, and therefore equipotential throughout the coupler length. In the folded version, the transmission and isolated ports are exchanged, so that the coupled and the transmission port are on the same side of the coupler; there is also a DC path between the upper and the lower side of the coupler. These characteristics make the folded version more common and convenient than the unfolded version.

To design a Lange coupler, we express, as a function of the centerband coupling C and of the closing impedance R_0, the even and odd mode impedances of the two-conductor coupled lines having the needed design parameters W and S. Inverting (4.12) and (4.13) so as to express the impedance ratio R as a function of the coupling C (the number of line k is given), and deriving the product $Z_{0o}(2)Z_{0e}(2)$ from (4.14) given the closing impedance R_0, we obtain that $Z_{0o}(2)$ and $Z_{0e}(2)$ can be expressed as:

$$Z_{0o}(2) = R_0 \sqrt{\frac{1-C}{1+C}} \frac{(k-1)(1+q)}{(C+q)+(k-1)(1-C)} \qquad (4.15)$$

$$Z_{0e}(2) = Z_{0o}(2) \frac{C+q}{(k-1)(1-C)}, \qquad (4.16)$$

where:

$$q = \sqrt{C^2 + (1-C^2)(k-1)^2}. \qquad (4.17)$$

In the most common case, the Lange coupler has four conductors ($k = 4$). An approximate design technique is as follows:

1. Starting from the centerband coupling and the closing impedance we derive the even and odd mode impedances of the two-conductor coupler having dimensions W and S, $Z_{0o}(2)$ and $Z_{0e}(2)$, using (4.15), (4.16) and (4.17).
2. According to the approximate method in [10], we derive the ratio W/h needed to obtain $Z_{0o}(2)/2$ and $Z_{0e}(2)/2$, respectively. To this purpose, we can use the approximations (2.44) and (2.45) for $Z \geq 44 - 2\epsilon_r$ Ω, and the approximations (2.46), (2.47), for $Z < 44 - 2\epsilon_r$ Ω (where $Z = Z_{0e}(2)/2$ or $Z_{0o}(2)/2$) yielding W/h as a function of the characteristic impedance Z.
3. The ratios W/h found are defined as $(W/h)_e$ and $(W/h)_o$. The real parameters S/h and W/h can be derived from the following equations, to be inverted numerically [10]:

$$\left(\frac{W}{h}\right)_e = f_e(W/h, S/h)$$

$$\left(\frac{W}{h}\right)_o = f_o(W/h, S/h),$$

where:

$$f_e = \frac{2}{\pi} \cosh^{-1}\left(\frac{2a - g + 1}{g + 1}\right)$$

and:

$$f_o = \frac{2}{\pi} \cosh^{-1}\left(\frac{2a - g - 1}{g - 1}\right) + \frac{4}{\pi(1 + \epsilon_r/2)} \cosh^{-1}\left(1 + 2\frac{W/h}{S/h}\right) \quad \epsilon_r \leq 6$$

$$f_o = \frac{2}{\pi} \cosh^{-1}\left(\frac{2a - g - 1}{g - 1}\right) + \frac{1}{\pi} \cosh^{-1}\left(1 + 2\frac{W/h}{S/h}\right) \quad \epsilon_r \geq 6.$$

The parameters g and a are defined as:

$$g = \cosh\left(\frac{\pi S}{2h}\right)$$

$$a = \cosh\left(\frac{\pi W}{h} + \frac{\pi S}{2h}\right).$$

The previous formulae allow the ratios W/h and S/h for the multiconductor line to be evaluated from the even and odd W/h.

The coupler centerband length can be approximated as the average of the quarter wavelengths for the even and odd modes of the two-conductor line.

Example 4.2 Design a 3 dB four-conductor Lange coupler on alumina at 10 GHz. The reference impedance is 50 Ω, the substrate dielectric constant is 9, the substrate thickness is $h = 25$ mils (0.635 mm).

Solution
We have $C = 0.707$, $k = 4$, $R_0 = 50$ Ω; from the design formulae (4.15) and (4.16) where:

$$q = \sqrt{C^2 + 9(1 - C^2)} = \sqrt{0.5 + 9 \cdot 0.5} = \sqrt{5} = 2.24,$$

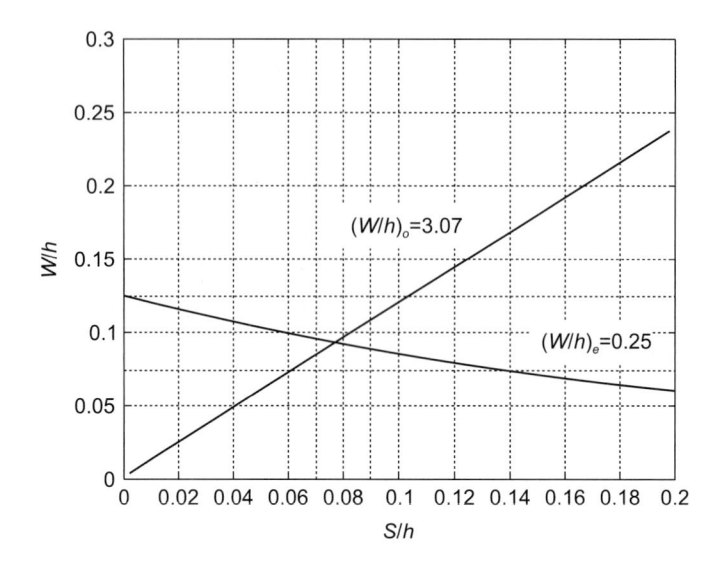

Figure 4.24 Graph of $f_e(W/h, S/h) = (W/h)_e = 0.25$ and of $f_o(W/h, S/h) = (W/h)_o = 3.07$, see Example 4.2; the solution corresponds to the crossing of the two curves.

we obtain $Z_{0o}(2) = 52.6\,\Omega$, $Z_{0e}(2) = 176.2\,\Omega$. The even mode W/h ratio therefore refers to $Z_{0e}(2)/2 = 88.1\,\Omega$, while the odd mode W/h ratio is derived from $Z_{0o}(2)/2 = 26.3\,\Omega$. Since $44 - 2\epsilon_r = 26\,\Omega$, we need to use (2.44) and (2.45); we obtain $(W/h)_e = 0.25$, $(W/h)_o = 3.07$. Inverting numerically or exploiting the chart in Fig. 4.24, we obtain $S/h = 0.076$, $W/h = 0.09$ from which $S \approx 46\,\mu$m, $W \approx 57\,\mu$m.

For the even and odd mode permittivities of the four-conductor line we have $\epsilon_{\text{eff}_o} \approx 5$ and $\epsilon_{\text{eff}_e} \approx 5.6$ from which:

$$\lambda_e/4 = 30/\sqrt{5.6}/4 = 3.17 \text{ mm}, \quad \lambda_e/4 = 30/\sqrt{5}/4 = 3.35 \text{ mm},$$

i.e., an average length $l \approx 3.26$ mm. As a first approximation, we note that $\epsilon_{\text{eff}} \approx (\epsilon_r + 1)/2 = 5$ for both modes, from which $l = 3.35$ mm.

4.5 Interference Couplers

Coupled microstrip line couplers allow for high coupling only in multiconductor form, as in the Lange coupler. Other coupler structures obtain power division and isolation through an interference principle; typical examples are the branch line and rat-race couplers.

4.5.1 Branch-Line Coupler

The analysis of the branch-line coupler, see Fig. 4.25 for the microstrip layout, can be carried out exploiting the structure quadrantal symmetry; see Example 4.3. For symmetry, the scattering matrix results:

4.5 Interference Couplers

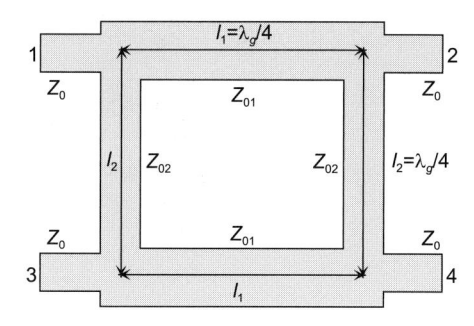

Figure 4.25 Branch-line coupler.

$$\mathbf{S} = \begin{pmatrix} S_{11} & S_{12} & S_{13} & S_{14} \\ S_{12} & S_{11} & S_{14} & S_{13} \\ S_{13} & S_{14} & S_{11} & S_{12} \\ S_{14} & S_{13} & S_{12} & S_{11} \end{pmatrix},$$

where, as shown in Example 4.3:

$$S_{11} = \frac{\Gamma_a + \Gamma_b + \Gamma_c + \Gamma_d}{4} \qquad (4.18a)$$

$$S_{21} = \frac{\Gamma_a + \Gamma_b - \Gamma_c - \Gamma_d}{4}$$

$$S_{31} = \frac{\Gamma_a - \Gamma_b + \Gamma_c - \Gamma_d}{4}$$

$$S_{41} = \frac{\Gamma_a - \Gamma_b - \Gamma_c + \Gamma_d}{4}.$$

The four indices refer to even and odd excitations with respect to the vertical and horizontal directions. Case (a) is even in both directions, case (d) is odd in both directions, (b) is even horizontally and odd vertically, (c) is odd horizontally and even vertically. At centerband, i.e., for $\theta_1 = \theta_2 = \pi/4$ (this means that the lengths of the two lines are a quarter wavelength) we obtain:

$$\Gamma_a = \frac{Y_0 - j Y_{01} - j Y_{02}}{Y_0 + j Y_{01} + j Y_{02}}$$

$$\Gamma_b = \frac{Y_0 - j Y_{01} + j Y_{02}}{Y_0 + j Y_{01} - j Y_{02}}$$

$$\Gamma_c = \frac{Y_0 + j Y_{01} - j Y_{02}}{Y_0 - j Y_{01} + j Y_{02}}$$

$$\Gamma_d = \frac{Y_0 + j Y_{01} + j Y_{02}}{Y_0 - j Y_{01} - j Y_{02}},$$

from which, substituting into (4.18a), we find for S_{11}:

$$S_{11} = \frac{Y_0^4 - (Y_{01}^2 - Y_{02}^2)^2}{Y_0^4 + 2Y_0^2(Y_{01}^2 + Y_{02}^2) + (Y_{01} - Y_{02})^4}.$$

Matching is obtained with respect to the reference impedance if:

$$Y_0^2 = |Y_{01}^2 - Y_{02}^2|. \tag{4.19}$$

Taking into account of this last expression and imposing for instance $Y_{01} > Y_{02}$, (4.19) becomes $Y_0^2 = Y_{01}^2 - Y_{02}^2$, i.e.:[1]

$$S_{21} = -j\frac{Y_0}{Y_{01}} \tag{4.20}$$

$$S_{31} = 0$$

$$S_{41} = -\frac{Y_{02}}{Y_{01}}. \tag{4.21}$$

Expressions (4.19), (4.20) and (4.21) can be used as design equations. They also imply that a phase relationship between port 2 (coupled) and port 4 (transmission or through) exists at centerband as in the coupled line coupler, i.e., a phase shift of 90 degrees. The centerband scattering matrix therefore is:

$$S = \begin{pmatrix} 0 & -j\dfrac{Y_0}{Y_{01}} & 0 & -\dfrac{Y_{02}}{Y_{01}} \\ -j\dfrac{Y_0}{Y_{01}} & 0 & -\dfrac{Y_{02}}{Y_{01}} & 0 \\ 0 & -\dfrac{Y_{02}}{Y_{01}} & 0 & -j\dfrac{Y_0}{Y_{01}} \\ -\dfrac{Y_{02}}{Y_{01}} & 0 & -j\dfrac{Y_0}{Y_{01}} & 0 \end{pmatrix},$$

that can be equivalently expressed in terms of impedances, since $Y_0/Y_{01} = Z_{01}/Z_0$ and $Y_{02}/Y_{01} = Z_{01}/Z_{02}$.

The centerband coupling between port 2 and 1 is now $C = Z_{01}/Z_0$ and, for power conservation in a lossless structure, we have $\sqrt{1 - C^2} = Z_{01}/Z_{02}$, from which:

$$Z_{01} = CZ_0$$

$$Z_{02} = \frac{CZ_0}{\sqrt{1 - C^2}}.$$

For a 3 dB coupler, $C = 1/\sqrt{2}$, i.e.:

$$Z_{02} = Z_0$$

$$Z_{01} = Z_0/\sqrt{2}.$$

Starting from an access line of 50 Ω, we find $Z_{01} = 35.35$ Ω and $Z_{02} = 50$ Ω, easily realizable in microstrip.

---

[1] If we impose instead $Y_{01} < Y_{02}$ or $Y_0^2 = Y_{02}^2 - Y_{01}^2$ we obtain by symmetry $S_{11} = 0$, $S_{21} = 0$, $S_{31} = -jY_0/Y_{02}$ and $S_{41} = -Y_{01}/Y_{02}$.

4.5 Interference Couplers

Example 4.3 Evaluate the scattering matrix of a branch-line coupler.

Solution

We start from the following remarks:

1. Due to symmetry and reciprocity, the scattering matrix is completely identified by the first row of the matrix.
2. Since the circuit is linear, we can apply the superposition principle, i.e., decompose the excitation required to evaluate the elements of the first row of the S-matrix into more convenient even and odd excitations.
3. Even and odd excitations can take the form of progressive voltages (or power waves) rather than of total voltages.

Let us impress (through a forward wave generator) the forward voltage V_n^+ entering port n ($n = 1, 2, 3, 4$) and denote with V_n^- the corresponding backward or reflected voltages. Consider now the following excitations at the four ports:

1. case (a): $V_{1a}^+ = V_{2a}^+ = V_{3a}^+ = V_{4a}^+ = V^+$;
2. case (b): $V_{1b}^+ = V_{2b}^+ = V^+$, $V_{3b}^+ = V_{4b}^+ = -V^+$;
3. case (c): $V_{1c}^+ = V_{3c}^+ = V^+$, $V_{2c}^+ = V_{4c}^+ = -V^+$;
4. case (d): $V_{1d}^+ = V_{4d}^+ = V^+$, $V_{2d}^+ = V_{3d}^+ = -V^+$.

The four excitations have the (already mentioned) even and/or odd character with respect to the vertical and horizontal plane. Superimposing, we have:

$$V_1^+ = 4V^+$$
$$V_2^+ = V_3^+ = V_4^+ = 0,$$

that are the excitation needed to evaluate the elements of the first row of the scattering matrix. Owing to the structure symmetry, we also have for the reflected waves:

1. $V_{1a}^- = V_{2a}^- = V_{3a}^- = V_{4a}^-$
2. $V_{1b}^- = V_{2b}^- = -V_{3b}^- = -V_{4b}^-$
3. $V_{1c}^- = -V_{2c}^- = V_{3c}^- = -V_{4c}^-$
4. $V_{1d}^- = -V_{2d}^- = -V_{3d}^- = V_{4d}^-$;

it follows that:

$$S_{11} = \frac{V_{1a}^- + V_{1b}^- + V_{1c}^- + V_{1d}^-}{V_{1a}^+ + V_{1b}^+ + V_{1c}^+ + V_{1d}^+} = \frac{V_{1a}^- + V_{1b}^- + V_{1c}^- + V_{1d}^-}{4V^+}$$

$$S_{21} = \frac{V_{2a}^- + V_{2b}^- + V_{2c}^- + V_{2d}^-}{V_{1a}^+ + V_{1b}^+ + V_{1c}^+ + V_{1d}^+} = \frac{V_{1a}^- + V_{1b}^- - V_{1c}^- - V_{1d}^-}{4V^+}$$

$$S_{31} = \frac{V_{3a}^- + V_{3b}^- + V_{3c}^- + V_{3d}^-}{V_{1a}^+ + V_{1b}^+ + V_{1c}^+ + V_{1d}^+} = \frac{V_{1a}^- - V_{1b}^- + V_{1c}^- - V_{1d}^-}{4V^+}$$

$$S_{41} = \frac{V_{4a}^- + V_{4b}^- + V_{4c}^- + V_{4d}^-}{V_{1a}^+ + V_{1b}^+ + V_{1c}^+ + V_{1d}^+} = \frac{V_{1a}^- - V_{1b}^- - V_{1c}^- + V_{1d}^-}{4V^+}$$

and, taking into account that $V_{1a}^+ = V_{1b}^+ = V_{1c}^+ = V_{1c}^+ = V^+$, we can also write:

$$S_{11} = \frac{1}{4}\left(\frac{V_{1a}^-}{V_{1a}^+} + \frac{V_{1b}^-}{V_{1b}^+} + \frac{V_{1c}^-}{V_{1c}^+} + \frac{V_{1d}^-}{V_{1c}^+}\right) = \frac{\Gamma_a + \Gamma_b + \Gamma_c + \Gamma_d}{4}$$

$$S_{21} = \frac{1}{4}\left(\frac{V_{1a}^-}{V_{1a}^+} + \frac{V_{1b}^-}{V_{1b}^+} - \frac{V_{1c}^-}{V_{1c}^+} - \frac{V_{1d}^-}{V_{1c}^+}\right) = \frac{\Gamma_a + \Gamma_b - \Gamma_c - \Gamma_d}{4}$$

$$S_{31} = \frac{1}{4}\left(\frac{V_{1a}^-}{V_{1a}^+} - \frac{V_{1b}^-}{V_{1b}^+} + \frac{V_{1c}^-}{V_{1c}^+} - \frac{V_{1d}^-}{V_{1c}^+}\right) = \frac{\Gamma_a - \Gamma_b + \Gamma_c - \Gamma_d}{4}$$

$$S_{41} = \frac{1}{4}\left(\frac{V_{1a}^-}{V_{1a}^+} - \frac{V_{1b}^-}{V_{1b}^+} - \frac{V_{1c}^-}{V_{1c}^+} + \frac{V_{1d}^-}{V_{1c}^+}\right) = \frac{\Gamma_a - \Gamma_b - \Gamma_c + \Gamma_d}{4},$$

where Γ_a, Γ_b, Γ_c and Γ_d are the reflection coefficients at port 1 obtained for the four excitations.

Application of the four excitations physically corresponds to introducing in the structure magnetic planes (planes of even symmetry implying zero current, i.e., an open circuit) or electric planes (planes of odd symmetry implying zero voltage, i.e., a short circuit) as shown in Fig. 4.26 (a), 4.26 (b), 4.26 (c), 4.26 (d). It follows that the reflection coefficients at port 1 corresponding to the four excitations can be simply derived from the circuits shown in Fig. 4.27 (a), 4.27 (b), 4.27 (c), 4.27 (d). To analyze the four configurations, we only need to recall the expression of a short circuit and open circuit line input admittances:

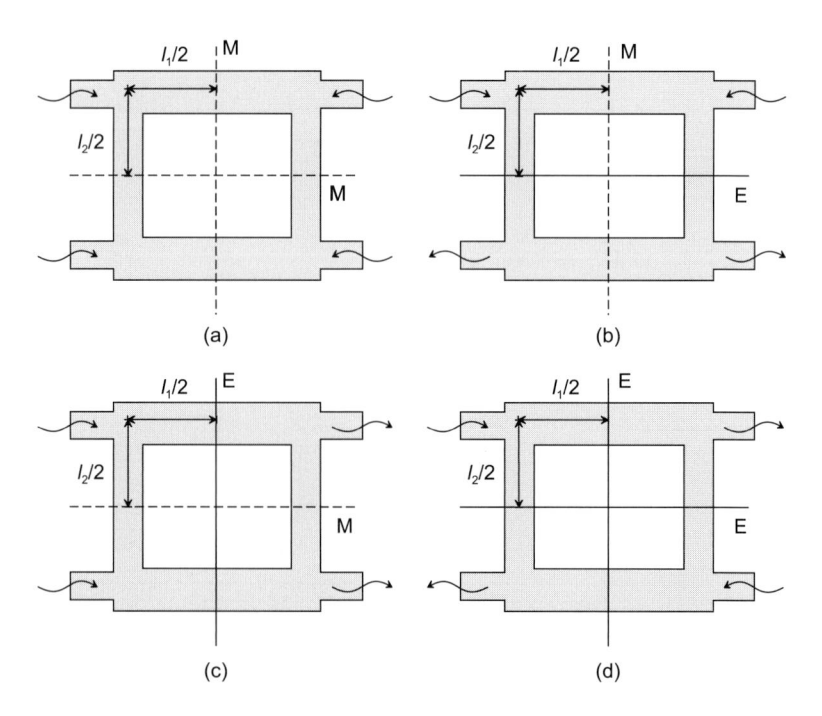

Figure 4.26 Even and/or odd excitation of a symmetric branch-line coupler.

4.5 Interference Couplers

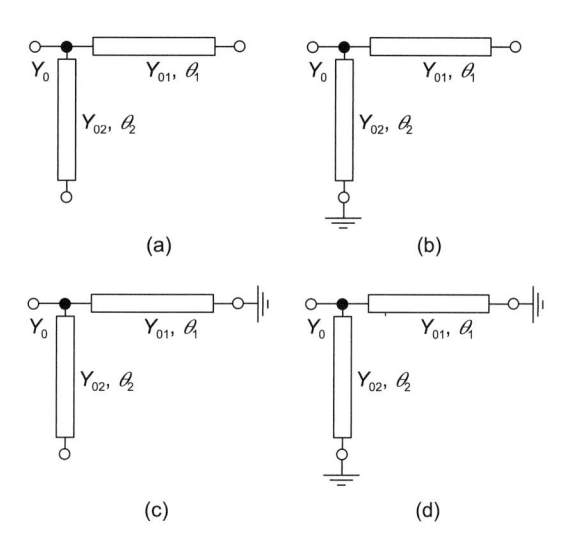

Figure 4.27 Equivalent circuit for the four excitations reported in Fig. 4.26.

$$Y_{in}^{sc} = j\, Y_0 \tan\theta$$
$$Y_{in}^{oc} = -j\, Y_0 \cot\theta,$$

where the electrical length is $\theta = 2\pi L/\lambda$; in the four cases we have that the input admittance can be obtained as the parallel of two open or short circuit admittances as follows:

$$Y_a = Y_{in1}^{sc} + Y_{in2}^{sc} = j\, Y_{01} \tan\theta_1 + j\, Y_{02} \tan\theta_2$$
$$Y_b = Y_{in1}^{sc} + Y_{in2}^{oc} = j\, Y_{01} \tan\theta_1 - j\, Y_{02} \cot\theta_2$$
$$Y_c = Y_{in1}^{oc} + Y_{in2}^{sc} = -j\, Y_{01} \cot\theta_1 + j\, Y_{02} \tan\theta_2$$
$$Y_d = Y_{in1}^{oc} + Y_{in2}^{oc} = -j\, Y_{01} \cot\theta_1 - j\, Y_{02} \cot\theta_2,$$

where Y_{01} and $\theta_1 = \pi l_1/\lambda$ (Y_{02} and $\theta_2 = \pi l_2/\lambda$) are the characteristic admittance and electrical length of the horizontal (vertical) lines. The line length considered is half of the total side length, see Fig. 4.25. Taking into account the definition of the reflection coefficient in terms of admittances:

$$\Gamma_k = \frac{Y_0 - Y_k}{Y_0 + Y_k}, \quad k = a, b, c, d,$$

we immediately have:

$$\Gamma_a = \frac{Y_0 - j\, Y_{01} \tan\theta_1 - j\, Y_{02} \tan\theta_2}{Y_0 + j\, Y_{01} \tan\theta_1 + j\, Y_{02} \tan\theta_2}$$

$$\Gamma_b = \frac{Y_0 - j\, Y_{01} \tan\theta_1 + j\, Y_{02} \cot\theta_2}{Y_0 + j\, Y_{01} \tan\theta_1 - j\, Y_{02} \cot\theta_2}$$

$$\Gamma_c = \frac{Y_0 + j\, Y_{01} \cot\theta_1 - j\, Y_{02} \tan\theta_2}{Y_0 - j\, Y_{01} \cot\theta_1 + j\, Y_{02} \tan\theta_2}$$

$$\Gamma_d = \frac{Y_0 + j\, Y_{01} \cot\theta_1 + j\, Y_{02} \cot\theta_2}{Y_0 - j\, Y_{01} \cot\theta_1 - j\, Y_{02} \cot\theta_2},$$

with centerband values (corresponding to $\theta_1 = \theta_2 = \pi/4$):

$$\Gamma_a(f_0) = \frac{Y_0 - j\,Y_{01} - j\,Y_{02}}{Y_0 + j\,Y_{01} + j\,Y_{02}}$$

$$\Gamma_b(f_0) = \frac{Y_0 - j\,Y_{01} + j\,Y_{02}}{Y_0 + j\,Y_{01} - j\,Y_{02}}$$

$$\Gamma_c(f_0) = \frac{Y_0 + j\,Y_{01} - j\,Y_{02}}{Y_0 - j\,Y_{01} + j\,Y_{02}}$$

$$\Gamma_d(f_0) = \frac{Y_0 + j\,Y_{01} + j\,Y_{02}}{Y_0 - j\,Y_{01} - j\,Y_{02}}.$$

4.5.2 Lumped-Parameter Branch-Line Couplers

Branch-line couplers can be implemented not only with the help of transmission lines, but also through lumped-parameter elements. Lumped couplers have an important advantage in terms of size when compared to distributed ones, but are limited in the operating frequency by losses and element resonances.

We start by demonstrating that a π network made of two parallel capacitors C and one series inductor L (see Fig. 4.28) shows, around the resonant frequency $\omega_0 = 1/\sqrt{LC}$, a behavior similar to a quarter-wave transmission line. In fact, imposing the resonance at ω_0, the capacitor admittance Y_c and inductor impedance Z_l can be written as:

$$Y_c = j\frac{\omega}{\omega_0}Y, \quad Z_l = j\frac{\omega}{\omega_0}\frac{1}{Y}, \quad Y = \sqrt{\frac{C}{L}}.$$

Suppose now that the π section is loaded by an impedance Z_L (admittance Y_L); the input admittance of the loaded section Y_{in} is:

$$Y_{in} = Y \frac{j\dfrac{\omega}{\omega_0} + \left(\dfrac{Y_L}{Y} + j\dfrac{\omega}{\omega_0}\right)\left[1 - \left(\dfrac{\omega}{\omega_0}\right)^2\right]}{\left[1 - \left(\dfrac{\omega}{\omega_0}\right)^2\right] + j\dfrac{\omega}{\omega_0}\dfrac{Y_L}{Y}}.$$

At the resonant frequency we have:

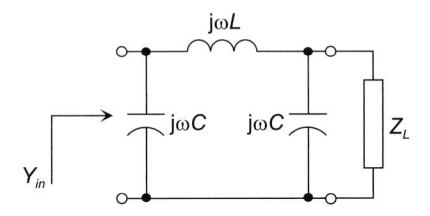

$$Y_{in}(\omega_0) = \frac{Y^2}{Y_L},$$

Figure 4.28 Lumped parameter π section replacing a quarter-wave transformer.

4.5 Interference Couplers

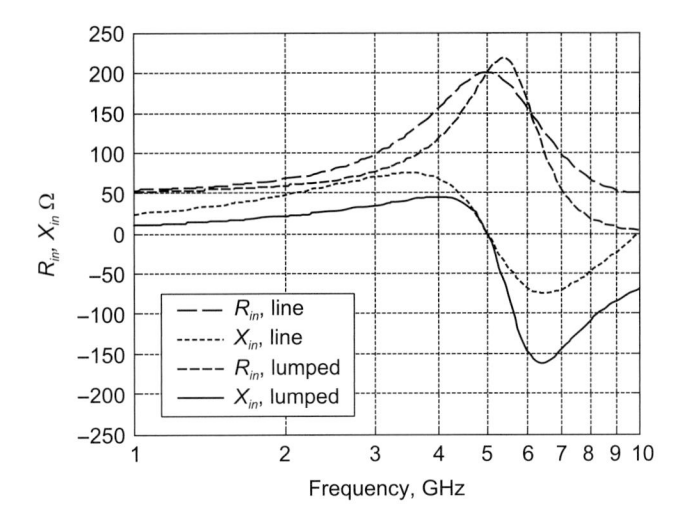

Figure 4.29 Comparison between the frequency behavior of a lumped and distributed quarter-wave transformer closed on 50 Ω and with a center frequency of 5 GHz; the equivalent line impedance is 100 Ω.

i.e., the input admittance is the same as for a quarter-wave transformer. The frequency behavior of the distributed and lumped transformers is compared in Fig. 4.29; the center frequency is 5 GHz, the load is 50 Ω and the equivalent characteristic impedance is $Z = 1/Y = 100$ Ω. While the frequency behavior of the lumped design is slightly less favorable than in the distributed case, the centerband behavior is exactly the same.

Taking into account that a branch-line coupler has (see Fig. 4.25) two horizontal quarter-wave lines of impedance Z_{01} and two vertical lines of impedance Z_{02}, we can mimic the centerband behavior of the coupler by replacing the quarterwave lines with a lumped equivalent section, having capacitances and inductances C_1, L_1 and C_2, L_2 respectively (see Fig. 4.30). Defining the susceptances at centerband (the resonant condition imposes that at centerband the inductor and capacitor susceptances must be equal and opposite):

$$j\omega_0 C_1 = jB_1, \quad j\omega_0 C_2 = jB_2$$
$$\frac{1}{j\omega_0 L_1} = \frac{1}{jX_1} = -jB_1, \quad \frac{1}{j\omega_0 L_2} = \frac{1}{jX_2} = -jB_2,$$

we can perform the same analysis already discussed for the distributed coupler and based on the superposition of even and odd excitations with respect to the horizontal and vertical planes. The four cases (a)–(d) analyzed in Example 4.3 now lead, for the input admittance at port 1, to the configurations shown in Fig. 4.31; in correspondence of a magnetic wall (even excitation) the inductive element is split in a series of two and terminated by an open, while in the presence of an electric wall (odd excitation) the element is split in the series of two and terminated by a short. Finally, the input admittances at port 1 and the related reflection coefficients with respect to the termination admittance Y_0 corresponding to the four cases are:

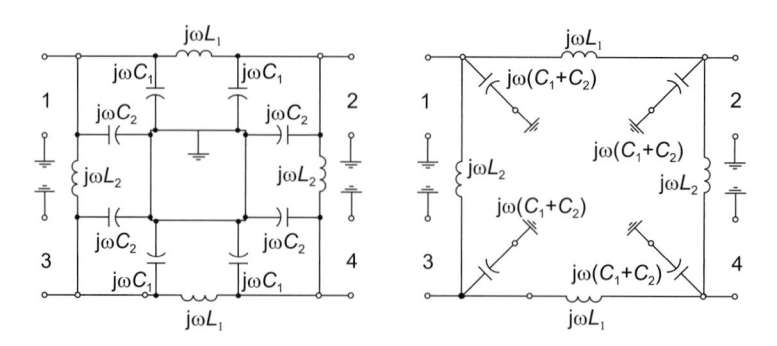

Figure 4.30 Lumped parameter branch-line coupler: left, implementation from π sections; right, practical implementation.

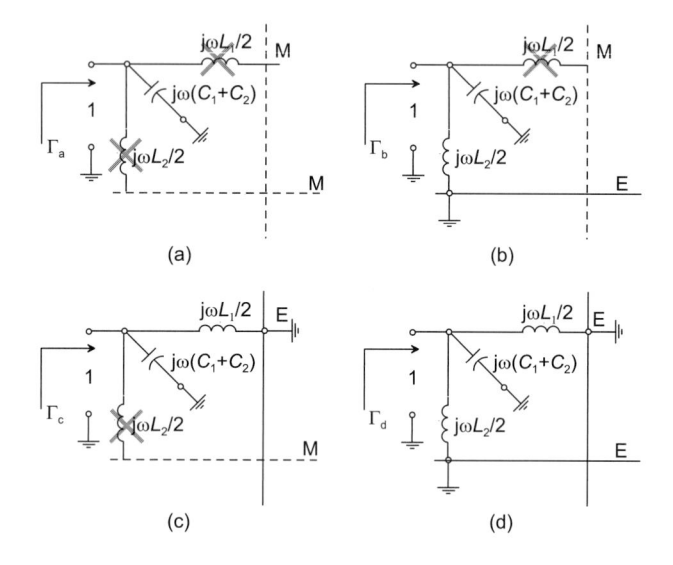

Figure 4.31 Equivalent circuits at port 1 resulting from odd and even excitations along the vertical and horizontal planes.

$$Y_{in}^a = j(B_1 + B_2) \rightarrow \Gamma_a = \frac{Y_0^2 - 2jY_0(B_1 + B_2) - (B_1 + B_2)^2}{Y_0^2 + (B_1 + B_2)^2}$$

$$Y_{in}^b = j(B_1 - B_2) \rightarrow \Gamma_b = \frac{Y_0^2 - 2jY_0(B_1 - B_2) - (B_1 - B_2)^2}{Y_0^2 + (B_1 - B_2)^2}$$

$$Y_{in}^c = -j(B_1 - B_2) \rightarrow \Gamma_c = \frac{Y_0^2 + 2jY_0(B_1 - B_2) - (B_1 - B_2)^2}{Y_0^2 + (B_1 - B_2)^2}$$

$$Y_{in}^d = -j(B_1 + B_2) \rightarrow \Gamma_d = \frac{Y_0^2 + 2jY_0(B_1 + B_2) - (B_1 + B_2)^2}{Y_0^2 + (B_1 + B_2)^2}.$$

Straightforward but lengthy computations lead to the centerband result:

4.5 Interference Couplers

$$S_{11} = \frac{\Gamma_a + \Gamma_b + \Gamma_c + \Gamma_d}{4} = \frac{Y_0^4 - (B_1^2 - B_2^2)^2}{[Y_0^2 + (B_1 + B_2)^2][Y_0^2 + (B_1 - B_2)^2]}.$$

Matching at port 1 ($S_{11} = 0$) yields the condition:

$$\left|B_1^2 - B_2^2\right| = Y_0^2$$

and, assuming for instance $B_1 > B_2$, this leads to:

$$B_1^2 - B_2^2 = Y_0^2.$$

Imposing this condition, the other elements of the first row of the scattering matrix assume the values:

$$S_{21} = \frac{\Gamma_a + \Gamma_b - \Gamma_c - \Gamma_d}{4} = \left(\frac{-j2Y_0 B_1}{Y_0^2 + B_1^2 + B_2^2}\right) = -j\frac{Y_0}{B_1}$$

$$S_{31} = \frac{\Gamma_a - \Gamma_b + \Gamma_c - \Gamma_d}{4} = 0$$

$$S_{41} = \frac{\Gamma_a - \Gamma_b - \Gamma_c + \Gamma_d}{4} = \frac{-2B_1 B_2}{Y_0^2 + B_1^2 + B_2^2} = -\frac{B_2}{B_1}.$$

Thus, port 3 is isolated while the phase difference between port 2 (coupled) and port 4 (transmission or through) is again 90 degrees as in the distributed branch-line coupler. For 3 dB coupling we have the conditions:

$$|S_{21}| = \frac{Y_0}{B_1} = \frac{1}{\sqrt{2}} \rightarrow B_1 = \sqrt{2} Y_0$$

$$|S_{41}| = \frac{B_2}{B_1} = \frac{1}{\sqrt{2}} \rightarrow B_2 = \frac{1}{\sqrt{2}} B_1 = Y_0.$$

Notice that this identically satisfies the matching condition $B_1^2 - B_2^2 = Y_0^2$.

Example 4.4 Design a lumped coupler with center frequency 10 GHz, 3 dB coupling, closed on 50 Ω.

Solution
We have:

$$B_1 = \sqrt{2} Y_0 = \sqrt{2}/50 = 2.8284 \times 10^{-2} \text{ S}$$
$$B_2 = Y_0 = 1/50 = 0.02 \text{ S}.$$

This corresponds to the element values:

$$C_1 = \frac{B_1}{\omega_0} = \frac{2.829 \times 10^{-2}}{2\pi \times 10^{10}} = 0.45 \text{ pF}$$

$$C_2 = \frac{B_2}{\omega_0} = \frac{2. \times 10^{-2}}{2\pi \times 10^{10}} = 0.318 \text{ pF}$$

$$L_1 = \frac{1}{\omega_0 B_1} = \frac{1}{2\pi \times 10^{10} \cdot 2.829 \times 10^{-2}} = 0.563 \text{ nH}$$

$$L_2 = \frac{1}{\omega_0 B_2} = \frac{1}{2\pi \times 10^{10} \cdot 2. \times 10^{-2}} = 0.795 \text{ nH}.$$

4.5.3 The Hybrid Ring

The hybrid ring is a four-port coupler shown in Fig. 4.32. For the sake of brevity we only discuss the centerband behavior; the analysis of the ring can be carried out by decomposing the excitation at port 1 into an even–odd pair [11, Sec. 7.8], or by extending to a four port the technique in Example 4.5. At centerband the scattering matrix can be shown to be:

$$\mathbf{S} = \begin{pmatrix} 0 & -j\dfrac{Z_0}{Z_{02}} & 0 & j\dfrac{Z_0}{Z_{01}} \\ -j\dfrac{Z_0}{Z_{02}} & 0 & -j\dfrac{Z_0}{Z_{01}} & 0 \\ 0 & -j\dfrac{Z_0}{Z_{01}} & 0 & -j\dfrac{Z_0}{Z_{02}} \\ j\dfrac{Z_0}{Z_{01}} & 0 & -j\dfrac{Z_0}{Z_{02}} & 0 \end{pmatrix}, \qquad (4.22)$$

where Z_0 is the matching impedance and the line impedances Z_{01} and Z_{02} are defined in Fig. 4.32. The impedances should satisfy the power conservation condition:

$$\frac{Z_0^2}{Z_{01}^2} + \frac{Z_0^2}{Z_{02}^2} = 1.$$

The coupler is matched at all ports at centerband; port 2 is coupled to port 1, port 3 is isolated, port 4 is transmitted but with 180° phase shift with respect to port 2; the hybrid ring (also called rat race coupler) is therefore a 180 degrees hybrid. The behavior with input at ports 2, 3 or 4 is defined by the structure scattering matrix (4.22).

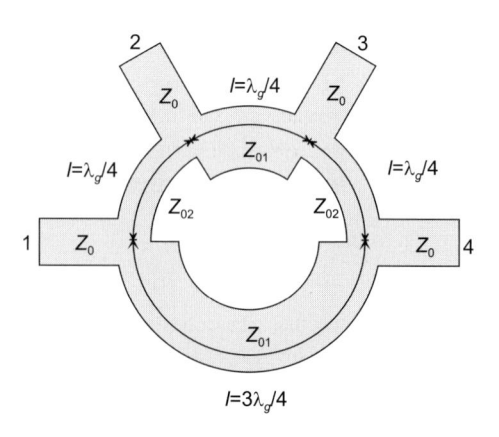

Figure 4.32 The hybrid ring.

Given Z_0, impedances Z_{01} and Z_{02} are immediately obtained from the desidered coupling. In particular, if we identify with port 2 the coupled port, we have $C = Z_0/Z_{02}$, $\sqrt{1 - C^2} = Z_0/Z_{01}$, where C is the (centerband) coupling, i.e.:

$$Z_{01} = \frac{Z_0}{\sqrt{1 - C^2}}$$
$$Z_{02} = \frac{Z_0}{C}.$$

In a 3 dB coupler we furthermore obtain $Z_{01} = Z_{02}$ and:

$$Z_{01} = Z_{02} = \sqrt{2} Z_0.$$

Thus, a 3 dB coupler matched to 50 Ω terminations requires 70.7 Ω microstrip impedances, that are easily realized. Notice that the 3 dB coupler has uniform impedances along the whole ring.

The 3 dB hybrid ring if often exploited in circuits where the sum and difference of two signals has to be evaluated. In fact, suppose to assign the incident wave a_1 at port 1 and a_3 at port 3. Since the circuit is linear, we can superimpose the result from the two excitations. We immediately have:

$$b_1 = b_3 = 0$$
$$b_2 = S_{21}a_1 + S_{23}a_3 = -j\frac{1}{\sqrt{2}}a_1 - j\frac{1}{\sqrt{2}}a_3 = -j\frac{1}{\sqrt{2}}(a_1 + a_3)$$
$$b_4 = S_{41}a_1 + S_{43}a_3 = j\frac{1}{\sqrt{2}}a_1 - j\frac{1}{\sqrt{2}}a_3 = j\frac{1}{\sqrt{2}}(a_1 - a_3),$$

i.e., ports 1 and 3 are matched and mutually isolated, port 2 has a sum signal (with an extra phase shift with respect to the input), port 4 has a difference signal (again with an extra phase shift).

4.6 Power Combiners and Dividers

In power amplifiers, the need often arises of dividing the input power among several active devices, and then to combine into a single load the device outputs. Series or parallel device combinations (that sum the output voltages and currents, respectively; similar remarks hold for voltage or current input dividers) of active devices do not clearly preserve the impedance level on which the single device is matched. Power dividing and combining should therefore be performed through structures able to preserve the impedance matching level, at least at centerband.

The most common case of power division and combination is the division by 2 and combination of two elements; powers of 2 can be obtained by properly cascading power dividers and combiners. Several power divider structures have been proposed in the past; we concentrate here on the Wilkinson divider [12]. As in interference couplers, this structure is easily introduced as distributed, but can be also implemented in concentrated form.

4.6.1 The Wilkinson Distributed Divider

The structure of a distributed Wilkinson divider is shown in Fig. 4.33. The divider, if properly designed, allows the power to be split from port 1 to ports 2 and 3 in equal parts, granting impedance matching at port 1 but also at port 2 and 3. Moreover, port 2 and 3 are isolated with each other (i.e., an excitation at port 2 leads to no signal at port 3 and *vice versa*). The analysis can be carried out by taking into account that the divider results from the connection of 3 two-ports, two transmission lines with characteristic impedance Z_{01} (admittance Y_{01}) and electrical length θ, and a π structure with zero parallel conductance and series resistance R (conductance G); see Fig. 4.34. As shown in Example 4.5 the admittance matrix of the divider is:

$$\mathbf{Y} = \begin{pmatrix} 2Y_{11}^l & Y_{12}^l & Y_{12}^l \\ Y_{12}^l & Y_{11}^l + Y_{11}^R & Y_{12}^R \\ Y_{12}^l & Y_{12}^R & Y_{11}^l + Y_{11}^R \end{pmatrix},$$

where the elements of the admittance matrix of the two lines of electrical length θ are (see Example 4.6):

$$Y_{11}^l = Y_{22}^l = -jY_{01}\cot\theta$$
$$Y_{12}^l = Y_{21}^l = jY_{01}/\sin\theta,$$

while the admittance matrix of the π structure is:

$$Y_{11}^R = Y_{22}^R = G$$
$$Y_{12}^R = Y_{21}^R = -G.$$

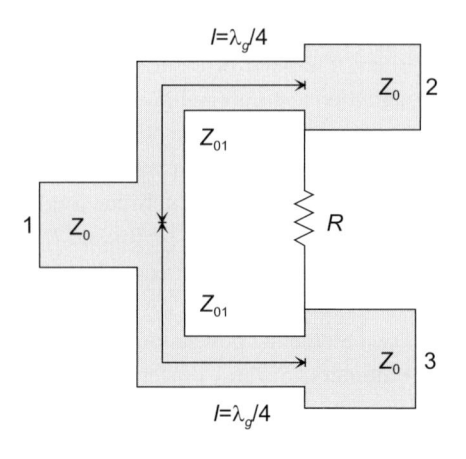

Figure 4.33 Wilkinson divider.

4.6 Power Combiners and Dividers

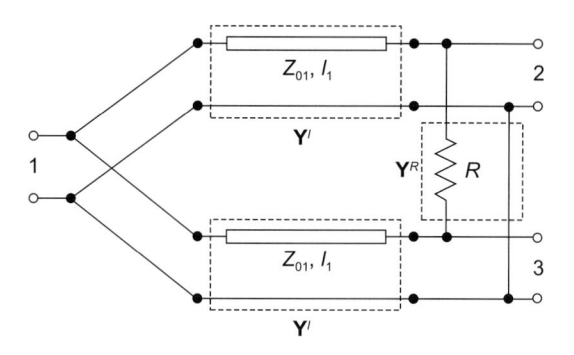

Figure 4.34 Scheme for the analysis of the Wilkinson divider.

We therefore obtain:

$$Y = \begin{pmatrix} -j2Y_{01}\cot\theta & jY_{01}/\sin\theta & jY_{01}/\sin\theta \\ jY_{01}/\sin\theta & G - j2Y_{01}\cot\theta & -G \\ jY_{01}/\sin\theta & -G & G - j2Y_{01}\cot\theta \end{pmatrix}.$$

At centerband $\theta = \pi/2$, $l = \lambda/4$; thus, the normalized matrix on Z_0 is:

$$\mathbf{y} = \mathbf{Y}Z_0 = \begin{pmatrix} 0 & jY_{01}Z_0 & jY_{01}Z_0 \\ jY_{01}Z_0 & GZ_0 & -GZ_0 \\ jY_{01}Z_0 & -GZ_0 & GZ_0 \end{pmatrix},$$

from which the scattering matrix:

$$\mathbf{S} = \begin{pmatrix} \dfrac{1 - 2Y_{01}^2 Z_0^2}{1 + 2Y_{01}^2 Z_0^2} & \dfrac{-2jY_{01}Z_0}{1 + 2Y_{01}^2 Z_0^2} & \dfrac{-2jY_{01}Z_0}{1 + 2Y_{01}^2 Z_0^2} \\ \dfrac{-2jY_{01}Z_0}{1 + 2Y_{01}^2 Z_0^2} & \dfrac{1 - 4Y_{01}^2 Z_0^3 G}{A} & \dfrac{-2(G - Y_{01}^2 Z_0) Z_0}{A} \\ \dfrac{-2jY_{01}Z_0}{1 + 2Y_{01}^2 Z_0^2} & \dfrac{-2(G - Y_{01}^2 Z_0) Z_0}{A} & \dfrac{1 - 4Y_{01}^2 Z_0^3 G}{A} \end{pmatrix},$$

where:

$$A = 1 + 2GZ_0 + 2Y_{01}^2 Z_0^2 + 4Y_{01}^2 Z_0^3 G.$$

Imposing matching at port 1 we obtain:

$$Y_{01} = \frac{1}{\sqrt{2}Z_0},$$

i.e.:

$$Z_{01} = \sqrt{2}Z_0,$$

from which we obtain the centerband scattering matrix:

$$S = \begin{pmatrix} 0 & -\frac{1}{\sqrt{2}}j & -\frac{1}{\sqrt{2}}j \\ -\frac{1}{\sqrt{2}}j & \frac{1-2GZ_0}{2+4GZ_0} & 2\left(G-\frac{1}{2Z_0}\right)\frac{Z_0}{2+4GZ_0} \\ -\frac{1}{\sqrt{2}}j & 2\left(G-\frac{1}{2Z_0}\right)\frac{Z_0}{2+4GZ_0} & \frac{1-2GZ_0}{2+4GZ_0} \end{pmatrix}.$$

Matching at ports 2 and 3 implies:

$$R = 1/G = 2Z_0,$$

from which, finally:

$$S = \begin{pmatrix} 0 & -\frac{1}{\sqrt{2}}j & -\frac{1}{\sqrt{2}}j \\ -\frac{1}{\sqrt{2}}j & 0 & 0 \\ -\frac{1}{\sqrt{2}}j & 0 & 0 \end{pmatrix}.$$

The structure of the scattering matrix implies matching at all ports, isolation between ports 2 and 3, transmission from 1 to 2 and from 1 to 3, with 3 dB power division, phase shift of $-\pi/2$ between the input and the output signals. Notice that, contrarily to the case of the 3 dB hybrids, the two outputs are in phase. The divider is narrowband, but the bandwidth can be increased by multisection structures.

Example 4.5 Consider three two-ports with admittance matrices $\mathbf{Y}^a$, $\mathbf{Y}^b$, $\mathbf{Y}^c$ combined in a triangle so that at port 1 the inputs of a and b are in parallel, at port 2 the output of a is in parallel with the input of c and at port 3 the output of c is in parallel with the output of b. Derive the admittance matrix of the three-port with ports 1, 2 and 3.

Solution
We have the following equations:

$$\begin{aligned} I_1^a &= Y_{11}^a V_1^a + Y_{12}^a V_2^a & I_2^a &= Y_{21}^a V_1^a + Y_{22}^a V_2^a \\ I_1^b &= Y_{11}^b V_1^b + Y_{12}^b V_2^b & I_2^b &= Y_{21}^b V_1^b + Y_{22}^b V_2^b \\ I_1^c &= Y_{11}^c V_1^c + Y_{12}^c V_2^c & I_2^c &= Y_{21}^c V_1^c + Y_{22}^c V_2^c, \end{aligned}$$

but $V_1^a = V_1^b = V_1$; $V_2^a = V_1^c = V_2$; $V_2^c = V_2^b = V_3$; moreover $I_1 = I_1^a + I_1^b$; $I_2 = I_2^a + I_1^c$; $I_3 = I_2^b + I_2^c$. Substituting:

$$\begin{aligned} I_1^a &= Y_{11}^a V_1 + Y_{12}^a V_2 & I_2^a &= Y_{21}^a V_1 + Y_{22}^a V_2 \\ I_1^b &= Y_{11}^b V_1 + Y_{12}^b V_3 & I_2^b &= Y_{21}^b V_1 + Y_{22}^b V_3 \\ I_1^c &= Y_{11}^c V_2 + Y_{12}^c V_3 & I_2^c &= Y_{21}^c V_2 + Y_{22}^c V_3 \end{aligned}$$

4.6 Power Combiners and Dividers

and then summing we have:

$$I_1 = I_1^a + I_1^b = \left(Y_{11}^a + Y_{11}^b\right) V_1 + Y_{12}^a V_2 + Y_{12}^b V_3$$
$$I_2 = I_2^a + I_1^c = Y_{21}^a V_1 + \left(Y_{22}^a + Y_{11}^c\right) V_2 + Y_{12}^c V_3$$
$$I_3 = I_2^b + I_2^c = Y_{21}^b V_1 + Y_{21}^c V_2 + \left(Y_{22}^b + Y_{22}^c\right) V_3,$$

which immediately yields the admittance matrix elements. In the distributed Wilkinson divider case, we have:

$$Y_{11}^a = Y_{22}^a = Y_{11}^b = Y_{22}^b = Y_{11}^l$$
$$Y_{12}^a = Y_{21}^a = Y_{12}^b = Y_{21}^b = Y_{12}^l$$
$$Y_{11}^c = Y_{22}^c = Y_{11}^R$$
$$Y_{12}^c = Y_{21}^c = Y_{12}^R,$$

leading to the parallel representation:

$$I_1 = 2Y_{11}^l V_1 + Y_{12}^l V_2 + Y_{12}^l V_3$$
$$I_2 = Y_{12}^l V_1 + \left(Y_{11}^l + Y_{11}^R\right) V_2 + Y_{11}^R V_3$$
$$I_3 = Y_{12}^l V_1 + Y_{12}^R V_2 + \left(Y_{11}^l + Y_{11}^R\right) V_3.$$

Example 4.6 Evaluate the admittance matrix of a line with electrical length θ and characteristic admittance Y_0.

Solution
A line is a symmetric and reciprocal two-port; from the definition of the admittance matrix we have;

$$Y_{11} = Y_{22} = \left.\frac{I_1}{V_1}\right|_{V_2=0}$$
$$Y_{21} = Y_{12} = \left.\frac{I_2}{V_1}\right|_{V_2=0}.$$

Taking into account the expression of the voltages and currents in terms of forward and backward propagating waves we have (we assume the line length to be L and the guided wavelength λ_g):

$$V(z) = V^+(z) + V^-(z)$$
$$I(z) = Y_0 \left[V^+(z) - V^-(z)\right]$$
$$V^+(z) = V^+(L) \exp\left(j2\pi \frac{L-z}{\lambda_g}\right)$$
$$V^-(z) = V^-(L) \exp\left(-j2\pi \frac{L-z}{\lambda_g}\right)$$

$$I^+(z) = Y_0 V^+(L) \exp\left(j2\pi \frac{L-z}{\lambda_g}\right)$$

$$I^-(z) = -Y_0 V^-(L) \exp\left(j2\pi \frac{L-z}{\lambda_g}\right),$$

where section $z = 0$ is port 1 and $z = L$ is port 2. If port 2 is shorted, then $V_2 = V(L) = 0$ or:

$$V^+(L) = -V^-(L)$$
$$I(L) = 2Y_0 V^+(L).$$

The port voltages and currents at port 1 and 2 can therefore expressed as:

$$V_1 = V(0) = V^+(0) + V^-(0) = V^+(L)\left[\exp(j\theta) - \exp(-j\theta)\right] = 2jV^+(L)\sin\theta$$
$$I_1 = I(0) = Y_0 V^+(L)(\exp(j\theta) + \exp(-j\theta)) = 2Y_0 V^+(L)\cos\theta$$
$$V_2 = V(L) = 0$$
$$I_2 = -I(L) = -2Y_0 V^+(L);$$

thus:

$$Y_{11} = Y_{22} = \left.\frac{I_1}{V_1}\right|_{V_2=0} = -jY_0 \cot\theta$$

$$Y_{21} = Y_{12} = \left.\frac{I_2}{V_1}\right|_{V_2=0} = jY_0/\sin\theta.$$

The centerband design equations of the Wilkinson divider (in particular, the value of the output resistor) can also be derived in a direct way, exploiting the matching and symmetry properties of the structure.

Assuming that the two branches with characteristic impedance Z_{01} are quarter-wave transformers, we have that, at centerband, the input impedance seen at port 1, $Z_{in,1}$, is the parallel of two equal quarterwave transformers loaded with Z_0:

$$Z_{in,1} = \frac{1}{2}\frac{Z_{01}^2}{Z_0} = Z_0, \qquad (4.23)$$

from which we immediately have $Z_{01} = \sqrt{2}Z_0$. Notice that if the divider is excited from port 1, due to the symmetry of ports 2 and 3 the output resistor connected between these ports does not dissipate any power. Since, thanks to condition (4.23), port 1 is matched and port 2 and 3 are closed to the matching resistor Z_0, power conservation and symmetry imply that:

$$|a_1|^2 = |b_2|^2 + |b_3|^2 = 2|b_2|^2,$$

i.e.:

$$2|S_{21}|^2 = 1.$$

Concerning the design of the output resistor, let us first demonstrate that isolation between ports 2 and 3 and matching of port 2 and 3 implicate each other. In fact, suppose that we excite the divider with two equal power waves $a_2 = a_3$ at ports 2 and 3, while port 1 is connected to Z_0; due to the symmetric excitation, again the output resistor does not dissipate any power. Power conservation and the matching condition at port 1 ($a_1 = 0$) yields:

$$|a_2|^2 - |b_2|^2 + |a_3|^2 - |b_3|^2 = 2\left(|a_2|^2 - |b_2|^2\right) = |b_1|^2. \quad (4.24)$$

However:

$$b_1 = S_{12}a_2 + S_{13}a_3 = 2S_{21}a_2,$$

since $S_{12} = S_{13}$ for symmetry and $S_{12} = S_{21}$ for reciprocity. Moreover:

$$b_2 = S_{22}a_2 + S_{23}a_3 = (S_{22} + S_{23})a_2.$$

We thus have from (4.24) that:

$$2\left(1 - |S_{22} + S_{23}|^2\right)|a_2|^2 = 4|S_{21}|^2|a_2|^2,$$

i.e.,

$$|S_{22} + S_{23}|^2 + 2|S_{21}|^2 = 1.$$

This in turn implies, since $2|S_{21}|^2 = 1$, that:

$$|S_{22} + S_{23}|^2 = 0 \rightarrow S_{23} = -S_{22}.$$

In other words, if ports 2 and 3 are matched ($S_{22} = 0$ and $S_{33} = 0$), they are also isolated ($S_{23} = 0$), and *vice versa*.

The above property allows us to easily confirm that a resistor R of value $2Z_0$ connected between ports 2 and 3 provides matching at both ports. For the sake of definiteness, assume that the divider is excited at port 2, while ports 3 and 1 are closed on Z_0. Since port 3 is matched, it will be isolated from port 2; therefore, no current will flow in port 3 due to the excitation at port 2, and $V_3 = 0$, i.e., port 3 will be shorted and R will be grounded at port 3. Consider now the quarterwave transformer from port 3 to port 1; since port 3 is shorted the input of the corresponding quarterwave line will be an open circuit. Therefore, the matching resistance Z_0 loading port 1 is unperturbed by the quarterwave line from port 3 to port 1. Furthermore, the resistance Z_0 loading port 1 will be seen from port 2 (through the quarterwave transformer) as $2Z_0$. The resistance $2Z_0$ will be in parallel to the output resistance $R = 2Z_0$ connected between port 2 and the (shorted) port 3. In conclusion, the total resistance seen at port 2 is Z_0, i.e., port 2 is matched.

4.6.2 Wilkinson Lumped Dividers

The structure of a lumped Wilkinson divider is shown in Fig. 4.35. The two transmission lines are replaced by lumped element quarterwave equivalents. We shall limit the analysis to centerband where:

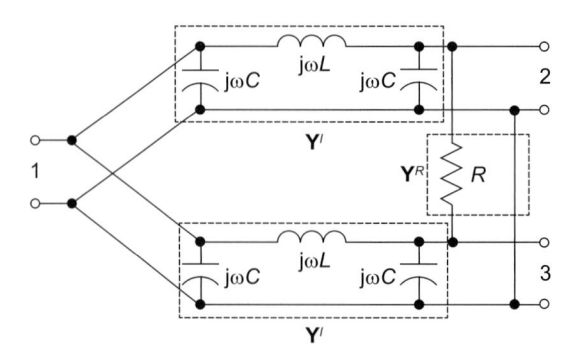

Figure 4.35 Lumped Wilkinson divider.

$$j\omega_0 C = jB$$
$$\frac{1}{j\omega_0 L} = \frac{1}{jX} = -jB,$$

leading to the admittance matrix elements of the lumped quarterwave equivalent:

$$Y^l_{11} = Y^l_{22} = 0$$
$$Y^l_{12} = Y^l_{21} = jB,$$

while the admittance matrix of the π resistive structure is always:

$$Y^R_{11} = Y^R_{22} = G$$
$$Y^R_{12} = Y^R_{21} = -G.$$

Exploiting again the result of Example 4.5, the centerband admittance matrix of the divider is:

$$\mathbf{Y} = \begin{pmatrix} 2Y^l_{11} & Y^l_{12} & Y^l_{12} \\ Y^l_{12} & Y^l_{11} + Y^R_{11} & Y^R_{12} \\ Y^l_{12} & Y^R_{12} & Y^l_{11} + Y^R_{11} \end{pmatrix} = \begin{pmatrix} 0 & jB & jB \\ jB & G & -G \\ jB & -G & G \end{pmatrix},$$

with normalized matrix:

$$\mathbf{y} = \mathbf{Y}Z_0 = \begin{pmatrix} 0 & jBZ_0 & jBZ_0 \\ jBZ_0 & GZ_0 & -GZ_0 \\ jBZ_0 & -GZ_0 & GZ_0 \end{pmatrix}$$

and scattering matrix:

$$S = \begin{pmatrix} \dfrac{1-2B^2Z_0^2}{1+2B^2Z_0^2} & \dfrac{-2jBZ_0}{1+2B^2Z_0^2} & \dfrac{-2jBZ_0}{1+2B^2Z_0^2} \\ \dfrac{-2jBZ_0}{1+2B^2Z_0^2} & \dfrac{1-4B^2Z_0^3G}{A} & \dfrac{-2(G-B^2Z_0)Z_0}{A} \\ \dfrac{-2jBZ_0}{1+2B^2Z_0^2} & \dfrac{-2(G-B^2Z_0)Z_0}{A} & \dfrac{1-4B^2Z_0^3G}{A} \end{pmatrix},$$

where:

$$A = 1 + 2GZ_0 + 2B^2Z_0^2 + 4B^2Z_0^3G.$$

Imposing matching at port 1 we obtain:

$$B = \frac{1}{\sqrt{2}Z_0},$$

while matching the ports 2 and 3 implies:

$$1 - 4B^2Z_0^3G = 1 - 2Z_0G = 0 \rightarrow R = 2Z_0, \qquad (4.25)$$

leading to the centerband scattering matrix of the matched coupler:

$$S = \begin{pmatrix} 0 & -\dfrac{1}{\sqrt{2}}j & -\dfrac{1}{\sqrt{2}}j \\ -\dfrac{1}{\sqrt{2}}j & 0 & 0 \\ -\dfrac{1}{\sqrt{2}}j & 0 & 0 \end{pmatrix},$$

exactly as in the distributed implementation.

4.7 Directional Couplers Summary

Fig. 4.36 reports the design rules and the centerband behavior of the distributed directional couplers and power dividers examined in this chapter. Their behavior can be summarized as follows:

- The coupled line and branch line couplers have a phase shift of 90 degrees between the two coupled and transmission ports, and are therefore called 90 degrees hybrid; the hybrid ring or rat race coupler introduces a phase shift of 180 degrees between the two outputs (180 degrees hybrid), while in the Wilkinson divider the two outputs are in phase.
- The branch line and hybrid ring couplers easily permit high coupling, while they are critical for low coupling; on the other hand, coupled line couplers do not allow for high coupling, apart from the multiconductor version (the Lange coupler).

Directional Couplers and Power Dividers

Coupling C	3dB coupler	Centerband behaviour
Two coupled lines		
$Z_{0p} = Z_0 \sqrt{\dfrac{1+C}{1-C}}$ $Z_{0d} = Z_0 \sqrt{\dfrac{1-C}{1+C}}$	$Z_{0e} = 2.4142 Z_0$ $Z_{0o} = 0.4142 Z_0$	
Branch line		
$Z_{02} = \dfrac{CZ_0}{\sqrt{1-C^2}}$ $Z_{01} = CZ_0$	$Z_{02} = Z_0$ $Z_{01} = \dfrac{Z_0}{\sqrt{2}}$	
Hybrid ring (rat race)		
$Z_{01} = \dfrac{Z_0}{\sqrt{1-C^2}}$ $Z_{02} = \dfrac{Z_0}{C}$	$Z_{01} = Z_0\sqrt{2}$ $Z_{02} = Z_0\sqrt{2}$	
Wilkinson divider		
	$Z_{01} = Z_0\sqrt{2}$ $R = 2Z_0$	

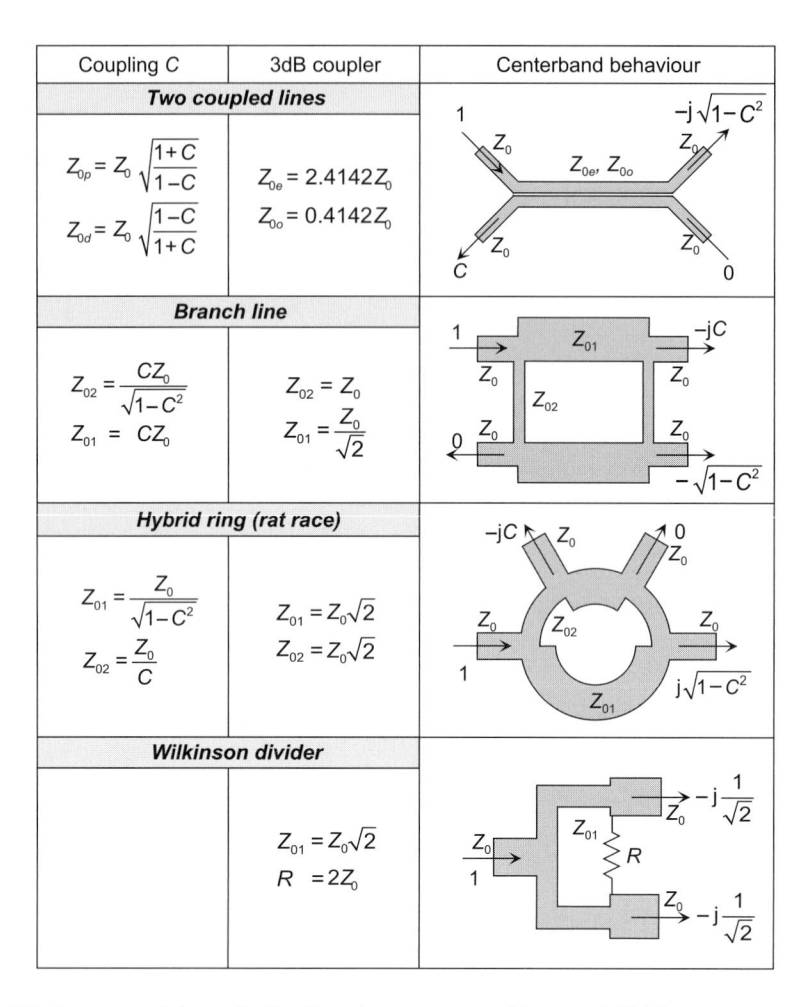

Figure 4.36 Summary of the main distributed parameter combiners and dividers.

- All distributed directional couplers have a rather large layout. Therefore, distributed couplers are seldom used in integrated circuits in the low microwave range, where lumped parameter couplers and dividers are preferred.

4.8 Questions and Problems

4.8.1 Questions

1. In a coupled two-conductor microstrip the even mode effective permittivity is 8, while the odd mode effective permittivity is 6. The odd and even mode impedances are 75 and 40 Ω, respectively. Are the previous data physically correct?
2. Sketch the layout of a Lange coupler and of a branch-line coupler, specifying the dimensions with respect to the centerband guided wavelength.

4.8 Questions and Problems

3. Is the hybrid ring a 90° or 180° coupler?
4. Explain the difference between a 90° and 180° hybrid.
5. Sketch the layout of a branch-line directional coupler.
6. Sketch the layout of a Wilkinson power divider.
7. Explain the difference between an unfolded and a folded Lange coupler.
8. Justify the fact that 3 dB couplers can be implemented through multiconductor coupled microstrips but not, in practice, through two-conductor coupled microstrips.
9. Explain why a low-coupling coupler (e.g., a 10 dB coupler) cannot easily be implemented through a branch-line coupler.
10. In two coupled microstrips:
    1. the even and odd modes are always velocity matched
    2. the even and odd modes are typically velocity mismatched
    3. the even mode velocity is larger than the odd mode velocity.
11. Discuss the opportunity of implementing Lange couplers with a large number of conductors (say greater than 6).
12. Velocity mismatch in a directional coupler is particularly critical for:
    1. input matching
    2. coupling and transmission
    3. isolation.
13. Discuss available compensation techniques for the coupler velocity mismatch.
14. Sketch a hybrid ring coupler, explaining its operation.

4.8.2 Problems

1. Imagine that an ideal 3dB, 90° coupler is fed with a 100 mW signal. What is the power on the coupled and the transmission port, respectively? What is the power on the insulated port? What is the phase difference between the coupled and transmission ports?
2. A Wilkinson divider on 50 Ω loads operates at 10 GHz. Assuming $\epsilon_{\text{eff}} = 4$ evaluate the lengths and characteristic impedance of the divider arms.
3. Design a 10 dB coupler on two-conductor coupled microstrips; the substrate is GaAs (permittivity 13) with thickness 0.3 mm; the centerband frequency is 10 GHz and the closing impedance 50 Ω.
4. Design a four-conductor Lange 3 dB coupler; the substrate is GaAs (permittivity 13) with thickness 0.3 mm; the centerband frequency is 10 GHz and the closing impedance 50 Ω.
5. Design (dimensions and impedances) a hybrid ring with 3 dB coupling on 50 Ω at 5 GHz. Assume that the line effective permittivity is 5.
6. Design (dimensions and impedances) a branch-line coupler with 3 dB coupling on 100 Ω at 20 GHz. Assume that the line effective permittivity is 5.
7. Design (dimensions and impedances) a Wilkinson divider on 70 Ω at 30 GHz. Assume that the line effective permittivity is 2.

References

[1] M. Kirschning and R. Jansen, "Accurate wide-range design equations for the frequency-dependent characteristic of parallel coupled microstrip lines," *Microwave Theory and Techniques, IEEE Transactions on*, vol. 32, no. 1, pp. 83–90, Jan. 1984.

[2] M. Kirschning and R. Jansen, "Accurate wide-range design equations for the frequency-dependent characteristics of parallel coupled microstrip lines (corrections)," *Microwave Theory and Techniques, IEEE Transactions on*, vol. 33, no. 3, p. 288, Mar. 1985.

[3] J. Helszajn, *The stripline circulators: theory and practice*. John Wiley & Sons, 2008, vol. 206.

[4] D. K. Linkhart, *Microwave circulator design*. Artech house, 2014.

[5] C. Buntschuh, "Octave-bandwidth, high-directivity microstrip couplers," DTIC Document, Tech. Rep., 1974.

[6] V. Rizzoli and A. Lipparini, "The design of interdigitated couplers for MIC applications," *Microwave Theory and Techniques, IEEE Transactions on*, vol. 26, no. 1, pp. 7–15, Jan. 1978.

[7] W. Ou, "Design equations for an interdigitated directional coupler," *IEEE Transactions on Microwave Theory and Techniques*, vol. 23, no. 2, pp. 253–255, Feb. 1975.

[8] A. Presser, "Interdigitated microstrip coupler design," *IEEE Transactions on Microwave Theory and Techniques*, vol. 26, no. 10, pp. 801–805, Oct. 1978.

[9] J. Lange, "Interdigitated strip-line quadrature hybrid," in *Microwave Symposium, 1969 G-MTT International*, May 1969, pp. 10–13.

[10] S. Akhtarzad, T. Rowbotham, and P. B. Johns, "The design of coupled microstrip lines," *IEEE Transactions on Microwave Theory and Techniques*, vol. 23, no. 6, pp. 486–492, Jun. 1975.

[11] D. M. Pozar, *Microwave engineering*. John Wiley & Sons, 2012.

[12] E. J. Wilkinson, "An n-way hybrid power divider," *IRE Transactions on Microwave Theory and Techniques*, vol. 8, no. 1, pp. 116–118, 1960.

5 Active RF and Microwave Semiconductor Devices

5.1 Active Microwave Components

Active devices for RF, microwave and millimeter wave circuits offer several functions, such as signal generation, signal amplification, mixing and frequency multiplication. Active components operating in switching (ON–OFF) mode are found in duplexers, switches and switching-mode power amplifiers. From the standpoint of the element circuit topology, active devices can be classified as one-ports or two-ports. One-ports typically are used for signal generation, while two-ports are the basis for oscillators but also for signal amplifiers.[1]

From a technological point of view, microwave active devices can be *vacuum tubes* or *semiconductor devices*. Examples of vacuum tubes for signal generation are the klystron and the magnetron; the traveling-wave tube (TWT) operates as a wideband amplifier. Unlike audio and video frequency vacuum components, microwave and millimeter wave vacuum tubes are far from being obsolete (for a review, see [1, 2, 3]) and still are a valid solution for very high-power sources. The concept of "high power" is however a relative one, since the available power of active devices exhibits a typical f^{-2} behavior, that becomes f^{-4} in the millimeter wave range. Fig. 5.1 presents a power-frequency diagram where the output power of a number of vacuum tubes is compared to the output power of some solid-state technologies. It is quite clear that in the high-power, high-frequency range vacuum tubes dominate,[2] while for lower powers the advantages of semiconductor sources in terms of small size and integrability can be fully exploited. During the last few years, a number of high-power semiconductor technologies (like gallium nitride, GaN, and silicon carbide, SiC) have appeared, competing with vacuum tubes in the medium power range.

5.1.1 Semiconductor Microwave Components

Signal generation through solid-state devices is preferably carried out through (bipolar) transistor-based oscillators thanks to their superior phase noise characteristics. Nevertheless, one-port semiconductor devices (often improperly called *microwave diodes*)

---

[1] Passive one-ports like the Schottky diode are used instead in passive mixers and frequency multipliers.
[2] Vacuum tubes are also widespread in certain consumer or industrial applications at RF and in the lower microwave range, see the 1 kW magnetrons exploited in domestic microwave ovens, operating around 2.4 GHz.

Active RF and Microwave Semiconductor Devices

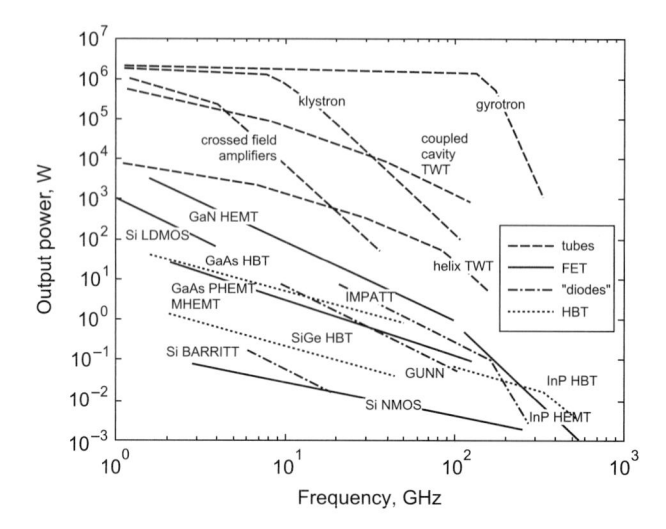

Figure 5.1 Frequency–power ranges of vacuum tubes (dashed lines) vs. solid-state transistor microwave sources (continuous line). The dashed-dotted line refers to the so-called microwave diodes, see text. Data are taken from [4, Fig. 1] (Copyright ©2004 IEEE. All rights reserved. Used with permission).

have been developed in the past for signal generation; such devices still are exploited in certain applications (e.g., radars). In microwave diodes, oscillations do not arise because of a gain block with positive feedback, but rather are caused by built-in mechanisms originating from the very device physics. Examples of microwave diodes are the Gunn diodes, also called TEDs (Transferred Electron Devices) and the transit-time effect devices (TTD, Transit Time Diodes). For a review see [5, Part IV].

Gunn diodes are junctionless devices made of a III-V semiconductor resistor with proper doping. Under DC bias, the electric field can reach values in the negative differential mobility region of the material, i.e., for GaAs, above 5 kV/cm, see Fig. 5.8. This leads to the generation of mobile charge dipoles, called *domains*, that travel though the device causing periodic current bursts in the terminals. The current waveform periodicity is dominated by the domain transit time through the diode length. In general, also devices exhibiting regions of small-signal negative differential conductivity are able to oscillate; an example is the *tunnel diode*, a *pn* diode with high doping levels, whose output power is limited to the mW range. The tunnel diode, however, exhibits very low noise.

In transit time diodes, the device driving voltage causes the generation of a charge pulse across, e.g., a *pn* junction; the pulse travels at the semiconductor saturation velocity through a drift region before reaching the device terminals. If the transit time is long enough to cause a π phase shift between the driving voltage and output current waveforms, the device behaves like a resistor having negative conductance. The generating mechanism of the charge burst can be charge injection through a barrier (BARrier Injection Transit Time or BARRITT diodes), or the avalanche generation in a reverse-biased junction (IMPATT and TRAPATT diodes). Transit time devices are able to provide medium- to high-power levels but typically are very noisy. For all devices considered, the effect of parasitic capacitances causes the available power to decrease with frequency at least as f^{-2}, i.e., $P_{av}(f) \approx P_{av}(f_0)(f_0/f)^2$ where, f_0 is a characteristic frequency.

Transistor-based oscillators provide the best noise performances, in particular when implemented through bipolar approaches, due to the lower low-frequency noise that is upconverted around the carrier due to a self-mixing effect. Whatever the technique, however, the generation of millimeter wave power for frequencies well above 100 GHz is increasingly difficult, as the overview of power-frequency characteristics in Fig. 5.1 shows.

A critical issue concerns today THz power generation, that is required by an increasing number of applications, e.g., in the field of security and remote sensing. Electronic power generation by means of transistors is limited by capacitive effects, that can be mitigated by device scaling down, but never eliminated. Vacuum tubes provide a higher-power solution but, ultimately, there is little possibility to generate practically useful EM power above 1 THz. Power generation is again efficient in the infrared region where lasers (semiconductor, gas, solid state) become available. Several technologies have been developed to bridge the gap, such as the Quantum Cascade Laser exploiting low-energy transitions, but the problem is far from being solved today. For an overview, see the IEEE T-MTT Special Issue [6].

For a comprehensive analysis of the competing technologies and transistor solutions exploited for microwave and millimeter wave circuits the reader can refer to the books [7] and [8] and to the review papers [9, 10]; here we briefly introduce the main solutions available.

The current solid-state transistor technologies exploited in signal generation, amplification and mixing can be classified in several ways. Field-effect transistors (FETs) have been the first devices to be used for microwave signal amplification under the form of GaAs Schottky-gate MESFETs (Metal Semiconductor FETs). The device was proposed by Mead in 1966 [11]; epitaxial and implanted MESFETs, both as discrete devices and as elements of GaAs-based MMICs, have dominated the application market till the mid nineties, covering frequencies up to the lower millimeter wave range.

The intrinsic limits posed by GaAs MESFET gate downscaling, implying a thinner active region with a larger doping, that in turn causes lower mobility in the channel due to impurity scattering, were overcome by the development of a new, heterojunction-based Schottky barrier FET family, the HEMT (High Electron Mobility Transistor) and its developments (pseudomorphic HEMT, PHEMT, and metamorphic HEMT, MHEMT). The basic building block of the HEMT conducting channel, the so-called modulation-doped AlGaAs/GaAs structure (grown by Molecular Beam Epitaxy, MBE, or Metal Organic Chemical Vapor Deposition, MOCVD, see [12]), was proposed in 1978 by Dingle et al. [13], while the HEMT as a device was demonstrated in 1979–1980 by T. Mimura et al. (Fujitsu) [14] and by D. Delagebeaudeuf et al. (Thomson-CSF) [15]. In the same years, the HEMT concept was also successfully applied to InP-based alloys, on which a large body of technological expertise had already been collected in the field of long-wavelength semiconductor lasers. Thanks to the superior transport properties of InP and related alloys, InP-based HEMTs showed significant improvements in the maximum frequency of operation, and still are the fastest transistors available, with maximum oscillation frequencies above the THz. GaAs-based HEMTs entered the application market starting from the mid-nineties, while, due to the high cost, InP-based devices have been always restricted to niche applications in the millimeter wave range.

Always in the field of compound semiconductors, the last two decades have seen a fast development of AlGaN/GaN-based HEMTs [16]; despite the problems related to the substrate (GaN is hexagonal and no native substrate exists), GaN has shown remarkable capabilities in terms of power density (one order of magnitude larger than for III-V devices) up to millimeter waves. A gradual market penetration of GaN-based HEMTs has taken place (in particular, for defense applications) during the last decade.

The steady progress in compound semiconductor FETs has been somewhat overshadowed, during the last few years, by a quiet revolution taking place in the field of Si MOSFETs [8]. The tremendous pressure exerted on the CMOS technology to downscale with the aim of obtaining, with nanoscale gate devices, faster and faster logical circuits, has led, in particular after 2000–2005, to a number of technology nodes having excellent properties in term of maximum operation frequency. Millimeter wave NMOS analog circuits have been demonstrated, the only limitation of the technology being in the achievable power levels. Already below 10 GHz, typical maximum output powers of MOSFET amplifiers are in the 100 mW range. Further developments in the CMOS technology leading to MISFETs with compound semiconductor channels may perhaps lead to the ultimate device, where the superior transport properties of compound semiconductors are coupled with the geometry and processing advantages of the CMOS approach.

In the above overview a number of FET technologies have not been mentioned: SiC-based devices, with excellent power capabilities but limited to the low microwave frequency range; the LDMOS, a MOSFET technology able to provide power levels up to 100 W for frequencies below 5 GHz, today dominating the market of cellular basestations (with some incoming competition from GaN), but with little possibility to substantially increase the operating frequency; the antimonide-based alloys, with low gap, restricted to niche application in the defense or aerospace field; finally, diamond MESFETs and MISFETs, that could be (thanks to the large gap and excellent thermal conductivity of diamond) the ultimate solution in terms of microwave and millimeter wave power devices, provided that a number of technological issues are solved, see [17, 18].

Bipolar devices with a conventional design (Si-based) are intrinsically limited to operation at a few GHz; in fact, in order to improve the speed, the base has to be made thinner, but also highly doped not to compromise the input resistance (affecting the maximum oscillation frequency and also the noise figure). Unfortunately, preservation of the emitter efficiency imposes a yet larger increase of the emitter doping, that leads in turn to emitter gap narrowing; this strongly counteracts the effect of increased doping, finally leading to a decrease of the emitter efficiency. To overcome this impasse, the concept of the heterojunction bipolar transistor (HBT) was proposed. For the sake of definiteness, consider a *npn* HBT; high emitter efficiency is obtained by blocking holes injected from the base into the emitter by means of the valence band discontinuity (hole barrier) generated by the emitter-base heterojunction. The barrier hole blocking effect

5.1 Active Microwave Components

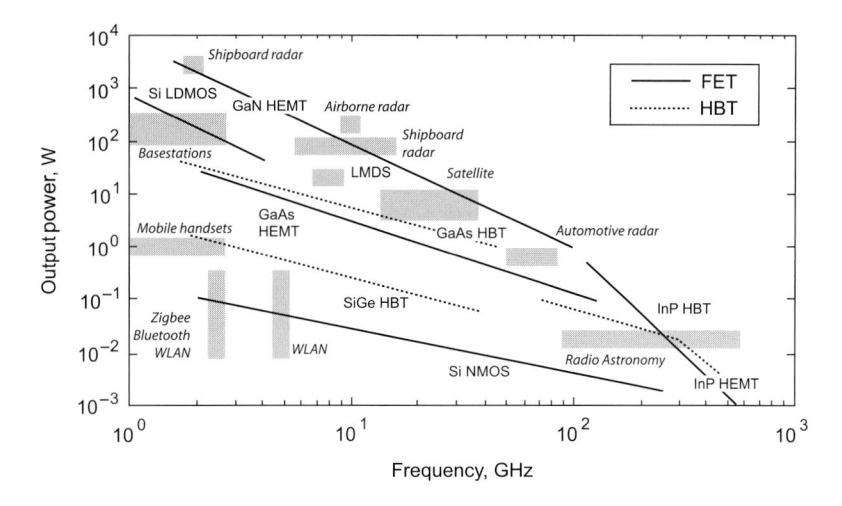

Figure 5.2 Power-frequency performances of a number of solid-state transistor technologies and power-frequency range of some RF, microwave and millimeter wave applications. LMDS is Local Multipoint Distribution System. System data are partly derived from [21] (Copyright ©2008 GAAS Association. All rights reserved. Reused with permission).

is so strong that the base doping can be made equal to the emitter doping, allowing to reduce the input base resistance.

The realization of the HBT had to wait for the development of a III-V based heterostructure technology; AlGaAs/GaAs HBTs were first demonstrated in 1972 [19] and have been present on the market since the middle of the nineties for aerospace but also consumer applications (e.g., the 1 W power stage of mobile phones). The HBT remains the device of choice in low phase noise oscillators, due to the favorable low-frequency noise typical of vertical conduction devices. However, HBTs can also be implemented within a Si-based technology through the SiGe HBT approach [20]. In such devices, the narrowgap SiGe alloy is the (strained) thin transistor base, while the collector and emitter are made with Si. SiGe HBTs have entered the market during the last 20 years in a number of millimeter-wave applications (also consumer, e.g., automotive radars); they exhibit more interesting power ranges than in the CMOS technology but a far more complex technological process. SiGe HBTs can be also used for extremely fast logical circuits based on the Emitter Coupled Logic (ECL) approach [20, Sec. 5.7].

A more detailed overview of the FET and bipolar technological options will be provided in Sec. 5.4, Sec. 5.5 and Sec. 5.7; we stress that, from the standpoint of circuit design, many of those technologies are interchangeable and anyway can be modeled (at least in small-signal conditions) with equivalent circuits having a very similar structure. To allow an easier introduction to the subject, we will address first the modeling of the GaAs MESFET and then extend the treatment to other FET families and to the HBTs.

Fig. 5.2 shows a synthetic overview of the power-frequency behavior of a number of solid-state technologies together with the power-frequency range of a number of RF, microwave and millimeter wave applications. Notice that for each application the power-frequency diagram stresses the potential for the transmitting state, while other technologies (e.g., better suited for low-noise amplifiers) are used in the receiver stage. In millimeter wave radio astronomy, for example, receivers rather than transmitters are used with the aim of detecting signals emitted by stars and other celestial bodies. A more detailed list of applications and enabling technologies is reported in [7, Table 1.14]; today, many low-power RF and microwave applications are implemented with silicon MOSFETs, rather than with compound semiconductor devices.

5.1.2 Device Modeling Approaches

The modeling of active semiconductor devices (for high-frequency applications, but also in general) can be carried out at different levels of complexity, computational efficiency, accuracy and generality. Most models are equation-based,[3] but a rough distinction can be made between *physics-based models* where the equations and their parameters are derived from theory or from the device geometry, and *measurement-based models* where a mathematical framework with some fitting parameters is exploited in order to simulate, with a prescribed accuracy, some electrical measurements. To make a simple example, theory predicts that a resistor is modeled by a linear equation of the kind $V = RI$ where R is the resistor resistance; a physics-based model of the resistor further suggests that $R = \rho l/A$ where ρ is the material conductivity (derived from the theory or from specific measurements), l and A (the resistor length and cross section) are geometrical parameters. On the other hand, starting from the same equation we can derive the resistance R as the slope of the measured IV characteristics, obtaining thus a measurement-based model of the resistor.

In measurement-based models the mathematical framework can be derived, up to a certain extent, from device physics, but in the *black box* modeling approach we make use of an entirely general system-level modeling framework, completely identified and parameterized on the basis of measurements. As an example, we know from theory that any linear two-port can be completely characterized in the frequency domain by its scattering parameters. A black box model in this case simply is an interpolation of the scattering parameters measured in a discrete set of frequencies.

Focusing on active semiconductor devices, the modeling frameworks available can be classified (non-univocally) into three main groups:

- *Physics-based models*: they solve the semiconductor equations (like a transport model coupled with the Poisson equation) at a microscopic level, analytically (approximately) or numerically (exactly); numerical models are in theory able to handle all possible operating conditions of the device and are potentially very accurate. However, they are computationally intensive. Most parameters are derived from physics and geometry, so that fitting on measurements is only possible up to some extent. They are not easily coupled to a circuit simulator.

---

[3] Also look-up table-based models are ultimately based on approximating equations with stored coefficients. A mathematical but not strictly speaking equation-based model is the neural network model, that has been sparsely used also for microwave device modeling.

5.1 Active Microwave Components

- *Equivalent circuit models* are approximate, but efficient, circuit representations of the device physics; the parameters of the model can be extracted from measurements or derived from physics-based approaches. They are easily coupled to circuit simulators.
- *Black-box models* are mathematical models entirely derived from measured data. They are numerically efficient, can be linked to circuit or system simulators, but require an extensive set of measurements and efficient optimization procedures to be implemented. They have little predictive capabilities outside the data range on which they were fitted.

Each class of active device models can be, moreover, more or less convenient when modeling a component in a specific operating regime. According to the kind of device operation (see Fig. 5.3) we can further classify the models according to a different perspective:

- *Nonlinear or large-signal* models aim at modeling the device in all possible operating conditions, independent of the input signal magnitude or waveform. In practice, a class of nonlinear models is optimized for *digital* or *switching* applications, see Fig. 5.3 (a); the signal is wideband, and time-domain simulations are typically used for analysis and design. Another class, the analog large-signal models, is optimized for *analog*, often *narrowband* operation, or, in the limit, periodic steady-state, see Fig. 5.3 (b); simulations are performed in the frequency-domain, e.g., through harmonic-balance nonlinear techniques (see Chapter 3). Such models are more critical, since they should accurately reproduce low-amplitude effects like harmonic distortion and intermodulation; they are essential in the modeling of power amplifiers, mixers, frequency multipliers, oscillators.
- *Small-signal* models are confined to modeling small deviations with respect to a DC working point; they are oriented to linear analog applications such as high-gain or low-noise amplifiers, see Fig. 5.3 (c).

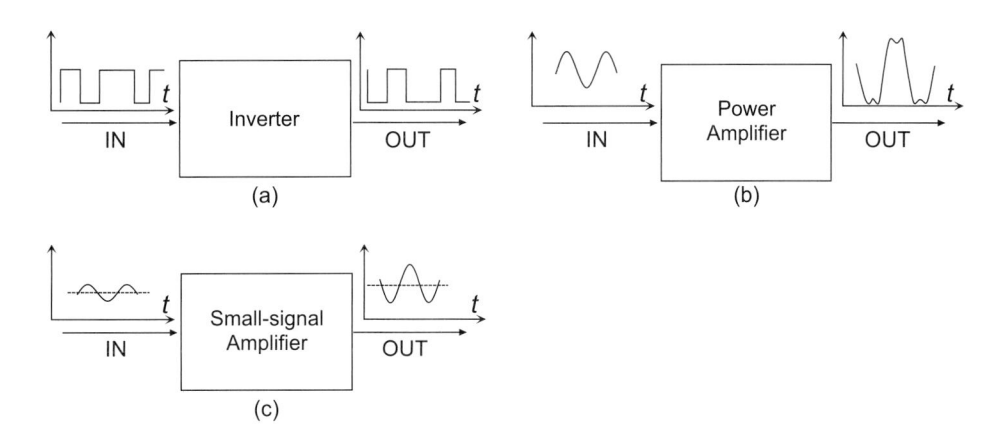

Figure 5.3 Operation modes for a device model (represented for simplicity like an input–output system): (a) switching large-signal operation; (b) large-signal analog narrowband, quasi-periodic operation; (c) small signal narrowband operation around a DC working point.

To this list, we can add *noise models* (see Chapter 7); since noise can be modeled as a small-amplitude fluctuation superimposed to a DC or periodic operating point, such models consists in a set of random generators added to either a small-signal or large-signal model. In Chapter 7 the treatment will concern the small-signal case, that is relevant to the analysis and design of low-noise amplifiers.

Both large-signal and small-signal models are dynamic (with memory) since they include reactive effects, that are essential in RF and microwave operation. An example of nonlinear memoriless model is the model of a transistor in DC, used to evaluate the DC working point of a circuit.

Physics-based numerical models are only marginally used in circuit simulation, due to their computational intensity. In large-signal analog operation, the most common modeling approach is perhaps the large-signal equivalent circuit, due to its simplicity and comparative ease of extraction from measured data. Black-box large-signal models, however, are increasingly popular in the modeling of entire components such as power amplifiers. Concerning small-signal modeling, although an exact black-box small-signal model is given by the measured scattering parameters on the frequency band of interest, small-signal equivalent circuits are popular in circuit design, even if they are approximate, because they allow for an easy separation of the intrinsic device from parasitics, and therefore device model scaling vs. the gate length, and wideband operation, also outside the bandwidth where measurements were carried out.

5.2 Semiconductors and Semiconductor Alloys for Microwave Transistors

As already recalled, the main actors in the field of RF, microwave and millimeter-wave solid-state field-effect and bipolar transistors are elementary semiconductors like Si (MOSFET) and compound semiconductors like GaAs, InP and GaN. SiGe (appearing in Si-based HBTs) is not a compound semiconductor, but rather a *semiconductor alloy*, that should be more properly written as Si_xGe_{1-x} where x is the Si fraction. Compound semiconductor alloys (like $Al_xGa_{1-x}As$) are also important building blocks in microwave FETs and HBTs.

Compound semiconductor families are generally classified according to the chemical nature of the metal and nonmetal component. If the metal component belongs to group III and the nonmetal to group V, we obtain a III-V compound. Examples of III-V compounds are GaAs, InP, GaSb, InAs, AlAs, GaP. III-V compounds with nitrogen such as GaN, InN, AlN are often referred to as III-N compounds. II-VI compounds include CdTe, HgTe, ZnS, CdSe, ZnO; the only important IV-IV compound is SiC (as already recalled SiGe is an alloy). From the standpoint of exploitation in semiconductor electronic devices, GaAs and InP, like Si, are general-purpose semiconductors, well suited to low-noise or medium-power transistors. Widegap semiconductors like SiC or GaN are particularly well suited for power transistors due to the large breakdown field. On the other hand, GaSb and its alloys are characterized by small gap and very high mobility, allowing for high-speed low-noise transistors or even extremely low-power logical circuits.

5.2.1 Semiconductor Properties: A Reminder

Single-crystal semiconductors have a particularly important place in RF and microwave electronics, as the starting material for high-quality active devices.

The electronic structure of crystals generally includes a set of allowed energy bands, populated by electrons according to quantum mechanics. The two topmost energy bands are the *valence* and *conduction* band, respectively, see Fig. 5.4. Above the conduction band energy, we find the *vacuum level*, i.e., the energy of an electron free to leave the crystal. Further details of the bandstructure are introduced in Fig. 5.4, such as the electron affinity $q\chi$, the distance between the conduction band edge and the vacuum level U_0, and the ionization I_0, i.e., the distance between the valence band edge and the vacuum level. In *insulators*, the valence band (which hosts the electrons participating to the chemical bonds) is separated from the conduction band by a large energy gap E_g, of the order of a few eV. A small number of electrons have enough energy to be promoted to the conduction band, where they could take part in electrical conduction, and the conductivity is extremely small. In *metals*, the valence and conduction bands overlap, so that all carriers belong to the conduction band, leading to a large conductivity. In *semiconductors*, the energy gap is typically of the order of 1–2 eV, so that some electrons have enough energy to reach the conduction band, leaving positively charged *holes* in the valence band. Holes and electrons participate to the current conduction. In pure, *intrinsic*, semiconductors, charge transport is *bipolar* (through electrons and holes), and the conductivity is low, exponentially dependent on the gap (the larger the gap, the lower the conductivity). However, impurities can be added (the *dopants*), able to provide a large amount of electrons to the conduction band (the *donors*) or of holes to the valence band (the *acceptors*). The resulting doped semiconductors are denoted as *n*-type and *p*-type, respectively; their conductivity can be artificially modulated by changing the amount of dopants; moreover, the dual doping option allows for the development of *pn* junctions, one of the basic building blocks of electronic devices.

Most high-quality electron devices are based on single-crystal semiconductors. Out of the possible crystal structures, only two are important at present, the *cubic* and the *hexagonal*. Most important semiconductors are cubic (examples are Si, Ge, GaAs,

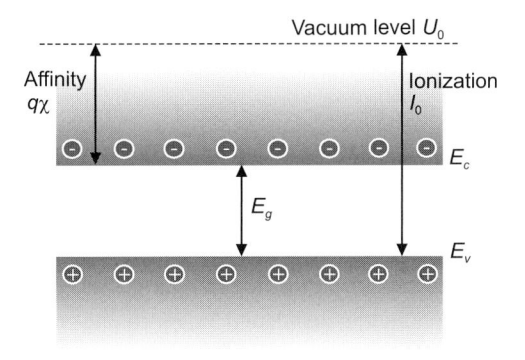

Figure 5.4 Main features of the semiconductor bandstructure.

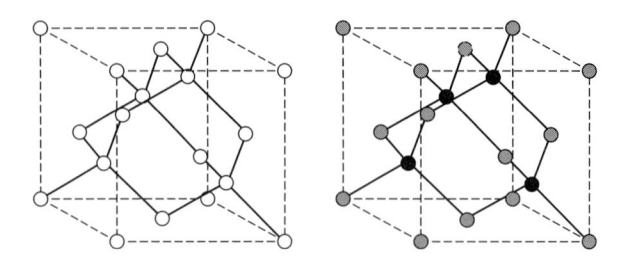

Figure 5.5 The diamond (left) and zinc-blende (right) lattices.

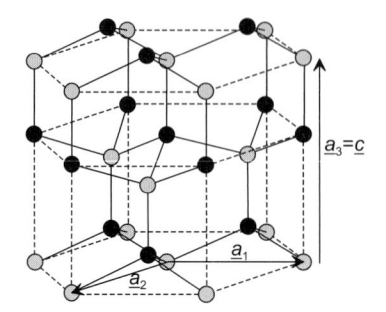

Figure 5.6 The hexagonal wurtzite cell. The c-axis corresponds to the direction of the vector $\underline{a}_3 = \underline{c}$.

InP...), but some are hexagonal (SiC, GaN). The basic cell of cubic semiconductors is shown in Fig. 5.5. In the diamond lattice, all atoms are the same (as in Si, Ge...), while in the *zinc-blende lattice* two different atoms are present (as in GaAs, InP, SiC...). In particular, the corner and face center atoms (in gray in Fig. 5.5, right) are metals (e.g., Ga) and the internal atoms (in black in Fig. 5.5, right) are nonmetals (e.g., As), or *vice versa*.

Some semiconductors, such as SiC and GaN, have the hexagonal *wurtzite* crystal structure. Hexagonal lattices admit many *polytypes* according to the stacking of successive atom layers; a large number of polytypes exists, but only a few have interesting semiconductor properties (e.g., 4H and 6H for SiC). The wurtzite cell is shown in Fig. 5.6, including 12 equivalent atoms. In the ideal lattice, one has:

$$|\underline{a}_3| = c, \quad |\underline{a}_1| = |\underline{a}_2| = a, \quad \frac{c}{a} = \sqrt{\frac{8}{3}} \approx 1.633.$$

Some properties of semiconductor lattices are shown in Table 5.1.[4] Wurtzite semiconductors are anisotropic (uniaxial) and have two dielectric constants, one parallel to the c axis, the other orthogonal to it.

An important feature of semiconductors is the energy–momentum relation for electrons in the conduction band and holes in the valence band, also called dispersion

---

[4] Semiconductor properties are documented in many textbooks; an online resource is provided by the Ioffe Institute of the Russian Academy of Sciences at the website [22].

5.2 Semiconductors and Semiconductor Alloys for Microwave Transistors

Table 5.1 Properties of some semiconductor lattices: the crystal is D (diamond), ZB (zinc-blende) or W (wurtzite); the bandgap is D (direct) or I (indirect); $\epsilon_\|$ is along the c axis, $\epsilon_\perp$ is orthogonal to the c axis for wurtzite materials. Permittivities are static to RF. Properties are at 300 K.

Material	Crystal	E_g	D/I	ϵ_r or $\epsilon_\|$	$\epsilon_\perp$	a	c	ρ
		(eV)	gap			(Å)	(Å)	(g/cm^3)
C	D	5.50	I	5.57		3.57		3.51
Si	D	1.12	I	11.9		5.43		2.33
SiC	ZB	2.42	I	9.72		4.36		3.17
Ge	D	0.66	I	16.2		5.66		5.32
GaAs	ZB	1.42	D	13.2		5.68		5.32
GaP	ZB	2.27	I	11.11		5.45		4.14
GaSb	ZB	0.75	D	15.7		6.09		5.61
InP	ZB	1.34	D	12.56		5.87		4.81
InAs	ZB	0.36	D	15.15		6.06		5.67
InSb	ZB	0.23	D	16.8		6.48		5.77
AlP	ZB	2.45	I	9.8		5.46		2.40
AlAs	ZB	2.17	I	10.06		5.66		3.76
AlSb	ZB	1.62	I	12.04		6.13		4.26
CdTe	ZB	1.47	D	10.2		6.48		5.87
GaN	W	3.44	D	10.4	9.5	3.17	5.16	6.09
AlN	W	6.20	D	9.14		3.11	4.98	3.25
InN	W	1.89	D	14.4	13.1	3.54	5.70	6.81
ZnO	W	3.44	D	8.75	7.8	3.25	5.21	5.67

relation or bandstructure. Due to the lattice periodicity in real space also the dispersion relation is periodic, the fundamental period being the Brillouin first zone, see [5]. The electron dispersion relation is typically complex, with many minima located for zero or nonzero momentum. Around the minimum the energy–momentum relation $E(\underline{p})$ can be approximated by a parabola, i.e., by the same relation holding for the kinetic energy of a free electron; the inverse of the curvature of the parabola is proportional to the *effective mass* rather than to the electron inertial mass. The effective mass accounts for the interaction between the electron and the periodic potential profile of the crystal and is often smaller than the inertial mass. In a semiconductor under the presence of a low applied field (in GaAs, indicatively much less than 5 kV/cm), most electrons have low energy and therefore are located near the lowest minimum, corresponding to the conduction band edge E_c; in most compound semiconductors the lowest minimum is located at $\underline{p} = 0$ (called the Γ point). With increasing applied field the electrons are scattered into highest minima (in GaAs and InP called the X and L minima) that have larger effective mass. On the other hand, in Si, Ge and some compound semiconductors (like AlAs and GaP) the lowest minimum is not located at $\underline{p} = 0$ but in a different point of the momentum space, close to the so-called X or L points, that lie on the boundary of the Brillouin first zone.

Concerning the valence band, the dispersion relation for holes is quite similar in all semiconductors, with a minimum hole energy E_v in the Γ point. Near this point two

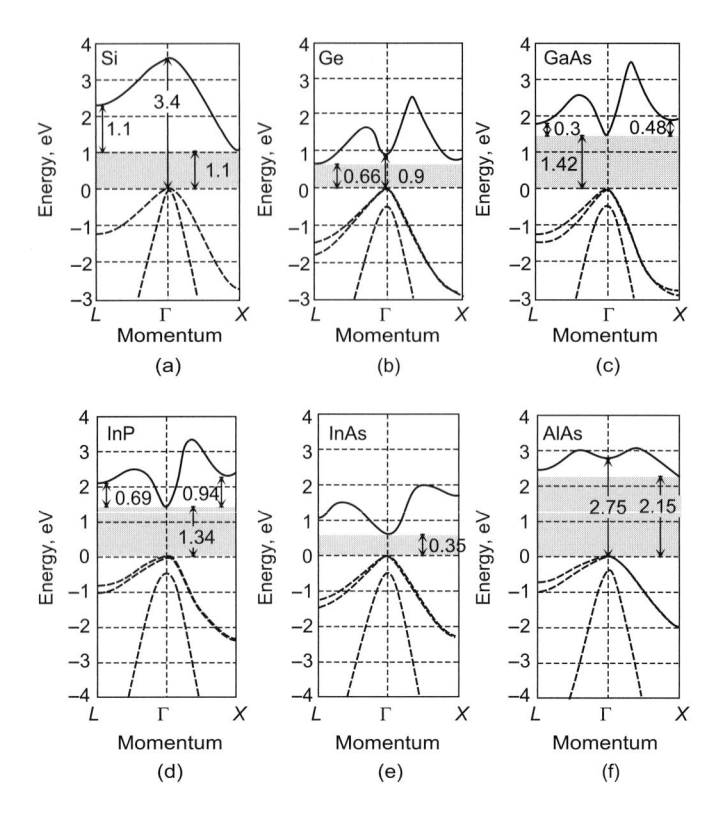

Figure 5.7 Simplified bandstructure of some important semiconductors: (a) Si; (b) Ge; (c) GaAs; (d) InP; (e) InAs; (f) AlAs. The energy–momentum relation for the electrons in the conduction band is shown as a continuous line, the same for holes (heavy, light, split-off) in the valence band as a dashed line.

kinds of holes exist, called the heavy and light holes with reference to their effective mass, so that the equivalent hole mass is an average of the two. A simplified sketch of the bandstructure of a few important semiconductors is reported in Fig. 5.7; notice that only a section of the dispersion relationship from point Γ to X and L is shown. Despite the simplification, this section allows the main features of the bandstructure to be readily appreciated.

Semiconductors in which the hole minimum energy and the electron minimum energy occur in the Γ point are called *direct bandgap*. If the valence and conduction band minima occur at different momenta, we have an *indirect bandgap* semiconductor. Examples of indirect bandgap semiconductors are Si, Ge, GaP, AlAs while GaAs, InP, GaN and are direct bandgap semiconductors. While the difference between direct and indirect bandgap semiconductors is particularly relevant to their ability to interact with photons – a feature that is not significant to the operation of high-speed transistors – it may be remarked that central (Γ) minima of the conduction band are characterized by a lower effective mass but also lower scattering from phonons, both leading to a large mobility, when compared with non-central minima. As a result, electrons are faster in most

5.2 Semiconductors and Semiconductor Alloys for Microwave Transistors

Table 5.2 Band properties of some important compound semiconductors. Mobility data are upper bounds referring to undoped material. m_n^* is the electron effective mass, m_{lh}^* and m_{hh}^* are the light and heavy hole effective masses, respectively; m_0 is the free electron mass. ϵ is the material dielectric permittivity, $\mathcal{E}_{br}$ thee breakdown electric field.

Property	In$_{0.53}$Ga$_{0.47}$As	GaAs	InP	AlAs	InAs
a (Å)	5.869	5.683	5.869	5.661	6.0584
E_g @300 K (eV)	0.717	1.424	1.34	2.168	0.36
$q\chi$ (eV)		4.07	4.37	3.50	4.90
m_n^*/m_0	0.041	0.067	0.077	0.150	0.027
m_{lh}^*/m_0	0.044	0.08	0.12	0.150	0.023
m_{hh}^*/m_0	0.452	0.45	0.6	0.76	0.60
$\epsilon(0)/\epsilon_0$	13.77	13.18	12.35	10.16	14.6
$\epsilon(\infty)/\epsilon_0$	11.38	10.9	9.52	8.16	12.25
$\mathcal{E}_{br}$ (kV/cm)	3.0	3.2	11		
μ_n (cm^2/Vs)	12000	8500	5500		40000
v_{max} (10^7 cm/s)	$\approx$ 2.5	$\approx$ 1.7	$\approx$ 2.7		
v_s (10^7 cm/s)	0.7	1			

compound semiconductors (like GaAs and InP) with respect to Si. Holes are typically slower and electrons, and their mobility is very similar in most semiconductors, unless light holes are made prevail at low energy by applying to the material proper strain, as done, e.g., in the base of SiGe heterojunction bipolar transistors (see Sec. 5.7).

In general, many III-V semiconductors have a bandstructure similar to GaAs. InP has a slightly lower bandgap, but a larger difference between the central and the lateral minima of the conduction band; this has important consequences on the transport properties, since it increases the electric field at which the electrons are scattered from the central minimum (characterized by lower effective mass and therefore high mobility, i.e., high electron velocity with the same applied electric field) to the lateral minima (with high effective mass and therefore low mobility). InAs has a very similar bandstructure, but with lower energy gap. For certain compound semiconductors, like AlAs the central minimum is higher than the lateral minima, thus making the material indirect-bandgap. The properties of important compound semiconductors are listed in Table 5.2, where v_s is the electron high-field saturation velocity (also denoted as $v_{n,\text{sat}}$), v_{max} is the maximum steady-state electron velocity, see Sec. 5.2.2; the values provided refer to an intrinsic (undoped) material.

5.2.2 Transport Properties

Under the application of an external electric field, electrons and holes experience a driving force which increases their average velocity. For low field the average velocity is proportional to the electric field:

$$\underline{v}_{n,av} = -\mu_n \underline{\mathcal{E}}, \quad \underline{v}_{h,av} = \mu_h \underline{\mathcal{E}},$$

where μ_n and μ_h are the electron and hole *mobilities*, measured in cm^2/Vs. The low-field mobility depends on the interaction with lattice vibrations (phonons), impurities etc., and typically decreases with increasing doping and increasing temperature (at least, at ambient temperature and above). For very large fields (values of the order of 10 kV/cm, depending on the semiconductor) the average velocity saturates:

$$v_{n,av} \to v_{n,\text{sat}}, \quad v_{h,av} \to v_{h,\text{sat}}.$$

In all semiconductors, the saturation velocities have magnitude around 10^7 cm/s.[5] The motion of electrons and holes due to the application of an electric field is called the *drift motion* and gives rise to the *drift (conduction) current density*:

$$\underline{J}_{n,\text{dr}} = -qn\underline{v}_{n,av} = qn\mu_n \underline{\mathcal{E}} \tag{5.1}$$

$$\underline{J}_{h,\text{dr}} = qp\underline{v}_{h,av} = qp\mu_h \underline{\mathcal{E}}. \tag{5.2}$$

While the average carrier velocity is limited by scattering with phonons and impurities, for extremely small time (ps) or space ($\ll 1$ μm, also depending on material) scales, electrons and holes experience *ballistic motion*, i.e., motion unaffected by collisions. In such conditions, the average carrier velocity can be substantially higher, in the presence of strong electric fields, than the static saturation velocity. Such effect is called *velocity overshoot* and plays a role in increasing the speed of transistors such as nanometer-gate FETs.

The electron velocity-field curves of a few relevant semiconductors are shown in Fig. 5.8. In compound semiconductors like GaAs, InP and InGaAs (lattice matched

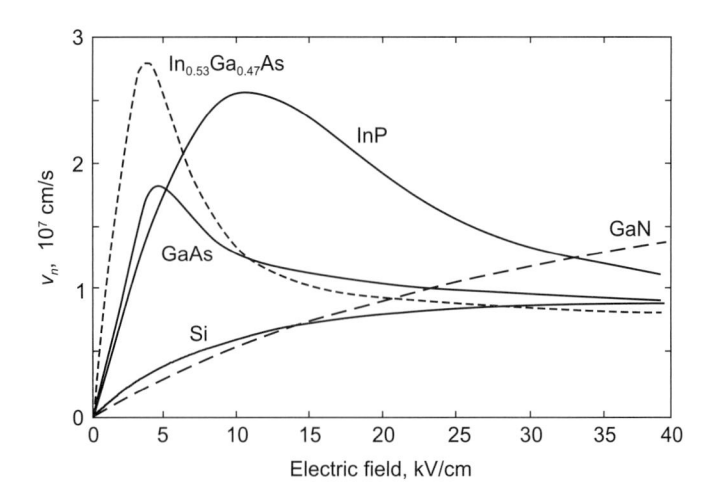

Figure 5.8 Electron drift velocity-field curves of Si, GaAs, InP, GaN, InGaAs lattice matched to InP. The GaN velocity has a peak towards 200 kV/cm and then saturates with a GaAs-like behavior. Adapted from [24], p. 13.

---

[5] Saturation velocities are difficult to measure and, from theory, related to the energy of the optical phonons emitted by carriers at high field, that is similar in all semiconductors. Expected saturation velocities range from one half to twice, approximately, of the value 10^7 cm/s.

to InP) the electron velocity-field curve is non-monotonic: at low electric field, electrons are mainly in the Γ minimum of the conduction band, but with increasing field they are ultimately scattered into the indirect-bandgap minima, where their velocity is lower. As a result, the average velocity decreases. This happens for electric fields of the order of 3 kV/cm in GaAs, where the energy difference is about 300 meV, but for larger fields (around 10 kV/cm) in InP where a larger energy difference (around 700 meV) is involved. In GaN, on the other hand, such energy difference is around 3 eV, leading to a peak field in excess of 200 kV/cm, see, e.g., [23]. The saturation velocity is however similar (within a factor of 2) in all semiconductors, including Si, which has a monotonically increasing velocity with a much lower initial (low field) mobility. The hole mobility in compound semiconductors is similar to (or even worse than) the hole mobility in Si, and therefore *n*-type (or *npn*) transistors are preferred for high-speed applications.

5.2.3 Heterostructures and Semiconductor Alloys

Crystals with different lattice constants grown on top of each other by epitaxial techniques are affected by interface defects called *misfit dislocations*. Such defects operate as electron or hole traps, and therefore the resulting structure is unsuited to the development of an electron device. However, if the lattice mismatch between the substrate and the heteroepitaxial overlayer is low or zero, an ideal or almost ideal crystal can be grown, made of two different materials. The resulting structure is called a *heterostructure*, and, since the electronic properties of the two layers are different, we also refer to it as a *heterojunction*. The material discontinuity arising in the heterojunction leads to important electronic properties, such as confinement of carriers, related to the discontinuity of the conduction or valence bands.

Heterostructures can be lattice-matched (if the two sides have the same lattice constant) or affected by a slight mismatch (less than 1 percent) which induces tensile or compressive strain. In this case, we talk about *pseudomorphic* or *strained* heterostructures, see Fig. 5.9. A little amount of strain in the heterostructure can lead in certain cases to an improvement of the material transport properties.

A double heterojunction made with a thin semiconductor layer (the thickness should be typically of the order of 100 nm) sandwiched between two layers (e.g., AlGaAs/GaAs/AlGaAs) creates a potential well in the conduction and/or valence band and is often referred to as a quantum well (QW).

The need to grow lattice-matched or strained heterostructures has fostered the development of semiconductor alloys. Semiconductor alloys are *artificial* semiconductors having intermediate properties with respect to the components. Among such properties are the lattice constant a and the energy gap E_g. In several material systems, both a and E_g approximately follow a linear law with respect to the individual component parameters. Tailoring the lattice constant by proper alloy composition enables to achieve lattice matching to the substrate; the resulting bandgap (and therefore conduction and valence band) discontinuity arising at the heterojunction allows, in electron devices, to create potential wells for electron and/or holes. Such a laterally confined carrier

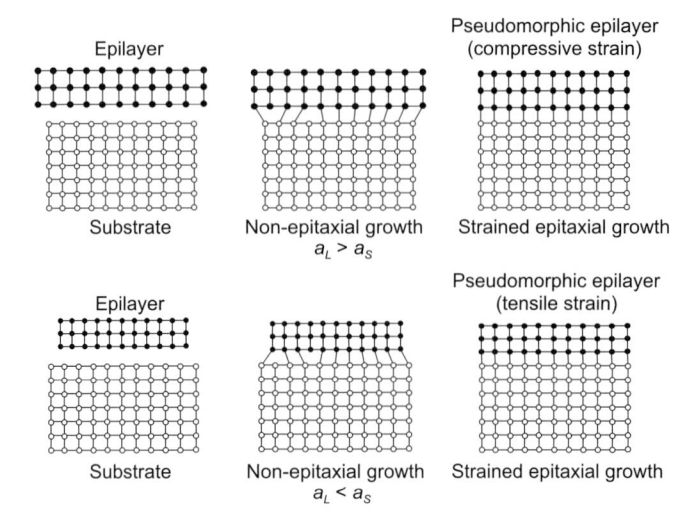

Figure 5.9 Pseudomorphic or strained growth: above, the epilayer lattice constant (a_L) is larger than the substrate's (a_S): non-epitaxial growth with interface misfit dislocations and strained epitaxy: below, the epilayer lattice constant is smaller than the substrate's.

population is the basic element for the conducting channel of heterojunction FETs (like the HEMT).

Alloys can be made of two components and three elements (*ternary alloys*: e.g., AlGaAs, alloy of GaAs and AlAs) or of four components and elements (*quaternary alloys*, e.g., InGaAsP, alloy of InAs, InP, GaAs, GaP). By properly selecting the alloy composition, semiconductor alloys matched to a specific substrate can be generated.

To quantitatively define an alloy, we have to consider that compound semiconductors (CS) are polar compounds with a metal M combined to a non-metal N in the form MN. Two different CS sharing the same metal or nonmetal give rise to a ternary alloy or compound:

$$(M_1N)_x (M_2N)_{1-x} = M_{1x}M_{2(1-x)}N \quad \text{e.g., } Al_xGa_{1-x}As$$
$$(MN_1)_y (MN_2)_{1-y} = MN_{1y}N_{2(1-y)} \quad \text{e.g., } GaAs_yP_{1-y},$$

where x and $1 - x$ denote the mole fraction of the two metal components, y and $1 - y$ the mole fraction of nonmetal components. Four different CS sharing two metal and two nonmetal components yield a quaternary alloy or compound. In the following formulae M and m are the metal components, N and n are the nonmetal components, and $\alpha + \beta + \gamma = 1$:

$$(MN)_\alpha (Mn)_\beta (mN)_\gamma (mn)_{1-\alpha-\beta-\gamma} = M_{\alpha+\beta}m_{1-\alpha-\beta}N_{\alpha+\gamma}n_{1-\alpha-\gamma}$$
$$= M_x m_{1-x} N_y n_{1-y} \quad \text{(e.g., } In_xGa_{1-x}As_yP_{1-y}).$$

Most alloy properties can be derived from the component properties through (global or piecewise) linear interpolation (Vegard law), often with second-order corrections (Abeles law); examples are the lattice constant, the energy gap, the inverse of the

5.2 Semiconductors and Semiconductor Alloys for Microwave Transistors

effective masses, and, in general, the bandstructure and related quantities. Varying the composition of a ternary alloy (one degree of freedom) changes the gap and related wavelength, but, at the same time, the lattice constant; in some cases (AlGaAs) the two components (AlAs and GaAs) are already matched, so that alloys with arbitrary Al content are lattice matched to the substrate (GaAs).

On the other hand, varying the composition of a quaternary alloy (two degrees of freedom) independently changes both the gap and the lattice constant, so as to allow for lattice matching to a specific substrate, e.g., InGaAsP on InP.

Alloys are often represented as a straight or curved segment (for ternary alloys) or as a quadrilateral region (for quaternary alloys) in a plane where the x coordinate is the lattice constant and the y coordinate is the energy gap, see Fig. 5.10. The segment extremes and the vertices of the quadrilateral are the semiconductor components. In Fig. 5.10 some important alloys are reported:

- AlGaAs, lattice matched for any composition to GaAs, direct bandgap up to an Al mole content of 0.45. The heterostructure AlGaAs/GaAs was the first one to be used in the development of High Electron Mobility field-effect transistors (HEMTs);
- InGaAsP, which can be matched either to GaAs or to InP substrates; the alloy is direct bandgap, apart from around the GaP corner, whose gap is indirect;
- InAlAs, than can be lattice matched to InP with composition $Al_{0.48}In_{0.52}As$; it is an important alloy in HEMTs based on InP substrates;

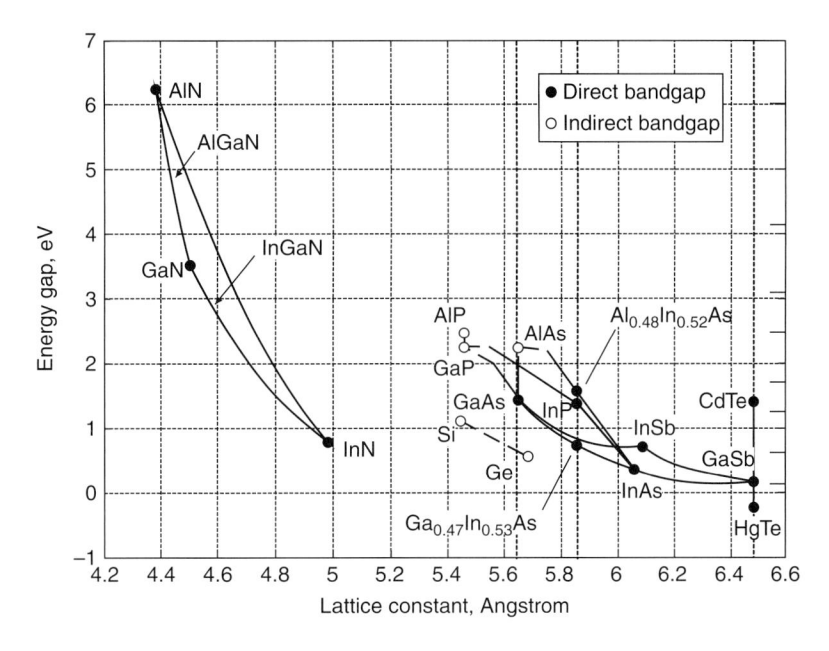

Figure 5.10 Some important alloys in the lattice constant – energy gap plane. In order of increasing gap and decreasing lattice constant: HgCdTe, InGaAsSb, InGaAsP, SiGe, AlGaAs, AlGaP, InGaN, AlGaN. For the widegap (wurtzite) nitrides (GaN, InN, AlN) the equivalent cubic lattice constant is shown. The InGaAsP, AlGaAs and InGaAsSb data are from [25]; the GaN, AlN and InN data are from [26, Fig. 3].

- InGaAs, a ternary alloy matched to InP with composition $Ga_{0.47}In_{0.53}As$; it is a subset of the quaternary alloy InGaAsP; it is again an important alloy in HEMTs based on InP substrates;
- InGaAsSb, the antimonide family, a possible material for low-noise and low-power electronics, with extremely low bias voltages;
- SiGe, an indirect bandgap alloy important for Si-based HBTs but also used in advanced MOSFET technologies to improve the channel transport properties through strain;
- III-N alloys, such as AlGaN and InGaN, with applications in RF and microwave power transistors. AlGaN can be grown by pseudomorphic epitaxy on a GaN virtual substrate; GaN has in turn no native substrate so far, but can be grown on SiC, sapphire (Al_2O_3) or Si.

Since GaN, AlN and InN have the wurtzite (hexagonal) crystal structure, an equivalent lattice constant $a_{C,eq}$ has to be defined for comparison to cubic crystal, so as to make the volume of the wurtzite cell V_H (per atom) equal to the volume of a cubic cell (per atom); taking into account that the wurtzite cell has 12 equivalent atoms, while the cubic cell has 8 equivalent atoms, we must impose:

$$\frac{1}{12}V_H = \frac{1}{12}\frac{3\sqrt{3}}{2}ca_H^2 = \frac{1}{8}a_{C,eq}^3 \rightarrow a_{C,eq} = \left(\sqrt{3}ca_H^2\right)^{1/3},$$

where V_H is the volume of the wurtzite cell prism of sides a and c. For GaN $a_H = 0.317$ nm, $c = 0.516$ nm; it follows that:

$$a_{C,eq} = \left(\sqrt{3}ca_H^2\right)^{1/3} = \left(\sqrt{3} \cdot 0.516 \cdot 0.317^2\right)^{1/3} = 0.448 \text{ nm}.$$

5.2.4 The Substrate Issue

Electronic devices require to be grown on a suitable (typically, semiconductor) substrate. In practice, the available semiconductor substrates have to be grown into monocrystal ingots through the Czochralsky or Bridgman method, see [12]. This limits the substrate choice (in order of decreasing quality and increasing cost) to Si, GaAs, InP, SiC and few others (GaP, GaSb, CdTe). Devices are grown so as to be either lattice matched to the substrate, or slightly (< 1 percent) mismatched (the so-called pseudomorphic approach). The use of graded buffer layers[6] allows the device to be grown on mismatched substrates, since it distributes the lattice mismatch over a larger thickness; the approach is often referred to as the *metamorphic* approach. Metamorphic devices often used to have reliability issues related to the migration of defects in graded buffer layers; however, high-quality metamorphic field-effect transistors with an InP active region on a GaAs substrate (Metamorphic HEMTs, MHEMTs) have been more recently developed with success.

Most undoped compound semiconductor substrates (in particular, GaAs, InP and SiC) are good insulators due to the very low intrinsic concentration of the intrinsic (or

---

[6] A graded buffer is a semiconductor alloy layer with a variable composition, e.g., AlGaAs with an Al fraction linearly increasing from 0 (lattice matched to GaAs) to 1 (lattice matched to AlAs).

compensated, i.e., intentionally doped with a dopant that compensates some unwanted doping resulting from the technological process). Such substrates are often referred to as SI or Semi Insulating substrates. Silicon substrates, on the other hand, exhibit higher losses also in the intrinsic variety, requiring more sophisticated device processes to allow separating the active devices from the substrates (for instance, the use of SOI, Silicon-on-Insulator, approaches).

Finally, gallium nitride FETs cannot be grown on a native substrate (no single-crystal GaN ingots are practically available at present). The present choice is to grow GaN FETs on a SiC substrate (high cost but high thermal conductivity) or on Si substrates (lower cost, but poorer electrical and thermal properties).

5.3 Heterojunctions and Reduced Dimensionality Structures

Heterojunctions are ideal, single-crystal junctions between semiconductors having different bandstructure. Lattice-matched or strained (pseudomorphic) junctions between different semiconductors or semiconductor alloys allow for carrier confinement (through potential wells in conduction or valence bands); if the carrier confinement takes place in a nanometer size region quantized structures such as quantum wells, quantum wires, quantum dots are obtained. Carrier confinement enables to create suitable conducting channels in FET devices; moreover, the strain induced by pseudomorphic heterostructures can lead to improvements in the transport properties of carriers. An example of heterostructure band diagram is shown in Fig. 5.11, where the band disalignment derives from application of the *affinity rule* (i.e., the conduction band discontinuity is the affinity difference, the valence band discontinuity is the difference in ionizations). In many practical cases, however, band disalignments are dominated by interfacial effects and do not follow the affinity rule exactly; for instance, in the AlGaAs-GaAs heterostructure one has:

$$|\Delta E_c| \approx 0.65 \Delta E_g, \quad |\Delta E_v| \approx 0.35 \Delta E_g. \tag{5.3}$$

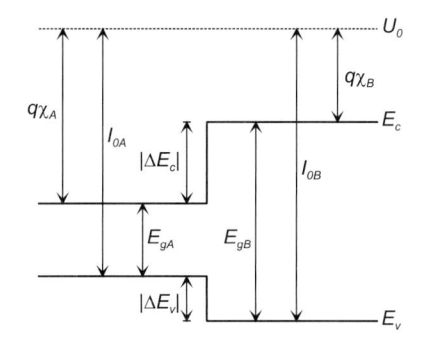

Figure 5.11 Heterostructure band alignment through application of the affinity rule to two materials having different bandstructure.

More specifically, the valence and conduction band discontinuities as a function of the Al fraction are (in eV) [22]:

$$|\Delta E_v| = 0.46x$$

$$|\Delta E_c| = \begin{cases} 0.79x, & x < 0.41 \\ 0.475 - 0.335x + 0.143x^2, & x > 0.41. \end{cases}$$

Heterojunctions can be made with two n-type or p-type materials (*homotype heterojunctions*) or so as to form a *pn* junction (*heterotype heterojunctions*). Often, the widegap material is conventionally denoted as N or P according to the type, the narrowgap material as n or p. According to this convention, a heterotype heterojunction is, e.g., *Np* or *nP* and a narrowgap intrinsic layer sandwiched between two widegap doped semiconductors is *NiP*.

As already recalled, single or double heterostructures create potential wells in the conduction and/or valence bands; the resulting carrier confinement leads to a conducting channel that is the main building block of a class of field-effect transistors.

Potential wells defined by heterojunctions can be narrow enough to possess quantum properties (i.e., quantized energy levels). In such case carriers are distributed in a set of subbands characterized by envelope wavefunctions with even or odd symmetry, see Fig. 5.12 for a simplified, qualitative example. The potential well can be defined in 1D (quantum well), 2D (quantum wire), 3D (quantum dot); these are collectively called *reduced dimensionality structures*. Assuming that the geometry be rectangular, we start from bulk (3D motion possible, no confinement) and obtain, by progressively restricting the degrees of freedom of the particle:

1. confinement in one direction (x): particles are confined along x by a potential well but are free to move along y and z (**quantum well**);
2. confinement in two directions (x and y): particles are confined along x and y, but they are free to move along z (**quantum wire**);
3. confinement in three directions (x, y, z): particles are entirely confined and cannot move (**quantum dot**).

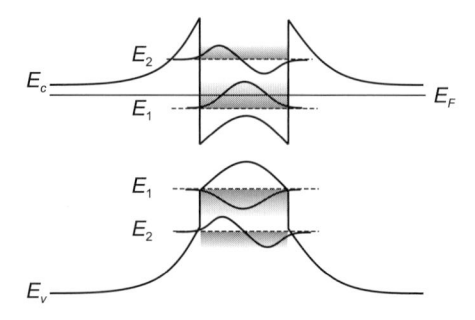

Figure 5.12 Qualitative example of quantization in a quantum well: subbands arise, in the conduction and valence bands, with minimum energies corresponding to the levels $E_1, E_2 \ldots$; the total electron and hole wavefunction is given by the product of the 3D (bulk) wavefunction and of the envelope wavefunction shown. For simplicity, only the heavy hole subbands are shown.

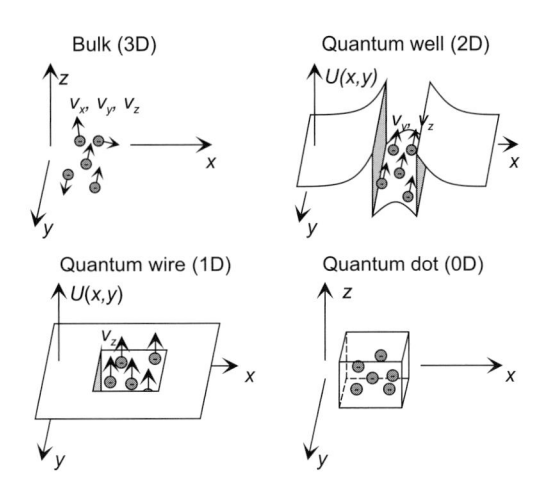

Figure 5.13 Degrees of freedom for electrons in bulk, in a quantum well, in a quantum wire and in a quantum dot. U is the potential profile.

Quantum wells and quantum wires allow for carrier transport. In quantum wells, transport takes place in a 2D thin layer, parallel to the potential barriers than confine the well; the electron population residing there is often called a *two-dimensional electron gas*, 2DEG. In quantum wires, transport occurs along a 1D channel, as shown in Fig. 5.13. Thus, both quantum wells and quantum wires can be exploited as conducting channels in field-effect transistors. On the other hand, quantum dots are a building block of optoelectronic devices only, since carriers can be injected into the quantum dot and radiatively recombine there, or be generated by photon absorption in the quantum dot and then released so as to generate a photocurrent. From the standpoint of electronic devices, carrier confinement is probably the most interesting property of potential barriers associated with single or double heterojunctions; also important is the possibility of using strained heterojunctions to remove the degeneracy of heavy and light holes in the valence band, making light holes dominate with significant improvements in their transport properties.

5.4 Microwave Schottky-Gate Field-Effect Transistors

High-speed field-effect transistors (FETs) are based on a conducting channel whose current (driven by the potential difference between the drain and source electrodes) is modulated by the potential applied to the control electrode (the gate). The gate electrode is isolated from the channel by a metal-oxide-semiconductor junction in MOSFETs, or by a reverse-biased Schottky junction in compound semiconductor FETs (see Fig. 5.14).

The first high-speed field-effect transistor able to operate at microwave frequencies was the GaAs-based MESFET (Metal Semiconductor FET), in which the channel is a layer of highly doped bulk semiconductor, obtained by ion implantation or epitaxy. In epitaxial MESFETs, the active channel is an epitaxial GaAs *n*-doped layer grown

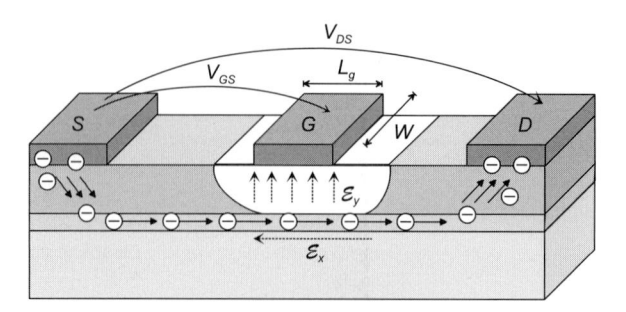

Figure 5.14 Schematic operation principle of an n-channel FET: electrons are driven by the channel electric field induced by the V_{DS} potential difference between drain and source, and the resulting channel is modulated by the control electrode (the gate, here a reverse-biased Schottky junction) through the application of a vertical control electric field. L_g is the gate length, W the gate periphery.

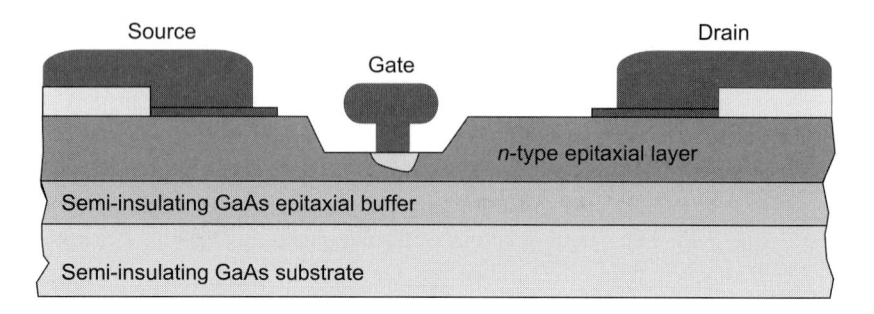

Figure 5.15 Epitaxial MESFET. The version shown is a power recessed-gate device, with breakdown voltages up to 20 V.

on top of a semi-insulating (often Cr-compensated) GaAs substrate, see Fig. 5.15 for the cross section. In MESFETs the control action of the gate takes place by depleting the active layer underneath the Schottky contact, thus reducing its cross section and, as a consequence, the total device current. MESFETs have been generally replaced by heterojunction-based FETs (HEMTs and PHEMTs), and are still produced for system compatibility reasons. Fig. 5.16 shows the layout of a MESFET (conceived for a satellite TV receiver) developed in Italy in 1983 by CISE, Milano within the framework of the MONOMIC project [27]; the side pads are the source, the central one is the drain, while two gate fingers are clearly visible in between. MESFET-based integrated circuits called for the development of ion-implanted (rather than epitaxial) devices, mainly normally on (depletion type), where current flows for zero gate-source bias, but, for some application, also normally off (enhancement type, the threshold voltage is positive). While E-type MESFETs are easily obtained, D-type devices are characterized by a restricted gate voltage range, due to the low built-in voltage of the Schottky gate junction (around 0.7–0.8 V). The cross section of D-type and E-type ion-implanted MESFETs is shown in Fig. 5.17.

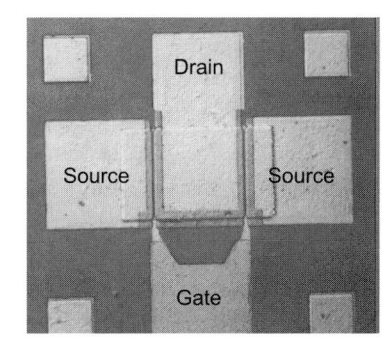

5.4 Microwave Schottky-Gate Field-Effect Transistors

Figure 5.16 Layout of an early epitaxial low-noise MESFET developed in Italy by CISE in 1983. Courtesy of Professor C. Naldi [28].

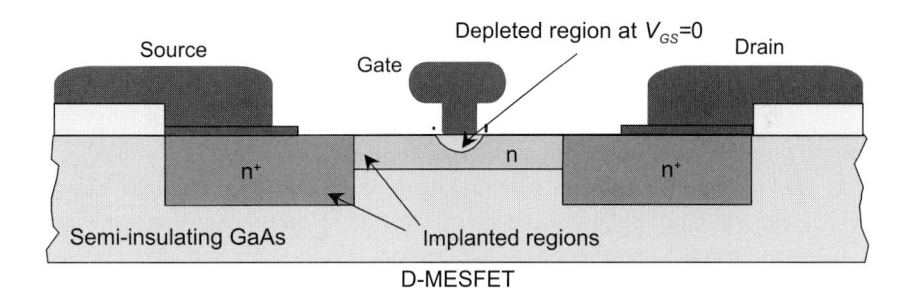

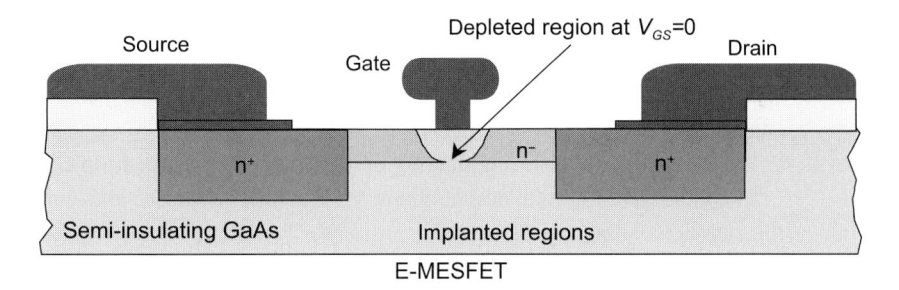

Figure 5.17 Cross section of depletion (D-MESFET, above) and enhancement (E-MESFET, below) ion-implanted integrated MESFET.

The qualitative behavior of the DC characteristics of the MESFET (similar to the standard FET) can be derived by inspection of Fig. 5.18. Suppose that most of the V_{DS} potential difference falls in the active channel located underneath the Schottky gate. For low V_{DS}, the potential difference between the gate and each point in the conducting channel is almost constant; as a consequence, the *channel potential* increases linearly from the grounded source to the drain, but the depletion region is almost uniform along the gate, and the channel cross section is constant. In such condition (Fig. 5.18 (a)), the device behaves as a variable resistor; the resistance depends on the applied V_{GS}. For $V_{GS} = V_{TH}$ (the threshold voltage) the conducting layer is completely depleted and

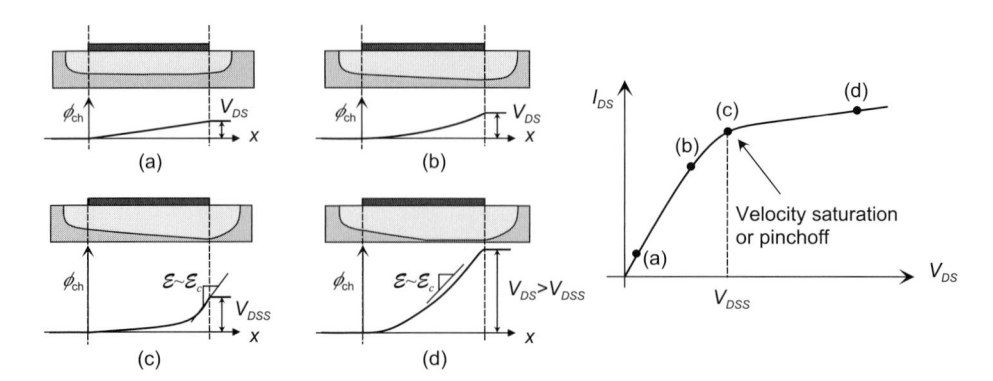

Figure 5.18 Qualitative explanation of the DC n-channel MESFET characteristics for constant V_{GS} and increasing V_{DS}.

the channel conductance is zero. Increasing V_{DS}, the local potential difference between the gate and the channel becomes increasingly nonuniform (Fig. 5.18 (b)) and, as a consequence, the channel cross section narrows from the source to the gate, leading to a nonlinear channel potential distribution. For large enough V_{DS} (Fig. 5.18 (c)) the electric field at the drain edge of the gate increases, causing velocity saturation of the carriers. Beyond this point, the channel current approximately saturates, and for larger V_{DS} the saturation point is displaced towards the source, leading to a small decrease in the length of the non-saturated (ohmic) part of the channel (Fig. 5.18 (d)). This causes in turn a small increase of the drain current I_D,[7] which is more evident in short-gate devices (e.g., with gate length $L_g < 0.5$ μm). Notice that in materials where the velocity saturation effect is less abrupt than in GaAs (e.g., in Si) and occurs at higher fields, current saturation is due to the channel pinchoff, i.e., it takes place at a voltage V_{DS} such as the cross section of the conducting channel vanishes at the drain edge of the gate.

In conclusion, the FET output characteristics (the input DC gate current is zero since the gate is isolated) exhibit a *linear* region for low V_{DS}, a *triode region* where the drain conductance begins to decrease, and finally, beyond a certain knee voltage, the *saturation region*, where the drain current is approximately independent from V_{DS}. In an n-channel FET, decreasing V_{GS} increases the amount of channel depletion and therefore leads to a lower current. For very large V_{DS} and/or large currents, breakdown occurs under the form of a sometimes catastrophic increase of the drain current. In high-speed Schottky gate FETs the breakdown voltage is low in the *on-state* (high current) and high in the *off-state* (low current, device at threshold). The resulting DC characteristics are qualitatively shown in Fig. 5.19; notice that the I_D corresponding to $V_{GS} = 0$ is often referred to as I_{DSS}. Since the gate is a Schottky barrier, the maximum V_{GS} is limited by the Schottky barrier turn-on voltage; this is lower than the barrier built-in voltage, whose approximate values are 0.6–0.7 V for GaAs, and 0.9–1 V for AlGaAs. The soft breakdown effect shown in Fig. 5.19 is a peculiarity of HEMTs and PHEMTs, due

---

[7] The drain current is also denoted with the symbol I_{DS}.

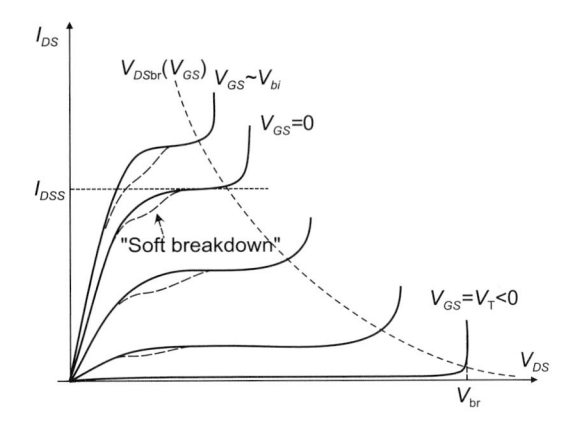

Figure 5.19 DC drain characteristics for Schottky gate FET. The kinks are typical of HEMT devices and are referred to as the *kink* or *soft breakdown* effect. V_{br} is the off-state breakdown voltage, corresponding to breakdown at threshold.

to a substrate charge injection mechanism; it is considerably alleviated by technology optimization.

5.4.1 FET DC Model

A simplified analytical model for the DC current of any FET can be easily developed assuming that saturation originates from channel pinch-off. Current saturation due to carrier velocity saturation will be discussed in Sec. 5.4.6.

Consider an *n*-type FET with channel along *x*; let us denote as $-Q_{ch}(x)$ the total mobile charge (per unit surface) in the channel; in a MESFET we simply have:

$$-Q_{ch}(x) = -q \int_{channel} n(x, y)\, dy,$$

where $n(x, y)$ is the channel electron density ($n \approx N_D$ in a MESFET); the channel drift current is:

$$I_D = W \mu_{n0} Q_{ch}(x) \frac{d\phi_{ch}}{dx}, \tag{5.4}$$

in which W is the gate periphery, μ_{n0} the low-field electron mobility, while the channel electric field (directed along *x*) is:

$$\mathcal{E}_{ch} = -\frac{d\phi_{ch}(x)}{dx};$$

ϕ_{ch} is the *channel potential*. The channel mobile charge $Q_{ch}(x)$ depends on the potential difference between the gate and section *x* of the channel:

$$Q_{ch}(x) = Q_{ch}(V_{GS} - \phi_{ch}(x)).$$

Integrating (5.4) from $x = 0$ to $x = L_g$ (corresponding to $\phi_{ch} = V_S = 0$ and $\phi_{ch} = V_D - V_S = V_{DS}$), we obtain:

$$\int_0^{L_g} I_D \, dx = W\mu_{n0} \int_0^{L_g} Q_{ch}(V_{GS} - \phi_{ch}) \frac{d\phi_{ch}}{dx} dx$$

$$= W\mu_{n0} \int_0^{V_{DS}} Q_{ch}(V_{GS} - \phi_{ch}) \, d\phi_{ch},$$

i.e.,

$$I_D = \frac{W\mu_{n0}}{L_g} \int_0^{V_{DS}} Q_{ch}(V_{GS} - \phi_{ch}) \, d\phi_{ch}. \tag{5.5}$$

We now approximate the function $Q_{ch}(V_{GS} - \phi_{ch})$ in a linear way (this is indeed almost exact in some FETs, like the MOSFET), as follows:

$$|Q_{ch}| = \begin{cases} C_{ch}(V_{GS} - \phi_{ch} - V_{TH}), & V_{GS} - \phi_{ch} - V_{TH} \geq 0 \\ 0, & V_{GS} - \phi_{ch} - V_{TH} < 0, \end{cases}$$

where C_{ch} has the dimension of a capacitance per unit surface. Substituting in (5.5) and integrating, we obtain:

$$I_D = \frac{W\mu_{n0}}{L_g} \int_0^{V_{DS}} C_{ch}(V_{GS} - \phi_{ch} - V_{TH}) \, d\phi_{ch}$$

$$= \frac{W\mu_{n0}C_{ch}}{L_g}\left[(V_{GS} - V_{TH})V_{DS} - \frac{1}{2}V_{DS}^2\right]. \tag{5.6}$$

For increasing V_{DS}, (5.6) yields a current that first increases linearly (linear region), then begins to saturate (triode region) to finally reach a maximum I_{Ds} for $V_{DS} = V_{DSS}$, corresponding to channel pinchoff at $x = L_g$, and therefore to current saturation. From (5.6) the saturation voltage V_{DSS} reads:

$$V_{DSS} = V_{GS} - V_{TH};$$

for $V_{DS} > V_{DSS}$ (5.6) does not hold any longer and the current is approximately constant with value $I_{Ds} = I_D(V_{DSS})$:

$$I_{Ds} = I_D(V_{DSS}) = \frac{W\mu_{n0}C_{ch}}{2L_g}(V_{GS} - V_{TH})^2. \tag{5.7}$$

Although saturation in compound semiconductor FETs is typically due to velocity saturation, the above model can be used as a simple analytical tool to explore some of the main features of different classes of high-speed FETs.

Two important FET characteristic parameters, directly connected (if no low-frequency dispersion effects are present) to the DC curves, are the transconductance g_m and the output resistance R_{DS}:

$$g_m = \frac{\partial I_{DS}}{\partial V_{GS}} \tag{5.8}$$

$$\frac{1}{R_{DS}} = \frac{\partial I_{DS}}{\partial V_{DS}}. \tag{5.9}$$

5.4 Microwave Schottky-Gate Field-Effect Transistors

The transconductance directly expresses the ability of the FET to amplify the input voltage into the output current (in a common source configuration); the output resistance should be ideally very large, since in a loaded device it is in parallel to the output load impedance.

The DC characteristics of the heterojunction FETs of the HEMT family are qualitatively similar to those of the MESFET, with some differences. First of all, the surface semiconductor layer in the HEMT is a widegap material, with built-in voltage of the Schottky barrier greater than 1 V; it follows that it is possible to bias the gate at positive voltages larger than in a MESFET. Moreover, the charge control in HEMTs is linear above the threshold but tends to saturate for voltages around 0 V, this leads to a compression of the transcharacteristics and to a bell-shape behavior in the transconductance vs. V_{GS} (note that in the simplified quadratic model in (5.7) the transconductance increases linearly from the pinchoff to the built-in voltage). As a consequence, the optimum g_m typically occurs in a MESFET for gate voltages close to the maximum, whereas in the HEMT the maximum occurs near $V_{GS} = 0$, i.e., in many devices almost midway between the threshold voltage and the maximum gate voltage.

5.4.2 Choice of the DC Working Point

The optimal bias point of a microwave FET depends on the device purpose within an amplifier. In maximum gain amplifiers the drain bias is located suitably above the knee voltage and the gate bias corresponds to the maximum transconductance. Low-noise amplifier have a gate bias allowing for the optimum (minimum) noise figure, that corresponds to relatively low drain voltages, low current; in the HEMT this requirement is relaxed, and good noise performance are obtained for higher currents, close to the condition of maximum transconductance. In the maximum power design the bias point has to allow for the maximum excursion of the drain voltage and current: drain bias midway between the knee (beginning of the saturation region) and breakdown voltages, and gate bias corresponding to one half of the maximum current.[8] The optimal bias points in the various conditions are shown qualitatively in Fig. 5.20.

5.4.3 FET Small-Signal Model and Equivalent Circuit

The analog operation of transistors often corresponds to small-signal conditions, i.e., a small-amplitude input signal source is superimposed to the DC bias. The small-signal device response can be approximated by a linear, frequency dependent model, e.g., under the form of the admittance or scattering matrix.[9]

It is however useful to associate the small-signal behavior with a circuit model (the *small-signal equivalent circuit*), whose elements can be physically mapped into the FET structure as shown Fig. 5.21. For simplicity, we refer to a MESFET, but the approach

---

[8] This bias point corresponds to the so-called Class A power amplifier design, in other power amplifier classes the gate bias could be different, e.g., at or below threshold.

[9] Remember that FETs mainly operate in the common source configuration, i.e., the source is signal grounded, the gate acts as the input and the drain as the output.

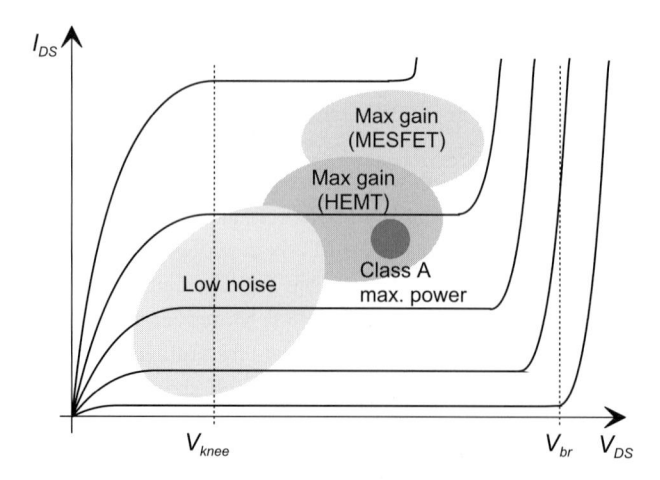

Figure 5.20 Optimum bias point for high-gain, low-noise, power amplifiers. The voltage V_{br} is the off-state breakdown voltage, corresponding to breakdown at threshold. The $I_{DS} - V_{DS}$ curves correspond to different values of the input voltage V_{GS}.

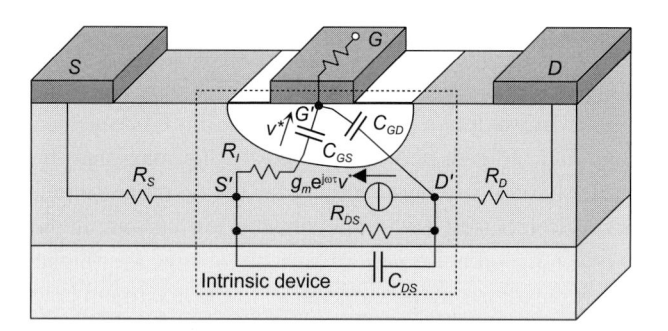

Figure 5.21 Physical mapping of a small-signal circuit into the MESFET cross section. The symbols G', D', S' denote the intrinsic gate, drain and source, respectively. The parasitic inductances of the gate, drain and source contacts are not shown.

holds for other FET families. The external parasitic resistances R_S, R_D, R_G appear here as access resistances of the source, drain and gate terminals, respectively, while the intrinsic resistance R_I is the resistance of the ohmic part of the channel. The capacitances C_{GS} and C_{GD} derive from partitioning the total capacitance associated with the gate depletion region, while C_{DS} is mostly a geometrical capacitance between drain and source. The other elements of the equivalent circuit of the intrinsic device (describing its specific transistor action) can be derived by differentiating the $I_D(V_{GS}, V_{DS})$ relation, see (5.8) and (5.9). The transconductance g_m is the device amplification between the input driving voltage (the voltage v^* applied between the intrinsic gate and the channel, across the depletion region) and the output (drain) current. The parameter τ describes an additional phase delay between the driving voltage v^* and the generated current $g_m v^*$; it is often associated with the delay that carriers accumulate drifting below the gate contact,

5.4 Microwave Schottky-Gate Field-Effect Transistors

Table 5.3 Indicative values of a Schottky-gate FET small signal parameters; the gate length is approximately between 0.25 and 1 μm and the gate with is $W \approx 0.5$ mm. For different gate widths the parameters scale according to the periphery scaling rules.

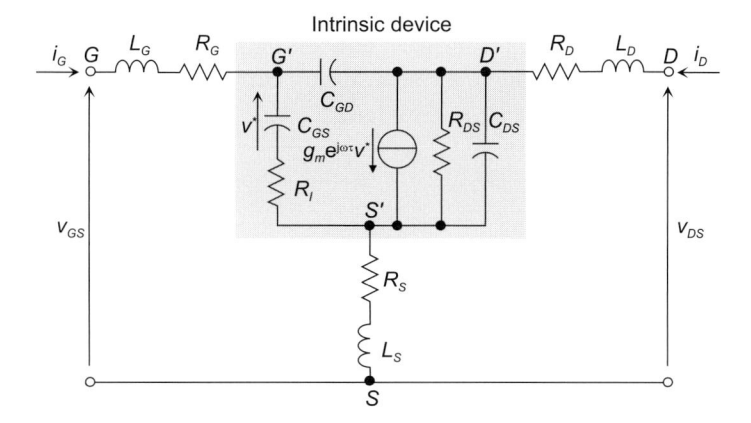

$R_G = 0.5 : 3$ Ω	$C_{DS} = 0.05 : 0.1$ pF
$R_S = 1 : 5$ Ω	$R_{DS} = 250 : 500$ Ω
$R_D = 1 : 5$ Ω	$g_m = 20 : 40$ mS
$C_{GS} = 0.15 : 0.4$ pF	$\tau = 0 : 5$ ps
$C_{GD} = 0.01 : 0.03$ pF	$R_I = 2 : 10$ Ω.

Figure 5.22 High-speed integrated FET equivalent circuit; the intrinsic part is enclosed in the box.

thus justifying the name *transit time*. Notice that the expression $g_m \exp(j\omega\tau)V^*$ is in the frequency domain (phasor notation), while in time domain it becomes $g_m v^*(t - \tau)$.

A more readable form of the equivalent circuit, in which also some external inductive parasitics (associated with the source and drain metallizations) have been added, is shown in Fig. 5.22. Table 5.3 presents some typical values of the small-signal MESFET parameters; the scaling vs. the *gate periphery* is comparatively straightforward, while reducing the *gate length* the small-signal parameters do not vary in a simple way: the gate capacitances decrease and the transconductance increases, but little can be said about the parasitic parameters.

The parameters of the equivalent circuit can be used to derive two figures of merit related to the device speed or, equivalently, to the maximum operating bandwidth:

- the *cutoff frequency f_T*;
- the *maximum oscillation frequency $f_{\max}$*.

The *cutoff frequency f_T* is the frequency at which the short-circuit current amplification (often somewhat improperly called *current gain*) of the transistor has unit magnitude. Such a parameter can be estimated from measurements, either directly, or from extrapolation. The short-circuit current gain can be evaluated from the intrinsic

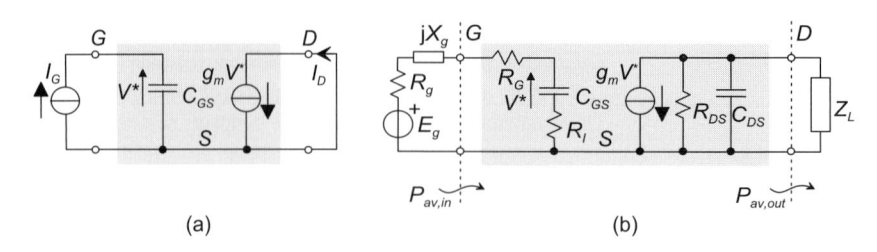

Figure 5.23 Evaluating the cutoff frequency (a) and the maximum oscillation frequency (b) of a FET.

equivalent circuit in Fig. 5.22. Taking into account that in output short-circuit conditions $v_2 = 0$ (port 1 is the gate, port 2 is the drain) and we have $i_1 = Y_{11}v_1$, $i_2 = Y_{21}v_1$, and that, by definition, the short-circuit current amplification is the element h_{21} of the hybrid matrix ($i_2 = h_{21}i_1$), we have:

$$h_{21} = \frac{Y_{21}}{Y_{11}} = \frac{g_m e^{j\omega\tau} - j\omega C_{GD}(1 + j\omega C_{GS}R_I)}{j\omega C_{GS} + j\omega C_{GD}(1 + j\omega C_{GS}R_I)}$$

$$\approx \frac{g_m e^{j\omega\tau}}{j\omega(C_{GS} + C_{GD})} \approx \frac{g_m e^{j\omega\tau}}{j\omega C_{GS}},$$

where the result in Sec. 5.4.5 was used, with reference to the intrinsic equivalent circuit in Fig. 5.27. Setting $|h_{21}| = 1$ we immediately obtain:

$$f_T = \frac{g_m}{2\pi(C_{GS} + C_{GD})} \approx \frac{g_m}{2\pi C_{GS}}. \tag{5.10}$$

The same result can be obtained by using, for the analysis, a simplified intrinsic equivalent circuit. By inspection of Fig. 5.23 (a) one has (we assume harmonic generators and phasor notation):

$$I_D = \frac{g_m}{j\omega C_{GS}} I_G,$$

i.e., imposing $|I_D/I_G| = 1$ at $\omega_T = 2\pi f_T$,

$$f_T = \frac{g_m}{2\pi C_{GS}}.$$

The *maximum oscillation frequency* is the frequency at which the maximum available power gain (MAG, the ratio between the available power at the device output and the generator available power, see Sec. 6.3) is unity. Since the MAG decreases with frequency, for $f > f_{max}$ the device becomes passive and cannot provide gain even if it is power matched at the input and output, i.e., if $Z_g = Z_{in}^*$, $Z_L = Z_{out}^*$ (the symbol Z_g was used instead of Z_G to avoid confusion with the index G denoting the gate). Again, f_{max} can be related to the parameters of a simplified equivalent circuit; some extra elements have to be added to avoid unphysical values in the output available power and in the input power. Assuming that the output is power matched, the output available power will be:

$$P_{av,out} = \frac{g_m^2 |V^*|^2}{4R_{DS}^{-1}};$$

however, if the input is power matched, the generator reactance compensates for the C_{GS} reactance, $R_g = R_G + R_I$ and the gate current is:

$$I_G = \frac{E_g}{R_g + R_G + R_I} = \frac{E_g}{2(R_G + R_I)} \rightarrow V^* = \frac{E_g}{2j\omega C_{GS}(R_G + R_I)};$$

thus, since $P_{av,in} = |E_g|^2/(4R_g)$ and $R_g = R_G + R_I$:

$$P_{av,out} = \frac{g_m^2 |E_g|^2 R_{DS}}{16\omega^2 C_{GS}^2 (R_G + R_I)^2} = \frac{g_m^2 R_{DS}}{4\omega^2 C_{GS}^2 (R_G + R_I)} P_{av,in}.$$

The available power gain $G_{av} = P_{av,out}/P_{av,in}$ corresponds to the MAG, because power matching is implemented at the input; moreover, it coincides (since also the output is matched) with the operational power gain $G_{op} = P_{out}/P_{in}$. The condition to be imposed to evaluate f_{max} is:

$$\text{MAG} = \frac{g_m^2 R_{DS}}{4(2\pi f_{max})^2 C_{GS}^2 (R_G + R_I)} = 1,$$

leading to:

$$f_{max} = f_T \sqrt{\frac{R_{DS}}{4(R_G + R_I)}}. \quad (5.11)$$

Since $R_{DS} \gg R_G + R_I$, typically by one order of magnitude, $f_{max} > f_T$ unless the input resistance is high. Eq. (5.11) also holds for bipolar transistors in common emitter configuration with the substitutions $R_G + R_I \rightarrow R_{BB'}$ (the base resistance), $R_{DS} \rightarrow R_{CE}$ (the collector–emitter output resistance), see (5.26).

Using the maximum oscillation frequency we can express the MAG as:

$$\text{MAG} = \left(\frac{f_{max}}{f}\right)^2, \quad (5.12)$$

i.e., the maximum available gain decreases as the square of the operating frequency. Notice that this behavior is in principle an approximation, since the device considered is unilateral (see Sec. 6.6.4) and external parasitics have been neglected. Nevertheless the MAG decrease with the square of the operating frequency is also experimentally observed both in field-effect and bipolar transistors up to the millimeter wave range, where a steeper decrease occurs, approximately as the fourth power of frequency, see Fig. 5.24. A more accurate expression for f_{max} is:

$$f_{max} = \frac{f_T}{2} \sqrt{\frac{R_{DS}}{R_I + R_G + R_S + 2\pi f_T R_{DS} R_G C_{GD}}}, \quad (5.13)$$

where the output and source resistances and feedback capacitance are taken into account.

Example 5.1 A microwave FET has the following parameters: $C_{GS} = 0.071$ pF; $R_I = 2.69$ Ω; $C_{GD} = 0.001$ pF; $g_m = 0.0152$ S; $\tau = 1.25$ ps; $C_{DS} = 0.025$ pF; $R_{DS} = 556$ Ω; $L_G = 0.05$ pH; $L_D = 0.01$ pH; $L_S = 0.01$ pH; $R_G = 4.46$ Ω; $R_S = 4.55$ Ω; $R_D = 6.7$

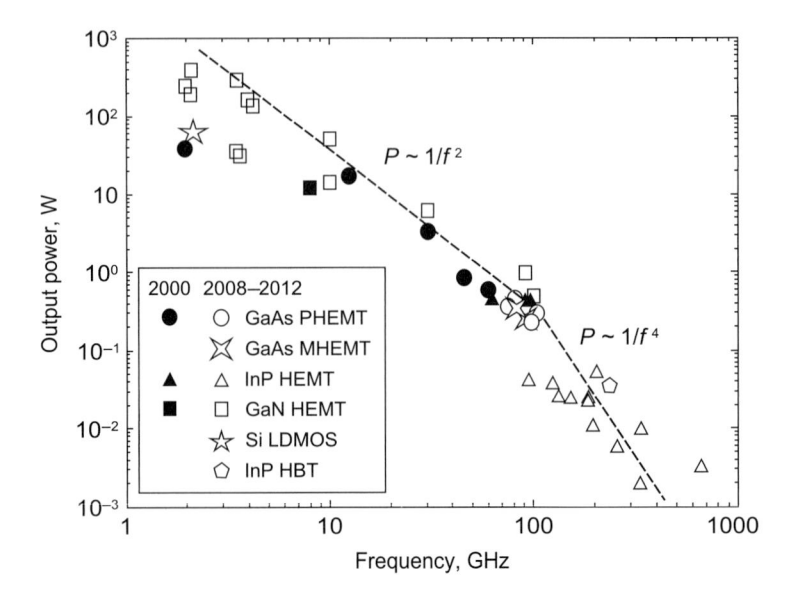

Figure 5.24 Output power vs. frequency for a number of transistor technologies. Adapted from [21] (Copyright ©2008 GAAS Association. All rights reserved. Reused with permission) with additional data from [29, Fig. 32] (Copyright ©2011 IEEE. All rights reserved. Reused with permission).

Ω. Estimate f_T and f_{max} (a) making use of the approximate expressions (5.10), (5.11), (5.13) and (b) exactly from the frequency behavior of the MAG.

Solution
From the approximate expressions we obtain $f_T = 34$ GHz, $f_{max} = 150$ GHz from (5.11), 97 GHz from (5.13). The device gains are shown in Fig. 5.25; the approximation for the cutoff frequency is accurate enough, while (5.11) overestimates the maximum oscillation frequency, that is reasonably foreseen by (5.13). The main cause for inaccuracy arises from neglecting the source parasitic resistance. Notice that the MAG (continuous line) is only defined in the region where the device is unconditionally stable and therefore the Linville parameter K defined in (6.52) is larger than unity, see Sec. 6.4.

Example 5.2 Evaluate the maximum oscillation frequency of the discrete microwave FET having the scattering parameters in Table 5.4.

Solution
The maximum and unilateral gains (MUG and MAG) from the scattering parameters are shown, together with the stability coefficient K, in Fig. 5.26. Extrapolating the theoretical slope of 6dB/octave of the MAG we obtain a maximum oscillation frequency around 45 GHz. Once again the device is unconditionally stable at high frequency, but potentially unstable at low frequency.

5.4 Microwave Schottky-Gate Field-Effect Transistors

Table 5.4 Measured scattering parameters between 2 and 22 GHz of the device considered in Example 5.2. The phase is in degrees.

f, GHz	$|S_{11}|$	ϕ_{11}	$|S_{21}|$	ϕ_{21}	$|S_{12}|$	ϕ_{12}	$|S_{22}|$	ϕ_{22}
2	.95	−26	3.57	157	.04	76	.66	−14
3	.93	−40	3.53	147	.05	69	.65	−20
4	.89	−52	3.23	136	.06	62	.63	−26
5	.86	−63	3.08	127	.07	56	.60	−31
6	.83	−71	2.98	121	.08	55	.59	−36
7	.80	−80	2.70	113	.09	50	.58	−40
8	.78	−87	2.58	108	.09	47	.57	−42
9	.75	−95	2.39	101	.09	44	.56	−46
10	.73	−102	2.27	96	.10	42	.54	−49
11	.68	−109	2.17	88	.10	41	.51	−52
12	.70	−116	2.11	84	.10	39	.48	−56
13	.70	−122	1.94	78	.10	37	.49	−64
14	.69	−126	1.79	77	.11	40	.51	−70
15	.68	−129	1.71	68	.11	36	.52	−74
16	.67	−130	1.62	68	.12	36	.53	−75
17	.66	−134	1.53	61	.12	38	.53	−76
18	.65	−136	1.51	57	.11	40	.54	−77
19	.64	−138	1.49	56	.11	40	.54	−78
22	.60	−144	1.30	40	.14	40	.56	−85

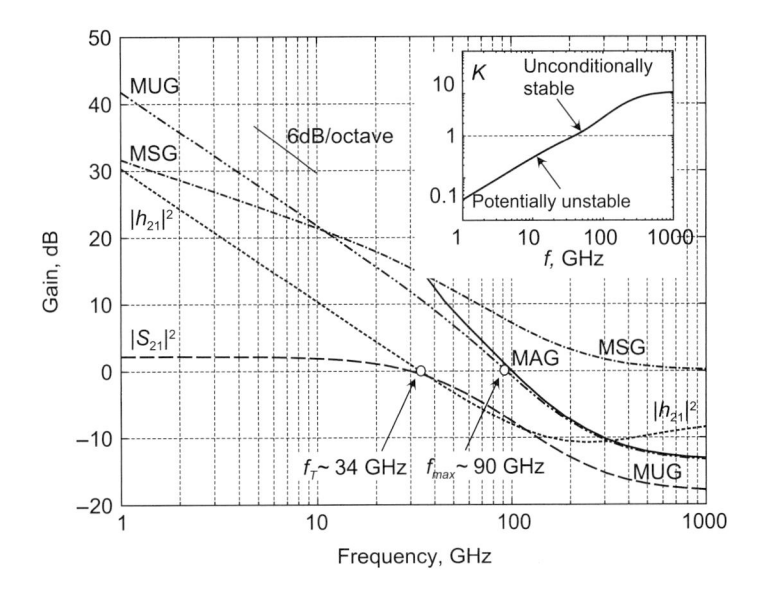

Figure 5.25 Gains and stability coefficient from Example 5.1.

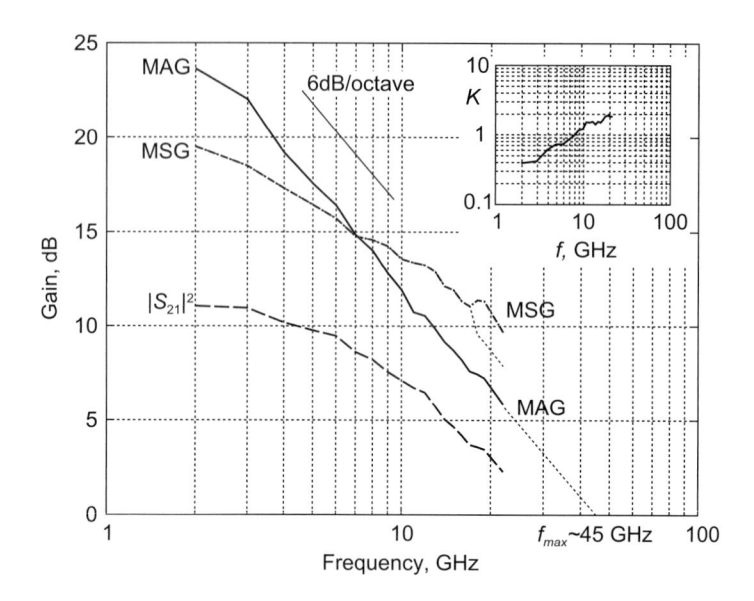

Figure 5.26 Gains and stability coefficient of the device in Example 5.2.

5.4.4 Scaling the Small-Frequency Parameters vs. the Gate Periphery

In the design of integrated circuits, devices can be (up to a certain extent) scaled with respect to the gate width or periphery W. From a circuit standpoint, increasing the gate periphery mainly amounts to making the parallel of several unit cells. This implies that the intrinsic parameters are scaled according to W if they are admittance-like, according to $1/W$ if they are impedance-like:

$$P_Y(W) = P_Y(W_0) \frac{W}{W_0}$$
$$P_Z(W) = P_Z(W_0) \frac{W_0}{W},$$

where P_Y and P_Z are admittance and impedance-like parameters, respectively. Concerning the parasitic parameters, some of them are in fact in series with the unit cell and therefore (P_Z case) they scale with W; an example is the gate resistance, mainly related to the gate metallization. Neglecting the inductive parameters we therefore have:

$$R_G \approx R'_G W, \quad R_S = R'_S/W, \quad R_D = R'_D/W, \quad R_{DS} = R'_{DS}/W, \quad R_I = R'_I/W$$
$$C_{GS} = C'_{GS} W, \quad C_{GD} = C'_{GD} W, \quad C_{DS} = C'_{DS} W, \quad g_m = g'_m W, \quad \tau = \tau',$$

where the primed parameters are per-unit-length. From the above formulae we see that, as a first approximation, the cutoff frequency is independent from the device periphery, while the maximum frequency of oscillation tends to deteriorate for large peripheries due to the increase of gate resistance. However this is just an approximation: in large devices, many gates are put in parallel, as propagation effects severely limit the length of each gate finger to a small fraction of the operating wavelength. Similarly, large powers cannot be obtained by increasing without limit the gate periphery putting several devices

5.4.5 Frequency Behavior of the Scattering Parameters

The purpose of the present section is to describe the *typical* behavior of a microwave FET S parameters when parametrically vs. frequency plotted on the Smith chart or on a polar plot. For simplicity we confine the treatment to the intrinsic device, see Fig. 5.27. This is a π section whose admittance parameters can be obtained by inspection (the derivation is left to the reader as an exercise):

$$Y = \begin{pmatrix} \dfrac{j\omega C_{GS}}{1+j\omega C_{GS}R_I} + j\omega C_{GD} & -j\omega C_{GD} \\ \dfrac{g_m e^{j\omega\tau}}{1+j\omega C_{GS}R_I} - j\omega C_{GD} & j\omega(C_{GD}+C_{DS}) + \dfrac{1}{R_{DS}} \end{pmatrix}. \quad (5.14)$$

The elements of the normalized admittance matrix are therefore:

$$y_{11} = Y_{11}R_0 = \left(j\omega\dfrac{C_{GS}}{1+j\omega C_{GS}R_I} + j\omega C_{GD}\right)R_0$$

$$y_{12} = Y_{12}R_0 = -j\omega C_{GD}R_0$$

$$y_{21} = Y_{21}R_0 = \left(\dfrac{g_m e^{j\omega\tau}}{1+j\omega C_{GS}R_I} - j\omega C_{GD}\right)R_0$$

$$y_{22} = Y_{22}R_0 = \left(j\omega(C_{GD}+C_{DS}) + \dfrac{1}{R_{DS}}\right)R_0.$$

The elements of the S-matrix can be directly derived by applying the conversion formulae in see Table 3.1.

The behavior of the S-matrix element is simple enough and can be easily understood by approximating the device with a unidirectional one since $C_{GD} \simeq 0$, so that $y_{12} \simeq 0$. The unidirectional admittance matrix elements are:

$$y_{11}^U = \dfrac{j\omega C_{GS}R_0}{1+j\omega C_{GS}R_I} \quad y_{21}^U = \dfrac{g_m R_0 e^{j\omega\tau}}{1+j\omega C_{GS}R_I}$$

$$y_{12}^U = 0 \quad y_{22}^U = \left(j\omega C_{DS} + \dfrac{1}{R_{DS}}\right)R_0,$$

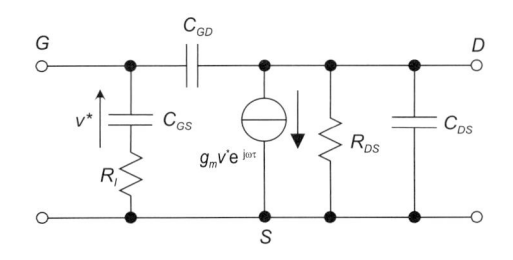

Figure 5.27 Intrinsic FET equivalent circuit used in estimating the scattering parameters.

and the unidirectional S-matrix elements:

$$S_{11}^U = -\frac{y_{11}^U - 1}{y_{11}^U + 1} = \frac{1 + j\omega C_{GS}(R_I - R_0)}{1 + j\omega C_{GS}(R_0 + R_I)}$$

$$S_{12}^U = 0$$

$$S_{21}^U = -\frac{2y_{21}^U}{(y_{11}^U + 1)(y_{22}^U + 1)} = -\frac{2R' g_m e^{j\omega\tau}}{(1 + j\omega C_{DS}R')[1 + j\omega C_{GS}(R_0 + R_I)]}$$

$$S_{22}^U = -\frac{y_{22}^U - 1}{y_{22}^U + 1} = -\frac{\dfrac{R_0 - R_{DS}}{R_0 + R_{DS}} + j\omega C_{DS}R'}{1 + j\omega C_{DS}R'},$$

where:

$$R' = \frac{R_0 R_{DS}}{R_0 + R_{DS}}.$$

The limits of the S-matrix elements at low frequency ($f \to 0$) are:

$$\lim_{\omega \to 0} S_{11} = 1 \qquad \lim_{\omega \to 0} S_{21} = -\frac{2g_m R_0 R_{DS}}{R_{DS} + R_0}$$

$$\lim_{\omega \to 0} S_{12} = 0 \qquad \lim_{\omega \to 0} S_{22} = \frac{R_{DS} - R_0}{R_{DS} + R_0},$$

while at high frequency ($f \to \infty$) we have:

$$\lim_{\omega \to \infty} S_{11} = \frac{R_I - R_0}{R_I + R_0} \qquad \lim_{\omega \to \infty} S_{21} = 0$$

$$\lim_{\omega \to \infty} S_{12} = 0 \qquad \lim_{\omega \to \infty} S_{22} = -1.$$

The behavior of the S-matrix elements of the device in Example 5.3 is shown in Fig. 5.28; both complete and unidirectional parameters are shown. S_{21} has two poles, that dominant (corresponding to the smallest characteristic frequency) is associated with the frequency:

$$f_{T1} = \frac{1}{2\pi C_{GS}(R_0 + R_I)}.$$

Example 5.3 A microwave FET has the following parameters: $C_{GS} = 0.3$ pF; $R_I = 2.5\ \Omega$; $C_{GD} = 0.02$ pF; $g_m = 40$ mS; $\tau = 0.5$ ps; $C_{DS} = 0.05$ pF; $R_{DS} = 600\ \Omega$ (parasitic inductances are neglected). Sketch the S parameters vs. $R_0 = 50\ \Omega$, for both the complete and the unilateral device.

Solution
The result is shown in Fig. 5.28; note that the S_{12} element is small, though being different from zero. The output reflectance is the parameter mostly affected by the unilateral approximation.

Example 5.4 Derive an approximate expression for the Maximum Stable Gain (MSG) of an intrinsic FET, see Sec. 6.6.2, valid at low frequency.

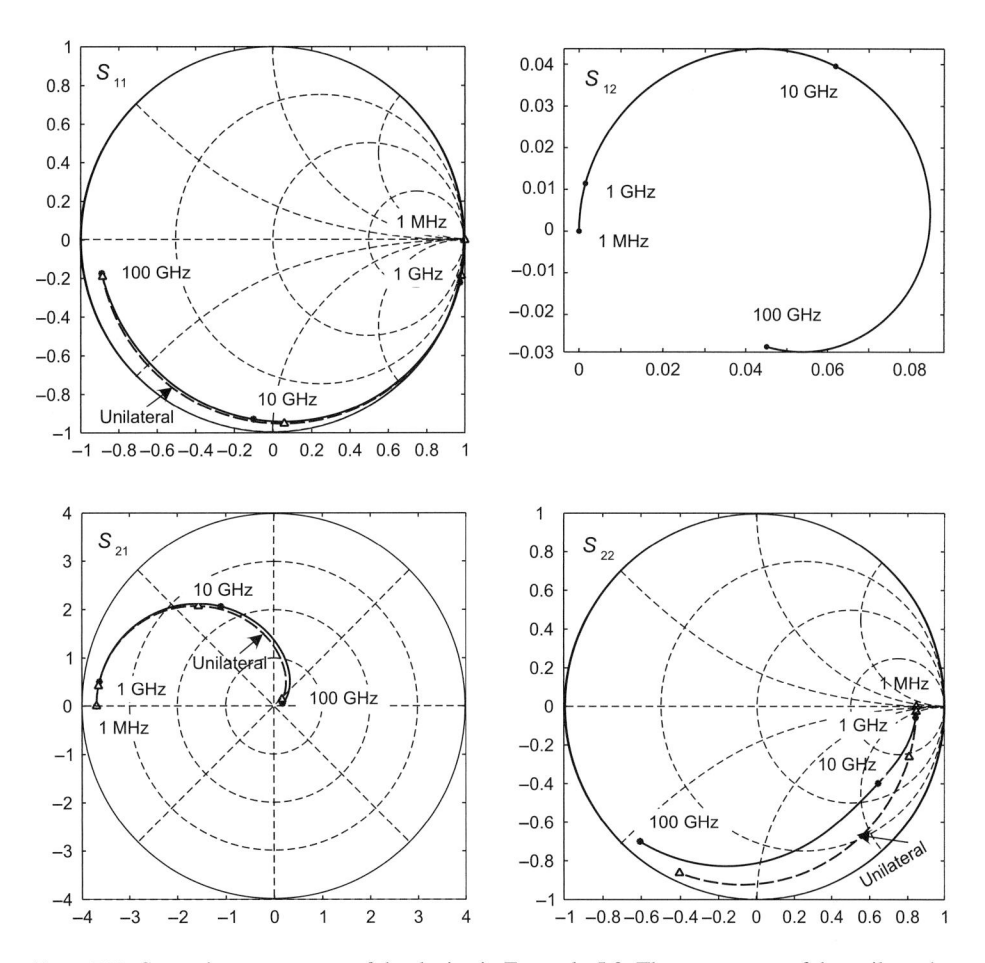

Figure 5.28 Scattering parameters of the device in Example 5.3. The parameters of the unilateral device are shown with a dashed line.

Solution
The MSG, corresponding to the maximum available gain of a device in the limit of instability ($K = 1$), is defined from (6.81) as:

$$\text{MSG} = \left|\frac{S_{21}}{S_{12}}\right| \equiv \left|\frac{Y_{21}}{Y_{12}}\right|,$$

where the second equality directly derives from the conversion formulae between the scattering and the admittance parameters, see Table 3.1. From (5.14) we have, at low frequency:

$$\text{MSG} = \left|\frac{Y_{21}}{Y_{12}}\right| \approx \frac{g_m}{\omega C_{GD}} = \frac{f_T}{f}\frac{C_{GS}}{C_{GD}}.$$

Contrarily to the MAG, decreasing like f^2, the MSG linearly decreases with frequency (at least within the approximations made). From (5.14), we also see that for $f \to \infty$ the MSG tends to unity (i.e., the device becomes passive and therefore reciprocal).

5.4.6 High-Speed Compound Semiconductor FETs: The HEMT Family

High-speed field-effect transistors (FET) today are implemented in a number of technologies based both on Si and on compound semiconductors. Deep scaling down to nanometer gate lengths of Si-based MOSFETs has allowed for transistors with cut-off frequencies in the millimeter wave range and comparable maximum oscillation frequencies. Heterostructure FETs are currently manufactured, based on several compound semiconductors, including GaAs, InP and also GaN. Representative examples are the GaAs- or InP-based Pseudomorphic High Electron Mobility Transistor (PHEMT), the GaAs-based metamorphic HEMT (MHEMT) and, for high-power or high-voltage applications, the GaN HEMT.

The basic building block of heterojunction FETs is the so-called *modulation doped heterojunction*. In its original form, this consists of a doped (uniformly or δ-doped) *n*-type widegap layer grown on top of an intrinsic (undoped) narrowgap layer. Fig. 5.29 shows an AlGaAs/GaAs modulation-doped structure; the conduction band and bandgap discontinuities are not to scale (the discontinuity as shown is larger than in reality). In equilibrium conditions, the donors in the doped layer (also called *supply layer*) are ionized and transfer their electrons into the conduction band potential well, located in correspondence of the bandgap and material discontinuity. A thin, confined conducting channel is thus generated in the interface well, in a way that is not dissimilar from the MOS surface inversion channel. Moreover, the electrons in the potential well suffer little or no impurity scattering, since the narrowgap material is undoped; this leads to an improvement in the channel electron mobility with respect to FETs (like the MESFET) that use a doped channel.[10] A last important advantage of the modulation doped

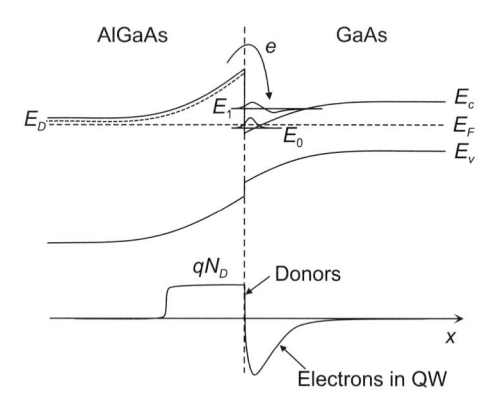

Figure 5.29 Example of band diagram for the AlGaAs/GaAs modulation-doped structure with triangular potential profile QW; the charge distribution (ionized donors in the supply layer, free electrons in the QW) is shown below. The envelope wavefunctions of the two first energy levels of the QW are also schematically shown. E_D is the shallow donor energy level in the AlGaAs supply layer.

---

[10] Huge mobility increases that led to the (historical) name of the HEMT, can however be detected only at cryogenic temperature, where phonon scattering becomes ineffective and the mobility is dominated by

5.4 Microwave Schottky-Gate Field-Effect Transistors

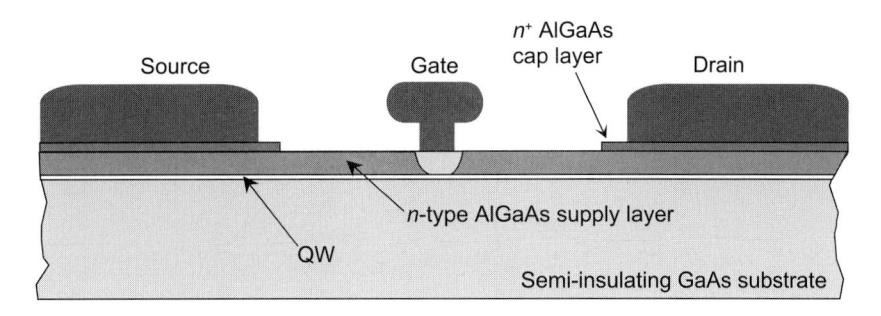

Figure 5.30 Cross section of the conventional AlGaAs/GaAs HEMT.

approach is the possibility to make the doped widegap supply layer thin, but highly doped, thus reducing the distance between the controlling gate and the channel. Since electrons in the channel are confined in the vertical direction, the modulation doping heterojunction typically is a QW for the channel electrons, which form a charge sheet, i.e., a *two-dimensional electron gas* (2DEG).

For the sake of definiteness we will refer to the HEMT based on the AlGaAs/GaAs modulation doped structure in Fig. 5.29 as the *conventional* HEMT. This is a legacy device in all material systems (with the exception of the AlGaN/GaN version) but is useful to introduce the subject. The conventional HEMT, shown in Fig 5.30, is a Schottky gate FET, where a reverse bias applied to the gate electrostatically modulates the 2DEG. The supply layer is AlGaAs (widegap), while the substrate is semi-insulating GaAs (narrowgap). The $Al_xGa_{1-x}As$ layer has an Al fraction x between 23 percent and 30 percent; larger Al fractions would be desirable to increase the conduction band discontinuity ΔE_c and the channel confinement; unfortunately, for $x > 30$ percent the AlGaAs supply layer develops a trap level (the so-called *DX centers*) which limits the effectiveness of the doping. The supply layer is anyway lattice matched to the GaAs substrate for any Al fraction x, see Fig. 5.31. In order to decrease the surface impurity scattering from the donors in the supply layer, a thin undoped AlGaAs *spacer layer* is epitaxially grown between the GaAs substrate and the doped supply layer.

State-of-the art HEMTs are based on a double heterostructure, wherein a narrowgap material is sandwiched between the widegap substrate and the widegap supply layer. The resulting rectangular (rather than triangular) conduction band QW has superior confinement properties and may ensure also better transport performances than in conventional HEMT. According to whether the narrowgap layer is lattice-matched or in a strained (tensile or compressive) condition we have the LMHEMT (Lattice Matched HEMT) or the PHEMT (Pseudomorphic HEMT).

Several material systems and modulation doping techniques in the supply layer (including δ doping instead of uniform doping) can be used in PHEMTs. Representative examples are:

> impurity scattering; at ambient temperature phonon scattering prevails and the mobility advantage is less significant.

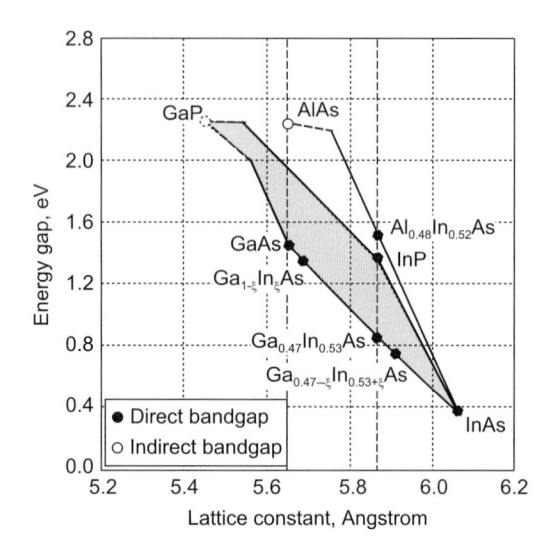

Figure 5.31 Details of the energy-gap/lattice constant diagram for AlGaAs, InGaAsP, and InAlAs alloys.

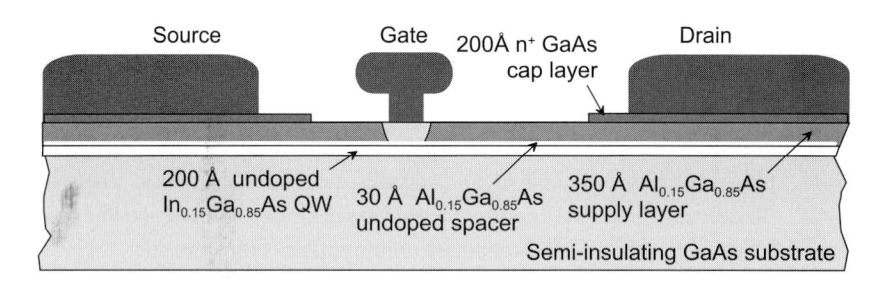

Figure 5.32 Cross section of asymmetrical (power) GaAs-based PHEMT.

- The GaAs-based PHEMT, where the supply layer is doped AlGaAs, the channel is In$_{1-x}$Ga$_x$As with low indium content (see the black dot close to GaAs in the energy-gap lattice-constant diagram in Fig. 5.31), and the substrate is GaAs. Due to the strained (pseudomorphic) channel, the conduction band discontinuity is increased, thus improving the confinement and sheet carrier density in the conducting channel. The potential barrier between the channel and the substrate is, however, low. An example of GaAs-based PHEMT structure is shown in Fig. 5.32; notice the asymmetric placement of the gate, typical of power devices and aimed at decreasing the maximum field in the drain-gate region, i.e., at increasing the breakdown voltage.
- The InP-based lattice-matched or pseudomorphic HEMT, for which several choices exist. Taking into account that Al$_{0.48}$In$_{0.52}$As and Ga$_{0.47}$In$_{0.53}$As are lattice matched to InP, a possible lattice-matched structure with an Al$_{0.48}$In$_{0.52}$As supply layer (doped), a Ga$_{0.47}$In$_{0.53}$As channel (undoped), and an InP substrate (undoped) is shown in Fig. 5.33. Implementing the channel through a pseudomorphic layer of

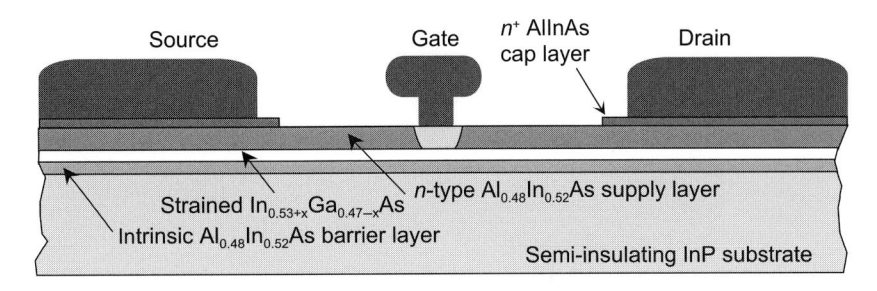

Figure 5.33 Cross section of InP-based PHEMT.

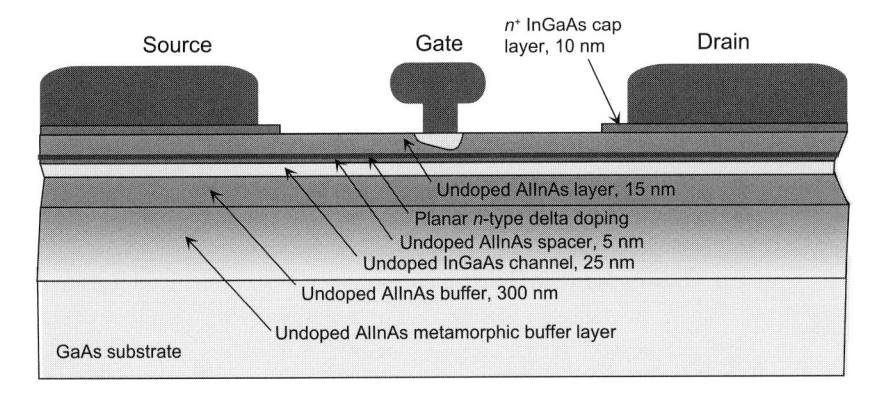

Figure 5.34 Cross section of GaAs-based metamorphic HEMTs. The lattice-matched layer structure is derived from [30, Fig. 10] (Copyright ©2001 Springer. All rights reserved. Reused with permission).

$Ga_{0.47-x}In_{0.53+x}As$, x small, we increase the bandgap discontinuity both towards the supply layer and the substrate, thus obtaining a PHEMT with improved performances.
- The GaAs-based metamorphic HEMT: an alternative to the direct growth on InP substrates, which would considerably decrease the device cost, has been proposed under the name of *metamorphic* HEMT (MHEMT), see the review paper [30]. Metamorphic devices are grown on substrates mismatched with respect to the active region (GaAs or even Si, see [31]); properly designed buffer layers have to be interposed. Mismatch defects are distributed in the buffer rather than concentrated on an abrupt interface and, moreover, the buffer technology can be optimized in order to avoid the migration of defects into the device active region. As a simple example, an HEMT structure that is typically grown on an InP substrate can be also grown on a GaAs substrate topped by a graded $Al_{1-x}In_xAs$ graded buffer that leads from the GaAs substrate (with lattice matching to AlAs, $x = 0$) to the value of the InP lattice constant, corresponding to the In fraction $x = 0.52$. The channel is made of narrow-gap $In_{0.53}Ga_{0.47}As$. A sketch of a possible δ-doped MHEMT structure is reported in Fig. 5.34. Notice that grading down to the InP lattice constant is not necessarily an optimum solution, since in principle a larger conduction band discontinuity and a better channel confinement can be achieved by using lattice-matched AlInAs (widegap) and InGaAs (narrowgap) layers

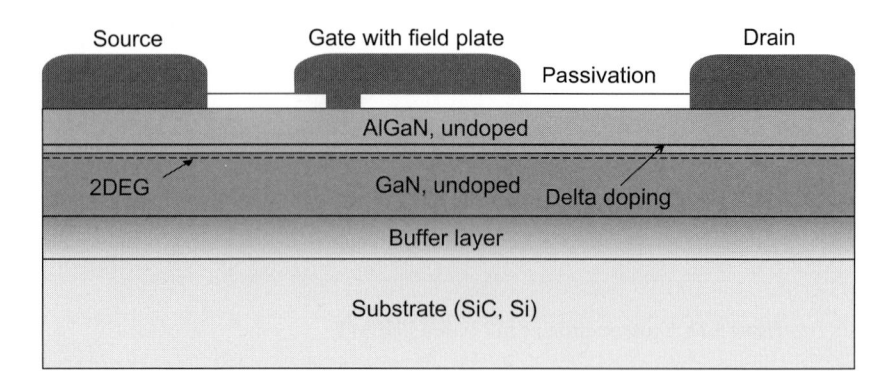

Figure 5.35 Cross section of GaN-based HEMT.

with proper composition, having a lattice constant larger than for InP. Despite the presence of a graded buffer introducing distributed defects, the recent MHEMT technology has shown good reliability and performances comparable to lattice-matched or pseudomorphic HEMTs.

- The GaN-based HEMT has a conventional HEMT structure made of a widegap supply layer of AlGaN grown on an undoped GaN substrate. A peculiar feature of the AlGaN/GaN system is the presence of a piezoelectrically induced electron charge in the interface QW, already without any supply layer doping. The AlGaN/GaN HEMT therefore has excellent properties in terms of channel charge, which can be further improved by intentional doping. A schematic cross section of a GaN HEMT is shown in Fig. 5.35; notice the gate field plate, aimed at further increasing the breakdown voltage thanks to a more favorable distribution of the electric field on the drain side of the gate. GaN HEMTs have reached record power densities per unit gate periphery when compared to other semiconductor technologies. While the maximum power density of III-V FETs typically is of the order of 1 W/mm, GaN-based devices can reach, up to millimeter waves, power densities one order of magnitude larger, i.e., around 10 W/mm.[11]

InP-based PHEMTs probably are the devices of this class offering the best performances in terms of high cutoff and maximum oscillation frequencies, with record values well into THz range. GaAs-based PHEMTs offer some advantages in terms of breakdown voltage (which can be pushed in the range 10–20 V) but are confined to applications below 50–60 GHz. However, the superior development of the GaAs technology and lower cost with respect to InP makes this the device of choice for many applications up to the millimeter wave range. In what follows we will develop an approximate physics-based analytical quasi-static model for the PHEMT channel charge control and current, also accounting for the carriers' velocity saturation.

---

[11] GaN-based HEMT typically are depletion (normally on) devices. The development of normally off (enhancement) devices, that is important in power switching applications, is underway, although a really convenient solution has not been found so far.

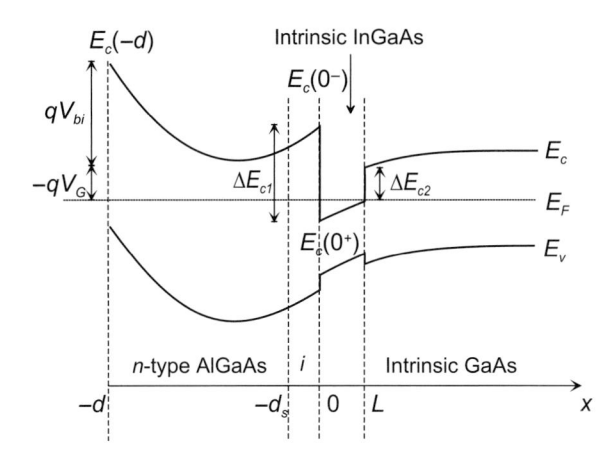

Figure 5.36 GaAs-based PHEMT band structure in equilibrium.

The carrier density in the interface potential well of a modulation doped structure can be evaluated taking into account the modified density of states of the QW. The complete analysis will be omitted here; we just remark that raising the Fermi level E_F with respect to $E_c(0)$ causes an increase of the E_c slope in 0^+ and therefore an increase of the QW surface electric field in 0^+, $\mathcal{E}_s$. However, from the Gauss law, see, e.g., [32], this is related to the total surface charge density n_s in the potential well:

$$\mathcal{E}_s = \frac{q}{\epsilon} n_s. \tag{5.15}$$

In general, n_s depends on $E_F - E_c(0)$ according to an implicit law, that, in a suitable range of energies and surface charge densities, can be linearly approximated as:

$$E_F - E_c(0) \approx qan_s, \tag{5.16}$$

where a is a fitting parameter of the order of 10^{-17} eV m^2; notice that (5.16) is expressed in MKS units.

The sheet carrier density n_s can be modulated by varying the bias of a reverse-biased Schottky junction connected to the supply layer. Be d the thickness of the supply layer, d_s the thickness of the spacer layer (meant to screen the channel carriers from surface donor scattering); if the Schottky barrier reverse bias is increased, the $E_c(0^+)$ slope decreases, leading to a decrease of n_s; conversely, if the Schottky negative bias is reduced, the slope increases, thus increasing the sheet carrier density n_s. The threshold voltage V_{TH} corresponds to the (flatband) condition in which $E_c(0^+)$, and therefore the channel charge, vanishes; the increase in n_s is, however, limited by the fact that, for positive Schottky contact (gate) bias, and well before the Schottky gate diode goes into direct conduction, the supply layer ultimately ceases to be depleted, thus decoupling the channel population of sheet density n_s from the gate control. Thus, n_s vanishes at threshold, then increases to saturate at a value n_{ss}; the behavior is qualitatively shown in Fig. 5.37.

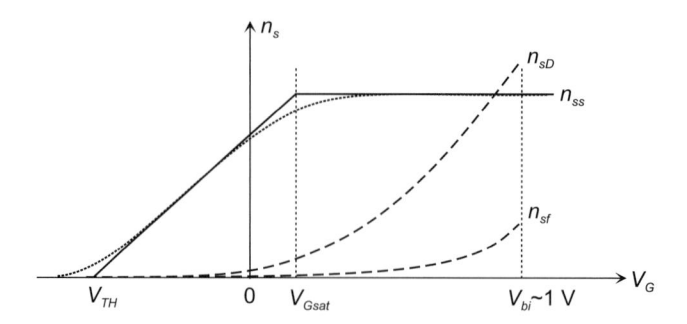

Figure 5.37 Behavior of sheet carrier density n_s for a PHEMT as a function of the gate bias. V_{TH} is the threshold voltage, V_{Gsat} is the gate voltage at which the sheet carrier density saturates at n_{ss}, V_{bi} is the Schottky gate barrier built-in voltage. The sheet carrier densities n_{sD} and n_{sf} refer to the donor trapped carriers and free carriers in the supply layer, respectively.

To relate the gate bias V_G to the sheet electron concentration n_s, the Poisson equation can be solved in the depleted supply layer and spacer. Taking into account the jump condition in 0:

$$E_c(0^-) = E_c(0^+) + \Delta E_c,$$

and the boundary condition in $-d$:

$$E_c(-d) = qV_{bi} - qV_G + E_F,$$

we have, from the solution of Poisson's equation:

$$qV_{bi} - qV_G + E_F = \frac{q^2 N_D}{2\epsilon}(d - d_s)^2 - q\mathcal{E}_s d + E_c(0^+) + \Delta E_c,$$

where N_D is the supply layer doping, ϵ the supply layer permittivity. Solving by the surface electric field we obtain:

$$\mathcal{E}_s = \frac{qN_D}{2d\epsilon}(d - d_s)^2 - \frac{V_{bi}}{d} + \frac{V_G}{d} + \frac{E_c(0^+) - E_F + \Delta E_c}{qd}. \quad (5.17)$$

Taking into account that, for $V_G = V_{TH}$, $n_s \propto \mathcal{E}_s \propto E_F - E_c(0^+) \approx 0$ by definition, we can define the threshold voltage V_{TH} as:

$$V_{TH} = -\frac{qN_D}{2\epsilon}(d - d_s)^2 + V_{bi} - \frac{\Delta E_c}{q}. \quad (5.18)$$

Finally, since from (5.15) $\epsilon \mathcal{E}_s = qn_s$ and, from (5.16):

$$n_s \approx \frac{E_F - E_c(0^+)}{aq},$$

we obtain from (5.17) and (5.18) the result:

$$\frac{1}{\epsilon}\left(d + \frac{\epsilon a}{q}\right) qn_s = (V_G - V_{TH}),$$

i.e., defining the equivalent thickness of the 2DEG, Δd, as:

$$\Delta d = \frac{\epsilon a}{q} \approx \frac{13 \cdot 8.86 \times 10^{-12} \cdot 10^{-17}}{1.69 \times 10^{-19}} \approx 7 \text{ nm},$$

we evaluate the surface mobile charge associated with the 2DEG, $Q_s = qn_s$, as:

$$Q_s = \frac{\epsilon}{d + \Delta d}(V_G - V_{TH}) = C_{eq}(V_G - V_{TH}), \tag{5.19}$$

where C_{eq} is the equivalent 2DEG capacitance. Eq. (5.19) provides a simple tool to analyze the PHEMT through the linear charge control approximation. Eq. (5.19) ceases to be valid when the 2DEG saturates to n_{ss}; the analysis, based again on the solution of the Poisson equation, leads to the result:

$$n_{ss} \approx \sqrt{\frac{\Delta E_c}{q} \frac{2N_D \epsilon}{q}},$$

which shows that the saturation density increases with increasing supply layer doping, but also with the conduction band discontinuity between the supply layer and the narrowgap channel. The saturation gate voltage can be derived, assuming $Q_s = qn_{ss}$ and $d + \Delta d \approx d$ in (5.19), as:

$$V_{Gsat} \approx V_{TH} + \frac{dq}{\epsilon} n_{ss}. \tag{5.20}$$

Starting from the charge control relation (5.19) and assuming, more realistically, that the drain current saturation is due to *velocity saturation* rather than to channel pinchoff, we can develop an approximate model for the drain current. To this purpose, we assume a simplified piecewise velocity-field curve for the electrons, in which abrupt saturation occurs for the threshold field $\mathcal{E}_{th} \approx v_{n,\text{sat}}/\mu_{n0}$, where μ_{n0} is the low-field mobility. The drain current can be obtained by taking into account that velocity saturation occurs at the drain edge of the gate ($x = L_g$) for $V_{DS} = V_{DSSv}$. In the velocity saturated region the current is:

$$I_{Ds} = Wv_{n,\text{sat}} q n_s = Wv_{n,\text{sat}} \frac{\epsilon}{d + \Delta d}[V_{GS} - V_{DSSv} - V_{TH}],$$

where W is the gate periphery, since $V_{GS} - V_{DSSv}$ is the potential difference between gate and channel at $x = L_g$ in velocity saturation conditions. As the current must be continuous from the ohmic to the velocity saturated part of the channel, we have:

$$Wv_{n,\text{sat}} \frac{\epsilon}{d + \Delta d}[V_{GS} - V_{DSSv} - V_{TH}]$$
$$= \frac{W\mu_{n0}}{L_g} \frac{\epsilon}{d + \Delta d} \left[(V_{GS} - V_{TH})V_{DSSv} - \frac{1}{2}V_{DSSv}^2\right],$$

where we used (5.6). We therefore obtain:

$$V_{DSSv} = (V_{GS} - V_{TH} + L_g\mathcal{E}_s) - \sqrt{(V_{GS} - V_{TH})^2 + (L_g\mathcal{E}_s)^2},$$

where $\mathcal{E}_s = v_{n,\text{sat}}/\mu_{n0}$, and the saturation current I_{Ds} due to velocity saturation:

$$I_{Ds} = Wv_{n,\text{sat}} \frac{\epsilon}{d + \Delta d}\left[\sqrt{(V_{GS} - V_{TH})^2 + (L_g\mathcal{E}_s)^2} - L_g\mathcal{E}_s\right]. \tag{5.21}$$

Eq. (5.21) defines the saturation transcharacteristics of the PHEMT; by differentiation, we obtain the device transconductance:

$$g_m = \frac{W\mu_{n0}}{L_g} \frac{\epsilon}{d+\Delta d} \frac{V_{GS} - V_{TH}}{\sqrt{\left(\frac{V_{GS} - V_{TH}}{L_g \mathcal{E}_s}\right)^2 + 1}}. \tag{5.22}$$

For short gate devices we can approximate the maximum transconductance by letting $L_g \to 0$ in (5.22), as:

$$g_m \approx \frac{W\mu_{n0}}{L_g} \frac{\epsilon}{d+\Delta d} L_g \mathcal{E}_s \approx W v_{n,\text{sat}} \frac{\epsilon}{d+\Delta d}, \tag{5.23}$$

and the maximum (angular) cutoff frequency approximately is:

$$2\pi f_T = \frac{g_m}{C_{GS}} \approx W v_{n,\text{sat}} \frac{\epsilon}{d+\Delta d} \times \frac{d+\Delta d}{WL_g \epsilon} = \frac{v_{n,\text{sat}}}{L_g} = \frac{1}{\tau_t}.$$

The expression of g_m in (5.23) suggests a major advantage of PHEMTs over MESFETs and other FETs, i.e., the fact that g_m can be increased by reducing the supply layer thickness d (and increasing its doping) without compromising the current (as it would happen in a MESFET due to the thinner active layer), nor introducing extra impurity scattering due to the increase in doping. Notice that to maintain control of the gate over the channel the channel-gate distance should be much lower than the gate length. This leads to the approximate condition $L_g d > 10$, where L_g is the gate length and d the gate-channel distance. The cutoff frequency can be readily interpreted in terms of the transit time of electrons below the gate, τ_t.

5.5 The RF MOSFET

The Metal Oxide Semiconductor FET (MOSFET) probably is the semiconductor device having experienced the most spectacular development since it was first introduced in the sixties. Nanometer-scale devices developed for mainstream digital applications can be, at the same time, high-performance analog RF devices, see [8], although process adjustments are needed to allow for passive RF devices, in particular inductors, for which ad-hoc multilevel low-resistance metallization processes have been developed together with proper insulation techniques from the Si substrate to minimize substrate losses. A synthesis of the nanometer MOSFET evolution is shown in Fig. 5.38. With the reduction of the gate length to the nanometer scale, the MOSFET technology underwent several major modifications, namely the development of strained Si channels to improve the Si (in particular hole) transport properties, obtained through SiGe drain and source regions that induce strain in the Si channel due to lattice mismatch; the replacement of the SiO_2 gate dielectric with the so-called high κ (high dielectric constant) materials, the aim being the possibility to increase the dielectric thickness, with the same capacitance, thus avoiding tunneling through the oxide; finally, the replacement of composite polysilicon plus salicide gates (silicides are metal-silicon compounds having

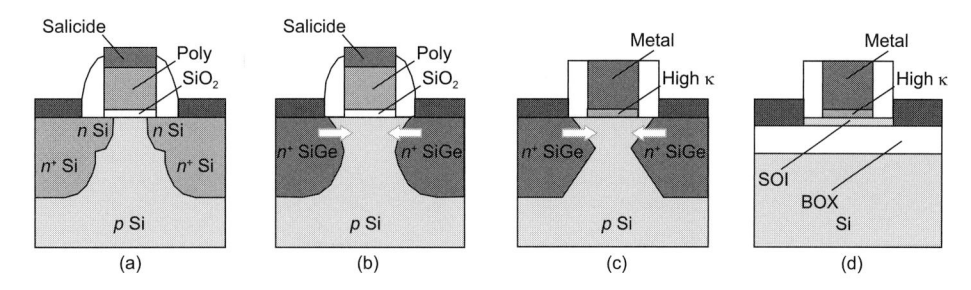

Figure 5.38 Technology evolution of nanometer MOSFETs: (a) conventional MOSFET with salicided polysilicon gate, low-doping drain double implant, 130 nm node; (b) strained Si channel MOSFET, 90–65 nm nodes; (c) strained Si channel, high-κ gate dielectric, metal gate MOSFET, 45–32 nm nodes; (d) Ultra thin body (UTB) SOI MOSFET, BOX is the Body Oxide, the gate length is 28 nm. The device section (a), (b), (c) are adapted from [33, Fig. 1] (Copyright ©2010 IEEE. All rights reserved. Reprinted with permission), (d) is adapted from [34, Fig. 2] (Copyright ©2012 IEEE. All rights reserved. Reprinted with permission).

better conductivity than polysilicon; the term *salicide* stands for Self Aligned Silicide, i.e., silicide deposed with a self-aligning process) with refractory metal gates, with a significant reduction in the gate resistance [33]. The final step is the introduction of 3D MOSFET structures called Tri-gate or FINFETs, with a perspective towards quantum wire MOSFETs where the gate entirely surrounds the Si or III-V quantum wire. FINFETs have shown interesting features as microwave analog devices, see [36] and references therein. Silicon-on-Insulator (SOI) approaches have also been followed to minimize the MOSFET parasitics [37, 34].

While nanometer MOSFET are able to achieve cutoff frequencies well above 100 GHz at least in n-channel devices (p-channel devices are inferior due to the lower hole mobility), the maximum oscillation frequency used to lag behind f_T due to the larger gate resistance when compared to Schottky-gate devices, where the metal gate cross section can be increased through mushroom shaped electrodes. In the latest technological nodes this issue was apparently overcome, thanks to the introduction of metal gates. Fig. 5.39 and Fig. 5.40 show the behavior of the cutoff frequency and of the maximum oscillation frequency vs. the drain bias, respectively, for a few technology nodes (28 nm for the Fully Depleted SOI (FDSOI) and 65, 55, 45 and 32 nm for the MOSFET); Fig. 5.39 and Fig. 5.40 also include, for comparison, an example of SiGe Heterojunction Bipolar Transistor (HBT). It can be seen that reducing the gate length leads to an increase of the maximum f_T or f_{max}; however the maximum f_{max} is achieved, for a slightly higher bias. For 32 nm gate devices the cutoff frequency has a maximum around 450 GHz, see Fig. 5.39, corresponding to bias currents around 1 mA/μm; such large cutoff frequencies make the MOSFET operation at millimeter wave possible, although a major limitation comes from the low output power of this technology, due to the low breakdown voltage and low current per unit gate periphery. Assuming for instance class A power operation (see Chapter 6) with a bias current of 1 A/mm and a breakdown voltage of 2 V the class A output power is $1 \times 2/8 = 0.25$ W/mm, one order of magnitude lower than the III-V limit. MOSFETs are therefore well suited for applications involving

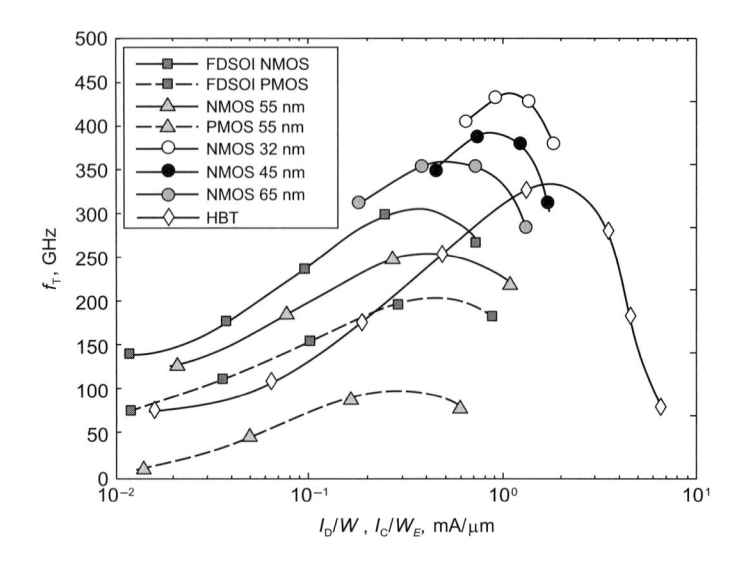

Figure 5.39 MOSFET p- and n-channel cutoff frequency vs. bias current for different technology nodes, including FDSOI devices. A SiGe HBT is also shown for comparison. The 55 nm gate length NMOS and PMOS are 40 gate devices with individual gate width of 500 nm, while the FDSOI 28 nm gate length devices are Ultra Thin Body and BOX (UTBB) FDSOI MOSFETs with individual gate width of 770 nm (data from [35, Fig. 2]. Copyright ©2015 IEEE. All rights reserved. Reused with permission). The NMOS with gate lengths of 32 nm, 45 nm, 65 nm are INTEL devices, data from [33, Fig. 8] (Copyright ©2010 IEEE. All rights reserved. Reused with permission).

mW power levels (such as for instance Bluetooth or WLAN systems) but cannot provide power levels needed, e.g., in wireless handsets or basestations. Medium power MOSFETs able to provide amplifiers with power levels up to 100 W up to 5 GHz have been introduced through the so-called LDMOS (Laterally Diffused MOS) technology, see Fig. 5.41. LDMOS are today the main component for wireless basestations, but are intrinsically limited in the microwave frequency operation to a few GHz.

5.6 Microwave FETs: A Comparison

A comparison between the trend lines for the cutoff and maximum oscillation frequencies of a number of competing Si-based and III-V-based FET technologies as a function of the gate length is shown in Fig. 5.42. In the f_{max} trend line for InP-based HEMTs, the record InP 32 nm gate length device with $f_{max} = 1.2$ THz presented in [38] was not considered, in order not to mask the overall trend of the majority of other devices.

Notice that the scaling trend for the MOS technology when reducing the gate length to nanometer scale is more effective than in III-V technologies, due in part to the greater effort in technology optimization that has been devoted to the mainstream Si MOSFETs. III-V technologies are superior, in particular when looking at the maximum frequency of oscillation; the best III-V devices are the InP-based HEMTs but also the

5.6 Microwave FETs: A Comparison 233

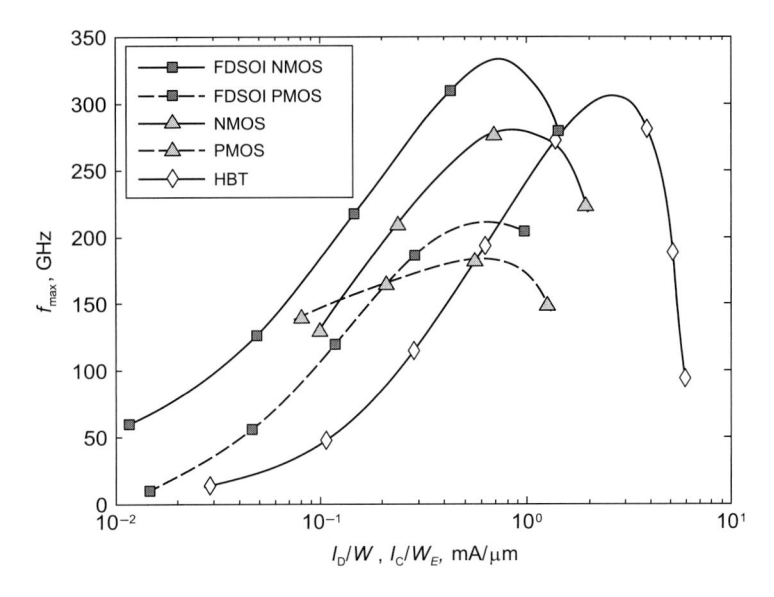

Figure 5.40 MOSFET p- and n-channel maximum oscillation frequency vs. bias current for different technology nodes, including FDSOI devices. A SiGe HBT is also shown for comparison. The 55 nm gate length NMOS and PMOS are 40 gate devices with individual gate width of 500 nm, while the FDSOI 28 nm gate length devices are Ultra Thin Body and BOX (UTBB) FDSOI MOSFETs with individual gate width of 770 nm (data from [35, Fig. 3]). Copyright ©2015 IEEE. All rights reserved. Reused with permission).

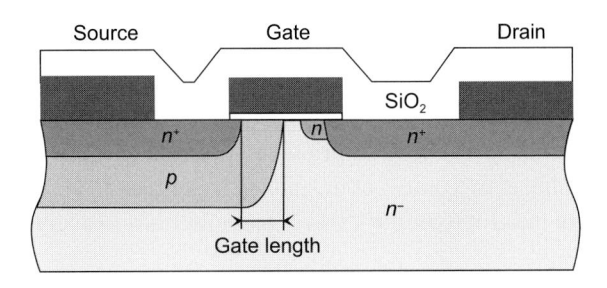

Figure 5.41 Example of Laterally Diffused MOS (LDMOS).

GaAs-based MHEMTs. Nanometer Si MOSFETs, however, perform in a somewhat surprisingly good way, considering that the Si mobility is much lower than the mobility in III-V devices. An interesting comparative appraisal of the strong points in the MOSFET vs. HEMT technologies is provided in [39]. First of all, the mobility penalty in MOSFET is not as large as it could appear, because (1) strained channels perform better; (2) nanoscale devices probably operate in velocity saturation conditions. Nevertheless, a further, less obvious asset of MOSFETs is provided by the fact that the device transconductance ultimately depends on the control action of the gate on the channel, that is inversely proportional (in HEMTs) to the thickness of the supply layer, see (5.23), but

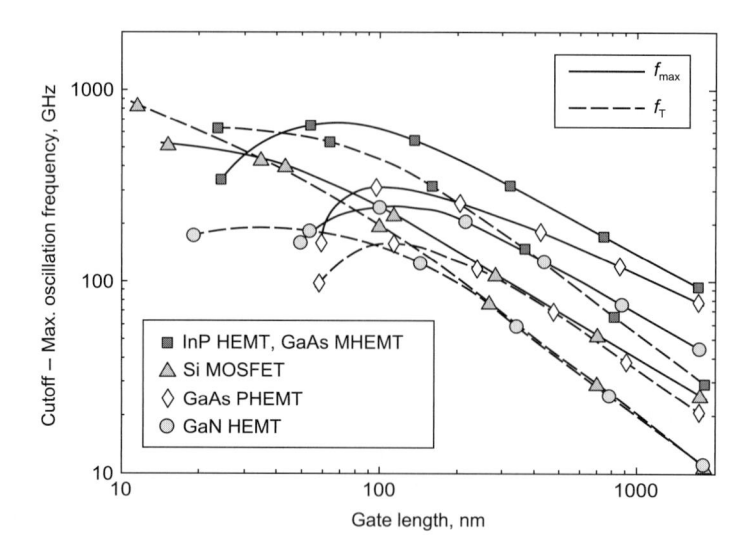

Figure 5.42 Trends in the cutoff frequency and maximum frequency of oscillation vs. gate length for a number of FET technologies (InP, GaAs, Si-based NMOS) with submicron gate length. The trend lines are derived from the data in [10, Fig. 10] for f_T and from the data in [10, Fig. 11] for f_{max}, respectively (Copyright ©2007 Elsevier. All rights reserved. Reused with permission). Also data from [40, Fig. 5] (Copyright ©2005 GAAS Association. All rights reserved. Reused with permission), [33, Fig. 9] (Copyright ©2010 IEEE. All rights reserved. Reused with permission), [35, Fig. 2–3] (Copyright ©2015 IEEE. All rights reserved. Reused with permission) were used.

in MOSFETs to the thickness of the oxide layer. Also taking into account the different values of the permittivities (III-V semiconductor vs. silicon oxide), this results in a kind of competitive geometrical advantage of the MOSFETs over Schottky gate FETs. Another factor that may contribute is the larger density of states in the conduction band of Si (almost two orders of magnitude larger than in GaAs, owing to the larger effective mass of electrons, see [5]), that potentially leads to a larger channel charge density.

5.7 Heterojunction Bipolar Transistors

In conventional bipolar junction transistors (BJT) two *pn* junctions are present with a common side; two dual solutions are possible, called the *pnp* and *npn* devices. In the *npn* transistor, that is preferred for analog applications due to the superior transport properties of electrons with respect to holes, the *np* emitter-base junction is forward biased and injects electrons into the thin *p*-type base. A small fraction of the injected electrons recombines in the base, attracting holes from the base and thus yielding a (small) base current I_B. Most of the electrons injected from the emitter do note recombine in the base and are instead swept by the electric field of the reverse-biased base-collector junction, to be ultimately collected by the collector contact as the collector current $I_C = -I_B - I_E = -\alpha I_E$ (we assume all currents as positive entering). The parameter

$\alpha < 1$ (but close to 1) is denoted as the *common base current gain*. We thus have, solving for I_C:

$$I_C = \frac{\alpha}{1-\alpha} I_B = \beta I_B, \quad \beta = \frac{\alpha}{1-\alpha} \gg 1,$$

where β is the *common emitter current gain*. The common base current gain α can be shown to be expressed as the product of two factors:

$$\alpha = \gamma b,$$

where $\gamma < 1$ is the *emitter efficiency*, $b < 1$ is the *base transport factor*. The emitter efficiency γ accounts for the fact that, in an *npn* transistor, the emitter current has two components, electrons injected from the emitter into the base, and holes injected from the base into the emitter. The latter component is useless, since it is not collected by the collector, and does not contribute to current gain. The emitter efficiency $\gamma < 1$ is the ratio between the useful component of I_E and the total emitter current. To maximize the current gain, it should be made as close as possible to unity. On the other hand, in an *npn* device the base transport factor b is the fraction of injected electrons that successfully travel through the base to reach the collector. To achieve a large current gain β, α (and therefore *both* γ and b) should be almost unity.

The base transport factor b can be optimized by making the base thickness (called base length) as small as possible with respect to the carrier diffusion length in the base, or, equivalently, by making the transit time of minority carriers in the base (electrons in an *npn*) much smaller than the minority carriers' lifetime in the base. Epitaxial growth allows the base thickness to be reduced to the nanometer scale. This however increases the base distributed resistance, unless the emitter width is reduced through self-aligning techniques similar to those adopted in submicron CMOS gate technology.

Optimization of the emitter efficiency γ in BJT requires the emitter-base junction to be strongly asymmetrical, with $N_{DE} \gg N_{AB}$. Assume for completeness that a bandgap difference $\Delta E_g = E_{gE} - E_{gB}$ may exist between the emitter and the base; it can be shown that [5]:

$$\gamma = \frac{1}{1 + \frac{N_{AB} D_{hE} W_B N_{vE} N_{cE}}{N_{DE} D_{nB} W_E N_{vB} N_{cB}} \exp\left(-\frac{\Delta E_g}{k_B T}\right)}, \quad (5.24)$$

where D_{hE} is the hole diffusivity in the emitter, D_{nB} the electron diffusivity in the base, W_B and W_E the base and emitter thickness, respectively, $N_{v\alpha}$ ($N_{c\alpha}$) the effective valence (conduction) band state densities in the emitter or base ($\alpha = E, B$). In a conventional BJT, $\Delta E_g = 0$ and $N_{\alpha E} = N_{\alpha B}$, $\alpha = c, v$; taking into account that $D_{hE} \approx D_{nB}$ (the hole diffusivity is in fact lower, but same order of magnitude as the electrons') and that $W_B < W_E$ (but again with similar orders of magnitude), the only way to have γ close to unity is to set:

$$\frac{N_{AB}}{N_{DE}} \ll 1,$$

i.e., to make the emitter doping much larger than the base doping. Since the base doping should be larger than the collector doping,[12] we finally have the BJT design rule $N_{DE} \gg N_{AB} > N_{DC}$. Assuming $b \approx 1$ and using the Einstein relation $D = (k_B T/q)\mu$ we have from (5.24):

$$\beta \approx \left(\frac{\mu_{nB} W_E N_{vB} N_{cB}}{\mu_{hE} W_B N_{vE} N_{cE}}\right) \frac{N_{DE}}{N_{AB}} \exp\left(\frac{\Delta E_g}{k_B T}\right). \tag{5.25}$$

As in field-effect transistors, two figures of merit can be introduced to characterize the bipolar transistor speed, the cutoff frequency f_T and the maximum oscillation frequency $f_{\max}$. Faster transistors can be generally obtained by scaling down the device geometry and scaling up the doping level. Increasing the doping level is mandatory in order to properly scale the junction depletion region sizes. Approximately, dopings scale as $l^{-1/2}$, where l is some characteristic dimension, so that a size scaling down of 100 approximately corresponds to an increase in the doping level of a factor 10. Conventional, homojunction bipolars are, unfortunately, affected by basic limitations if the cutoff frequency has to be pushed beyond a few GHz. In fact, the increase in the doping level in the emitter, that is the region with the highest doping in the whole device, leads to the so-called *bandgap narrowing* effect, whereby the material gap slightly decreases for high doping according to the Lanyon–Tuft model [41], [5]:

$$\Delta E_{gE} = \frac{3q^3}{16\pi \epsilon_s^{3/2}} \sqrt{\frac{N_{\alpha E}}{k_B T}},$$

where ϵ_s is the semiconductor permittivity, $N_{\alpha E}$ is the donor or acceptor emitter doping, T the absolute temperature, and ΔE_{gE} is the emitter gap decrease. For Si at ambient temperature one has:

$$\Delta E_{gE} \approx 22.5 \sqrt{\frac{N_{\alpha E}}{1 \times 10^{18}}} \quad \text{meV},$$

with the doping in cm^{-3} units. At $N_{DE} \approx 1 \times 10^{19}$ cm^{-3} the bandgap narrowing is of the order of 100 meV; since, from (5.25):

$$\beta(N_{DE}) \approx \beta(0) \exp\left(-\frac{\Delta E_{gE}}{k_B T}\right),$$

where $\beta(0)$ is the low-doping current gain, we see that even a small bandgap narrowing leads to a significant decrease in the common emitter current gain. Another important point to consider is the fact that the low base doping required by the condition $N_{DE} \gg N_{AB}$ implies a high value for the input base distributed resistance, which in turn compromises the transistor $f_{\max}$. In conclusion, homojunction bipolars have a limited space for optimization to achieve operation above a few GHz, because this would need increasing the base doping to reduce the input resistance, which would imply extremely large emitter dopings leading, in turn, to emitter bandgap narrowing.

---

[12] This is needed for two purposes: decreasing the width of the base depletion layer in the base-collector junction and its sensitivity to the applied V_{CE}, thus minimizing the so-called Early effect and as a consequence the small-signal output resistance of the transistor; increasing the breakdown voltage of the base-collector junction and therefore the transistor maximum output power.

5.7 Heterojunction Bipolar Transistors

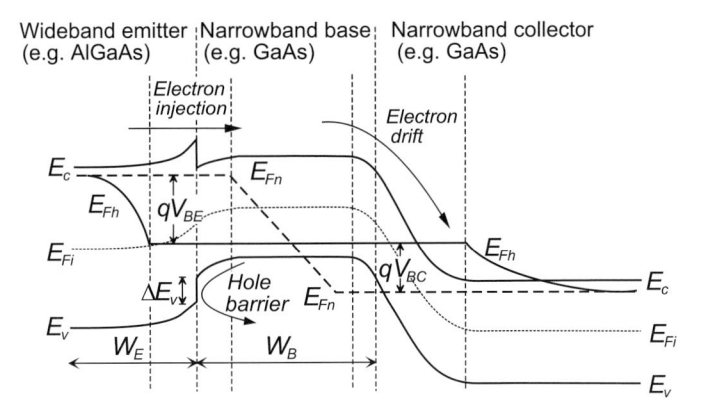

Figure 5.43 Bandstructure of a *npn* heterojunction bipolar in direct active region. In the example shown the emitter is widegap, the base and collector narrowgap.

A possible way out of this stalemate is obtained by a bipolar design where the bandgap of the base is different (in particular, smaller) than the emitter bandgap; the idea was already suggested in the 1951 BJT patent by William Shockley, see [42, Fig. 12]. In this case, we can satisfy the condition:

$$\frac{N_{AB}}{N_{DE}} \exp\left(-\frac{\Delta E_g}{k_B T}\right) \ll 1$$

by exploiting a suitably large ΔE_g, even if $N_{AB} \approx N_{DE}$. This enables to decrease the base spreading resistance and, as a consequence, to increase $f_{\max}$. Of course, also the emitter width has to be scaled down to submicron size to reduce the base resistance. The band diagram of a heterojunction bipolar transistor in the *forward active region*, i.e., when the emitter-base junction is in forward bias and the base-collector junction is in reverse bias, is shown in Fig. 5.43. The increase in the emitter efficiency can be readily interpreted also in terms of a potential barrier opposing the hole back-diffusion into the emitter; in fact, the global effect of the heterojunction is related to the entire bandgap difference ΔE_g, that also has an influence on the relative intrinsic concentrations. In the example shown, the base and collector are narrowgap and the base-emitter heterojunction is abrupt; alternative designs can be obtained through a graded base-emitter heterojunction and by using a widegap collector. The resulting device is called the Heterojunction Bipolar Transistor (HBT); if only the emitter-base heterojunction is present the device is referred to as Single HBT (SHBT); if also the base-collector junction is a heterojunction we talk about the Double HBT (DHBT).

5.7.1 HBT Equivalent Circuit

Bipolar transistors (both conventional and heterojunction) are junction-based devices for which a few fundamental ideal relations hold, at least approximately. In the direct active region (we refer to a *npn* device) the emitter current follows the junction law vs.

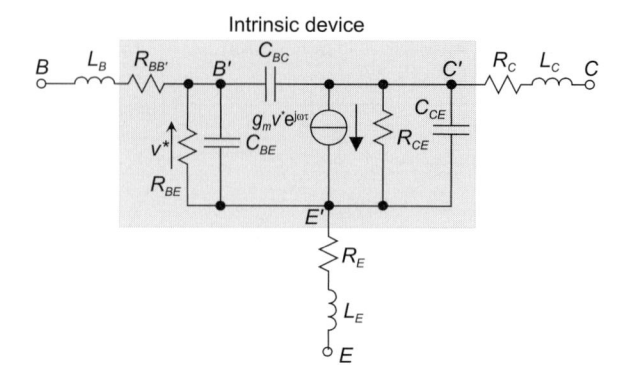

Figure 5.44 Small-signal equivalent circuit of a bipolar transistor. The gray box is the intrinsic device.

the driving voltage V_{BE}:

$$I_E \approx I_{E0}\left(e^{V_{BE}/V_T} - 1\right),$$

where $V_T = k_B T/q$ (26 mV at 300 K). Since $I_E \approx I_C$ we also have:

$$I_C \approx I_{C0}\left(e^{V_{BE}/V_T} - 1\right) \approx I_{C0} e^{V_{BE}/V_T},$$

so that the intrinsic bipolar transconductance g_m can be expressed as:

$$g_m = \frac{\partial I_C}{\partial V_{BE}} = \frac{I_C}{V_T}.$$

On the other hand, $I_B = I_C/\beta$; thus, the input differential conductance of the bipolar can be expressed as:

$$G_{B'E} = \frac{\partial I_B}{\partial V_{BE}} \approx \frac{I_B}{V_T} = \frac{I_C}{\beta V_T} = \frac{g_m}{\beta}.$$

The intrinsic base has been denoted as B', as opposed to the external (extrinsic) base contact. Due to the input junction structure, the input capacitance of the bipolar, C_{BE}, is the capacitance of a forward-biased pn junction, and therefore appears in the equivalent small-signal circuit in parallel to $G_{B'E} = R_{B'E}^{-1}$. Addition of other parasitic capacitances and inductances, together with the output resistance R_{CE} arising from the weak dependence of I_C on V_{CE} (the so-called Early effect) finally leads to the equivalent circuit in Fig. 5.44. The distributed base resistance $R_{BB'}$ models the resistive path between the transistor input and the intrinsic base (the base current flows in the narrow base layer orthogonal to the collector and emitter current densities); due to the very small base thickness such a resistance tends to be large (therefore negatively affecting the device $f_{\max}$) unless the base is suitably doped and the emitter is very narrow.

The short-circuit current gain and therefore the *cutoff frequency* of a bipolar transistor (in the unilateral approximation) can be/ evaluated from the simplified intrinsic equivalent circuit in Fig. 5.45 (a). Assuming that for $f \approx f_T$ we have the condition:

$$\omega_T C_{BE} \gg \frac{1}{R_{BE}},$$

5.7 Heterojunction Bipolar Transistors

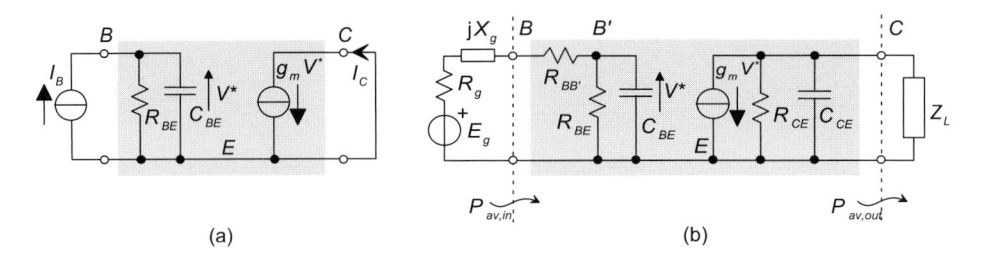

Figure 5.45 Evaluating the cutoff frequency (a) and the maximum oscillation frequency (b) of a BJT or HBT.

we obtain the short-circuit current gain as:

$$h_{21} = \frac{I_C}{I_B} \approx \frac{g_m}{j\omega C_{BE}}.$$

Setting $|h_{21}| = 1$ we derive the cutoff frequency as:

$$f_T \approx \frac{g_m}{2\pi C_{BE}}.$$

The *maximum oscillation frequency* f_{max} is again the frequency at which the maximum available power gain (MAG) is unity. With reference to Fig. 5.45 (b), assuming that the output is power matched, the output available power will be:

$$P_{av,out} = \frac{g_m^2 |V^*|^2 R_{CE}}{4};$$

however, if the input is power matched, the generator reactance compensates for the C_{BE} reactance (again we neglect R_{BE}), $R_g = R_{BB'}$ and the base current is:

$$I_B = \frac{E_g}{R_g + R_{BB'}} = \frac{E_g}{2R_{BB'}} \rightarrow V^* = \frac{E_g}{2j\omega C_{BE} R_{BB'}};$$

thus, since $P_{av,in} = |E_g|^2/(4R_g)$ and $R_g = R_{BB'}$:

$$P_{av,out} = \frac{g_m^2 |E_g|^2 R_{CE}}{16\omega^2 C_{BE}^2 R_{BB'}^2} = \frac{g_m^2 R_{CE}}{4\omega^2 C_{BE}^2 R_{BB'}} P_{av,in}.$$

The available power gain $G_{av} = P_{av,out}/P_{av,in}$ corresponds to the MAG, because power matching is implemented at the input; moreover, it coincides (since also the output is matched) with the operational power gain $G_{op} = P_{out}/P_{in}$. The condition to be imposed to evaluate f_{max} is:

$$\text{MAG} = \frac{g_m^2 R_{CE}}{4(2\pi f_{max})^2 C_{BE}^2 R_{BB'}} = 1,$$

leading to:

$$f_{max} = f_T \sqrt{\frac{R_{CE}}{4R_{BB'}}}. \tag{5.26}$$

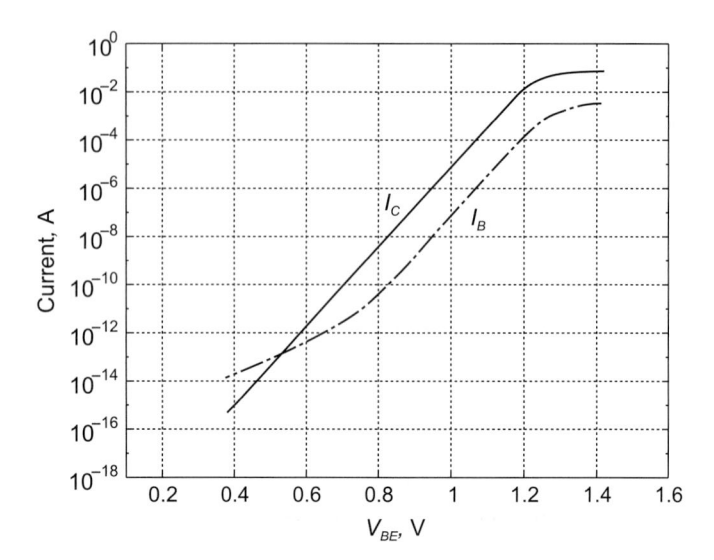

Figure 5.46 Qualitative Gummel plot of heterojunction bipolar transistor.

Since $R_{CE} \gg R_{BB'}$ (typically by one order of magnitude), $f_{max} > f_T$ unless the input resistance is high. The bipolar cutoff frequency:

$$f_T = \frac{g_m}{2\pi C_{BE}} = \frac{I_C}{2\pi C_{BE} V_T}$$

increases with the collector bias current, but for high values of I_C, saturation occurs due to *high-injection effects*, that causes a drop of the transistor current gain β, see Fig. 5.39 and Fig. 5.40. In conclusion, the cutoff frequency typically exhibits a maximum vs. I_C, and decreases for low and high I_C values. Plotting I_C and I_B as a function of V_{BE} in semilog scale we obtain the result shown in Fig. 5.46: for low currents, I_B is mainly associated with leakage and/or carrier generation effects, and the base current may be larger than the collector current. In an intermediate range $I_C = \beta I_B$ as foreseen by ideal transistor operation; finally, for large V_{BE} both currents saturate and β decreases. The diagram in Fig. 5.46 is referred to as the *Gummel plot* of the transistor.[13] InP-based HBTs currently reach cutoff frequencies of the order of several hundreds GHz and maximum oscillation frequencies of the order of 1 THz.

5.7.2 HBT Choices and Material Systems

Heterojunction bipolar transistors can be implemented in a variety of material systems. III-V-based HBTs are grown on either GaAs or InP substrates. For the GaAs-based device, the widegap emitter is obtained through lattice-matched AlGaAs; in AlGaAs the Al content must be kept below 30 percent since above this value a trap level (the *DX centers*) appears in the widegap material. The base can be either GaAs (lattice matched)

---

[13] From Hermann Gummel, a pioneer in the semiconductor industry.

5.7 Heterojunction Bipolar Transistors

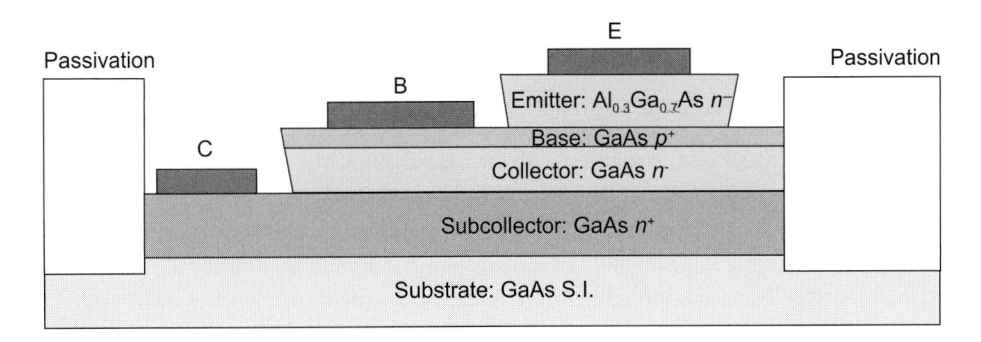

Figure 5.47 Qualitative cross section of epitaxial GaAs-based HBT.

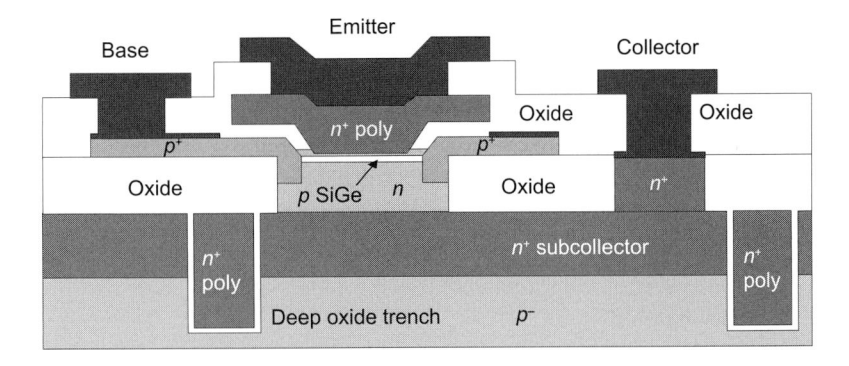

Figure 5.48 Cross section of a SiGe HBT.

or InGaAs (pseudomorphic), while the collector, subcollector and substrate are GaAs. A simplified example of GaAs-based epitaxial HBT structure is shown in Fig. 5.47.

Although the GaAs-based technology is well consolidated, still better performances can be obtained with InP-based devices, thanks to the larger bandgap difference between the emitter and the base. Typical devices have an InAlAs emitter, a lattice-matched or pseudomorphic InGaAs base, and an InP collector, subcollector and substrate.

Of particular interest are HBTs grown on Si substrates. The SiGe alloy is a narrow-band material, which can be epitaxially grown over a Si substrate in pseudomorphic form. In fact, the $Si_{1-x}Ge_x$ alloy exhibit a lattice mismatch with respect to Si equal to $\approx 4x$ percent, where x is the Ge fraction. An example of SiGe HBT is shown in Fig. 5.48; the only SiGe layer is the narrowgap base; the device is made with a self-centered polysilicon emitter process and deep trench isolation between neighboring devices. SiGe HBTs are an interesting alternative to other high-speed transistors for low-power analog circuits and digital circuits, although the process is more complex than for CMOS-based circuits. SiGe bipolars can reach millimeter wave operation, but with low breakdown voltages (below 3 V); the breakdown voltage decreases with increasing cut-off frequency, while GaAs-based (and, to a lesser extent, InP based) transistors exhibit larger breakdown voltages and power densities, see [7, Fig. 8.41].

5.8 Measurement-Based Microwave FET Small- and Large-Signal Models

5.8.1 Small- and Large-Signal Circuit Models

In small-signal conditions the device behavior is completely described by its frequency-dependent small-signal parameters (scattering, admittance, impedance parameters). Most commercially available circuit simulators allow a linear n-port to be described through a look-up table of the scattering parameters sampled as a function of frequency. Another common approach consists in extracting, from the measured S parameters, a suitable small-signal lumped equivalent circuit. Although such model is an approximation of measurements, it has some advantages: the equivalent circuit is valid also outside the measured frequency range, and due to the physical meaning of its elements it can be conveniently scaled with respect to the gate periphery. Finally, some time-domain circuit simulators do not support look-up table models.

The situation is much more involved if a large-signal model is required. This should be able to mimic the device behavior under arbitrary operating conditions. The nonlinear behavior that the device exhibits can be accounted for through different approaches. From a circuit standpoint, the large signal model is an equivalent circuit, whose topology is similar to the small-signal model, but whose elements are nonlinear.[14] With respect to the small-signal case, the nonlinear equivalent circuit should also account for effects such as gate direct conduction or device breakdown, that are not included in the small-signal topology.

5.8.2 Extracting the Small-Signal Equivalent Circuit from Measurements

Extracting a small-signal equivalent circuit from measurements means to identify, given a certain circuit topology (see Fig. 5.22), the element values in order to mimic, with some prescribed accuracy, the frequency behavior of the measured small-signal parameters.

The equivalent circuit has a set of frequency-independent parameters that can be ordered in vector $\underline{P}$. For instance in the equivalent circuit in Fig. 5.22 we have:

$$\underline{P} = (R_I, R_{DS}, g_m, \tau, C_{GS}, C_{CD}, C_{DS}; R_S, R_G, R_D, L_S, L_G, L_D).$$

Given the measured scattering parameters sampled in a set of N frequencies:

$$\underline{S}_{11} = [S_{11}(f_1), S_{11}(f_2), \ldots, S_{11}(f_N)]$$
$$\underline{S}_{12} = [S_{12}(f_1), S_{12}(f_2), \ldots, S_{12}(f_N)]$$
$$\underline{S}_{21} = [S_{12}(f_1), S_{12}(f_2), \ldots, S_{12}(f_N)]$$
$$\underline{S}_{22} = [S_{22}(f_1), S_{22}(f_2), \ldots, S_{22}(f_N)]$$

and the simulated scattering parameters from the equivalent circuit $\underline{S}_{ij}^s(\underline{P})$, $i, j = 1, 2$, the identification of the parameter set $\underline{P}$ is, in general, an optimization (or fitting) problem

---

[14] Nonlinear resistors, capacitors, current or voltage dependent sources are defined by constitutive equations where the output variable y is a nonlinear function of the input variable x. For example, in a linear resistor $y = Rx$ where x is the current and y the voltage, in a nonlinear resistor $y = f(x)$.

5.8 Measurement-Based Microwave FET Small- and Large-Signal Models

involving the minimization of some norm of the difference between the measured and simulated scattering parameters. Adopting for instance an L^2 norm we should minimize the following functional:

$$\min\left[\sum_{i,j}\left|\underline{S}_{ij} - \underline{S}^s_{ij}(\underline{P})\right|^2\right] = \min\left[\sum_{i,j}\sum_{k=1}^{N}\left|\underline{S}_{ij}(f_k) - \underline{S}^s_{ij}(f_k;\underline{P})\right|^2\right].$$

While many optimization techniques are available in circuit simulators to address the fitting problem, some difficulties arise in practice. First of all, the solution of the problem is related to the weight given to the individual scattering parameters and to the frequency samples (some parameters are large, some small, and their magnitude is frequency dependent). Moreover, the number of parameters involved is already large enough to lead the fitting process to secondary minima and spurious solutions, also related to the choice of the fitting starting point.

In order to increase the reliability of the fitting process techniques have been developed to directly extract some of the parameters from measurements and to use the resulting circuit elements as a starting point of a final optimization procedure. This technique is generally referred to as *de-embedding* [43]. For microwave FETs, the following de-embedding procedure, fist introduced in [44], is well established:

1. The extrinsic or parasitic elements are measured directly. Such measurements can be carried out on the so-called *cold FET* consisting in the unbiased device (drain and source are DC short circuited and the gate is in DC direct bias). Cold FET measurements allow us to directly evaluate R_S, R_G, R_D, L_S, L_G, L_D. The cold FET with DC shorted drain and source operates like a three-terminal passive circuit, where the gate junction is (approximately) a short circuit or anyway a low resistance (a diode in direct bias); the resulting small-signal equivalent circuit is shown in Fig. 5.49. Neglecting (i.e., approximating with a short-circuit) the elements in the intrinsic device, i.e., assuming that r_{DS}, r_{DD}, R_{ch} be much smaller than the extrinsic parameters, the impedance matrix of the cold FET can be approximated as:

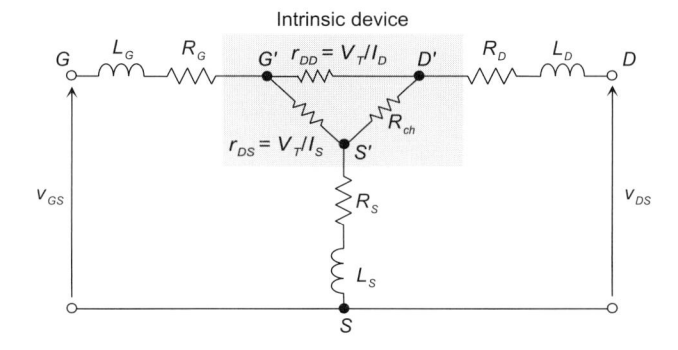

$$Z_{11} = (R_G + R_S) + j\omega(L_G + L_S)$$

Figure 5.49 Small-signal cold FET T equivalent circuit.

$$Z_{12} = Z_{21} = R_S + j\omega L_S$$
$$Z_{22} = (R_D + R_S) + j\omega(L_D + L_S).$$

i.e.:

$$R_S = \text{Re}\,[Z_{21}]$$
$$R_G = \text{Re}\,[Z_{11} - Z_{21}]$$
$$R_D = \text{Re}\,[Z_{22} - Z_{21}]$$
$$L_S = \text{Im}\,[Z_{21}]/\omega$$
$$L_G = \text{Im}\,[Z_{11} - Z_{21}]/\omega$$
$$L_D = \text{Im}\,[Z_{22} - Z_{21}]/\omega.$$

The cold FET impedance matrix is derived from the measured scattering matrix through circuit transformation, and the constant parasitic elements result from a simple straight line (e.g., least square) fitting; in fact, the real parts of the Z matrix are almost constant while the imaginary parts are approximately linear with frequency.

2. The extrinsic parameters are *de-embedded* from the complete device Z-matrix (derived from the complete scattering matrix in the working point). In fact, the parasitic elements are almost independent from the bias point. Be **Z** the biased FET impedance matrix and $\mathbf{Z}_{int}$ the matrix of the intrinsic FET, we immediately have from Fig. 5.50:

$$Z_{11} = Z_{11,int} + Z_G + Z_S$$
$$Z_{12} = Z_{12,int} + Z_S$$
$$Z_{21} = Z_{21,int} + Z_S$$
$$Z_{22} = Z_{22,int} + Z_D + Z_S,$$

where $Z_G = R_G + j\omega L_G$, $Z_D = R_D + j\omega L_D$, $Z_S = R_S + j\omega L_S$. Solving, we have:

$$Z_{11,int} = Z_{11} - Z_G - Z_S$$
$$Z_{12,int} = Z_{12} - Z_S$$

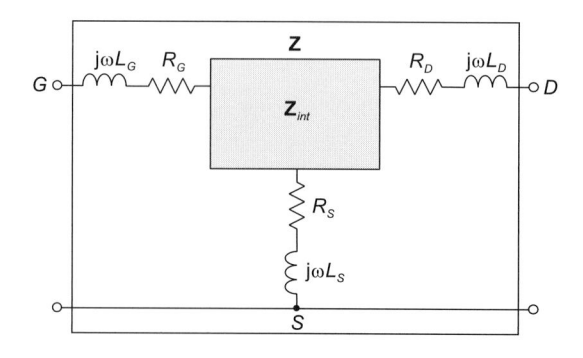

Figure 5.50 De-embedding the intrinsic impedance matrix from the total one measured at the bias point.

$$Z_{21,int} = Z_{21} - Z_S$$
$$Z_{22,int} = Z_{22} - Z_D - Z_S.$$

3. The intrinsic impedance matrix is now converted numerically, at each frequency, into the intrinsic admittance matrix. However, from (5.14) the intrinsic admittance matrix derived from the equivalent circuit is:

$$Y_{11,int} = \left(\frac{j\omega C_{GS}}{1 + j\omega C_{GS} R_I} + j\omega C_{GD} \right)$$

$$Y_{12,int} = -j\omega C_{GD}$$

$$Y_{21,int} = \left(\frac{g_m e^{j\omega \tau}}{1 + j\omega C_{GS} R_I} - j\omega C_{GD} \right)$$

$$Y_{22,int} = \left(j\omega (C_{GD} + C_{DS}) + \frac{1}{R_{DS}} \right).$$

Taking into account that the intrinsic circuit has seven parameters, it is quite obvious that, for each measured frequency, we can derive those parameters from the measured intrinsic matrix (that yield four complex equations, i.e., eight real ones). The inversion process can be carried out analytically as follows. For the capacitances we have:

$$C_{GD} = -\text{Im}\left[Y_{12,int}\right]/\omega$$
$$C_{DS} = \text{Im}\left[Y_{22,int}\right]/\omega - C_{GD}$$
$$C_{GS} = -\left\{ \text{Im}\left[\frac{1}{Y_{11,int} - j\omega C_{GD}} \right] \omega \right\}^{-1} \approx \text{Im}\left[Y_{11,int}\right]/\omega - C_{GD}.$$

The last equation is an approximation holding for frequencies $f \ll (C_{GS} R_I)^{-1}$. Resistances can be derived as:

$$R_{DS} = \left\{ \text{Re}\left[Y_{22,int}\right] \right\}^{-1}$$
$$R_I = \text{Re}\left[\frac{1}{Y_{11,int} - j\omega C_{GD}} \right].$$

Finally:

$$g_m e^{j\omega \tau} = (Y_{21,int} + j\omega C_{GD})(1 + j\omega C_{GS} R_I),$$

from which:

$$g_m = |(Y_{21,int} + j\omega C_{GD})(1 + j\omega C_{GS} R_I)|$$
$$\tau = \arg[(Y_{21,int} + j\omega C_{GD})(1 + j\omega C_{GS} R_I)]/\omega.$$

The parameters derived this way are in principle frequency-dependent. However, their frequency variation is weak, so that an average value can be exploited either directly or as a starting point for a global fitting process.

Notice that the above procedure works well for integrated devices with simple layout and low gate periphery. Packaged devices require equivalent circuit with additional elements (package parasitics) that can be estimated by separately characterizing a dummy package (i.e., a package with a short circuit replacing the device) or by characterizing the packaged device in proper bias conditions.

5.8.3 Large-Signal FET Circuit Model Principles

In establishing the topology of large-signal FET equivalent circuit models a physics-based approach is followed to decide what elements are nonlinear (and therefore depending on the instantaneous working point). The device physics suggests that parasitic elements (resistances and inductances) are almost independent from the instantaneous working point and therefore can be approximated through linear elements. This means that the nonlinear modeling effort can be reduced to the intrinsic device. For the sake of definiteness, let us address first the nonlinear modeling problem in DC conditions, where the device is nonlinear but memoryless (no reactive elements). Besides being a starting point for the general modeling exercise, the DC model is used in evaluating the DC bias point through a circuit simulator.

First of all, the DC intrinsic gate-source and drain-source voltages can be de-embedded from the measured DC characteristic taking into account that:

$$V_{GS,i} = V_{GS} - I_D R_S$$
$$V_{DS,i} = V_{DS} - I_D (R_S + R_D).$$

De-embedding the parasitic resistances is also important because the control variables of the intrinsic device model are the intrinsic voltages, not those external.

Thus, the nonlinear DC model reduces to a nonlinear dependent generator expressing the drain current as a function of the intrinsic voltages (the gate DC current is zero); see Fig. 5.51:

$$I_D = \hat{I}_D(V_{GS,i}, V_{DS,i}). \tag{5.27}$$

The small-signal DC (or low-frequency, neglecting reactive elements) representation can be directly obtained by linearly approximating (5.27) around a DC working point $V_{GS,i}, V_{DS,i}$, i.e.:

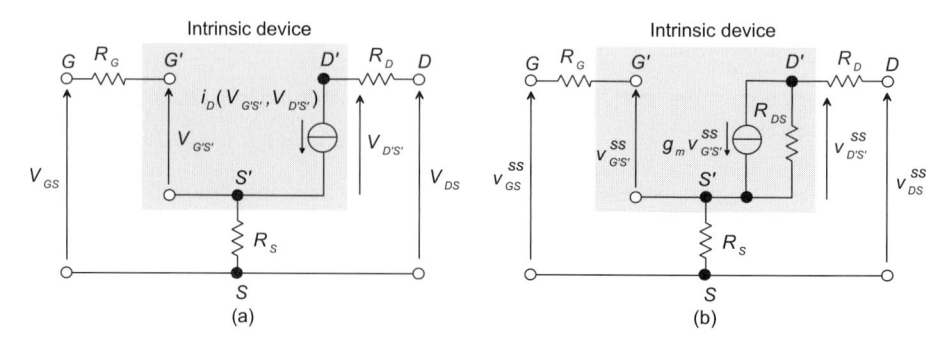

Figure 5.51 DC large-signal equivalent circuit (a) and the resulting small-signal circuit (b).

$$i_D(t) = I_D + i_D^{ss}(t) \approx I_D + \frac{\partial \hat{I}_D}{\partial v_{GS,i}}(v_{GS,i} - V_{GS,i}) + \frac{\partial \hat{I}_D}{\partial v_{DS,i}}(v_{GS,i} - V_{GS,i}),$$

i.e., defining:

$$g_m = \frac{\partial \hat{I}_D}{\partial v_{GS,i}} \tag{5.28}$$

$$R_{DS}^{-1} = \frac{\partial \hat{I}_D}{\partial v_{DS,i}}, \tag{5.29}$$

we obtain:

$$i_D^{ss} = \frac{\partial \hat{I}_D}{\partial v_{GS,i}} v_{GS,i}^{ss} + \frac{\partial \hat{I}_D}{\partial v_{DS,i}} v_{GS,i}^{ss} = g_m v_{GS,i}^{ss} + R_{DS}^{-1} v_{GS,i}^{ss}$$

corresponding to the circuit in Fig. 5.51, right. Interestingly, the DC model $\hat{I}_D(V_{GS,i}, V_{DS,i})$ can also be obtained, from (5.28) and (5.29), by integrating the parameters g_m and R_{DS}^{-1} derived, in small-signal conditions, from small-signal measurements carried out in different working points [45]. We refer to those characteristics as *pseudo-DC* ones, since they are not generally observable, as such, under real DC measurements. An important caveat, however, must be introduced. First of all the two small-signal parameters should obey the *consistency condition*:

$$\frac{\partial g_m}{\partial v_{DS,i}} = \frac{\partial R_{DS}^{-1}}{\partial v_{GS,i}} = \frac{\partial^2 \hat{I}_D}{\partial v_{DS,i} \partial v_{GS,i}}.$$

This condition is not necessarily forced when extracting the above parameters from small-signal measurements. More importantly, the pseudo-DC characteristics extracted from small-signal parameters measured at microwaves are *not necessarily* coincident with the true DC parameters (in particular, the output resistance can be much smaller). The physical reason for this discrepancy between the DC characteristics and the so-called dynamic DC curves measured with a fast sweep (sweep frequency in excess of 100 MHz typically, when reactive effects still are negligible) comes from *low-frequency dispersion* effects associated with surface and substrate traps acting in true DC conditions but not under a fast sweep, as well as to self-heating effects. While the evolution of technology has made those effects less and less important, there are ways to take into account this dual behavior in the model, although not in a simple manner.

Concerning the implementation of the model for the function $\hat{I}_D(v_{GS,i}, v_{DS,i})$, two main strategies are possible, a look-up table-based numerical interpolation, or the use of some approximating analytical function whose parameters should be properly fitted. Both strategies are currently implemented in circuit simulators, although the analytical function approach is computationally more efficient and better conditioned when the instantaneous working point is outside the set used for the model identification. Many circuit simulators allow for user-defined models where the model equations are directly entered, in some programming language, by the user.

As a last remark, the model in Fig. 5.51 (a) still is incomplete: a model for the breakdown behavior (essential in simulating power amplifiers) and for the direct gate conduction should be added, typically under the form of diodes connected between the gate and the other terminals. Finally, in some applications the subthreshold current model is important and the simplification consisting in stating that this current simply is zero is not enough.

Reactive elements such as junction capacitors associated with the gate junction are, again, nonlinear (i.e., the capacitor charge is a nonlinear function of the driving voltage or voltages). In the simplest case the charge depends on one driving voltage; e.g., the gate-source capacitance approximately depends on $V_{GS,i}$ only. In this case we can postulate a diode-like charge–voltage relation:

$$Q_G = \hat{Q}_G(v_{GS,i}),$$

from which, in time-varying conditions:

$$i_G(t) = \frac{dQ_G}{dt} = \frac{d\hat{Q}_G}{dv_{GS,i}} \frac{dv_{GS,i}}{dt} = C_{GS}(v_{GS,i}) \frac{dv_{GS,i}}{dt},$$

where $C_{GS}(v_{GS,i})$ coincides with the small-signal capacitance $C_{GS}(V_{GS,i})$ extracted from measurements in the DC (intrinsic) bias point $V_{GS,i}$. The same approach can be extended to the other capacitances.

Additional complexities arise when taking into account that the gate charge actually depends both on the gate-source and the gate-drain voltages $V_{GS,i}$, $V_{DS,i}$. The small-signal capacitances should again abide by a consistency relation ultimately implying charge conservation [46]. In static conditions we have:

$$Q_G = \hat{Q}_G(V_{GS,i}, V_{GD,i}),$$

but then, in time-varying conditions:

$$i_G = \frac{dQ_G}{dt} = \frac{\partial \hat{Q}_G}{\partial v_{GS,i}} \frac{dv_{GS,i}}{dt} + \frac{\partial \hat{Q}_G}{\partial v_{GD,i}} \frac{dv_{GD,i}}{dt} =$$
$$= C_{GS}(v_{GS,i}, v_{GD,i}) \frac{dv_{GS,i}}{dt} + C_{GD}(v_{GS,i}, v_{GD,i}) \frac{dv_{GD,i}}{dt}.$$

Since:

$$v_{GD,i} = v_{GS,i} - v_{DS,i},$$

we can also write:

$$C_{GS} = C_{GS}(v_{GS,i}, v_{DS,i})$$
$$C_{GD} = C_{GD}(v_{GS,i}, v_{DS,i}).$$

However, the consistency relationship arises:

$$\frac{\partial C_{GS}}{\partial v_{GD,i}} = \frac{\partial^2 \hat{Q}_G}{\partial v_{GS,i} \partial v_{GD,i}} = \frac{\partial C_{GD}}{\partial v_{GS,i}},$$

that, again, the small-signal capacitances extracted from multi-bias S parameter measurements do not necessarily follow. In conclusion, if the model has to include a charge depending on two controlling voltages, the model implementation has better be carried out on a charge control basis than on a capacitance basis (many circuit simulators use charges as state variables).

As a final remark, it should be noticed that analytical large-signal models should be in turn fitted (in order to identify their parameters) on a set of small- and large-signal performances. This task is particularly hard since accuracy in small-signal conditions for several working points does not automatically grant that the model is also accurate in typical large-signal conditions such as the one-tone test (where the benchmark is harmonic generation) or the two-tone test (where the benchmark is the ability to simulate intermodulation distortion). A particularly difficult task is the identification of accurate wideband large-signal models.

In the following sections, some examples will be provided of large-signal analytical FET models. Sec. 5.8.4 presents some representative MESFET models belonging to the so-called Curtice family, while HEMT state-of-the art models (in particular, the Chalmers model [47, 48]) are described in Sec. 5.8.5.

5.8.4 MESFET LS Models

The *quadratic Curtice* FET model [49] is derived from the ideal quadratic long-channel FET trancharacteristics. The model topology is shown in Fig. 5.22; the model neglects the intrinsic resistance and the gate delay τ. The two diodes connected to the gate aim at simulating direct gate conduction and breakdown conditions; drain breakdown is simulated by exploiting a standard SPICE-like diode breakdown model [50].

The drain current is approximated by the following empirical relation:

$$i_D = \begin{cases} \beta(v_{GS,i} - V_{TO})^2(1 + \lambda v_{DS,i})\tanh(\alpha v_{DS,i}) & \text{if } v_{GS,i} > V_{TO} \\ 0 & \text{if } v_{GS,i} \leq V_{TO}. \end{cases} \quad (5.30)$$

V_{TO} is the FET threshold voltage; an example of the resulting DC curves is shown in Fig. 5.53. The λ parameter is related to the slope, in the saturation region, of the drain

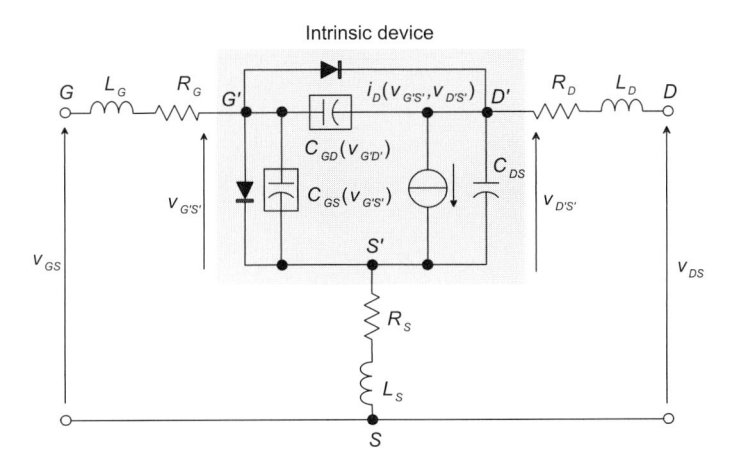

Figure 5.52 Quadratic Curtice large-signal equivalent circuit [49].

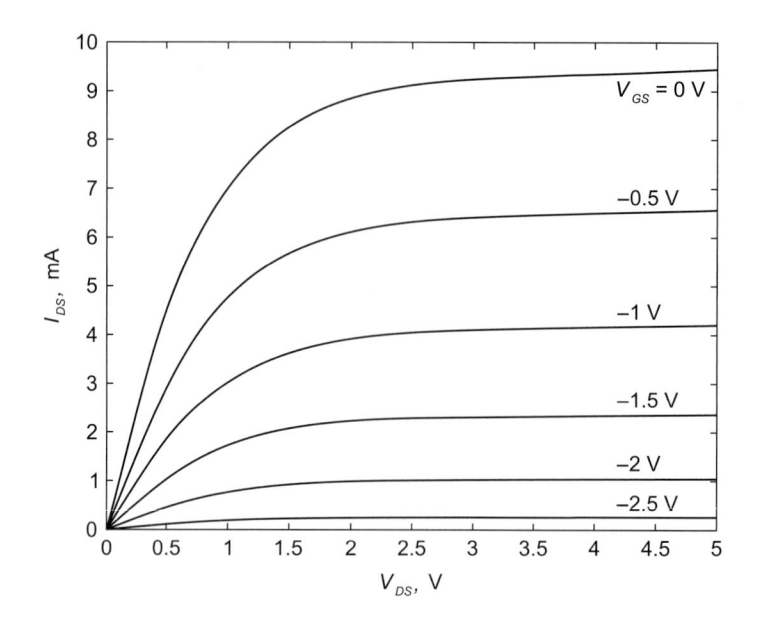

Figure 5.53 Quadratic Curtice DC model behavior. Notice that the breakdown model is not activated in the drain voltage range shown.

current vs. the drain voltage, i.e., to the device output resistance. From (5.28), (5.29) and (5.30), we obtain:

$$g_m = \begin{cases} \dfrac{2i_D}{v_{GS,i} - V_{TO}} & \text{if } v_{GS,i} > V_{TO} \\ 0 & v_{GS,i} \leq V_{TO} \end{cases} \quad (5.31)$$

and:

$$R_{DS}^{-1} = \begin{cases} \beta(v_{GS,i} - V_{TO})^2 \left(\dfrac{\alpha(1 + \lambda v_{DS,i})}{\cosh^2(\alpha v_{DS,i})} + \lambda \tanh(\alpha v_{DS,i}) \right) & v_{GS,i} > V_{TO} \\ 0 & v_{GS,i} \leq V_{TO} \end{cases} \quad (5.32)$$

and, in saturation:

$$g_m = 2\beta(v_{GS,i} - V_{TO})(1 + \lambda V_{DS,i})$$
$$R_{DS}^{-1} = \beta(V_{GS,i} - V_{TO})^2 \lambda.$$

The capacitance model is derived from the junction capacitance SPICE model:

$$C_{GS}(v_{GS,i}) = \begin{cases} \dfrac{C_{GS0}}{\sqrt{1 - v_{GS,i}/V_{bi}}} & \text{if } v_{GS,i} < F_C V_{bi} \\[2mm] \dfrac{C_{GS0}}{\sqrt{1 - F_C}} \left[1 + \dfrac{v_{GS,i} - F_C V_{bi}}{2V_{bi}(1 - F_C)} \right] & \text{if } v_{GS,i} \geq F_C V_{bi}, \end{cases} \quad (5.33)$$

5.8 Measurement-Based Microwave FET Small- and Large-Signal Models

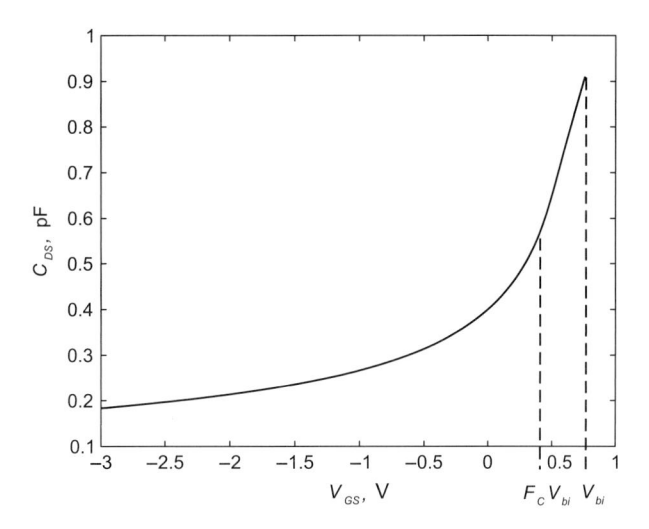

Figure 5.54 Quadratic Curtice model for the gate-source capacitance.

and, for the gate-drain capacitance:

$$C_{GD}(v_{GD,i}) = \begin{cases} \dfrac{C_{GD0}}{\sqrt{1 - v_{GD,i}/V_{bi}}} & \text{if } v_{GD,i} < F_C V_{bi} \\ \dfrac{C_{GD0}}{\sqrt{1 - F_C}} \left[1 + \dfrac{v_{GD,i} - F_C V_{bi}}{2V_{bi}(1 - F_C)} \right] & \text{if } v_{GD,i} \geq F_C V_{bi}, \end{cases} \quad (5.34)$$

where $v_{GD,i}$ is the intrinsic gate-drain voltage $v_{GS,i} - v_{DS,i}$.

In (5.33) and (5.34) the parameter V_{bi} is related to the built-in voltage of the gate Schottky junction. The empirical parameter F_C allows for a smooth behavior of the model. An example of the behavior of the gate-source capacitance is shown in Fig. 5.54. The drain-source capacitance C_{DS} is mainly a geometrical capacitance and therefore turns out to be weakly dependent on the driving voltages; thus it is approximated by a constant value. While the parameters of the DC model can be fitted on the DC (also pulsed) characteristics, the capacitive part can be fitted on the scattering parameters measured in several bias points. The diodes modeling direct conduction and breakdown can be directly derived from the direct conduction gate measurement and from the breakdown voltage measured in DC or pulsed conditions.

A simple evolution of the quadratic Curtice model is the so-called *cubic Curtice* or *Curtice–Ettenberg* model [51]. This model overcomes one of the main limitations of the quadratic model, i.e., the quadratic characteristics. Typical microwave FETs (in particular MESFETs) do not exhibit a linear transconductance vs. the input (gate-source) voltage. The cubic model introduces a polynomial transcharacteristics, with additional degrees of flexibility. Other differences in the model are the inclusion of the intrinsic resistance R_I as a linear resistor and the introduction of a delay time related to the small signal parameter τ. The resulting equivalent circuit is shown in Fig. 5.55.

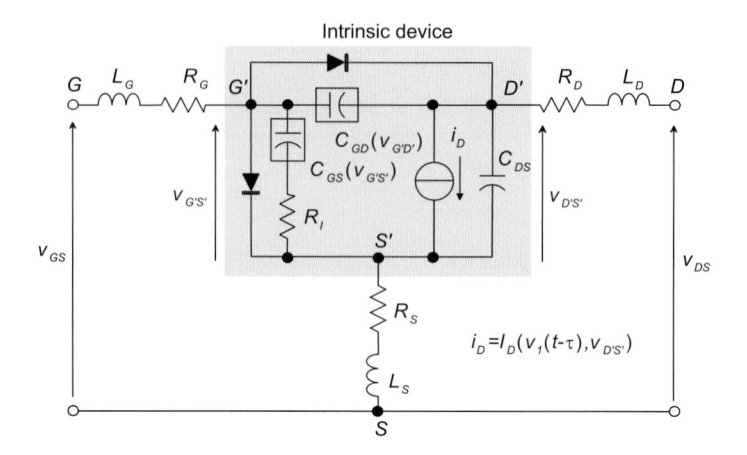

Figure 5.55 Cubic Curtice equivalent circuit model [51]; v_1 is defined in (5.35).

The DC current model for i_D is as follows:

$$i_D = \begin{cases} f(v_1, v_{DS,i}) & \text{if } f(v_1, v_{DS,i}) \geq 0 \\ 0 & \text{if } f(v_1, v_{DS,i}) < 0, \end{cases}$$

where:

$$f(v_1, v_{DS,i}) = (A_0 + A_1 v_1 + A_2 v_1^2 + A_3 v_1^3)(1 + \lambda v_{DS,i}) \tanh(\alpha v_{DS,i})$$
$$v_1 = v_{GS,i}[1 + \beta(v_{DS0} - v_{DS,i})], \quad (5.35)$$

and $v_1(t - \tau)$ is the time-delayed v_1. The i_D parameters to be determined by fitting are $A_0, A_1, A_2, A_3, \beta, \lambda$ and α. For the capacitances C_{GS} and C_{GD} the models are again in (5.33) and (5.34); C_{DS} is again taken as constant.

5.8.5 HEMT LS Models

The Curtice models (and many others) were basically developed for MESFET devices (the quadratic one is in fact a general-purpose FET model also applying to other devices, like the MOSFET). However, such models are not well fitted to simulating the HEMT due to the peculiar bell-shaped behavior of the device transconductance vs. the gate-source voltage. A generalized model able to accurately reproduce this behavior is the so-called Angelov or Chalmers model [47, 48]. Today this model and its modifications are perhaps the most widespread microwave FET models in circuit simulators. The Chalmers model is reported here, see Fig. 5.56, in its original implementation [47] where in fact the resistor R_{GD} was not present. The drain current is expressed as:

$$i_D = I_{pk}(1 + \tanh \psi)(1 + \lambda v_{D'S'}) \tanh \alpha v_{D'S'},$$

where:

$$\psi = P_1(v^* - V_{pk}) + P_2(v^* - V_{pk})^2 + P_3(v^* - V_{pk})^3 + \ldots$$
$$V_{pk} = V_{pk0} + \gamma v_{D'S'}.$$

5.8 Measurement-Based Microwave FET Small- and Large-Signal Models

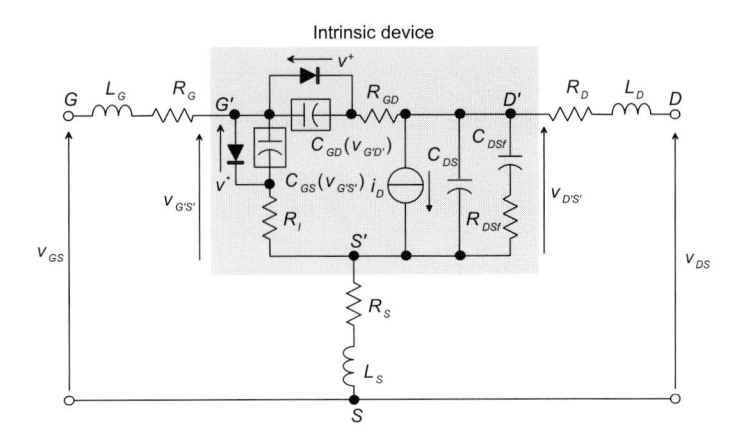

Figure 5.56 An implementation of the Chalmers model equivalent circuit.

The hyperbolic tangent functional dependence on the intrinsic input voltage v^* allows for the typical bell-shaped behavior of the HEMT g_m. In fact one has:

$$\frac{\partial i_D}{\partial v_{G'S'}} \approx \frac{\partial i_D}{\partial v^*} = I_{pk}\left(1 - \tanh^2 \psi\right)\frac{\partial \psi}{\partial v^*} =$$
$$= I_{pk}\left(1 - \tanh^2 \psi\right)\left[P_1 + 2P_2\left(v^* - V_{pk}\right) + 3P_3\left(v^* - V_{pk}\right)^2\right]. \quad (5.36)$$

The additional output series RC group takes into account low-frequency dispersion effects, i.e., discrepancies between the pulsed DC measurements and the true DC measurements originating from slow phenomena like trapping and heating. The capacitance models also exploit a tanh dependence:

$$C_{GS} = C_{GS0}\left(1 + \tanh \psi_1\right)\left(1 + \tanh \psi_2\right)$$
$$C_{GD} = C_{GD0}\left(1 + \tanh \psi_3\right)\left(1 - \tanh \psi_4\right),$$

where:

$$\psi_1 = P_{0GSg} + P_{1GSg}v^* + P_{2GSg}v^{*2} + \ldots$$
$$\psi_2 = P_{0GSd} + P_{1GSd}v_{D'S'} + P_{2GSg}v_{D'S'}^2 + \ldots$$
$$\psi_3 = P_{0GDg} + P_{1GDg}v^* + P_{2GDg}v^{*2} + \ldots$$
$$\psi_4 = P_{0GDd} + \left(P_{1GDd} + P_{1cc}v^*\right)v_{D'S'} + P_{2GDd}v_{D'S'}^2 + \ldots.$$

Since its initial version [47], the model has undergone evolutions and modifications where, in particular, the nonlinear input capacitors have been replaced by charge sources [48]. An example of DC characteristics and of the transconductance behavior is reported in Fig. 5.57 and Fig. 5.58, respectively. Notice that the model does not assume as a parameter the threshold voltage; indeed, V_{pk} is the voltage corresponding to the peak transconductance. The model well reproduces the bell-shaped behavior of the HEMT transconductance, see Fig. 5.58.

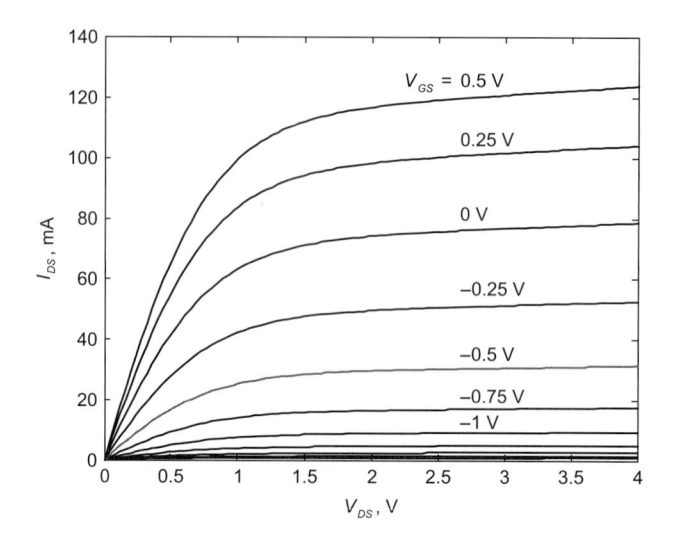

Figure 5.57 The Chalmers model DC characteristics. The parameters are: $I_{pk} = 69$ mA, $\lambda = 0.025$ V^{-1}, $\alpha = 1.3$ V^{-1}, $\gamma = 0$, $V_{pk} = -0.025$ V, $P_1 = 1.42$ V^{-1}, $P_2 = 0$, $P_3 = -0.02$ V^{-3} (data from [47, Fig. 4, caption]). Note that the drain voltage range considered is well below the breakdown voltage.

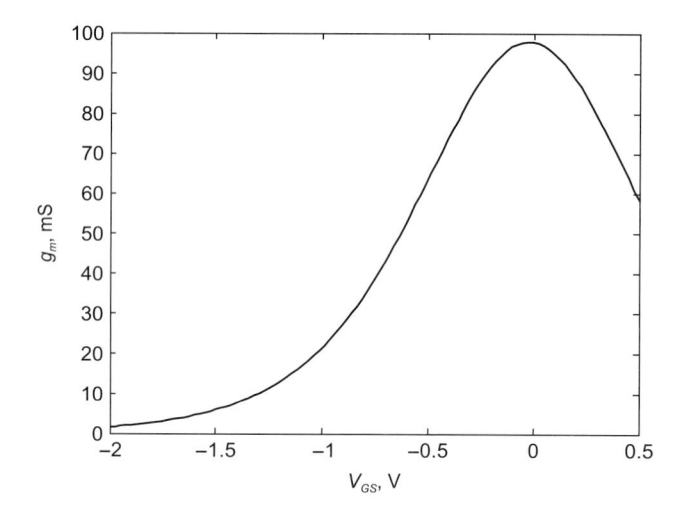

Figure 5.58 The Chalmers model DC transconductance from (5.36). The parameters are: $I_{pk} = 69$ mA, $\lambda = 0.025$ V^{-1}, $\alpha = 1.3$ V^{-1}, $\gamma = 0$, $V_{pk} = -0.025$ V, $P_1 = 1.42$ V^{-1}, $P_2 = 0$, $P_3 = -0.02$ V^{-3} (data from [47, Fig. 4, caption]).

5.9 Measurement-Based Bipolar Small- and Large-Signal Models

The extraction of small-signal bipolar models follows step quite similar to the those outlined in the FET case. Also for bipolar devices a "cold bipolar" measurement can be carried out if the device is driven in saturation (i.e., both the emitter-base and the base-collector junctions are in direct bias) in order to extract the parasitic elements, and the intrinsic device elements can be again fitted from the scattering parameters through a procedure that allows again a direct extraction frequency by frequency.

The situation is somewhat different concerning large-signal models, since bipolar devices allow for a modeling framework based on the device physics, i.e., the Gummel–Poon and Ebers–Moll approaches. Several proprietary device models, moreover, have been developed, like the MEXTRAM model; more details are available in the literature (see [52, 53] and references therein).

5.10 Questions and Problems

5.10.1 Questions

1. Sketch the equilibrium band diagram (E_c, E_v, E_F) of an intrinsic semiconductor assuming $N_c = N_v$. Repeat for an n-doped and p-doped semiconductor. Repeat for a semiconductor under high injection conditions (sketch the position of the quasi-Fermi levels).
2. Quote some compound semiconductors: wide-gap, narrow-gap, general purpose. What are wide-gap semiconductors for?
3. Explain the composition of the AlGaAs and InGaAsP alloys.
4. Explain the composition of the AlGaN alloy.
5. What is the difference between SiC and SiGe? Which is a compound semiconductor and which is an alloy?
6. Explain the purpose of compound semiconductor alloys.
7. Suppose you want to grow a lattice-matched layer of InGaAs on an InP substrate. What is the In composition? Repeat for InAlAs.
8. Explain the meaning of lattice-matched and pseudomorphic heterostructures.
9. List the following substrates in order of availability and maturity (one of the materials is not an available substrate today!): GaAs, GaN, SiC, Si, InP.
10. Explain why a GaAs substrate is semi-insulating, while a Si substrate is not.
11. Qualitatively explain why the velocity-field characteristic of electrons in a compound semiconductor first increases, then decreases and finally saturates at high electric field.
12. Define a heterostructure and sketch the related band diagram. Suppose to consider, for the sake of definiteness, a heterostructure between a widegap and a narrowgap material.
13. Explain how a quantum well can be used as the conducting channel of a field-effect transistor.
14. Sketch the structure of a GaAs MESFET and the DC characteristics of the device.
15. Sketch the structure of a GaAs-based PHEMT and the DC characteristics of the device.
16. Sketch the structure of an InP-based PHEMT and the DC characteristics of the device.

17. Sketch the cross section of a GaN-based HEMT. Quote a few advantages of widegap semiconductors.
18. What is a metamorphic HEMT?
19. Sketch the maximum gain, low-noise, maximum power working point of a FET.
20. List in order of performance (cutoff frequency) the following devices: InP-based PHEMT, GaAs-based conventional HEMT, Si-based MOSFET, GaAs-based MESFET, GaAs-based PHEMT, GaAs-based MHEMT.
21. What is the heterostructure bipolar transistor? Why are the frequency performances of this device better than the ones of a conventional bipolar transistor?
22. In a SiGe HBT-based circuit, where do you find Ge?
23. Explain what are physics-based models, circuit models, black-box models.
24. Sketch a MESFET small-signal equivalent circuit and separate the extrinsic parasitics from the intrinsic model.
25. Sketch the behavior of the scattering parameters of a microwave FET.
26. Define the cutoff frequency of a FET and evaluate it from the intrinsic FET parameters.
27. Define the maximum oscillation frequency of a FET. How is this related to the cutoff frequency?
28. Derive the $f_{\max}$ of a unilateral FET from the intrinsic parameters.
29. What is the typical behavior of the MAG as a function of frequency?
30. Comment on the power density (W/mm) in GaN HEMTs as compared to III-V HEMTs and MOSFETs.
31. What are the reasons for the good performances of microwave nanometer MOSFETs despite the inferior transport properties of electrons in Si?
32. Explain how the FET equivalent circuit can be extracted from measured S parameters.
33. Sketch the large-signal equivalent circuit of a FET in DC and explain the relationship of the output generator to the device transconductance and output resistance.
34. Explain how a nonlinear capacitor can be derived from small-signal measurements of the same.
35. Sketch a quadratic Curtice model and justify the presence of diodes in this model.
36. Sketch some DC and DC pulsed VI curves exhibiting low-frequency dispersion effects.
37. To what device does the Chalmers model specially apply?

5.10.2 Problems

1. A HEMT has gate length of 50 nm, thickness of the supply layer $d + \Delta d = 10$ nm, gate width 100 μm. The relative dielectric constant of the supply layer is $\epsilon_r = 13$. Evaluate the maximum device transconductance and cutoff frequency assuming an equivalent electron saturation velocity $v_n = 2 \times 10^7$ cm/s.
2. The cutoff frequency of a HEMT is 400 GHz while the maximum oscillation frequency is 900 GHz. The gate peryphery is $W = 200$ μm while the transconductance

per unit length is 800 mS/mm. The gate and intrinsic resistances are $R_G = 5\,\Omega$ and $R_I = 4\,\Omega$. Estimate the gate-source capacitance and the output resistance R_{DS}.
3. A heterojunction bipolar transistor has a base to emitter capacitance $C_{BE} = 5$ pF. The DC collector current is $I_C = 100$ mA. Estimate the ideal cutoff frequency.
4. A FET has $R_{DS} \to \infty$, $V_{T0} = -2$ V, drain current at $v_{GS} = 0$ equal to $I_{DSS} = 100$ mA, output conductance $\partial I_D/\partial V_{DS} = 100$ mS for $v_{GS} = 0$, $v_{DS} \to 0$. Evaluate the values of the parameters of the quadratic Curtice model β, α, V_{T0}, λ. (Neglect the difference between intrinsic and extrinsic voltages.)
5. Consider a simplified small-signal equivalent circuit of a bipolar transistor in the common emitter configuration, where only the intrinsic circuit is considered, the input includes the base-emitter capacitance C_{BE} and the base-emitter resistance R_{BE}, while the output has the current generator βI_B in parallel with the output resistance R_{CE}. Evaluate the maximum available power gain (see Sec. 6.3.1) of the stage and the optimum input and output matching condition.
6. Consider a simplified small-signal equivalent circuit of a FET in the common source configuration, where only the intrinsic circuit is considered, the input includes the gate resistance R_G and the gate-source capacitance C_{GS}, while the output has the transconductance generator $g_m V^*$, where V^* is the voltage across C_{GS} in parallel with the output resistance R_{DS}. Evaluate the maximum available power gain (see Sec. 6.3.1) of the stage, together with the optimum input and output matching conditions.

References

[1] J. X. Qiu, B. Levush, J. Pasour, A. Katz, C. M. Armstrong, D. R. Whaley, J. Tucek, K. Kreischer, and D. Gallagher, "Vacuum tube amplifiers," *IEEE Microwave Magazine*, vol. 10, no. 7, pp. 38–51, Dec. 2009.

[2] J. H. Booske, R. J. Dobbs, C. D. Joye, C. L. Kory, G. R. Neil, G. S. Park, J. Park, and R. J. Temkin, "Vacuum electronic high power terahertz sources," *IEEE Transactions on Terahertz Science and Technology*, vol. 1, no. 1, pp. 54–75, Sep. 2011.

[3] J. Whitaker, *Power vacuum tubes handbook*. CRC Press, 2012.

[4] R. J. Trew, "Wide bandgap transistor amplifiers for improved performance microwave power and radar applications," in *Microwaves, Radar and Wireless Communications, 2004. MIKON-2004. 15th International Conference on*, vol. 1, May 2004, pp. 18–23 Vol. 1.

[5] S. Sze and K. K. Ng, *Physics of semiconductor devices*. Wiley Online Library, 2007.

[6] P. H. Siegel, T. Loffler, D. Mittleman, K. Mizuno, and X. C. Zhang, "Guest editorial: terahertz technology: bridging the microwave-to-photonics gap," *IEEE Transactions on Microwave Theory and Techniques*, vol. 58, no. 7, pp. 1901–1902, Jul. 2010.

[7] F. Schwierz and J. J. Liou, *Modern microwave transistors: theory, design, and performance*. Wiley-Interscience, 2003.

[8] F. Schwierz, H. Wong, and J. J. Liou, *Nanometer CMOS*. Pan Stanford Publishing, 2010.

[9] F. Schwierz and C. Schippel, "Performance trends of Si-based RF transistors," *Microelectronics Reliability*, vol. 47, no. 2–3, pp. 384–390, 2007.

[10] F. Schwierz and J. J. Liou, "RF transistors: recent developments and roadmap toward terahertz applications," *Solid-State Electronics*, vol. 51, no. 8, pp. 1079–1091, 2007.

[11] C. A. Mead, "Schottky barrier gate field effect transistor," *Proceedings of the IEEE*, vol. 54, no. 2, pp. 307–308, Feb. 1966.

[12] S. M. Sze, *Semiconductor devices: physics and technology*. John Wiley & Sons, 2008.

[13] R. Dingle, H. L. Stoermer, A. C. Gossard, and W. Wiegmann, "Electron mobilities in modulation-doped semiconductor heterojunction superlattices," *Applied Physics Letters*, vol. 33, no. 7, pp. 665–667, 1978.

[14] T. Mimura, S. Hiyamizu, T. Fujii, and K. Nanbu, "A new field-effect transistor with selectively doped GaAs/n-Al_xGa_{1-x} as heterojunctions," *Japanese Journal of Applied Physics*, vol. 19, no. 5, p. L225, 1980.

[15] D. Delagebeaudeuf, P. Delescluse, P. Etienne, M. Laviron, J. Chaplart, and N. T. Linh, "Two-dimensional electron gas m.e.s.f.e.t. structure," *Electronics Letters*, vol. 16, no. 17, pp. 667–668, Aug. 1980.

[16] M. Asif Khan, A. Bhattarai, J. N. Kuznia, and D. T. Olson, "High electron mobility transistor based on a GaN/Al_xGa_{1-x} heterojunction," *Applied Physics Letters*, vol. 63, no. 9, pp. 1214–1215, 1993.

[17] H. Kawarada, "Diamond field effect transistors using H-terminated surfaces," in *Thin Film Diamond, Chapter 7*, C. E. Nebel and J. Ristein, Eds. San Diego, USA: Elsevier Inc., 2004.

[18] M. Kasu, "Diamond field-effect transistors as microwave power amplifiers," *NTT Tech. Rev*, vol. 8, pp. 1–5, 2010.

[19] W. Dumke, J. Woodall, and V. Rideout, "GaAs/GaAlAs heterojunction transistor for high frequency operation," *Solid-State Electronics*, vol. 15, no. 12, pp. 1339–1343, 1972.

[20] J. D. Cressler and G. Niu, *Silicon-Germanium heterojunction bipolar transistors*. Artech House, 2002.

[21] F. Schwierz, "Global trends in microwave and millimeter-wave power devices," in *Presentation in Workshop "Microwave power devices – the european perspective," European Microwave Week 2008*, 2008.

[22] "Ioffe Institute of the Russian Academy of Sciences web site on semiconductors," www.ioffe.ru/SVA/NSM/Semicond/.

[23] M. Farahmand, C. Garetto, E. Bellotti, K. F. Brennan, M. Goano, E. Ghillino, G. Ghione, J. D. Albrecht, and P. P. Ruden, "Monte Carlo simulation of electron transport in the III-nitride wurtzite phase materials system: binaries and ternaries," *IEEE Transactions on Electron Devices*, vol. 48, no. 3, pp. 535–542, Mar. 2001.

[24] F. Schwierz, "Wide bandgap and other non-III-V RF transistors: trends and prospects," in *CSSER 2004 Spring Lecture Series – ASU Tempe*, Mar. 2004.

[25] T. P. Pearsall, *GaInAsP alloy semiconductors*. John Wiley & Sons, 1982.

[26] I. Vurgaftman and J. Meyer, "Band parameters for nitrogen-containing semiconductors," *Journal of Applied Physics*, vol. 94, no. 6, pp. 3675–3696, 2003.

[27] E. M. Bastida, E. C. D'Oro, G. P. Donzelli, N. Fanelli, G. Fazzini, and G. Simonetti, "A monolithic 800 mhz bandwidth dbs front-end receiver for mass production," in *1984 14th European Microwave Conference*, Sep. 1984, pp. 755–760.

[28] C. Naldi, private communication.

[29] L. A. Samoska, "An overview of solid-state integrated circuit amplifiers in the submillimeter-wave and THz regime," *IEEE Transactions on Terahertz Science and Technology*, vol. 1, no. 1, pp. 9–24, Sep. 2011.

[30] S. Bollaert, Y. Cordier, M. Zaknoune, T. Parenty, H. Happy, and A. Cappy, "HEMT's capability for millimeter wave applications," *Annales Des Télécommunications*, vol. 56, no. 1, pp. 15–26, 2001.

[31] K. M. Lau, C. W. Tang, H. Li, and Z. Zhong, "AlInAs/GaInAs mHEMTs on silicon substrates grown by MOCVD," in *2008 IEEE International Electron Devices Meeting*, Dec. 2008, pp. 1–4.

[32] S. Ramo, J. R. Whinnery, and T. Van Duzer, *Fields and waves in communication electronics*, 3rd ed. New York: John Wiley & Sons, 1994.

[33] C. H. Jan, M. Agostinelli, H. Deshpande, M. A. El-Tanani, W. Hafez, U. Jalan, L. Janbay, M. Kang, H. Lakdawala, J. Lin, Y. L. Lu, S. Mudanai, J. Park, A. Rahman, J. Rizk, W. K. Shin, K. Soumyanath, H. Tashiro, C. Tsai, P. VanDerVoorn, J. Y. Yeh, and P. Bai, "RF CMOS technology scaling in high-k/metal gate era for RF SoC (system-on-chip) applications," in *Electron Devices Meeting (IEDM), 2010 IEEE International*, Dec. 2010, pp. 27.2.1–27.2.4.

[34] N. Planes, O. Weber, V. Barral, S. Haendler, D. Noblet, D. Croain, M. Bocat, P. O. Sassoulas, X. Federspiel, A. Cros, A. Bajolet, E. Richard, B. Dumont, P. Perreau, D. Petit, D. Golanski, C. Fenouillet-Branger, N. Guillot, M. Rafik, V. Huard, S. Puget, X. Montagner, M. A. Jaud, O. Rozeau, O. Saxod, F. Wacquant, F. Monsieur, D. Barge, L. Pinzelli, M. Mellier, F. Boeuf, F. Arnaud, and M. Haond, "28 nm FDSOI technology platform for high-speed low-voltage digital applications," in *VLSI Technology (VLSIT), 2012 Symposium on*, Jun. 2012, pp. 133–134.

[35] S. P. Voinigescu, S. Shopov, and P. Chevalier, "Millimeter-wave silicon transistor and benchmark circuit scaling through the 2030 ITRS horizon," in *Millimeter Waves (GSMM), 2015 Global Symposium On*, May 2015, pp. 1–3.

[36] G. Crupi, D. M.-P. Schreurs, J.-P. Raskin, and A. Caddemi, "A comprehensive review on microwave FinFET modeling for progressing beyond the state of art," *Solid-State Electronics*, vol. 80, pp. 81–95, 2013.

[37] J.-P. Colinge, V. S. Lysenko, and A. Nazarov, *Physical and technical problems of SOI structures and devices*. Springer Science & Business Media, 2012, vol. 4.

[38] R. Lai, X. B. Mei, W. R. Deal, W. Yoshida, Y. M. Kim, P. H. Liu, J. Lee, J. Uyeda, V. Radisic, M. Lange, T. Gaier, L. Samoska, and A. Fung, "Sub 50 nm InP HEMT Device with Fmax greater than 1 THz," in *2007 IEEE International Electron Devices Meeting*, Dec. 2007, pp. 609–611.

[39] F. Schwierz, "The frequency limits of field-effect transistors: MOSFET vs. HEMT," in *9th International Conference on Solid-State and Integrated-Circuit Technology, 2008. ICSICT 2008.*, Oct. 2008, pp. 1433–1436.

[40] A. R. Barnes, A. Boetti, L. Marchand, and J. Hopkins, "An overview of microwave component requirements for future space applications," in *European Gallium Arsenide and Other Semiconductor Application Symposium, GAAS 2005*, Oct. 2005, pp. 5–12.

[41] H. P. D. Lanyon and R. A. Tuft, "Bandgap narrowing in moderately to heavily doped silicon," *IEEE Transactions on Electron Devices*, vol. 26, no. 7, pp. 1014–1018, Jul. 1979.

[42] W. Shockley, "Circuit element utilizing semiconductive material," Sep. 1951, US Patent 2,569,347.

[43] G. Crupi and D. Schreurs, *Microwave de-embedding: from theory to applications*. Academic Press, 2013.

[44] G. Dambrine, A. Cappy, F. Heliodore, and E. Playez, "A new method for determining the FET small-signal equivalent circuit," *IEEE Transactions on Microwave Theory and Techniques*, vol. 36, no. 7, pp. 1151–1159, Jul. 1988.

[45] D. E. Root, S. Fan, and J. Meyer, "Technology independent large signal non quasi-static FET models by direct construction from automatically characterized device data," in *1991 21st European Microwave Conference*, vol. 2, Sep. 1991, pp. 927–932.

[46] D. E. Root and B. Hughes, "Principles of nonlinear active device modeling for circuit simulation," in *32nd ARFTG Conference Digest*, vol. 14, Dec. 1988, pp. 1–24.

[47] I. Angelov, H. Zirath, and N. Rosman, "A new empirical nonlinear model for HEMT and MESFET devices," *IEEE Transactions on Microwave Theory and Techniques*, vol. 40, no. 12, pp. 2258–2266, Dec. 1992.

[48] I. Angelov, L. Bengtsson, and M. Garcia, "Extensions of the Chalmers nonlinear hemt and mesfet model," *IEEE Transactions on Microwave Theory and Techniques*, vol. 44, no. 10, pp. 1664–1674, Oct. 1996.

[49] W. R. Curtice, "A MESFET model for use in the design of GaAs integrated circuits," *IEEE Transactions on Microwave Theory and Techniques*, vol. 28, no. 5, pp. 448–456, May 1980.

[50] P. Antognetti and G. Massobrio, *Semiconductor device modeling with SPICE*. McGraw-Hill, Inc., 1993.

[51] W. R. Curtice and M. Ettenberg, "A nonlinear GaAs FET model for use in the design of output circuits for power amplifiers," *IEEE Transactions on Microwave Theory and Techniques*, vol. 33, no. 12, pp. 1383–1394, Dec. 1985.

[52] J. Mcmacken, S. Nedeljkovic, J. Gering, and D. Halchin, "HBT Modeling," *IEEE Microwave Magazine*, vol. 9, no. 2, pp. 48–71, Apr. 2008.

[53] M. Rudolph, "Current trends and challenges in III-V HBT compact modeling," in *Microwave Integrated Circuit Conference, 2008. EuMIC 2008. European*, Oct. 2008, pp. 278–281.

6 Microwave Linear Amplifiers

6.1 Introduction

The task of RF amplifiers is to transfer the input signal to the load with an increased power, without overly corrupting the signal through added noise and linear or nonlinear distortion (harmonic or intermodulation). From the energy standpoint, the amplifier converts DC power from the power supply into RF power to the load with a certain *efficiency*. The output signal should be as faithful as possible a replica of the input signal, or, in a digital context, should replicate with an acceptable error the symbols in the input signal. This requires constant gain over the frequency bandwidth and the input signal dynamic range, linear phase relationship between input and output, and, in general, low enough nonlinear distortion. Such requirements should be satisfied for the whole amplifier frequency bandwidth and for a range of the input signal amplitude defining the amplifier *dynamics* or Spurious Free Dynamic Range (SFDR); see Sec. 8.2.3. Because of such requirements, amplifiers should behave, within the SFDR, as linear or quasi-linear components, i.e., nonlinearity is a factor that adversely affects the amplifier performance.[1]

As already discussed, three main amplifier classes can be found in transceivers, the Low-Noise Amplifier (LNA) whose main purpose is to amplify weak input signals to the receiver chain with an acceptable compromise between gain and noise; the high-gain amplifier (present in the RX and TX stages, often in the IF section), where noise is not the main concern but the primary purpose of design is to maximize the amplifier gain; the Power Amplifier (PA), whose main purpose is to deliver an output signal of the TX chain with a power level adequate for the specific system. The PA maximum power is limited by signal distortion and, ultimately, by power saturation. The maximum power that a transistor can deliver is in fact approximately proportional to the product of the maximum output current and of the breakdown voltage, and distortion increases with increasing output power. Indeed, increasing the transistor periphery or area, thus increasing the saturation power, and keeping the output signal level low with respect to the saturation power (the so-called output backoff, see Chapter 8) is a way to reduce (in class A amplifiers) distortion, and therefore to increase the amplifier dynamics. However, since the DC power dissipated in a class A amplifier is independent from

---

[1] A nonlinear model is of course needed to investigate the limitations of the amplifier dynamics caused by distortion.

the signal level, working in backoff with a "large" device negatively affects the amplifier efficiency. Since distortion should be under control also in a low-noise or high-gain linear amplifier, the efficiency–linearity tradeoff (to be discussed in detail in Chapter 8) is an issue in the design of *any* amplifier. In the present analysis we will however neglect nonlinear effects (distortion, power saturation) and efficiency issues and confine the treatment to the purely linear design, i.e., to the *small-signal* approximation.

The chapter starts from the analysis of power transfer between a generator and a load through a two-port (Sec. 6.2) to define the loaded two-port power gains (Sec. 6.3). Sec. 6.4 is devoted to a discussion on the issue of two-port stability; stability criteria are introduced in Sec. 6.5. The connection between amplifier gains and stability issues is finally reviewed in Sec. 6.6. The main classes of linear, maximum gain amplifiers are introduced in Sec. 6.8, while Sec. 6.9 is devoted to the classical narrowband open-loop amplifier design. Stabilization and bias issues are discussed in Sec. 6.10. Approaches to design wideband amplifiers (in particular, the balanced architecture) are introduced in Sec. 6.11; finally, Sec. 6.12 and Sec. 6.13 are devoted to extremely wideband amplifiers based on the feedback and distributed techniques, respectively.

6.2 Generator-Load Power Transfer

6.2.1 Generator Directly Connected to Load

Consider the real generator connected to a load in Fig. 6.1; the power absorbed by the load impedance $Z_L = R_L + jX_L$, P_L, is:

$$P_L = \text{Re}\left[V_L I_L^*\right] = |V_0|^2 \frac{R_L}{|Z_G + Z_L|^2}, \tag{6.1}$$

where $Z_G = R_G + jX_G$ is the generator impedance. The maximum power transfer (maximum power on the load) occurs in power matching conditions, i.e., when:

$$Z_L = Z_G^*. \tag{6.2}$$

(see Example 6.1); the maximum load power, also called *generator available power*, is:

$$P_{\text{av}} = \frac{|V_0|^2}{4R_G}. \tag{6.3}$$

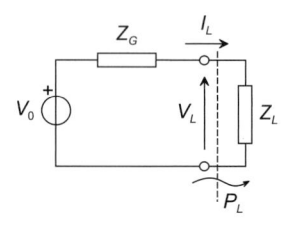

Figure 6.1 Evaluating the power transfer on a load.

Similarly, for a generator with short-circuit current I_0 and internal admittance $Y_G = G_G + jB_G$, the generator available power can be expressed as:

$$P_{av} = \frac{|I_0|^2}{4G_G}, \qquad (6.4)$$

where G_G is the source internal conductance.

Example 6.1 Obtain (6.2) and (6.3).

Solution

The maximum of P_L vs. R_L and X_L corresponds to a zero of the partial derivatives of P_L vs. the two variables, i.e., to conditions:

$$\begin{cases} \dfrac{\partial P_L}{\partial R_L} = |V_0|^2 \dfrac{|Z_G + Z_L|^2 - 2R_L(R_L + R_G)}{|Z_G + Z_L|^4} = 0 \\ \dfrac{\partial P_L}{\partial X_L} = -|V_0|^2 \dfrac{2R_L(X_L + X_G)}{|Z_G + Z_L|^4} = 0; \end{cases}$$

from the second equation we find $X_L = -X_G$; substituting in the first we obtain $R_L = R_G$ and therefore the load corresponding to the maximum power transfer from a generator with internal impedance Z_G is $Z_L = Z_G^*$. (The value $R_L = R_G$ clearly corresponds to a maximum since the P_L vanishes for $R = 0$ and for $R_L \to \infty$.) Substituting, we find the generator maximum or available power in (6.3). A similar procedure holds for a generator in its parallel (Norton) form, leading to (6.4).

The same result derived by exploiting voltages and currents can be obtained by describing the circuit in terms of power waves. The circuit in Fig. 6.1 results from the connection of two one-ports, the voltage generator with internal impedance Z_G and the load with impedance Z_L. A power-wave version of the same circuit is shown in Fig. 6.2; for simplicity, the normalization resistance is the same for both one-ports. The scattering parameter of the generator and load are the reflection coefficients Γ_G and Γ_L, respectively; these, and the forward wave generator b_0, are expressed from (3.31) and (3.13) as:

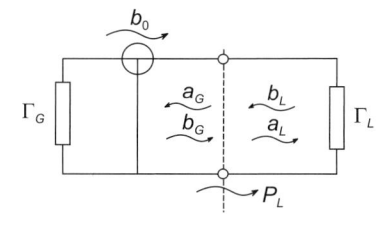

Figure 6.2 Evaluating the power on a load through power waves.

$$\begin{cases} \Gamma_G = \dfrac{Z_G - R_0}{Z_G + R_0} \\ \Gamma_L = \dfrac{Z_L - R_0}{Z_L + R_0} \\ b_0 = V_0 \dfrac{\sqrt{R_0}}{Z_G + R_0}. \end{cases} \qquad (6.5)$$

The power on the load can be evaluated by taking into account that the circuit in Fig. 6.2 implies two topological relationships in terms of power wave continuity and two constitutive relations (generator and load):

$$\begin{cases} a_L = b_G \\ a_G = b_L \\ b_G = b_0 + \Gamma_G a_G \\ b_L = \Gamma_L a_L. \end{cases}$$

Solving, the power waves on the load are:

$$\begin{cases} a_L = \dfrac{b_0}{1 - \Gamma_G \Gamma_L} \\ b_L = \dfrac{b_0 \Gamma_L}{1 - \Gamma_G \Gamma_L}. \end{cases}$$

Thus, from (3.5) the power on the load is:

$$P_L = |a_L|^2 - |b_L|^2 = |a_L|^2 (1 - |\Gamma_L|^2) = |b_0|^2 \frac{1 - |\Gamma_L|^2}{|1 - \Gamma_G \Gamma_L|^2}. \qquad (6.6)$$

P_L is maximum for $\Gamma_G = \Gamma_L^*$ (i.e., for $Z_L = Z_G^*$); the maximum power (or generator available power) is:

$$P_{\text{av}} = |b_0|^2 \frac{1}{1 - |\Gamma_G|^2} = |V_0|^2 \frac{R_0}{|Z_G + R_0|^2} \frac{1}{1 - |\Gamma_G|^2}, \qquad (6.7)$$

that is equivalent, with proper substitutions, to (6.3).[2] From (6.6) and (6.7) we obtain:

$$P_L = P_{\text{av}} \frac{(1 - |\Gamma_G|^2)(1 - |\Gamma_L|^2)}{|1 - \Gamma_G \Gamma_L|^2}. \qquad (6.8)$$

Notice that for $\Gamma_L = 0$ we do not have maximum power transfer; in fact in that case we obtain from (6.8):

$$P_L = P_{\text{av}} (1 - |\Gamma_G|^2) \leq P_{\text{av}}$$

and $P_L = P_{\text{av}}$ only if $\Gamma_G = 0$, implying that the load and source impedances are the normalization resistance.

---

[2] To obtain (6.7) from (6.6) with the condition $\Gamma_G = \Gamma_L^*$ remember that $1 - \Gamma_G \Gamma_L = 1 - |\Gamma_L|^2 \geq 0$ ($|\Gamma_L| \leq 1$ since passive) so that $(1 - |\Gamma_L|^2)/|1 - |\Gamma_L|^2|^2 = 1/(1 - |\Gamma_L|^2) = 1/(1 - |\Gamma_G|^2)$.

6.2 Generator-Load Power Transfer

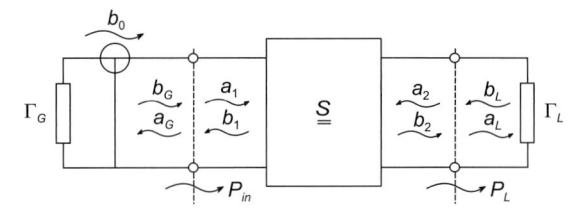

Figure 6.3 Power transfer between generator and load through a two-port.

6.2.2 Power Transfer in Loaded Two-Ports

A loaded linear two-port is a general model for a linear, small-signal amplifier. The purpose of the analysis is to understand under which conditions the power transfer from the generator to the load is maximized. Contrarily to the simple direct interconnection case considered in Sec. 6.2.1, we will find here that power transfer maximization through simultaneous conjugate matching at the input and output ports is possible if and only if the two-port is *unconditionally stable*, with significant impact on the design strategy of linear amplifiers.

The purpose of the initial analysis is to evaluate, in the circuit in Fig. 6.3, the load power P_L as a function of the S-parameters of the two-port and of the load and generator reflection coefficients Γ_L and Γ_G. For simplicity, we assume the same normalization resistance at all ports. The power wave continuity and constitutive relationships read:

$$\begin{cases} \text{topological relations:} & \begin{cases} a_1 = b_G \\ b_1 = a_G \\ b_2 = a_L \\ a_2 = b_L \end{cases} \\ \text{constitutive relations:} & \begin{cases} b_G = b_0 + \Gamma_G a_G \\ b_1 = S_{11} a_1 + S_{12} a_2 \\ b_2 = S_{21} a_1 + S_{22} a_2 \\ b_L = \Gamma_L a_L. \end{cases} \end{cases}$$

Eliminating a_G, b_G, a_L and b_L from the power wave continuity equations we obtain the reduced system:

$$\begin{pmatrix} 1 & -\Gamma_G & 0 & 0 \\ -S_{11} & 1 & -S_{12} & 0 \\ -S_{21} & 0 & -S_{22} & 1 \\ 0 & 0 & 1 & -\Gamma_L \end{pmatrix} \begin{pmatrix} a_1 \\ b_1 \\ a_2 \\ b_2 \end{pmatrix} = b_0 \begin{pmatrix} 1 \\ 0 \\ 0 \\ 0 \end{pmatrix}, \qquad (6.9)$$

with solution:

$$a_1 = b_G = b_0 \frac{1 - S_{22} \Gamma_L}{\Delta} \qquad (6.10a)$$

$$b_1 = a_G = b_0 \frac{S_{12} S_{21} \Gamma_L + S_{11}(1 - S_{22} \Gamma_L)}{\Delta} = b_0 \frac{S_{11} - \Delta_S \Gamma_L}{\Delta} \qquad (6.10b)$$

$$a_2 = b_L = b_0 \frac{\Gamma_L S_{21}}{\Delta} \qquad (6.10c)$$

$$b_2 = a_L = b_0 \frac{S_{21}}{\Delta}, \qquad (6.10d)$$

where Δ_S is the determinant of the S-matrix and Δ is the determinant of the system matrix (6.9):

$$\Delta = (1 - S_{11}\Gamma_G)(1 - S_{22}\Gamma_L) - S_{12}S_{21}\Gamma_G\Gamma_L.$$

From the system solution (6.10) we can derive a number of parameters, that are now discussed.

The **input reflection coefficient** of the loaded two-port seen from the input generator (see Fig. 6.4, above) can be obtained as the ratio b_1/a_1:

$$\Gamma_{in} = \frac{b_1}{a_1} = S_{11} + \frac{S_{12}S_{21}\Gamma_L}{1 - S_{22}\Gamma_L} = \frac{S_{11} - \Delta_S \Gamma_L}{1 - S_{22}\Gamma_L}. \qquad (6.11)$$

Conversely, the two-port, loaded by the input generator, has to be modeled, at the output port, as a non-autonomous one-port, see Fig. 6.4 (below), due to the presence of the input generator. The **two-port output equivalent circuit** will be described by the following, non-homogeneous constitutive equation:

$$b_2 = \bar{b}_0 + \Gamma_{out} a_2, \qquad (6.12)$$

where the **output reflection coefficient** Γ_{out} can be derived by symmetry exchanging ports 1 and 2 and Γ_L with Γ_G in (6.11):

$$\Gamma_{out} = S_{22} + \frac{S_{12}S_{21}\Gamma_G}{1 - S_{11}\Gamma_G} = \frac{S_{22} - \Delta_S \Gamma_G}{1 - S_{11}\Gamma_G}. \qquad (6.13)$$

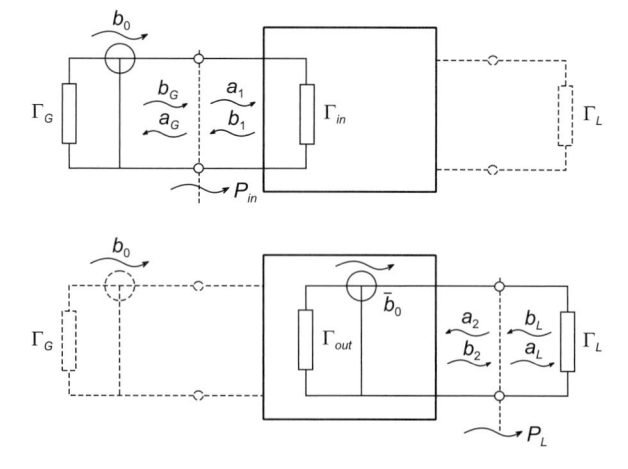

Figure 6.4 Loaded two-port: equivalent circuit at port 1 (above), equivalent circuit at port 2 (below).

6.2 Generator-Load Power Transfer

Moreover, we have from (6.12) that $b_2 = \bar{b}_0$ when port 2 is loaded by the normalization resistance, i.e., when $\Gamma_L = 0$. In this case we have from the solution for b_2:

$$\bar{b}_0 = b_0 \frac{S_{21}}{1 - S_{11}\Gamma_G}. \tag{6.14}$$

The power entering the two-port and delivered to the load can be expressed directly from the input and load power waves. For the **two-port input power** we have:

$$P_{in} = |a_1|^2 - |b_1|^2 = |b_0|^2 \frac{|1 - S_{22}\Gamma_L|^2 - |S_{11} - \Delta_S \Gamma_L|^2}{|(1 - S_{11}\Gamma_G)(1 - S_{22}\Gamma_L) - S_{12}S_{21}\Gamma_G\Gamma_L|^2}. \tag{6.15}$$

An alternative expression can be derived by introducing the input reflection coefficient; the analysis reduces to a one-port (the generator) loaded by Γ_{in}, and therefore:

$$a_1 = b_0 \frac{1}{1 - \Gamma_G \Gamma_{in}},$$

from which:

$$P_{in} = |a_1|^2(1 - |\Gamma_{in}|^2) = |b_0|^2 \frac{1 - |\Gamma_{in}|^2}{|1 - \Gamma_G \Gamma_{in}|^2}. \tag{6.16}$$

Expressions (6.15) and (6.16) are equivalent. The **input available power** is the available power of the generator; it can be obtained from (6.16) assuming input conjugate matching as:

$$P_{av,in} = |b_0|^2 \frac{1}{1 - |\Gamma_G|^2}. \tag{6.17}$$

The **power on the load** can be also expressed in several equivalent ways, either directly:

$$P_L = |a_L|^2 - |b_L|^2 = |b_0|^2 \frac{|S_{21}|^2(1 - |\Gamma_L|^2)}{|(1 - S_{11}\Gamma_G)(1 - S_{22}\Gamma_L) - S_{12}S_{21}\Gamma_G\Gamma_L|^2}, \tag{6.18}$$

or by means of the equivalent circuit at port 2, cfr. Fig. 6.4 (below):

$$P_L = |\bar{b}_0|^2 \frac{1 - |\Gamma_L|^2}{|1 - \Gamma_{out}\Gamma_L|^2} = |b_0|^2 \frac{|S_{21}|^2(1 - |\Gamma_L|^2)}{|1 - \Gamma_L \Gamma_{out}|^2 |1 - S_{11}\Gamma_G|^2}, \tag{6.19}$$

or, finally, with reference to the input reflection coefficient:

$$P_L = |b_0|^2 \frac{|S_{21}|^2(1 - |\Gamma_L|^2)}{|1 - \Gamma_G\Gamma_{in}|^2 |1 - S_{22}\Gamma_L|^2}. \tag{6.20}$$

In fact, from the expression of a_L:

$$a_L = b_0 \frac{S_{21}}{(1 - S_{11}\Gamma_G)(1 - S_{22}\Gamma_L) - S_{12}S_{21}\Gamma_G\Gamma_L},$$

we have, collecting the term $(1 - S_{22}\Gamma_L)$ and taking into account the expression of Γ_{in} (6.11):

$$a_L = b_0 \frac{S_{21}}{(1 - S_{22}\Gamma_L)(1 - \Gamma_G\Gamma_{in})},$$

from which, using $P_L = |a_L|^2(1 - |\Gamma_L|^2)$, we obtain (6.20).

6.3 Power Gains of a Loaded Two-Port

Inserting a two-port between a generator and a load has an influence on the power exchange, introducing gain (as in an amplifier) or loss (as in attenuator).[3] We quantify this effect introducing a set of *gains* in terms of ratios of load- and generator-referred powers:

1. The *operational gain* G_{op} (also called the operating gain): the ratio between the power on the load P_L and the power P_{in} entering the input port of the two-port. As will be shown later, the operational gain depends on Γ_L but does not depend on Γ_G; changing Γ_G, in fact, we change the input available power and the input matching condition, thus the input power, but, in the same way, also the output power is changed in such a way that their ratio is constant;
2. The *available power gain* G_{av}: the ratio between the input (generator) available power $P_{av,in}$ and the available power at the output port of the two-port $P_{av,L}$. It depends on Γ_G but does not depend on Γ_L; in fact, the output available power is the power on the load when the load is power matched to the output port of the two-port, independent of the actual value of Γ_L;
3. The *transducer gain* G_t: the ratio between the power on the load P_L and the input available power $P_{av,in}$; it depends both on Γ_G and on Γ_L.

Note that a fourth possible gain, expressed as the ratio of the output available power and the input power, is not exploited in practice. In considering the use (and popularity) of G_{op}, G_{av} and G_t, several remarks are in order. As discussed in Sec. 6.3.1, the maximization of the power transfer between generator and load only corresponds to the maximization of the transducer gain, that provides, for an unconditionally stable two-port, the condition for simultaneous conjugate matching at both ports. Optimization of the operational and available power gains is however more straightforward from a mathematical standpoint, since these only depend on the load or generator reflectances; the maximum gain is in fact the same for all considered gains. From a measurement standpoint, the transducer gain is particularly easy to characterize, since only the measurement of the load power is required (the input available power is imposed by the generator). Conversely, the operational gain requires a full two-port network analyzer, since also a fraction of the input incident wave has to be recovered through a directional coupler to evaluate the input power. Furthermore, the operational gain is not overly significant in devices whose input power is low, as in FETs due to the capacitive input circuit. Finally, the available gain is important in the theory of low-noise amplifiers, where noise power is typically introduced in terms of available power.

6.3.1 Maximum Gain and Maximum Power Transfer

The maximum power transfer between a generator and a load, connected through a two-port, occurs when two conditions are met, i.e., the input power is the generator (input)

---

[3] We will concentrate on the amplifier case, although the loss of an attenuator simply is the inverse of its gain.

6.3 Power Gains of a Loaded Two-Port

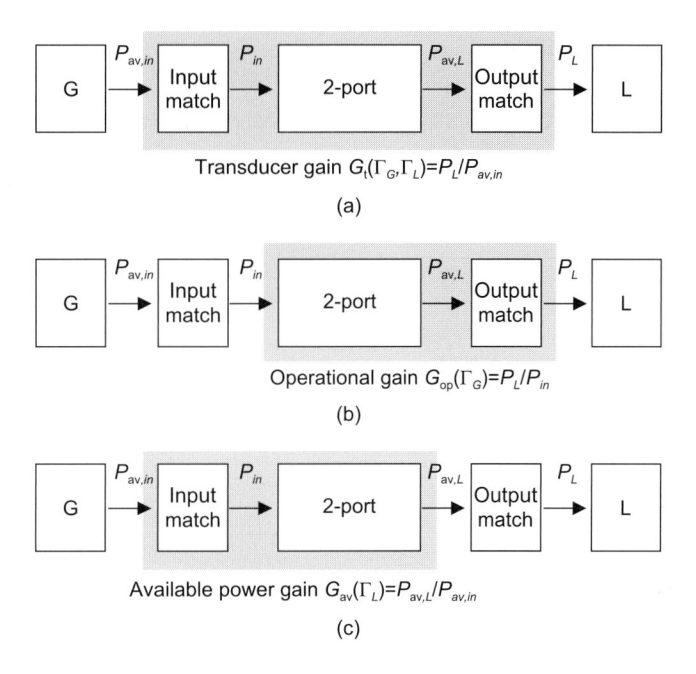

Figure 6.5 Block diagram describing the flow from the generator available power to the load power and the definitions of (a) operational gain, (b) available power gain, (c) transducer gain.

available power, and the power on the load is the load (output) available power, see Fig. 6.5 (a). This condition of maximum power transfer implies, when possible, simultaneous power (conjugate) impedance matching at the two-ports. While maximum power transfer implies maximum gain, the opposite is true only for the transducer gain. In fact, a maximum in the operational gain and in the available power gain imply maximum power transfer only if a second condition is met, corresponding to the input or output matching, respectively. However, it is quite clear that the maxima of all gains coincide. In fact, we can write:

$$G_{t,M} = \left.\frac{P_L}{P_{av,in}}\right|_M \equiv \frac{P_L^M}{P_{av,in}^M},$$

where P_L^M and $P_{av,in}^M$ are the power on the load and the input available power, respectively, in maximum power transfer conditions, and $G_{t,M}$ is the maximum transducer gain, obtained by properly selecting the generator and load reflectances. However, in maximum power transfer conditions, the input power coincides with the source available power, implying that conjugate matching is achieved at the input by properly selecting the source reflectance. We thus have:

$$G_{t,M} = \frac{P_L^M}{P_{av,in}^M} = \frac{P_L^M}{P_{in}^M} \equiv G_{op,M},$$

where P_{in}^M is the input power in maximum power transfer conditions. Thus, the maximum of the operational gain (obtained by properly selecting the load reflectance) coincides with the maximum transducer gain, but the maximum power transfer condition additionally requires the conjugate input matching to be achieved (see Fig. 6.5 (b)). Similarly, in maximum power transfer conditions, $P_L^M = P_{av,L}^M$ because if the output is power matched (see Fig. 6.5 (c)) then the load power coincides with the output available power. Therefore we have:

$$G_{t,M} = \frac{P_L^M}{P_{av,in}^M} = \frac{P_{av,L}^M}{P_{av,in}^M} \equiv G_{av,M},$$

i.e., the maximum available power gain coincides with the maximum transducer gain; notice that $P_{av,in}^M$ derives from the gain optimization with respect to the source impedance, while the condition $P_L^M = P_{av,L}^M$ is obtained through an additional constraint on the load impedance.

In conclusion, while we have:

$$G_{t,M} = G_{op,M} = G_{av,M},$$

the optimization of the whole chain leading to $P_{av,in}$ to P_L, see Fig. 6.5 (a), by proper selection of the optimum input and output reflectances yields maximum power transfer, while optimizing the chain from P_{in} to P_L by selection of the optimum load reflectance (maximum operational gain), or the chain from $P_{av,in}$ to $P_{av,L}$ by selection of the optimum source reflectance (maximum available gain) do lead to maximum gain, but not to maximum power transfer, unless a second condition is met on the output or input power matching, respectively. For historical reasons, however, the maximum gain is often referred to as MAG, i.e., the Maximum Available Gain.

Achieving simultaneous power matching at the input and output ports is not, as already recalled, always possible. Such a condition may be implemented only if the two-port is *unconditionally stable*, i.e., stable for any load and generator impedances having positive real part. If this is not the case, the two-port is *potentially unstable* and the maximum gain tends to infinity. The potentially unstable condition was regarded in the early stages of the development of electronics as a useful tool to boost the amplifier gain, and was even artificially caused through positive feedback; today, however, the preferred design strategy, at least for integrated circuits,[4] is based on the stabilization of active devices within the design bandwidth. Out-of-band stabilization (in particular at low frequency) is mandatory anyway to suppress spurious oscillations that may desensitize and saturate the amplifier also in the operating bandwidth.

6.3.2 Operational Gain

The expression for the operational gain is obtained from (6.15) and (6.18):

$$G_{op} = \frac{P_L}{P_{in}} = |S_{21}|^2 \frac{1 - |\Gamma_L|^2}{|1 - S_{22}\Gamma_L|^2 - |S_{11} - \Delta_S \Gamma_L|^2}, \qquad (6.21)$$

---

[4] Stabilization however may be incompatible with noise requirements; for this reason low-noise amplifiers are sometimes implemented with not unconditionally stable devices.

i.e., developing:

$$G_{\text{op}} = |S_{21}|^2 \frac{1 - |\Gamma_L|^2}{1 - |S_{11}|^2 + |\Gamma_L|^2(|S_{22}|^2 - |\Delta_S|^2) + 2\text{Re}\left[\Gamma_L(S_{11}^*\Delta_S - S_{22})\right]}. \quad (6.22)$$

Note from (6.21) that the operational gain is a real function of the complex variable Γ_L; as shown in Example 6.2, the constant gain curves are non-concentric circles in the Γ_L plane. Moreover, the operational gain vanishes on the unit circle of the Γ_L Smith chart, i.e., for $|\Gamma_L| = 1$; in such a case the load is reactive and $P_L = 0$.

Example 6.2 Show that the constant operational gain curves are circles in the Γ_L plane, and find their center and radius.

Solution
Reworking (6.22) we obtain:

$$|\Gamma_L|^2(G_{\text{op}}(|S_{22}|^2 - |\Delta_S|^2) + |S_{21}|^2)$$
$$+ G_{\text{op}} 2\text{Re}\left[\Gamma_L(S_{11}^*\Delta_S - S_{22})\right] = |S_{21}|^2 - G_{\text{op}}(1 - |S_{11}|^2)$$

and, rearranging:

$$|\Gamma_L|^2 - 2\text{Re}\left[\Gamma_L \frac{G_{\text{op}}(S_{22} - S_{11}^*\Delta_S)}{G_{\text{op}}(|S_{22}|^2 - |\Delta_S|^2) + |S_{21}|^2}\right] = \frac{|S_{21}|^2 - G_{\text{op}}(1 - |S_{11}|^2)}{G_{\text{op}}(|S_{22}|^2 - |\Delta_S|^2) + |S_{21}|^2}. \quad (6.23)$$

Adding to both sides of (6.23) the parameter α, defined as:

$$\alpha = \frac{\left|G_{\text{op}}(S_{22} - S_{11}^*\Delta_S)\right|^2}{\left[G_{\text{op}}(|S_{22}|^2 - |\Delta_S|^2) + |S_{21}|^2\right]^2}$$

and applying the equality $|a - b|^2 = |a|^2 + |b|^2 - 2\text{Re}\left[ab^*\right]$, Eq. (6.23) can be rewritten as:

$$|\Gamma_L - C|^2 = \frac{|S_{21}|^2 - G_{\text{op}}(1 - |S_{11}|^2)}{G_{\text{op}}(|S_{22}|^2 - |\Delta_S|^2) + |S_{21}|^2} + \alpha = R^2,$$

that describes a circle with center:

$$C = \left[\frac{G_{\text{op}}(S_{22} - S_{11}^*\Delta_S)}{G_{\text{op}}(|S_{22}|^2 - |\Delta_S|^2) + |S_{21}|^2}\right]^* = \frac{G_{\text{op}}(S_{22}^* - S_{11}\Delta_S^*)}{G_{\text{op}}(|S_{22}|^2 - |\Delta_S|^2) + |S_{21}|^2}. \quad (6.24)$$

and radius R, defined by its square:

$$R^2 = \frac{|S_{21}|^2 - G_{\text{op}}(1 - |S_{11}|^2)}{G_{\text{op}}(|S_{22}|^2 - |\Delta_S|^2) + |S_{21}|^2} + \alpha \geq 0.$$

Thus, the constant operational gain curves are circles with center C and radius R. Introducing the factor $C_2 = S_{22} - S_{11}^*\Delta_S$, and expanding the expression for $|C_2|^2$ as follows:

$$|C_2|^2 = |S_{22}|^2 + |S_{11}|^2|\Delta_S|^2 - 2\text{Re}\left[S_{22}S_{11}\Delta_S^*\right]$$

$$=|S_{22}|^2 + |S_{11}|^2|\Delta_S|^2 - |S_{11}|^2|S_{22}|^2 - |\Delta_S|^2 + |S_{21}|^2|S_{12}|^2$$
$$=(|S_{22}|^2 - |\Delta_S|^2)(1 - |S_{11}|^2) + |S_{21}|^2|S_{12}|^2,$$

we obtain for the radius the explicit expression:

$$R = |S_{21}| \frac{\sqrt{|S_{21}|^2 - 2K|S_{21}||S_{12}|G_{op} + |S_{12}|^2 G_{op}^2}}{|G_{op}(|S_{22}|^2 - |\Delta_S|^2) + |S_{21}|^2|}, \qquad (6.25)$$

where we have introduced the real parameter K defined in (6.52), called the *Linville* or *Rollet* coefficient, which plays a fundamental role in assessing the two-port stability. From (6.24) we also find that, varying G_{op}, the centers of the circles in the Γ_L plane lie on a straight line, crossing the origin of the Γ_L plane (for $G_{op} = 0$), with slope:

$$\arg\left[C_2^*\right] = -\arg\left[C_2\right] = -\tan^{-1}\frac{\operatorname{Im}\left[C_2\right]}{\operatorname{Re}\left[C_2\right]}.$$

In conclusion, if the radius R defined in (6.25) is real, the constant gain curves in plane Γ_L are non-concentric circles, whose centers lie on a straight line. The circle $G_{op} = 0$ correspond to the unit circle of the Smith chart, with center in $\Gamma_L = 0$. A qualitative plot of the constant level circles is provided in Fig. 6.6.

The radius R defined by (6.25) must be real, and therefore the second-order polynomial in G_{op} under square root in the numerator of (6.25) should be positive. Since the second-order term in G_{op} has positive coefficient, the polynomial is positive for values of G_{op} either smaller or larger than the two roots of the polynomial. Solving for the polynomial's zeros we have

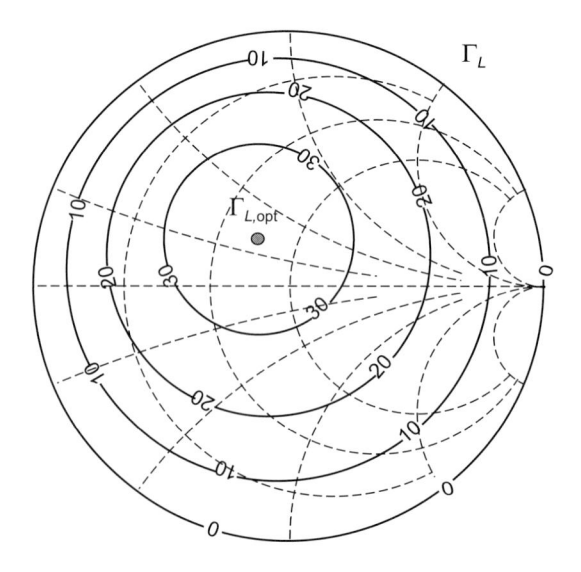

Figure 6.6 Constant operational gain circles in plane Γ_L. The curve labels are in natural units.

6.3 Power Gains of a Loaded Two-Port

$$G_{op} \leq \left|\frac{S_{21}}{S_{12}}\right|\left(K - \sqrt{K^2 - 1}\right) = G_{op,M} \tag{6.26}$$

$$G_{op} \geq \left|\frac{S_{21}}{S_{12}}\right|\left(K + \sqrt{K^2 - 1}\right) = G_{op,m}, \tag{6.27}$$

where $G_{op,m} \geq G_{op,M}$. Taking into account that $K > 1$ is a necessary condition for unconditional stability (see Sec. 6.5.1), $G_{op,m}$ and $G_{op,M}$ are both positive, and therefore $G_{op,m} \geq G_{op,M} > 0$ where the equal sign in the limiting case $K = 1$.

When $G_{op} = G_{op,M}$ and $G_{op} = G_{op,m}$ we have from (6.25) that $R = 0$, i.e., the operational gain circle collapses to a point. The two corresponding values of Γ_L can be found, after some algebraic manipulations, to be:

$$\Gamma_{L,M} = \frac{B_2 - \sqrt{B_2^2 - 4|C_2|^2}}{2C_2} \tag{6.28}$$

$$\Gamma_{L,m} = \frac{B_2 + \sqrt{B_2^2 - 4|C_2|^2}}{2C_2}, \tag{6.29}$$

where:

$$B_2 = 1 + |S_{22}|^2 - |S_{11}|^2 - |\Delta_S|^2 \tag{6.30}$$

$$C_2 = S_{22} - \Delta_S S_{11}^*. \tag{6.31}$$

As shown in Sec. 6.6.1, see the discussion before (6.79), in an unconditionally stable two-port we will have $|\Gamma_{L,m}| \geq 1$ (outside the Smith chart) and $|\Gamma_{L,M}| \leq 1$ (within the Smith chart). The equal sign again holds in the limiting case $K = 1$. This means that only the load termination leading to $G_{op,M}$ corresponds to a passive load reflection coefficient. We will define such an optimum termination as:

$$\Gamma_{L,opt} \equiv \Gamma_{L,M} = \frac{B_2 - \sqrt{B_2^2 - 4|C_2|^2}}{2C_2}. \tag{6.32}$$

In conclusion, within the Smith chart the operational gain takes values ranging from $G_{op} = 0$ on the unit circle to $G_{op} = G_{op,M}$ in correspondence of $\Gamma_{L,opt}$. The (single-valued) surface describing the operational gain, whose constant-level curves are circles, has therefore, within the Smith chart, an absolute maximum $G_{op} = G_{op,M}$ in $\Gamma_{L,opt}$, with value:

$$G_{op,M} = \left|\frac{S_{21}}{S_{12}}\right|(K - \sqrt{K^2 - 1}). \tag{6.33}$$

Such a maximum only exists in an unconditionally stable two-port. In the limiting condition $K = 1$ we finally have:[5]

$$G_{op,M}\big|_{K=1} = \left|\frac{S_{21}}{S_{12}}\right|$$

---

[5] Notice that under such a condition (see Sec. 6.6) the constant gain circles are tangential in a point of the unit circle of the Smith chart and the optimum termination corresponding to MSG is in the limit on the unit circle itself.

called the Maximum Stable Gain (MSG). From (6.27) and (6.26) it is also easy to show that:

$$\text{MSG} = \sqrt{G_{\text{op},M} G_{\text{op},m}},$$

since:

$$K - \sqrt{K^2 - 1} = \frac{1}{K + \sqrt{K^2 - 1}}.$$

6.3.3 Available Power Gain

The available power gain is the ratio between the load available power ($P_{\text{av},L}$) and the input available power, see (6.17), while the output available power is equal to the load power when $\Gamma_L = \Gamma_{out}^*$, see (6.13). In such conditions (6.19) becomes:

$$P_{\text{av},L} = |b_0|^2 |S_{21}|^2 \frac{1}{(1 - |\Gamma_{out}|^2)|1 - S_{11}\Gamma_G|^2},$$

from which, substituting Γ_{out} from (6.13), we obtain:

$$G_{\text{av}} = \frac{P_{\text{av},L}}{P_{\text{av},in}} = |S_{21}|^2 \frac{1 - |\Gamma_G|^2}{|1 - S_{11}\Gamma_G|^2 - |S_{22} - \Delta_S \Gamma_G|^2}, \quad (6.34)$$

i.e.:

$$G_{\text{av}} = |S_{21}|^2 \frac{1 - |\Gamma_G|^2}{1 - |S_{22}|^2 + |\Gamma_G|^2(|S_{11}|^2 - |\Delta_S|^2) + 2\text{Re}\left[\Gamma_G(S_{22}^* \Delta_S - S_{11})\right]}. \quad (6.35)$$

G_{av} only depends on Γ_G and, in particular, it vanishes on the unit circle of the Γ_G plane. In such conditions, in fact, the generator available power becomes infinity, leading to zero gain.

Maximization of the available power gain can be done with analogy to the case of the operational gain. In fact, comparing $G_{\text{av}}/|S_{21}|^2$ to $G_{\text{op}}/|S_{21}|^2$ we immediately notice that the two terms correspond to each other by replacing the generator with the load and exchanging port 1 with port 2. It follows that the constant level curves of G_{av} in the plane Γ_G will be again circles with centers lying on a straight line; if again the Linville coefficients satisfies $K > 1$ the radius and centers of the circles can be derived from (6.24) and (6.25) by exchanging the indices 1 and 2. Moreover, for an unconditionally stable two-port the available power gain will show a maximum $G_{\text{av},M}$ that coincides with the operational gain maximum in (6.33). In fact we obtain $G_{\text{av}}/|S_{21}|^2$ from $G_{\text{op}}/|S_{21}|^2$ by exchanging ports 1 and 2, but the maximum of $G_{\text{op}}/|S_{21}|^2$, $G_{\text{op},M}/|S_{21}|^2$, is invariant with respect to such an exchange; thus $G_{\text{av},M}/|S_{21}|^2 = G_{\text{op},M}/|S_{21}|^2$ and therefore $G_{\text{av},M} = G_{\text{op},M}$, i.e.:

$$G_{\text{av},M} = \left|\frac{S_{21}}{S_{12}}\right|(K - \sqrt{K^2 - 1}) \equiv G_{\text{op},M}. \quad (6.36)$$

The optimum Γ_G leading to maximum available gain is:

$$\Gamma_{G,\text{opt}} = \frac{B_1 - \sqrt{B_1^2 - 4|C_1|^2}}{2C_1}, \quad (6.37)$$

where:

$$B_1 = 1 + |S_{11}|^2 - |S_{22}|^2 - |\Delta_S|^2 \qquad (6.38)$$
$$C_1 = S_{11} - \Delta_S S_{22}^*. \qquad (6.39)$$

B_1, C_1 and $\Gamma_{G,\text{opt}}$ can be derived from B_2, C_2 and $\Gamma_{L,\text{opt}}$ by exchanging ports.

6.3.4 Transducer Gain

The transducer gain G_t is the ratio between the power on the load and the input available power. From (6.17) and (6.18) we immediately obtain:

$$G_t = \frac{P_L}{P_{\text{av},in}} = |S_{21}|^2 \frac{(1-|\Gamma_L|^2)(1-|\Gamma_G|^2)}{|(1-\Gamma_L S_{22})(1-\Gamma_G S_{11}) - S_{12}S_{21}\Gamma_G\Gamma_L|^2}. \qquad (6.40)$$

An equivalent alternative expression can be derived as the ratio of (6.20) and of (6.17) as:

$$G_t = \frac{P_L}{P_{\text{av},in}} = \frac{|S_{21}|^2(1-|\Gamma_L|^2)\left(1-|\Gamma_G|^2\right)}{|1-\Gamma_G\Gamma_{in}|^2|1-S_{22}\Gamma_L|^2}. \qquad (6.41)$$

As already noticed, the transducer gain depends both on Γ_G and on Γ_L. The maximum transducer gain corresponds to the simultaneous conjugate matching at both ports (assuming this is feasible), i.e., to the coupled equations:

$$\begin{cases} \Gamma_G = \Gamma_{in}^*(\Gamma_L) \\ \Gamma_L = \Gamma_{out}^*(\Gamma_G). \end{cases} \qquad (6.42)$$

For an unconditionally stable two-port, the optimum load and generator reflectances implicitly defined in (6.32) and (6.37) are the unique solution to the system (6.42). Direct substitution of (6.37) and (6.32) in (6.41) (see, e.g., [1, Appendix F]) allows us to demonstrate again that the maximum transducer gain coincides with the maximum operational and available power gain:

$$G_{t,M} = G_{\text{op},M} = G_{\text{av},M} \equiv \text{MAG} = \frac{|S_{21}|}{|S_{12}|}(K - \sqrt{K^2 - 1}). \qquad (6.43)$$

6.3.5 Power Matching and Unconditional Stability

As discussed in the previous sections, simultaneous power matching is the key for gain optimization and therefore for the design of maximum gain amplifiers. However, as discussed in Sec. 6.4, this condition exists if and only if the two-port is unconditionally stable. In practice, given a device operating in small-signal conditions and represented as a two-port, two different conditions can exist.

If the two-port is *unconditionally stable* within the design bandwidth, simultaneous power matching at the two-ports is possible, the input power is the source available power, the power on the load is the output available power, and all gains (operational, available, transducer) have the same maximum, with value given by (6.33). Proper design of matching networks allows, at least on a narrow enough bandwidth, to ensure the two-port the optimum terminations. As already recalled, out-of-band stabilization is anyway mandatory to avoid spurious, typically low-frequency oscillations that would saturate the amplifier.

If the two-port is *potentially unstable* within the design bandwidth, simultaneous power matching at the two-ports is not possible and there is a set of generator and load impedances (with positive real part) for which the gain tends to infinity, thus leading to the onset of oscillations. While out-of-band stabilization is mandatory to avoid spurious low-frequency oscillations, a design with an in-band potentially unstable device is possible, provided that the terminating impedances are chosen so as to be far enough away from the set of potentially unstable terminations.

6.4 Two-Port Stability

The stability issue is important in the design of amplifiers, mixers and oscillators (a mixer can be seen like an amplifier also providing frequency conversion). Typically we want stable, non-self-oscillating behavior from amplifiers and mixers. On the other hand, instability is sought in the design of oscillators: in the linear approximations an unstable circuit generates oscillations with infinite amplitude, but the amplitude is ultimately limited by nonlinear saturation effects that are present in all active devices.

A two-port loaded with generator and load impedances with *positive real part* (also defined as passive or "physically realizable") is *unconditionally stable* if (1) the input impedance has positive real part for *any* value of the load impedance and (2) the output impedance has positive real part for *any* value of the generator impedance. The same condition can be expressed through reflectances: for any value of Γ_L (with $|\Gamma_L| < 1$) we have $|\Gamma_{in}| < 1$, and for any value of Γ_G (with $|\Gamma_G| < 1$) we have $|\Gamma_{out}| < 1$. Since the two-port parameters are in general frequency dependent, stability depends on frequency. Moreover, stability is a global property of a circuit, unless this is made of subcircuits that are isolated with respect to each other.

On the other hand, we say that a two-port is *conditionally stable* or *potentially unstable* if there is a set of passive impedances at port 1 or 2 such as the output or input reflection coefficient of the two-port has magnitude larger than one. Notice that the fact that, e.g., $|\Gamma_{in}| > 1$ does not automatically imply that the circuit will oscillate, since the oscillation condition is in fact $\Gamma_{in}\Gamma_G = 1$. Such a condition is more clearly stated in terms of impedances (Z_{in} and Z_G) or admittances (Y_{in} and Y_G); it is straightforward to show that the condition $\Gamma_{in}\Gamma_G = 1$ implies, independent of the choice of the normalization resistance R_0, $R_{in} + R_G = 0$ and $X_{in} + X_G = 0$ (or, equivalently, $G_{in} + G_G = 0$ and $B_{in} + B_G = 0$). If a generator is connected to a load having negative internal resistance, the total circuit can operate without violating the oscillation condition if the total loop resistance (generator plus load) is positive, see Example 6.3; however such a circuit is rather an example of *reflection amplifier* exploiting devices characterized by a negative small-signal differential resistance. The dual case of a parallel generator connected to a load admittance is similar.

An important warning has however to be given about using the value of $|\Gamma_{in}\Gamma_G|$ as a marker for the total loop resistance or admittance being positive or negative; while $|\Gamma_{in}\Gamma_G| = 1$ implies zero total resistance or admittance, it is *not* true that $|\Gamma_{in}\Gamma_G| < 1$ implies positive total resistance, $|\Gamma_{in}\Gamma_G| > 1$ negative total resistance, as discussed in Example 6.4. The only proper choice for design is therefore to select, in a potentially unstable two-port, a set of terminations such as the input and output reflectances of the two-port have *magnitude less than one*, i.e., correspond to immittances (i.e., impedances or admittances) with positive real part.

As already stressed, for an unconditionally stable two-port a well defined maximum power transfer condition exists corresponding to conjugate matching at both ports. For a potentially unstable devices in the operating bandwidth, on the other hand, the linear gain is theoretically unbounded, and the choice of terminations should be made so as to ensure stability, a large enough gain, but also a termination not too close to the instability boundary, to avoid self-oscillations induced by process variations. An important result is that unconditional stability can be detected by a set of simple equivalent (two- or one-parameter stability conditions) tests to be made on the two-port scattering parameters. If the two-port is potentially unstable graphical tools like the *stability circles* can be exploited in order to make sure that the terminations ensure stable behavior with good enough margins.

Example 6.3 A sinusoidal generator with open-circuit voltage $E_G = 1$ V has internal resistance $R_G = 50$ Ω. The generator is connected with a small-signal load $R_L = -25$ Ω (provided, e.g., by a device with a local negative differential resistance). Analyze the power transfer within the circuit.

Solution

The current flowing in the circuit (positive out of the generator and positive entering the load) is:

$$I = \frac{E_G}{R_G + R_L} = \frac{1}{50 - 25} = 40 \text{ mA}.$$

The voltage on the internal generator resistance will be $V_G = R_G I = 50 \cdot 0.04 = 2$ V while the voltage on the load resistance will be $V_L = R_L I = -25 \cdot 0.04 = -1$ V. Thus, the power transfer within the circuit is as follows: the generator supplies a power $P_G = E_G I = 1 \cdot 0.04 = 40$ mW; the internal generator resistance absorbs a power $P_{R_G} = 2 \cdot 0.04 = 80$ mW, and the load resistance absorbs a negative power $P_L = -1 \cdot 0.04 = -40$ mW, i.e., supplies 40 mW. In other words the circuit is formally stable because the loop resistance is not zero (it is in fact positive), but the load supplies power rather than absorbing it. (From the energetic standpoint the power supplied by the negative resistance is actually converted from the DC bias of the device providing the negative differential resistance.) If the load is the input of an amplifier, clearly this is not the operation corresponding to amplification as we commonly understand – it is rather the principle of the so-called reflection amplifiers.

Microwave Linear Amplifiers

Example 6.4 Consider, as in Example 6.3, a generator with internal resistance $R_G > 0$ connected to a load R_L. Investigate the behavior of the product $|\Gamma_G \Gamma_L|$ (1) when R_L assumes positive and negative values; (2) for different values of the normalization resistance R_0.

Solution
We have:

$$|\Gamma_G \Gamma_L| = \left| \frac{R_G - R_0}{R_G + R_0} \frac{R_L - R_0}{R_L + R_0} \right| = \left| \frac{r_G - 1}{r_G + 1} \frac{r_L - 1}{r_L + 1} \right|.$$

Investigation of the values assumed by the product $\Gamma_G \Gamma_L$ (real in this case) allows us to clarify that when $\Gamma_G \Gamma_L = 1$, then, correctly, $r_G + r_L = 0$ corresponding to the oscillation condition. However $r_G + r_L > 0$ does not imply necessarily $|\Gamma_G \Gamma_L| < 1$, nor $r_G + r_L < 0$ implies $|\Gamma_G \Gamma_L| > 1$. This is clearly seen from the result shown in Fig. 6.7. In fact, imagine that $R_G = R_0$ and therefore $\Gamma_G = 0$; in such case $|\Gamma_G \Gamma_L| = 0 < 1$ independent from the value of R_L. In Example 6.3 we had the condition $r_L = -r_G/2$ which always implies of course $r_G + r_L > 0$ but may correspond to $|\Gamma_G \Gamma_L| > 1$ or to $|\Gamma_G \Gamma_L| < 1$ according to the choice of the normalization resistance. In conclusion, while $\Gamma_G \Gamma_L = 1$ implies $r_G + r_L = 0$ and *vice versa* the condition $|\Gamma_G \Gamma_L| < 1$ does not necessarily imply $r_G + r_L > 0$.

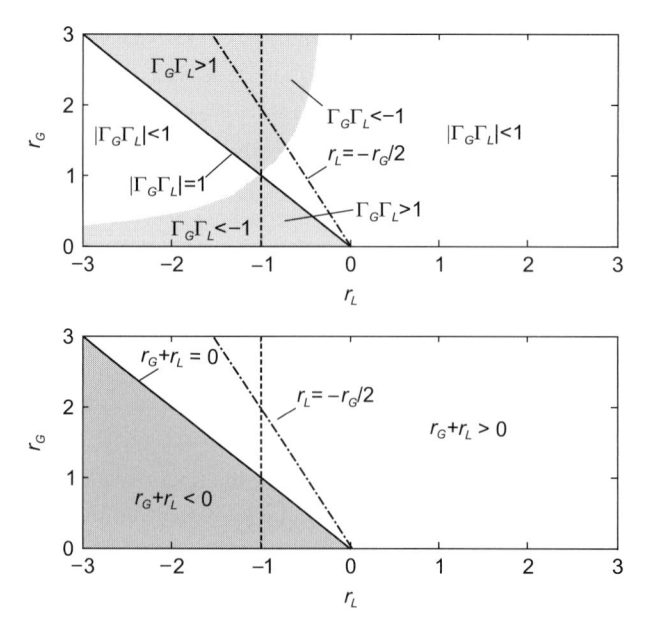

Figure 6.7 Above: Values of the product $\Gamma_G \Gamma_L$ (see Example 6.4) in the normalized $r_L - r_G$ plane. Below: regions of the $r_L - r_G$ plane corresponding to positive or negative total loop resistance. The dashed-dotted line corresponds to the values assumed in Example 6.3.

6.4.1 Analysis of Stability Conditions

Let us consider again (6.9); the system admits a nonzero solution when $b_0 = 0$ only if the system matrix determinant is zero, i.e.:

$$\begin{vmatrix} 1 & -\Gamma_G & 0 & 0 \\ -S_{11} & 1 & -S_{12} & 0 \\ -S_{21} & 0 & -S_{22} & 1 \\ 0 & 0 & 1 & -\Gamma_L \end{vmatrix} = (1 - S_{11}\Gamma_G)(1 - S_{22}\Gamma_L) - S_{12}S_{21}\Gamma_G\Gamma_L = 0.$$

Taking into account (6.11) and (6.13) the condition can be written in one of the two following ways:

$$(1 - S_{11}\Gamma_G)(1 - \Gamma_L\Gamma_{out}) = 0 \qquad (6.44a)$$

$$(1 - S_{22}\Gamma_L)(1 - \Gamma_G\Gamma_{in}) = 0. \qquad (6.44b)$$

Therefore, *at least one* of the following conditions should be met:

$$S_{11}\Gamma_G = 1 \qquad (6.45a)$$

$$S_{22}\Gamma_L = 1 \qquad (6.45b)$$

$$\Gamma_L\Gamma_{out} = 1 \qquad (6.45c)$$

$$\Gamma_G\Gamma_{in} = 1. \qquad (6.45d)$$

We immediately remark that a two-port having $|S_{11}| \geq 1$ or $|S_{22}| \geq 1$ has little practical interest, since they cannot be safely measured when terminated on the normalization impedances (however this condition could occur in real devices, thus requiring, e.g., a change of the normalization impedance from, e.g., 50 Ω to another value, as it happens in large periphery power transistors). Unless stated differently, we will therefore assume that $|S_{11}| < 1$ and $|S_{22}| < 1$; therefore, if we take into account that Γ_L and Γ_G refer to passive terminations and have magnitude lower than 1, the first two equations of (6.45) are never satisfied.

We therefore reduce to the analysis of the behavior of the term including $\Gamma_G\Gamma_{in}$ or $\Gamma_L\Gamma_{out}$ as a function of Γ_L and Γ_G. The condition $\Gamma_L\Gamma_{out} = 1$ or $\Gamma_G\Gamma_{in} = 1$ implies linear instability, since it is equivalent to the Kurokawa criterion for oscillation (zero total loop impedance or zero total node admittance, see [2]). This condition is never met if $|\Gamma_{in}\Gamma_G| < 1$ or $|\Gamma_{out}\Gamma_L| < 1$. Since Γ_G and Γ_L are passive, i.e., with subunitary magnitude, the two-port certainly is *unconditionally stable* if:

- for any passive Γ_L, $|\Gamma_{in}| < 1$; or
- for any passive Γ_G, $|\Gamma_{out}| < 1$.

Using the Smith chart and taking into account that the transformation between Γ_L (Γ_G) and Γ_{in} (Γ_{out}) is a conformal mapping between complex planes (also called linear fractional transformation or Möbius transformation) transforming circles into circles, we have the following interpretation:

1. The circles $|\Gamma_L| < 1$ ($|\Gamma_G| < 1$) are transformed into an image in the plane Γ_{in} (Γ_{out}) consisting in a circle plus its interior or exterior (see Fig. 6.8); if the image

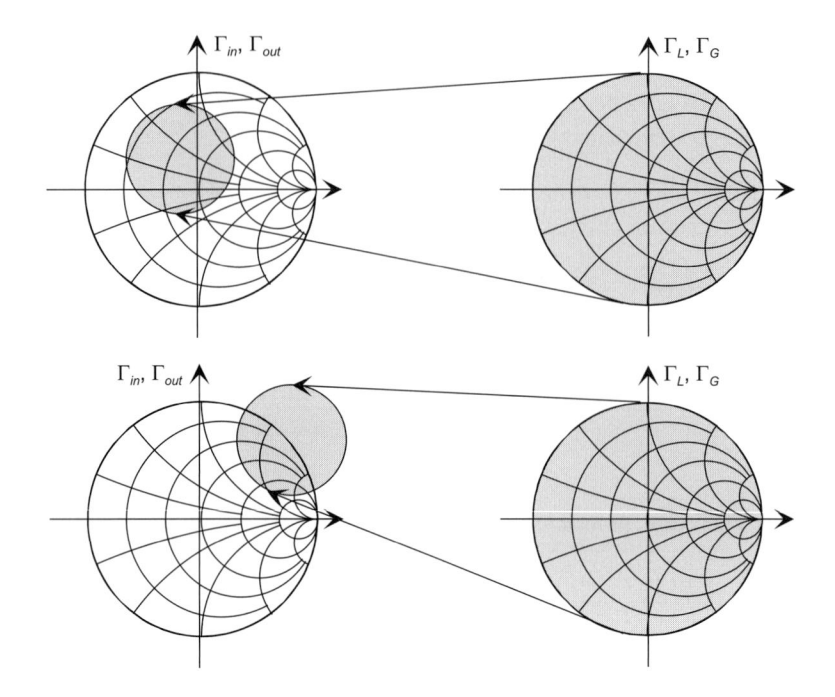

Figure 6.8 Example of unconditional stability (above) and conditional stability (below) in the input (output) reflection coefficient plane.

falls within the Smith chart, we have unconditional stability (Fig. 6.8, above); if it partially falls outside the Smith chart, we have potential instability (Fig. 6.8, below).
2. Alternatively, the circle $|\Gamma_{in}| < 1$ ($|\Gamma_{out}| < 1$) has as a counterimage a circle of the Γ_L (Γ_G) plane and its interior or exterior. If the counterimage includes the whole Γ_L (Γ_G) Smith chart, we have unconditional stability because certainly all Γ_L (Γ_G) within the unit circle will yield a Γ_{in} (Γ_{out}) within the unit circle, see Fig. 6.9, above. If on the other hand the counterimage does not include the whole Γ_L (Γ_G) Smith chart, we have potential instability (see Fig. 6.9, below).

The counterimage of $|\Gamma_{out}| = 1$ in the plane Γ_G is denoted as the *input stability circle*, whereas with *output stability circle* we denote the counterimage of $|\Gamma_{in}| = 1$ in plane Γ_L.[6]

As shown in Example 6.5, the center and the radius of the output and input stability circles (plane Γ_L and Γ_G, respectively) are given by:

$$\Gamma_{LC} = \frac{S_{11}\Delta_S^* - S_{22}^*}{|\Delta_S|^2 - |S_{22}|^2} \quad (6.46a)$$

$$R_{LC} = \frac{|S_{12}S_{21}|}{\left||\Delta_S|^2 - |S_{22}|^2\right|} \quad (6.46b)$$

---

[6] Notice that the stability circle only identifies the boundary of the stable region, since the image of the stable region is not necessarily a circle and its interior, i.e., a disk, but may be a circle and its exterior.

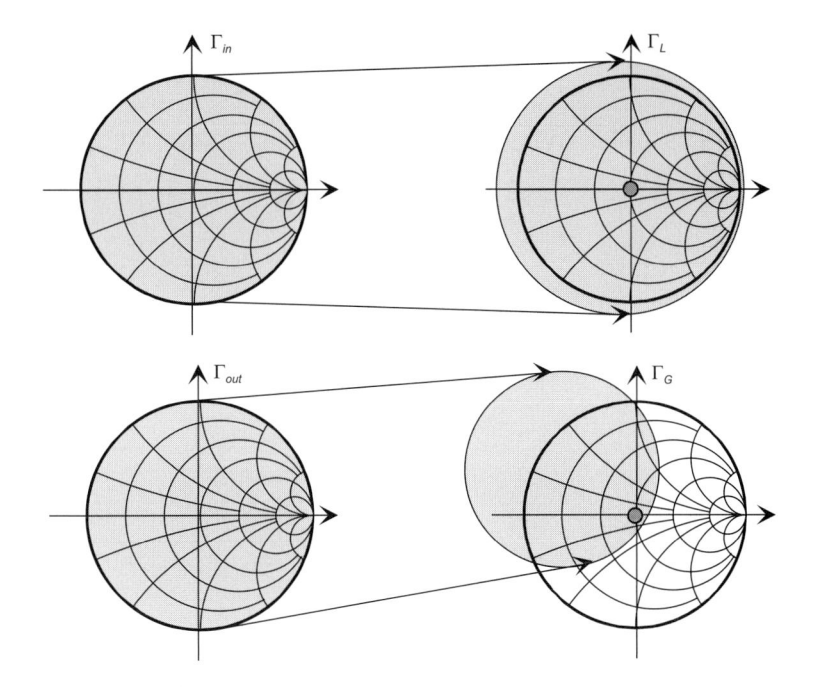

Figure 6.9 Above: input stability circle, example of unconditional stability. Below: output stability circle, example of conditional stability.

and by:

$$\Gamma_{GC} = \frac{S_{22}\Delta_S^* - S_{11}^*}{|\Delta_S|^2 - |S_{11}|^2} \quad (6.47a)$$

$$R_{GC} = \frac{|S_{12}S_{21}|}{\left||\Delta_S|^2 - |S_{11}|^2\right|}. \quad (6.47b)$$

Supposing that $|S_{11}| < 1$ and $|S_{22}| < 1$, we can immediately understand whether the stable region identified by the stability circle corresponds to the region internal or external to the circle. In fact, the origin of the Γ_G (Γ_L) plane corresponds in the Γ_{in} (Γ_{out}) plane to the point S_{11} (S_{22}), that we have assumed to lie within the unit circle. The rationale is that the two-port, closed on the normalization resistances, should be stable, otherwise the measurement itself of the scattering parameters would be impossible. Therefore, the output (input) stability circle is the region (disk) of the plane Γ_L (Γ_G) delimited by the circles described by (6.46) and (6.47) *including the origin*. According to whether the stability circle is internal, external or partially overlapped to the Smith chart, we can have the six cases shown in Fig. 6.10.

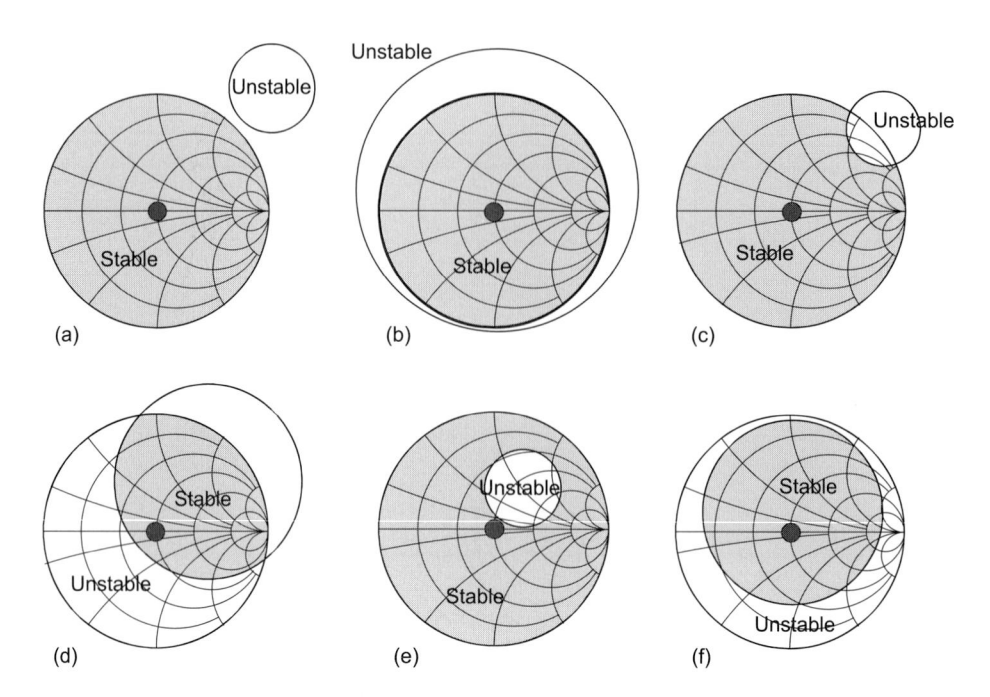

Figure 6.10 Stability cases (a) and (b): unconditional stability; (c), (d), (e) and (f): conditional stability. The Smith chart refers to Γ_G (output stability circles) or Γ_L (input stability circles). The center of the Smith chart (black dot) marks the stable region within the Smith chart (gray shading).

Example 6.5 Demonstrate Eqs. (6.46) and (6.47).

Solution
The relation (6.11) yielding Γ_{in} as a function of Γ_L is a linear fractional transformation between complex variables of the kind:

$$w = \frac{az+b}{cz+d}, \qquad (6.48)$$

that transforms the circles of z plane into circles of w plane. The unit circle in w plane will therefore correspond to the condition:

$$\left|\frac{az+b}{cz+d}\right|^2 = \frac{(az+b)(a^*z^*+b^*)}{(cz+d)(c^*z^*+d^*)} = 1$$

and thus:

$$|z|^2 + z\frac{(ab^* - cd^*)}{|a|^2 - |c|^2} + z^*\frac{(a^*b - c^*d)}{|a|^2 - |c|^2} = \frac{|d|^2 - |b|^2}{|a|^2 - |c|^2}. \qquad (6.49)$$

Eq. (6.49) is the equation of a circle in the z plane, as it is clear if we sum and subtract the factor

$$\frac{|c^*d - a^*b|^2}{(|a|^2 - |c|^2)^2},$$

to the left-hand side of (6.49), that becomes:

$$\left| z - \frac{c^*d - a^*b}{|a|^2 - |c|^2} \right|^2 = \frac{|d|^2 - |b|^2}{|a|^2 - |c|^2} + \frac{|c^*d - a^*b|^2}{(|a|^2 - |c|^2)^2} = \frac{|ad - cb|^2}{(|a|^2 - |c|^2)^2}. \quad (6.50)$$

From (6.50) we immediately obtain that the center C and radius R of the afore mentioned circle are given by:

$$\begin{cases} C = \dfrac{c^*d - a^*b}{|a|^2 - |c|^2} \\ R = \left| \dfrac{ad - cb}{|a|^2 - |c|^2} \right|. \end{cases} \quad (6.51)$$

From (6.11), comparing with (6.48), we obtain:

$$\begin{cases} a = \Delta_S \\ b = -S_{11} \\ c = S_{22} \\ d = -1, \end{cases}$$

that, after substitution into (6.51), yield the first two equations (6.46). If we exchange Γ_L with Γ_G and Γ_{in} with Γ_{out} we similarly obtain Eq. (6.47).

6.5 Two-Port Stability Criteria

To decide whether a two-port is unconditionally stable or potentially unstable *stability criteria* have been developed. Such criteria exploit inequalities on a set of two parameters being function of the S-parameters of the two-port (two-parameter stability criteria) or a based on just one parameter (one-parameter stability criteria).

6.5.1 Two-Parameter Criteria

Suppose a two-port is stable when closed by its normalization resistances (i.e., $|S_{11}| < 1$ and $|S_{22}| < 1$). A set of necessary and sufficient conditions for unconditional stability is given by the necessary (but not sufficient) condition on the Linville or stability coefficient K [3], [4]:

$$K = \frac{1 - |S_{22}|^2 - |S_{11}|^2 + |\Delta_S|^2}{2|S_{21}S_{12}|} > 1 \quad (6.52)$$

together with **one** of the following conditions:

$$|S_{12}S_{21}| < 1 - |S_{11}|^2 \quad (6.53a)$$

$$|S_{12}S_{21}| < 1 - |S_{22}|^2 \qquad (6.53\text{b})$$
$$|\Delta_S| < 1 \qquad (6.53\text{c})$$
$$B_1 = 1 - |S_{22}|^2 + |S_{11}|^2 - |\Delta_S|^2 > 0 \qquad (6.53\text{d})$$
$$B_2 = 1 + |S_{22}|^2 - |S_{11}|^2 - |\Delta_S|^2 > 0. \qquad (6.53\text{e})$$

The parameters B_1 and B_2 were already defined in (6.38) and (6.30); their definition is repeated here for clarity. The most popular set probably is $K > 1$, $\Delta_S < 1$. Most CAD tools show the frequency behavior of K; if $K < 1$ the two-port is potentially unstable, but also in regions where $K > 1$ potential instability can arise if the determinant of the scattering matrix is larger than one, see Example 6.6. The stability criteria are demonstrated in Sec. 6.5.2.

Example 6.6 Show an example of a two-port having $K > 1$ but being potentially unstable.

Solution
Consider a two-port with the scattering matrix:

$$\mathbf{S} = \begin{pmatrix} 0.5292 - j0.6643 & 0.1375 - j0.1346 \\ -5.3756 + j2.9848 & 0.5918 - j0.5800 \end{pmatrix}.$$

The Linville coefficient is $K = 1.2787 > 1$; however, the two-port is not unconditionally stable. For example, close port 2 with the passive load $\Gamma_L = 0.3762 + j0.5264$; we have $\Gamma_{in} = -1.0464 - j0.4481$ with magnitude larger than one. This is confirmed by the fact that $|\Delta_S| = 1.8528 > 1$, contrarily to what requested by condition (6.53c).

6.5.2 Proof of Two-Parameter Stability Criteria

Output stability criterion. The output stability circle is the region (disk) of plane Γ_L delimited by the circle with center Γ_{LC} and radius R_{LC}, see (6.46), including the origin. Two cases are possible:

1. $\Gamma_L = 0$ lies outside the disk, and the stable region is external to it (Fig. 6.11). We have unconditional stability if the region defined by (6.46) lies completely outside the Smith chart, i.e., when $|\Gamma_{LC}| > 1 + R_{LC}$, implying *a fortiori* $|\Gamma_{LC}|^2 > |R_{LC}|^2$. Substituting the relevant expressions (6.46) the previous inequality becomes:

$$\left|S_{11}\Delta_S^* - S_{22}^*\right|^2 = |S_{12}S_{21}|^2 + \left(1 - |S_{11}|^2\right)\left(|S_{22}|^2 - |\Delta_S|^2\right) > |S_{12}S_{21}|^2, \quad (6.54)$$

that in turn implies:

$$\left(1 - |S_{11}|^2\right)\left(|S_{22}|^2 - |\Delta_S|^2\right) > 0,$$

that is satisfied only if $|S_{22}| > |\Delta_S|$. This result allows the sign of the denominator of R_{LC} to be chosen correctly, eliminating the absolute value; see (6.46). We start again

6.5 Two-Port Stability Criteria

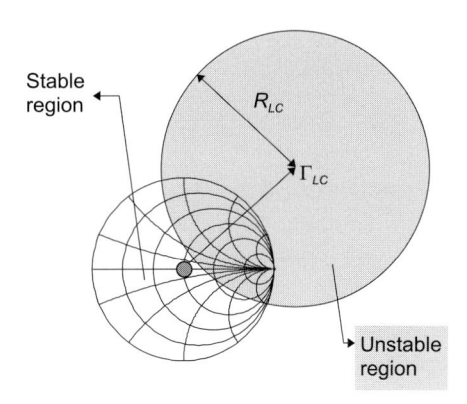

Figure 6.11 Case (1): the stable region in the load reflection coefficient is external to the output stability circle.

from condition $|\Gamma_{LC}| > 1 + R_{LC}$; taking the square of both members and substituting the value of R_{LC} we obtain:

$$\left|S_{11}\Delta_S^* - S_{22}^*\right|^2 > \left(|S_{22}|^2 - |\Delta_S|^2 + |S_{12}S_{21}|\right)^2$$

and, exploiting (6.54):

$$|S_{12}S_{21}|^2 + \left(1 - |S_{11}|^2\right)\left(|S_{22}|^2 - |\Delta_S|^2\right) > \left(|S_{22}|^2 - |\Delta_S|^2 + |S_{12}S_{21}|\right)^2. \tag{6.55}$$

This relation can be rewritten by using the Linville coefficient K defined in (6.52), and repeated here for clarity:

$$K = \frac{1 - |S_{22}|^2 - |S_{11}|^2 + |\Delta_S|^2}{2|S_{21}S_{12}|} > 1. \tag{6.56}$$

We have therefore shown that the output unconditional stability implies $K > 1$; thus, this condition is *necessary*.

2. $\Gamma_L = 0$ is internal to the circle, and therefore the stable region lies within the circle (Fig. 6.12). Thus, we have unconditional stability if the circle defined in (6.47) is completely overlaps with the Smith chart, i.e., if $|\Gamma_{LC}| < R_{LC} - 1$. This implies *a fortiori* $|\Gamma_{LC}|^2 < R_{LC}^2$. With analogy to case (1), we find that, this time, the following condition should be verified:

$$|S_{22}| < |\Delta_S|;$$

therefore condition $|\Gamma_{LC}| < R_{LC} - 1$ is equivalent to the pair:

$$|\Gamma_{LC}|^2 < (R_{LC} - 1)^2 \tag{6.57a}$$

$$R_{LC} > 1. \tag{6.57b}$$

Developing (6.57a) we have:

$$\left[\left|\left(|\Delta_S|^2 - |S_{22}|^2\right)\right| - |S_{12}S_{21}|\right]^2 > |S_{12}S_{21}|^2 + \left(1 - |S_{11}|^2\right)\left(|S_{22}|^2 - |\Delta_S|^2\right),$$

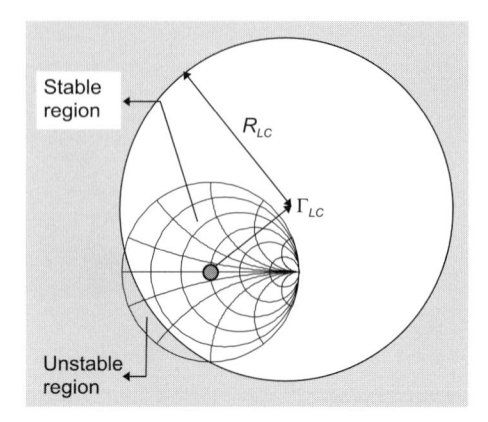

Figure 6.12 Case (2): the stable region of the load reflection coefficient is internal to the input stability circle.

from which we get back condition $K > 1$. To impose (6.57b) we start from the expression of R_{LC}; since $|\Delta_S| > |S_{22}|$, we have:

$$|S_{12}S_{21}| > |\Delta_S|^2 - |S_{22}|^2. \tag{6.58}$$

From condition $K > 1$ we have:

$$\left(1 - |S_{11}|^2\right) + \left(|\Delta_S|^2 - |S_{22}|^2\right) > 2|S_{12}S_{21}|,$$

that is:

$$1 - |S_{11}|^2 - |S_{12}S_{21}| > |S_{12}S_{21}| - \left(|\Delta_S|^2 - |S_{22}|^2\right),$$

which, to satisfy (6.58), yields:

$$1 - |S_{11}|^2 > |S_{12}S_{21}|. \tag{6.59}$$

Thus (6.59) must be verified together with condition $K > 1$. In case (1) ($\Gamma_L = 0$ lies outside the disk, and the stable region is external to it) the condition is implicitly satisfied if $K > 1$.

By running backward the above proof, conditions $K > 1$ and $1 - |S_{11}|^2 > |S_{12}S_{21}|$ can be easily shown to be not only necessary, but also sufficient.

Input stability criterion. By exchanging port 1 and port 2 and the input and output reflection coefficients we immediately obtain:

$$\begin{cases} K & > & 1 \\ 1 - |S_{22}|^2 & > & |S_{12}S_{21}|. \end{cases}$$

Input and output stability. Since stability is global, the input and output stability criteria should be equivalent. Assembling the stability conditions obtained in the previous sections we obtain that a necessary and sufficient set for the unconditional stability of a two-port (with $|S_{11}| < 1$ and $|S_{22}| < 1$) turns out to be:

$$K > 1 \qquad (6.60)$$
$$|S_{12}S_{21}| < 1 - |S_{11}|^2 \qquad (6.61)$$
$$|S_{12}S_{21}| < 1 - |S_{22}|^2. \qquad (6.62)$$

If $K > 1$ we can show that (6.61) implies (6.62) and *vice versa*. In fact, suppose that (6.61) is verified and that $K > 1$. In this case, we have unconditional output stability, i.e., for every Γ_L with magnitude < 1 we have that Γ_{in} has magnitude < 1. Since $|S_{11}| < 1$, $\Gamma_L = 0$ (in correspondence of which $\Gamma_{in} = S_{11}$) certainly falls within the unit circle. This implies that the image circle $|\Gamma_L| = 1$ in plane Γ_{in} must have radius < 1 (if it were not so there would be loads that make potentially unstable the two-port at the input). Proceeding as discussed in Example 6.2, we obtain that the image circle $\Gamma_{in}(|\Gamma_L| = 1)$ has radius:

$$R_{in} = \frac{|S_{12}S_{21}|}{|1 - |S_{22}|^2|}. \qquad (6.63)$$

We have $R_{in} < 1$ if the following condition holds:

$$|S_{12}S_{21}| < |1 - |S_{22}|^2|,$$

but we also have $|S_{22}| < 1$; therefore it follows:

$$|S_{12}S_{21}| < 1 - |S_{22}|^2$$

i.e., (6.62) holds. Similarly, we show that if (6.60) holds, (6.62) implies (6.61).

Stability criterion based on Δ_S. A further condition, alternative to (6.62) or to (6.61) is obtained by summing (6.62) and (6.61); we obtain:

$$|S_{12}S_{21}| < 1 - \frac{1}{2}|S_{11}|^2 - \frac{1}{2}|S_{22}|^2.$$

Taking into account that:

$$|\Delta_S| = |S_{11}S_{22} - S_{21}S_{12}| < |S_{11}S_{22}| + |S_{21}S_{12}|$$

and exploiting the previous equation we find:

$$\begin{aligned}|\Delta_S| &< |S_{11}S_{22}| + 1 - \frac{1}{2}|S_{11}|^2 - \frac{1}{2}|S_{22}|^2 \\ &= 1 - \frac{1}{2}(|S_{11}| - |S_{22}|)^2 < 1.\end{aligned}$$

If (6.61) and (6.62) hold, then $|\Delta_S| < 1$; it follows that if a two-port is unconditionally stable we also have:

$$\begin{cases} K & > \quad 1 \\ |\Delta_S| & < \quad 1. \end{cases} \qquad (6.64)$$

Inversely, if $|\Delta_S| < 1$ and $K > 1$, we have:

$$2|S_{21}S_{12}| < \left(1 - |S_{22}|^2\right) + \left(1 - |S_{11}|^2\right),$$

from which we obtain that at least one of the two expressions must be true:

$$|S_{21}S_{12}| < 1 - |S_{11}|^2$$
$$|S_{21}S_{12}| < 1 - |S_{22}|^2,$$

but, since if $K > 1$ they imply each other, both must be true. The set (6.64) can be therefore used to test the unconditional stability of the two-port.

Stability criterion based on B_1 and B_2. In an unconditionally stable two-port $K > 1$, implies:
$$\left(1 - |S_{22}|^2 - |S_{11}|^2 + |\Delta_S|^2\right)^2 > 4|S_{21}S_{12}|^2.$$

Summing and subtracting the term:
$$4\left(1 - |S_{22}|^2\right)\left(|S_{11}|^2 - |\Delta_S|^2\right),$$

we obtain:
$$\left(1 - |S_{22}|^2 + |S_{11}|^2 - |\Delta_S|^2\right)^2 > 4|S_{21}S_{12}|^2 + 4\left(1 - |S_{22}|^2\right)\left(|S_{11}|^2 - |\Delta_S|^2\right). \tag{6.65}$$

Taking into account (6.54), rewritten exchanging port 1 and 2:
$$\left|S_{22}\Delta_S^* - S_{11}^*\right|^2 = |S_{12}S_{21}|^2 + \left(1 - |S_{22}|^2\right)\left(|S_{11}|^2 - |\Delta_S|^2\right),$$

we obtain that (6.65) is equivalent to:
$$\left(1 - |S_{22}|^2 + |S_{11}|^2 - |\Delta_S|^2\right)^2 > 4\left|S_{22}\Delta_S^* - S_{11}^*\right|^2.$$

The two members of the above equation clearly are always positive. Taking the square root we obtain:
$$\frac{|B_1|}{2|C_1|} = \frac{\left|1 - |S_{22}|^2 + |S_{11}|^2 - |\Delta_S|^2\right|}{2|S_{22}\Delta_S^* - S_{11}^*|} > 1, \tag{6.66}$$

where B_1 and C_1 were defined in (6.38) and (6.39), respectively. Exchanging again ports we obtain that if $K > 1$ then:
$$\frac{|B_2|}{2|C_2|} > 1, \tag{6.67}$$

where B_2 and C_2 are defined in (6.30) and (6.31), respectively. We will now show that if the two-port is unconditionally stable then $B_1 > 0$; notice that if this occurs the stability circle lies completely outside the Smith chart, or covers it completely (see cases (a) and (b) of Fig. 6.10). In the first case, we certainly have $|S_{11}| > |\Delta_S|$ and therefore $2(|S_{11}|^2 - |\Delta_S|^2) > 0$; summing this equation to (6.52) we immediately obtain $B_1 > 0$. In the second case $|S_{11}| < |\Delta_S|$ and the stability circle has radius larger than one; thus, from the second equation in (6.47), we obtain $|\Delta_S|^2 - |S_{11}|^2 < |S_{12}S_{21}|$. Since in the case of unconditional stability (6.53b) holds, $1 - |S_{22}|^2 > |S_{12}S_{21}|$, we obtain:
$$|\Delta_S|^2 - |S_{11}|^2 < 1 - |S_{22}|^2$$

and therefore:
$$B_1 = 1 - |S_{22}|^2 - |\Delta_S|^2 + |S_{11}|^2 > 0$$

Similarly, we can show that unconditional stability implies $B_2 > 0$.

6.5.3 One-Parameter Stability Criterion

The classical stability criterion based on $K > 1$ plus one additional condition, as summarized above, still is widely exploited in CAD tools; it is, however, somewhat inconvenient, since the simple condition $K > 1$ is necessary but not sufficient to ensure stability. In 1992 Edwards and Sinksky [5] proved that a condition on a single parameter is sufficient to assess unconditional stability; namely, they showed that the two-port unconditional stability conditions can be put into bi-unique correspondence with one of the following conditions:

$$\mu_1 = \frac{1 - |S_{11}|^2}{|S_{22} - S_{11}^* \Delta| + |S_{12} S_{21}|} > 1 \tag{6.68}$$

$$\mu_2 = \frac{1 - |S_{22}|^2}{|S_{11} - S_{22}^* \Delta| + |S_{12} S_{21}|} > 1. \tag{6.69}$$

6.5.4 Proof of the One-Parameter Criterion

The demonstration exploits a proof of the equivalence of the two conditions $K > 1$ and $B_1 > 1$ with (6.68) [6, 7]. We first notice that, from (6.68), it is trivial to see that $\mu_1 < 1$ implies $|S_{11}| > 1$ and *vice versa* so that, when the two-port cannot be unconditionally stable because $|S_{11}| > 1$, the parameter μ_1 correctly predicts conditional stability. We can therefore focus on the case $|S_{11}| < 1$ or $\mu_1 > 0$.

We now prove that μ_1 implies $K > 1$ and $B_1 > 1$. In fact, the condition $\mu_1 > 1$ can be expressed as:

$$\left| S_{22} - S_{11}^* \Delta \right| < 1 - |S_{11}|^2 - |S_{12} S_{21}| = |S_{21}||S_{12}|(B_1 - 1), \tag{6.70}$$

from which we immediately find that $B_1 > 1$. Then, squaring (6.70), and using the identity:

$$|S_{22} - S_{11}^* \Delta|^2 = (|S_{22}|^2 - |\Delta|^2)(1 - |S_{11}|^2) + |S_{12}|^2 |S_{21}|^2, \tag{6.71}$$

we find that:

$$\left(|S_{22}|^2 - |\Delta|^2\right)\left(1 - |S_{11}|^2\right) + |S_{12} S_{21}|^2$$
$$< \left(1 - |S_{11}|^2\right)^2 - 2|S_{12} S_{21}|\left(1 - |S_{11}|^2\right) + |S_{12} S_{21}|^2. \tag{6.72}$$

Since $|S_{11}| < 1$, and therefore $1 - |S_{11}|^2 > 0$, it follows that:

$$|S_{22}|^2 - |\Delta|^2 < 1 - |S_{11}|^2 - 2|S_{12} S_{21}|, \tag{6.73}$$

and therefore:

$$1 - |S_{11}|^2 - |S_{22}|^2 + |\Delta|^2 > 2|S_{12} S_{21}|, \tag{6.74}$$

i.e., from the definition of the Linville coefficient, $K > 1$.

By reversing the above proof, we show now that $B_1 > 1$ and $K > 1$ implies $\mu_1 > 1$. Starting from (6.74), we obtain (6.73) and then (6.72) by multiplying (6.73) by the factor $1 - |S_{11}|^2$ (positive because $B_1 > 1$), and adding $|S_{12}S_{21}|^2$ to both sides. Taking the square root of the two terms, and using (6.71), we have:

$$\left|S_{22} - S_{11}^* \Delta\right| \leq \left|1 - |S_{11}|^2 - |S_{12}S_{21}|\right|,$$

but, since $B_1 > 1$, we have $1 - |S_{11}|^2 - |S_{12}S_{21}| > 0$ and finally obtain (6.70), which is equivalent to condition $\mu_1 > 1$.

By means of the same procedure, we obtain that $\mu_2 > 1$ implies $K > 1$, $b_2 > 1$ and *vice versa*. The reciprocal implication of the conditions $\mu_1 > 1$ and $\mu_2 > 1$ can be stated equivalently by showing that the condition $K > 1$, $B_1 > 1$ implies $b_2 > 1$. This demonstration has already been carried out in [5], but a simpler proof will be given here, which also has the advantage of immediately showing that if $K > 1$ conditions $B_1 > 1$ and $b_2 > 1$ mutually imply each other.

We start by evaluating the product $(B_1 - 1)(b_2 - 1)$; a direct computation shows that:

$$\begin{aligned}
(B_1 - 1)(b_2 - 1) \\
= \frac{1 - |S_{11}|^2 - |S_{22}|^2 + |S_{11}|^2|S_{22}|^2 + |S_{21}|^2|S_{12}|^2}{|S_{12}|^2|S_{21}|^2} \\
- \frac{2|S_{21}||S_{12}| - |S_{21}||S_{12}|\left(|S_{11}|^2 + |S_{22}|^2\right)}{|S_{12}|^2|S_{21}|^2} \\
= \frac{1 - |S_{11}|^2 - |S_{22}|^2 + |\Delta|^2 - 2|S_{21}||S_{12}|}{|S_{12}|^2|S_{21}|^2} \\
+ \frac{|S_{21}||S_{12}|\left(|S_{11}|^2 + |S_{22}|^2\right) + 2\mathrm{Re}\left(S_{11}S_{22}S_{21}^*S_{12}^*\right)}{|S_{12}|^2|S_{21}|^2} \\
= 2\frac{K-1}{|S_{21}||S_{12}|} + \frac{\left|S_{11}\sqrt{S_{21}^*S_{12}^*} + S_{22}^*\sqrt{S_{21}S_{12}}\right|^2}{|S_{12}|^2|S_{21}|^2}.
\end{aligned}$$

This implies that, if $K > 1$, $(B_1 - 1)(b_2 - 1) > 0$, i.e., B_1 and b_2 are both either larger or smaller than unity. Therefore condition $K > 1$, $B_1 > 1$ implies $b_2 > 1$, and thus $\mu_1 > 1$ implies $\mu_2 > 1$.

6.6 Two-Port Stability and Power Matching

6.6.1 Unconditional Stability and Simultaneous Power Matching

We show now that the simultaneous power matching problem is uniquely solved only if the two-port in unconditionally stable. In fact, to maximize the power transfer between port 1 and 2, we must impose simultaneous matching at the two-ports, see (6.42):

$$\begin{cases} \Gamma_{in} = \Gamma_G^* \\ \Gamma_{out} = \Gamma_L^*. \end{cases} \quad (6.75)$$

6.6 Two-Port Stability and Power Matching

Replacing the expressions of Γ_{in} (6.11) and Γ_{out} (6.13) we obtain the following nonlinear system:

$$\begin{cases} \Gamma_G^* = \Gamma_{in} = \dfrac{S_{11} - \Delta_S \Gamma_L}{1 - S_{22} \Gamma_L} \\ \Gamma_L = \Gamma_{out}^* = \dfrac{S_{22}^* - \Delta_S^* \Gamma_S^*}{1 - S_{11}^* \Gamma_G^*} \end{cases}.$$

Substituting the second equation into the first one we obtain a second-order equation in Γ_G yielding its optimum value; similarly, we can derive a second-order equation for the optimum Γ_L. Solving we have:

$$\Gamma_{G,\text{opt}} = \frac{1}{2C_1} \left[B_1 \pm \sqrt{B_1^2 - 4|C_1|^2} \right] \tag{6.76}$$

$$\Gamma_{L,\text{opt}} = \frac{1}{2C_2} \left[B_2 \pm \sqrt{B_2^2 - 4|C_2|^2} \right], \tag{6.77}$$

where the coefficients B_1, C_1, B_2 and C_2 were already defined in (6.38), (6.39), (6.30) and (6.31).

The choice between signs $+$ and $-$ should grant that the optimum reflection coefficients have magnitude less than one. This is possible if and only if the two-port is unconditionally stable. If the two-port is unconditionally stable we have in fact $B_1 > 0$, $B_2 > 0$. Therefore we can write (6.76) and (6.77) as:

$$\left| \Gamma_{G,\text{opt}} \right| = \frac{B_1}{2|C_1|} \left(1 \pm \sqrt{1 - \frac{4|C_1|^2}{B_1^2}} \right)$$

$$\left| \Gamma_{L,\text{opt}} \right| = \frac{B_2}{2|C_2|} \left(1 \pm \sqrt{1 - \frac{4|C_2|^2}{B_2^2}} \right).$$

From the previous formulae, taking into account (6.66) and (6.67), we have that the two reflectances have magnitude less than one if we select the solutions with minus sign in the formulae; also the choice of the sign in (6.76) and (6.77) is forced, and the optimum reflectances maximizing gain are:

$$\Gamma_{G,\text{opt}} = \frac{1}{2C_1} \left[B_1 - \sqrt{B_1^2 - 4|C_1|^2} \right] \tag{6.78}$$

$$\Gamma_{L,\text{opt}} = \frac{1}{2C_2} \left[B_2 - \sqrt{B_2^2 - 4|C_2|^2} \right]. \tag{6.79}$$

Eqs. (6.78) and (6.79) coincide with (6.37) and (6.32), respectively. Therefore, if the two-port is unconditionally stable the optimum terminations exist and are uniquely defined and the maximum gain is given by the expression, already introduced:

$$\text{MAG} = \frac{|S_{21}|}{|S_{12}|} (K - \sqrt{K^2 - 1}). \tag{6.80}$$

This corresponds to the maxima of the transducer, available, and operating gain, as already stated for simultaneous power matching the three conditions coincide.

6.6.2 Managing Conditional Stability

If the two-port is not unconditionally stable we should identify the regions of plane Γ_G and Γ_L granting values of Γ_{out} and Γ_{in} within the Smith chart, i.e., the stability of the two-port with a given set of loads. The regions sought for clearly are the intersection of the stability circles with the Smith chart. With reference, e.g., to Γ_G we should:

1. identify the circle limiting the input stability region;
2. decide if the stability circle is the external or internal region to the circle; remember that the stable region includes the origin of the Γ_G plane;
3. identify the intersection of the above region with the Smith chart.

We similarly proceed to find the region of the Γ_L plane corresponding to stable behavior.

In the case of conditional stability, the conditions corresponding to conjugate matching at both ports and maximum gain do not exist any more; within the stable region the gain is finite (it is zero on the unit circle of the Smith chart), and tends to infinity on the boundary between the stable and the unstable region. The already introduced Maximum Stable Gain (MSG), whose definition is recalled here,

$$\text{MSG} = \left| \frac{S_{21}}{S_{12}} \right|, \qquad (6.81)$$

corresponds to the MAG when $K = 1$. Since K depends on frequency and is, in transistors, typically smaller than one at low frequency and larger than one (but asymptotically tending to one) at high frequency, there is at least one frequency in which $K = 1$ exactly; when plotting the MAG as a function of frequency, the MSG is usually shown as a figure of merit in the frequency range where $K \leq 1$ (and therefore the MAG is not defined). Since from (6.33) $G_{op,M} \leq \text{MSG}$, if a potentially unstable device is stabilized by inserting a dissipative element in series or parallel to the device input or output (see Sec. 6.10.1), the MSG is the maximum gain that can be achieved when the stabilization is marginal ($K = 1$); if K is made larger than unity, the corresponding MAG will be smaller than the MSG. Under this respect, the MSG can be conveniently exploited as figure of merit for a potentially unstable device.[7]

6.6.3 Stability Circles and Constant Gain Contours

There is a close relationship between the stability circles and the constant gain contours of a loaded two-port. In particular:

1. the constant operational gain contours in plane Γ_L are related to the output stability circle ($|\Gamma_{in}| = 1$ circles in plane Γ_L);
2. the constant available power gain contours in plane Γ_G are related to the input stability circle ($|\Gamma_{out}| = 1$ circles in plane Γ_G).

---

[7] It can be shown that the MSG is invariant upon cascading a two-port with passive reciprocal networks, as done in the stabilization schemes in Sec. 6.10.1, excluding the parallel resistive feedback.

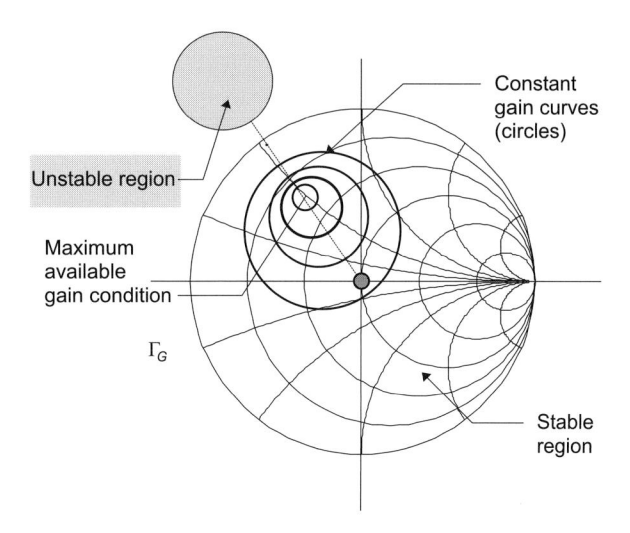

Figure 6.13 Constant gain level curves and input stability circle for an unconditionally stable two-port.

For the sake of definiteness, let us refer to the level curves of the available power gain in plane Γ_G. For an unconditionally stable two-port (Fig. 6.13) the gain has a maximum within the Smith chart and is singular outside it (i.e., for active terminations) on the boundary of the stability circle, that is completely outside the unit circle.

In the previous case $K > 1$; if $K = 1$ we are in a limiting condition, the stability circle is tangent to the unit circle of the Smith chart and the constant gain level curves are tangent to the same tangent point (Fig. 6.14). In this situation the maximum gain occurs in the limit on the tangent point and corresponds to the MSG.

Finally, if the two-port is conditionally stable the gain goes to infinity in the part of the stability circle internal to the Smith chart (Fig. 6.15). In theory, this gain amplification could be exploited in circuit design but the choice of a load too close to the unstable region is dangerous due to possible technological fluctuations that may lead the circuit to oscillate. The load and generator should be therefore chosen by allowing enough stability margin.

6.6.4 Unilateral Two-Port

In a unilateral device the internal feedback between port 2 and port 1 is zero, i.e., $S_{12} = 0$. In many semiconductor transistors this condition is almost verified, and some devices can be (as a first approximation) considered as unilateral. Besides, the unilateral case allows for an easy design of the matching networks, also considering that unilateral devices typically are unconditionally stable. Before the advent of computers, unilateral design was a suitable tool to obtain an approximate design by hand.

The unilateral nature of a device can be quantified by the unilaterality index U defined as:

294   Microwave Linear Amplifiers

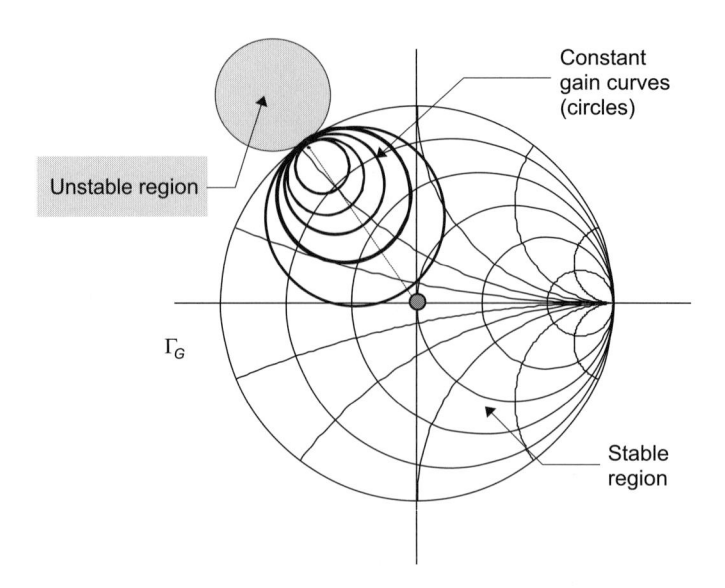

Figure 6.14 Constant gain level curves and input stability circle for a two-port in the limit of stability.

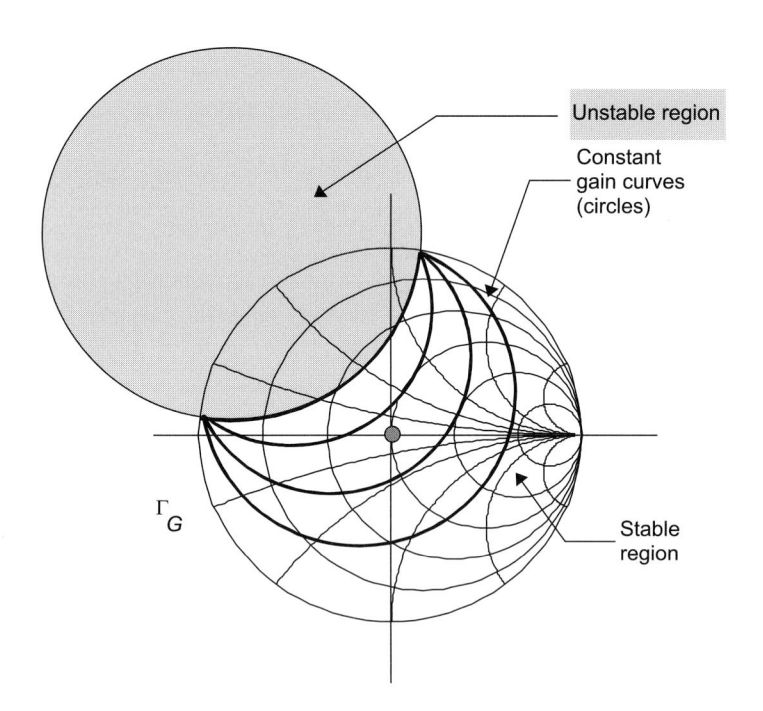

Figure 6.15 Constant gain level curves and input stability circle for a two-port that is conditionally stable.

$$U = \frac{|S_{11}S_{12}S_{21}S_{22}|}{(1-|S_{11}|^2)(1-|S_{22}|^2)}.$$

For an exactly unilateral device $U = 0$.

Let us define now a new parameter, the *unilateral transducer gain* G_u. This is the transducer gain of a two-port having $S_{12} = 0$. If $S_{12} = 0$ we obtain from (6.40):

$$G_u = G_t|_{S_{12}=0} = |S_{21}|^2 \frac{(1-|\Gamma_L|^2)(1-|\Gamma_G|^2)}{|1-\Gamma_L S_{22}|^2|1-\Gamma_G S_{11}|^2}. \tag{6.82}$$

In this case G_u is clearly maximum when conjugate matching is simultaneously achieved at both ports:

$$\begin{cases} \Gamma_G = S_{11}^* \\ \Gamma_L = S_{22}^*, \end{cases} \tag{6.83}$$

with maximum unilateral gain (MUG):

$$G_{u,M} = \frac{|S_{21}|^2}{(1-|S_{11}|^2)(1-|S_{22}|^2)}. \tag{6.84}$$

We can show that the ratio between the MAG and the maximum unilateral gain (6.84) satisfies the inequality:

$$(1+U)^{-2} < \text{MAG}/\text{MUG} < (1-U)^{-2},$$

i.e., for small U, the error introduced by assuming the device unilateral is of the order of $4U$.

We should however stress that the unilateral approximation neglects stability problems; in fact, for a device with $|S_{11}| < 1$ and $|S_{22}| < 1$ (and $\Delta = |S_{11}S_{22}| < 1$) the unilateral device always is unconditionally stable (the Linville parameter K tends to infinity in this case), unless S_{11} or S_{22} have magnitude larger than one. Therefore, the unilateral approximation is meaningful only if the original device is unconditionally stable.

6.7 Examples of FET Stability Behavior

6.7.1 Stability and Gains at Constant Frequency

In this section we analyze some two-ports whose scattering parameters are given, see Table 6.1. Out of the nine cases considered, the last three are unidirectional devices. Examples deliberately include cases not common in practice or anomalous.

Consider beginning with cases 1–6, referring to non-unidirectional two-ports. Table 6.2 shows the values of the center (modulus and phase) and the radius of the input and output stability circles. Note that in case 4, $|\Delta_S| = |S_{22}|$ and the radius of the output stability circle is infinity. Table 6.3 shows the values of K and $|\Delta_S|$ for the cases in which there is unconditional stability (denoted by "st.", the remaining "unst."). We also report the optimal value of the source and load terminations and the value of maximum gain. The following comments apply:

Table 6.1 Scattering parameters of the analyzed two-ports.

	$|S_{11}|$	$\angle S_{11}$, degrees	$|S_{12}|$	$\angle S_{12}$, degrees	$|S_{21}|$	$\angle S_{21}$, degrees	$|S_{22}|$	$\angle S_{22}$, degrees
1	0.2	20	0.05	120	3	30	0.5	−50
2	0.75	−60	0.3	70	6	90	0.5	60
3	1.05	20	0.05	120	3	40	0.5	−50
4	0.5	0	0.025	180	2	0	0.1	0
5	0.95	−22	0.04	80	3.5	165	0.61	−13
6	0.69	−123	0.11	48	1.29	78	0.52	−77
7	0.1	0	0	0	0	0	0.3	0
8	1.2	0	0	0	0	0	0.3	0
9	0.1	0	0	0	0	0	1.3	0

Table 6.2 Values of the center and radius of the stability circles for the non-unidirectional two-ports in Table 6.1.

	$|\Gamma_{GC}|$	$\angle \Gamma_{GC}$, degrees	R_{SC}	$|\Gamma_{LC}|$	$\angle \Gamma_{LC}$, degrees	R_{LC}
1	3.33	160	6.70	2.40	50	0.80
2	0.10	107	0.44	0.26	−36	0.41
3	1.10	−19	0.23	1.02	42	0.74
4	2.04	0	0.21	undef.	undef.	∞
5	1.07	30	0.24	3.45	71	3.12
6	1.37	127	0.34	1.74	86	0.69

- In case 1, $K > 1$ and $|\Delta_S| < 1$, the device is unconditionally stable.
- In cases 2 and 3 we have potential instability because at least one of the stability conditions is violated. In case 3 we also have $|S_{11}| > 1$. The parameters of the maximum gain and simultaneous power matching are therefore not defined.
- In case 4 we have unconditional stability, since $K > 1$ and $|\Delta_S| < 1$.
- In case 5 we have potential instability because $K < 1$, the gain and power matching parameters are not defined.
- Finally, in case 6 $K > 1$ and $|\Delta_S| < 1$, thus the two-port is unconditionally stable.

Cases 7–9 refer to unidirectional two-ports, always unconditionally stable if $|S_{22}| < 1$ and $|S_{11}| < 1$. Therefore, case 7 is unconditionally stable, cases 8 and 9 potentially unstable. Note that in the latter two cases we have $K < 1$, and obviously we always have $|\Delta_S| = |S_{11}S_{22}| < 1$ since $|S_{11}| < 1$ and $|S_{22}| < 1$. Note that in case 7, the device has no gain, i.e., its MAG is, in dB, equal to $-\infty$ and 0 in natural units (shown in the table).

6.7 Examples of FET Stability Behavior

Table 6.3 Coefficients for the calculation of the stability for the bi-directional two-ports in Table 6.1; cases 1, 4, 6 are unconditionally stable; cases 2, 3, 5 are conditionally stable.

	K	$|\Delta_S|$	Type	$|\Gamma_{G,opt}|$	$\angle\Gamma_{G,opt}$, degrees	$|\Gamma_{L,opt}|$	$\angle\Gamma_{L,opt}$, degrees	MAG, dB
1	2.57	0.249	st.	0.10	−20	0.48	50	10.8
2	1.34	2.156	unst.	undef.	undef.	undef.	undef.	undef.
3	0.34	0.673	unst	undef.	undef.	undef.	undef.	undef.
4	7.50	0.1	st.	0.50	0	0.07	0	7.3
5	0.19	0.572	unst	undef.	undef.	undef.	undef.	undef.
6	1.12	0.254	st.	0.88	127	0.82	86	8.6

Table 6.4 Coefficients for the stability evaluation of unidirectional two-ports in Table 6.1.

	K	$|\Delta_S|$	Type	$|\Gamma_{G,opt}|$	$\angle\Gamma_{G,opt}$, degrees	$|\Gamma_{L,opt}|$	$\angle\Gamma_{L,opt}$, degrees	MAG
7	$+\infty$	0.03	st.	0.1	0	0.3	0	0
8	$-\infty$	0.36	unst.	undef.	undef.	undef.	undef.	undef.
9	$-\infty$	0.13	unst.	undef.	undef.	undef.	undef.	undef.

Table 6.5 Scattering parameters from 1 to 10 GHz of an active microwave device; the phase is in degrees.

f, GHz	$|S_{11}|$	ϕ_{11}	$|S_{21}|$	ϕ_{21}	$|S_{12}|$	ϕ_{12}	$|S_{22}|$	ϕ_{22}
1.000	0.949	−29.8	4.825	151.1	0.038	72.1	0.781	−14.4
2.000	0.821	−59.8	4.531	123.8	0.070	56.0	0.696	−28.9
3.000	0.648	−94.2	4.092	97.6	0.092	41.4	0.600	−42.4
4.000	0.512	−133.0	3.516	73.9	0.102	30.5	0.518	−51.8
5.000	0.472	−165.2	3.025	54.7	0.108	25.3	0.444	−57.8
6.000	0.464	176.0	2.714	38.4	0.118	23.7	0.367	−65.4
7.000	0.441	158.2	2.505	22.1	0.134	20.2	0.302	−80.8
8.000	0.411	127.5	2.321	4.0	0.151	15.0	0.281	−105.9
9.000	0.454	91.4	2.093	−15.1	0.168	7.0	0.300	−134.2
10.000	0.551	66.6	1.836	−34.5	0.181	−2.8	0.328	−169.8

6.7.2 Stability and Gains as a Function of Frequency

Table 6.5 shows the frequency behavior of a MESFET measured at intervals of 1 GHz. Although the device is almost unilateral (i.e., S_{12} is small), the stability parameters (K and Δ_S) evaluated as a function of frequency (Fig. 6.16) show that the device is potentially unstable for frequencies below 5 GHz. Such potential instability at low frequency is typical of devices operating in the microwave field, and is related to the decrease of $|S_{21}|$ with frequency.

In Fig. 6.17 we show (in semi-logarithmic scale) the maximum, unilateral and maximum stable gains (MAG, MUG, MSG), and the squared modulus of S_{21}, which

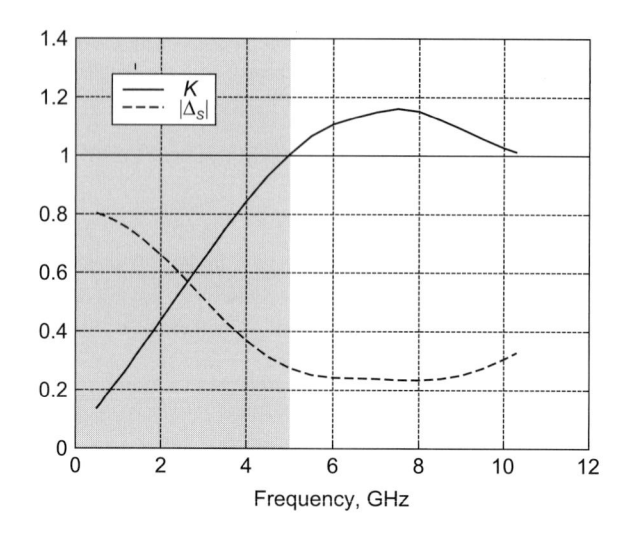

Figure 6.16 Frequency behavior of K and Δ_S for the device in Table 6.5. The potentially unstable region is shown in gray.

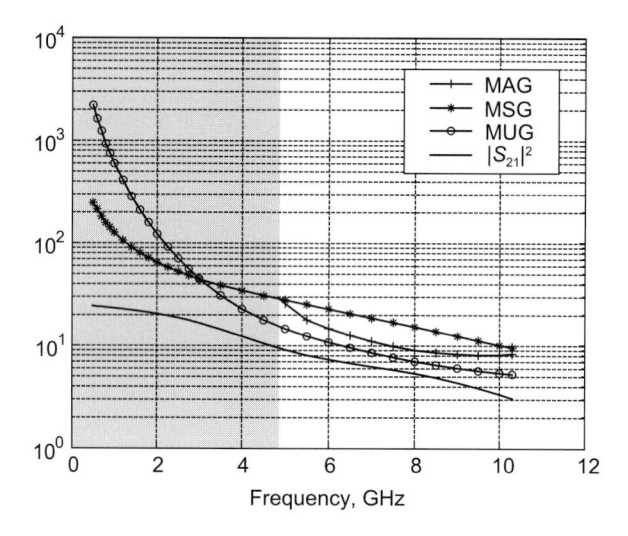

Figure 6.17 Frequency behavior of the gain of the devices in Table 6.5. The unstable region is in gray. Notice that in that region the MAG is not defined.

represents the operational gain when the device is closed on its reference impedances. Notice that the MAG coincides with the MSG at the limit frequency corresponding to the transition between unconditional and conditional stability; beyond that frequency, the MAG is undefined and the MSG is shown instead. Also note that, in the frequency range in which the device is stable, the MSG is greater than the MAG; this was expected since the MSG refers to a device brought to the edge of instability. The unilateral gain MUG is always quite different from the MAG, showing that the device is not, actually, unilateral. Finally, the gain on the reference impedances, which do not correspond, in the stable frequency range, to the optimum condition, is always lower than the MAG,

6.7 Examples of FET Stability Behavior

as expected. In the unstable region the maximum gain is obviously infinite, so the MSG should be understood as a figure of merit of the device, not the maximum gain that may be actually achieved.

Suppose now that we modify the device so as to make it more unilateral, for example by dividing S_{12} by 10. The new device thus obtained, while retaining all other parameters unchanged, presents different characteristics. The region of instability moves to lower frequencies, see Fig. 6.18, while the unilateral gain comes very close to the MAG, see Fig. 6.19. Although in this case the device is, at high frequency, virtually unilateral, we cannot neglect its potential instability at low frequency. In conclusion, the unilateral

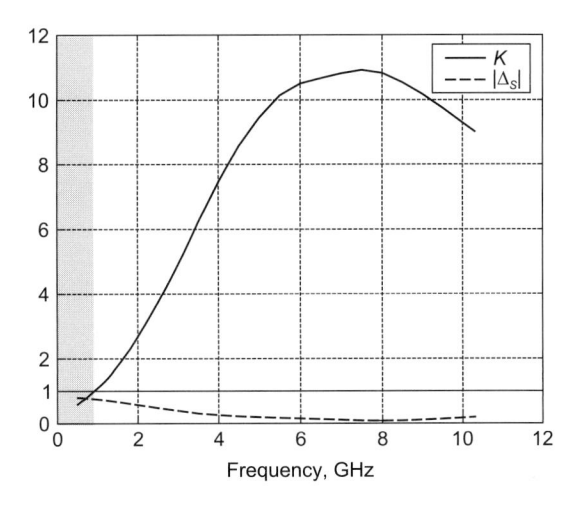

Figure 6.18 Frequency behavior of K and Δ_S for the device in Table 6.5, but with S_{12} made 10 times smaller. The potentially unstable region is in gray.

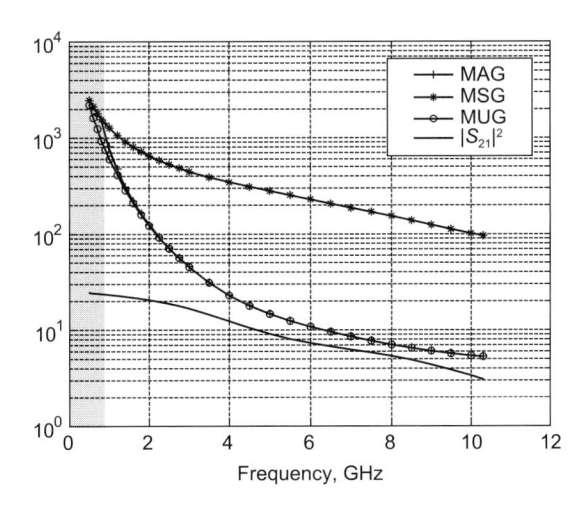

Figure 6.19 Frequency behavior of the gain of the devices in Table 6.5, with S_{12} modified. The unstable region is in gray. Notice that in that region the MAG is not defined.

approximation is sometimes acceptable in the design bandwidth, but quite dangerous on a wide bandwidth including low frequencies.

6.8 Linear Microwave Amplifier Classes

Linear amplifiers can be divided into two classes, according to the main design goal:

- low-noise amplifiers, whose goal is to provide a low-noise figure together with an acceptable associated gain;
- high-gain amplifiers, maximizing the power gain, the noise figure being a secondary design goal.

Another classification concerns the amplifier bandwidth Δf vs. the central frequency f_0; we have:

- very narrowband amplifiers, with $\Delta f / f_0 < 1$ percent; such amplifiers can often be designed at the centerband frequency;
- narrowband amplifiers, with $\Delta f / f_0 < 10$ percent; such amplifiers require some amount of equalization but can be designed with extensions of the very narrowband approach;
- wideband amplifiers, with $\Delta f / f_0 > 10$ percent; such amplifiers require proper design strategies to allow for flat gain and input matching;
- extremely wideband amplifiers, with bandwidths ranging, e.g., from DC to an upper limit; they require proper specialized topologies to be implemented.

A further classification concerns the amplifier topology. The most common topology makes use of a number of single-transistor stages in cascade; within this framework we have open loop vs. feedback amplifiers (parallel or series feedback) single stage vs. multi-stage amplifiers.[8] Particular topologies are derived from cascaded stages (the distributed amplifiers) or by a different stage setup (like in matrix [8] and stacked amplifiers, see [9] and references therein).

As already remarked, open-loop amplifiers, see Fig. 6.20, are a cascade of two-ports connecting the RF generator to the load. In feedback amplifiers, stages have an individual feedback (Fig. 6.22), or a single feedback can involve more than one stage or the whole amplifier chain. A limitation in the use of feedback amplifiers in the microwave range is the typically low gain of stages at high frequency.

Open-loop amplifiers are a classical solution above all in hybrid integrated circuits, according to the nature of the matching networks:

- LMA, Lossy Matched Amplifiers, usually with *RC* matching networks; they have favorable stability and input reflection coefficient properties but are noisier and cannot provide the intrinsic device MAG due to losses;

---

[8] Single-stage amplifiers are uncommon, a typical stage number is 3 while more than three stages are again uncommon (unless in distributed amplifiers) because the device technology is unable to provide enough gain. Notice that in LNAs only the first stage is properly low noise, while in PAs only the last stage is the power stage.

- RMA, Reactively Matched Amplifiers, with matching networks exploiting transmission lines or reactive elements.

Matching networks should in fact provide matching in narrowband amplifiers, but they should also equalize the amplifier gain vs. frequency in wideband amplifiers. In a wideband RMA however we have an intrinsic limitation of the achievable bandwidth in the presence of a specified level of input matching; in other words, we cannot obtain a wideband amplifier with good matching exploiting reactive matching networks. Such theoretical limitations, also referred to as *Bode–Fano's limits* [10, 11], are discussed in Example 6.7. Moreover, the requirement of good matching in a wideband, reactively matched amplifier is in conflict with the fact that the amplifier MAG decreases with frequency (ideally, like $1/f^2$); as a consequence, the input matching network should provide a larger reflection at low frequency to obtain a globally flat gain vs. frequency. In order to allow for simultaneous input matching and flat gain a peculiar solution is sometimes exploited, the *balanced amplifier*, properly combining the output of two amplifiers in tandem. The balanced amplifier is, moreover, an example of power combination between amplifier stages.

Finally, a peculiar solution for extremely wideband amplifiers operating at microwaves is given by the *distributed amplifier*; this structure combines the input and output of cascaded stages through the help of an input and output transmission line, obtaining wideband operation with a comparatively low gain.

While the design of a narrowband or single-frequency amplifier can be carried out, up to a certain extent, by analytical techniques or, for a single stage case, virtually by hand, the design of multi-stage and moderately wideband solutions usually requires optimization through CAD tools.

Example 6.7 Discuss Fano's limit to the input and output wideband matching of a transistor (FET or bipolar).

Solution
Consider the simplified equivalent circuit of a field-effect or bipolar transistor. The input and output equivalent circuits can be approximated as:

- for the FET, the input as a series *RC* network (gate plus intrinsic resistance and gate-source capacitance), the output as a parallel *RC* network (drain-source resistance and capacitance);
- for the bipolar, the input as a parallel *RC* network (base-emitter resistance and capacitance), the output as a parallel *RC* network (collector-emitter resistance and capacitance).

In both cases we only need to discuss the possibility to match, over a prescribed bandwidth and to a prescribed reflection coefficient Γ, a series or parallel *RC* network. According to Fano's theory [11], the following inequality holds independent of the complexity of the reactive matching network for a series *RC* network:

$$\int_0^\infty \frac{1}{\omega^2} \log |\Gamma^{-1}(\omega)| d\omega \leq \pi RC,$$

where Γ is the reflection coefficient of the matching network loaded by the series RC load. Similarly, for a parallel RC load we have:

$$\int_0^\infty \log|\Gamma^{-1}(\omega)|d\omega \le \frac{\pi}{RC}.$$

For the sake of definiteness, suppose that the reflection coefficient be constant (Γ_0) on the bandwidth B centered around the centerband frequency f_0, and equal to 1 outside such a bandwidth. In the parallel RC case we obtain:

$$\int_0^\infty \frac{1}{\omega^2} \log\left|\frac{1}{\Gamma(\omega)}\right| d\omega = \log\left|\frac{1}{\Gamma_0}\right| \int_{f_0-\frac{B}{2}}^{f_0+\frac{B}{2}} \frac{1}{\omega^2} d\omega$$
$$= \frac{B}{f_0^2 - \left(\frac{B}{2}\right)^2} \frac{1}{2\pi} \log\frac{1}{|\Gamma_0|} \le \pi RC,$$

i.e., in the case where the inverse of the reflection coefficient is maximum (and therefore the reflection coefficient is minimum):

$$|\Gamma_0| = \exp\left[-\frac{2\pi^2 RC}{B}\left(f_0^2 - \frac{B^2}{4}\right)\right].$$

In the parallel RC case we have instead:

$$\int_0^\infty \log|\Gamma^{-1}(\omega)|d\omega = -2\pi B \log|\Gamma_0| \le \frac{\pi}{RC},$$

i.e., in the best case:

$$|\Gamma_0| = \exp\left(-\frac{1}{2BRC}\right).$$

As an example, suppose that in a HEMT $R_G + R_I = 5\,\Omega$ and $C_{GS} = 0.2$ pF; we have $RC = 5 \cdot 0.2 \times 10^{-12} = 1$ ps. Assuming $f_0 = 10$ GHz the minimum reflection coefficient on a 10 percent bandwidth ($B = 1$ GHz) is:

$$|\Gamma_0| = \exp\left[-\frac{2 \cdot \pi^2 \cdot 1 \times 10^{-12}}{1 \times 10^9}\left(\left(10 \times 10^9\right)^2 - \frac{\left(1 \times 10^9\right)^2}{4}\right)\right] = 0.14.$$

but for a 100 percent bandwidth ($B = 10$ GHz, i.e., from 5 to 15 GHz) we obtain:

$$|\Gamma_0| = \exp\left[-\frac{2 \cdot \pi^2 \cdot 1 \times 10^{-12}}{10 \times 10^9}\left(\left(10 \times 10^9\right)^2 - \frac{\left(10 \times 10^9\right)^2}{4}\right)\right] = 0.86.$$

In other words, in the second case the amplifier will be severely mismatched all over the bandwidth. In the limiting case of a bandwidth from DC to 20 GHz we have $B = 20$ GHz and the minimum reflection coefficient will be 1. Notice that with a real matching network the input reflection coefficient is not constant, but is typically oscillating; the above conclusions can however be referred to the average reflection coefficient on the band.

6.8.1 The Linear Amplifier – Design Steps

In the design of a linear amplifier a number of design steps have to be followed:

1. Choose an active device. The choice of a specific technology (bipolar, FET, Si or compound semiconductor) depends on the operating frequency and the desired output power level; in fact, the maximum power that a device can yield can be shown to be:

$$P_{max} \approx \frac{I_{max} V_{max}}{8},$$

   where the maximum voltage is the output breakdown voltage and the maximum current the maximum output current. Supposing that the amplifier final stage has to operate with a certain maximum power, the constraint on small-signal operation suggests to choose a device able to provide a significantly larger maximum or saturation power. This implies a proper selection of the FET gate periphery or of the bipolar emitter area. A more realistic analysis can be carried out on the basis of distortion specifications and requires the use of nonlinear device simulation tools. In practice, technological choices are also influenced by cost, availability and reliability of a specific process.

2. Select an amplifier topology (number of stages, matching networks, open loop or with feedback). The choice is typically suggested by experience, meaning that for each specific application, the microwave cookbook suggests a preferred recipe (the term *cookbook* is intentional, see [12]). In a multi-stage amplifier, the device periphery usually scales up from the first to the last stage, since, due to amplification, the power level is also increasing correspondingly.

3. Choose the bias point and bias network for each stage. The bias point changes according to the device choice and the application (low-noise, high-gain, power).

4. Obtain (from measurements, data sheets, equivalent circuits) a small-signal model for all active devices.

5. Take care of device stabilization for each stage both out-of-band and in-band. Since the device gain is decreasing with frequency, device tend to be unconditionally stable at high frequency and the out-of-band stabilization is typically confined to the low-frequency range. Packaged devices used in hybrid circuits often have built-in stabilization networks; in integrated circuit design, those have to be realized by the designer. In-band stabilization can be, of course, avoided if the designer is prepared to cope with the design of a potentially unstable stage; this choice is sometimes convenient (since it allows us to profit from a larger gain), but has to be handled with care. Since the amount of stabilization needed is a function of frequency, stabilization networks typically combine resistive and reactive elements, see Sec. 6.10.

6. Design the matching input, output and interstage networks by CAD optimization. The choice of lumped element network is common in MMICs while in hybrid circuits distributed solutions are more customary.

7. Design the amplifier layout with a CAD tool, often integrated with the design suite exploited for circuit optimization.

8. And of course, realize and characterize the design!

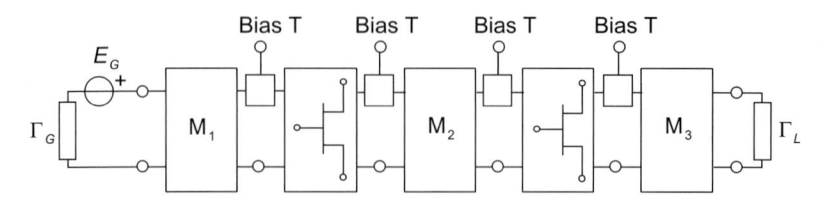

Figure 6.20 Open-loop amplifier.

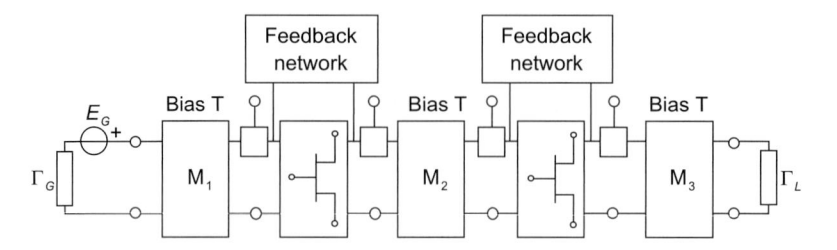

Figure 6.21 Feedback stage amplifier.

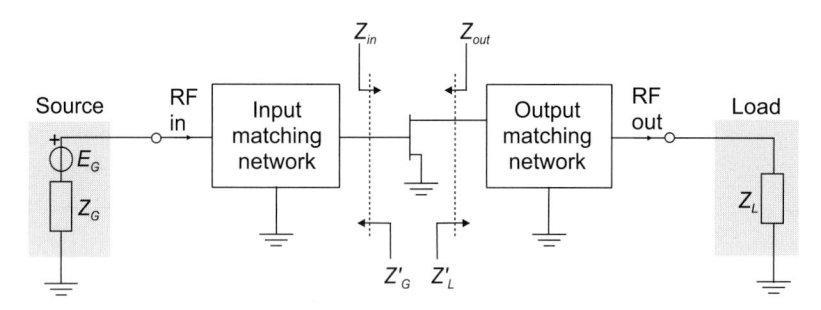

Figure 6.22 Single-stage amplifier with input and output matching networks. The circuit shown is small-signal, neglecting bias networks.

6.9 The Open-Loop Narrowband Amplifier

In narrowband amplifiers the project can be carried out mainly at the centerband frequency, plus some extra optimization. For a multi-stage cascaded amplifier we need to design the overall input and output terminations but also the interstage matching networks. In a single-stage amplifier, on the other hand, we only need to design the input and output terminations according to the project specifications. Namely:

- In the maximum gain design, the input and output matching networks should grant the simultaneous conjugate matching at both ports, i.e., $Z'_G = Z^*_{in}$, $Z'_L = Z^*_{out}$ (see Fig. 6.22).
- In the low-noise design, the input termination should minimize the noise figure, as discussed in Chapter 7; the output termination should grant output conjugate matching once the input is closed on the optimum noise source impedance.

6.9 The Open-Loop Narrowband Amplifier

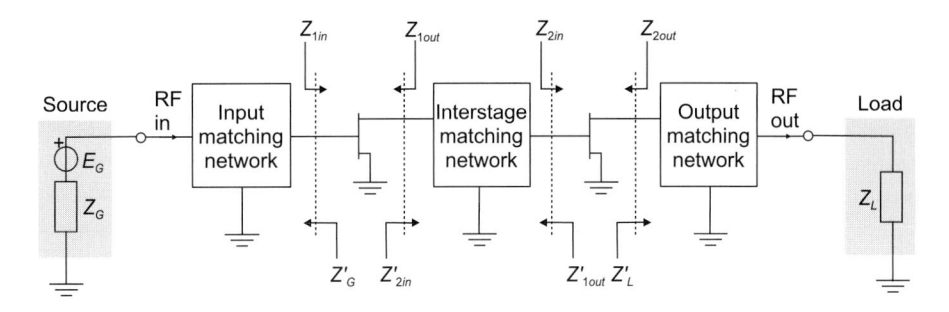

Figure 6.23 Two-stage amplifier with input, output and interstage matching sections. The circuit shown is small-signal, neglecting bias networks.

In both cases, a convenient choice for narrowband design is using reactive matching networks, thus minimizing noise and losses. In multi-stage design the problem is slightly more involved due to the need of simultaneously power matching throughout the whole stage chain.

For the single stage case, thus, the design of low-noise and high-gain amplifiers reduce to the synthesis of the input an output matching network transforming the nominal, e.g., 50 Ω loads into the required optimum admittances. Reactive matching stages are frequency selective and therefore operate as bandpass filters, allowing for matching at centerband but exhibiting an increasingly large out-of-band mismatch. Broadbanding can be obtained, up to a certain extent (remember Fano's limits), by increasing the complexity of the matching sections.

The discussion of multi-stage maximum gain amplifiers is similar. For the sake of definiteness, consider a two-stage amplifier; the design requirements are summarized in Fig. 6.23. The high-gain design is comparatively straightforward, at least as a matter of principle. In fact, the maximum gain of the chain is obtained when each stage is closed on the impedances granting the maximum available gain (MAG). In other words, Z'_G and Z'_{2in} should be the optimum impedances for the first stage, Z'_{1out} and Z'_L the optimum impedances for the second stage. Such terminations are known from the stage scattering parameters at the design frequency, and allow for conjugate matching at the input and output ports of the two stages. Now the input, output and interstage matching sections should be designed with the goals:

1. the input matching stage transforms Z_G into Z'_G;
2. the output matching stage transforms Z_L into Z'_L;
3. the interstage matching stage transforms $Z_{1out} = Z^*_{2in}$ into $Z'_{1out} = Z'^*_{2in}$. This also allows Z_{2in} to be transformed automatically into Z'_{2in}, taking into account that this condition corresponds to conjugate matching at both ports. However, the design of the matching stage may be more involved due to the need of transforming a complex impedance into another one rather than into a real impedance.

6.9.1 Design of Matching Sections

Matching sections can be implemented, at a single frequency, using lumped or distributed approaches.[9] Many solutions are available according to whether the input and output impedances of the matching section are real or complex. Minimal solutions (in terms of network complexity) are available at a single frequency, that can be made more complex if broadbanding is required. While broadband matching between two resistive loads is a well-known problem, admitting solution in terms of equal-ripple (Chebyshev) matching sections, matching a complex to a real impedance is a well defined problem at a single frequency, but is subject to the Fano limits if a broadband matching is looked for.

A typical matching problem is the transformation of a 50 Ω impedance into the optimum input or output impedance of a transistor, Z^*, where Z is the transistor input or output impedance; often it is easier to transform $Z \to 50$ Ω rather than 50 Ω $\to Z^*$, see Fig. 6.24 (b) and Fig. 6.24 (a), respectively. Notice that, in general, a reactive two-port transforming Z_2 into Z_1^* also transforms Z_1 into Z_2^*, see Fig. 6.24 (c). In fact, a reactive reciprocal two-port is described, e.g., by the series representation:

$$V_1 = jX_{11}I_1 + jX_m I_2$$
$$V_2 = jX_m I_1 + jX_{22}I_2;$$

loading with Z_2 implies $V_2 = -Z_2 I_2$; it follows:

$$-Z_2 I_2 = jX_m I_1 + jX_{22}I_2 \to I_2 = -\frac{jX_m}{Z_2 + jX_{22}} I_1,$$

i.e., from the first equation we can impose the input power matching condition as:

$$Z_{in} = \frac{V_1}{I_1} = jX_{11} + \frac{X_m^2}{Z_2 + jX_{22}} = Z_1^*.$$

This equation, however, can be rewritten as:

$$Z_2 = -jX_{22} + \frac{X_m^2}{Z_1^* - jX_{11}},$$

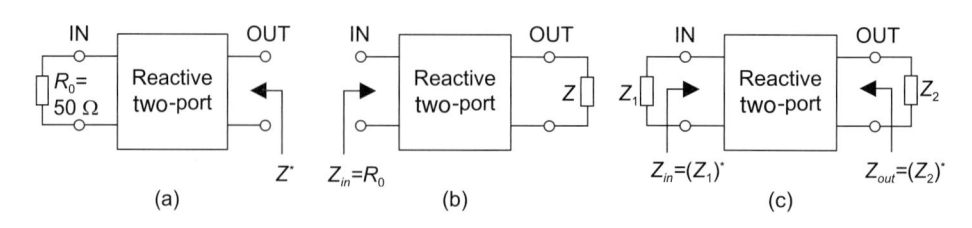

Figure 6.24 Matching network seen as an impedance transformer (a) or a network providing conjugate matching (b). Equivalence of conjugate matching at the two-ports (c).

---
[9] An exhaustive treatment of this topic can be found in [13].

6.9 The Open-Loop Narrowband Amplifier

i.e., taking into account that, exchanging the input and output ports:

$$Z_{out} = jX_{22} + \frac{X_m^2}{Z_1 + jX_{11}} = \left(-jX_{22} + \frac{X_m^2}{Z_1^* - jX_{11}}\right)^* = Z_2^*,$$

we clearly see that the power matching condition at port 1 implies the same at port 2, which is obvious, considering that a reactive two-port does not dissipate any power. The above result can be exploited to formulate in a different way the matching problem - instead to look for the transformation from a *real* impedance R_0 into a *complex* impedance Z we can start from the *complex* impedance Z^* (notice the complex conjugate!) and design the matching section in order to obtain the *real* impedance R_0, a procedure that is sometimes easier to carry out.

For the sake of definiteness let us call Z_M the goal impedance (the output impedance of the matching section) and Z_0 the starting impedance (the generator impedance of the matching section). In other words, the matching section transforms Z_M into Z_0.

Lumped parameter matching sections are compact and convenient, above all in an integrated implementation. At very high frequency (e.g., above 50 GHz) losses may be too high to allow for this choice. The simplest topology allowing for a lumped parameter matching at a single frequency are the L sections shown in Fig. 6.25 (a) and Fig. 6.25 (b). Be $Y_M = G_M + jB_M$ the admittance to be synthesized (the impedance is $Z_M = R_M + jX_M$), starting from the impedance $Z_0 = R_0 + jX_0$ (or $Y_0 = G_0 + jB_0$). For the circuit in Fig. 6.25 (a) the matching conditions are:

$$Y_M = G_M + jB_M = (Z_0 + jX_2)^{-1} + (jX_1)^{-1}$$
$$= \frac{R_0}{R_0^2 + (X_2 + X_0)^2} - j\left(\frac{1}{X_1} + \frac{X_2 + X_0}{R_0^2 + (X_2 + X_0)^2}\right),$$

from which we obtain:

$$X_1 = \left[-B_M \mp \sqrt{\frac{G_M}{R_0}(1 - G_M R_0)}\right]^{-1} \quad (6.85)$$

$$X_2 = -X_0 \pm \sqrt{\frac{R_0}{G_M}(1 - G_M R_0)}. \quad (6.86)$$

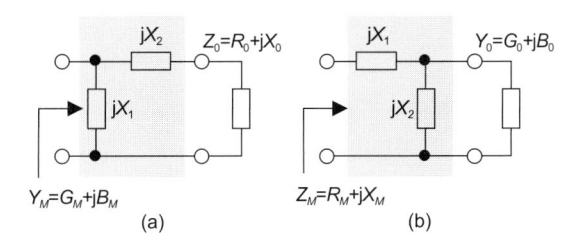

Figure 6.25 Lumped L-section matching networks, cases A and B ((a) and (b) respectively).

From (6.85) and (6.86) we see that the solution is acceptable (with real X_1 and X_2) only if (case A):

$$G_M R_0 < 1 \rightarrow G_M < \frac{1}{R_0}.$$

If the condition is not met we can exploit the circuit in Fig. 6.25 (b) leading to the following dual expressions for X_1 and X_2:

$$X_1 = X_M \pm \sqrt{\frac{R_M}{G_0}(1 - R_M G_0)} \qquad (6.87)$$

$$X_2 = \left[B_0 \mp \sqrt{\frac{G_0}{R_M}(1 - R_M G_0)} \right]^{-1}. \qquad (6.88)$$

The solution is now physically acceptable if (case B):

$$R_M G_0 \leq 1 \rightarrow R_M \leq \frac{1}{G_0}.$$

Notice that at least one of the two solutions is always acceptable since we cannot have at the same time $R_0 G_M > 1$ and $R_M G_0 \leq 1$. From a graphical standpoint, consider two Smith charts, the chart of Z_0 assuming as normalization resistance $1/G_M$ (Fig. 6.26, left), and the chart of Y_0, assuming as the normalization conductance $1/R_M$ (Fig. 6.26, right), redrawn on the Z_0 Smith chart plane. For Case A the relation $G_M < 1/R_0$ or $R_0 < 1/G_M$ or $R_0/(1/G_M) = r_0 < 1$ identifies the Smith chart of the Z_0 plane, excluding the $r_0 > 1$ circle. For Case B the relation $R_M \leq 1/G_0$ or $G_0/(1/R_M) = g_0 \leq 1$ identifies the Smith chart of the Y_0 plane, excluding the $g_0 \leq 1$ circle; taking into account that the voltage and current reflection coefficients have opposite sign the image of the circle in the Z_0 plane is shown in Fig. 6.26, right. We clearly see that, two sets of impedances exist for which both solutions (A and B) are possible, while for the two sets in the white circle only one solution (A or B) can be applied. Notice that if both structures were unacceptable we would have at the same time:

$$G_M > \frac{1}{R_0}, \quad R_M \geq \frac{1}{G_0} \rightarrow G_M R_M G_0 R_0 > 1,$$

i.e.:

$$\frac{R_M^2}{R_M^2 + X_M^2} \frac{R_0^2}{R_0^2 + X_0^2} > 1,$$

that is never verified.

To realize the transformer design we finally have to assign to the reactances a component value, given the centerband frequency f_0. If $X > 0$ we can synthesize it with an inductance $L = X_1/(2\pi f_0)$, while if $X < 0$ a capacitor is needed with value $C = |X_1|^{-1}/(2\pi f_0)$. Summarizing, an L section can always transform, at a given frequency, an arbitrary impedance Z_0 into an arbitrary impedance Z_M.

6.9 The Open-Loop Narrowband Amplifier

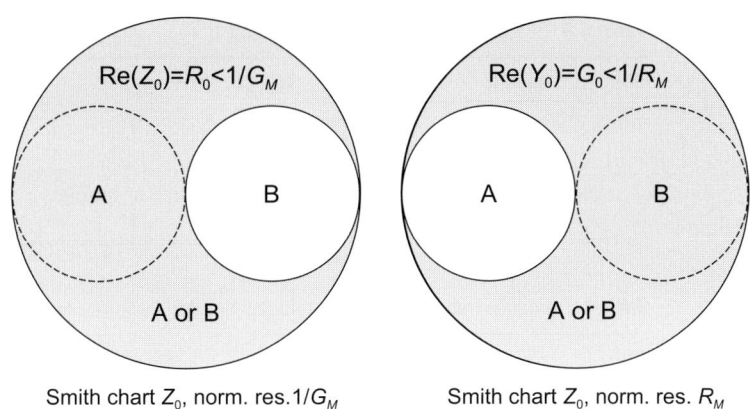

Figure 6.26 Choice between the two L matching sections (cases A and B) to transform impedance Z_0 (admittance Y_0) into impedance Z_M (admittance Y_M).

Example 6.8 Design a matching network to synthesize the two reflection coefficients (with respect to $Z_0 = 50\,\Omega$): (1) $\Gamma_G = 0.062 + j0.198$, (2) $\Gamma_L = 0.049 + j0.793$, starting from and impedance of 50 Ω, at the frequency $f_0 = 4$ GHz. Exploit a lumped parameter L section.

Solution
From the relationship:

$$Z = R_0 \frac{1+\Gamma}{1-\Gamma},$$

we obtain:

$$Z_G = 52.06 + j \cdot 21.54\ \Omega, \quad Y_G = \frac{1}{Z_G} = 16.4 - j \cdot 6{,}79\ \text{mS}$$

$$Z_L = 12.0 + j \cdot 51.7\ \Omega, \quad Y_L = \frac{1}{Z_L} = 4.26 - j \cdot 18{,}3\ \text{mS}.$$

Using the previous notation we have $R_0 = 50\,\Omega$, $G_0 = 20$ mS, and

$$R_{ML} = 52.06\ \Omega, \quad X_{ML} = 21.54\ \Omega$$
$$G_{ML} = 16.4\ \text{mS}, \quad B_{ML} = -6.79\ \text{mS}$$
$$R_{MG} = 12.0\ \Omega, \quad X_{MG} = 51.7\ \Omega$$
$$G_{MG} = 4.26\ \text{mS}, \quad B_{MG} = -18.3\ \text{mS}.$$

Concerning Z_L and Z_G, we have:

$$G_{ML} R_0 = 16.4 \times 10^{-3} \cdot 50 = 0.82 \leq 1$$
$$G_0 R_{ML} = 20 \times 10^{-3} \cdot 52.06 = 1.041 > 1$$
$$G_{MG} R_0 = 4.26 \times 10^{-3} \cdot 50 = 0.213 \leq 1$$
$$G_0 R_{MG} = 20 \times 10^{-3} \cdot 12 = 0.24 < 1.$$

Table 6.6 Results from Example 6.8, matching of Z_G.

Sol.	X_{1_G}, Ω	X_{2_G}, Ω	C_1, pF	C_2, pF	L_1, nH	L_2, nH
1	−98.51	−96.04	0.40	0.41	-	-
2	−37.69	96.04	1.06	-	-	3.82
3	−73.09	28.13	0.54	-	-	1.12
4	−30.35	−28.13	1.31	1.41	-	-

Table 6.7 Results from Example 6.8, matching of Z_L.

Sol.	X_{1_L}, Ω	X_{2_L}, Ω	C_1, pF	C_2, pF	L_1, nH	L_2, nH
1	−1113.9	23.43	0.036	-	-	0.93
2	69.1	−23.43	-	1.7	2.75	-

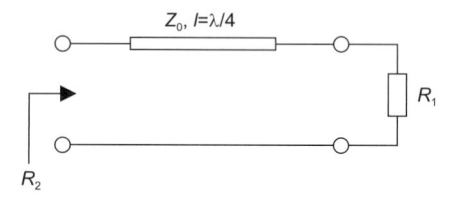

Figure 6.27 Quarter-wavelength ($\lambda/4$) matching section.

For Z_G both the topologies in Fig. 6.25 (a) and Fig. 6.25 (b) can be exploited. From (6.85), (6.86), (6.87) and (6.88) using the double sign and taking into account that:

$$C = \frac{1}{2\pi f |X_C|}, \quad L = \frac{X_L}{2\pi f}$$

for capacitive and inductive reactances, respectively, we find the four solutions listed in Table 6.6. Concerning Z_L, only Case A applies, see Fig. 6.25 (a). From (6.85), (6.86) we obtain the results listed in Table 6.7.

Distributed matching sections are commonly used in hybrid integrated circuits, while (as already mentioned) in integrated circuits the preferred solution, at least up to millimeter waves, is the concentrated one. Distributed matching sections are typically made of transmission lines and transmission line stubs in short or open circuit. In what follows, we will consider the single frequency design; wideband designs can be carried out by numerical optimization or according to analytic approaches. For the sake of definiteness (and with the exception of the quarterwave transformer) we will always refer to the transformation from an arbitrary complex impedance to a real impedance of value R_0. The following choices for distributed matching sections are available:

1. **Quarterwave transformer.** Performs the transformation $R_1 \to R_2$ with the structure in Fig. 6.27. From the input impedance of a lossless line loaded with an impedance Z_L:

$$Z_{in} = Z_0 \frac{Z_L \cos\beta l + jZ_0 \sin\beta l}{Z_0 \cos\beta l + jZ_L \sin\beta l},$$

we have, for $l = \lambda/4$, that $\beta l = \pi/2$; it follows: $Z_{in} = Z_0^2/Z_L$. For a resistive load $Z_L = R_1$ we have $Z_{in} = R_2$ if the condition holds:

$$Z_0 = \sqrt{R_1 R_2}.$$

A generalization of the quarterwave transformer is presented in Example 6.9.

2. **Stub and quarterwave transformer.** The quarterwave transformer can be extended to the case $Z_1 \to R_2$ where Z_1 is complex by compensating the imaginary part of Z_1 with a series stub or the imaginary part of Y_1 by a parallel stub, see Fig. 6.28. After compensation the load impedance is real and can be transformed to R_2. As a further generalization, we can obtain the transformation $Z_1 \to Z_2$ where both Z_1 and Z_2 are complex by adding a second stub in series or parallel. The quarterwave transformer impedance is related to the transformer ratio while the stub impedances (Z_{0s}) are arbitrary. Of course, for the series or parallel stub, two minimum length solutions are available corresponding to the short-circuit (sc) and open circuit (oc) case; we have:

$$X_{sc} = Z_{0s} \tan\beta l_s, \quad B_{sc} = -Y_{0s} \cot\beta l_s$$
$$X_{oc} = -Z_{0s} \cot\beta l_s, \quad B_{sc} = Y_{0s} \tan\beta l_s.$$

As is well known, a short stub (i.e., with length less than $\lambda_g/4$) in *short circuit* is inductive, a short stub in *open circuit* is capacitive. Due to technological constraints the minimum length solution is not always available.

3. **Line and stub adapter.** A possible scheme is shown in Fig. 6.29: a line of impedance Z_{02} and length l_2 is connected to the load Z_L, that is transformed into the admittance

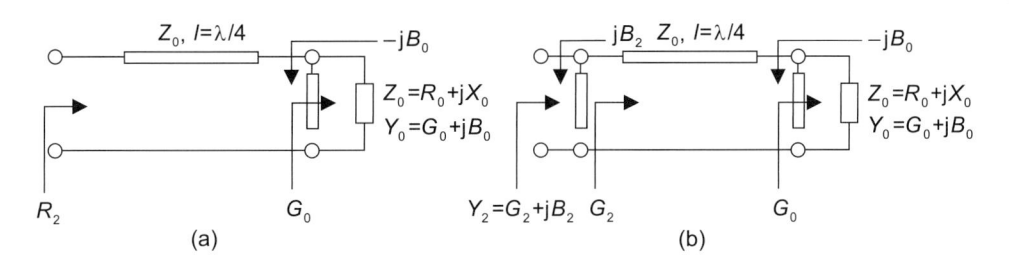

Figure 6.28 Stub and quarterwave transformers: from complex to real impedance (a), from complex to complex impedance (b).

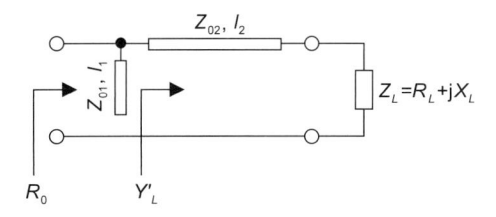

Figure 6.29 Distributed L matching section.

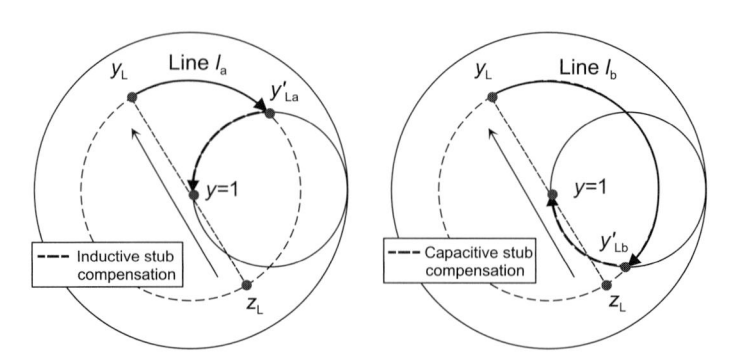

Figure 6.30 Line and stub adapter: possible solutions with minimum line length.

$Y'_L = G'_L + jB'_L$ such as $G'_L = 1/R_0$. The stub (impedance Z_{01} and length l_1, in parallel or series, in short or open circuit) has (for the parallel case) an input admittance $-jB'_L$ compensating the imaginary part of Y'_L; the final resulting input admittance is therefore G'_L coinciding with the wanted resistance R_0. The adapter has four degrees of freedom but typically the characteristic impedances Z_{01} and Z_{02} will be the same and will coincide with some convenient value (e.g., 50 Ω, sometimes coinciding with R_0); thus, the adapter design requires to obtain the two lengths l_1 and l_2. The line design becomes particularly straightforward on the Smith chart. Suppose for simplicity that $Z_{01} = Z_{02} = R_0$ and define $z_L = Z_L/R_0$ as the normalized load impedance. Suppose also that a parallel stub is exploited; as shown in Fig. 6.30 we identify from z_L the corresponding admittance y_L (opposite reflection coefficient) and then we rotate the reflection coefficient on the Smith chart till it intersects the circle $\text{Re}(y'_L) = 1$; two alternatives are possible (referred to as a and b, respectively), on the left the resulting admittance has a positive (capacitive) imaginary part, on the right a negative (inductive) imaginary part. In case a we need to synthesize an (e.g., open) stub with inductive input susceptance, in case b with capacitive input susceptance. In case a the line is short but the stub longer, in case b the line is long but the stub shorter. The choice is often a matter of layout, unless a shorted stub is available, in which case the optimum solution is the first. The evaluation of the line length can be, on the other hand, carried out also analytically; we have the constraint:

$$y'_L = \frac{1 + jz_L \tan \beta l_2}{z_L + j \tan \beta l_2} = \frac{1 + j(r_L + jx_L) \tan \beta l_2}{(r_L + jx_L) + j \tan \beta l_2}$$
$$= \frac{(1 - x_L \tan \beta l_2) + jr_L \tan \beta l_2}{r_L + j(x_L + \tan \beta l_2)},$$

i.e., imposing $\text{Re}(y'_L) = 1$ we obtain a quadratic equation in $\tan \beta l_2$ yielding the two minimum length solutions:

$$r_L \frac{1 - 2x_L \tan \beta l_2 - \tan^2 \beta l_2}{r_L^2 + (x_L + \tan \beta l_2)^2} = 1.$$

4. **Cascaded line adapter.** The circuit is reported in Fig. 6.31; it is able to transform $Z_L \to R_0$ without stubs exploiting two fixed length lines ($\lambda/4$ and $\lambda/8$) with

6.9 The Open-Loop Narrowband Amplifier

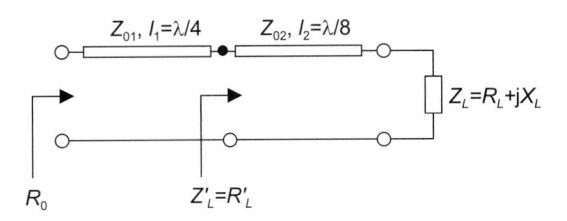

Figure 6.31 Cascaded line adapter.

impedances Z_{01} and Z_{02} to be determined. The $\lambda/8$ line transforms $Z_L = R_L + jX_L$ into a purely resistive load. Taking into account that in this case $\beta_2 l_2 = \pi/4$ we have:

$$Z_{in} = Z_0 \frac{Z_L \cos \beta l + jZ_0 \sin \beta l}{Z_0 \cos \beta l + jZ_L \sin \beta l},$$

$$Z'_L = R'_L = Z_{02} \frac{R_L + j(X_L + Z_{02})}{(Z_{02} - X_L) + jR_L} = Z_{02} \frac{2Z_{02}R_L + j\left(Z_{02}^2 - R_L^2 - X_L^2\right)}{(Z_{02} - X_L)^2 + R_L^2},$$

i.e., imposing zero imaginary part:

$$Z_{02} = \sqrt{R_L^2 + X_L^2}.$$

Therefore:

$$R'_L = \sqrt{R_L^2 + X_L^2} \frac{2\sqrt{R_L^2 + X_L^2}R_L}{\left(\sqrt{R_L^2 + X_L^2} - X_L\right)^2 + R_L^2} = \frac{R_L}{1 - \frac{X_L}{\sqrt{R_L^2 + X_L^2}}} = \frac{R_L}{1 - \frac{X_L}{Z_{02}}}.$$

To obtain an input R_0, the quarter-wave transformer impedance will be:

$$Z_{01} = \sqrt{R'_L R_0} = \sqrt{\frac{R_0 R_L}{1 - \frac{X_L}{Z_{02}}}}.$$

Example 6.9 Show that a load impedance Z_L can, under certain conditions, be matched to an impedance Z_i using a lossless transmission line with real characteristic impedance R_0 and electrical length θ to be found. Derive the conditions under which this matching is possible.

Solution
Let us consider the input impedance Z_i of a lossless line loaded with Z_L and with electrical length θ, the characteristic impedance is real and denoted as R_0. We have:

$$Z_i = R_0 \frac{Z_L + jZ_0 \tan \theta}{Z_0 + jZ_L \tan \theta}.$$

Solving we obtain:

$$j\tan\theta = R_0 \frac{Z_i - Z_L}{R_0^2 - Z_L Z_i} = R_0 \frac{R_i + jX_i - R_L - jX_L}{R_0^2 - (R_i + jX_i)(R_L + jX_L)}. \quad (6.89)$$

However, the term $j\tan\theta$ must be imaginary. Equating to 0 the real part in (6.89) we obtain:

$$R_0^2 (R_i - R_L) - R_L \left(R_i^2 + X_i^2\right) + R_i \left(R_L^2 + X_L^2\right) = 0,$$

i.e.:

$$R_0^2 (R_i - R_L) - R_L |Z_i|^2 + R_i |Z_L|^2 = 0,$$

from which:

$$R_0^2 = |Z_i|^2 |Z_L|^2 \frac{R_L/|Z_L|^2 - R_i/|Z_i|^2}{R_i - R_L} = |Z_i|^2 |Z_L|^2 \frac{G_L - G_i}{R_i - R_L},$$

where G_L and G_i are the load and generator conductances, respectively. To obtain a realizable R_0 we must have:

$$\frac{G_L - G_i}{R_i - R_L} > 0, \quad (6.90)$$

i.e., either $G_L - G_i > 0$ and $R_i - R_L > 0$ or $G_L - G_i < 0$ and $R_i - R_L < 0$. Given Z_L, the allowed input impedances Z_i and reflectances can be identified on the Smith chart by considering constant resistance and constant conductance circles.

If condition (6.90) is verified, the electrical length of the line can be obtained from:

$$\tan\theta = \frac{R_0 (R_L X_i + R_i X_L)}{\left(R_0^2 - R_L R_i + X_L X_i\right)^2 + (R_L X_i + R_i X_L)^2}, \quad (6.91)$$

where the imaginary part of the left-hand side of (6.89) has been explicited. Notice that if $Z_L = R_L$ and $Z_i = R_i$ condition (6.90) becomes:

$$R_0^2 = (R_L R_i)^2 \frac{1/R_L - 1/R_i}{R_i - R_L} = R_L R_i > 0, \quad (6.92)$$

i.e., the condition on R_0 is always satisfied, while (6.91), taking into account (6.92), yields $\tan\theta \to \infty$, corresponding to a quarter-wavelength line. On the other hand, condition $R_0 = \sqrt{R_L R_i}$ also corresponds to the characteristic impedance of a quarter-wavelength transformer. Thus, the proposed matching section can be considered as a generalization of the quarter-wave transformer.

6.10 Active Device Stabilization and Bias

6.10.1 Stabilization

Open-loop amplifiers should always been stabilized off-band (i.e., at low frequency) and in-band stabilization offers advantages in terms of design strategies for high-gain and low-noise amplifiers. Stabilization of transistors can be achieved by adding proper

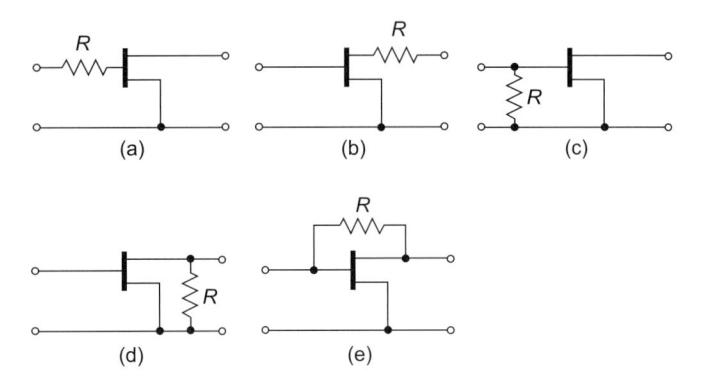

Figure 6.32 Resistive stabilization of a transistor.

dissipative elements in the input/output or in feedback, see Fig. 6.32. The principle of resistive stabilization is indeed straightforward [14]: if the device input resistance or conductance is negative at a certain frequency, a compensating series or parallel resistor can be connected at the input, with a convenient value leading to a positive input resistance or conductance. The same considerations hold for the device output, taking into account that stabilization can be carried out using both the device input and output. In particular the device becomes unconditionally stable if the resistance in series or the conductance in parallel are able to compensate for the negative input/output resistance or conductance for any passive output/input termination, respectively. The effect of a series resistance or parallel conductance connected at the device input or output on the stability factor K can be immediately seen by taking into account that K can be equivalently expressed (using the conversion formulae in Table 3.1) in terms of admittances or (similarly) of impedances as:

$$K = \frac{2\operatorname{Re}[Y_{11}]\operatorname{Re}[Y_{22}] - \operatorname{Re}[Y_{12}Y_{21}]}{|Y_{12}Y_{21}|}$$
$$= \frac{2\operatorname{Re}[Z_{11}]\operatorname{Re}[Z_{22}] - \operatorname{Re}[Z_{12}Z_{21}]}{|Z_{12}Z_{21}|}. \tag{6.93}$$

By inspection, connecting in series to the device input a resistance R_{in} and in series to the device output a resistance R_{out} the elements of the device impedance matrix Z_{12} and Z_{21} are unaffected while we have $Z_{11} \to Z_{11} + R_{in}$, $Z_{22} \to Z_{22} + R_{in}$. It is therefore obvious that the positive term in the numerator of K increases, while the negative term in the numerator is unaffected and so is the denominator. We conclude that a large enough resistance R_{in} and/or R_{out} connected in series to the device input and/or output is always able to make K larger than unity. A similar conclusion holds for a conductance G_{in} in parallel to the device input and/or G_{out} in parallel to the device output.[10] The effect of a parallel or series feedback resistance connected between the gate and the drain or

---

[10] Notice, however, that if the real part of the input (output) immittance is negative, an increase of the output (input) immittance real part would lead to worse stability.

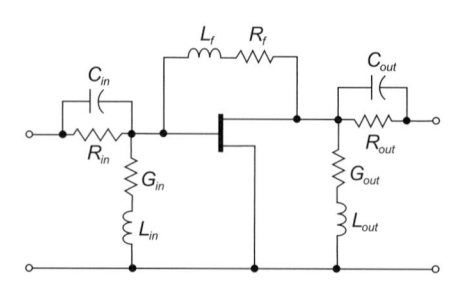

Figure 6.33 Stabilizing a transistor through frequency-selective networks.

between the source and the ground, respectively, is less straightforward because feedback affects all the elements of the admittance or impedance matrix. Series feedback is typically not considered because it decreases the device transconductance. On the other hand, the parallel feedback solution is typically avoided in open-loop amplifiers and requires anyway a DC block not to DC connect the input and output bias.

In general, since the amount of added stabilization is usually large at low frequencies but lower in-band, reactive elements are integrated in the stabilization network so as to allow for a variable degree of stabilization vs. frequency. As shown in Example 6.10, active devices are increasingly unstable when decreasing the operating frequency (the context is a simplified intrinsic equivalent circuit but the conclusion holds true also in many real devices). A typical solution is connecting RC parallel blocks in series, see Fig. 6.33, or RL series blocks in parallel. In both cases the stabilizing effect vanishes at high frequency. Similarly, negative (resistive) parallel feedback can stabilize a device, but has a negative impact on gain and is exploited, often as an auxiliary tool, in feedback amplifiers. Resistive input stabilization is finally inappropriate in low-noise amplifiers since dissipative elements at the input have a negative impact on the noise figure. Since stabilization through output resistive networks is not necessarily able to reach the goal within the amplifier bandwidth, low-noise amplifiers often are designed, to avoid unacceptable penalties on the minimum noise figure, through out-of-band stabilization only, while the in-band design follows the strategy used for potentially unstable devices.

Concerning the practical realization of RL parallel and RC series blocks, RC blocks are of course much more compact and do not require, as RL parallel blocks do, a large inductor connected to ground through a via hole. Nevertheless RL blocks can be sometimes integrated within the bias network.

Stabilizing networks can be designed by optimization requiring $K > 1$ at low frequency and $1 < K < 1 + \epsilon$ ($\epsilon > 0$) in the frequency band, where ϵ is a suitably small number. At the same time we should ensure that $|\Delta| < 1$. These design goals can be imposed as others, more conventional, such as the flat gain over the bandwidth and the input and output matching.

Example 6.10 Discuss the stability factor of an intrinsic FET. For simplicity, neglect the intrinsic resistance and the transit time τ.

6.10 Active Device Stabilization and Bias

Solution
From (6.93) and the admittance parameters in (5.14) we immediately have that Y_{11} is purely imaginary. Thus:

$$K = -\frac{\text{Re}\,[Y_{12}Y_{21}]}{|Y_{12}||Y_{21}|} = \frac{1}{\sqrt{1+\left(\dfrac{g_m}{\omega C_{GD}}\right)^2}} \leq 1.$$

The stability factor is therefore $K = 0$ in DC and tends to 1 for $f \to \infty$.

Example 6.11 A FET has the following lumped equivalent circuit parameters: $C_{GS} = 0.071 \times 10^{-12}$ F, $C_{GD} = 0.001 \times 10^{-12}$ F, $C_{DS} = 0.025 \times 10^{-12}$ F, $g_m = 15.2$ mS, $\tau = 1.25 \times 10^{-12}$ s, $R_I = 2.69\,\Omega$, $R_{DS} = 556\,\Omega$, $L_G = 70 \times 10^{-12}$ H, $L_D = 10 \times 10^{-12}$ H, $L_S = 10 \times 10^{-12}$ H, $R_G = 1.46\,\Omega$, $R_S = 1.55\,\Omega$, $R_D = 1.7\,\Omega$. Evaluate the unconditional stability of the device and the effect of stabilizing schemes at the input and output of the device. Also consider the effect of reactive elements with the aim of making the stabilizing effect frequency-dependent.

Solution
From Fig. 6.34 we see that the transistor is unstable at low frequency, as expected. The four stabilization schemes considered are an *RC* or *RL* group able to disconnect the resistive element at high frequency. In detail we have:
1. input series resistance with capacitive bypass:

$$Z_{in} = \frac{R_{in}}{1 + j\omega R_{in} C_{in}} = \frac{R_{in}}{1 + j\omega \tau};$$

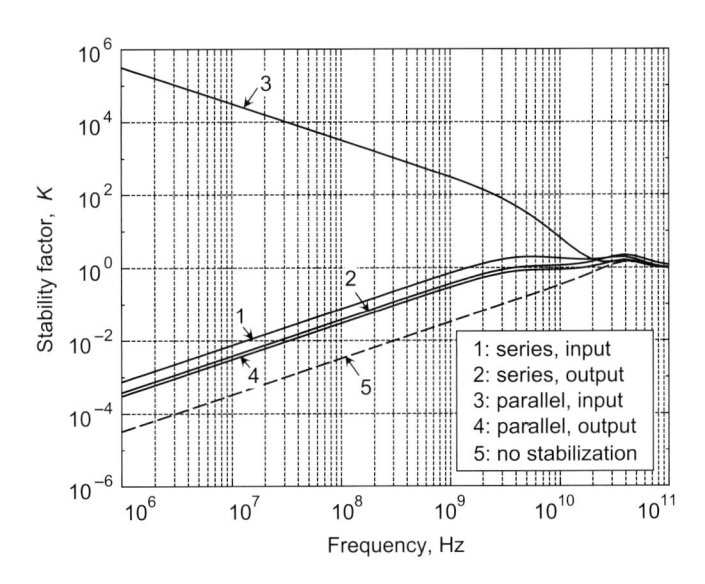

Figure 6.34 Stability coefficient of the device in Example 6.11 vs. frequency.

2. output series resistance with capacitive bypass:

$$Z_{out} = \frac{R_{out}}{1 + j\omega R_{out} C_{out}} = \frac{R_{out}}{1 + j\omega \tau};$$

3. input parallel admittance with inductive block:

$$Y_{in} = \frac{G_{in}}{1 + j\omega G_{in} L_{in}} = \frac{G_{in}}{1 + j\omega \tau};$$

4. output parallel admittance with inductive block:

$$Y_{out} = \frac{G_{out}}{1 + j\omega G_{out} L_{out}} = \frac{G_{out}}{1 + j\omega \tau}.$$

Note that we did not consider the feedback resistance scheme. Since the device becomes stable above 20 GHz we will design a cutoff frequency lower than 20 GHz for the stabilization blocks. As a first guess, we start from a cutoff frequency f_T (i.e., a frequency where the real and imaginary parts of the impedance or admittance of the stabilization block are equal, $2\pi f_T \tau = 1$) of 5 GHz for the isolated block; of course this is inaccurate since the block will be loaded by the rest of the circuit, but the guess can be improved by optimization of the whole circuit. From $f_T = 5$ GHz we obtain a time constant $\tau = 1/(2\pi f_T) = 1/(2\pi \cdot 5 \cdot \times 10^9) = 0.32 \times 10^{-12}$ s; imposing for each of the stabilization schemes this time constant, we obtain:

1. $R_{in} = 100\ \Omega$, $C_{in} = \tau/R_{in} = 3.2 \times 10^{-13}$ F;
2. $R_{out} = 1000\ \Omega$, $C_{out} = \tau/R_{out} = 3.2 \times 10^{-14}$ F;
3. $G_{in} = 1/100\ \Omega$, $L_{in} = \tau/G_{in} = 32 \times 10^{-9}$ H;
4. $G_{out} = 1/100\ \Omega$, $L_{out} = \tau/G_{out} = 32 \times 10^{-9}$ H.

The resistance or conductance values are indicative only, the aim being to stress the effect of each element on the stability factor. The four schemes are separately applied and the resulting stability factor is shown in Fig. 6.34. Only the third scheme (parallel input conductance) is able to stabilize on the whole low-frequency range; the other schemes may allow this result to be obtained only with a large increase of the resistance or conductance, leading to a large gain penalty, and fail anyway at low frequency. Notice the decrease of the S_{21} (at least out of band) that follows the application of the stabilization schemes, see Fig. 6.35.

6.10.2 DC Bias

The DC bias of microwave active devices requires considerable care. The DC bias networks must be decoupled from the RF circuit through proper circuit blocks (called *bias Ts*). The DC voltages that are applied to the gate and the drain of a transistor are obtained from one or two DC sources through a biasing circuit, which must be separated electrically from the RF circuit. Separation is needed (a) because the DC current should not flow in the passive part of the RF circuit, causing power dissipation and possibly damage; (b) because the DC bias circuit should not load the RF circuit, as its electrical behavior at microwaves is completely unpredictable.

6.10 Active Device Stabilization and Bias

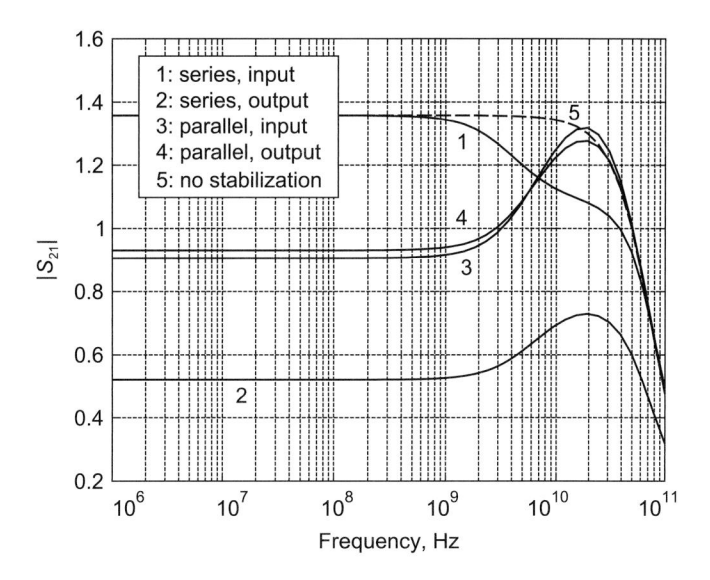

Figure 6.35 Behavior of the magnitude of the S_{21} of the device in Example 6.11 vs. frequency.

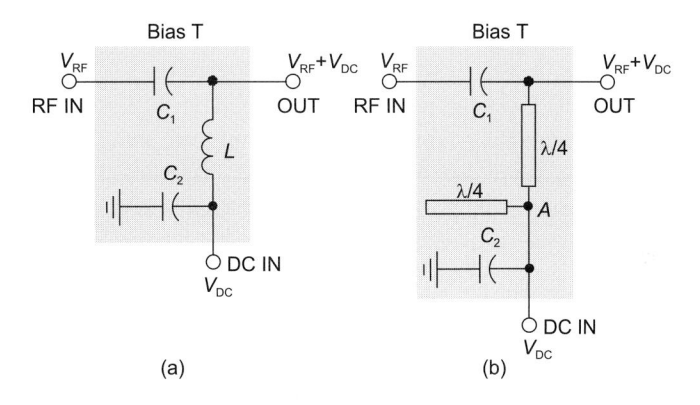

Figure 6.36 Lumped-parameter (a) and distributed parameter (b) bias Ts.

A possible scheme of DC–RF decoupling network referred to as the *bias T* is shown in Fig. 6.36 (a). The capacitor C_1 is a DC block, but it is designed so as to present a low impedance to the RF signal; the inductor L is a DC short but almost an open circuit at RF. The decoupling capacitor C_2 is a further RF ground connected to the DC source. In conclusion, the bias T allows the DC source to be directly connected (through a short) to the device input or output, while exhibiting a large impedance (almost an open) at RF from the input or output connecting node. At high frequency, the implementation of inductors may be difficult, suggesting alternative topologies based on distributed elements. The bias T shown in Fig. 6.36 (b) uses a lumped capacitor C_1 as a RF short; at centerband the open $\lambda/4$ stub RF grounds node A, that becomes an open in the second quarterwave section. Thus, the RF impedance from the bias node is infinity. Out of the centerband frequency the RF is anyway grounded through the blocking capacitor C_2.

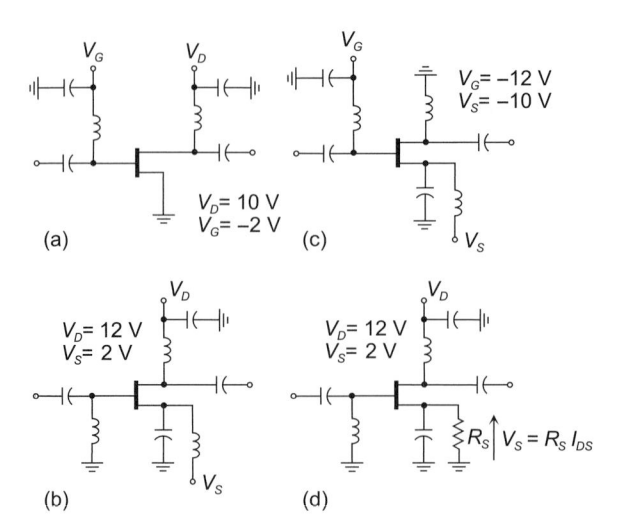

Figure 6.37 Examples of FET passive bias networks: (a), (b), (c) with two DC supplies; (d) with a single positive DC supply.

Most practical bias circuits aim at keeping constant the bias point of the transistor vs. changes in the transistor parameters and/or the operating temperature. Active bias networks can be exploited whenever temperature stability is a main concern.

Some common FET passive bias networks are shown in Fig. 6.37. The networks in Fig. 6.37 (a)–(c) require two bias sources, while that in Fig. 6.37 (d) only needs one bias source. In the network in Fig. 6.37 (a) the source is directly connected to ground, thus minimizing the parasitic source inductance, while in the networks in Fig. 6.37 (b)–(c) the source is DC separated from the ground; since the decoupling capacitor has some parasitic inductance this leads to an increase of the source inductive loading. In the network in Fig. 6.37 (a) the sequence in applying the DC source is important: first the gate voltage has to be applied (i.e., $V_G < 0$) and then the drain voltage ($V_D > 0$). In switching off the bias the inverse sequence has to be followed. Similar criteria can be applied in the bias circuits in Fig. 6.37 (b)–(c); also in this case the gate bias has to be applied first in order to avoid a large DC current to flow on the application of the drain bias. Finally, the network in Fig. 6.37 (d) exploits a resistor connected to the source, that grants to the device a protection for the DC bias; however, the use of a source resistor can deteriorate the noise properties and introduce instabilities. The resistor R_S implements a DC negative feedback that decreases the variation of I_D due to variation in the operating temperature and device parameters (I_{DSS} and V_T). For further details see, e.g., [1].

6.11 The Open-Loop Wideband Amplifier

In the design of maximum gain broadband amplifiers, the design goal is to achieve flat gain over the operation bandwidth with low input and output reflection coefficients (e.g., better than -10 dB). We immediately see that imposing conjugate matching over a wide bandwidth is not only impossible due to Fano's limit, but would also lead to an amplifier operating with MAG at each of the operating frequencies, i.e., to a gain decreasing with

6.11 The Open-Loop Wideband Amplifier

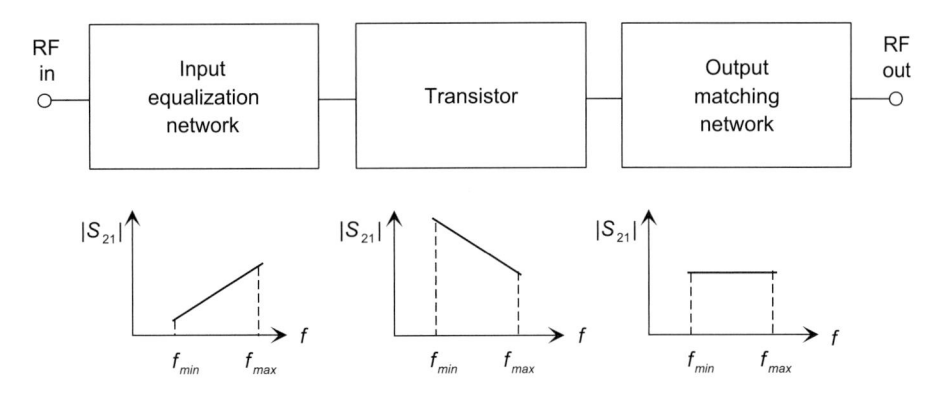

Figure 6.38 Scheme of an open-loop single-stage wideband amplifier with reactive input equalization and reactive output matching.

the square of the frequency, see (5.12). Therefore, in the wideband design some amount of input or output equalization is needed, see Fig. 6.38, to compensate for the decreasing gain by attenuating the signal (typically at the input: the output equalization would lead to either dissipating a large power or to reflecting back some output power into the device, both being inconvenient solutions) at the lowest frequency. The overall gain of the amplifier is then dominated by the gain achieved in the upper limit of the bandwidth.

Equalization networks can be either reactive or resistive. Resistive input equalization yields the possibility of dissipating the reflected power, and therefore allows for a low input reflection coefficient. Reactive input equalization operates by reflecting back part of the input power in the lower part of the operating bandwidth, and is therefore incompatible with a low input reflection coefficient. However, proper combining solutions, such as the balanced amplifier, can be exploited in order to simultaneously achieve low input reflection coefficient and gain equalization.

6.11.1 The Balanced Amplifier

The balanced amplifier [15] is a tandem combination of two equal amplifier stages by means of one input and one output 3 dB 90° directional couplers. The two amplifiers have a flat gain over a certain bandwidth thanks to the reactive input equalization, but at the same time their input reflection coefficients may be large in the lower part of the operating bandwidth. However, thanks to the action of the directional couplers (power splitting and phase delay) the power reflected from the input of the two stages is dissipated in the resistor R_0 connected to the input coupler, while none reaches the amplifier input port. The same principle holds for reflections from the amplifier output. The amplifier bandwidth is limited either by the equalization and matching networks of the two stages or, anyway, by the coupler bandwidth, of the order of one octave (i.e., from f to $2f$). The balanced amplifier scheme is shown in Fig. 6.39. Notice that the two amplifiers in tandem require an input power that is twice the one of the single amplifier; since also the output power is doubled, the amplifier gain is the same. The output saturation power, however, is doubled.

Microwave Linear Amplifiers

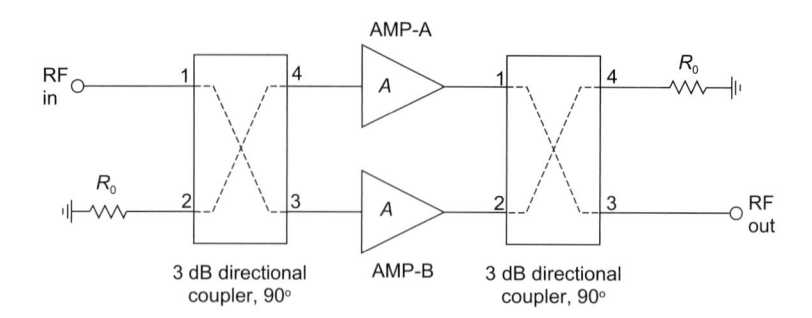

Figure 6.39 Scheme of a balanced amplifier.

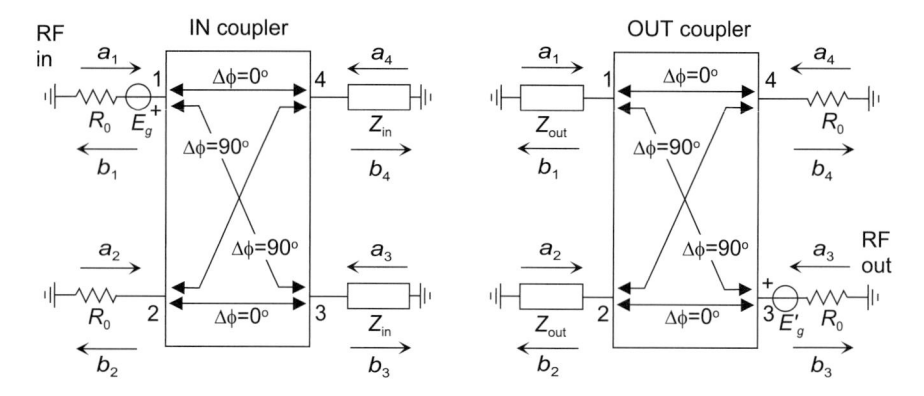

Figure 6.40 Evaluating the input impedances of the input and output directional couplers loaded by the amplifiers inputs (left) and outputs (right).

In order to understand the operation of the balanced amplifier, one should consider how the 3 dB directional coupler behaves when it is loaded by the normalization resistances at ports 1 and 2 and by the amplifier identical input impedances Z_{in} at ports 3 and 4, see Fig. 6.40 (left), or, conversely, by the normalization resistances at ports 3 and 4 and by the amplifier identical output impedances Z_{out} at ports 1 and 2, see Fig. 6.40 (right). We are going to show that port 1 of the input coupler is matched – and so is port 4 of the output coupler – even if the coupler is not closed at all ports on R_0. Let us consider first the centerband frequency and suppose that the coupler is ideal concerning input matching and isolation. Consider first the input coupler. The centerband scattering matrix reads:

$$\begin{pmatrix} b_1 \\ b_2 \\ b_3 \\ b_4 \end{pmatrix} = \begin{pmatrix} 0 & 0 & -j/\sqrt{2} & 1/\sqrt{2} \\ 0 & 0 & 1/\sqrt{2} & -j/\sqrt{2} \\ -j/\sqrt{2} & 1/\sqrt{2} & 0 & 0 \\ 1/\sqrt{2} & -j/\sqrt{2} & 0 & 0 \end{pmatrix} \begin{pmatrix} a_1 \\ a_2 \\ a_3 \\ a_4 \end{pmatrix},$$

i.e.:

$$b_1 = -\frac{j}{\sqrt{2}} a_3 + \frac{1}{\sqrt{2}} a_4$$

6.11 The Open-Loop Wideband Amplifier

$$b_2 = \frac{1}{\sqrt{2}}a_3 - \frac{j}{\sqrt{2}}a_4.$$

Moreover, since port 2 is loaded by R_0, we have $a_2 = 0$ and therefore:

$$b_3 = -\frac{j}{\sqrt{2}}a_1, \quad b_4 = \frac{1}{\sqrt{2}}a_1.$$

However, due to the loading condition at ports 3 and 4, we also have:

$$a_3 = \Gamma_{in}b_3 = \Gamma_{in}\left(-\frac{j}{\sqrt{2}}a_1\right)$$

$$a_4 = \Gamma_{in}b_4 = \Gamma_{in}\left(\frac{1}{\sqrt{2}}a_1\right),$$

where Γ_{in} is the reflectance corresponding to the amplifier input impedance Z_{in}. Thus, substituting:

$$b_1 = -\frac{j}{\sqrt{2}}\Gamma_{in}\left(-\frac{j}{\sqrt{2}}a_1\right) + \frac{1}{\sqrt{2}}\Gamma_{in}\left(\frac{1}{\sqrt{2}}a_1\right) = 0$$

$$b_2 = \frac{1}{\sqrt{2}}\Gamma_{in}\left(-\frac{j}{\sqrt{2}}a_1\right) - \frac{j}{\sqrt{2}}\Gamma_{in}\left(\frac{1}{\sqrt{2}}a_1\right) = -j\Gamma_{in}a_1.$$

Port 1 is therefore matched while the transmission coefficient between port 1 and 2 is $-j\Gamma_{in}$, i.e., the amplifier input reflection coefficient with a phase delay of $-\pi/2$. In a similar way we can show that the output coupler is matched at port 4 while the transmission coefficient between port 4 and 3 is $-j\Gamma_{out}$, i.e., the amplifier output reflectance.

As a further step, let us analyze the input and output loading impedances seen by the two tandem amplifiers, see Fig. 6.41. For the sake of definiteness let us consider the

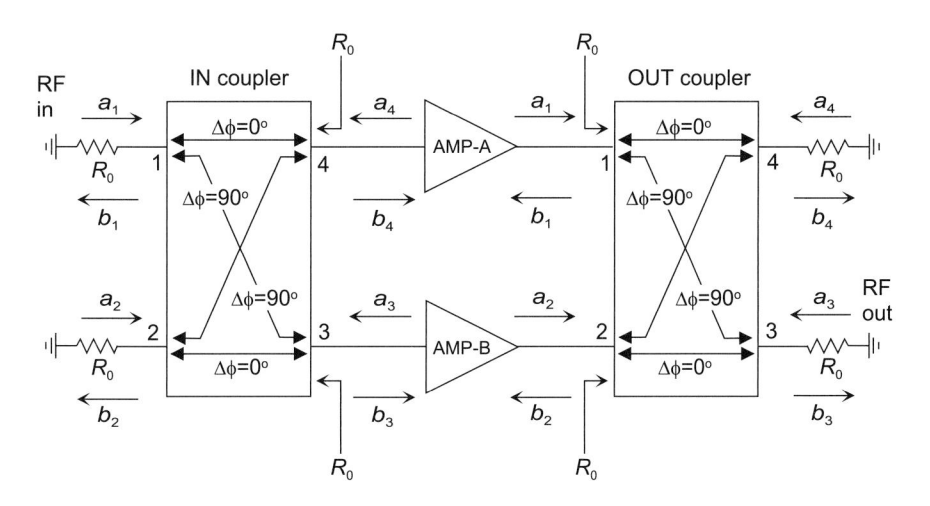

Figure 6.41 Evaluating the input and output loading impedances of the two amplifier stages in tandem.

input coupler. Since ports 1 and 2 are loaded with R_0 we have $a_1 = a_2 = 0$. It follows, from the scattering matrix definition (at centerband and outside it), that:

$$\begin{pmatrix} b_1 \\ b_2 \\ b_3 \\ b_4 \end{pmatrix} = \begin{pmatrix} 0 & 0 & s_{13} & s_{14} \\ 0 & 0 & s_{14} & s_{13} \\ s_{13} & s_{14} & 0 & 0 \\ s_{14} & s_{13} & 0 & 0 \end{pmatrix} \begin{pmatrix} a_1 \\ a_2 \\ a_3 \\ a_4 \end{pmatrix}$$

and thus:

$$b_3 = 0, \quad b_4 = 0,$$

i.e., independent of frequency, the two amplifiers are closed at the input on the normalization resistances. The same conclusion can be reached concerning the output. This means that the relevant power waves are related by the scattering parameters, and in particular $\Gamma_{in} = S_{11}$, $\Gamma_{out} = S_{22}$ with reference to the amplifier scattering matrix (denoted with a capital S).

We can now conclude with the analysis of the balanced amplifier at centerband, from the scheme in Fig. 6.42 where S_{ij} are the scattering parameters of the two amplifiers in tandem referred to the resistance R_0. Assuming that a unit power wave enters port 1 no reflection is seen, while the power reflected from the amplifiers is deviated on the matched resistance in port 2. From the power wave flow we see that the power wave on the amplifier output port leads to a global transmission parameter equal to jS_{21}. With similar arguments we see that the amplifier is also matched (ideally at centerband) at the output port, see Fig. 6.43.

Deviating from the centerband frequency the couplers' transmission and coupling become unbalanced, but the $\pi/2$ phase shift is preserved. This implies that, if the S_{21} of the two amplifiers is equalized, the total balanced amplifier S_{21} well be comparatively flat also out of centerband. From Fig. 6.44, defining the input amplifier port with the subscript i and the output with the subscript o, we find that the scattering matrix of the whole balanced amplifier has elements in the first column:

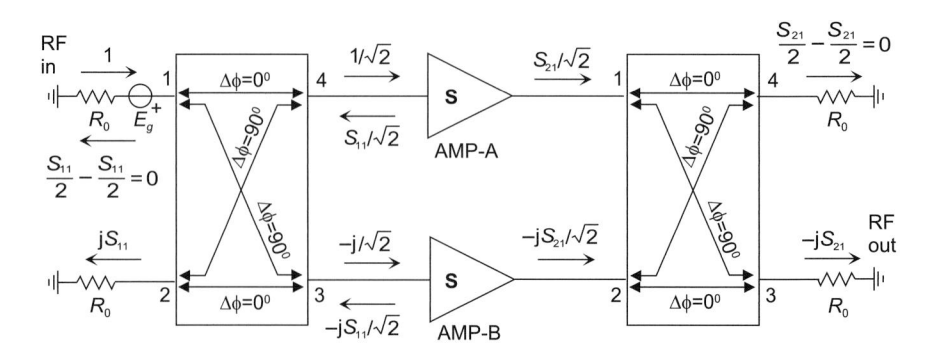

Figure 6.42 Operation of the balanced amplifier at centerband: input matching and gain.

6.11 The Open-Loop Wideband Amplifier

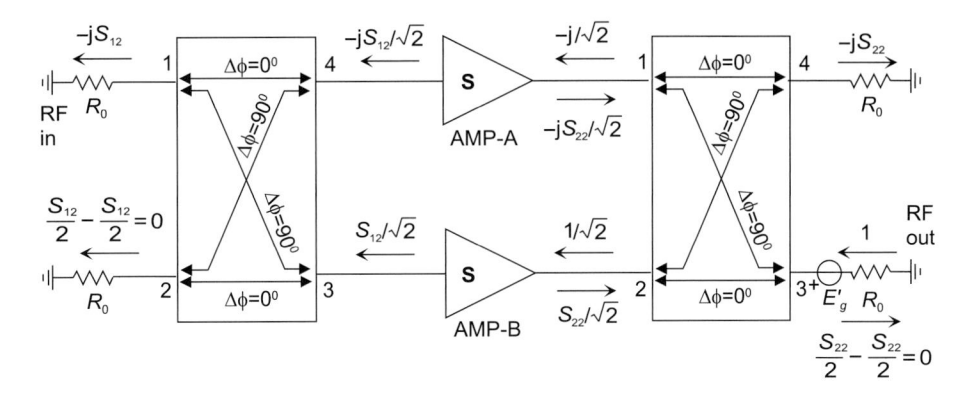

Figure 6.43 Output matching of the balanced amplifier at centerband.

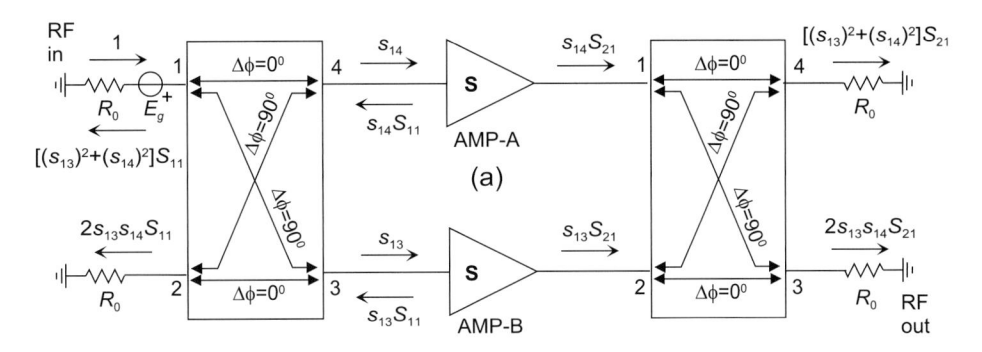

Figure 6.44 Operation of the balanced amplifier out of centerband: input matching and gain.

$$S_{ii} = \left(s_{13}^2 + s_{14}^2\right) S_{11}$$

$$S_{oi} = 2s_{13}s_{14}S_{21}.$$

Taking into account that for a 3 dB coupler we have, see (4.8):

$$s_{14} = \frac{\Gamma\left(1 - \exp\left(-2j\theta\right)\right)}{1 - \Gamma^2 \exp\left(-2j\theta\right)}$$

$$s_{13} = \frac{\left(1 - \Gamma^2\right) \exp\left(-j\theta\right)}{1 - \Gamma^2 \exp\left(-2j\theta\right)}$$

$$\Gamma = \frac{\sqrt{Z_e} - \sqrt{Z_o}}{\sqrt{Z_e} + \sqrt{Z_o}},$$

where θ is the electrical length ($\pi/2$ at centerband) and that, for a 50 Ω closure the even and odd mode impedances (we assume for simplicity a two-conductor coupler) are:

$$Z_e = 120.73 \ \Omega, \quad Z_o = 20.71 \ \Omega,$$

we obtain $\Gamma = 0.414$. The behavior of the absolute value of the normalized transmittance $2s_{13}s_{14}$ and input reflectance $s_{13}^2 + s_{14}^2$ are shown in Fig. 6.45. Notice that if the

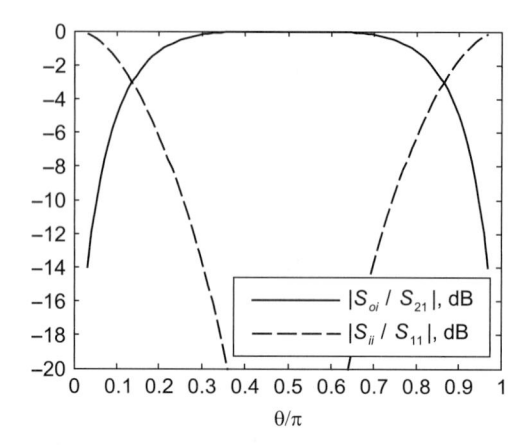

Figure 6.45 Normalized input–output transmittance and input reflectance for an ideal balanced amplifier.

amplifier S_{21} is flat, the 3 dB bandwidth of the balanced amplifier is much wider than one octave; it is in fact around two octaves; however, the amount of the reflectance suppression provided on the same bandwidth can be insufficient. In fact, on a two-octave bandwidth the amplifier MAG decreases from the lower to the upper limit by a factor of 16 (12 dB) requiring the input reflectance to increase accordingly from the upper to the lower limit. If, for instance, the reflectance is −20 dB in the upper limit of the bandwidth, it will be −8 dB in the upper limit, requiring 12 dB suppression by the coupler, than can be achieved on a narrower bandwidth. Of course the actual design depends on the actual reflectance level that can be allowed by design.

The balanced amplifier architecture has enjoyed a constant popularity both in hybrid and monolithic implementations, due to its general ability to correct on a moderately wide bandwidth input or output mismatches. Along this line, it has been used not only for broadbanding purposes, but also to improve the input matching of low-noise amplifiers with input noise matching (see Sec. 10.10) or the output matching of power amplifiers designed for maximum output power. A hybrid implementation of the balanced amplifier was already presented in Fig. 1.16.

In Fig. 6.46 we show a more recent example, a MMIC three-stage millimeter-wave cascode balanced amplifier. The circuit is fabricated using an InAlAs/InGaAs MHEMT technology on GaAs substrate from IAF [16]. The input and output hybrid coupler is implemented using stacked conductors, thus forming a broadside-coupled line coupler that is able to provide 3 dB coupling (the Lange coupler solution is critical at this high frequency due to the parasitic parameters of the airbridges). The amplifier exhibits and average 20 dB gain between 150 and 240 GHz and input and output reflection coefficient below −10 dB on the same bandwidth.

6.12 The Wideband Feedback Amplifier

Negative feedback amplifiers are commonly used in electronics; their main purpose is to equalize the device gain making it independent from the device parameters and only

6.12 The Wideband Feedback Amplifier

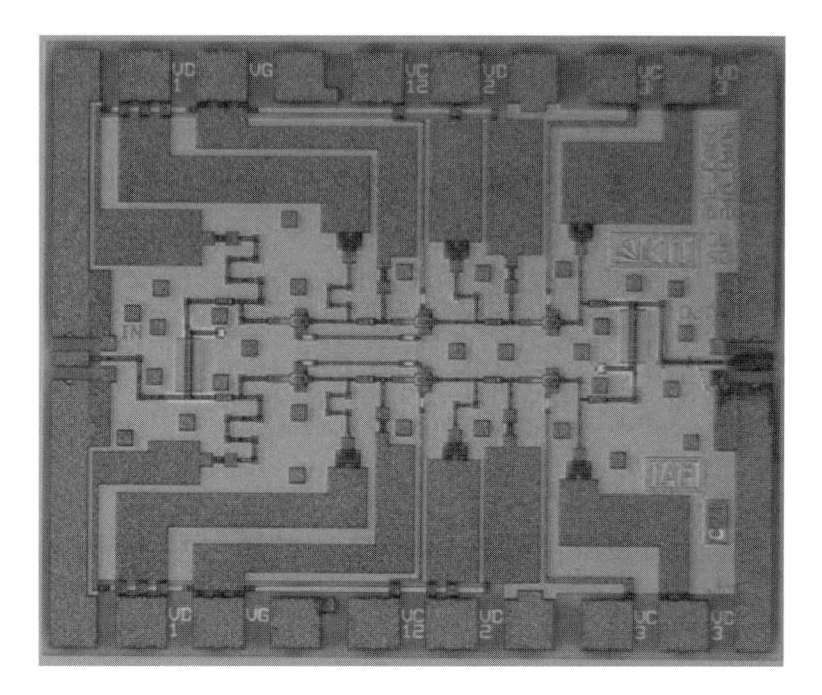

Figure 6.46 Millimeter-wave integrated balanced amplifier. Copyright ©2013 IEEE. All rights reserved. Reprinted, with permission, from [16, Fig. 5].

dependent on the values of passive (feedback) elements. However, the feedback principle operates correctly, as well known, only if the open-loop gain is much larger than the gain with feedback. This constraint makes the use of feedback amplifiers rather difficult at microwaves, since in this region the device gain typically decreases like the square of the frequency. Therefore, the feedback principle is effective only at comparatively low frequency vs. f_T. Moreover, the use of parallel feedback with reactive elements (namely, *peaking inductors*) is often exploited to equalize the amplifier gain on the operating band [17].

Microwave feedback amplifiers fall into two classes. In a first class, we have *narrowband* amplifiers were feedback is exploited mainly to make the device gain stable and independent from the device parameters; this is a typical solution for integrated implementation exhibiting large parameter fluctuations associated with the technological process. To a second class belong the *wideband* feedback amplifiers where feedback is exploited to implement a flat gain over a large frequency range (ideally from DC to an upper cutoff frequency). This second design approach is somewhat similar to the low-frequency feedback amplifier concept. However, it can be stressed that, from a high-frequency analog standpoint, resistive parallel feedback also allows, at least ideally, to achieve good input and output matching together with flat gain.

As a general example of resistive feedback amplifier, let us consider a feedback amplifier with both parallel and series feedback: R_s is the series feedback resistance while R_p is the parallel feedback resistance. The purpose of the additional series feedback resistor

Microwave Linear Amplifiers

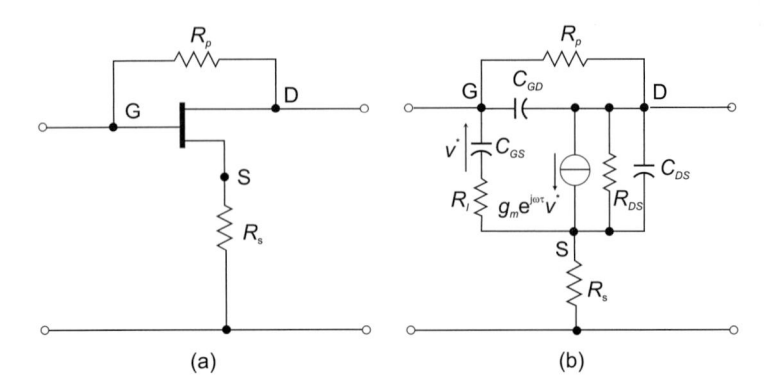

Figure 6.47 FET with series and parallel feedback (a); the FET is replaced by its intrinsic equivalent circuit (b).

will be clarified in what follows. In the analysis, we will exploit an intrinsic FET equivalent circuit; the design will be carried out at low frequency, where reactive effects are negligible.

Let us evaluate the scattering matrix of the FET with parallel and series feedback. From the intrinsic admittance of the FET, see (5.14):

$$\mathbf{Y}_{in} = \begin{pmatrix} \dfrac{j\omega C_{GS}}{1+j\omega C_{GS}R_I} + j\omega C_{GD} & -j\omega C_{GD} \\ \dfrac{g_m e^{j\omega\tau}}{1+j\omega C_{GS}R_I} - j\omega C_{GD} & j\omega(C_{GD}+C_{DS}) + \dfrac{1}{R_{DS}} \end{pmatrix},$$

in the low-frequency limit we obtain:

$$\mathbf{Y}_{in}(0) = \begin{pmatrix} 0 & 0 \\ g_m & \dfrac{1}{R_{DS}} \end{pmatrix}.$$

Finally, summing the admittance matrix of the parallel feedback network we have:

$$\mathbf{Y}_1 = \mathbf{Y}_{in}(0) + \begin{pmatrix} \dfrac{1}{R_p} & -\dfrac{1}{R_p} \\ -\dfrac{1}{R_p} & \dfrac{1}{R_p} \end{pmatrix} = \begin{pmatrix} \dfrac{1}{R_p} & -\dfrac{1}{R_p} \\ g_m - \dfrac{1}{R_p} & \dfrac{1}{R_p} + \dfrac{1}{R_{DS}} \end{pmatrix}.$$

The impedance matrix with parallel feedback $Z_1 = Y_1^{-1}$ is obtained by inversion as:

$$\mathbf{Z}_1 = \begin{pmatrix} \dfrac{R_p^2}{R_p - R_p' + R_p'R_p g_m} & \dfrac{R_p'R_p}{R_p - R_p' + R_p'R_p g_m} \\ \dfrac{R_p'R_p - R_p'R_p^2 g_m}{R_p - R_p' + R_p'R_p g_m} & \dfrac{R_p'R_p}{R_p - R_p' + R_p'R_p g_m} \end{pmatrix},$$

where $R'_p = R_p // R_{DS}$. The impedance matrix with series feedback is finally obtained by summing the impedance matrix of the feedback network:

$$\mathbf{Z}_2 = \mathbf{Z}_1 + \begin{pmatrix} R_s & R_s \\ R_s & R_s \end{pmatrix}$$

$$= \begin{pmatrix} \dfrac{R_p^2}{R_p - R'_p + R'_p R_p g_m} + R_s & \dfrac{R'_p R_p}{R_p - R'_p + R'_p R_p g_m} + R_s \\ \dfrac{R'_p R_p - R'_p R_p^2 g_m}{R_p - R'_p + R'_p R_p g_m} + R_s & \dfrac{R'_p R_p}{R_p - R'_p + R'_p R_p g_m} + R_s \end{pmatrix}.$$

If $R_{DS} \gg R_p$ we have $R'_p \approx R_p$ and:

$$\mathbf{Z}_2 = \begin{pmatrix} \dfrac{1}{g_m} + R_s & \dfrac{1}{g_m} + R_s \\ \dfrac{1}{g_m} - R_p + R_s & \dfrac{1}{g_m} + R_s \end{pmatrix},$$

so that the low-frequency scattering matrix elements with feedback (approximated at zero frequency) vs. R_0 become:

$$S_{11f}(0) = \dfrac{\left(R_p R_s - R_0^2\right) g_m + R_p}{\left(2R_0 R_s + R_p R_s + R_0^2\right) g_m + 2R_0 + R_p}$$

$$S_{12f}(0) = 2R_0 \dfrac{1 + R_s g_m}{\left(2R_0 R_s + R_p R_s + R_0^2\right) g_m + 2R_0 + R_p}$$

$$S_{21f}(0) = -2R_0 \dfrac{\left(R_p - R_s\right) g_m - 1}{\left(2R_0 R_s + R_p R_s + R_0^2\right) g_m + 2R_0 + R_p}$$

$$S_{22f}(0) = \dfrac{\left(R_p R_s - R_0^2\right) g_m + R_p}{\left(2R_0 R_s + R_p R_s + R_0^2\right) g_m + 2R_0 + R_p}.$$

Taking into account that $S_{11f} = S_{22f}$ at low frequency we can *simultaneously match* the amplifier both at the input and at the output imposing:

$$R_s = \dfrac{R_0^2}{R_p} - \dfrac{1}{g_m}. \tag{6.94}$$

Substituting into S_{12f} and S_{21f} we obtain, independent of the device transconductance:

$$S_{12f}(0) = \dfrac{R_0}{R_0 + R_p}$$

$$S_{21f}(0) = \dfrac{R_0 - R_p}{R_0}.$$

Notice that, due to the presence of the parallel feedback resistor, the S_{12f} parameter becomes large, i.e., the device with feedback in not any more unidirectional. The value of the parallel feedback resistance can be now derived from the gain required at low-frequency $S_{21f}(0)$:

$$R_p = R_0 \left(1 - S_{21f}(0)\right) = R_0 \left(1 + |S_{21f}(0)|\right) \tag{6.95}$$

(remember that the low-frequency S_{21} is real negative). Eqs. (6.94) and (6.95) yield the parallel and series feedback resistor values allowing for input and output matching and a certain amount of gain with feedback. However, (6.94) suggests that, since R_s cannot be negative, the transconductance should be at least:

$$g_m \geq \frac{R_p}{R_0^2}. \tag{6.96}$$

Taking into account that the low-frequency open-loop S_{21} approximately is $S_{21} \approx -2g_m R_0$, we can express the above condition as a condition on the minimum open-loop S_{21}; since, from (6.96) and (6.95):

$$g_m = \frac{|S_{21}|}{2R_0} \geq \frac{R_p}{R_0^2} = \frac{1 + |S_{21f}(0)|}{R_0},$$

we obtain:

$$|S_{21}| \geq 2\left(1 + |S_{21f}(0)|\right), \tag{6.97}$$

i.e., the open-loop S_{21} must be at least about twice the S_{21} with feedback. Notice that the equality sign holds when the series feedback resistance is zero.

Example 6.12 Consider a FET with $g_m = 0.152$ S. Find the series and parallel feedback resistances needed to obtain $S_{21f} = 15$ dB at low frequency with matching on $R_0 = 50\ \Omega$.

Solution
Without feedback we have $S_{21} = -2R_0 g_m = -15.2$. The required $S_{21f}|_{dB} = 15$ dB corresponds, in natural units, to $|S_{21f}(0)| = 10^{15/20} = 5.623$. Since the relation $|S_{21}| \geq 2\left(1 + |S_{21f}(0)|\right)$ here corresponds to the values $15.2 > 2(1+5.623)$, the design is possible and:

$$R_p = R_0\left(1 + |S_{21f}(0)|\right) = 331.17\ \Omega.$$

The transconductance must satisfy the relation:

$$g_m = 0.152 > \frac{R_p}{R_0^2} = 0.132\ \text{S};$$

since g_m is larger than the minimum theoretical value required we must introduce a series feedback resistor of value:

$$R_s = \frac{R_0^2}{R_p} - \frac{1}{g_m} = 0.970\ \Omega.$$

In practice, in an integrated implementation where the device periphery (and therefore the transconductance) can be scaled, it is convenient to avoid using the series feedback resistance ($R_s = 0$) using for g_m the minimum value:

6.12 The Wideband Feedback Amplifier

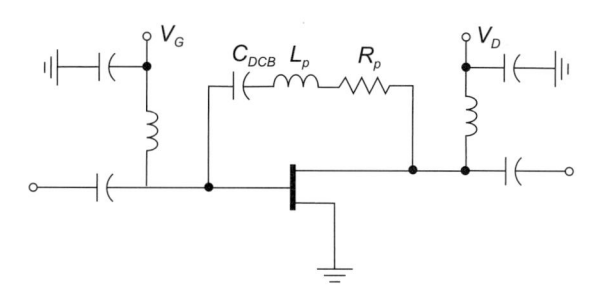

Figure 6.48 FET with reactive feedback and bias Ts.

$$g_m = \frac{R_p}{R_0^2}.$$

Concerning the frequency behavior, the design was carried out at low frequency, but, increasing the frequency, the input and output matching deteriorate and the open-loop S_{21} decreases; the parallel feedback network can be modified introducing reactive elements (a DC blocking capacitor and a peaking inductor) to partly compensate for the decrease in the open-loop gain as shown in Fig. 6.48. In other words we have:

$$R_p \rightarrow Z_p = j\omega L_p + R_p + \frac{1}{j\omega C_{DCB}}.$$

The series peaking inductor has to be designed in order to partly compensate the decrease of the open-loop S_{21}; its value can be determined by optimization.

A warning should however be mentioned about integrated feedback amplifier design. Apparently, the constraint on the transconductance can be always achieved by scaling the device periphery. However, increasing this way the transconductance, while it keeps the cutoff frequency constant, correspondingly decreases the output resistance to values that violate the condition $R_{DS} \gg R_p$. Besides, the effect of parasitics typically decreases the amplifier performances vs. the ideal case. This point is stressed in Example 6.13 (where an ideal design exercise is carried out) and in Example 6.14 (where the design is implemented with a real high-performance FET).

Example 6.13 We want to design a resistive feedback amplifier having a low-frequency $S_{21f} = 20$ dB for a FET with $R_0 = 50\ \Omega$ matching. Estimate the minimum g_m needed to implement it without series feedback and the resulting S_{21}.

Solution
We have $|S_{21f}(0)| = 10^{20/20} = 10$, from which:

$$R_p = R_0 \left(1 + |S_{21f}(0)|\right) = 550\ \Omega,$$

i.e., the minimum transconductance is:

$$g_m = \frac{R_p}{R_0^2} = 220\ \text{mS}.$$

Table 6.8 Rescaled elements of the small-signal equivalent circuit in Example 6.14.

$C'_{GS} = \alpha C_{GS} = 1.62 \times 10^{-13}$ F	$C'_{GD} = \alpha C_{GD} = 1.1 \times 10^{-15}$ F
$C'_{DS} = \alpha C_{DS} = 2.2 \times 10^{-14}$ F	$g'_m = \alpha g_m = 220$ mS
$R'_I = \alpha^{-1} R_I = 1.45$ Ω	$R'_{DS} = \alpha^{-1} R_{DS} = 869$ Ω
$L'_G = \alpha^{-1} L_G = 6.36 \times 10^{-12}$ H	$L'_D = \alpha^{-1} L_D = 9.09 \times 10^{-13}$ H
$L'_S = \alpha^{-1} L_S = 9.09 \times 10^{-13}$ H	$R'_G = \alpha^{-1} R_G = 1.33$ Ω
$R'_S = \alpha^{-1} R_S = 1.41$ Ω	$R'_D = \alpha^{-1} R_D = 1.45$ Ω

The low-frequency open-loop transmittance is $S_{21} = -2R_0 g_m = -22$, equal, as it should be, to the limit value $2(1 + |S_{21f}(0)|) = 22$ (we have $R_s = 0$ in fact).

Example 6.14 Implement the design in Example 6.13 by exploiting a properly rescaled (with respect to the device periphery W) and integrated nanometer gate length device with $W = 0.25$ mm and small-signal parameters $C_{GS} = 1.47 \times 10^{-14}$ F, $C_{GD} = 0.1 \times 10^{-14}$ F, $C_{DS} = 0.2 \times 10^{-13}$ F, $g_m = 200$ mS, $\tau = 0.05 \times 10^{-12}$ s, $R_I = 1.6$ Ω, $R_{DS} = 956$ Ω; $L_G = 7 \times 10^{-12}$ H, $L_D = 1 \times 10^{-12}$ H, $L_S = 1 \times 10^{-12}$ H, $R_G = 1.46$ Ω, $R_S = 1.55$ Ω, $R_D = 1.6$ Ω.

Solution
To obtain a 220 mS transconductance we need a rescaling ratio α:

$$\alpha = \frac{W'}{W} = \frac{220}{200} = 1.1,$$

where W is the original gate periphery, W' the modified one, $W' = 0.25 \cdot 1.1 = 0.275$ mm. The new parameters (series elements scale according to $1/W$, parallel elements according to W, $\tau' = \tau = 0.05 \times 10^{-12}$ s is unchanged) are listed in Table 6.8.

The (intrinsic and ideal) cutoff and maximum oscillation frequency of the rescaled device are:

$$f_T = \frac{g'_m}{2\pi C'_{GS}} = \frac{0.220}{2\pi \cdot 1.62 \times 10^{-13}} = 216 \text{ GHz}$$

$$f_{\max} = \frac{f_T}{2} \sqrt{\frac{R'_{DS}}{R'_I + R'_G + R'_S + 2\pi f_T R'_{DS} R'_G C'_{GD}}} = 1.3 \text{ THz}.$$

Notice that the theoretical value of $f_{\max}$ is somewhat unrealistically high, since parasitics are only partially accounted for in (5.13). However, the figures of merit obtained point out that the device is a high performance one.

With the parallel feedback of 550 Ω derived in Example 6.13, we obtain the results in Fig. 6.49 (stability coefficient) and in Fig. 6.50 (scattering parameters). The RL feedback results are obtained with an additional feedback peaking inductance $L_p = 2$ nH. We notice the following points.

The device is unstable without feedback, and becomes stable with a resistive feedback. This points out the stabilizing effect of the resistive feedback. The use of a RL feedback, however, makes the device again potentially unstable at very high frequency

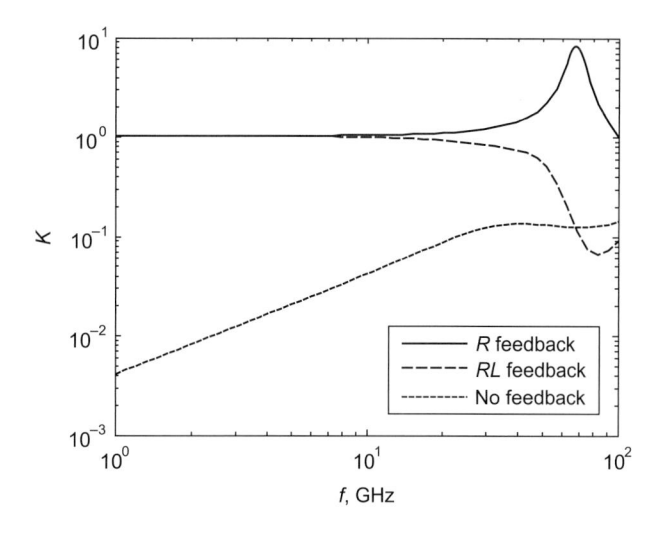

Figure 6.49 Stability coefficient without feedback, with resistive feedback, with RL feedback, see Example 6.14.

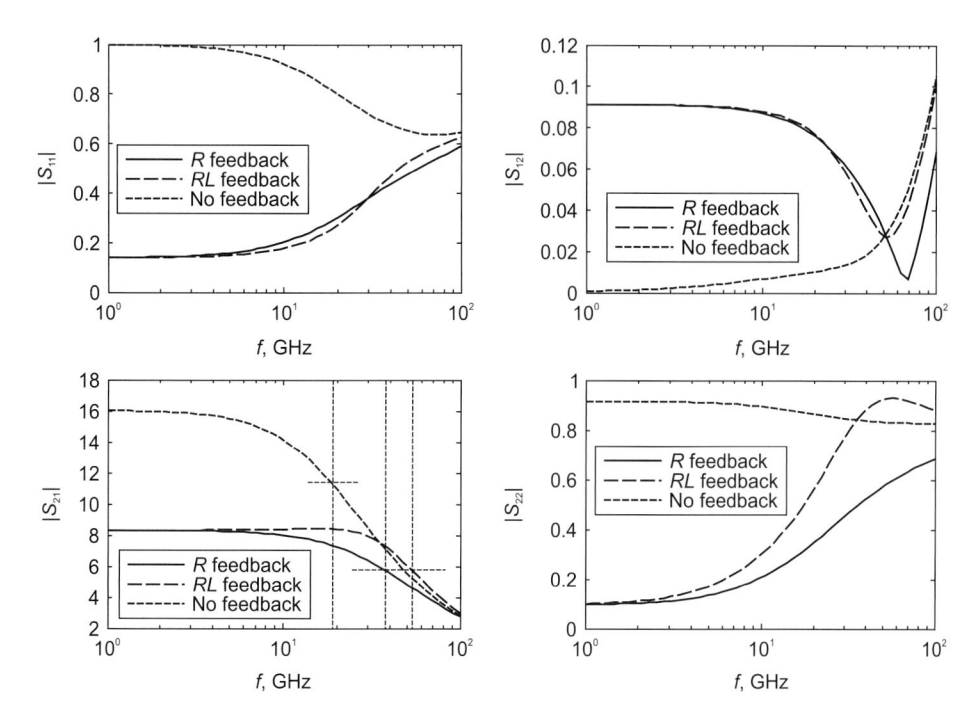

Figure 6.50 Scattering parameters without feedback, with resistive feedback, with RL feedback, see Example 6.14.

(notice that this is still well below the device $f_{\max}$), thus requiring additional out-of-band stabilization. In practice also a low-frequency additional stabilization is needed since a DC blocking capacitor must be inserted in the feedback loop, that opens the loop at low frequency thus canceling the stabilizing effect of feedback.

The values of S_{21} and S_{21f} are lower than the theoretical ones (22 and 10, respectively) as evaluated on the basis of the ideal intrinsic equivalent circuit, see Example 6.13. This is due to the effect of parasitics (in particular the source resistance, that lowers the effective transconductance) and of the output resistance R_{DS}. The introduction of feedback clearly reduces the gain while increasing the amplifier bandwidth, as expected. Without feedback the 3 dB bandwidth is slightly below 20 GHz while it becomes above 35 GHz with resistive feedback and above 50 GHz with RL feedback. Notice that the amount of gain that can be obtained (on the basis of the scattering parameters) could in fact be improved by optimizing the closing resistance, chosen here as 50 Ω (reactive terminations are virtually impossible to implement in a very broadband structure). The peaking inductor leads to a significant amount of broadbanding, while decreasing at high frequency the input matching. Finally, the device matching is improved vs. the original (no feedback) condition, and, as expected, the feedback device is less unidirectional than the original one.

In conclusion, parallel feedback is a tool to obtain extremely broadband amplifiers with fairly good input and output matching. The theoretical values derived can be exploited as a starting point for further optimization. Parallel feedback also improves the device stability. Further broadbanding can be obtained through the use of reactive elements (peaking inductors) in the feedback loop. However, the price to be paid is the availability of a device with a cutoff frequency much larger than the upper operating frequency of amplifier. As discussed in Sec. 6.13, another approach to extremely broadband amplifiers, the distributed amplifier, is less demanding in terms of device performances.

6.13 The Distributed Amplifier

In the discussion of the wideband feedback amplifier we saw how we can trade gain for bandwidth: the open-loop gain is large but on a narrower bandwidth, feedback allows lower gain on a wider bandwidth to be achieved. According to a somewhat simplified model of the amplifier response we can show that the gain-bandwidth product of a device is approximately constant and related to the cutoff frequency of the device. Consider a simplified single-stage FET amplifier closed on the normalization impedance R_0 at the input and output, see Fig. 6.51. Evaluating the voltage amplification between the generator and the load voltage we obtain:

$$A_V(f) = \frac{V_L}{E_g} = -\frac{R_0 g_m}{1 + j\omega C_{GS} R_0}.$$

Defining the bandwidth as the frequency f_β at which the voltage amplification drops by 3 dB vs. the DC value (i.e., by a factor $1/\sqrt{2}$ in natural units, corresponding to a factor 1/2 in terms of power) we immediately have:

$$f_\beta = \frac{1}{2\pi C_{GS} R_0}.$$

6.13 The Distributed Amplifier

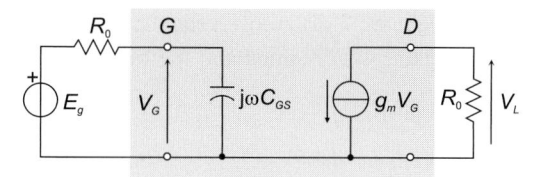

Figure 6.51 Evaluation of the gain-bandwidth product of an ideal single-stage FET amplifier.

We see that increasing the load impedance R_0 the low-frequency gain increases but the bandwidth decreases in such a way that their product (GBP, Gain Bandwidth Product) remains constant:

$$\text{GPB} = A_V(0) f_\beta = \frac{g_m}{2\pi C_{GS}} \equiv f_T.$$

This is of course a simplified result referring to an ideal device topology, but yields an interesting suggestion when it comes to designing a wideband amplifier: we can trade bandwidth for gain, but if we ask for extremely wideband behavior maintaining a reasonable gain (e.g., $G = 10$ or 10 dB), the bandwidth we can achieve is of the order of f_T/G. The GBP cannot be however increased with conventional combining structures; in particular, this is independent from the device periphery (since g_m and the input capacitance scale in the same way) and does not improve by connecting devices in series.

In fact, consider a series arrangement of FETs modeled by a simplified equivalent circuit, and suppose that for each stage the input capacitance is C_{GS}, the output resistance is R_{DS} and the transconductance is g_m. For each stage, the output circuit can be given a series representation, with an open-circuit voltage generator $g_m R_{DS} v^*$ (where v^* is the voltage on C_{GS}) and a series internal resistance R_{DS}. Connecting N stages in series, with an input voltage v_{in}, we have that the total input capacitance is $C'_{GS} = C_{GS}/N$ but the output series circuit has a total resistance NR_{DS} and a total open-circuit voltage generator $N \cdot g_m R_{DS} v_{in}/N$. Getting back to the parallel output circuit, we obtain that the total transconductance current generator will be:

$$g'_m v_{in} = \frac{N \cdot g_m R_{DS} v_{in}}{N} \frac{1}{NR_{DS}} = \frac{g_m}{N},$$

and therefore the cutoff frequency of the series stage f'_T is:

$$f'_T = \frac{g'_m}{2\pi C'_{GS}} = \frac{g_m}{N} \frac{1}{2\pi C_{GS}/N} = \frac{g_m}{2\pi C_{GS}} = f_T,$$

i.e., the GBP is unchanged.

As a first approximation, also device cascading is not increasing the gain-bandwidth product (GBP). In fact, consider a cascade of stages with single-pole response; the amplification will follow the frequency behavior:

$$A_t(\omega) = \prod_{i=1,N} \frac{A(0)}{1 + j\omega\tau} \approx \frac{NA(0)}{1 + j\omega N\tau}$$

and therefore, from the 3 dB cutoff frequency of the single stage, the approximate total 3 dB cutoff frequency is obtained and the total gain–bandwidth product is found to be unchanged:

$$f_\beta = \frac{1}{2\pi\tau}$$

$$f_{\beta t} = \frac{1}{2\pi N\tau}$$

$$\text{GBP}_t = A_t(0) f_{\beta t} = \frac{NA(0)}{2\pi N\tau} = \frac{A(0)}{2\pi\tau} = \text{GBP}.$$

As already discussed, broadband amplifiers can be obtained through conventional feedback circuit approaches. Resistive feedback applied to a high-gain amplifier is a way to achieve flat gain over a broad frequency band; however, the open loop gain of the amplifier should be suitably larger than the gain with feedback at the maximum operating frequency, a difficult requirement at microwaves.

An interesting alternative approach, allowing broadband operation up to a frequency which can, in theory, exceed the device cutoff frequency (but is, in practice, only somewhat larger than $f_T/2$), consists in turning the amplifier into a distributed, traveling-wave structure. This solution is called the *distributed amplifier* (DAMP) and can, in theory, achieve a GBP that is larger than the GBP of the devices used. The distributed amplifier is also interesting from the standpoint of the operating principle, since it is an example of electronic amplifier exploiting a distributed or quasi-distributed interaction, yielding wideband operation.

In the distributed amplifier, devices are not cascaded but rather connected (gates with gates, drains with drains) through two transmission lines (gate and drain line, these can be replaced by proper delay blocks). The input signal travels on the gate line and is progressively amplified into the drain line, in such a way that, if the gate (input) and drain (output) travel in a synchronous way, a gradual signal buildup (caused by constructive interference) takes place on the drain line. In practice, the structure is made of a discrete number of cells connected by delay blocks, as shown in Fig. 6.52. The analysis carried out on a more idealized, truly distributed structure, see Fig. 6.53, shows that, in the absence of losses and for synchronous coupling between the drain and gate line, the bandwidth is theoretically infinite. In practice a discrete-cell amplifier is limited by losses and by the finite bandwidth of the delay blocks (that can be made by lumped cells or transmission line blocks), so that the overall amplifier bandwidth rarely exceeds f_T/n with $n \approx 2 \div 3$ and with a gain typically between 10 and 20 dB. We can however state that, in such structures, the gain-bandwidth product is actually improved vs. the single-stage solution.

Sec. 6.13.1 will be devoted to the analysis of the idealized continuous distributed amplifier shown in Fig. 6.53, while the analysis of the discrete-cell version is carried out in Sec. 6.13.2.

6.13.1 The Continuous Distributed Amplifier

A somewhat idealized structure for the distributed FET (e.g., PHEMT) amplifier is made of an input (gate) transmission line connecting, in a continuous way, the input of

6.13 The Distributed Amplifier

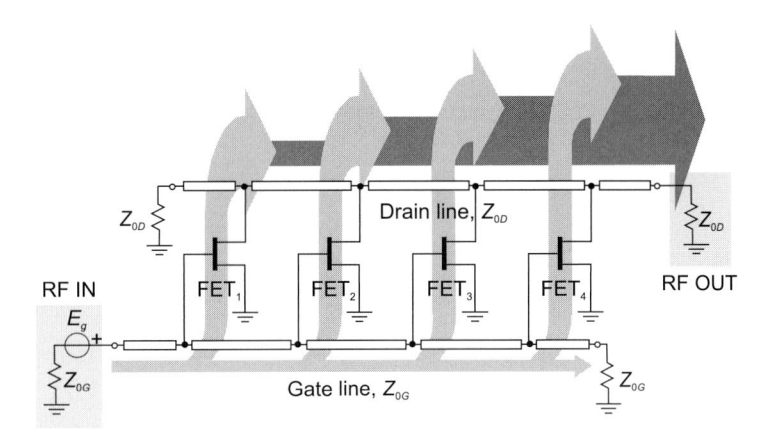

Figure 6.52 Example of discrete cell distributed amplifier with four cells.

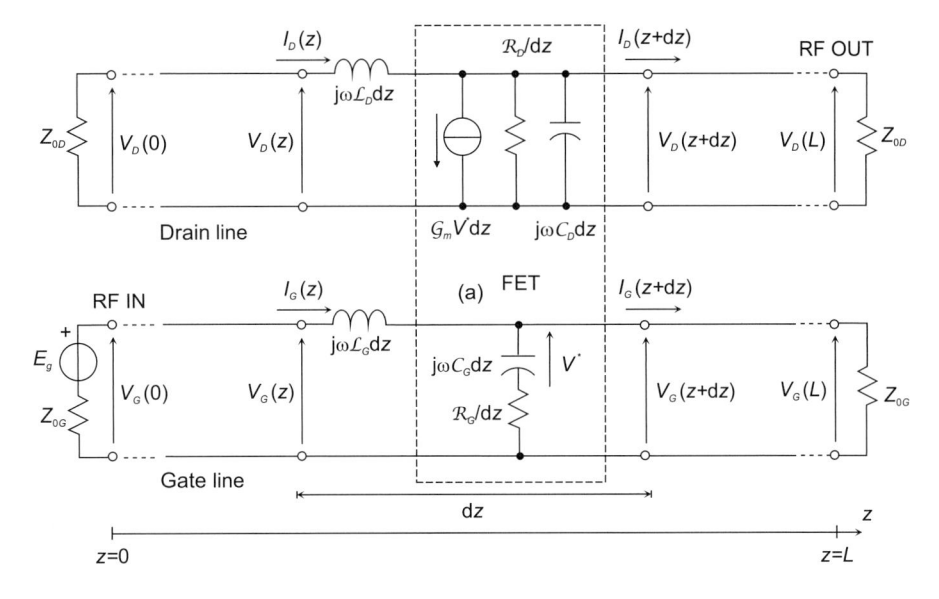

Figure 6.53 Continuous distributed amplifier.

each infinitesimal device cell. The output (drain) transmission line collects the current injected by the transconductance generator. In the simplified model shown in Fig. 6.53, the input line includes a distributed per-unit-length (p.u.l.) inductance $\mathcal{L}_G$, a distributed p.u.l. capacitance $\mathcal{C}_G$ and a distributed p.u.l. resistance $\mathcal{R}_G$ including the gate and intrinsic resistances. The output line includes the transconductance current generators ($\mathcal{G}_m = g_m/L$ is the p.u.l. device transconductance, g_m being the total transconductance and L the gate periphery) of p.u.l. current $\mathcal{G}_m V_G(z)$, and a p.u.l. inductance and capacitance $\mathcal{L}_D$ and $\mathcal{C}_D$ (including the drain-source device capacitance per unit length). The gate line is coupled to the drain line only in the forward direction, since the device is supposed unidirectional. Apart from the gate resistance, all parasitics are neglected. The gate and drain lines are closed on the characteristic impedances; the RF signal is injected at the input of the gate line and collected at the output of the drain line.

The gate line is characterized by the p.u.l. series impedance and parallel admittance:

$$\mathcal{Z}_G = j\omega \mathcal{L}_G$$
$$\mathcal{Y}_G = \frac{j\omega \mathcal{C}_G}{1 + j\omega \tau_G},$$

where $\tau_G = \mathcal{R}_G \mathcal{C}_G$. For the drain line we have instead:

$$\mathcal{Z}_D = j\omega \mathcal{L}_D$$
$$\mathcal{Y}_D = j\omega \mathcal{C}_D + \frac{1}{\mathcal{R}_D} = j\omega \mathcal{C}_D \left(1 + \frac{1}{j\omega \tau_D}\right),$$

where $\tau_D = \mathcal{R}_D \mathcal{C}_D$. The gate and drain line characteristic impedances are therefore:

$$Z_{0G} = \sqrt{\frac{\mathcal{Z}_G}{\mathcal{Y}_G}} = \sqrt{1 + j\omega \tau_G} \sqrt{\frac{\mathcal{L}_G}{\mathcal{C}_G}}$$

$$Z_{0D} = \sqrt{\frac{\mathcal{Z}_D}{\mathcal{Y}_D}} = \sqrt{\frac{j\omega \tau_D}{1 + j\omega \tau_D}} \sqrt{\frac{\mathcal{L}_D}{\mathcal{C}_D}},$$

while the complex propagation constants are:

$$\gamma_G = \sqrt{\mathcal{Z}_G \mathcal{Y}_G} = j\omega \sqrt{\mathcal{L}_G \mathcal{C}_G} \frac{1}{\sqrt{1 + j\omega \tau_G}} = \alpha_G + j\beta_G$$

$$\gamma_D = \sqrt{\mathcal{Z}_D \mathcal{Y}_D} = j\omega \sqrt{\mathcal{L}_D \mathcal{C}_D} \sqrt{\frac{1 + j\omega \tau_D}{j\omega \tau_D}} = \alpha_D + j\beta_D.$$

Notice that the synchronous coupling condition that, as we will see, maximizes the amplifier bandwidth, together with the condition that the input impedance equals the output one ($Z_{0G} \approx Z_{0D}$) implies $\mathcal{L}_G \approx \mathcal{L}_D$ and $\mathcal{C}_G \approx \mathcal{C}_D$. Since, however, $\mathcal{R}_D \gg \mathcal{R}_G$, we have $\tau_D \gg \tau_G$. In such conditions, a frequency range exists where $\omega \tau_G \ll 1$ while $\omega \tau_D \gg 1$; in such a frequency range the lines behave like LC lines with losses, with parameters:

$$Z_{0G} \approx \sqrt{\frac{\mathcal{L}_G}{\mathcal{C}_G}}, \quad Z_{0D} \approx \sqrt{\frac{\mathcal{L}_D}{\mathcal{C}_D}}$$

$$\gamma_G \approx j\omega \sqrt{\mathcal{L}_G \mathcal{C}_G} \left(1 - j\frac{\omega \tau_G}{2}\right)$$

$$= \frac{\omega^2 \mathcal{R}_G \mathcal{C}_G \sqrt{\mathcal{L}_G \mathcal{C}_G}}{2} + j\omega \sqrt{\mathcal{L}_G \mathcal{C}_G} = \alpha_G + j\beta_G$$

$$\gamma_D \approx j\omega \sqrt{\mathcal{L}_D \mathcal{C}_D} \left(1 - j\frac{1}{2\omega \tau_D}\right)$$

$$= \frac{1}{2\mathcal{R}_D} \sqrt{\frac{\mathcal{L}_D}{\mathcal{C}_D}} + j\omega \sqrt{\mathcal{L}_D \mathcal{C}_D} = \alpha_D + j\beta_D.$$

6.13 The Distributed Amplifier

Let us evaluate first the voltage $V_G(z)$ on the gate line, that is closed on both ends on its characteristic impedance. Since the line is matched we have:

$$V_G(z) = \frac{E_g}{2} e^{-\gamma_G z}.$$

The drain line includes distributed transconductance current generators of value:

$$\mathcal{G}_m V^* dz = \mathcal{G}_m \frac{V_G}{1+j\omega\tau_G} dz.$$

From the Kirchhoff laws applied to an infinitesimal drain line section we obtain:

$$V_D(z+dz) = V_D(z) - \mathcal{Z}_D dz I_D(z)$$

$$I_D(z+dz) = I_D(z) - \mathcal{Y}_D dz V_D(z+dz) - \mathcal{G}_m \frac{V_G}{1+j\omega\tau_G} dz,$$

from which the modified telegraphers' equations for the drain line:

$$\frac{dV_D}{dz} = -\mathcal{Z}_D I_D(z)$$

$$\frac{dI_D}{dz} = -\mathcal{Y}_D V_D(z) - \tilde{\mathcal{G}}_m V_G = -\mathcal{Y}_D V_D(z) - \frac{\tilde{\mathcal{G}}_m E_g}{2} e^{-\gamma_G z},$$

with:

$$\tilde{\mathcal{G}}_m = \frac{\mathcal{G}_m}{1+j\omega\tau_G}.$$

Differentiating the first equation and substituting into the second we obtain the second-order differential equation:

$$\frac{d^2 V_D}{d^2 z} = \gamma_D^2 V_D + \frac{\tilde{\mathcal{G}}_m \mathcal{Z}_D E_g}{2} e^{-\gamma_G z}.$$

We can express the solution as $V_D(z) = V_1 + V_2$, where V_1 is the solution of the homogeneous equation (no forcing term), V_2 is a particular solution of the forced equation. We have:

$$V_1(z) = V_{D0}^+ e^{-\gamma_D z} + V_{D0}^- e^{\gamma_D z}.$$

We seek V_2 under the form $K \exp(-\gamma_G z)$, K to be determined; substituting we have:

$$V_D(z) = V_{D0}^+ e^{-\gamma_D z} + V_{D0}^- e^{+\gamma_D z} + \frac{E_g}{2} \left(\frac{\tilde{\mathcal{G}}_m \mathcal{Z}_D}{\gamma_G^2 - \gamma_D^2} \right) e^{-\gamma_G z}.$$

For the drain current we obtain:

$$I_D(z) = -\frac{1}{\mathcal{Z}_D} \frac{dV_D}{dz} = \frac{V_{D0}^+}{Z_{0D}} e^{-\gamma_D z} - \frac{V_{D0}^-}{Z_{0D}} e^{+\gamma_D z} + \frac{E_g}{2} \left(\frac{\tilde{\mathcal{G}}_m \gamma_G}{\gamma_G^2 - \gamma_D^2} \right) e^{-\gamma_G z}.$$

The integration constants V_{D0}^+ and V_{D0}^- are obtained from the boundary conditions:

$$V_D(0) = Z_{0D}(-I_D(0))$$

$$V_D(L) = Z_{0D} I_D(L),$$

leading to:

$$V_{D0}^+ = -\frac{E_g}{4}\left[\frac{\tilde{G}_m(Z_{0D}\gamma_G + Z_D)}{\gamma_G^2 - \gamma_D^2}\right]$$

$$V_{D0}^- = \frac{E_g}{4}\left[\frac{\tilde{G}_m(Z_{0D}\gamma_G - Z_D)}{\gamma_G^2 - \gamma_D^2}\right]e^{-(\gamma_G + \gamma_D)L}.$$

The drain line output voltage finally reads:

$$V_D(L) = \frac{E_g}{4}\left[\frac{\tilde{G}_m(Z_{0D}\gamma_G + Z_D)}{\gamma_G^2 - \gamma_D^2}\right]\left(-e^{-\gamma_D L} + e^{-\gamma_G L}\right)$$

$$= -\frac{E_g}{2}\left[\frac{\tilde{G}_m L Z_{0D}}{2}\right]e^{-\left(\frac{\gamma_D + \gamma_G}{2}\right)L}\frac{\sinh\left(\frac{\gamma_D - \gamma_G}{2}\right)L}{\left(\frac{\gamma_D - \gamma_G}{2}\right)L}.$$

Since the input voltage is $V_i = V_G(0) = E_g/2$, the voltage amplification of the amplifier is:

$$A_V = \frac{V_D(L)}{V_G(0)} = -\frac{g_m Z_{0D}}{2(1 + j\omega\tau_G)}e^{-\left(\frac{\gamma_D + \gamma_G}{2}\right)L}\frac{\sinh\left(\frac{\gamma_D - \gamma_G}{2}\right)L}{\left(\frac{\gamma_D - \gamma_G}{2}\right)L}.$$

For negligible losses we have:

$$|A_V| = \frac{g_m Z_{0D}}{2}\frac{\sin\left(\frac{\beta_D - \beta_G}{2}\right)L}{\left(\frac{\beta_D - \beta_G}{2}\right)L},$$

that is maximum in synchronous conditions $\beta_D - \beta_G = 0$:

$$\mathcal{L}_D \mathcal{C}_D = \mathcal{L}_G \mathcal{C}_G.$$

If, moreover, we impose:

$$Z_{0G} = Z_{0D} \rightarrow \sqrt{\frac{\mathcal{L}_G}{\mathcal{C}_G}} = \sqrt{\frac{\mathcal{L}_D}{\mathcal{C}_D}},$$

we finally obtain the condition:

$$\mathcal{L}_D = \mathcal{L}_G, \quad \mathcal{C}_D = \mathcal{C}_G.$$

This ideally leads to constant, frequency independent voltage amplification:

$$|A_V| = \frac{Z_{0D} g_m}{2}. \tag{6.98}$$

In a similar way we can obtain the reverse amplification $V_D(L)$; in the absence of losses and with synchronous coupling we have:

$$|A_V^B| = \frac{V_D(0)}{V_G(0)} = A_V \frac{\sin\beta_D L}{\beta_D L}.$$

Since the term in $\sin(x)/x$ is small, the reverse amplification is negligible.

From (6.98) we obtain an infinite GBP. In fact, the GBP is anyway limited by the effect of losses. In synchronous conditions we have in fact:

$$|A_V| = \frac{\mathcal{G}_m L Z_{0D}}{2\sqrt{1+\omega^2 \tau_G^2}} e^{-\left(\frac{\alpha_D+\alpha_G}{2}\right)L} \frac{\sinh\left[\left(\frac{|\alpha_D-\alpha_G|}{2}\right)L\right]}{\left(\frac{|\alpha_D-\alpha_G|}{2}\right)L}$$

$$= \frac{\mathcal{G}_m Z_{0D}}{2\sqrt{1+\omega^2 \tau_G^2}} \frac{|e^{-\alpha_G L} - e^{-\alpha_D L}|}{|\alpha_D - \alpha_G|}, \qquad (6.99)$$

implying that the amplification decreases with frequency. Moreover, in the presence of drain and gate losses there is an optimum length, given by:

$$L_{\text{opt}} = \frac{\log(\alpha_D/\alpha_G)}{\alpha_D - \alpha_G}.$$

The optimum length is of course also frequency dependent but, since both α_D and α_G increase with frequency (the drain attenuation due to the skin effect is associated not with R_{DS} but with some additional series losses, not considered in the simplified model) the limitation is more critical at the upper operating frequency.

In conclusion, from the analysis of the continuous distributed amplifier we obtain that the bandwidth is maximized in synchronous conditions, but is also limited by losses. Moreover, the amount of gain that can be achieved with a certain transconductance per unit length is limited and corresponds to an amplifier with optimum length. Finally, since perfect synchronous coupling cannot be achieved in practice, a bandwidth limitation always appears, that is more critical for a very long distributed amplifier.

Example 6.15 A FET with gate periphery $W = 300$ μm has the following intrinsic parameters: $C_{GS} = 0.1 \times 10^{-12}$F, $C_{GD} \approx 0$, $C_{DS} = 0.002 \times 10^{-12}$F, $R_I + R_G = 5$ Ω, $g_m = 0.02$ S, $R_{DS} = 556$ Ω. Moreover, $L_D = 50 \times 10^{-12}$H, $L_G = 100 \times 10^{-12}$H. Derive the parameters per unit length of the device, exploited to design an ideal, continuous distributed amplifier. Evaluate the additional drain capacitance needed to achieve synchronous coupling between the drain and gate line, and the frequency behavior of the voltage amplification as a function of the device length (periphery) L.

Solution
The FET specific parameters are:

$$\mathcal{C}_G \approx \mathcal{C}_{GS} = C_{GS}/W = 3.33 \times 10^{-10} \text{ F/m}$$
$$\mathcal{C}_{GD} \approx 0.$$
$$\mathcal{C}_{DS} = C_{DS}/W = 6.67 \times 10^{-12} \text{ F/m}$$
$$\mathcal{R}_G = (R_I + R_G) W = 0.0015 \text{ } \Omega \cdot \text{m}$$
$$\mathcal{G}_m = g_m/W = 66.67 \text{ S/m}$$
$$\mathcal{R}_D \approx \mathcal{R}_{DS} = R_{DS} W = 0.167 \text{ } \Omega \cdot \text{m}$$
$$\mathcal{L}_D = L_D/W = 1.67 \times 10^{-7} \text{ H/m}$$
$$\mathcal{L}_G = L_G/W = 3.33 \times 10^{-7} \text{ H/m}.$$

To obtain synchronous coupling we should add to the drain a p.u.l. capacitance C_a such that:

$$\mathcal{L}_G \mathcal{C}_{GS} = \mathcal{L}_D (C_a + \mathcal{C}_{DS}) \to C_a = 6.57 \times 10^{-10} \text{ F/m},$$

leading to a total drain p.u.l. capacitance:

$$\mathcal{C}_D = C_a + \mathcal{C}_{DS} = 6.64 \times 10^{-10} \text{ F/m}.$$

We then have:

$$\tau_G = 4.995 \times 10^{-13} \text{ s}, \quad \tau_D = 1.1076 \times 10^{-10} \text{ s},$$

with related cutoff frequencies $f_D = 1.44$ GHz, $f_G = 318$ GHz; therefore in the microwave range the lossy LC approximation holds for the characteristic impedances of the gate and drain line:

$$Z_{0G} = \sqrt{\frac{\mathcal{L}_G}{\mathcal{C}_G}} = 22.394 \ \Omega, \quad Z_{0D} = \sqrt{\frac{\mathcal{L}_D}{\mathcal{C}_D}} = 15.859 \ \Omega$$

and for the losses:

$$\alpha_G = \frac{\omega^2 \mathcal{R}_G \mathcal{C}_G \sqrt{\mathcal{L}_G \mathcal{C}_G}}{2} = 1.0372 \times 10^{-19} f^2 = 0.1 f_{GHz}^2 \text{ Np/m}$$

$$\alpha_D = \frac{1}{2} \sqrt{\frac{\mathcal{L}_D}{\mathcal{R}_D}} = 5 \times 10^{-4} \text{ Np/m}.$$

Since the dominant attenuation is strongly frequency dependent, so is the optimum length. This parameter is therefore hardly useful in this case, since if a wideband response is intended the length should be of the order of the optimum one close to the maximum design frequency. For this reason it is more convenient to plot the amplifier response for different amplifier lengths as a function of frequency from (6.99), see Fig. 6.54. We see that losses imply a decrease of the gain at high frequency. However, taking into account that the original cutoff frequency of the device was $f_T = 32$ GHz, we have with $L = 5$ mm a low-frequency gain of 8 dB with a 3 dB bandwidth of around 40 GHz, with a GBP around 100 GHz; with $L = 10$ mm a low-frequency gain of 14.5 dB with a 3 dB bandwidth of around 27 GHz, with a GBP around 140 GHz. We see therefore an effective increase of the GPB with the conventional device. As a last remark, notice that the gain drop at high frequency is dominated by the ω^2 behavior of the gate line losses.

6.13.2 The Discrete Cell Distributed Amplifier

In practice, distributed amplifiers are made with a number of discrete devices connected on the inputs and outputs by delay lines, as shown in Fig. 6.52. Although a real continuous distributed amplifier, according to the layout shown in Fig. 6.55, has been actually attempted in the past, the approach gives, in practice, poor gain and bandwidth performances. The main motivations for the alternative discrete-cell design are the difficulty of achieving synchronous coupling in a continuous device (due to the fact that the device

6.13 The Distributed Amplifier

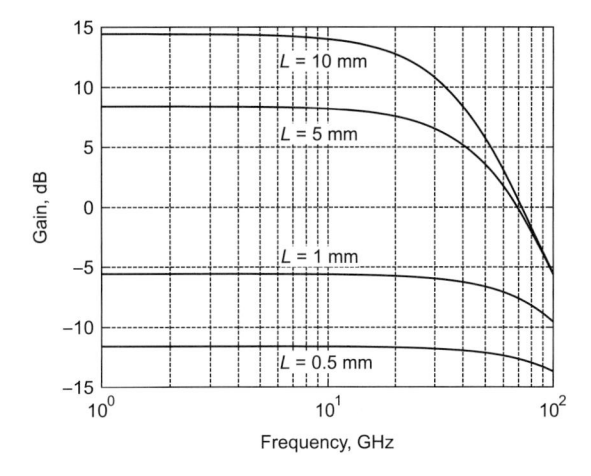

Figure 6.54 Frequency behavior of the voltage amplification of the distributed amplifier in Example 6.15.

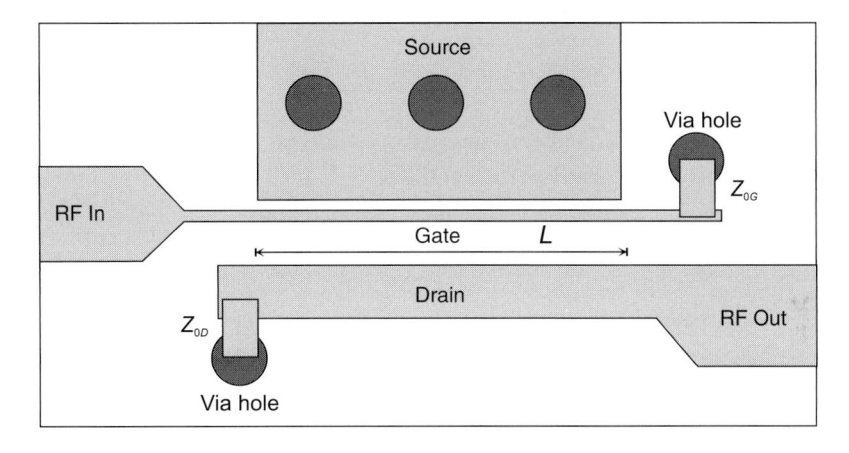

Figure 6.55 Possible layout of a continuous distributed amplifier obtained by properly connecting a single-finger FET with suitable periphery.

input capacitance C_{GS} typically is much larger than the output capacitance C_{DS}), and the large losses induced in the input line by the extremely thin gate electrode.

With the discrete setup, on the other hand, the output line can be capacitively loaded so as to improve velocity matching (or the line length can be properly increased to compensate delays), and losses can be decreased due to the wider conductors used (typically 100 μm against less than 1 μm as in the gate fingers).

In practice, the discrete-cell amplifier exploits, to implement the synchronous condition between the device inputs and outputs, delay blocks made of lumped cells or transmission line sections. We discuss for example a delay structure made of LC cells known as an *artificial line*, see Fig. 6.56. The characteristic impedance of such a structure is defined as the input impedance of an infinite line cascade, and is therefore

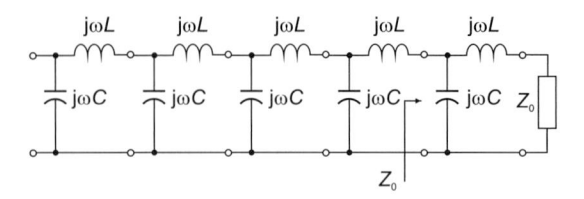

Figure 6.56 Artificial line with LC lowpass sections: evaluating the characteristic impedance as the iterative impedance.

equivalent to the so-called cell *iterative impedance* (i.e., the impedance such as if a cell is loaded with it, it also coincides with the input impedance of the loaded cell). We have to impose:

$$Z_0 = \frac{1}{j\omega C + \dfrac{1}{Z_0 + j\omega L}},$$

from which the second-order equation:

$$Z_0^2 + j\omega L Z_0 - \frac{L}{C} = 0.$$

Solving for Z_0 and choosing the square root with positive sign (we need to obtain a positive value for $\omega = 0$) we obtain:

$$Z_0 = -j\omega\frac{L}{2} + \sqrt{-\frac{\omega^2 L^2}{4} + \frac{L}{C}}.$$

The behavior of $Z_0(\omega)$ is shown in Fig. 6.57. At low frequency the artificial line has characteristic impedance $Z_0 = \sqrt{L/C}$, like a classical transmission line. However, the artificial line has a finite bandwidth f_C corresponding to the frequency at which the impedance becomes purely imaginary (i.e., the argument of the square root vanishes):

$$f_C = \frac{Z_0(0)}{\pi L} = \frac{1}{\pi}\sqrt{\frac{L}{C}\frac{1}{L}} = \frac{1}{\pi Z_0(0) C}.$$

This imposes a further limitation to the bandwidth of the distributed amplifier also in the ideal, lossless case.

The analysis of the discrete-cell structure suggests again that the synchronous coupling condition is:

$$L_G C_G \approx L_G C_{GS} = L_D C_D.$$

which implies, assuming also $Z_0 = Z_{0G} = Z_{0D}$, that $L_G = L_D$, $C_G = C_D$. Due to the capacitance unbalance between the gate and drain line the drain line has to be capacitively loaded or must exploit longer line sections than the gate line. In all cases, the additional degrees of freedom allow the synchronous condition to be achieved in an easier way than with the continuous, ideal structure. It can be shown that in a synchronous structure with n cells the low-frequency gain is:

6.13 The Distributed Amplifier

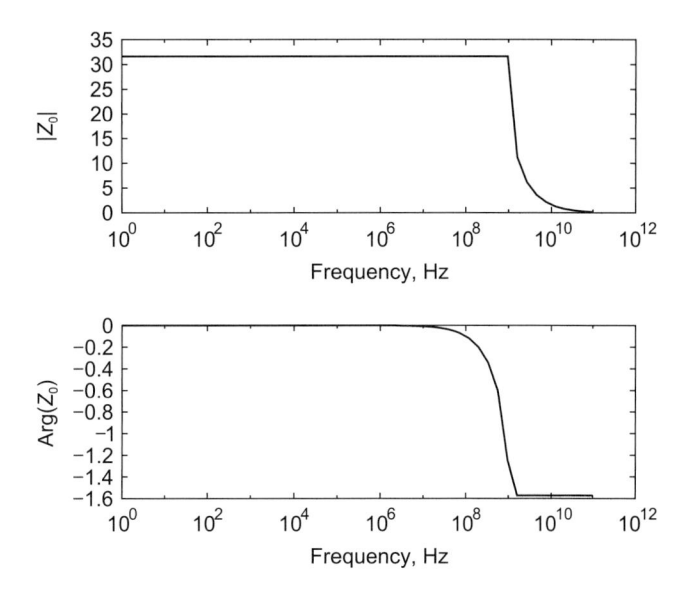

Figure 6.57 Frequency behavior of the characteristic impedance of an LC lowpass artificial line.

$$|A_V(0)| = \frac{V_{D,out}}{V_{G,in}} = \frac{Z_0}{2} n g_m, \qquad (6.100)$$

where g_m is the cell transconductance. This result can be directly obtained taking into account that in DC all cells are in parallel at the input, with an input voltage $V^* = E_g/2 = V_{G,in}$. The total output impressed current on the drain line is therefore $ng_m V^* = ng_m V_{G,in}$; however, this current is uniformly split into the two drain line terminations, both equal to Z_0. The current in the amplifier output termination is therefore half of the impressed current, i.e., $ng_m V_{G,in}/2$ with a load voltage $V_{D,out} = Z_0 n g_m V_{G,in}/2$. Eq. (6.100) immediately follows.

Having equalized the artificial line capacitances, those are approximately given by C_{GS}, with cutoff frequency (that we identify with the lossless amplifier 3 dB bandwidth f_β):

$$f_C = f_\beta = \frac{1}{\pi Z_0 C_{GS}},$$

with gain-bandwidth product:

$$f_T = f_\beta |A_V(0)| = \frac{n g_m}{2\pi C_{GS}} = n f_{T1},$$

where f_{T1} is the gain-bandwidth product of the single stage.

Again, the possibility of increasing without limit the gain-bandwidth product is made difficult in practice by the presence of losses. From the analysis of a discrete-cell structure like the one in Fig. 6.58 we obtain, assuming synchronous conditions and $Z_0 = Z_{0G} = Z_{0D}$ [18]:

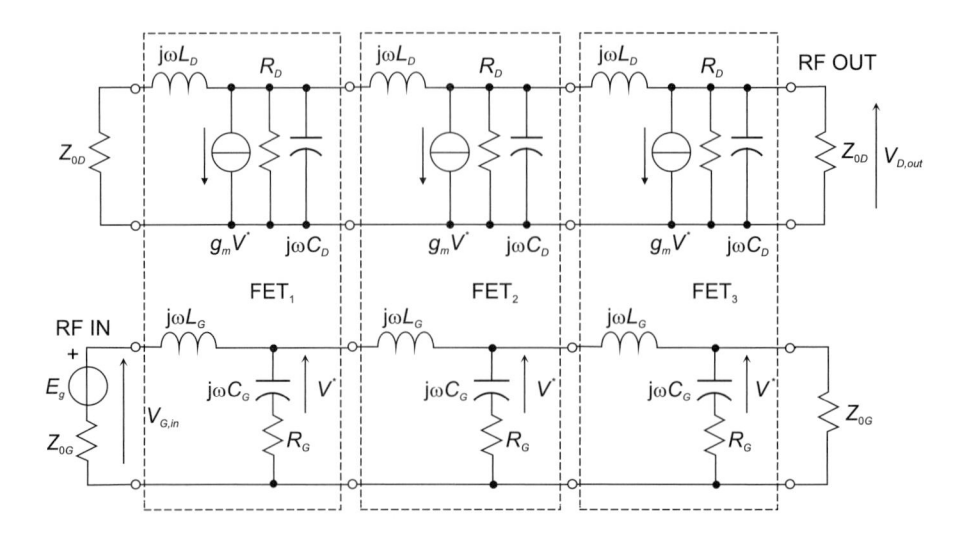

Figure 6.58 Equivalent circuit of a discrete-cell distributed amplifier with three cells.

$$|A_V| = \frac{g_m Z_0}{2\sqrt{\left[1+\left(\frac{\omega}{\omega_G}\right)^2\right]\left[1-\left(\frac{\omega}{\omega_C}\right)^2\right]}} e^{-n\frac{A_D+A_G}{2}} \frac{\sinh\left(n\frac{A_D-A_G}{2}\right)}{\sinh\left(\frac{A_D-A_G}{2}\right)}, \qquad (6.101)$$

where:

$$A_G = \frac{\left(\frac{\omega_C}{\omega_G}\right)\left(\frac{\omega}{\omega_C}\right)^2}{\sqrt{1-\left[1-\left(\frac{\omega_C}{\omega_G}\right)^2\right]\left(\frac{\omega}{\omega_C}\right)^2}}$$

$$A_D = \frac{\frac{\omega_D}{\omega_C}}{\sqrt{1-\left(\frac{\omega}{\omega_C}\right)^2}}$$

$$\omega_G = \frac{1}{R_G C_{GS}}, \quad \omega_D = \frac{1}{R_D C_{DS}}$$

$$\omega_c = \frac{2}{\sqrt{L_G C_G}} = \frac{2}{\sqrt{L_D C_D}}.$$

The parameters A_G and A_D are the cell attenuations of the drain and gate line, respectively. Also in this case, we can derive an optimum cell number:

$$n_{ott} = \frac{\log(A_D/A_G)}{A_D - A_G}.$$

Practical amplifiers have a low number of cells (typically less than ten).

Owing to the need of broadbanding as much as possible the available devices, a quite popular configuration in the design of distributed amplifiers is the *cascode* cell configuration. The cascode transistor configuration (Fig. 6.59) is, for a bipolar, a common emitter stage connected to a common base stage. The same configuration can be

6.13 The Distributed Amplifier

Figure 6.59 Cascode configuration of (a) bipolar transistors (right) including a Common Base (CB) and a Common Emitter (CE) stage, compared to the conventional CE (left) configuration; (b), same for FETs, where CS stands for common-source, CG for common-gate.

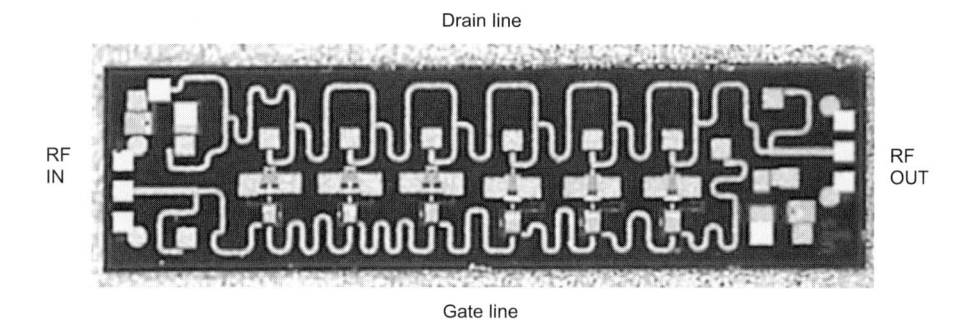

Figure 6.60 Example of six-stage discrete-cell DAMP layout. Copyright ©2005 GAAS Association. All rights reserved. Reprinted, with permission, from [21, Fig. 6].

implemented in FETs with a common source and common gate stage. The cascode configuration has the same transconductance as the CE or CS transistor and the same Y_{11}, but the internal feedback capacitance (C_{GD} or C_{BC}) is strongly decreased, leading to a suppression of the Miller capacitance as seen from the input. Due to this the cascode stage exhibits moderate broadbanding and a better stability vs. the CE and CS stages.

Distributed amplifiers have a long story (they were proposed in 1936 with a vacuum tube implementation [19], see also [20]) and still are a significant circuit solution whenever extremely large bandwidth with high upper frequencies are needed. Typical examples of systems where this solution is commonly exploited are certain wideband communication systems exploiting spread spectrum techniques or frequency hopping, instrumentation amplifiers (where, however, medium-power solutions still require vacuum components like the traveling wave tubes, itself a distributed amplifier from a physical standpoint), and, finally, in high-speed optoelectronic communication systems, whenever a high-speed (e.g., 40 Gbps) baseband bit sequence has to be amplified. Electro-optic modulator drivers are a typical application requiring a driving voltage with peak-to-peak values around 5 V, out of reach of logical SiGe driving stages.

An example of distributed amplifier layout is shown Fig. 6.60 [21]. The amplifier has six stages and includes drain and gate delay lines having different length to achieve synchronous coupling. The chip technology is power PHEMT with chip size 3.75×1 mm^2. The average gain is 7 dB up to around 15 GHz and the DAMP was designed as a 10 Gbps electro-optic modulator driver. In other DAMP realizations the amplifier can exploit

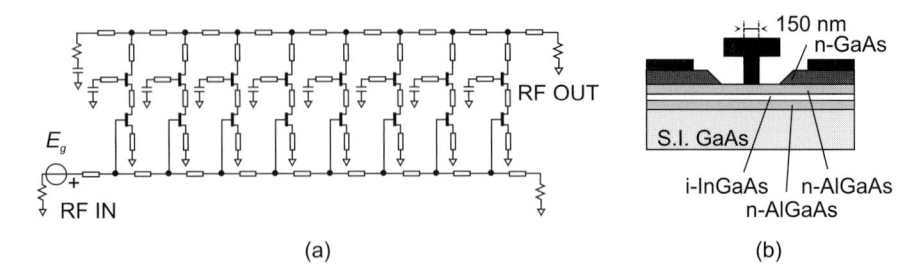

Figure 6.61 Eight-stage PHEMT-based cascode distributed amplifier designed as 40 Gbps modulator driver (a) and GaAs PHEMT cross section (b). Reprinted, with permission, from [22], Fig. 2 (adapted) and Fig. 8. Copyright ©2002 IEEE. All rights reserved.

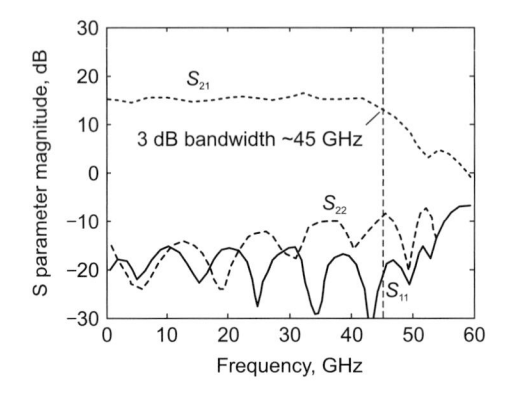

Figure 6.62 Scattering parameters of the eight-stage cascode amplifier with 6 V peak-to-peak output. Reprinted, with permission, from [22, Fig. 10]. Copyright ©2002 IEEE. All rights reserved.

nonuniform cells or delay lines with nonuniform cross section, also to accommodate for larger currents near the output of the drain line.

Another example of high-speed electro-optic modulator driver distributed amplifier for 40 Gbps systems developed by Fujitsu, using this time a cascode configuration, is shown in Fig. 6.61 [22]. The technology is a 0.15 μm InGaAs/GaAs PHEMT; the amplifier provides 6 V peak-to-peak output voltage with a gain around 14 dB; the 3 dB electrical bandwidth is 45 GHz. The electrical response is shown in Fig. 6.62.

In conclusion, distributed amplifiers are a microwave wideband amplifier solution that is able to exploit devices up to a frequency theoretically close or larger than the cutoff frequency. They exhibit low gain (10–20 dB) on an extremely wide bandwidth ranging from almost DC to an upper limit. In some cases DAMPs are also designed as wideband power amplifiers, with a different design strategy optimizing the power delivered by each section. Weak points of the DAMPs are the large layout, the large DC current consumption related to a high number of stages operating in class A, and in general a rather low efficiency.

6.14 Questions and Problems

6.14.1 Questions

1. Maximum power transfer between generator and load through a two-port implies simultaneous power matching at both ports. Is this condition always possible?
2. A two-port has $K = 2.5$, $|\Delta_S| = 1.5$. Is the two-port unconditionally stable?
3. Discuss the stability (according to the one- and two-parameter criteria) of the unilateral two-port with scattering matrix:

$$S = \begin{pmatrix} j1.1 & 0 \\ 5 & 0.1 \end{pmatrix}.$$

4. Is a unilateral device always unconditionally stable?
5. Suppose a device in unconditionally stable above f_0 and potentially unstable below f_0. Qualitatively sketch the behavior of the device MAG and MSG as a function of frequency.
6. Consider two passive two-ports, one reactive (lossless), the other resistive (lossy). What kind of property do we expect from their stability factors?
7. Explain the difference between a maximum gain, a low-noise and a power amplifier.
8. Define a narrowband, a wideband and an ultrawideband amplifier.
9. Explain the design flow of a narrowband maximum gain RF amplifier.
10. Explain the operation and structure of a *bias-T*.
11. Sketch some possible FET stabilization circuits and explain why those circuits include reactive elements together with resistive ones.
12. Explain why the narrowband design strategy with reactive matching sections cannot be extended to wideband design.
13. What is the impact of Fano's limit on the design of wideband amplifiers?
14. Justify why a wideband amplifier with input reactive matching cannot have at the same time flat gain and input matching.
15. Sketch a balanced amplifier and explain why this circuit is able to provide flat gain and good input matching.
16. Justify the fact that a balanced amplifier has the same gain of a single stage but twice as the maximum power.
17. Explain the purpose of an RF parallel/series feedback amplifier in terms of gain flatness and input and output matching.
18. Explain the operation of a distributed amplifier. What happens if the phase velocities on the input and output transmission lines are different? What limits the amplifier bandwidth?
19. Explain why practical distributed amplifiers are discrete-cell rather than continuous.

6.14.2 Problems

1. A real generator with $\Gamma_G = 0.2$ and $b_0 = 1 \text{ W}^{1/2}$ is connected to a load with $\Gamma_L = 0.5$. Evaluate the power delivered to the load and the maximum available power of the generator.

2. A loaded two-port has the following characteristics: $P_{in} = 10$ mW; $P_{av,in} = 20$ mW; $P_L = 100$ mW; $P_{av,out} = 300$ mW. Evaluate the two-port gains G_{op}, G_{av}, G_t.

3. A two-port has the following scattering matrix ($R_0 = 50\ \Omega$):

$$S = \begin{pmatrix} 0 & 0 \\ 10 & 0 \end{pmatrix}.$$

Evaluate the two-port MAG. Is the two-port unilateral?

4. A two-port has the following scattering matrix ($R_0 = 50\ \Omega$):

$$S = \begin{pmatrix} 0.1 & 0.01 \\ 10 & 0.1 \end{pmatrix}.$$

Compute the input and output reflection coefficients when the two-port is loaded on $100\ \Omega$.

5. A two-port has $K = 2$, $S_{21} = 15(1 + j)$ and $S_{12} = 0.1$. Evaluate the two-port MAG and MSG. Assume the two-port is unconditionally stable.

6. Consider the parameter $|S_{21}|^2$. To what power gain (and in which loading conditions) does it correspond?

7. Discuss the stability (according to the one- and two-parameter criteria) of the two-port with scattering matrix:

$$S = \begin{pmatrix} j0.1 & 10 \\ 0.1 & 0.1 \end{pmatrix}.$$

Suppose now that you exchange ports 1 and 2; the new scattering matrix becomes:

$$S' = \begin{pmatrix} 0.1 & 0.1 \\ 10 & j0.1 \end{pmatrix}.$$

Does the two-port stability change?

8. Discuss the stability (according to the one- and two-parameter criteria) of the unilateral two-port with scattering matrix:

$$S = \begin{pmatrix} j1.1 & 0 \\ 5 & 0.1 \end{pmatrix}.$$

9. We want to design a 10 dB amplifier with parallel/series feedback. What is the minimum device $|S_{21}|$?

References

[1] G. Gonzalez, *Microwave transistor amplifiers: analysis and design*. New Jersey: Prentice Hall, 1997.

[2] K. Kurokawa, *An introduction to the theory of microwave circuits*. Academic Press, 1969.

[3] J. Rollett, "Stability and power-gain invariants of linear twoports," *IRE Transactions on Circuit Theory*, vol. 9, no. 1, pp. 29–32, Mar. 1962.

[4] D. Woods, "Reappraisal of the unconditional stability criteria for active 2-port networks in terms of S parameters," *IEEE Transactions on Circuits and Systems*, vol. 23, no. 2, pp. 73–81, Feb. 1976.

[5] M. L. Edwards and J. H. Sinsky, "A new criterion for linear 2-port stability using a single geometrically derived parameter," *IEEE Transactions on Microwave Theory and Techniques*, vol. 40, no. 12, pp. 2303–2311, Dec. 1992.

[6] P. Bianco, G. Ghione, and M. Pirola, "New simple proofs of the two-port stability criterium in terms of the single stability parameter μ_1 (μ_2)," *IEEE Transactions on Microwave Theory and Techniques*, vol. 49, no. 6, pp. 1073–1076, Jun. 2001.

[7] M. Pirola and G. Ghione, "Immittance and S-parameter-based criteria for the unconditional stability of linear two-ports: relations and invariance properties," *IEEE Transactions on Microwave Theory and Techniques*, vol. 57, no. 3, pp. 519–523, Mar. 2009.

[8] K. B. Niclas and R. R. Pereira, "The matrix amplifier: a high-gain module for multioctave frequency bands," *IEEE Transactions on Microwave Theory and Techniques*, vol. 35, no. 3, pp. 296–306, Mar. 1987.

[9] P. Asbeck, "Stacked Si MOSFET strategies for microwave and mm-wave power amplifiers," in *Silicon Monolithic Integrated Circuits in RF Systems (SiRF), 2014 IEEE 14th Topical Meeting on*, Jan. 2014, pp. 13–15.

[10] H. W. Bode, *Network analysis and feedback amplifier design*. van Nostrand, 1945.

[11] R. M. Fano, "Theoretical limitations on the broadband matching of arbitrary impedances," *Journal of the Franklin Institute*, vol. 249, no. 1, pp. 57–83, 1950.

[12] S. A. Maas, *The RF and microwave circuit design cookbook*. Artech House, 1998.

[13] G. L. Matthaei, L. Young, and E. Jones, *Microwave filters, impedance-matching networks, and coupling structures*. Artech House, 1964, vol. 1.

[14] L. Besser, "Avoiding RF oscillations," *Applied Microwave and Wireless*, pp. 44–55, Spring 1995.

[15] K. B. Niclas, W. T. Wilser, R. B. Gold, and W. R. Hitchens, "Application of the two-way balanced amplifier concept to wide-band power amplification using GaAs MESFET's," *IEEE Transactions on Microwave Theory and Techniques*, vol. 28, no. 3, pp. 172–179, Mar. 1980.

[16] J. Laengst, S. Diebold, H. Massler, S. Wagner, A. Tessmann, A. Leuther, T. Zwick, and I. Kallfass, "Balanced medium power amplifier MMICs from 200 to 270 GHz," in *2013 38th International Conference on Infrared, Millimeter, and Terahertz Waves (IRMMW-THz)*, Sep. 2013, pp. 1–3.

[17] K. B. Niclas, W. T. Wilser, R. B. Gold, and W. R. Hitchens, "The matched feedback amplifier: ultrawide-band microwave amplification with GaAs MESFET's," *IEEE Transactions on Microwave Theory and Techniques*, vol. 28, no. 4, pp. 285–294, Apr. 1980.

[18] J. B. Beyer, S. N. Prasad, R. C. Becker, J. E. Nordman, and G. K. Hohenwarter, "MESFET distributed amplifier design guidelines," *IEEE Transactions on Microwave Theory and Techniques*, vol. 32, no. 3, pp. 268–275, Mar. 1984.

[19] W. S. Percival, "Thermionic valve circuits," *British patent*, vol. 460562, p. 25, 1937.

[20] E. L. Ginzton, W. R. Hewlett, J. H. Jasberg, and J. D. Noe, "Distributed amplification," *Proceedings of the IRE*, vol. 36, no. 8, pp. 956–969, Aug. 1948.

[21] J. Shohat, I. D. Robertson, and S. J. Nightingale, "High efficiency 10 Gb/s optical modulator driver amplifier using a power pHEMT technology," in *European Gallium Arsenide and Other Semiconductor Application Symposium, GAAS 2005*, Oct. 2005, pp. 129–132.

[22] H. Shigematsu, M. Sato, T. Hirose, and Y. Watanabe, "A 54-GHz distributed amplifier with 6-VPP output for a 40-Gb/s LiNbO$_3$ modulator driver," *IEEE Journal of Solid-State Circuits*, vol. 37, no. 9, pp. 1100–1105, Sep. 2002.

7 Low-Noise Amplifier Design

7.1 Introduction

Noise is a random unwanted signal, typically of small amplitude, superimposed on the ideal, deterministic voltages and currents of the circuit. In electronics, the term *noise* has two main interpretations.

Noise can be an interfering signal caused by a set of deterministic signals generated by an external circuit (or by another part of the circuit under consideration). Interference is caused by electromagnetic coupling of the interfering signal source with the circuit interconnects. While this kind of noise has an ultimately deterministic cause, it is often characterized in a statistical way. Electromagnetic compatibility analyzes and models this kind of noise, and develops circuit design approaches having low sensitivity to interferers.

On the other hand, noise (also called *intrinsic noise*) is a random signal generated by the very elements of the circuit (typically resistors, diodes, transistors; reactive elements are ideally noiseless). Such noise is intrinsically associated with the charge transport and generation-recombination processes in semiconductors and conductors, and cannot be eliminated, though its effect may be, as we shall see, alleviated through proper circuit design. Intrinsic noise is therefore an ultimate limit to the performance of the circuit in dealing with signals of very small amplitude. In fact, when the signal power is comparable with the noise power, the signal over noise ratio (S/N ratio) tends to unity, becoming incompatible with the detection of a signal in the receiver stage.

Due to the presence of intrinsic noise, the open circuit voltage (or short-circuit current) observed at the ports of any electronic device is affected by stochastic fluctuations having zero mean value, but nonzero mean square value (and therefore nonzero available electrical power). Since intrinsic noise is a random phenomenon, it should be characterized as a stochastic process, see Sec. 7.2 for a review.

This chapter is devoted to the basic principles of electrical noise in circuits, to noise device models (active and passive), and to the design of low-noise amplifiers, starting from the analysis of circuits including random noise generators (Sec. 7.3), and including a short discussion on the physical origin of electrical noise (Sec. 7.5). The minimization of the noise figure in a loaded two-port is addressed in Sec. 7.7, while the noise models of passive and active devices are discussed in Sec. 7.6 and Sec. 7.8, respectively. Sec. 7.10 is devoted to the design of low-noise amplifiers both according to traditional noise figure minimization approaches and using ad hoc topologies.

7.2 A Review of Random Processes

Intrinsic noise is associated with a random, time dependent electrical variable, such as a voltage, current, power wave, or also power. Such a variable can be generally modeled as a *random process*, that we will consider first as adimensional signals.[1]

7.2.1 First- and Second-Order Statistics

A random process $x(t)$ is a time-dependent random variable x depending on time t. The random process can be interpreted as the collection of its *realizations*, each realization $x_\alpha(t)$ being a function of time that can be randomly selected. For $t = t_0$ the random process $x(t_0)$ is a random variable providing the value of the process sampled at t_0.

The *mean value* or *average* of $x(t)$ can be defined in two ways, as the *ensemble average*, i.e., the expected value $E[\cdot]$ of the realizations of the process sampled at time t:

$$\bar{x} = E[x(t)],$$

or as the *time average* of each of the realizations:

$$\langle x_\alpha \rangle = \lim_{T \to \infty} \frac{1}{T} \int_{-T/2}^{T/2} x_\alpha(t) dt.$$

While the ensemble average is, in general, a function of t, the time average in principle depends on the realization chosen. In a *stationary* and *ergodic* process the ensemble average is time-independent and coincides with the time average, independent of the realization chosen. This happens (at least up to second-order statistical averages) in the process involved in noise in small-signal amplifiers; therefore we have:

$$\bar{x} = \langle x_\alpha \rangle.$$

We now take into examination second-order averages, i.e., averages involving the product of a process by itself or the product of two processes. The *self-correlation* of $x(t)$ (for generality we assume that $x(t)$ is complex) is the joint average of $x(t)$ evaluated in t_1, t_2:

$$R_{xx}(t_1, t_2) = \overline{x(t_1) x^*(t_2)}.$$

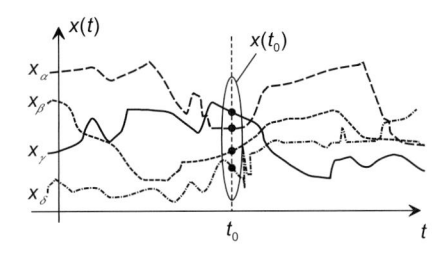

Figure 7.1 Random process and its realizations.

[1] As a comprehensive reference for this section the reader can refer to [1].

If the process is stationary and ergodic, R_{xx} only depends on $\tau = t_1 - t_2$ and can be defined both in terms of the ensemble average and in terms of the time average:

$$R_{xx}(t_1, t_2) = \overline{x(t)x^*(t+\tau)} \equiv R_{xx}(\tau) = \langle x(t)x^*(t+\tau)\rangle$$

$$= \lim_{T\to\infty} \frac{1}{T} \int_{-T/2}^{T/2} x(t)x^*(t+\tau)\,dt.$$

The self-correlation (or autocorrelation) function $R_{xx}(\tau)$ is a measure of the statistical correlation of the process with itself delayed by time τ. If the process rapidly varies between t_1 and t_2, the random variable $x(t_1)$ and $x(t_2)$ are almost completely uncorrelated and therefore the self-correlation function rapidly decreases as a function of $|\tau|$. In slow processes the self-correlation function decreases slowly as a function of $|\tau|$, see Fig. 7.2 (a) and (b).

For stationary processes we also have:

$$R_{xx}(\tau) = \langle x(t)x^*(t+\tau)\rangle = \langle x^*(t')x(t'-\tau)\rangle = R_{xx}^*(-\tau), \qquad (7.1)$$

where $t' = t + \tau$. Taking into account (7.1) we have:

$$R_{xx}(\tau) = R'_{xx}(\tau) + jR''_{xx}(\tau) = R_{xx}^*(-\tau) = R'_{xx}(-\tau) - jR''_{xx}(-\tau),$$

i.e., the real part of R_{xx} is even, the imaginary part is odd. If the process is real we obviously have:

$$R_{xx}(\tau) = R_{xx}(-\tau).$$

The *power spectrum* of x is the Fourier transform of the self-correlation function $R_{xx}(\tau)$:

$$\hat{S}_{xx}(\omega) = \int_{-\infty}^{\infty} R_{xx}(\tau)\exp(j\omega\tau)\,d\tau.$$

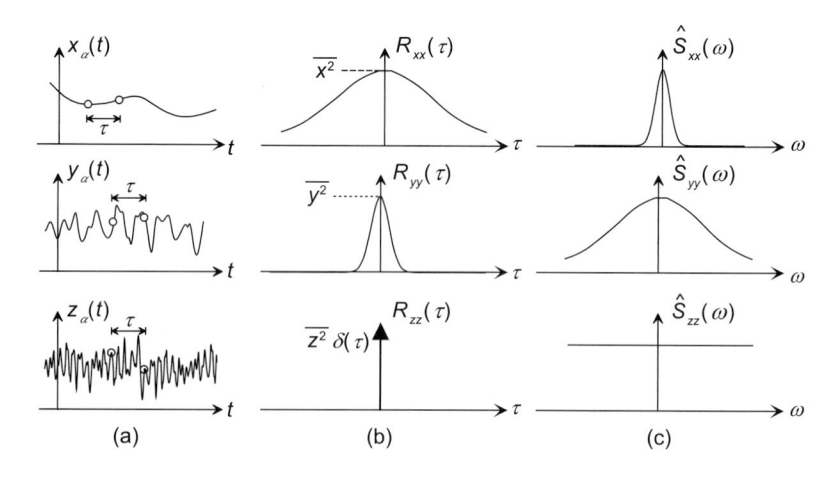

Figure 7.2 Correlation function of a slowly varying vs. a rapidly varying vs. white process (first, second and third row, respectively): (a) example of a realization of the process; (b) autocorrelation function; (c) bilateral power spectrum. The white process realization is qualitative only.

The function $\hat{S}_{xx}(\omega)$ is the *bilateral spectrum* defined for positive and negative angular frequencies. Taking into account the even/odd properties of the real and imaginary parts of the self-correlation function we have:

$$\hat{S}_{xx}(\omega) = \int_{-\infty}^{\infty} \left[R'_{xx}(\tau) + jR''_{xx}(\tau) \right] \cdot \left[\cos(\omega\tau) + j\sin(\omega\tau) \right] d\tau$$

$$= \int_{-\infty}^{\infty} R'_{xx}(\tau) \cos(\omega\tau) d\tau - \int_{-\infty}^{\infty} R''_{xx}(\tau) \sin(\omega\tau) d\tau,$$

i.e., the power spectrum of a process is always real. Moreover, if $x(t)$ is real, the self-correlation is real and even, and the power spectrum reduces to an even function of the angular frequency:

$$\hat{S}_{xx}(\omega) = \int_{-\infty}^{\infty} R_{xx}(\tau) \cos(\omega\tau) d\tau.$$

Since for a real process $x(t)$ (like a voltage or a current) the bilateral power spectrum is even, i.e., $\hat{S}_{xx}(\omega) = \hat{S}_{xx}(-\omega)$, in circuit theory often positive frequencies only are considered; to this aim, we define the *unilateral spectrum* $S_{xx}(\omega)$, in which only the range $\omega > 0$ is considered but the positive frequencies are counted twice:

$$S_{xx}(\omega) = 2\hat{S}_{xx}(\omega) = 2 \int_{-\infty}^{\infty} R_{xx}(\tau) \exp(j\omega\tau) d\tau, \quad \omega > 0.$$

The power spectrum provides a frequency-domain interpretation of the random process. Slow processes have a power spectrum concentrated in the low-frequency range, fast processes have a broad power spectrum, see Fig. 7.2 (c). A limiting case of a fast process is the *white process*, that is only correlated with itself at $\tau = 0$ (i.e., $R_{xx}(\tau) = 0$ for $\tau \neq 0$):

$$R_{xx}(\tau) = \overline{x^2} \delta(\tau),$$

where δ is the Dirac delta function and $\overline{x^2}$ is the *mean square value* of x, defined in (7.2). The power spectrum of white noise is thus frequency independent. Notice that the white process model is a convenient model wherever the spectrum of a noise source is only limited by the finite bandwidth of the embedding circuit or system.

The *quadratic mean* of a process is the expected value of the square of the process, $\overline{x^2}$ (or $\overline{x^2} \equiv \overline{|x|^2}$ for complex processes). We immediately obtain from the definition of the self-correlation function:

$$\overline{x^2} = \langle x(t)x^*(t) \rangle = \overline{x(t)x^*(t)} = R_{xx}(0). \tag{7.2}$$

A spectral domain expression is obtained taking into account that:

$$R_{xx}(\tau) = \frac{1}{2\pi} \int_{-\infty}^{\infty} \hat{S}_{xx}(\omega) \exp(-j\omega\tau) d\omega.$$

We thus have, with $\tau = 0$:

$$\overline{x^2} = R_{xx}(0) = \frac{1}{2\pi} \int_{-\infty}^{\infty} \hat{S}_{xx}(\omega) d\omega = \frac{1}{2\pi} \int_{0}^{\infty} S_{xx}(\omega) d\omega = \int_{0}^{\infty} S_{xx}(f) df. \tag{7.3}$$

From (7.3) we obtain that the *average power* associated with the signal $x(t)$ (represented by $\overline{x^2}$) can be obtained by superimposing the spectral contribution at all frequencies. Finally, the *effective value* of a process (also called *root mean square, r.m.s., value* or *effective value*) is defined as:

$$x_{rms} = \sqrt{\overline{x^2}}.$$

In circuit theory the r.m.s. value of a voltage or current is the equivalent DC voltage or current providing the same power dissipation; the definition can also applied to evaluating the effective value of a deterministic signal, as common in circuit theory. Notice that for a zero-average signal (like the current and voltage fluctuations associated with electrical noise) the r.m.s. value coincides with the signal *standard deviation*, σ_x, defined as:

$$\sigma_x = \sqrt{\overline{x^2} - \overline{x}^2}.$$

Second-order averages can be applied to two different processes. Given two ergodic and stationary processes x and y (complex in general) the *mutual correlation function* is:

$$R_{xy}(\tau) = \overline{x(t)y^*(t+\tau)} = \lim_{T \to \infty} \frac{1}{T} \int_{-T/2}^{T/2} x(t)y^*(t+\tau) dt,$$

while the *correlation spectrum* is the Fourier transform:

$$S_{xy}(\omega) = 2 \int_{-\infty}^{\infty} R_{xy}(\tau) \exp(j\omega\tau) d\tau.$$

Exchanging x and y we obtain:

$$R_{yx}(\tau) = \overline{y(t)x^*(t+\tau)} = \lim_{T \to \infty} \frac{1}{T} \int_{-T/2}^{T/2} y(t)x^*(t+\tau) dt$$

$$= \lim_{T \to \infty} \frac{1}{T} \int_{-T/2}^{T/2} y(t'-\tau)x^*(t') dt' = R_{xy}^*(-\tau).$$

For real processes we simply have $R_{xy}(\tau) = R_{yx}(-\tau)$. Concerning the correlation spectrum, replacing $xy \to yx$ we have:

$$S_{xy}(\omega) = S_{yx}^*(\omega).$$

In fact:

$$S_{xy}(\omega) = 2 \int_{-\infty}^{\infty} R_{xy}(\tau) \exp(j\omega\tau) d\tau = 2 \int_{-\infty}^{\infty} R_{yx}^*(-\tau) \exp(j\omega\tau) d\tau,$$

or, with $\tau' = -\tau$:

$$S_{xy}(\omega) = -2 \int_{\infty}^{-\infty} R_{yx}^*(\tau') \exp(-j\omega\tau') d\tau'$$

$$= 2 \int_{-\infty}^{\infty} R_{yx}^*(\tau') \exp(-j\omega\tau') d\tau' = S_{yx}^*(\omega).$$

If $x = y$ we again have that the self-correlation spectrum is real. To keep the notation light, we often define $R_{xx} \equiv R_x$ and similarly $S_{xx} \equiv S_x$.

The *spectral correlation coefficient*, defined as:

$$C_{xy}(\omega) = \frac{S_{xy}(\omega)}{\sqrt{S_{xx}(\omega)S_{yy}(\omega)}},$$

is a normalized form of the correlation spectrum $S_{xy}(\omega)$. $C_{xy}(\omega)$ is a complex function of frequency that becomes a complex number if the processes under consideration are white and therefore have constant power and correlation spectra. In general, if $|C_{xy}| = 0$, x and y are uncorrelated; if $|C_{xy}| = 1$ the two processes are completely correlated.

7.2.2 Random Process Through Linear Systems

In a linear circuit, the effect of a noise generator (i.e., a generator whose open circuit voltage or short-circuit current is a random process) on some output voltage or current can be described through an input-output linear system where the generator process is the input and the resulting voltage or current is the output. For analog circuits, the most convenient representation of such linear system is in the frequency domain.

Let us consider a random process $x(t)$ as the input of a time-invariant linear system of pulse response $h(t)$ and transfer function $H(\omega)$; the power spectra of the input $x(t)$ and output $y(t)$ are related as:

$$S_{yy}(\omega) = |H(\omega)|^2 S_{xx}(\omega).$$

The above relation can be readily extended to multiple-input multiple-output system with n inputs and m outputs, see Fig. 7.3; having defined the input and output random process vectors:

$$\underline{x} = (x_1, x_2, \ldots, x_n)^T, \qquad \underline{y} = (y_1, y_2, \ldots, y_m)^T,$$

with their Fourier transforms:

$$\underline{X} = (X_1, X_2, \ldots, X_n)^T, \qquad \underline{Y} = (Y_1, Y_2, \ldots, Y_m)^T.$$

The system is defined, in the frequency domain, by the matrix transfer function $\mathbf{H}(\omega)$ such as:

$$\underline{Y}(\omega) = \mathbf{H}(\omega)\underline{X}(\omega).$$

Notice that we have introduced the Fourier transforms $\underline{Y}$ and $\underline{X}$ of the processes $\underline{y}$ and $\underline{x}$, respectively. In principle such transforms are frequency-domain random processes that we will only exploit symbolically, i.e., by formally defining through their help the process power spectrum. This is the basis for the symbolic phasor-like treatment of linear circuits with random generators, discussed in Sec. 7.3.

We can now define the correlation matrix of $\underline{x} = (x_1, x_2, \ldots, x_n)^T$, $\mathbf{R}_{\underline{xx}}$, with elements:

$$R_{ij}(\tau) = R_{x_i x_j}(\tau) = \langle x_i(t) x_j^*(t+\tau) \rangle,$$

and the correlation spectrum $\mathbf{S}_{\underline{xx}}$ defined as:

$$\mathbf{S}_{\underline{xx}} = 2 \int_{-\infty}^{\infty} \mathbf{R}_{\underline{xx}}(\tau) \exp(j\omega\tau) d\tau.$$

Low-Noise Amplifier Design

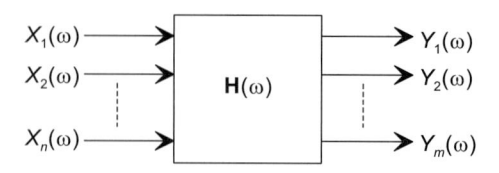

Figure 7.3 Multiple *n*-input *m*-output linear system.

The diagonal of the correlation matrix includes all power spectra of the processes in vector $\underline{x}$, while its out-of-diagonal elements are the correlation spectra:

$$\mathbf{S}_{\underline{xx}} = \begin{pmatrix} S_{x_1 x_1} & S_{x_1 x_2} & \cdots & S_{x_1 x_N} \\ S_{x_2 x_1} & S_{x_2 x_2} & & \vdots \\ \vdots & & \ddots & \vdots \\ S_{x_N x_1} & \cdots & \cdots & S_{x_N x_N} \end{pmatrix}.$$

Since $S_{xy} = S_{yx}^*$ we also have:

$$\mathbf{S}_{\underline{xx}} = \mathbf{S}_{\underline{xx}}^{T*} = \mathbf{S}_{\underline{xx}}^{\dagger},$$

where † denotes the Hermitian conjugate, i.e., the transpose of the complex conjugate, so that for an *n* element process the correlation spectrum is an $n \times n$ matrix with n real elements and $n(n-1)/2$ complex elements, i.e., n^2 independent real elements. The input correlation spectrum $\mathbf{S}_{\underline{xx}}$ and the output correlation spectrum $\mathbf{S}_{\underline{yy}}$ are related by the expression:

$$\mathbf{S}_{\underline{yy}} = \mathbf{H}(\omega) \cdot \mathbf{S}_{\underline{xx}} \cdot \mathbf{H}^{T*}(\omega),$$

that is consistent with the symbolic notation introduced in Sec. 7.2.3.

7.2.3 Symbolic Frequency-Domain (Phasor) Notation

We introduce a symbolic notation of the power spectrum of a process, based on the process Fourier transform, as:

$$S_{xx}(\omega) = \overline{X(\omega)X^*(\omega)}$$
$$S_{yy}(\omega) = \overline{Y(\omega)Y^*(\omega)},$$

while for the correlation spectrum we have:

$$S_{xy}(\omega) = \overline{X(\omega)Y^*(\omega)}.$$

The symbolic definition is consistent with the usual theory of linear systems with random input processes; in fact we have, for a single-input single-output process:

$$S_{yy}(\omega) = \overline{Y(\omega)Y^*(\omega)} = \overline{(HX)(H^*X^*)} = HH^* \overline{XX^*} = |H(\omega)|^2 S_{xx}(\omega),$$

that coincides with the standard result. For a system with *n* inputs and *m* outputs the symbolic definition yields:

$$\mathbf{S}_{\underline{xx}} = \overline{\underline{X} \cdot \underline{X}^{*T}}$$
$$\mathbf{S}_{\underline{yy}} = \overline{\underline{Y} \cdot \underline{Y}^{*T}}$$

and, since:

$$\underline{Y}(\omega) = \mathbf{H}(\omega) \cdot \underline{X}(\omega)$$
$$\underline{Y}^{T*}(\omega) = \underline{X}^{T*}(\omega) \cdot \mathbf{H}^{T*}(\omega),$$

we obtain:

$$\overline{\underline{Y} \cdot \underline{Y}^{T*}} = \overline{\mathbf{H}(\omega) \cdot \underline{X}(\omega) \cdot \underline{X}^{T*}(\omega) \cdot \mathbf{H}^{T*}(\omega)}$$

and therefore, taking out of the average sign the deterministic transfer function matrix:

$$\mathbf{S}_{\underline{yy}} = \mathbf{H}(\omega) \cdot \mathbf{S}_{\underline{xx}} \cdot \mathbf{H}^{T*}(\omega).$$

The symbolic notation allows a straightforward phasor notation to be applied to the analysis of linear circuits with random noise generators, as described in Sec. 7.3.

7.3 Analyzing Circuits with Random Generators: The Symbolic Technique

Consider a linear circuit with random generators, typically associated with component noise. Such generators are current or voltage sources and their power and correlation spectra are assumed as known. We want to evaluate the power and correlation spectra of a set of voltages or currents in the circuit, selected as output variables; from these we can easily derive the noise power dissipated by a component or the available noise power of a generator with random open circuit voltage or short circuit current.

The symbolic procedure proposed is a direct consequence of the linear system theory in the frequency domain, and provides a straightforward approach based on the symbolic phasor frequency-domain circuit analysis technique. The steps are as follows:

1. Suppose that in the circuit there are M random voltage sources and N random current sources. Associate, with each source, a voltage or current phasor: $e_{ni} \rightarrow E_{ni}$, $i = 1\ldots M$, $i_{nj} \rightarrow I_{nj}$, $j = 1\ldots N$. Phasors are denoted as uppercase variables.
2. Solve the circuit in the frequency domain with the phasor technique for the desired m voltages and n current phasors (hereafter *output variables*): V_k, $k = 1\ldots m$, I_l, $l = 1\ldots n$. Since the circuit is linear, the output variable phasors will be a linear combination of the sources phasors through complex, frequency dependent coefficients.
3. The power and correlation spectra of the output variables are now evaluated through the symbolic definitions in Sec. 7.2.3 as a function of the (known) power and correlation spectra of the noise sources. As an example, assume that the output variable is V_k and that only random current sources are present; we have:

$$V_k = \sum_{j=1}^{N} Z_{kj} I_{nj},$$

where Z_{kj} is a transimpedance; the power spectrum of v_k can be now expressed as:

$$S_{v_k} = \overline{V_k V_k^*} = \overline{\sum_{j=1}^{N} Z_{kj} I_{nj} \sum_{l=1}^{N} Z_{kl}^* I_{nl}^*} = \sum_{j,l=1}^{N} Z_{kj} Z_{kl}^* \overline{I_{nj} I_{nl}^*},$$

where the summation can be easily shown to be real (terms with $j = l$ have a real coefficient and are proportional to the power spectrum of the current generators, terms with $j \neq l$ appear in complex conjugate pairs and are proportional to the correlation spectrum of two current generators). In a similar way, we can evaluate the correlation spectrum of two output variables, e.g., $S_{v_i v_j} = \overline{V_i V_j^*}$; this will be in general complex.

In order to evaluate the power dissipated by an element, the square of the effective value of the current or voltage (i.e., the quadratic mean) can be recovered by integrating the relevant power spectra on the system bandwidth, see (7.3). For example, in a resistor with resistance R and current i the average dissipated power will be:

$$P_d = R\overline{i^2} = R \int_0^\infty S_i(f) df \equiv \int_0^\infty p_{,d}(f) df,$$

where $p_d(f) = RS_i(f)$ is the dissipated power spectral density (W/Hz). In many cases the power spectrum $S_i(f)$ is white and the spectral integration is carried out on the system bandwidth, often identified by some passband filter inserted in the system chain. Similarly, for a random voltage generator with open circuit voltage e and internal resistance R_g the average available power is:

$$P_{av} = \frac{\overline{e^2}}{4R_g} = \frac{1}{4R_g} \int_0^\infty S_e(f) df \equiv \int_0^\infty p_{av}(f) df,$$

where $p_{av}(f) = \overline{e^2}/4R_g$ is the spectral density of the generator available power. The same treatment holds with parallel components and generators. An example of the symbolic technique is provided in Example 7.1.

Example 7.1 Evaluate the power spectrum of the output voltage v_n in the circuit in Fig. 7.4, assuming that the power spectrum of the noise source e_n, S_{e_n}, is known. Evaluate the total power dissipated in R assuming a narrow system bandwidth B located around the frequency f_0. With the same assumption, evaluate the average available power at the output port.

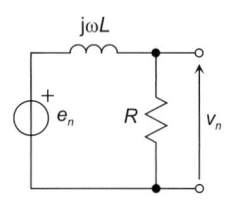

Figure 7.4 Circuit in the Example 7.1.

7.3 Analyzing Circuits with Random Generators: The Symbolic Technique

Solution
Let us first operate with a formal linear system approach. Assume that the system input is the noise voltage e_n, and the system output the output voltage v_n. In the frequency domain, the transfer function between the input and the output is readily evaluated as:

$$H(\omega) = \frac{V_n}{E_n} = \frac{R}{R + j\omega L}$$

and therefore:

$$S_{v_n} = |H(\omega)|^2 S_{e_n} = \left|\frac{R}{R + j\omega L}\right|^2 S_{e_n} = \frac{R^2}{R^2 + (\omega L)^2} S_{e_n}. \qquad (7.4)$$

Having associated with e_n and v_n the phasors E_n and V_n, circuit analysis immediately yields:

$$V_n = \frac{R}{R + j\omega L} E_n.$$

Hence, applying the symbolic definition of the power spectrum (Sec. 7.2.3) we obtain:

$$S_{v_n} = \overline{V_n V_n^*} = \frac{R}{R + j\omega L} E_n \frac{R}{R - j\omega L} E_n^* = \frac{R^2}{R^2 + (\omega L)^2} \overline{E_n E_n^*} = \frac{R^2}{R^2 + (\omega L)^2} S_{e_n},$$

that coincides with the result in (7.4). To evaluate the power dissipated on the output resistance R we derive the effective value of v_n (i.e., the root mean square value):

$$v_{n,\text{eff}}^2 \equiv \overline{v_n^2} = \int_{f_0 - \frac{B}{2}}^{f_0 + \frac{B}{2}} S_{v_n}(f) df = \int_{f_0 - \frac{B}{2}}^{f_0 + \frac{B}{2}} \frac{R^2 S_{e_n}(f)}{R^2 + (2\pi f L)^2} df \approx \frac{R^2 B S_{e_n}(f_0)}{R^2 + (2\pi f_0 L)^2}.$$

The power dissipated in R finally is:

$$P_d = \frac{v_{n,\text{eff}}^2}{R} = \frac{R B S_{e_n}(f_0)}{R^2 + (2\pi f_0 L)^2} \equiv \frac{R \cdot e_{n,\text{eff}}^2}{R^2 + (2\pi f_0 L)^2}.$$

The average available power can be obtained by considering that the input resistance R_{in} (i.e., the real part of the input impedance) at the port is:

$$R_{in}(\omega) = \frac{R(\omega L)^2}{R^2 + (\omega L)^2};$$

thus the available power is:

$$P_{av} = \frac{v_{n,\text{eff}}^2}{4 R_{in}} = \frac{R^2 B S_{e_n}(f_0)}{R^2 + (2\pi f_0 L)^2} \cdot \frac{R^2 + (2\pi f_0 L)^2}{4R(2\pi f_0 L)^2} = \frac{R B S_{e_n}(f_0)}{4(2\pi f_0 L)^2}.$$

Notice that if the narrowband condition $B \ll f_0$ is violated a proper integration over the system bandwidth is needed.

7.4 Equivalent Circuit of Noisy Linear *N*-Ports

Let us consider a linear *N*-port (linear operation can also be the result of small-signal linearization of a nonlinear device around a DC working point). We assume that noise is *additive*, i.e., that noisy signals can be modeled as the superposition of an ideal, noiseless signal, and of a zero-average stochastic process. Since noise has nonzero r.m.s. value, it is associated, at a circuit level, with power exchange; as already discussed, the noise power depends on the second-order statistical properties of the signal fluctuations, such as the autocorrelation function, the power spectrum, the r.m.s. value or the quadratic mean.

7.4.1 Noisy One-Ports

Let us consider a noisy one-port; we assume that the input voltage and current can be expressed by superimposing a deterministic signal ($v(t)$ and $i(t)$) to a random signal whose statistical properties are independent from the deterministic signal.[2] Let us denote the random process associated with voltage noise as e_n and the random process associated with current noise as i_n. Suppose that the noiseless one-port, i.e., the ideal device obtained by suppressing all noise sources, has short-circuit current i_0 (taken positive outgoing from the + pole of the voltage open-circuit generator) and open-circuit voltage e_0. If the processes i_n and e_n are independent from the deterministic (instantaneous) operating condition of the one-port, $e_0 + e_n$ is the total voltage across the device port when $i = 0$ (open circuit conditions) while $i_0 + i_n$ is the total short-circuit current ($v = 0$). If we associate with e_n (i_n) the meaning of *fluctuations with respect of the average value* of voltage and currents, this obviously implies that $\overline{e_n} = \overline{i_n} = 0$, while the quadratic average of the two processes will be generally nonzero.

For simplicity, suppose that the one-port be a resistor, with $e_0 = i_0 = 0$. Applying the Thévenin and Norton theorems, respectively, the equivalent circuit in Fig. 7.5 results. Notice that the resistor noise is extracted under the form of a series or parallel voltage or current generator, while the equivalent circuit includes, as the passive part, a noiseless, ideal resistor. From circuit theory we moreover obtain $i_n = e_n/R$. The instantaneous available power of the noisy resistor can be immediately derived as:

$$p_d(t) = \frac{e_n^2(t)}{4R}.$$

The function p_d is itself a random process whose time or ensemble average (for typical noise generators the two values coincide since the process is second-order stationary and ergodic) is the average available power, from now on denoted for simplicity as the available power:

---

[2] In some cases (e.g., in a resistor) noise is independent from the instantaneous working point of the device. In transistors, however, noise typically is modulated by the instantaneous device working point and may therefore become a non-stationary random process. However, in circuit operating in small-signal condition we may assume that noise only depends on the DC working point and is therefore a stationary, though bias-dependent, process. In such cases the deterministic signal has to be interpreted under the small-signal approximation.

7.4 Equivalent Circuit of Noisy Linear N-Ports

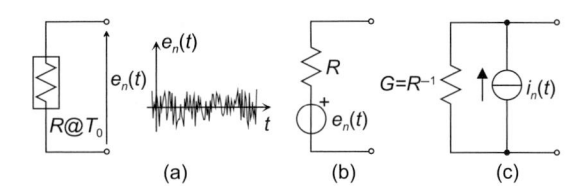

Figure 7.5 (a) Noisy resistor (denoted with a box) and qualitative behavior of its open-circuit voltage; series (b) and parallel (c) equivalent circuit of the noisy resistor.

$$P_d = \frac{\overline{e_n^2}}{4R} = \frac{\overline{i_n^2}}{4G},$$

where $G = 1/R$ is the resistor conductance; P_d is therefore proportional to the process quadratic mean or to the square of the process effective value.

We can immediately associate with the noise available power the available power unilateral spectral density, since:

$$P_d = \frac{1}{4R} R_{e_n e_n}(0) = \int_0^\infty \frac{S_{e_n e_n}(f)}{4R} df = \int_0^\infty p_d(f) df.$$

$S_{e_n e_n}(f)$ is the process power spectrum and the *available power spectral density* reads:

$$p_d(f) = \frac{S_{e_n e_n}(f)}{4R} = \frac{S_{i_n i_n}(f)}{4G}. \tag{7.5}$$

The second expression in (7.5) results from the definition of the available power in terms of the short-circuit current. The total available power therefore results from the superposition, extended to the whole spectrum, of its spectral density $p_d(f)$, having the dimension of W/Hz. Remember that the power spectrum of a voltage has dimension V^2/Hz, the power spectrum of a current A^2/Hz.

For a general linear passive one-port including reactive elements we have the frequency-domain series or parallel representation:

$$V = Z(\omega)I + E_n, \qquad I = Y(\omega)V - I_n,$$

where E_n is the Fourier transform of e_n, I_n the Fourier transform of i_n (positive outgoing), Z and Y are the one-port impedance and admittance, respectively. Finally, a power-wave (see Sec. 3.2.2) based representation of noise is also possible:

$$b = Sa + B_n,$$

where B_n is the Fourier transform of the random process $b_n(t)$. Contrarily to the treatment of gain and stability, the treatment of noise will be mainly based on standard circuit theory. The one-port representations are shown in Fig. 7.6.

7.4.2 Noisy N-Ports

The same approach exploited for the case of noisy one-ports can be easily extended to two-ports and N-ports. A linear noisy non-autonomous two-port or tripole (i.e., with

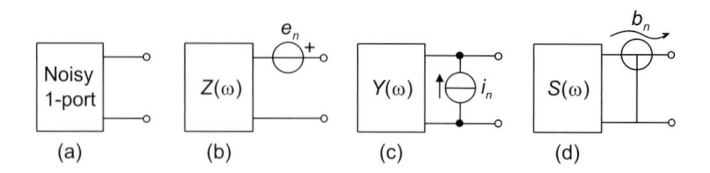

Figure 7.6 Noisy one-port (a) and series (b), parallel (c) and power wave (d) equivalent circuits. The open-circuit voltages and short-circuit currents are denoted as time-domain random processes.

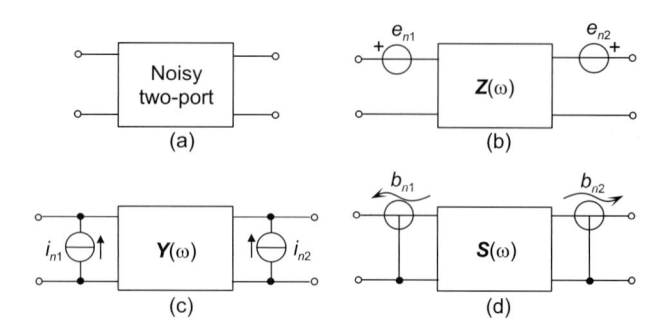

Figure 7.7 Noisy two-port (a) and series (a), parallel (b) and scattering matrix (c) equivalent circuits.

zero open circuit voltages or short circuit currents) is characterized, in the frequency domain, by a series or parallel constitutive relation:

$$\underline{V} = \mathbf{Z} \cdot \underline{I} + \underline{E}_n \tag{7.6}$$

$$\underline{I} = \mathbf{Y} \cdot \underline{V} - \underline{I}_n, \tag{7.7}$$

where $\underline{E}_n$ is the Fourier transform of the open circuit voltage vector at ports 1 and 2, respectively, $\underline{e}_n = [e_{n1}, e_{n2}]^T$, while $\underline{I}_n$ is the Fourier transform of the short circuit current (positive outcoming) vector at ports 1 and 2, respectively, $\underline{i}_n = [i_{n1}, i_{n2}]^T$. Other voltage and current-based representations (like the hybrid representations) can be treated in a similar way. Finally, according to a power wave representation:

$$\underline{b} = \mathbf{S} \cdot \underline{a} + \underline{b}_n, \tag{7.8}$$

where noise sources are random forward or backward wave generators. The relevant equivalent circuits are shown in Fig. 7.7, where the generators e_{n1}, e_{n2}, i_{n1}, i_{n2}, b_{n1} and b_{n2} are random processes, in general correlated.

For the sake of definiteness, let us consider a series (impedance) representation. The noise model of the two-port consists of the impedance matrix, the real power spectra of the open-circuit noise voltages at ports 1 and 2 and their complex correlation spectrum. Such parameters can be assembled into the *correlation matrix* of the two open-circuit noise voltage generators:

$$\mathbf{S}_{\underline{e}_n \underline{e}_n} = \begin{pmatrix} S_{e_{n1} e_{n1}} & S_{e_{n1} e_{n2}} \\ S_{e_{n2} e_{n1}} & S_{e_{n2} e_{n2}} \end{pmatrix},$$

where, by definition, $S_{e_{n2}e_{n1}} = S^*_{e_{n1}e_{n2}}$. Similarly, for the parallel representation we have the admittance matrix $\mathbf{Y}$ and the short-circuit noise current correlation matrix:

$$\mathbf{S}_{\underline{i}_n\underline{i}_n} = \begin{pmatrix} S_{i_{n1}i_{n1}} & S_{i_{n1}i_{n2}} \\ S_{i_{n2}i_{n1}} & S_{i_{n2}i_{n2}} \end{pmatrix}.$$

Given a structure including noise generators (for instance, associated with the equivalent circuit of a noisy resistor or active element), the series or parallel model can be evaluated by applying the rules of circuit theory according to the symbolic technique discussed in Sec. 7.3. An example of the procedure is provided in Example 7.2.

Example 7.2 Evaluate the impedance matrix and the correlation spectrum of the open circuit noise voltages of the circuit in Fig. 7.8.

Solution
To evaluate the impedance matrix we consider the circuit with all independent generators deactivated. The structure is a reciprocal T tripole, thus we immediately have:

$$Z_{11} = j\omega L + R$$
$$Z_{12} = Z_{21} = R$$
$$Z_{22} = 2R.$$

Supposing that the statistical properties of the e_1 and e_2 generators are known ($S_{e_1e_1}$, $S_{e_1e_2}$, $S_{e_2e_2}$) we want to evaluate the power and correlation spectra of the open-circuit voltage generators $S_{e_{n1}e_{n1}}$, $S_{e_{n1}e_{n2}}$ and $S_{e_{n2}e_{n2}}$. Let us define the Fourier transforms $e_1(t) \to E_1(\omega)$, $e_2(t) \to E_2(\omega)$. The open-circuit voltages (E_{n1} and E_{n2}) of the tripole are:

$$E_{n1} = E_1$$
$$E_{n2} = E_1 - E_2;$$

according to the symbolic technique we immediately get the correlation matrix of the open-circuit noise generators:

$$S_{e_{n1}e_{n1}} = \overline{E_{n1}E^*_{n1}} = \overline{E_1 E^*_1} = S_{e_1e_1}$$
$$S_{e_{n1}e_{n2}} = \overline{E_{n1}E^*_{n2}} = \overline{E_1(E^*_1 - E^*_2)} = \overline{E_1 E^*_1} - \overline{E_1 E^*_2} = S_{e_1e_1} - S_{e_1e_2}$$
$$S_{e_{n2}e_{n2}} = \overline{E_{n2}E^*_{n2}} = \overline{(E_1 - E_2)(E^*_1 - E^*_2)} = \overline{E_1 E^*_1} - \overline{E_1 E^*_2} - \overline{E^*_1 E_2} + \overline{E_2 E^*_2}$$
$$= S_{e_1e_1} - S_{e_1e_2} - S_{e_2e_1} + S_{e_2e_2}.$$

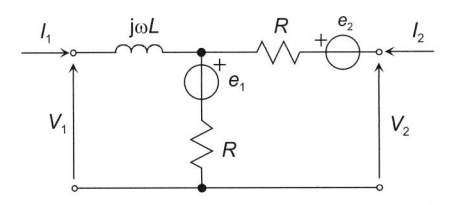

Figure 7.8 Circuit of the Example 7.2.

From the above equations we see that the open circuit generators can be correlated even if the internal generators are uncorrelated ($S_{e_1 e_2} = 0$). Correlation between noise generators can therefore be introduced by the circuit itself, or, in a device, by the fact that certain noise sources induce noise on two different device ports.

Eqs. (7.6), (7.7) and (7.8) can be further generalized to the case of linear, passive (i.e., without internal generators), noisy N−ports. Also in the general case the second-order statistical properties of $\underline{e}_n$ and $\underline{i}_n$ are associated with the correlation matrices $\mathbf{S}_{\underline{e}_n,\underline{e}_n}(\omega)$ and $\mathbf{S}_{\underline{i}_n,\underline{i}_n}(\omega)$, whose elements are defined, respectively, as:

$$(\mathbf{S}_{\underline{e}_n,\underline{e}_n})_{i,j} = S_{e_{ni},e_{nj}}$$
$$(\mathbf{S}_{\underline{i}_n,\underline{i}_n})_{i,j} = S_{i_{ni},i_{nj}},$$

i.e., as the power and correlation spectra of the voltage (current) port noise generators. Since the correlation matrix is Hermitian, a linear, noisy N-port is characterized by a set of N (real) power spectra and of $N(N-1)/2$ (complex) correlation spectra that generally depend on the operating frequency. Taking into account that (7.6) and (7.7) yield:

$$\underline{I}_n = -\mathbf{Y} \cdot \underline{E}_n$$
$$\underline{E}_n = -\mathbf{Z} \cdot \underline{I}_n,$$

by exploiting, e.g., the symbolic definition of power spectra the correlation spectra of the voltage and current noise generators are related as follows:

$$\mathbf{S}_{\underline{i}_n,\underline{i}_n} = \mathbf{Y} \cdot \mathbf{S}_{\underline{e}_n,\underline{e}_n} \cdot \mathbf{Y}^\dagger \tag{7.9a}$$
$$\mathbf{S}_{\underline{e}_n,\underline{e}_n} = \mathbf{Z} \cdot \mathbf{S}_{\underline{i}_n,\underline{i}_n} \cdot \mathbf{Z}^\dagger, \tag{7.9b}$$

where $\mathbf{A}^\dagger = \mathbf{A}^{*T}$.

As already pointed out, the circuit models developed for linear, passive, noisy N-ports can be readily extended to nonlinear N-ports operating under small-signal conditions. In this case voltages and currents are small-signal values (i.e., small perturbations with respect to a DC operating point), while the (differential) impedance and admittance matrices originate from linearization, around the DC operating point, of the current or voltage-controlled constitutive relationships of the nonlinear N-port. In turn, the noise generators will generally depend on the DC operating point. Since fluctuations have small amplitude, noise generators are *compatible with the small-signal regime*.

7.5   The Physical Origin of Noise

Current and voltage random fluctuations associated with intrinsic noise are caused by two fundamental mechanisms operating at a microscopic level: the carrier average velocity fluctuations (thermal and diffusion noise) and the carrier population fluctuations (generation-recombination noise). Other noise mechanisms lead to so-called $1/f$

7.5 The Physical Origin of Noise

or flicker noise, with reference to an ubiquitous frequency dependence whose physical origin probably is not unique. The spectrum of RG and $1/f$ noise decreases with frequency; therefore, both are negligible at microwave frequencies, unless frequency upconversion takes place due to mixing and nonlinearities – this is not the case in small-signal operation. Thermal noise related to velocity fluctuations, with an almost white (i.e., frequency independent) spectrum, is therefore the main noise process to be considered in microwave circuits operating above a few GHz. For a review on the physical noise processes the reader can refer to [2].

7.5.1 Thermal and Diffusion Noise

In conductors, charge carriers experience a motion consisting in free flights and scattering events. Scattering takes mainly place with impurities (doping included) and lattice vibrations (phonons). Because of scattering, the instantaneous velocity of each carrier undergoes huge fluctuations vs. its average value. Nevertheless, due to the extremely large number of carriers present in a conductor such fluctuations are averaged out, so that the fluctuations in the ensemble average velocity of a set of carriers becomes extremely small. Noise related to velocity fluctuations is denoted as *diffusion noise* (out of thermodynamic equilibrium) or *thermal noise* (at or near thermodynamic equilibrium). Velocity fluctuations immediately imply fluctuations in the current density; for electrons we have, e.g.:

$$\delta \underline{J}_n = -q \cdot n \cdot \delta \underline{v}_n,$$

where n is the electron density, $\delta \underline{v}_n$ the average velocity fluctuation on the ensemble of carriers. The power spectrum of diffusion or thermal noise is white up to frequencies of the order of the mean time between collisions (typically above 100 GHz); in microwave operation, therefore, thermal or diffusion noise can be considered as white.

7.5.2 RG Noise

Generation-Recombination (RG) noise is associated with population fluctuations following generation and recombination events associated with different microscopic mechanism, such as band-to-band or direct RG, Shockley–Read–Hall trap-assisted RG, Auger recombination and impact generation [3]. RG noise arises from the fact that RG events take place at random instants, and therefore the total amount of mobile charge decreases or increases by fixed amounts in random instants. Population fluctuations ultimately lead to current density fluctuations; for electrons we have, e.g.:

$$\delta \underline{J}_n = -q \cdot \delta n \cdot \underline{v}_n,$$

where δn is the electron population fluctuation. Notice that at or near thermodynamic equilibrium the average velocity is zero and therefore the current fluctuation induced by population fluctuations also vanishes. RG noise is therefore proportional to the current; for this reason, it is also denoted as *excess noise*. Since the time constants associated with RG events are slow (the order of magnitude is the carriers' lifetime, which, according to the mechanism, is in the approximate range 1 ms – 1 ns) the RG noise dynamics is low pass, with cutoff frequencies ranging from a few kHz to a few GHz. The power spectrum typically is Lorenzian:

$$S_{\delta J \delta J}(\omega) \approx \frac{1}{1+\omega^2 \tau^2},$$

where τ is the carrier lifetime. RG noise again is often negligible in microwave linear operation.

7.5.3 1/f or Flicker Noise

Flicker or $1/f$ noise is an ubiquitous noise present in many physical systems and associated with a variety of basic causes. The power spectrum of the current density fluctuations follows the law:

$$S_{\delta J \delta J} = \frac{\alpha_H}{f^\beta},$$

where α_H is the Hooge constant [2] and $\beta \cong 1$. Low-frequency noise is often associated with an energy distribution of RG noise mechanisms with different time constant, but a unified theory still is missing. In many devices the overall noise arises from a superposition of $1/f$ and thermal noise, see Fig. 7.9. The *corner frequency* usually is around 1 GHz, so that, again, flicker noise is irrelevant to microwave operation, unless upconversion mechanisms play a role, as in oscillators. Bipolar devices typically have a lower corner frequency than FETs, and are therefore preferred in applications (like oscillators) where the low-frequency noise is upconverted.

7.6 Noise Models of Passive Devices

As discussed, semiconductor noise is associated with fluctuations of the carrier velocity and population. In many cases, noise models exist (at least for thermal or diffusion

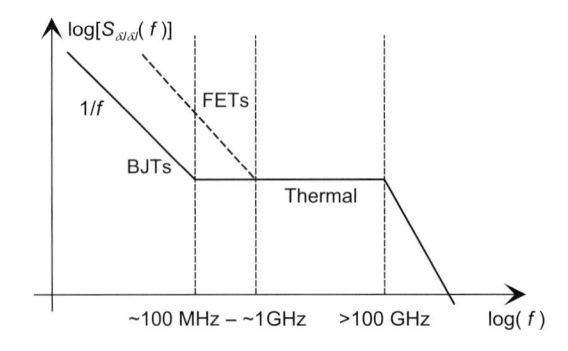

Figure 7.9 Qualitative behavior of the total current fluctuation spectrum derived from the superposition of thermal and $1/f$ noise for bipolar and FET devices. The crossing point between low-frequency and thermal noise is referred to as the corner frequency; this is larger in FETs than in bipolar devices.

noise) relating the noise behavior to the DC or small-signal parameters of the device. Examples are passive circuits, diodes, and, with some added complexity, bipolar and field-effect transistors. Since the physically meaningful quantity at a circuit level is the noise power exchanged by the elements, circuit noise characterizations are entirely based on second-order statistical properties, in particular power or correlation spectra of the external noise sources. Such spectra are directly related to the small-signal response only if operation is at or near equilibrium, in which case the Nyquist theorem holds. Junction devices can be often described by the so-called shot noise model, where the noise power spectrum is proportional to the DC current. Finally, in field-effect transistors semi-empirical models can be applied, in which noise is related, through fitting factors, to the device small-signal parameters, see Sec. 7.8. Remember that noise generators associated with different devices are always uncorrelated, while noise generators related to a single device typically are correlated.

7.6.1 Noise in Passive Devices: The Nyquist Law

According to the *Nyquist law* [4, 2], the (white, i.e., frequency independent) power spectrum of the open-circuit noise voltage or short-circuit noise current of a resistor at temperature T are:

$$S_{v_n}(f) = 4k_B TR \ \text{V}^2/\text{Hz}, \quad S_{i_n}(f) = 4k_B TG \ \text{A}^2/\text{Hz},$$

with available power spectral density:

$$p_n(f) = \frac{S_{v_n}(f)}{4R} = \frac{S_{i_n}(f)}{4G} = k_B T \ \text{W/Hz}.$$

The r.m.s. or effective values of the open-circuit voltage and short-circuit current are, respectively:

$$v_{n,\text{rms}} = \sqrt{\overline{v_n^2}} = \sqrt{4k_B TRB} \ \text{V}, \quad i_{n,\text{rms}} = \sqrt{\overline{i_n^2}} = \sqrt{4k_B TGB} \ \text{A},$$

where B is the system bandwidth. For a passive *RLC* one-port, the generalized Nyquist law holds:

$$S_{v_n}(\omega) = 4k_B T \text{Re}\,[Z(\omega)] \ \text{V}^2/\text{Hz}, \quad S_{i_n}(\omega) = 4k_B T \text{Re}\,[Y(\omega)] \ \text{A}^2/\text{Hz},$$

where Z and Y are the one-port impedance and admittance, respectively.

The Nyquist law can be extended to passive multiport circuits. Following a definition that is customary in circuit theory, albeit not entirely rigorous, we shall define as *linear passive multi-port* any multi-port operating at or near thermodynamic equilibrium. Assume that the multi-port is characterized through its admittance matrix **Y**. The correlation spectrum of the short-circuit noise currents can be shown to take the form:

$$\mathbf{S}_{\underline{i},\underline{i}} = 2k_\text{B} T \left(\mathbf{Y} + \mathbf{Y}^\dagger \right) = 4k_B T \text{Re}[\mathbf{Y}(\omega)]. \tag{7.10}$$

The generalized Nyquist theorem (7.10) can be readily extended to evaluating the correlation spectrum of the open-circuit noise voltage taking into account (7.9); one immediately has:

$$\mathbf{S}_{\underline{e},\underline{e}} = 2k_B T \left(\mathbf{Z} + \mathbf{Z}^\dagger\right) = 4k_B T \text{Re}[\mathbf{Z}(\omega)], \tag{7.11}$$

where $\mathbf{Z}$ is the multi-port impedance matrix.

A proof of (7.10) is presented in [5, Sec. 8]. The generalized Nyquist theorem obviously yields the same result as the direct computation, see Example 7.3.

Example 7.3 Evaluate the series noise representation for the noisy tripole in Fig. 7.10; all resistors are at the same temperature T.

Solution
The impedance matrix can be derived by inspection as:

$$\mathbf{Z} = \begin{pmatrix} R_1 + R_2 & R_2 \\ R_2 & R_2 + R_3 \end{pmatrix},$$

with open circuit voltages:

$$E_{01} = E_1 + E_2, \quad E_{02} = E_2 + E_3,$$

where the power spectra of the resistor open-circuit noise voltages are given according to Nyquist law by:

$$S_{e_1 e_1} = 4k_B T R_1, \quad S_{e_2 e_2} = 4k_B T R_2, \quad S_{e_3 e_3} = 4k_B T R_3.$$

Remember that all resistor noise sources are uncorrelated. Applying the symbolic method for evaluating the power and correlation spectra of the open-circuit tripole voltages and taking into account that the resistor noise sources are uncorrelated we obtain:

$$S_{e_{01} e_{01}} = \overline{E_{01} E_{01}^*} = S_{e_1 e_1} + S_{e_2 e_2} = 4k_B T (R_1 + R_2)$$
$$S_{e_{01} e_{02}} = \overline{E_{01} E_{02}^*} = S_{e_2 e_2} = 4k_B T R_2$$
$$S_{e_{02} e_{02}} = \overline{E_{02} E_{02}^*} = S_{e_3 e_3} + S_{e_2 e_2} = 4k_B T (R_1 + R_3).$$

The source correlation matrix turns out to be consistent with the generalized Nyquist law:

$$\mathbf{S}_{\underline{e}_0 \underline{e}_0} = 4k_B T \begin{pmatrix} R_1 + R_2 & R_2 \\ R_2 & R_2 + R_3 \end{pmatrix} = 4k_B T \text{Re}[\mathbf{Z}].$$

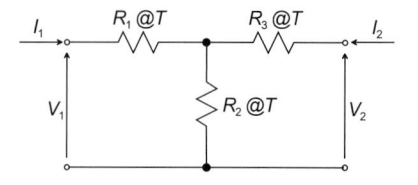

Figure 7.10 Circuit in the Example 7.3.

Notice that if resistors are at different temperatures (T_i for resistor i) the direct computation method is still valid, with result:

$$S_{\underline{e_0}\underline{e_0}} = 4k_B \begin{pmatrix} T_1 R_1 + T_2 R_2 & T_2 R_2 \\ T_2 R_2 & T_2 R_2 + T_3 R_3 \end{pmatrix},$$

but of course the Nyquist theorem does not apply. In fact, a circuit characterized by a nonuniform temperature is not in thermodynamic equilibrium. In practice, the Nyquist theorem holds whenever the resistor self-heating is negligible.

7.6.2 Noise in Junction Diodes

Diffusion and RG noise from junction devices (like diodes and bipolar transistors) can be modeled through a *shot noise model* according to which noise is described by a Poissonian process, whose power spectrum is proportional to the process mean (DC) value according to Campbell's theorem [1]. Notice that the Poissonian model, originally inspired by the physics of vacuum tubes where electrons travel in vacuum with a ballistic motion, does not properly describe the actual physics of semiconductor junction devices, where carrier transport is limited by scattering. Nevertheless, a microscopic analysis of junction noise does indeed shows that the shot noise model also applies to semiconductor junction devices, see, e.g., [6]. For a diode, the power spectrum of the short-circuit noise current is [2]:

$$S_{i_n}(\omega) = 2q(I_D + 2I_S) + 4k_B T \mathrm{Re}\left[Y_D(\omega) - Y_D(0)\right], \tag{7.12}$$

where I_D is the total diode current, I_0 the reverse saturation current and Y_D the small-signal diode admittance. At low frequency:

$$S_{i_n}(\omega) \approx 2q\left(I_D + 2I_0\right) \approx 2q I_D,$$

where the last approximation holds in forward bias. Notice that in reverse bias $I_D = -I_0$ and $S_{i_n}(\omega) = 2qI_0$. The bipolar transistor noise is also described by a shot noise model, as discussed in Sec. 7.8.2. Close to thermodynamic equilibrium (i.e., at zero bias) the diode noise does follow the Nyquist law; in fact we have, for $I_D = 0$:

$$S_{i_n}(\omega) = 4qI_0,$$

but, from the diode characteristic:

$$I = I_0 \exp(V/V_T) - I_0.$$

The diode differential conductance results as:

$$g_d = \frac{I_0}{V_T} \exp(V/V_T)$$

and, for $V = 0$, we have $g_d = I_0/V_T$, with Nyquist noise current spectrum:

$$S_{i_n i_n} = 4k_B T g_d = 4k_B T I_0 / V_T = 4qI_0,$$

that coincides with the shot noise result. Notice that the small-signal differential conductance is exploited in the Nyquist formula.

7.7 System-Oriented Device Noise Parameters: The Noise Figure

Short-circuit current or open-circuit voltage fluctuation power and correlation spectra are a convenient device-level model, which enables us to evaluate the noise power exchanged in a network through circuit analysis. Nevertheless, these parameters are difficult or impossible to measure directly and do not provide straightforward information on system noise performances. Thus, other, mathematically equivalent, noise parameters are customarily introduced to express the noise properties of a (usually two-port or three-pole) device.

Let us consider a noisy two-port connected at port 1 (the input) to a noisy generator and at port 2 (the output) to a load, see Fig. 7.11. From a system design standpoint, we want to assess the relative effect of the noisy two-port and of the noisy generator on the noise power delivered to the load, through the ratio, called the two-port *noise figure F*:

$$F = \frac{P_{n,\text{avL}}}{P'_{n,\text{avL}}},$$

where $P_{n,\text{avL}}$ is the total *noise available power* on the load originating from both the noisy generator and the noisy two-port, while $P'_{n,\text{avL}}$ is the total *noise available power* that would be delivered to the load if the generator were noisy, but the two-port were noiseless. In the standard definition of the noise figure F the generator noise must be thermal, at a reference temperature $T_0 = 290$ K, close to ambient temperature, see [7].[3] We will often also exploit for F the notation NF (sometimes restricted to the expression of F in log units). If the noise available powers are interpreted per unit frequency, F, called in this case the *spot noise figure*, is defined as the ratio of noise available power spectral densities:

$$F(f) = \frac{p_{n,\text{avL}}(f)}{p'_{n,\text{avL}}(f)}.$$

Since the noise available power can also be expressed in terms of the power spectrum of the open-circuit voltage fluctuations v_{2n} or short-circuit current fluctuations i_{2n} at port 2, we have the alternative definitions for the spot noise figure:

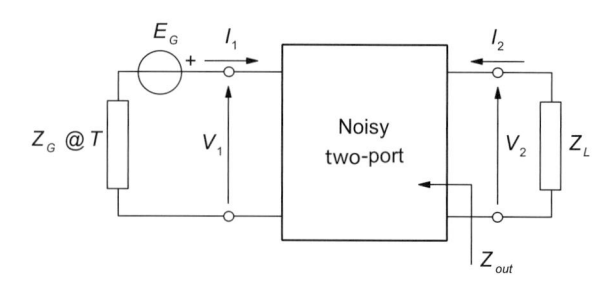

Figure 7.11 Loaded noisy two-port.

---
[3] The choice of the value 290 K rather than the usual ambient temperature 300 K may be historically justified by the fact that $k_B T_0$ is almost exactly equal to 25 meV (the value is 24.990257990408068 meV).

$$F(f) = \frac{\frac{S_{v_{2n}}(f)}{4\text{Re}(Z_{out})}}{\frac{S'_{v_{2n}}(f)}{4\text{Re}(Z_{out})}} = \frac{S_{v_{2n}}(f)}{S'_{v_{2n}}(f)} = \frac{\frac{S_{i_{2n}}(f)}{4\text{Re}(Y_{out})}}{\frac{S'_{i_{2n}}(f)}{4\text{Re}(Y_{out})}} = \frac{S_{i_{2n}}(f)}{S'_{i_{2n}}(f)},$$

where the unprimed spectra refer to the total power spectrum, the primed one to the fluctuations caused by the generator noise only.

Supposing, as physically obvious, that the generator and two-port noise are uncorrelated, we have $P_{n,\text{avL}} = P'_{n,\text{avL}} + P''_{n,\text{avL}}$, where the second term is the noise available power on the load due to the two-port only; therefore:

$$F = \frac{P'_{n,\text{avL}} + P''_{n,\text{avL}}}{P'_{n,\text{avL}}} = 1 + \frac{P''_{n,\text{avL}}}{P'_{n,\text{avL}}}.$$

Notice that F is a relative, rather than absolute, measure of the two-port noise, since the noise generated by the two-port is compared to the noise generated by the generator. In fact, the ideal value $F = 1$ corresponding to a noisy two-port is also obtained in the limit when the generator noise tends to infinity; for this reason a standard noise figure is defined with a source providing thermal noise at a reference temperature $T_0 = 290$ K. Given the available power gain of the two-port G_{av}, we have (notice that we are using the spot definition of the noise figure) $p'_{n,\text{avL}} = G_{\text{av}} p_{n,\text{avG}}$, $p''_{n,\text{avL}} = G_{\text{av}} p''_{n,\text{avG}}$, where $p''_{n,\text{avG}}$ is the *source-referred* available noise power generated by the two-port. We thus obtain, since $p_{n,\text{avG}} = k_B T_0$:

$$F = \frac{p'_{n,\text{avL}} + p''_{n,\text{avL}}}{p'_{n,\text{avL}}} = \frac{G_{\text{av}} p_{n,\text{avG}} + G_{\text{av}} p''_{n,\text{avG}}}{G_{\text{av}} p_{n,\text{avG}}} = 1 + \frac{p''_{n,\text{avG}}}{p_{n,\text{avG}}} = 1 + \frac{p''_{n,\text{avG}}}{k_B T_0}. \quad (7.13)$$

If the input-referred noise from the two-port were generated by a thermal source at temperature T_n (called the two-port *noise temperature*) we would have from (7.13):

$$p''_{n,\text{avG}} = k_B T_0 (F - 1) \equiv k_B T_n$$

and therefore:

$$T_n = T_0 (F - 1). \quad (7.14)$$

For extremely low-noise two-ports the noise figure is very close to unity and the noise temperature T_n is preferred as a figure of merit.

An alternative definition of the noise figure is based on the generator and load (input and output) signal over noise ratios. Let us denote as S_{in} and S_{out} the input and output signal available power and by N_{in} and N_{out} the input and output available powers. We have:

$$F = \frac{N_{out}}{G_{\text{av}} N_{in}} = \frac{\frac{N_{out}}{S_{out}}}{\frac{G_{\text{av}} N_{in}}{S_{out}}} = \frac{\frac{N_{out}}{S_{out}}}{\frac{G_{\text{av}} N_{in}}{G_{\text{av}} S_{in}}} = \frac{\frac{S_{in}}{N_{in}}}{\frac{S_{out}}{N_{out}}} \equiv \frac{(S/N)_{in}}{(S/N)_{out}}.$$

In other words, the noise figure of a two-port is the ratio between the input and the output signal over noise ratios. We clearly have $(S/N)_{in} > (S/N)_{out}$ and therefore $F > 1$. The system definition is particularly helpful in defining parameters like the dynamics of an amplifier.

A second important issue is whether an *optimum generator termination* (impedance Z_{Go} or admittance Y_{Go}) exists for which the noise figure has a minimum value F_{min}. Since the noise figure is a ratio of available output powers, it is independent from the load impedance. The analysis will show that such a minimum indeed exists for a finite value of the source impedance, although in some simplified circuits the optimum corresponds to short- or open-circuit conditions.

Finally, the circuit designer is interested in the sensitivity of the noise figure to changes in the input impedance or admittance with respect to Z_o (Y_o), which will be shown to be proportional to a parameter called the series noise conductance g_n (parallel noise resistance R_n).[4]

Example 7.4 Evaluate the noise figure and noise temperature of a resistive attenuator of available power attenuation $L = 1/G_{av}$ operating at temperature T_0.

Solution
We can directly evaluate the noise figure from the noise available power at the device output. Assuming a resistive attenuator operating (together with the generator) at a temperature T_0, the total output noise available power will be thermal noise at T_0, i.e., $p_{n,avL} = k_B T_0$. However, the output available power due to the generator noise only will be:

$$p'_{n,avL} = k_B T_0 \frac{1}{L}.$$

Therefore:

$$F = \frac{p_{n,avL}}{p'_{n,avL}} = \frac{k_B T_0}{k_B T_0 \frac{1}{L}} = L,$$

i.e., the noise figure is the attenuator loss. The noise temperature will be:

$$T_n = T_0 (F - 1) = (L - 1)T.$$

7.7.1 Evaluating the Noise Figure

The analysis follows the classical method in [9]. As shown in Example 7.5, the noise figure of a two-port with series model **Z** and open-circuit noise generators e_{n1} and e_{n2} (with spectral representation E_{n1} and E_{n2}), with correlation matrix $\mathbf{S}_{\underline{e}_n \underline{e}_n}$ is:

---

[4] Introducing four parameters, the series noise resistance and conductance (r_n and g_n, respectively) and the parallel noise resistance and conductance (R_n and G_n, respectively) can be somewhat confusing for the reader; unfortunately, such notation has become a standard. To make things worse, sometimes the series parameters are introduced with uppercase notation. A weak point in the parameter R_n and g_n is that, while the minimum noise figure is invariant under lossless embedding and the optimum source immittance transforms according to simple circuit laws, the transformation of R_n (g_n) under lossless embedding is complex, thus making their experimental evaluation more involved. Julius Lange suggested in 1967 [8] a new parameter $N = R_n \text{Re}\,[Y_o]$ that is invariant under lossless embedding. Notice that the term *lossless embedding* means that the two-port under examination is cascaded (at the input or output) with a lossless two-port. A lossless feedback indeed changes the minimum noise figure; see Example 7.10.

7.7 System-Oriented Device Noise Parameters: The Noise Figure

$$F = 1 + \frac{r_n}{R_G} + \frac{g_n}{R_G}\left[(R_c + R_G)^2 + (X_c + X_G)^2\right],$$

where $R_G + jX_G = Z_G$ is the source impedance, while $R_c + jX_c = Z_c$ is the *correlation impedance*:

$$Z_c = Z_{11} - Z_{21}\frac{S_{e_{n1}e_{n2}}}{S_{e_{n2}e_{n2}}}. \tag{7.15}$$

The parameters r_n (the *series noise resistance*) and g_n (the *series noise conductance*) are defined by the following equations:

$$S_{e_{nc}e_{nc}} = 4k_B T_0 \cdot r_n \tag{7.16}$$

$$S_{ii} = 4k_B T_0 \cdot g_n, \tag{7.17}$$

where $S_{e_{nc}e_{nc}}$ and S_{ii} are the spectra of the *uncorrelated input voltage generator* and of the *input current generator* respectively, defined by the relations:

$$S_{e_{nc}e_{nc}} = S_{e_{n1}e_{n1}} - \frac{|S_{e_{n1}e_{n2}}|^2}{S_{e_{n2}e_{n2}}} \tag{7.18}$$

$$S_{ii} = \frac{S_{e_{n2}e_{n2}}}{|Z_{21}|^2}. \tag{7.19}$$

From (7.7.1), the noise figure is independent from the load, but, as expected, only depends on the generator impedance.

A dual expression can be derived from the parallel equivalent circuit, see Fig. 7.12 (b). As shown in Example 7.6, the noise figure of a two-port with parallel model **Y**, and short circuit noise generators i_{n1} and i_{n2} (with Fourier transform I_{n1} and I_{n2}), having correlation matrix $\mathbf{S}_{i_n i_n}$, is:

$$F = 1 + \frac{G_n}{G_G} + \frac{R_n}{G_G}\left[(G_c + G_G)^2 + (B_c + B_G)^2\right], \tag{7.20}$$

where $G_G + jB_G = Y_G$ is the generator admittance and $G_c + jB_c = Y_c$ is the *correlation admittance*, expressed as:

$$Y_c = Y_{11} - Y_{21}\frac{S_{i_{n1}i_{n2}}}{S_{i_{n2}i_{n2}}}; \tag{7.21}$$

the parameters R_n (the parallel noise resistance) and G_n (the parallel noise conductance) are defined by the relations:

$$S_{i_{nc}i_{nc}} = 4k_B T_0 \cdot G_n$$

$$S_{ee} = 4k_B T_0 \cdot R_n,$$

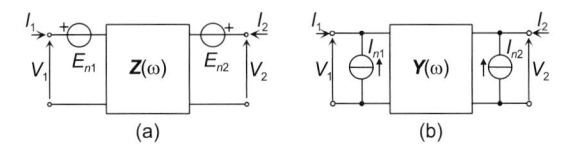

Figure 7.12 Series (a) and parallel (b) representation of a noisy two-port.

where $S_{i_{nc}i_{nc}}$ and S_{ee} are the spectra of the *uncorrelated input current generator* and of the *input voltage generator*, respectively, defined by the relations:

$$S_{i_{nc}i_{nc}} = S_{i_{n1}i_{n1}} - \frac{|S_{i_{n1}i_{n2}}|^2}{S_{i_{n2}i_{n2}}}$$

$$S_{ee} = \frac{S_{i_{n2}i_{n2}}}{|Y_{21}|^2}.$$

The series and parallel representation parameters are related as follows:

$$Y_c = \frac{Z_c^*}{|Z_c|^2 + \frac{r_n}{g_n}}$$

$$G_n = \frac{r_n}{|Z_c|^2 + \frac{r_n}{g_n}}$$

$$R_n = r_n + g_n |Z_c|^2.$$

Notice that $Y_c \neq Z_c^{-1}$.

Example 7.5 Demonstrate (7.7.1).

Solution
According to a classical procedure, the noise generators of the two-port in Fig. 7.12 are moved to the two-port input, as shown in Fig. 7.13. The two configurations are equivalent if they yield the same open-circuit voltages V_{01} and V_{02}, however we have:

$$V_{01} = E + Z_{11}I = E_{n1}$$
$$V_{02} = Z_{21}I = E_{n2},$$

from which:

$$I = \frac{E_{n2}}{Z_{21}} \tag{7.22}$$

$$E = E_{n1} - \frac{Z_{11}}{Z_{21}}E_{n2}. \tag{7.23}$$

Since e_{n1} and e_{n2} are correlated, also i and e will be correlated. It is convenient to decompose the process e into the sum of two processes, the first completely correlated to i (e_c), the second completely uncorrelated (e_{nc}). Exploiting the symbolic notation we set:

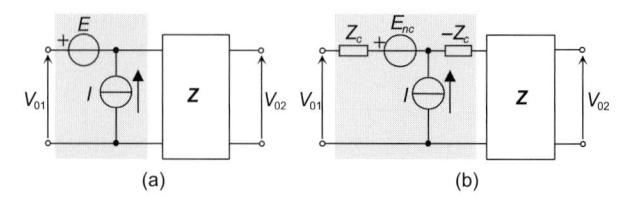

Figure 7.13 (a) Representation with correlated input-referred generators; (b) same, but with uncorrelated generators. The two circuits should exhibit the same open-circuit voltages.

7.7 System-Oriented Device Noise Parameters: The Noise Figure

$$E = E_c + E_{nc} \equiv -Z_c I + E_{nc}, \qquad (7.24)$$

i.e., we express the completely correlated part of E as $-Z_c I$, where Z_c is a constant called the *correlation impedance*. From a circuit standpoint the decomposition $E = -Z_c I + E_{nc}$ is equivalent to transforming the circuit in Fig. 7.13 (a) into the circuit in Fig. 7.13 (b).

Consider now the two tripoles, the one formed by the two input correlated generators and the one including the uncorrelated generators and the correlation impedances, see the gray box in Fig. 7.13 (a) and Fig. 7.13 (b), respectively. The two tripoles are equivalent if their representation is the same; by inspection, however, we find that only their hybrid representation is defined (the series and parallel representation are singular). The equivalence of the passive part is quite obvious, since, eliminating all generators, the two tripoles reduce to a short circuit between the input and output port. Then, we must impose that the open circuit voltage at port 1 and the short circuit current at port 2 be equal in the circuits shown in Fig. 7.14 (a) and (b).

We then have $I_2 = I$ for both circuits, while $E_1 = E$ for circuit (a), $E_1 = E_{nc} - Z_c I$ for circuit (b); it follows that:

$$E_{nc} = Z_c I + E, \qquad (7.25)$$

as initially postulated in (7.24). The parameter Z_c can be derived by imposing that e_{nc} and i are uncorrelated, i.e., in symbolic notation:

$$\overline{E_{nc}I^*} = \overline{(Z_c I + E)I^*} = \overline{Z_c II^*} + \overline{EI^*} = 0,$$

from which:

$$Z_c = -\frac{\overline{EI^*}}{\overline{II^*}}. \qquad (7.26)$$

Substituting into E (7.23) and I (7.22) their expressions, we moreover obtain:

$$\overline{EI^*} = \overline{\left(E_{n1} - \frac{Z_{11}}{Z_{21}}E_{n2}\right) \cdot \left(\frac{E_{n2}^*}{Z_{21}^*}\right)}$$

$$\overline{II^*} = \frac{\overline{E_{n2}E_{n2}^*}}{|Z_{21}|^2},$$

and thus, substituting in (7.26):

$$Z_c = Z_{11} - Z_{21}\frac{\overline{E_{n1}E_{n2}^*}}{\overline{E_{n2}E_{n2}^*}}, \qquad (7.27)$$

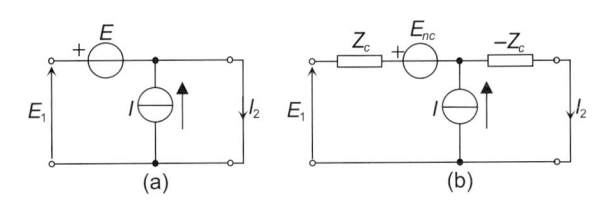

Figure 7.14 Comparison between the open circuit voltage at port 1 and the short circuit current at port 2 for the two structures with correlated (a) and uncorrelated (b) generators.

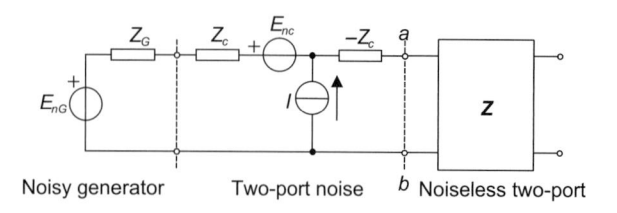

Figure 7.15 Final structure with input referred noise generators.

that coincides with (7.15). Moreover, from (7.25) and (7.23) we obtain:

$$E_{nc} = E_{n1} - E_{n2} \frac{\overline{E_{n1}E_{n2}^*}}{\overline{E_{n2}E_{n2}^*}}, \qquad (7.28)$$

with power spectrum:

$$S_{e_{nc}e_{nc}} = \overline{E_{nc}E_{nc}^*} = \overline{E_{n1}E_{n1}^*} - \frac{\left|\overline{E_{n1}E_{n2}^*}\right|^2}{\overline{E_{n2}E_{n2}^*}}. \qquad (7.29)$$

It is often useful to express a noise power spectrum in terms of the thermal noise associated with a proper noise resistance (for voltage noise generators) or noise conductance (for current noise generators). To this aim, we define r_n (the series noise resistance) and g_n (the series noise conductance) such as:

$$S_{e_{nc}e_{nc}} = 4k_B T_0 \cdot r_n \qquad (7.30)$$
$$S_{ii} = 4k_B T_0 \cdot g_n. \qquad (7.31)$$

Adding the input generator impedance and voltage noise source e_{nG} (e_G is the signal generator), we finally obtain from Fig. 7.13 (b) the circuit in Fig. 7.15. Supposing that the generator noise is thermal (Nyquist law) we have:[5]

$$S_{e_{nG}e_{nG}} = 4k_B T_0 R_G, \qquad (7.32)$$

and the related available power spectral density is $p_{n,\text{avG}} = k_B T_0$. The total open-circuit voltage at section a-b will therefore be:

$$E_{nG} + E_{2p} = E_{nG} - E_{nc} + (Z_c + Z_G)I,$$

where we have denoted as E_{2p} the total input-referred noise contribution of the two-port; the internal impedance of the generator referred to section a-b is Z_G. We thus have:

$$S_{e_{2p}e_{2p}} = \overline{E_{2p}E_{2p}^*} = S_{e_{nc}e_{nc}} + |Z_c + Z_G|^2 S_{ii} = 4k_B T_0 r_n + |Z_c + Z_G|^2 \cdot 4k_B T_0 g_n, \qquad (7.33)$$

---

[5] The assumption of thermal noise is well verified in RF or microwave receivers, but does not apply to optoelectronic receivers where the source noise is shot noise from a photodetector, and therefore does not depend on the generator impedance. The noise optimization carried out here fails if R_G is not, as in this case, influencing both the generator noise and the generator impedance matching with the two-port.

7.7 System-Oriented Device Noise Parameters: The Noise Figure

accounting for the fact that e_{nc} and i are uncorrelated. Since the equivalent generator representing the two-port noise at section a-b has internal resistance R_G, its available noise power spectral density will be:

$$p''_{n,\text{av}G} = \frac{S_{e_{2p}e_{2p}}}{4R_G}.$$

Thus, we immediately have from (7.13):

$$F = \frac{G_{\text{av}}p_{n,\text{av}G} + G_{\text{av}}p''_{n,\text{av}G}}{G_{\text{av}}p_{n,\text{av}G}} = 1 + \frac{p''_{n,\text{av}G}}{p_{n,\text{av}G}} = 1 + \frac{S_{e_{2p}e_{2p}}}{4k_B T_0 R_G}$$

$$= 1 + \frac{r_n + |Z_c + Z_G|^2 g_n}{R_G} = 1 + \frac{r_n}{R_G} + \frac{g_n}{R_G}\left[(R_c + R_G)^2 + (X_c + X_G)^2\right], \quad (7.34)$$

coinciding with (7.7.1).

Example 7.6 Demonstrate (7.20).

Solution
The demonstration is dual with respect with the one of (7.7.1) in Example 7.5, and will be only summarily outlined. The starting point is the parallel equivalent circuit, leading to the circuit with input-referred generators, see Fig. 7.16 (a). The equivalence condition yields:

$$I_{n2} = Y_{21}E, \quad I_{n1} = I + Y_{11}E,$$

from which:

$$E = \frac{I_{n2}}{Y_{21}} \qquad (7.35)$$

$$I = I_{n1} - \frac{Y_{11}}{Y_{21}} I_{n2}. \qquad (7.36)$$

Process i is now decomposed in two parts, one fully correlated to e (called i_c), the other completely uncorrelated (called i_{nc}). From the equivalence of the two circuits in the gray box of Fig. 7.16 (a) and (b) we have:

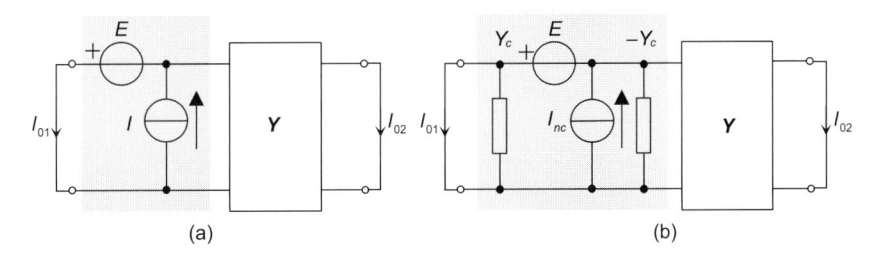

$$I_{nc} = Y_c E + I, \qquad (7.37)$$

Figure 7.16 Noise representations starting from the parallel circuits (a) with uncorrelated input referred generators (b). The two circuits should yield the same short-circuit currents.

where Y_c is the correlation admittance. Imposing the uncorrelation condition between e_{nc} and i we obtain:

$$Y_c = -\frac{\overline{IE^*}}{\overline{EE^*}} \tag{7.38}$$

and, from the expression of E (7.35) and of I (7.36), we obtain:

$$\overline{IE^*} = \overline{\left(I_{n1} - \frac{Y_{11}}{Y_{21}}I_{n2}\right) \cdot \left(\frac{I_{n2}^*}{Y_{21}^*}\right)}$$

$$\overline{II^*} = \frac{\overline{I_{n2}I_{n2}^*}}{|Y_{21}|^2},$$

or, introducing in (7.38):

$$Y_c = Y_{11} - Y_{21}\frac{\overline{I_{n1}I_{n2}^*}}{\overline{I_{n2}I_{n2}^*}},$$

coinciding with (7.21) in the symbolic notation. From Y_c we further obtain:

$$I_{nc} = I_{n1} - I_{n2}\frac{\overline{I_{n1}I_{n2}^*}}{\overline{I_{n2}I_{n2}^*}},$$

and the power spectrum of i_{nc} results:

$$S_{i_{nc}i_{nc}} = \overline{I_{nc}I_{nc}} = \overline{I_{n1}I_{n1}^*} - \frac{\left|\overline{I_{n1}I_{n2}^*}\right|^2}{\overline{I_{n2}I_{n2}^*}}.$$

Also in this case, it is convenient to express the power spectra in terms of pseudo-thermal noise, i.e.:

$$S_{ee} = 4k_BT_0G_n$$

$$S_{i_{nc}i_{nc}} = 4k_BT_0R_n,$$

where G_n and R_n are the *parallel noise conductance* and *parallel noise resistance*, respectively. Assuming thermal noise from the generator conductance, we have that the total noise current at the two-port input is:

$$I_{nG} + I_{2p} = I_{nG} - I_{nc} + (Y_c + Y_G)E,$$

where I_{2p} is the input-referred two-port contribution, I_{nG} the generator contribution. The power spectrum of the noise current generator associated with the two-port noise is:

$$S_{i_{2p}i_{2p}} = \overline{I_{2p}I_{2p}^*} = S_{i_{nc}i_{nc}} + |Y_c + Y_G|^2 S_{ee} = 4k_BT_0G_n + |Y_c + Y_G|^2 \cdot 4k_BT_0R_n,$$

where we have taken into account that i_{nc} and e are uncorrelated. Since the associated available noise power spectral density is now:

$$P''_{n,\text{av}G} = \frac{S_{i_{2p}i_{2p}}}{4G_G}$$

and the generator available power spectral density is $p_{n,\text{av}G} = k_B T_0$, we again have:

$$F = \frac{G_{\text{av}} p_{n,\text{av}G} + G_{\text{av}} p''_{n,\text{av}G}}{G_{\text{av}} p_{n,\text{av}G}} = 1 + \frac{p''_{n,\text{av}G}}{p_{n,\text{av}G}} = 1 + \frac{S_{i_{2p} i_{2p}}}{4k_B T_0 G_G}$$

$$= 1 + \frac{G_n + |Y_c + Y_G|^2 R_n}{G_G} = 1 + \frac{G_n}{G_G} + \frac{R_n}{G_G}\left[(G_c + G_G)^2 + (B_c + B_G)^2\right],$$

coinciding with (7.20).

7.7.2 The Minimum Noise Figure

The noise figure in (7.7.1) or (7.20), depends on the generator impedance and admittance. The analysis easily shows that F exhibits an absolute minimum vs. Z_G or Y_G, the *minimum noise figure* $F_{\min}$ occurring for the optimum generator impedance $Z_{Go} = R_{Go} + jX_{Go} = Y_{Go}^{-1}$. We will assume that the optimum source impedance (and the associated two-port load, typically corresponding to output matching) are compatible with the device stability condition.[6]

From (7.7.1), we have by inspection that the minimum of F vs. X_G occurs for:

$$X_{Go} = -X_c. \qquad (7.39)$$

Imposing this condition in (7.7.1) we have:

$$F = 1 + \frac{r_n}{R_G} + \frac{g_n(R_c + R_G)^2}{R_G} = 1 + \frac{r_n}{R_G} + \frac{g_n R_c^2}{R_G} + \frac{2g_n R_c R_G}{R_G} + \frac{g_n R_G^2}{R_G}$$

$$= 1 + \frac{1}{R_G}\left[r_n + g_n R_c^2\right] + g_n R_G + 2g_n R_c.$$

Since F is large for small R_G and again for large R_G, a minimum should exist in corresponding of an optimum R_{Go} that can be found by imposing $dF/dR_G = 0$ as:

$$R_{Go} = \sqrt{\frac{r_n}{g_n} + R_c^2}. \qquad (7.40)$$

With $Z_G = Z_{Go}$ the minimum noise figure is:

$$F_{\min} = 1 + 2g_n R_c + 2g_n \sqrt{\frac{r_n}{g_n} + R_c^2}. \qquad (7.41)$$

Similarly, for the parallel representation one has $Y_{Go} = G_{Go} + jB_{Go}$; we obtain:

$$B_{Go} = -B_c \qquad (7.42)$$

$$G_{Go} = \sqrt{\frac{G_n}{R_n} + G_c^2}, \qquad (7.43)$$

---

[6] This implies that either the device is unconditionally stable or that generator and load lie in the region of the Smith chart where the device is conditionally stable.

with minimum noise figure:

$$F_{\min} = 1 + 2R_n G_c + 2R_n \sqrt{\frac{G_n}{R_n} + G_c^2}. \tag{7.44}$$

The optimum source impedance and the minimum noise figure are (obviously) unique, independent of the series or parallel representation, see Example 7.7.

Example 7.7 Show that the series and parallel representation lead to the same optimum source impedance and to the same minimum noise figure.

Solution
The relation between the parallel and series parameters is as follows:

$$Y_c = \frac{Z_c^*}{|Z_c|^2 + \frac{r_n}{g_n}}, \quad G_n = \frac{r_n}{|Z_c|^2 + \frac{r_n}{g_n}}, \quad R_n = r_n + g_n |Z_c|^2.$$

Thus, the minimum noise figure in the parallel form becomes:

$$F_{\min} = 1 + 2R_n G_c + 2R_n \sqrt{\frac{G_n}{R_n} + G_c^2}$$

$$= 1 + \frac{2R_c \left(r_n + g_n |Z_c|^2\right)}{|Z_c|^2 + \frac{r_n}{g_n}} + 2\left(r_n + g_n |Z_c|^2\right)$$

$$\times \sqrt{\frac{r_n}{\left(|Z_c|^2 + \frac{r_n}{g_n}\right)\left(r_n + g_n |Z_c|^2\right)} + \frac{R_c^2}{\left(|Z_c|^2 + \frac{r_n}{g_n}\right)^2}}$$

$$= 1 + 2g_n R_c + 2\left(r_n + g_n |Z_c|^2\right) \sqrt{\frac{g_n r_n + g_n^2 R_c^2}{\left(r_n + g_n |Z_c|^2\right)^2}}$$

$$= 1 + 2g_n R_c + 2g_n \sqrt{\frac{r_n}{g_n} + R_c^2}.$$

Similarly, we have:

$$Y_{Go} = \sqrt{\frac{G_n}{R_n} + G_c^2} - jB_c = \frac{g_n}{r_n + g_n |Z_c|^2} \sqrt{\frac{r_n}{g_n} + R_c^2} + j\frac{g_n X_c}{r_n + g_n |Z_c|^2}$$

$$Y_{Go}^{-1} = \left(\frac{r_n}{g_n} + |Z_c|^2\right) \frac{1}{\sqrt{\frac{r_n}{g_n} + R_c^2 + jX_c}}$$

$$= \left(\frac{r_n}{g_n} + |Z_c|^2\right) \frac{\sqrt{\frac{r_n}{g_n} + R_c^2} - jX_c}{\frac{r_n}{g_n} + R_c^2 + X_c^2} = \sqrt{\frac{r_n}{g_n} + R_c^2} - jX_c = Z_{Go},$$

that confirms the result.

7.7 System-Oriented Device Noise Parameters: The Noise Figure

The set consisting of the minimum noise figure, the optimum impedance and the noise series resistance or parallel conductance:

$$F_{\min}, Z_c, g_n \quad (7.45)$$

$$F_{\min}, Y_c, R_n \quad (7.46)$$

can be exploited as a measurable, frequency-dependent characterization of a two-port, from which the correlation matrix of the noise generators can be derived. Besides, the parameters g_n and R_n have an important physical meaning. Using (7.7.1), (7.41), (7.39), (7.40) and (7.20), (7.44), (7.42), (7.43), respectively, we can easily show that:

$$F = F_{\min} + \frac{g_n}{R_G} |Z_G - Z_{Go}|^2 \quad (7.47)$$

$$F = F_{\min} + \frac{R_n}{G_G} |Y_G - Y_{Go}|^2. \quad (7.48)$$

Therefore, the noise figure is a quadratic function of the deviation with respect of the optimum source impedance; in fact, setting $Z_G - Z_{Go} = \Delta Z_G$ we have, around the optimum:

$$F - F_{\min} \approx \frac{g_n}{R_{Go}} |\Delta Z_G|^2,$$

i.e., g_n (R_n) are the sensitivity of the noise figure vs. the minimum value when mismatching the impedance or admittance vs. the optimum. From a measurement standpoint, the identification of the optimum impedance, minimum noise figure and g_n (R_n) is better carried out on the Smith chart exploiting a Γ-plane representation. Introducing:

$$\Gamma_G = \frac{Z_G - R_0}{Z_G + R_0}, \quad \Gamma_{Go} = \frac{Z_{Go} - R_0}{Z_{Go} + R_0},$$

we have:

$$|Z_G - Z_{Go}|^2 = 4R_0^2 \frac{|\Gamma_G - \Gamma_{Go}|^2}{|1 - \Gamma_G|^2 |1 - \Gamma_{Go}|^2}$$

$$R_G = R_0 \frac{1 - |\Gamma_G|^2}{|1 - \Gamma_G|^2}.$$

Substituting in (7.47) we obtain:

$$F = F_{\min} + \frac{4g_n R_0^2}{R_0} \frac{|1 - \Gamma_G|^2}{1 - |\Gamma_G|^2} \frac{|\Gamma_G - \Gamma_{Go}|^2}{|1 - \Gamma_G|^2 |1 - \Gamma_{Go}|^2}$$

$$= F_{\min} + 4g_n R_0 \frac{|\Gamma_G - \Gamma_{Go}|^2}{\left(1 - |\Gamma_G|^2\right) |1 - \Gamma_{Go}|^2}. \quad (7.49)$$

The constant noise figure curves is the Γ_G plane are non-concentric circles called *noise circles*, see [10, Sec. 4.3], whose centers lie on a straight line. Notice that $F \to \infty$ on the unit circle of the Smith chart; in such a case, in fact the generator becomes reactive and therefore noiseless; thus the two-port noise is compared to zero generator noise, and the noise figure diverges.

Low-Noise Amplifier Design

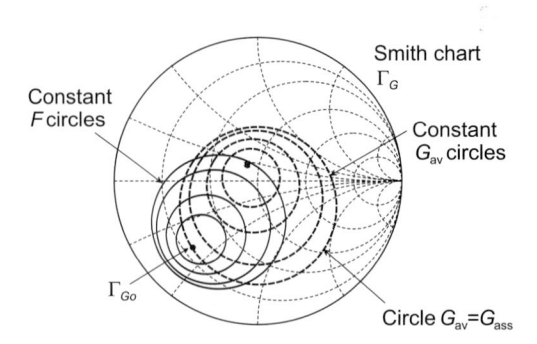

Figure 7.17 Constant noise figure circles ("noise circles") and constant available power gain circles on the Γ_G Smith chart.

Since the optimum condition for noise and the optimum condition form maximizing the available power gain do not typically correspond to the same source impedance, the available power gain at Γ_{Go} is lower than the MAG. The *associated gain*, G_{ass}, is the available power gain when $\Gamma_G = \Gamma_{Go}$:

$$G_{\text{ass}} = G_{\text{av}}(\Gamma_{Go}).$$

When designing a low-noise amplifier a compromise can be sought between the optimum noise and the optimum gain conditions. An example of noise circles and constant available gain circles is shown in Fig. 7.17; according to the device and bias point chosen the tradeoff between gain and noise can be more or less favorable. The situation in Fig. 7.17 of course refers to an unconditionally stable device.

Example 7.8 Evaluate the noise figure of a passive two-port as a function of its scattering matrix. Suppose that the two-port is at the reference temperature T_0.

Solution
In such conditions the noise figure is equal to the loss, $F = L = 1/G_{\text{av}}$; from the expression of the available power gain in (6.35) we immediately have:

$$F = \frac{|1 - S_{11}\Gamma_G|^2 - |S_{22} - \Delta_S\Gamma_G|^2}{(1 - |\Gamma_G|^2)|S_{21}|^2}.$$

Notice that the result does not depend on T_0, since for a passive two-port thermal noise depends on temperature according to the same law followed by the generator reference noise. By exploiting the properties of the scattering matrix of a reactive two-port, the associated gain and noise figure can be shown to be equal to one. Notice that this result does not contradict the fact that this structure can be operating as a filter by reflecting back part of the incident power. In fact, the available power gain corresponds with the operating gain only when the input and output ports are matched – a condition that a filter not in the passband clearly violates.

7.8 Noise Models of Microwave Transistors

7.8.1 Noise Models of Field-Effect Transistors

Bipolar and field effect transistors have similar noise performances at high frequency for devices having comparable cutoff and maximum oscillation frequency. However, the low-frequency noise is larger in surface devices like the FETs when compared to BJTs or HBTs due to the larger $1/f$ noise of the former, see Fig. 7.18; as already remarked, BJTs are thus preferred in devices where frequency upconversion is present, like oscillators. On the other hand FETs have an advantage at high frequency since the high-frequency noise figure increases linearly with f in FETs but quadratically with f in BJTs [11].

HEMTs typically have better noise performances than MESFETs with the same gate length, and also a better compromise between noise and gain. An approximate formula relates, for both devices, the minimum noise figure at a given frequency to the small-signal parameters (the *Fukui formula* [11], originally derived from noise compact models, and in fair agreement with experimental data):

$$F_{\min} = 1 + K_f \frac{f}{f_T} \sqrt{g_m (R_G + R_S)}, \tag{7.50}$$

where K_f is an empirical factor with value around 2.5 (MESFETs) but generally ranging between 1.5 and 3. Notice that the minimum noise figure increases monotonically with increasing frequency, at least for frequencies above the corner frequency where the high-frequency noise equals the $1/f$ noise. HEMTs typically exhibit, with the same gate lengths, larger cutoff frequencies and lower parasitic resistances when compared to MESFETs, and have therefore globally superior noise performances.

As shown in Fig. 7.19, HEMTs also exhibit a better compromise between noise and gain as a function of the bias current. For HEMTs the optimum low-noise bias current can be as large as 50 percent of the maximum, while it is 10–20 percent for MESFETs.

For field-effect transistors (FETs), compact noise models exist relating the short circuit gate and drain current fluctuation power ($S_{i_{Dn}}$, $S_{i_{Gn}}$) and correlation ($S_{i_{Gn}i_{Dn}}$) spectra to a set of small-signal parameters. An example is the so-called *PRC* high-frequency noise model for the intrinsic FET noise, see [12, 13] and references therein, yielding:

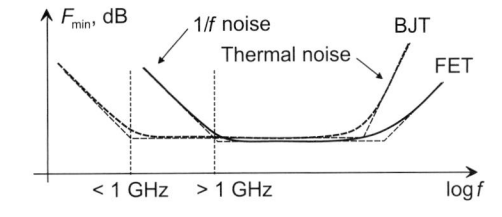

Figure 7.18 Qualitative example of noise figure behavior as a function of frequency for bipolar and field-effect transistors. The corner frequency for $1/f$ noise can be for bipolar devices as low as a few kHz and, for field-effect devices, in excess of a few GHz. Notice the different high-frequency theoretical slope of the noise figures.

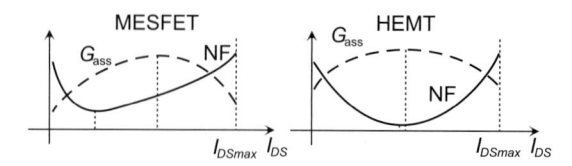

Figure 7.19 Noise figure and associated gain vs. the drain bias current for MESFETs and HEMTs.

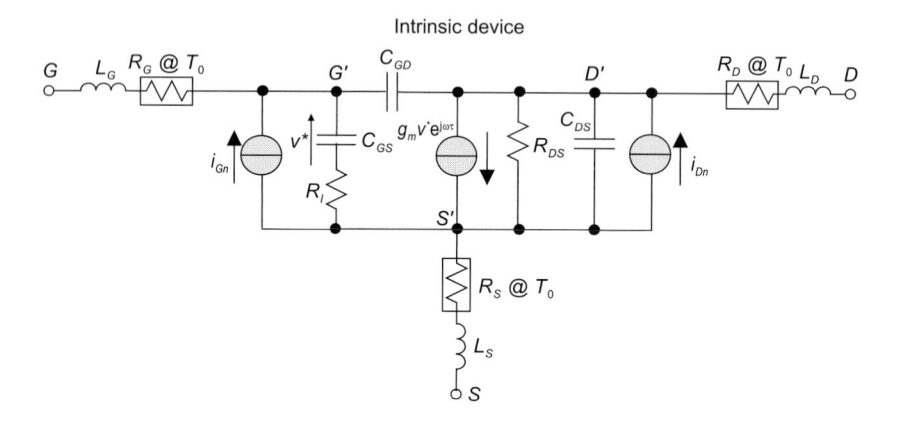

Figure 7.20 FET noise equivalent circuit.

$$S_{i_{Dn}} \approx 4k_B T_0 g_m P \tag{7.51}$$

$$S_{i_{Gn}} \approx 4k_B T_0 \frac{\omega^2 C_{GS}^2}{g_m} R \tag{7.52}$$

$$S_{i_{Gn} i_{Dn}} \approx jC\sqrt{S_{i_{Dn}} S_{i_{Gn}}}, \tag{7.53}$$

where g_m is the device transconductance, C_{GS} the input capacitance, according to the equivalent circuit shown in Fig. 7.20, T_0 is a reference temperature at which the parameters P, R and C are measured (as already recalled, the default value is $T_0 = 290$ K). The circuit also include thermal noise models for the parasitic resistances. The parameters P, R and C can be considered as fitting factors, to be derived from theory or measurements; the ideal, long-channel values of the *PRC* parameters (see [6, pp. 64–66]):

$$P = \frac{2}{3} \approx 0.67, \quad R = \frac{16}{135} \approx 0.12, \quad C = \frac{\sqrt{10}}{8} \approx 0.4. \tag{7.54}$$

Notice that P is also called γ in MOSFET modeling. Realistic P values for short-gate devices typically are larger; C expresses the correlation magnitude (typically $C \approx 0.6 - 0.7$ in short-gate devices). The imaginary correlation spectrum implies that short-circuit gate and drain current fluctuations are in quadrature due to the capacitive coupling of channel current fluctuations to the gate. At low frequency, the gate noise spectrum $S_{i_{Gn}}$ and the correlation spectrum $S_{i_{Gn} i_{Dn}}$ become negligible, and only the output drain noise (white) is significant.

7.8 Noise Models of Microwave Transistors

Taking into account of the effect (typically not at all negligible) of the gate resistance, the minimum noise figure, the series noise conductance and the optimum generator impedance can be derived from the *PRC* model (neglecting the drain parasitic resistance and the gate-drain capacitance) as in [13]:

$$F_{\min} \approx 1 + 2K_1\sqrt{g_m(R_S + R_G) + K_2}\,\frac{f}{f_T} \tag{7.55}$$

$$g_n = g_m K_1 \left(\frac{f}{f_T}\right)^2 \tag{7.56}$$

$$Z_{Go} = \frac{\sqrt{g_m(R_S + R_G) + K_2}}{K_1}\frac{1}{\omega C_{GS}} + j\frac{P - C\sqrt{PR}}{K_1^2}\frac{1}{\omega C_{GS}}, \tag{7.57}$$

where $f_T = g_m/2\pi C_{GS}$ is the FET cutoff frequency, R_S is the source resistance and:

$$K_1 = \sqrt{P + R - 2C\sqrt{PR}}$$
$$K_2 = PR(1 - C^2)/K_1^2.$$

Notice that the optimum source impedance has an inductive imaginary part with a capacitor-like frequency behavior, potentially able to compensate for the capacitive input reactance of the FET. The real part is instead a resistance with a capacitor-like frequency behavior. It is therefore clear that the input noise matching partly coincides (for the imaginary part) with the input power matching, while little can be said for the real part. Moreover, wideband noise matching is difficult due to the need to implement a proper frequency-dependent behavior of the generator resistance.

Neglecting the effect of intrinsic noise sources (expressed by the parameter K_2 under square root) with respect to the extrinsic noise sources (related to the resistances R_S and R_G) we finally obtain the Fukui formula (7.50) where $K_f = 2K_1$ can be considered as a fitting factor derived from measurements. Assuming the ideal values in (7.54) we obtain $K_f \approx 1.5$.

A different approach to the measurement-oriented noise modeling was proposed by Pospieszalski [14] and Hughes [15], whereby the noise properties of the two-port are reduced to two frequency-independent noise temperatures, the gate noise temperature T_G associated with the device intrinsic resistance R_I, and the drain noise temperature T_D associated with the output conductance $G_{DS} = 1/R_{DS}$, see Fig. 7.21. At low frequency, the following expressions result for the short-circuit current power and correlation spectra:

$$S_{i_{Dn}} \approx 4k_B(T_G g_m^2 R_I + T_D G_{DS})$$
$$S_{i_{Gn}} \approx 4k_B T_G (\omega C_{GS})^2 R_I$$
$$S_{i_{Gn} i_{Dn}} \approx -j4k_B T_G \omega C_{GS} g_m R_I.$$

Measurements carried out on several state-of-the-art MESFETs and HEMTs [14, 15] suggest that the two-temperature model is able to fit accurately enough the experimental data over a wide frequency band; moreover, T_G turns out to be close to the ambient temperature T, while typically $T_D \gg T$. Furthermore, Hughes [15] shows that a simple

Low-Noise Amplifier Design

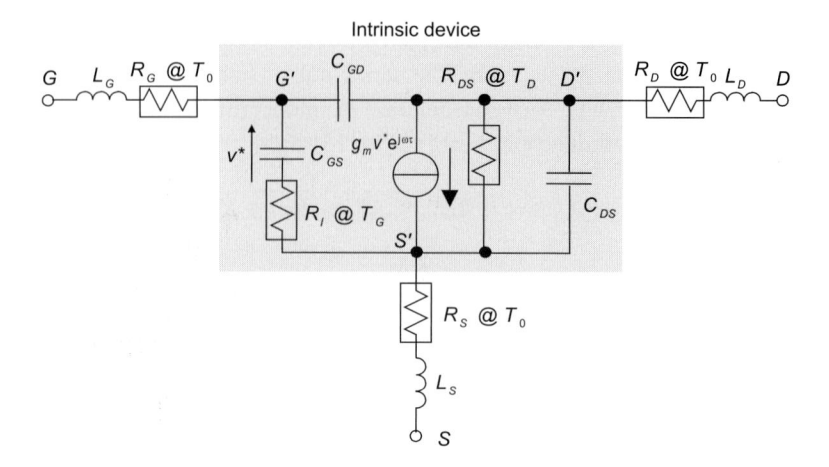

Figure 7.21 Two-temperature FET noise model.

approximate relationship exists (see Example 7.9) between the noise figure and the optimum associated gain G_{ass} of a microwave FET with low $F_{\min}$:

$$G_{\text{ass}}(F_{\min} - 1) \approx \frac{T_D}{T_0}. \qquad (7.58)$$

Example 7.9 Give a proof of (7.58).

Solution
To derive (7.58) we have to make a number of simplifying assumptions. The device parasitics are neglected, and so is the internal feedback gate-drain capacitance; moreover, also the gate noise is not considered. Finally, we assume that the generator impedance is the optimum source impedance. In such conditions the available power gain is the associate gain G_{ass}. The output noise spectral density results from the superposition of drain noise (related to the drain temperature) and of the generator noise amplified by the stage:

$$p_{nL} = k_B T_D + G_{\text{ass}} k_B T_0,$$

while the contribution of the generator noise only is:

$$p'_{nL} = G_{\text{ass}} k_B T_0.$$

Therefore the noise figure (minimum in this case, since we suppose that the generator termination is the optimum one) will be:

$$F_{\min} = \frac{p_{nL}}{p'_{nL}} = 1 + \frac{T_D}{G_{\text{ass}} T_0},$$

i.e.:

$$G_{\text{ass}}(F_{\min} - 1) \approx \frac{T_D}{T_0}.$$

7.8.2 Noise Models for Bipolar Transistors

The noise model of bipolar and heterojunction bipolar transistors makes use of the shot noise approach already described with reference to *pn* junction diode noise models [2, 6]. For the sake of brevity, we will only present a bipolar transistor noise model in the common emitter configuration in the forward active region, see Fig. 7.22. The relevant intrinsic noise generators are:

$$S_{i_{Bn}} \approx 2q\,|I_B| + 4k_B T \text{Re}\,[Y_{EE}(\omega) - Y_{EE}(0)]$$
$$\qquad - 4k_B T \text{Re}\,[Y_{CE}(\omega) - Y_{CE}(0)] \qquad (7.59a)$$
$$S_{i_{Cn}} \approx 2q\,|I_C| \qquad (7.59b)$$
$$S_{i_{Bn} i_{Cn}} \approx 2k_B T \left\{ Y^*_{CE}(\omega) - Y_{CE}(0) \right\}, \qquad (7.59c)$$

where the small-signal admittances Y_{EE} and Y_{CE} are evaluated at the bias point. At low frequency therefore the base and collector current exhibit completely uncorrelated shot noise. It may be interesting to compare the intrinsic noise sources pertaining to a bipolar and to a FET. Working at low frequency we have for the output noise current:

$$S_{i_{Dn}} \approx 4k_B T_0 g_m P \approx 3k_B T_0 g_m$$
$$S_{i_{Cn}} \approx 2q\,|I_C| = 2q \frac{V_T\,|I_C|}{V_T} = 2k_B T_0 g_m,$$

where we have approximated the *P* parameter with its ideal, long gate value. Similar output noises therefore arise for devices having the same transconductance, or, to put it in a different way, the ideal bipolar *P* is $P = 1/2$. On the other hand the input noise exhibits different behaviors, white for the bipolar and instead quadratically increasing with frequency for the FET. The different frequency dependence of the input noise generators, together with the different structure of the input (series *RC* for the FET, parallel *RC* for the bipolar) makes the comparative behavior of the minimum noise figure quite different as a function of frequency. As shown in [16], an approximate expression for the bipolar minimum noise figure is:

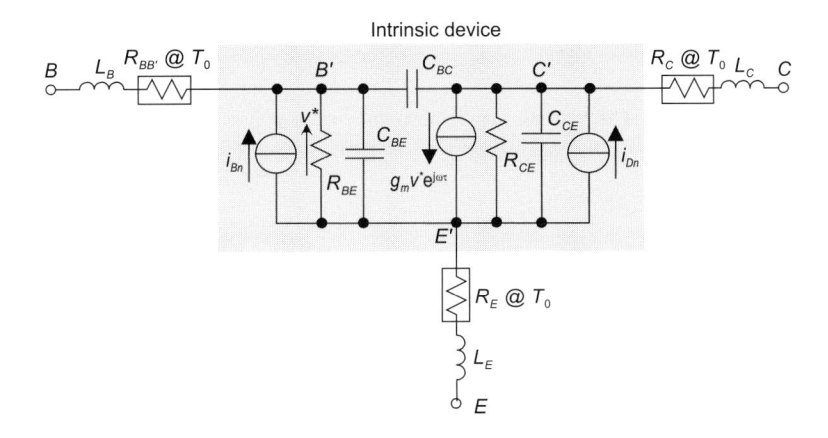

Figure 7.22 Noise model for a high-frequency bipolar transistor.

$$F_{\min} \approx 1 + g_m R_B \left(\frac{f}{f_T}\right)^2 \left[1 + \sqrt{1 + \frac{2}{g_m R_B}\left(\frac{f_T}{f}\right)^2}\right],$$

where $g_m = I_C/V_T$ is the transistor transconductance, $R_B \equiv R_{BB'}$ is the total external base resistance and f_T the cutoff frequency [16]. For typical bias currents and base resistances the frequency-dependent term in square root is small vs. unity; for instance, assume $f/f_T = 0.5$, $I_C = 100$ mA, $R_B = 5\ \Omega$; we have:

$$\frac{2}{g_m R_B}\left(\frac{f_T}{f}\right)^2 = \frac{2}{\frac{100 \times 10^{-3}}{26 \times 10^{-3}} \cdot 5}\left(\frac{1}{0.5}\right)^2 = 0.416$$

and of course this term is decreasing with frequency. It follows that the dominating term will indeed be the quadratic one, leading to the conclusion that the minimum noise figure in a bipolar grows faster with frequency than in a field-effect transistor.

7.9 Noise Figure of Cascaded Two-Ports

Many transmission systems are made of several stages in cascade; imagine a receiver including a band selection filter, a low-noise amplifier, some image frequency filter, a mixer, an intermediate frequency section, another mixer and the baseband section. The overall noise figure of a cascaded series of stages can be simply evaluated from the available power gains and noise figures of the single stages if certain conditions are met. The approach, leading to the so-called *Friis formula* [17], is useful at a system level, while it cannot be typically applied in the circuit-level analysis of multi-stage circuits, as justified at the end of the section.

As a starting point, suppose to connect in cascade two two-ports, with available power gains $G_1(\Gamma_G)$ and $G_2(\Gamma_{out})$ and noise figures $F_1(\Gamma_G)$ and $F_2(\Gamma_{out})$, see Fig. 7.23. Notice that the available power gains and noise figures are evaluated in operational conditions, i.e., taking into account the reflection coefficient of the input, see Fig. 7.24. The source is at the reference temperature T_0. From the definition of the noise figure we have:

$$F_1 = 1 + \frac{p_{n1}}{p_{nG} G_1},$$

where p_{n1} is the noise available power at port 2 of the first two-port due to the internal noise generators only, and $p_{nG} = k_B T_0$ is the available noise power of the generator

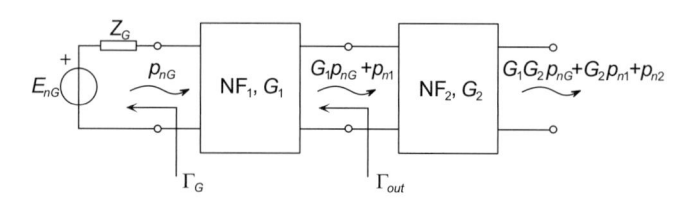

Figure 7.23 Noise figure of two cascaded stages. For each stage the input and output noise available power spectral densities are shown.

7.9 Noise Figure of Cascaded Two-Ports

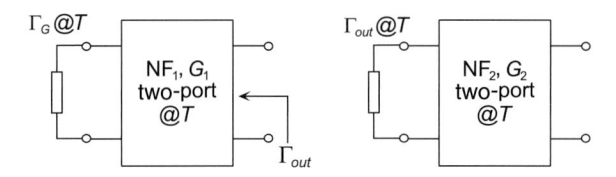

Figure 7.24 Condition in which the noise figure and available gains of the cascaded stages are measured (see Fig. 7.23).

connected to port 1 of first two-port. Notice that we use the spot definition (per unit bandwidth) of the noise figure. Therefore:

$$p_{n1} = (F_1 - 1)G_1 k_B T_0.$$

Similarly, for the noise available power of port 2 of the second two-port due to internal generators only we have:

$$p_{n2} = (F_2 - 1)G_2 k_B T_0.$$

Therefore, the available noise power density at the output port of the cascade is:

$$p_{nout} = G_2 p_{n1} + p_{n2} = (F_1 - 1)G_1 G_2 k_B T_0 + (F_2 - 1)G_2 k_B T_0,$$

while the output available power spectral density due to the noisy generator connected to the cascade input port is:

$$p'_{nout} = G_1 G_2 k_B T_0.$$

Therefore, the total noise figure is:

$$F_{tot} = 1 + \frac{p_{nout}}{p'_{nout}} = 1 + (F_1 - 1) + \frac{(F_2 - 1)}{G_1}. \tag{7.60}$$

Eq. (7.60) can be directly extended to the N−stage case as:

$$F_{tot} = F_1 + \frac{F_2 - 1}{G_1} + \frac{F_3 - 1}{G_1 G_2} + \frac{F_4 - 1}{G_1 G_2 G_3} \cdots.$$

The main conclusion from the Friis formula is that, if the first stage has a large enough gain, the noise figure of all stages beyond the first does not have a significant impact on the overall noise figure of the chain. In other words, in a receiver chain we have to place the low-noise amplifier as soon as possible at the input of the chain in order to decrease the effect of the noise coming from the following stages. In particular we should avoid placing a lossy structure at the beginning of the chain (with available power gain < 1) because this would amplify the noise from the following stage. In practice, however, we cannot avoid placing a filter (i.e., a lossy stage) at the beginning of a receiver chain (with a low loss however in the passband), and the main contributor to the noise figure of the chain is often not the first stage, but the mixer stage, due to its typically very high noise figure.

The application of the Friis formula has a main caveat, all gains and noise figures refer to the source impedance of the previous stage, which is not necessarily coinciding with the optimum source impedance or with the impedance (typically 50 Ω) at which the noise figure has been measured. In a multi-stage circuit, moreover, it is difficult to clearly identify the input impedance of a single stage. The situation is different at a system level, where stages are well defined two-ports, often designed to be matched on a 50 Ω impedance. In this case we can also (at least in principle) exchange two stages without making the Friis formula invalid.

7.9.1 Noise Measure

Suppose to consider two cascaded stages designed on the same input impedance R_0 (i.e., each stage has output impedance R_0 when its input impedance is R_0). The available power gains and noise figures are $G_1(R_0)$, $G_2(R_0)$ and $F_1(R_0)$, $F_2(R_0)$. We evaluate the noise figure of two configurations, 1-2 (case a) and 2-1 (case b). Notice that the two stages are interchangeable since both are designed on R_0. From the Friis formula we have:

$$F_{tot,a} = 1 + (F_1 - 1) + \frac{(F_2 - 1)}{G_1}$$
$$F_{tot,b} = 1 + (F_2 - 1) + \frac{(F_1 - 1)}{G_2}.$$

The problem is now to understand when the noise figure in case a is better (i.e., lower) than the noise figure in case b. We easily find that:

$$F_{tot,a} < F_{tot,b} \rightarrow F_1 + \frac{F_2 - 1}{G_1} < F_2 + \frac{F_1 - 1}{G_2} \rightarrow M_1 < M_2,$$

where the parameter M, defined as:

$$M = \frac{F - 1}{1 - G^{-1}}$$

is called the *noise measure* of the two-port. The noise measure is therefore a mixed figure of merit addressing a compromise between the noise figure and the gain.

7.10 Low-Noise Amplifiers

The classical design of low-noise amplifiers is based on obtaining the minimum noise figure F_{min} by terminating the amplifier input with a source impedance equal to the optimum source impedance Z_{Go}. The optimum source impedance can be obtained from the source impedance (typically a resistance R_0) through a proper matching section, as shown in Fig. 7.25 (b).

It is important to notice that the matching section should be designed by transforming R_0 into the amplifier optimum input impedance Z_{Go} and **not**, as possible in the maximum gain design (Fig. 7.25 (a)), by transforming the amplifier input impedance Z_{in} into R_0. In fact, the input impedance Z_{in} of the input matching section seen from the generator will be, in general, $Z_{in} \neq R_0$, i.e., input noise matching does not allow for power or resistance

7.10 Low-Noise Amplifiers

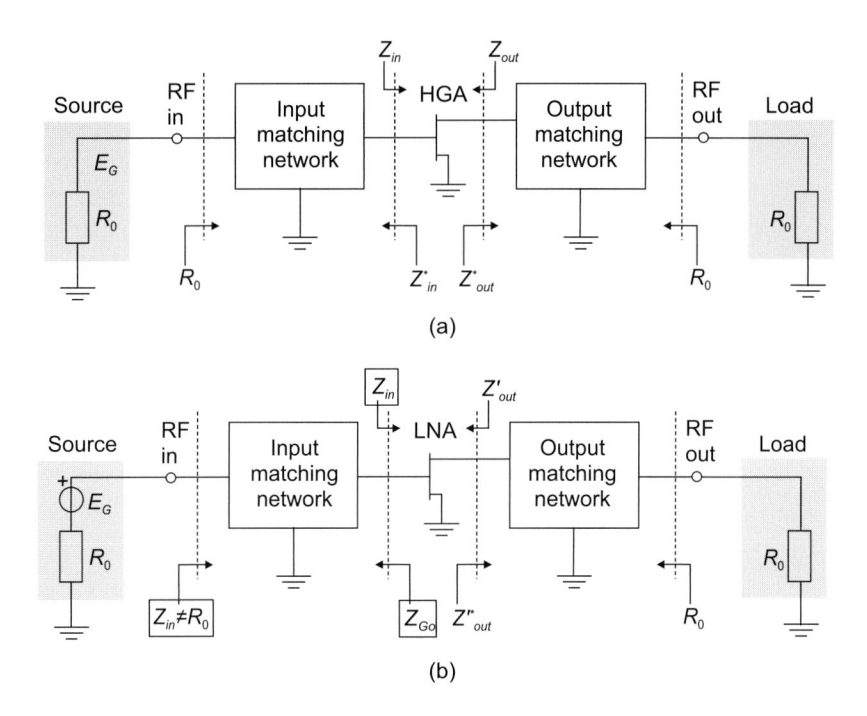

Figure 7.25 Maximum gain and minimum noise design for a single-stage amplifier: (a) maximum gain matching (input and output conjugate matching); (b) minimum noise matching (input noise matching, output conjugate matching). HGA is a high-gain amplifier stage while LNA is a low-noise amplifier stage.

matching of the generator. As a consequence, the generator and amplifier input will be *mismatched*.

Concerning the amplifier output, since the noise figure does not depend on the loading conditions we can choose to power match the output device impedance Z'_{out}, the same strategy adopted in maximum gain design.

In conclusion, while for a real generator and load R_0 the single-frequency maximum gain design yields, at the same time, the device MAG and input/output resistance matching (i.e., zero input and output reflection coefficients), the minimum noise figure design leads to a gain (the associated gain) G_{ass} <MAG, but, at the same time, grants that the amplifier is matched only at the output (and not at the input) to the generator and load resistance R_0.

Practical low-noise amplifier design has to face a number of issues:

- The input matching section should be reactive, otherwise input losses would deteriorate the noise figure. Moreover, parasitic losses should be low, suggesting a distributed implementation at high frequency and making the realization of a lumped integrated matching section difficult if the noise figure constraints are severe.
- Even if the gain penalty vs. MAG can be accepted, input mismatch is often unacceptable within a receiver chain. For example, the input band selection filter is typically designed on 50 Ω and does not correctly work if it is severely mismatched. In general,

little can be said on the amount of input mismatching that is introduced by the minimum noise figure strategy, since it also depends on technological issues. Circuit solutions like the use of input isolators or circulators allow the input reflection to be eliminated by dissipating it on an internal or external load, but the technological implementations are bulky and not integrable.[7]

- Device stabilization is an issue since the use of resistive input stabilization in-band is ruled out, due to the negative impact on the noise figure. Alternative strategies to improve stability can be, e.g., the use of cascode configurations that reduce the device internal feedback, thus potentially improving the device stability. In many cases the in-band stabilization is avoided not to compromise the noise figure and only the out-of-band stabilization is enforced [10]; for the design strategy in the non-unconditionally stable case see, e.g., [18].
- In some systems (e.g., a satellite receiver) the constraints on the noise figure are severe, while in others (e.g., a mobile phone receiver) the acceptable noise figure is much larger (e.g., 2 dB) while cost and size problems dominate, excluding sophisticated solutions devised to suppress the input reflections.

In conclusion, alternative (vs. the classical minimum noise figure strategy) design approaches for LNAs are based on a number of circuit solutions that:

- allow for input matching without introducing dissipative elements at the amplifier input, and at the same time provide a gain close or equal to the MAG;[8]
- without achieving the minimum noise figure, can be suitably optimized to yield a noise figure acceptable for a specific application.

In the following sections, two widespread LNA alternative schemes are described, the common gate LNA stage and the LNA stage with inductive series source feedback. Another possibility, not described here in detail, is the adoption of a balanced configuration. This solution, first proposed in 1965 by Engelbrecht Kurokawa [19] and still popular also in integrated design, see, e.g., [20], allows good input matching to be achieved at the expense of a greater circuit complexity and a worse noise figure (since noisy elements are added at the LNA inputs); it can also help in broadbanding the LNA.

7.10.1 Common-Gate (Base) LNA Stage

A first example of low-noise stage allowing for input matching to some specified value of R_0 is the *common gate* (or common base for bipolar implementation) stage, first proposed by Estreich in [21]. In this stage the input is on the source, the output on the drain, while the gate is shorted in small-signal conditions. The stage is shown in Fig. 7.26; we suppose that an input adapter is present and for simplicity we introduce

---

[7] The idea is not dissimilar for the one in the balanced amplifier. Indeed, this configuration can be exploited to obtain an input matched LNA, of course with increased circuit complexity.

[8] Notice that in some cases we compare the gain of the LNA with the MAG of a different configuration of the same active element, for instance the gain of a common base stage with the one of a common source stage, that is typically larger.

7.10 Low-Noise Amplifiers

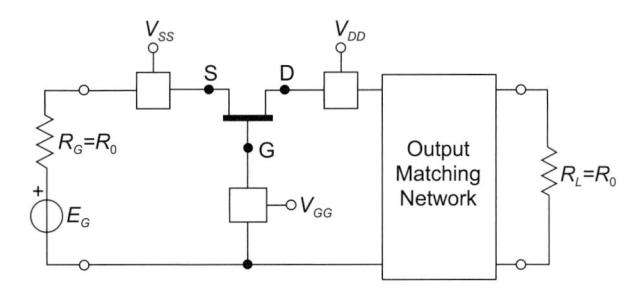

Figure 7.26 Common gate low-noise amplifier stage.

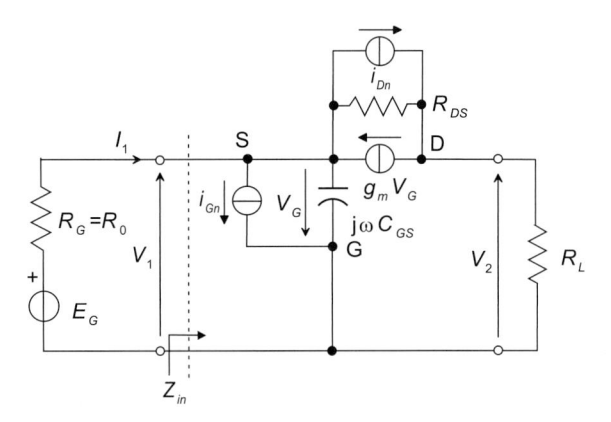

Figure 7.27 Simplified small-signal equivalent circuit of common-gate low-noise stage amplifier.

three T-bias for the DC polarization. The simplified small-signal equivalent circuit is shown in Fig. 7.27.

As a first step, we identify the stage input impedance. For simplicity we will neglect the effect of the output resistance $R_{DS} = 1/G_{DS}$. Since $V_G = -V_1$, we have that the input current is given by:

$$I_1 = j\omega C_{GS} V_1 - g_m V_G = (g_m + j\omega C_{GS}) V_1$$

and therefore:

$$Z_{in} = \frac{V_1}{I_1} = \frac{1}{g_m + j\omega C_{GS}} \xrightarrow{\omega \to 0} \frac{1}{g_m}.$$

The input matching condition at low frequency is therefore:

$$R_0 = \frac{1}{g_m}.$$

In particular, for 50 Ω matching we need a transconductance of $g_m = 1/50 = 20$ mS that can be obtained by a small periphery high-frequency FET (HEMTs have specific transconductance of the order of 200 mS/mm or more). In order to adjust g_m two degrees of freedom are available:

- given the device periphery, we can adjust the gate bias in order to comply with the input matching requirement;
- given the gate bias, we can change (in an integrated device) the device periphery W in order to comply with the input matching requirement.

Once matching is imposed, the input impedance has a frequency behavior that limits the amplifier bandwidth:

$$Z_{in} = \frac{R_0}{1 + j\omega R_0 C_{GS}}.$$

An estimate of the available power gain of the stage can be obtained as follows. To this purpose, we evaluate the open-circuit voltage V_{02} by transforming the output parallel current generator into the series equivalent, see Fig. 7.28. We have for the open-circuit voltage:

$$V_1 = \frac{E_G}{1 + j\omega C_{GS} R_0}$$

$$V_{02} = V_1 + g_m R_{DS} V_1 = \frac{1 + g_m R_{DS}}{1 + j\omega C_{GS} R_0} E_G,$$

while for the output impedance:

$$Z_{out} = R_{out} + jX_{out} = \frac{R_0}{1 + j\omega R_0 C_{GS}} + R_{DS}$$

$$= \frac{R_0 + R_{DS}\left[1 + (\omega R_0 C_{GS})^2\right]}{1 + (\omega R_0 C_{GS})^2} - \frac{j\omega R_0^2 C_{GS}}{1 + (\omega R_0 C_{GS})^2}.$$

Thus, the available power gain can be expressed as:

$$G_{av} = \frac{\frac{|V_{02}|^2}{4R_{out}}}{\frac{|E_G|^2}{4R_0}} \approx \frac{\frac{(1 + g_m R_{DS})^2}{1 + (\omega R_0 C_{GS})^2} \frac{1 + (\omega R_0 C_{GS})^2}{R_0 + R_{DS}\left[1 + (\omega R_0 C_{GS})^2\right]} \frac{|E_G|^2}{4}}{\frac{|E_G|^2}{4R_0}}$$

$$= \frac{R_0 (1 + g_m R_{DS})^2}{R_0 + R_{DS}\left[1 + (\omega R_0 C_{GS})^2\right]}.$$

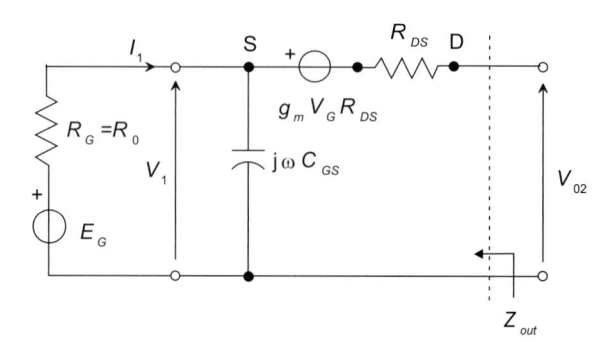

Figure 7.28 Evaluating the available power gain of common-gate low-noise stage amplifier.

7.10 Low-Noise Amplifiers

At low frequency and assuming $R_{DS} \gg R_0$ we have:

$$G_{\text{av}} \underset{\omega \to 0}{\to} \frac{R_0 (1 + g_m R_{DS})^2}{R_0 + R_{DS}} \approx g_m^2 R_{DS} R_0 \approx \frac{R_{DS}}{R_0},$$

where the last approximation holds in matching conditions. The available gain has a cutoff frequency related to the input time constant. Since the output resistance is larger than R_0 typically (above all in a small device) we can expect an available power gain larger than unity but, typically, less than 10 dB.

We now estimate the low-frequency noise figure of the stage according to the circuit in Fig. 7.29. Thanks to the low-frequency assumption, we neglect the gate noise and the effect of the input capacitance. The short-circuit output current I_2 can be immediately expressed as:

$$I_2 = I_{Dn} - g_m V_G,$$

but:

$$E_{Gn} = R_0 I_2 - V_G = R_0 (I_{Dn} - g_m V_G) - V_G$$
$$= R_0 I_{Dn} - (g_m R_0 + 1) V_G$$
$$V_G = \frac{-E_{Gn} + R_0 I_{Dn}}{1 + g_m R_0},$$

from which:

$$I_2 = \frac{I_{Dn} + g_m E_{Gn}}{1 + g_m R_0}.$$

Define now as I_{n2} the total output noise current and as I'_{n2} the output noise current due to the input generator only. We have:

$$I_{n2} = \frac{I_{Dn} + g_m E_{Gn}}{1 + g_m R_0}, \quad I'_{n2} = \frac{g_m E_{Gn}}{1 + g_m R_0}.$$

The noise figure can be now defined as:

$$F = \frac{\overline{I_{n2} I_{n2}^*}}{\overline{I'_{n2} I'^*_{n2}}} = \frac{\overline{I_{Dn} I_{Dn}^*} + g_m^2 \overline{E_{Gn} E_{Gn}^*}}{g_m^2 \overline{E_{Gn} E_{Gn}^*}} \frac{4k_B T g_m P + 4k_B T R_0 g_m^2}{4k_B T R_0 g_m^2}$$

$$= 1 + \frac{P}{R_0 g_m} \approx 1 + P,$$

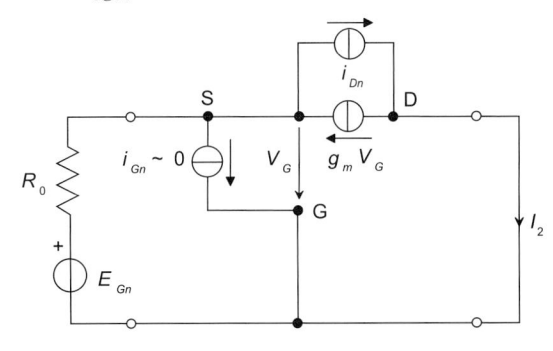

Figure 7.29 Simplified noise equivalent circuit of common-gate low-noise stage amplifier. E_{Gn} is the noise generator associated with the generator resistance.

where we have again imposed the input matching condition. Taking into account long-gate parameters, see (7.54), we have:

$$F \approx 1 + \frac{2}{3} \rightarrow F_{\text{dB}} = 2.2 \text{ dB}.$$

In the bipolar case the result is the same but the equivalent $P \approx 1/2$ leading to a slightly better noise figure of 1.76 dB. The value found may be not good enough for microwave applications where a very small noise figure is required but could be appropriate in systems (e.g., a portable phone receiver) where low-noise requirements are not so stringent.

Besides allowing for input matching without any additional complexity and with a moderate available power gain, the input common gate (or common base) low-noise stage exhibit a reduction of the internal feedback capacitance when compared to the common source (emitter) stage, i.e., a better stability. The main disadvantage of the stage is of course the moderately high noise figure, that does not exhibit the typical decreasing behavior with decreasing frequency ($F \approx 1 + Kf/f_T$), typical of common-source stages.

7.10.2 LNA with Inductive Source/Emitter Series Feedback

A popular topology for the first stage of high-performance microwave integrated LNAs is the low-noise stage with inductive source series feedback (Fig. 7.30); the same topology applies to bipolar with inductive collector feedback. It was probably proposed for the first time by Nevin and Wong in 1978 [22]. The stage is also referred to as *inductively degenerated* stage.[9] The solution allows for input matching without introducing noisy element exploiting a proper combination of reactive and transconductance blocks. Moreover, the resulting noise figure, albeit potentially larger (but also smaller, see Example 7.10) than the device minimum one,[10] can be conveniently optimized by

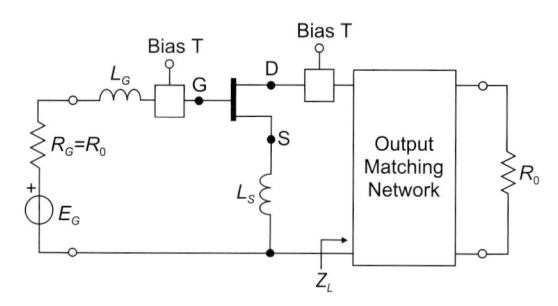

Figure 7.30 Low-noise amplifier with source inductive feedback.

---
[9] The terms "regenerative" and "degenerative" feedback refer to positive or negative feedback, respectively. An emitter degenerated amplifier is therefore an amplifier with a negative (series) feedback on the emitter, and a source inductively degenerated stage is a FET with a series (negative) feedback on the source. The therm "degenerative" to mean a "negative" feedback can be found, e.g., in the 1959 book [23], but it is clearly much older.

[10] The inductive feedback indeed changes the minimum noise figure of a two-port. This variation can be negligible, see (7.67) and the following discussion, in a proper frequency and parameter value range.

7.10 Low-Noise Amplifiers

playing on the device size, also reducing the stage power consumption. Stability issues can be alleviated by the use of cascode stage, even though the series inductor feedback itself is leading to improved stability, as discussed in [24], at least in a range of source inductor values.

Variations on the topology including rather than two uncoupled inductors (the source feedback inductor and the input gate inductor) a single transformer have been proposed by Long et al., see [25, 26]. The resulting topology is more compact (since the two coupled inductors are often stacked) and allows for an additional degree of freedom in the design (the mutual inductance). The inductively degenerated stage is often implemented with cascode transistors to improve the input-output isolation and therefore the stability. For a bipolar cascode implementation see, e.g., [27].

The source inductive series feedback will be discussed starting from the equivalent circuit in Fig. 7.31 without noise generators and where parasitics and internal feedback are neglected. Also, the output capacitance will be neglected together with the intrinsic resistance R_I. Starting from the intrinsic device admittance:

$$\mathbf{Y}_{in} = \begin{pmatrix} j\omega C_{GS} & 0 \\ g_m & 1/R_{DS} \end{pmatrix},$$

we obtain the series representation of the device with inductive feedback and input gate inductor (whose purpose will be clarified soon):

$$\mathbf{Z} = \mathbf{Y}_{in}^{-1} + \begin{pmatrix} j\omega(L_S + L_G) & j\omega L_S \\ j\omega L_S & j\omega L_S \end{pmatrix}$$

$$= \begin{pmatrix} -\dfrac{j}{\omega C_{GS}} + j\omega(L_S + L_G) & j\omega L_S \\ j\dfrac{g_m R_{DS}}{\omega C_{GS}} + j\omega L_S & R_{DS} + j\omega L_S \end{pmatrix}, \quad (7.61)$$

from which the input impedance seen by the generator with load equal to Z_L:

$$Z_{in} = Z_{11} - \dfrac{Z_{12}Z_{21}}{Z_{22} + Z_L}$$

$$= -\dfrac{j}{\omega C_{GS}} + j\omega L_S + j\omega L_G - \dfrac{j\omega L_S \left(j\dfrac{g_m R_{DS}}{\omega C_{GS}} + j\omega L_S \right)}{R_{DS} + j\omega L_S + Z_L}.$$

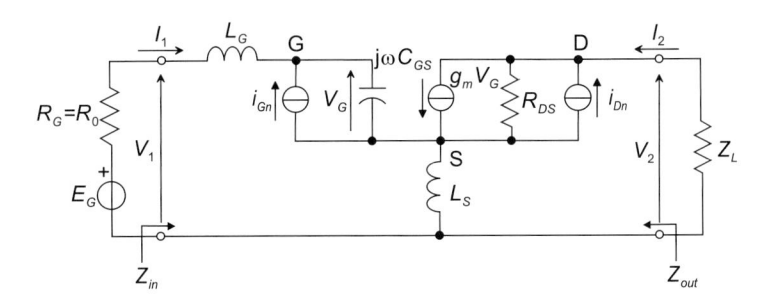

Figure 7.31 Simplified equivalent circuit of low-noise amplifier with series source inductive feedback.

Assuming that the output resistance R_{DS} of the device is suitably large ($R_{DS} \gg |j\omega L_S + Z_L|$ and $g_m R_{DS} \gg \omega^2 L_S C_{GS}$) we can approximate:

$$Z_{in} \approx -\frac{j}{\omega C_{GS}} + j\omega L_S + j\omega L_G + \frac{g_m L_S}{C_{GS}},$$

i.e., for large enough R_{DS} the input impedance is independent from the load impedance.

Input matching to R_0 can now be obtained by imposing that the real part of Z_{in} be equal to R_0 (this imposes the value of L_S) and by canceling the imaginary part through resonance with the gate inductor L_G:

$$\frac{g_m L_S}{C_{GS}} = R_0 \rightarrow L_S = \frac{C_{GS} R_0}{g_m} = \frac{R_0}{\omega_T}$$

$$\omega(L_G + L_S) = \frac{1}{\omega C_{GS}},$$

where $\omega_T = 2\pi f_T$ is the device cutoff angular frequency (only dependent on technology). Since the condition $\omega_T L_S = R_0$ holds, L_S typically is a "small" inductor, e.g., a 50 Ω inductor but at a frequency f_T much larger than the operating frequency. Moreover, this suggests that the resonance with C_{GS} at ω will usually require the gate inductance (i.e., that L_S does not overcompensate C_{GS}). It is worth noticing that the resistance matching is wideband, while the reactance matching is resonant; the input resistance of the stage does not derive from a real resistor, but from the combination of two reactances and a transconductance generator.

Let us now evaluate the available power gain of the stage with feedback. Since the input is matched, this coincides with the MAG. We have from Fig. 7.31:

$$V_1 = E_G - R_0 I_1 = Z_{11} I_1 + Z_{12} I_2 \tag{7.62}$$

$$V_2 = Z_{21} I_1 + Z_{22} I_2, \tag{7.63}$$

from which, solving (7.62) for I_1:

$$I_1 = -\frac{Z_{12}}{Z_{11} + R_0} I_2 + \frac{E_G}{Z_{11} + R_0},$$

where, at the design frequency:

$$Z_{11} + R_0 = -\frac{j}{\omega C_{GS}} + j\omega L_S + j\omega L_G + R_0 = R_0$$

$$I_1 = -\frac{Z_{12}}{R_0} I_2 + \frac{E_G}{R_0}$$

thanks to the input matching condition; substituting in (7.63), we obtain:

$$V_2 = \left(Z_{22} - \frac{Z_{21} Z_{12}}{R_0}\right) I_2 + \frac{Z_{21} E_G}{R_0} = Z_{out} I_2 + V_{02},$$

where the output impedance and open-circuit voltage are:

$$Z_{out} = \frac{Z_{22} R_0 - Z_{21} Z_{12}}{R_0}$$

$$V_{02} = \frac{Z_{21} E_G}{R_0}.$$

7.10 Low-Noise Amplifiers

Thus, the available power gain is:

$$G_{av} = \frac{\frac{|V_{02}|^2}{4\text{Re}(Z_{out})}}{\frac{|E_G|^2}{4R_0}} = \frac{\frac{|Z_{21}|^2 |E_G|^2}{4\text{Re}(Z_{out}) R_0^2}}{\frac{|E_G|^2}{4R_0}} = \frac{|Z_{21}|^2}{\text{Re}(Z_{22}R_0 - Z_{21}Z_{12})}. \quad (7.64)$$

Taking into account the input matching conditions we have:

$$\frac{g_m}{\omega C_{GS}} = \frac{R_0}{\omega L_S}$$

$$\omega L_S - \frac{1}{\omega C_{GS}} = -\omega L_G,$$

so that the impedance matrix in (7.61) becomes:

$$\mathbf{Z} = \begin{pmatrix} 0 & j\omega L_S \\ j\frac{R_0 R_{DS}}{\omega L_S} + j\omega L_S & R_{DS} + j\omega L_S \end{pmatrix}. \quad (7.65)$$

Thus:

$$G_{av} = \frac{|Z_{21}|^2}{\text{Re}(Z_{22}R_0 - Z_{21}Z_{12})} =$$

$$= \frac{\left(\frac{R_0 R_{DS}}{\omega L_S} + \omega L_S\right)^2}{\text{Re}\left[(R_{DS} + j\omega L_S) R_0 - j\omega L_S \left(j\frac{R_0 R_{DS}}{\omega L_S} + j\omega L_S\right)\right]}$$

$$= \frac{\left(\frac{R_0 R_{DS}}{\omega L_S} + \omega L_S\right)^2}{2R_0 R_{DS} + \omega^2 L_S^2} \approx \frac{R_0 R_{DS}}{2\omega^2 L_S^2} = \frac{R_{DS} \, \omega_T^2}{2R_0 \, \omega^2}. \quad (7.66)$$

Let us now evaluate the noise figure obtained in such conditions. We first evaluate the correlation spectrum of the series (open circuit voltage) noise generators. Since the feedback inductive network is noiseless, and so is the input gate inductor, the open-circuit noise voltages of the device without feedback and without gate inductor will coincide with those of the device with feedback. We start from the impedance matrix of the device without feedback, obtained from (7.65) by putting $L_S = 0$:

$$\mathbf{Z}' = \mathbf{Y}_{in}^{-1} = \begin{pmatrix} -\frac{j}{\omega C_{GS}} & 0 \\ j\frac{g_m R_{DS}}{\omega C_{GS}} & R_{DS} \end{pmatrix},$$

and from the correlation matrix of the short-circuit current generators given, according to the *PRC* representation, by (see (7.51)–(7.53)):

$$\mathbf{S}_{i_G i_D} = \begin{pmatrix} 4k_B T \frac{\omega^2 C_{GS}^2}{g_m} R & jC \cdot 4k_B T \omega C_{GS} \sqrt{PR} \\ -jC \cdot 4k_B T \omega C_{GS} \sqrt{PR} & 4k_B T g_m P \end{pmatrix}.$$

We obtain:

$$S_{e_Ge_D} \equiv S_{v_{n1}v_{n2}} = \mathbf{Z}' \cdot S_{i_Gi_D} \cdot \mathbf{Z}'^{T*}$$

$$= \begin{pmatrix} 4k_BT\dfrac{R}{g_m} & -4k_BT\left(R - C\sqrt{PR}\right)R_{DS} \\ -4k_BT\left(R - C\sqrt{PR}\right)R_{DS} & 4k_BTg_m\left(R - 2C\sqrt{PR} + P\right)R_{DS}^2 \end{pmatrix}.$$

Taking now into account the full impedance matrix $\mathbf{Z}$ of the device with source inductive feedback and input gate inductance in (7.65) we derive the correlation impedance (see (7.27)–(7.31)):

$$Z_c = Z_{11} - Z_{21}\frac{(S_{e_Ge_D})_{12}}{(S_{e_Ge_D})_{22}}$$

$$= \frac{j}{\omega C_{GS}}\left(\frac{R - C\sqrt{PR}}{R - 2C\sqrt{PR} + P}\right) + \frac{j\omega L_S}{g_m R_{DS}}\left(\frac{R - C\sqrt{PR}}{R - 2C\sqrt{PR} + P}\right)$$

$$\approx \frac{j}{\omega C_{GS}}\left(\frac{R - C\sqrt{PR}}{R - 2C\sqrt{PR} + P}\right), \quad (7.67)$$

where we have assumed that the output resistance R_{DS} is large and/or the frequency is low:

$$\frac{g_m R_{DS}}{\omega C_{GS}} \gg \omega L_S.$$

This yields for the correlation impedance:

$$X_c = R_0 Q_L \left[\frac{R - C\sqrt{PR}}{R - 2C\sqrt{PR} + P}\right], \quad R_c = 0. \quad (7.68)$$

From (7.16), (7.17), (7.18) and (7.19) we obtain the noise conductances and resistances:

$$r_n = \frac{\omega}{\omega_T}\frac{Q_L R_0\left(1 - C^2\right)PR}{R - 2C\sqrt{PR} + P} \quad (7.69)$$

$$g_n \approx \frac{\omega}{\omega_T}\frac{R - 2C\sqrt{PR} + P}{Q_L R_0}, \quad (7.70)$$

where we have set:

$$Q_L = \frac{\omega(L_G + L_S)}{R_0} = \frac{1}{\omega C_{GS} R_0} = \frac{\omega_T}{\omega}\frac{1}{g_m R_0}, \quad (7.71)$$

and we have assumed $R_{DS} \to \infty$. From (7.7.1) with $R_G = R_0$ and $X_G = 0$ the noise figure is:

$$F = 1 + \frac{r_n}{R_0} + \frac{g_n}{R_0}\left[(R_c + R_0)^2 + X_c^2\right]$$

$$= 1 + \frac{\omega}{\omega_T}\left[Q_L R + \frac{1}{Q_L}\left(R - 2C\sqrt{PR} + P\right)\right]. \quad (7.72)$$

Since the generator is designed according to the input matching condition, (7.72) does not define the minimum noise figure. However, a further optimization can be performed with respect to the parameter Q_L. From (7.71) we see that, given the operating frequency, the generator resistance R_0 and the technological process cutoff frequency f_T, we have one more degree of freedom, at least in an integrated circuit, that is $g_M \propto W$. Minimizing vs. Q_L leads to an *optimum* quality factor and noise figure:

$$Q_{Lo} = \sqrt{\frac{R - 2C\sqrt{PR} + P}{R}} \tag{7.73}$$

$$F_o = 1 + 2\frac{\omega}{\omega_T}\sqrt{R\left(R - 2C\sqrt{PR} + P\right)}. \tag{7.74}$$

Notice that this does not lead to the classical minimum noise figure $F_{\min}$ but rather to an optimum noise figure F_o. With the approximations made, we have $F_o > F_{\min}$, although, since the inductive feedback may indeed lower the noise figure, in practical cases the optimum noise figure can also be smaller or comparable with $F_{\min}$.

We can obtain an estimate of the optimum quality factor and minimum noise figure using a consistent set of the *PRC* parameters from [13, Fig. 5], i.e., $P \approx 1$, $R \approx 0.25$, $C \approx 0.8$ (the values correspond to $V_{DS} \approx 1.5$ V). We obtain:

$$Q_{Lo} = \sqrt{\frac{0.25 - 2 \cdot 0.8\sqrt{1 \cdot 0.25} + 1}{0.25}} \approx 1.3416$$

$$F_o = 1 + 2\frac{f}{f_T}\sqrt{0.25 \cdot \left(0.25 - 2 \cdot 0.8\sqrt{1 \cdot 0.25} + 1\right)} \approx 1 + 0.67\frac{f}{f_T}.$$

Since typically $R < P$ the optimum quality factor tends to be large, but a large Q_{Lo} corresponds to a small g_m and therefore to a small periphery, small DC consumption device. Under this respect the quality factor optimization is also often referred to as a *power optimization* [28] of the low-noise amplifier.

We can compare the optimum noise figure thus obtained with the (approximate) minimum noise figure of the device without inductive feedback. From (7.57), neglecting R_S and R_G we have:

$$R'_{Go} = \frac{\sqrt{PR(1-C^2)}}{\omega C_{GS}\left(R - 2C\sqrt{PR} + P\right)}, \quad X'_{Go} = \frac{1}{\omega C_{GS}}\left[\frac{P - C\sqrt{PR}}{R - 2C\sqrt{PR} + P}\right], \tag{7.75}$$

while the minimum noise figure is, from (7.57):

$$F_{\min} = 1 + 2\frac{\omega}{\omega_T}\sqrt{(1 - C^2)PR}. \tag{7.76}$$

Comparing (7.76) with (7.74), since:

$$R\left(R - 2C\sqrt{PR} + P\right) - \left(1 - C^2\right)PR = \left(R - C\sqrt{PR}\right)^2 > 0,$$

we find that the optimum noise figure of the series inductive feedback amplifier is larger, according to the present approximation, than the theoretical minimum noise figure.

However the optimum impedance is now, with the parameters $P \approx 1$, $R \approx 0.25$, $C \approx 0.8$ and with the same device periphery (such as $Q_L \omega C_{GS} R_0 = 1$ with $Q_L \approx 1.3416$):

$$R'_{Go} = \frac{0.6667}{\omega C_{GS}} = 1.3416 \cdot R_0 \cdot 0.6667 = 0.8944 R_0$$

$$X'_{Go} = \frac{1.3333}{\omega C_{GS}}.$$

The optimum generator impedance is close to R_0 but an inductive reactance has to put in series to it, almost compensating for the input capacitance of the device. In a real device with a finite gate and source resistance, this means that the generator is connected to a very low resistance, almost a short circuit; it is therefore severely mismatched. On the other hand, the optimum noise figure is very close to the minimum one. In fact, for an ideal, long-gate device we would have:

$$F_{\min} = 1 + 2 \frac{f}{f_T} \sqrt{(1 - 0.8^2) \cdot 1 \cdot 0.25} = 1 + 0.6 \frac{f}{f_T} < 1 + 0.671 \cdot \frac{f}{f_T} = F_o.$$

We see therefore that the device with inductive source feedback is close to its own noise optimum conditions.

In conclusion, for an integrated device the optimum design procedure (leading to an initial approximation to be refined with the help of CAD techniques) is:

1. From R_0 (e.g., 50 Ω) we evaluate, given the process ω_T, the matching feedback inductor L_S. This does not depend on the gate periphery.
2. At the design frequency, we derive the optimum Q_L. In practice we can simulate the noise figure for different values of the device periphery W through CAD tools, adjusting each time L_G so as to achieve input matching. From the optimum noise figure achieved as a function of W we derive W_o and Q_{Lo}; the other parameters like C_{GS} follow.

A variation in the optimum Q_L design is discussed in [29]. The idea is to exploit for the design a device having a Q_L larger than the optimum one; the quality factor is then adjusted to the optimum value through an external capacitor C'_{GS} connected between source and drain. Of course this strategy can be successfully applied only if the device parasitics are not important (the extra capacitor should be connected, in theory, between the intrinsic source and the intrinsic drain).

Example 7.10 Show that the minimum noise figure does not change if a series inductor is connected at the input or output of a tripole, while it changes if the inductor is connected in series feedback to the common pole of the tripole.

Solution
As a first remark, note that any element in series to the input, output or common pole of a tripole does not change the correlation matrix of the open circuit noise generators. This is quite obvious, considering that in open-circuit conditions the voltage drop on the

added series elements is zero. Thanks to this, the power spectrum of the uncorrelated input voltage generator $S_{e_{nc}e_{nc}}$ and therefore r_n do not change either.

Suppose now to add a reactive series element, jX_{in} and jX_{out}, at the input and output, respectively, of the tripole. We have:

$$Z_{11} \to Z'_{11} = Z_{11} + jX_{\text{in}}$$
$$Z_{22} \to Z'_{22} = Z_{22} + jX_{\text{out}},$$

while the elements Z_{12} and Z_{21} are unchanged. Therefore also S_{ii} (depending on Z_{21}) and g_n do not change. Concerning the correlation resistance, we have:

$$Z_c = Z_{11} - Z_{21}\frac{S_{e_{n1}e_{n2}}}{S_{e_{n2}e_{n2}}} \to Z'_c = Z_c + jX_{\text{in}}$$
$$R'_c = R_c$$

and also the correlation resistance is invariant. In conclusion, series input or output reactive elements have no influence on the minimum noise figure. Using the $F_{\min}$ expression in terms of the parallel parameters, it is straightforward to show that also reactive elements in parallel to the device input or output do not change the minimum noise figure.

On the other hand, a series feedback reactive element jX_{sf} alters all the elements of the impedance matrix, i.e.:

$$Z_{ij} \to Z'_{ij} = Z_{ij} + jX_{\text{sf}}, \ i,j = 1,2.$$

Again, the variation of Z_{11} and Z_{22} has no effect. However, we have:

$$S_{ii} = \frac{S_{e_{n2}e_{n2}}}{|Z_{21}|^2} \to S'_{ii} = \frac{S_{e_{n2}e_{n2}}}{|Z_{21} + jX_{\text{sf}}|^2}$$

leading, for inductive feedback, to a decrease of $g_n \to g'_n$. Moreover:

$$Z_c = Z_{11} - Z_{21}\frac{S_{e_{n1}e_{n2}}}{S_{e_{n2}e_{n2}}} \to Z'_c = Z_c + jX_{\text{sf}}\left(1 + \frac{S_{e_{n1}e_{n2}}}{S_{e_{n2}e_{n2}}}\right)$$
$$R'_c = R_c + \operatorname{Re}\left(\frac{jX_{\text{sf}}S_{e_{n1}e_{n2}}}{S_{e_{n2}e_{n2}}}\right) = R_c - \frac{X_{\text{sf}}\operatorname{Im}\left(S_{e_{n1}e_{n2}}\right)}{S_{e_{n2}e_{n2}}}.$$

Taking into account the expression of the minimum noise figure, reported here for convenience, we have:

$$F'_{\min} = 1 + 2g'_n R'_c + 2\sqrt{g'_n r_n + \left(g'_n R'_c\right)^2}.$$

The resulting minimum noise figure can be smaller or larger than the original one according to the sign of $\operatorname{Im}\left(S_{e_{n1}e_{n2}}\right)$; if this term is positive both R_c and g_n decrease, leading to a decrease of $F_{\min}$. However, if this term is negative, R_c will increase and g_n decrease, and the minimum noise figure could decrease, increase or remain constant according to the values involved. The same remarks hold, starting from the parallel model, for the effect of a reactive parallel feedback element.[11]

---

[11] As discussed in [30] any lossless embedding does not vary the *minimum noise measure* of the device; this implies that if the minimum noise figure improves, the associated gain in fact decreases.

Example 7.11 Consider a FET with cutoff frequency $f_T = 20$ GHz and *PRC* noise parameters $P = 2$, $R = 0.25$, $C = 0.7$. The p.u.l. transconductance is 200 mS/mm and $R_{DS} = 500$ Ω. Design an inductive feedback LNA at 10 GHz on a 50 Ω generator. Compare the noise figure and available power gain of the amplifier realized with the same device, according to the conventional low-noise design.

Solution
We have:

$$L_S = \frac{C_{GS}R_0}{g_m} = \frac{R_0}{\omega_T} = \frac{50}{2\pi \cdot 20 \times 10^9} = 0.398 \quad \text{nH}.$$

Thus, from (7.73) and (7.74) we immediately obtain for the optimum quality factor and noise figure $Q_{Lo} = 2.245$ and $F_o = 1.5613$, respectively. Then:

$$Q_{Lo} = \frac{\omega(L_G + L_S)}{R_0} = \frac{1}{\omega C_{GS}R_0},$$

from which:

$$C_{GS} = \frac{1}{\omega R_0 Q_{Lo}} = \frac{1}{2\pi \cdot 10 \times 10^9 \cdot 50 \cdot 2.245} = 0.1418 \quad \text{pF}$$

$$L_G + L_S = \frac{1}{\omega^2 C_{GS}} = \frac{1}{(2\pi \cdot 10 \times 10^9)^2 \cdot 0.1418 \times 10^{-12}} = 1.7863 \quad \text{nH}$$

$$L_G = 1.7863 - 0.398 = 1.3883 \quad \text{nH}.$$

For the transconductance corresponding to the optimum noise figure we obtain the device periphery:

$$g_m = \omega_T C_{GS} = 2\pi \cdot 20 \times 10^9 \cdot 0.1418 \times 10^{-12} = 17.82 \quad \text{mS},$$
$$W = 17.82/200 \approx 90 \quad \mu\text{m}.$$

In such conditions:

$$Z_{11} = -\frac{j}{\omega C_{GS}} + j\omega(L_S + L_G) = 0$$

$$Z_{12} = j\omega L_S = j25 \; \Omega$$

$$Z_{21} = j\frac{g_m}{\omega C_{GS}G_{DS}} + j\omega L_S = j1022.4 \; \Omega$$

$$Z_{22} = R_{DS} + j\omega L_S = 500 + j25 \; \Omega,$$

and the available power gain is, from (7.66)

$$G_{\text{av}} = \frac{|Z_{21}|^2}{\text{Re}(Z_{22}R_0 - Z_{21}Z_{12})} = 20.67.$$

We can now evaluate the minimum noise figure according to the classical minimum noise design. We have from (7.76) $F_{\min} = 1.505$, only marginally better than the optimum one. Taking into account that the optimum source impedance is, from (7.75):

$$Z_{Go} = R_{Go} + jX_{Go} = 44.981 \; \Omega + 134.06j \; \Omega.$$

The available gain of the conventional stage can be expressed as:

$$G'_{av} = \frac{|g_m V^*|^2 R_{DS}}{4} \times \left(\frac{|E_G|^2}{4R_{Go}}\right)^{-1},$$

where V^* is the voltage on the input capacitor of the device:

$$V^* = E_G \frac{\frac{1}{j\omega C_{GS}}}{R_{Go} + jX_{Go} + \frac{1}{j\omega C_{GS}}}.$$

We thus have:

$$G'_{av} = \frac{g_m^2 R_{DS} R_{Go}}{(\omega C_{GS})^2} \frac{1}{R_{Go}^2 + \left(X_{Go} - \frac{1}{\omega C_{GS}}\right)^2} = \frac{\omega_T^2}{\omega^2} \frac{R_{DS} R_{Go}}{R_{Go}^2 + \left(X_{Go} - \frac{1}{\omega C_{GS}}\right)^2}$$

$$= \frac{20^2}{10^2} \cdot \frac{500 \cdot 44.981}{44.981^2 + \left(134.06 - \frac{1}{2\pi \cdot 10 \times 10^9 \cdot 0.1418 \times 10^{-12}}\right)^2} \approx 36.$$

In conclusion, inductor degeneration leads to a strong improvement of the device input matching without overly compromising the device noise figure vs. the minimum one. The available power gain of the series inductive feedback solution is decreased but remains acceptable if the cutoff frequency is substantially larger than the operating frequency. Notice that the example is somewhat ideal, since the gate and source resistances are neglected, and so are the feedback capacitance and intrinsic resistance.

7.11 Questions and Problems

7.11.1 Questions

1. Explain why noise causes power transfer in a circuit even if it is a zero-average random signal.
2. Sketch the series and parallel equivalent circuits for a noisy one-port and for a two-port.
3. Explain the physical cause of electrical noise from microscopic fluctuations.
4. Sketch the *PRC* noise model for a FET.
5. Define the noise figure of a two-port.
6. What is the behavior of the noise figure with respect to the minimum when the generator impedance is varied with respect to the optimum value?
7. In a minimum noise amplifier, what is the associated gain G_{ass}? Can the associated gain be larger than the MAG?
8. Compare the noise behavior of FETs and bipolar transistors in terms of (a) output noise source spectra, (b) input noise source spectra, (c) minimum noise figure vs. frequency.

Low-Noise Amplifier Design

9. Discuss minimum noise vs. maximum gain amplifier design.
10. In a minimum noise amplifier, what is the associated gain G_{ass}? Can the associated gain be larger than the MAG?
11. What are the advantages and disadvantages of the common base (gate) low-noise amplifier stage? Is that well suited for a high-performance low-noise amplifier?
12. Explain the purpose of the inductive series feedback LNA design vs. the conventional LNA approach through input noise matching.
13. Derive the parameters of the *PRC* model from those of the two-temperature model T_G and T_D.

7.11.2 Problems

1. A noisy tripole has the following admittance matrix:

$$\mathbf{Y} = \begin{pmatrix} j\omega C & -j\omega C \\ -j\omega C & \dfrac{1}{R} + j\omega C \end{pmatrix}.$$

Derive an equivalent circuit of the tripole and evaluate the short-circuit noise current correlation matrix.

2. A voltage noise source has a power spectrum of 1 (nV)^2/Hz[12]. Assuming a bandwidth of 500 MHz, evaluate the mean square value of the noise voltage and the noise available power on 50 Ω. Evaluate the noise available power spectral density of the generator.

3. A resistor with $R = 1$ kΩ operates with a bandwidth of 5 GHz. Evaluate the power spectral density at 300 K. Evaluate the spectral density of the resistor noise voltage and the r.m.s. noise voltage value over the specified bandwidth.

4. In the circuit in Fig. 7.32 (a) assume $Z_1 = 50+\text{j}50$ Ω, $Z_L = 50-\text{j}50$ Ω; the two noise generators are the thermal noise (Nyquist law) generators of the two impedances, respectively (i.e., i_{n1} is associated with Z_1, e_{n2} to Z_L). Assuming 1 GHz bandwidth, evaluate at 300 K the total power on the load.

5. In the circuit in Fig. 7.32 (b) compute the minimum noise figure and optimum generator impedance of the two-port in the gray box assuming $Z_1 = 10$ Ω,

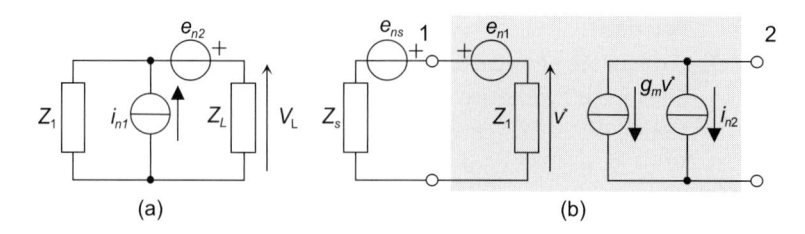

Figure 7.32 Circuits from Problem 4 (a) and Problem 5 (b).

---

[12] Spectral units like V^2/Hz, A^2/Hz etc., when the V or A unit is associated with a multiplier, like in μV^2/Hz, must be interpreted as $(\mu\text{V})^2$/Hz, that is 10^{-12}V/Hz. This meaning has been explicited in the text whenever possible.

$g_m = 500$ mS. The two (uncorrelated) noise generators e_{n1} and i_{n2} are white, with spectral density equal to 100 (pV)^2/Hz and 100 (pA)^2/Hz, respectively. The system bandwidth is 100 MHz. (Hint: the noise figure is a ratio of available noise powers at the output port, that is independent from the output conductance of the two-port and therefore reduces to a ratio of short-circuit noise current spectra at port 2.)

6. A noisy two-port has the optimum source impedance $Z_{Go} = 25 + j32$ Ω, minimum noise figure NF$_{min}$ = 2 dB, and series noise conductance $g_n = 50$ mS. Supposing that the source reactance is the optimum one, estimate the variation of the noise figure when the source resistance varies between 5 and 50 Ω.
7. Two amplifiers are cascaded (50 Ω design) with $G_{av,1} = 10$ dB, $G_{av,2} = 20$ dB, NF$_1$ = 1 dB, NF$_2$ = 6 dB. Evaluate the total noise figure according to the Friis formula.
8. A resistive attenuator designed on 50 Ω has 3 dB loss. What is the noise figure?
9. A common-gate LNA has to be designed on a 50 Ω generator with a device having a specific transconductance of 400 mS/mm. Evaluate the device periphery needed and the low-frequency noise figure, assuming $P = 0.9$.
10. An inductive series feedback amplifier is designed on 50 Ω at 10 GHz. Assuming $C_{GS} = 0.2$ pF and $g_m = 200$ mS, evaluate L_S and L_G. Evaluate the noise figure, assuming $P = 0.7$ and neglecting the gate noise source (and therefore the correlation).

References

[1] A. Papoulis and S. U. Pillai, *Probability, random variables, and stochastic processes*. McGraw-Hill, 1985.
[2] A. Van der Ziel, *Noise in solid state devices and circuits*. Wiley-Interscience, 1986.
[3] S. Sze and K. K. Ng, *Physics of semiconductor devices*. Wiley Online Library, 2007.
[4] H. Nyquist, "Thermal agitation of electric charge in conductors," *Phys. Rev.*, vol. 32, pp. 110–113, Jul. 1928.
[5] P. Russer and S. Müller, "Noise analysis of linear microwave circuits," *International Journal of Numerical Modelling: Electronic Networks, Devices and Fields*, vol. 3, no. 4, pp. 287–315, 1990.
[6] F. Bonani and G. Ghione, *Noise in semiconductor devices: modeling and simulation*. Springer Science & Business Media, 2001, vol. 7.
[7] D. E. Meer, "Noise figures," *IEEE Transactions on Education*, vol. 32, no. 2, pp. 66–72, May 1989.
[8] J. Lange, "Noise characterization of linear twoports in terms of invariant parameters," *IEEE Journal of Solid-State Circuits*, vol. 2, no. 2, pp. 37–40, Jun. 1967.
[9] H. Rothe and W. Dahlke, "Theory of noisy fourpoles," *Proceedings of the IRE*, vol. 44, no. 6, pp. 811–818, Jun. 1956.
[10] G. Gonzalez, *Microwave transistor amplifiers: analysis and design*. New Jersey: Prentice Hall, 1997.
[11] H. Fukui, "Optimal noise figure of microwave GaAs MESFET's," *IEEE Transactions on Electron Devices*, vol. 26, no. 7, pp. 1032–1037, Jul. 1979.
[12] R. A. Pucel, H. A. Haus, and H. Statz, "Signal and noise properties of gallium arsenide microwave field-effect transistors," *Advances in Electronic & Electron Physics*, vol. 38, pp. 195–265, 1975.

[13] A. Cappy, "Noise modeling and measurement techniques," *IEEE Transactions on Microwave Theory and Techniques*, vol. 36, no. 1, pp. 1–10, Jan. 1988.

[14] M. W. Pospieszalski, "Modeling of noise parameters of MESFETs and MODFETs and their frequency and temperature dependence," *IEEE Transactions on Microwave Theory and Techniques*, vol. 37, no. 9, pp. 1340–1350, Sep. 1989.

[15] B. Hughes, "A temperature noise model for extrinsic FETs," *IEEE Transactions on Microwave Theory and Techniques*, vol. 40, no. 9, pp. 1821–1832, Sep. 1992.

[16] H. Fukui, "The noise performance of microwave transistors," *IEEE Transactions on Electron Devices*, vol. ED-13, no. 3, pp. 329–341, Mar. 1966.

[17] H. T. Friis, "Noise figures of radio receivers," *Proceedings of the IRE*, vol. 32, no. 7, pp. 419–422, Jul. 1944.

[18] M. L. Edwards, S. Cheng, and J. H. Sinsky, "A deterministic approach for designing conditionally stable amplifiers," *IEEE Transactions on Microwave Theory and Techniques*, vol. 43, no. 7, pp. 1567–1575, Jul. 1995.

[19] R. S. Engelbrecht and K. Kurokawa, "A wide-band low noise l-band balanced transistor amplifier," *Proceedings of the IEEE*, vol. 53, no. 3, pp. 237–247, Mar. 1965.

[20] Y. Xue, Y. Hao, Z. Haiying, Z. Xinnian, D. Zhiwei, L. Zhiqiang, and D. Zebao, "A monolithic 60 GHz balanced low noise amplifier," *Journal of Semiconductors*, vol. 36, no. 4, p. 045003, 2015.

[21] D. B. Estreich, "A monolithic wide-band GaAs IC amplifier," *IEEE Journal of Solid-State Circuits*, vol. 17, no. 6, pp. 1166–1173, Dec. 1982.

[22] L. Nevin and R. Wong, "L-band GaAs FET amplifier," in *Microwave Conference, 1978. 8th European*, Sep. 1978, pp. 140–145.

[23] H. A. Haus and R. B. Adler, *Circuit theory of linear noisy networks*. The MIT Press, 1959, no. 2.

[24] D. D. Henkes, "LNA design uses series feedback to achieve simultaneous low input VSWR and low noise," *Applied Microwave and Wireless*, vol. 10, pp. 26–33, Oct. 1998.

[25] A. Tasic, W. A. Serdijn, and J. R. Long, "Concept of transformer-feedback degeneration of low-noise amplifiers," in *Circuits and Systems, 2003. ISCAS '03. Proceedings of the 2003 International Symposium on*, vol. 1, May 2003, pp. I–421–I–424 vol.1.

[26] D. J. Cassan and J. R. Long, "A 1-V transformer-feedback low-noise amplifier for 5-GHz wireless LAN in 0.18-μm CMOS," *IEEE Journal of Solid-State Circuits*, vol. 38, no. 3, pp. 427–435, Mar. 2003.

[27] G. Girlando and G. Palmisano, "Noise figure and impedance matching in RF cascode amplifiers," *IEEE Transactions on Circuits and Systems II: Analog and Digital Signal Processing*, vol. 46, no. 11, pp. 1388–1396, Nov. 1999.

[28] T.-K. Nguyen, C.-H. Kim, G.-J. Ihm, M.-S. Yang, and S.-G. Lee, "CMOS low-noise amplifier design optimization techniques," *IEEE Transactions on Microwave Theory and Techniques*, vol. 52, no. 5, pp. 1433–1442, May 2004.

[29] P. Andreani and H. Sjoland, "Noise optimization of an inductively degenerated cmos low noise amplifier," *IEEE Transactions on Circuits and Systems II: Analog and Digital Signal Processing*, vol. 48, no. 9, pp. 835–841, Sep. 2001.

[30] H. A. Haus and R. B. Adler, "Optimum noise performance of linear amplifiers," *Proceedings of the IRE*, vol. 46, no. 8, pp. 1517–1533, Aug. 1958.

8 Power Amplifiers

8.1 Introduction

The purpose of RF and microwave power amplifiers [1, 2, 3] is to obtain, rather than the maximum power gain, the maximum output power compatible with a given device, with acceptable efficiency and linearity. The output power of active devices is, in fact, limited by the maximum voltage and current swing, in turn related to the drain or collector breakdown voltage and to the maximum current, respectively. Moreover, while in small-signal operation the device is approximately linear, power amplifiers, while being ultimately limited by output power saturation, exhibit nonlinear effects (i.e., distortion) that become increasingly important for large input power. Nonlinearities yield signal distortion due to the generation of *harmonics* and *intermodulation products* (IMPs); such a nonlinear distortion has to be kept under control to satisfy system requirements.

Power amplifiers are traditionally divided in *classes*. Class A (Sec. 8.4) amplifiers are quasi-linear since, at least ideally, the device output voltage and current swings corresponding to a single-tone sinusoidal inputs are again single-tone (neglecting distortion). In class B (Sec. 8.6.1) and C (Sec. 8.6.2) amplifiers (to confine ourselves to traditional amplifier classes) the output device current swing corresponding to a sinusoidal input is a set of sine pulses; this harmonic-rich, wideband waveform is converted on the load into a narrowband one through filtering (the so-called *tuned load* approach).[1] The rationale behind introducing strongly nonlinear amplifiers is the increase of the *amplifier efficiency*, i.e., the ratio between the RF output power and the power from the DC supply (Sec. 8.2.4). Typically, there is a trade-off between efficiency and distortion; in class A amplifiers, low distortion can be achieved by reducing the input and output powers, at the expense of the amplifier efficiency. In traditional amplifier classes, high efficiency can be only achieved at the expense of a large gain penalty.

More advanced amplifier classes, like the *harmonic loading* amplifiers (class F, Sec. 8.7.1) or the *switching amplifiers* (e.g., class E, Sec. 8.7.2) are able to achieve a theoretical maximum 100 percent efficiency with acceptable gain. Other strategies (like the Doherty amplifier, Sec. 8.7.3) aim at providing an acceptable efficiency also when the input signal power is not constant, but covers wide dynamics, as needed in many real-life last-generation communication systems. A final issue concerns the efficiency–linearity trade-off. Highly efficient amplifier classes typically exhibit poor distortion

---

[1] A discussion on filter design criteria can be found in Example 8.4.

properties. A different approach to comply with the very stringent distortion requirements of many communication systems is to reduce distortion by properly modifying the input waveform (the so-called pre-distortion) or by compensating for the output distortion by cancellation techniques (the so-called feedforward amplifier), see Sec. 8.9.

8.2 Characteristics of Power Amplifiers

8.2.1 Power, Gain, Distortion

According to the input power level and to the amplifier structure, power amplifiers (PAs) can be considered as quasi-linear systems (class A from small signal to compression), or as nonlinear systems; examples of the latter condition are the class A amplifier under power saturation, or all other amplifier classes for most operating conditions. The PA input typically is a narrowband signal centered around f_0, with total input power P_{in}. The DC bias supplies the amplifier with a power P_{DC}, and both the DC power and the input RF power contribute to yield, on the load, the output power P_{out}. From an energy standpoint, the power amplifier can be therefore seen as a DC–RF converter controlled by the input RF signal. Contrarily to linear amplifiers, where the output signal is, in the ideal case, an amplified and delayed replica of the input signal, or, at worst, in the presence of linear distortion, a filtered version of it, PAs exhibit nonlinear effects, such as the generation of harmonics and intermodulation products, and power saturation. The output signal spectrum is therefore generally different from the input signal spectrum, and the amplifier behavior significantly depends on the input power level, but also on the specific spectrum of the input signal.

The characterization of power amplifiers (or of active devices operating under large-signal, LS, conditions) can be carried out at different levels of complexity of the input signal. We will introduce here the following amplifier-level or device-level classical large-signal test procedures:

- The *single-tone test*, where the input signal is a sinusoid at f_0 with power P_{in}. The input is an unmodulated signal having constant average power. In the test, the output power P_{out} is measured at the carrier frequency f_0 and at the harmonics nf_0, sweeping the input power from small-signal conditions to power saturation. The $P_{out}(P_{in})$ curve, typically in log units, is called the single-tone $P_{in} - P_{out}$ characteristic. Notice that there is a significant difference between performing the single-tone test on a device (transistor) or on an amplifier. In fact, while in device testing the load can be resistive at the fundamental frequency and at all harmonics, most power amplifiers exploit a tuned load that is resistive at the fundamental and reactive (typically a short or an open circuit) at the harmonics; this is to avoid power dissipation in the load at higher harmonics, thus increasing the amplifier efficiency.[2] If the measurement system supports this feature, also the input-output phase delay as a function of the input power can be obtained, giving a measure of the so-called AM–PM characteristic of the amplifier.

---

[2] The measurement of harmonics can be made difficult anyway by the limited instrumentation bandwidth.

8.2 Characteristics of Power Amplifiers

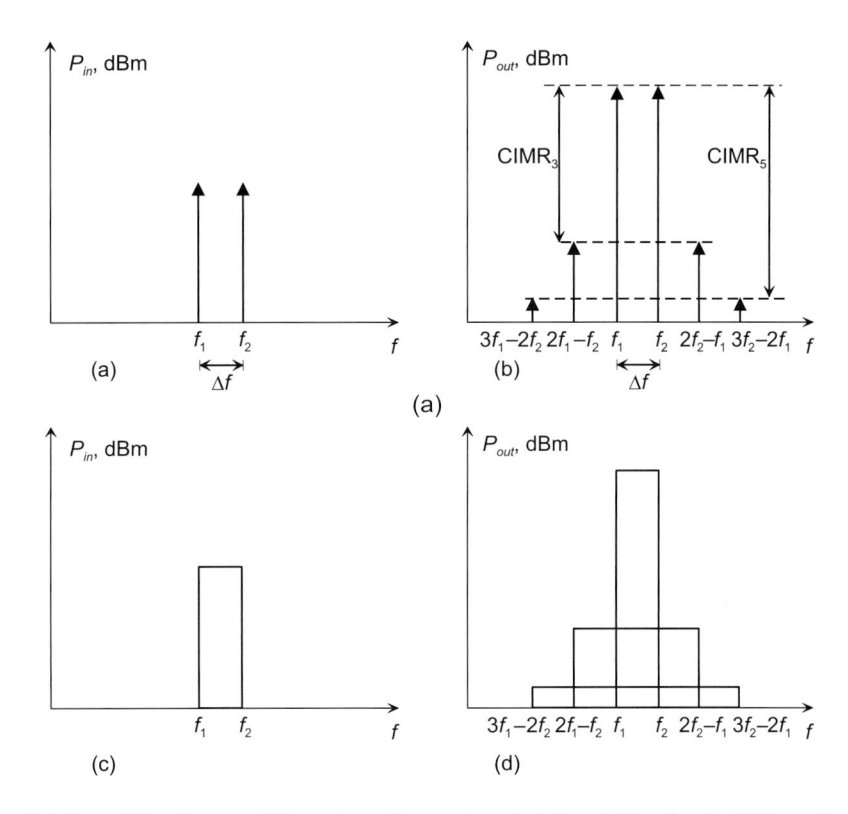

Figure 8.1 Input (a) and output (b) spectrum for two tone test; input (a) and output (b) spectrum for an input modulated signal text. Only the spectrum region close to the input channel is shown, including the third- and fifth-order intermodulation products, but neglecting out-of-band harmonics and IMPs.

- The *two-tone test*, where the input signal is a set of two closely spaced tones (f_1 and f_2) around f_0, with spacing $\Delta f \ll f_0$ and usually equal power $P'_{in} = P_{in}/2$, see Fig. 8.1 (a), where P_{in} is the total power of the two tones. The input signal is now sinusoidally modulated with frequency Δf; contrarily to the single-tone test, the peak signal amplitude varies with time. In the test, the output power $P_{out} = 2P'_{out}$ (including the two tones at the fundamentals, each with approximately equal power P'_{out}) is measured by sweeping the input power from small-signal conditions to power saturation. Measurements may include the fundamentals f_1 and f_2, their harmonics, and the intermodulation products $|mf_1 \pm nf_2|$. While the same remarks hold concerning the generation of harmonics on the load in a device or amplifier context as in the single-tone test case, the measured signal in the two-tone test is often limited to the output fundamentals (f_1 and f_2) and to the third-order intermodulation products $2f_1 - f_2$ and $2f_2 - f_1$ falling close to the input frequencies. According to the instrumentation available, the measurement can also include further odd-order in-band IMPs. An example of output power spectrum of the signal around $f_{1,2}$ is shown in Fig. 8.1 (b). The $P_{out}(P_{in})$ curve, typically in log units, is called the two-tone $P_{in} - P_{out}$ characteristic;

above compression it differs from the single-tone curve because a modulated signal exhibits, with the same average power, a larger peak power than a constant envelope signal, thus causing the earlier compression of the output power. On the same $P_{in} - P_{out}$ characteristic also the lower and upper IMP$_3$ (third-order intermodulation products) are reported as a function of P_{in}. The two-tone test can be extended by using n input tones, with the purpose of approximating a real modulated signal; also in this case, information on the AM–PM characteristic can be derived from phase measurements.

- The *modulated signal test*, where the input is a modulated signal according to some communication standard. A qualitative example is shown in see Fig. 8.1 (c). The signal is randomly generated by a waveform generator in order to accurately reproduce the modulation characteristics of the standard, like the signal constellation and the instantaneous power statistics. For a given signal, we define the Peak to Average Power Ratio (PAPR) as the ratio between the maximum signal instantaneous power and the average signal power. The PAPR is also directly related to the Crest Factor C_F defined as the ratio between the maximum and the average signal amplitude; we have PAPR $= C_F^2$ and the value in logarithmic units is the same. The peak-to-average power ratio (PAPR) is 2 (3 dB) in the single tone test and becomes 4 (6 dB) in the two-tone test, assuming equal tone amplitude, see Example 8.1. For realistic modulation schemes the PAPR depends on the type and order of the modulation scheme; the QAM case is discussed in detail in Example 8.2.

Typically, the spectrum of the output signal is measured around the carrier, see Fig. 8.1 (d), allowing to appreciate the effect of the amplification but also of the intermodulation distortion. A parameter describing the intermodulation of modulated signals is the Adjacent Channel Power Ratio (ACPR), expressing the total power introduced at the output by the effect of the intermodulation in the frequency bands close to the input channel. For a digital signal modulation, the output constellation resulting from a statistically significant input bit sequence is measured, and the total error (called the Error Vector Magnitude, EVM) with respect to the ideal output can be derived. The modulated signal test is always carried out at an amplifier level rather than at a device level.

We will now provide a quantitative definition of the parameters relevant to the above amplifier or device tests. Consider first the **single-tone test**. As in small-signal amplifiers, we define the *operational gain*:

$$G_{\text{op}} = \frac{P_{out}(f_0)}{P_{in}(f_0)}$$

and the *transducer gain*:

$$G_t = \frac{P_{out}(f_0)}{P_{av,in}(f_0)}.$$

The available power gain is not used in power amplifiers, also because the output matching condition that is optimal for power amplifiers is different from the optimum small-signal value. The definition can be extended to the two-tone test. For low values of P_{in}, the gains coincide with their small-signal values, i.e., $G_{op} = G_{op}^{ss}$ and $G_t = G_t^{ss}$. For large values of the input power, the amplifier gain is decreased (compressed) and ultimately the output power saturates to $P_{sat,out}$. According to the instrumentation available, all PA tests can be performed by using as the independent variable the input power P_{in} or the input available power $P_{av,in}$; the latter solution is easier to implement, since it does not require a full network analyzer. To keep the treatment light, in the present section we will always refer to the input power P_{in} and to the operational power and conversion gain; the extension to the transducer case is straightforward.

Harmonics in the single or two-tone test can be shown (see Sec. 8.5.1) to approximately behave, for low input power, according to the law:

$$P_{out}(nf_0) = K_n \left[P_{in}(f_0)\right]^n,$$

where the coefficient K_n is the (operational) *conversion gain*. The power amplifier load typically is tuned at the fundamental, and the harmonic power cannot be observed on the load.[3] An example of the $P_{in} - P_{out}$ single-tone characteristics (including fundamental and harmonics) is shown in Fig. 8.2. The graph is in log scale, with powers defined in dBm (i.e., dB vs. 1 mW):

$$P|_{dBm} = 10 \log_{10}\left(\frac{P}{1 \text{ mW}}\right).$$

In log scale we have:

$$P_{out}(f_0)|_{dBm} = G_{op}|_{dB} + P_{in}(f_0)|_{dBm}$$
$$P_{out}(nf_0)|_{dBm} = -30(n-1) + 10\log_{10} K_n + n P_{in}(f_0)|_{dBm}.$$

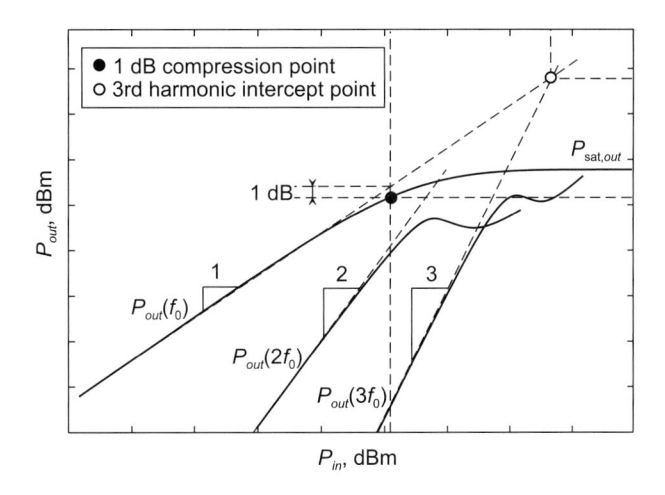

Figure 8.2 $P_{in} - P_{out}$ curves of a power amplifier or transistor under single-tone test. The load is resistive rather then tuned, thus allowing also the harmonics to be measured. The 1 dB compression point and the third harmonic intercept are shown.

---
[3] In frequency multipliers, on the other hand, the load is tuned on the n-th harmonic and the device is designed so as to maximize the conversion gain.

In log scale therefore the slope of the small-signal fundamental is 1 while the slope of the harmonics is n.

Important points of the $P_{in} - P_{out}$ single-tone characteristics in Fig. 8.2 are:

- The *1 dB compression point*: the input power for which the output power is:

$$P_{out}(f_0)|_{dBm} = G_{op}^{ss}\big|_{dB} + P_{in}(f_0)|_{dBm} - 1 \text{ dB}$$

and similarly for the available input power case. Notice that the input and output coordinates of the point are related through the small-signal gain.

- The *n-th harmonic intercept* is the input power for which:

$$G_{op}^{ss} P_{in}(f_0) = K_n \left[P_{in}(f_0)\right]^n,$$

(and similarly for the transducer gain case) i.e.:

$$P_{in}(f_0) = \sqrt[n-1]{\frac{G_{op}^{ss}}{K_n}}.$$

The intercept point can be used as a linearity figure of merit, and is the result of an extrapolation from the small-signal or low-power behavior, see Fig. 8.2. Also in this case, the intercept point can be referred to the input or to the output power; the two values are related by the small-signal gain. The third harmonic intercept has particular relevance, since it depends on the third-order device nonlinearity, in turn the main responsible for third-order intermodulation.

The $P_{in} - P_{out}$ single-tone characteristic may also report the gain as a function of the input power.

In the **two-tone test** the input signal is a set of two single-tone signals at frequencies f_1 and f_2, centered around $f_0 = (f_1 + f_2)/2$ (see Fig. 8.1 (a)), typically having the same amplitude and a low (relative) spacing $\Delta f/f_0$, where $\Delta f = |f_2 - f_1|$. The $P_{in} - P_{out}$ two-tone characteristic, shown in Fig. 8.3, usually includes, apart from the power at fundamentals, also the upper and lower third-order intermodulation products (IMP$_3$) at frequencies:

$$f_a = 2f_1 - f_2$$
$$f_b = 2f_2 - f_1.$$

Those products are the main cause for distortion, since they fall close to the input band (or even, for an input signal with a continuous spectrum, within the input band). Other odd-order intermodulation products (IMP$_5$, IMP$_7$...) have a similar behavior and lead to an output spectrum made of an equispaced comb with spacing Δf, see Fig. 8.1 (b). It can be shown (Sec. 8.5.1) that for low input power the IMP$_n$ power is approximated by:

$$P_{out}(\text{IMP}_n) = H_n \left[P_{in}\right]^n,$$

where the input power may refer to the two-tone total power, assumed to be the same, or to the power of a single-tone, and H_n is a proper (operational) conversion gain. In logarithmic scale we have, for $n = 3, 5$:

8.2 Characteristics of Power Amplifiers

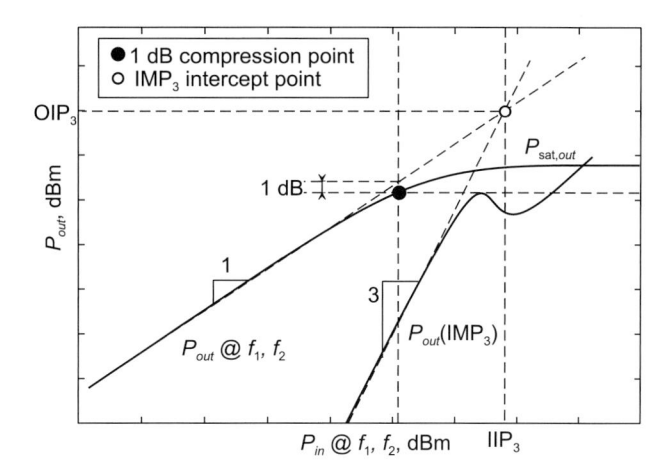

Figure 8.3 $P_{in} - P_{out}$ curves of a power amplifier or transistor under two-tone test. The 1 dB compression point and the third-order intermodulation intercept are shown.

$$P_{out}(\text{IMP}_3)|_{\text{dBm}} = -60 + 10 \log_{10} H_3 + 3 P_{in}|_{\text{dBm}}$$
$$P_{out}(\text{IMP}_5)|_{\text{dBm}} = -120 + 10 \log_{10} H_5 + 5 P_{in}|_{\text{dBm}},$$

i.e., the slope of the IMP_n output power is n. From the power at fundamentals and the IMP_n (with n odd), the n-th-order carrier-to-intermodulation ratio CIMR_n is defined as:

$$\text{CIMR}_n = \frac{P_{out}}{P_{out}(\text{IMP}_n)}$$
$$\text{CIMR}_n|_{\text{dB}} = P_{out}|_{\text{dBm}} - P_{out}(\text{IMP}_n)|_{\text{dBm}},$$

see Fig. 8.1 (b) where the CIMR_3 and CIMR_5 are shown.

As a figure of merit of intermodulation distortion, the intercept point is used between the fundamental low-power $P_{in} - P_{out}$ straight line and the IMP_3 low-power straight line with slope 3. Let us define as IIP_3 (Input Intercept Point) and OIP_3 (Output Intercept Point) the x and y coordinates of the intercept point. In general, the third-order intermodulation power can be expressed as:

$$P_{out}(\text{IMP}_3) = \frac{G_{op}^{ss} P_{in}^3}{\text{IIP}_3^2}.$$

The above equation is correct since it has the desired input power behavior and satisfies the intercept condition for $P_{in} = \text{IIP}_3$:

$$\text{OIP}_3 = G_{op}^{ss} \text{IIP}_3.$$

Solving we obtain:

$$\text{IIP}_3 = \sqrt{\frac{G_{op}^{ss} P_{in}^3}{P_{out}(\text{IMP}_3)}} \tag{8.1}$$

$$\text{OIP}_3 = \sqrt{\frac{\left(G_{op}^{ss} P_{in}\right)^3}{P_{out}(\text{IMP}_3)}} = \sqrt{\frac{P_{out}^3}{P_{out}(\text{IMP}_3)}},$$

or, in log scale:

$$\text{OIP}_3|_{\text{dBm}} = \frac{3}{2} P_{out}|_{\text{dBm}} - \frac{1}{2} P_{out}(\text{IMP}_3)|_{\text{dBm}}$$

$$\text{IIP}_3|_{\text{dBm}} = \frac{3}{2} P_{out}|_{\text{dBm}} - \frac{1}{2} P_{out}(\text{IMP}_3)|_{\text{dBm}} - G^{ss}|_{\text{dB}}.$$

The output intercept can be therefore measured by performing, for small input power, a measurement of the fundamental and intermodulation powers. A plot of the two-tone $P_{in} - P_{out}$ characteristics is reported in Fig. 8.3, where also the third-order intermodulation intercept is shown.

In the **modulated signal test** the amplifier input has a continuous spectrum, typically limited to a narrow band around f_0, see Fig. 8.1 (c). The output spectrum will also be continuous and the generation of odd-order intermodulation products will lead to the so-called spectral regrowth, i.e., the broadening of the input spectrum to the nearby channels, as shown in Fig. 8.1 (d). Notice that the IMP_3 level also affects the main channel and should be regarded as a kind of deterministic noise. The Adjacent Channel Power Ratio (ACPR) is defined as:

$$\text{ACPR}_k = \frac{\int_{\text{MC}} p_{out}(f) df}{\int_{C_k} p_{out}(f) df},$$

where MC is the main channel, C_k the k-th adjacent channel. An example of the shape of the output spectrum is shown in Fig. 8.4.

Finally, the Error Vector Magnitude (EVM), specific to a digital modulation format, can be evaluated as the signal power associated with the difference between each measured constellation point and the ideal value, normalized vs. the ideal value. In other words, suppose that $N \times N$ constellation points are defined with position in the complex (Q, I) plane (referring the in-phase and quadrature components of the baseband signal,

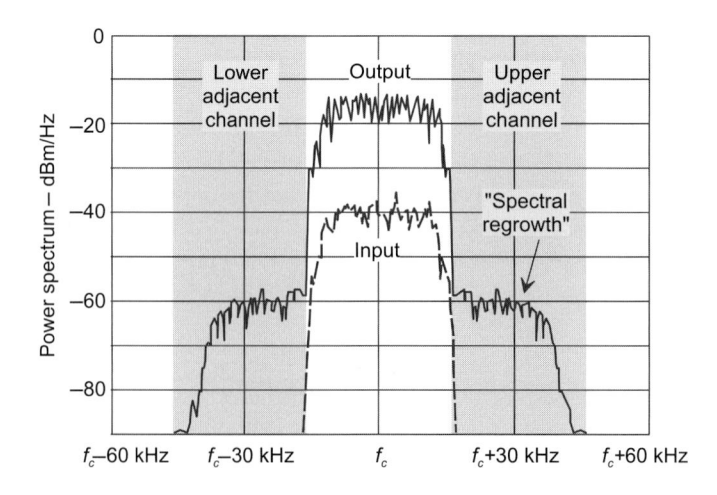

Figure 8.4 Input modulated signal and output power spectrum after amplification, affected by spectral regrowth.

see Sec. 1.2.1) given by $I_i + jQ_j$ and ideal values $\hat{I}_i + j\hat{Q}_j$; we have:

$$\text{EVM} = \sum_{i=1,j=1}^{i=N,j=N} \frac{\left(\hat{I}_i - I_i\right)^2 + \left(\hat{Q}_j - Q_j\right)^2}{\hat{I}_i^2 + \hat{Q}_j^2}.$$

Example 8.1 Evaluate the PAPR of a single-tone and of a two-tone signal, assuming that the two tones have the same amplitude.

Solution
In a sinusoidal signal of peak value x_p, the effective value is $x_p/\sqrt{2}$; since power is proportional to the square of the signal, the ratio between the peak power and the average power, i.e., the PAPR, is 2. For a two-tone signal with peak value y_p and equal tone amplitude, the effective value for each tone is $y_p/\sqrt{2}$ and the total average power of the two tones is $2y_p^2/2 = y_p^2$. However, for the two-tone signal, the peak value is $2y_p$, leading to a peak power $4y_p^2$. Thus, the peak to average power ratio is 4.

Example 8.2 Consider a QAM (Quadrature Amplitude Modulation) constellation of $N = M^2$ signals, M even, regularly deployed on a $X\,Y$ Cartesian grid. Any point of the constellation has amplitude $i - 1/2$ and $j - 1/2$ along the X or Y axis, respectively, where $-M/2 + 1 < i < M/2$ and $-M/2 + 1 < j < M/2$ are integer numbers. Evaluate the PAPR for such a signal.

Solution
First of all, notice that a scale factor common to all constellation points has no influence on the PAPR or the crest factor. The RMS signal value is defined as:

$$\text{RMS}^2 = \frac{1}{N} \cdot 4 \cdot \sum_{i=1}^{M/2} \sum_{j=1}^{M/2} \left[\left(i - \frac{1}{2}\right)^2 + \left(j - \frac{1}{2}\right)^2 \right]$$

$$= \frac{1}{N} \cdot 4 \cdot \left[\frac{M}{2} \sum_{i=1}^{M/2} \left(i - \frac{1}{2}\right)^2 + \frac{M}{2} \sum_{j=1}^{M/2} \left(j - \frac{1}{2}\right)^2 \right]$$

$$= \frac{1}{M^2} \cdot 4 \cdot 2 \cdot \frac{M}{2} \sum_{k=1}^{M/2} \left(k - \frac{1}{2}\right)^2 = \frac{1}{M^2} \cdot 4 \cdot 2 \cdot \frac{M}{2} \cdot \frac{M}{24} \cdot \left(M^2 - 1\right),$$

where the factor $1/N^2$ normalizes the power of the symbols, assumed here equiprobable, while the factor 4 accounts for the number of quadrants. We thus have:

$$\text{RMS}^2 = \frac{M^2 - 1}{6} = \frac{N - 1}{6}.$$

The peak value is:

$$\text{PEAK} = \sqrt{2} \cdot \left(\frac{M}{2} - \frac{1}{2}\right) = \frac{M - 1}{\sqrt{2}}.$$

The crest factor C_F in term of RMS value is therefore:

$$C_F = \frac{\text{PEAK}}{\text{RMS}} = \sqrt{\frac{6}{M^2-1}\frac{M-1}{\sqrt{2}}} = \sqrt{3}\sqrt{\frac{M-1}{M+1}} = \sqrt{3}\frac{\sqrt{N}-1}{\sqrt{N-1}}.$$

For example we have: 4QAM, $N = 4 \rightarrow C_F = 1$, 0 dB (trivial since the four points are on a circle, hence constant power, non-varying envelope, no PAPR); 16QAM, $N = 16 \rightarrow C_F = 3/\sqrt{5}$, 2.6 dB; 64QAM, $N = 64 \rightarrow C_F = \sqrt{7/3}$, 3.7 dB; 1024QAM, $N = 1024 \rightarrow C_F = 31/\sqrt{341}$, 4.5 dB; ∞QAM, $N = \infty \rightarrow C_F = \sqrt{3}$, 4.8 dB.

8.2.2 AM–AM and AM–PM Characteristics

Consider a power amplifier undergoing a single-tone test and suppose, for simplicity, that the PA is described as an input–output system with input x and output y, where y is the output waveform *at the fundamental*. Assume that y is measured both in amplitude (i.e., power) and phase. Since both x and y are sinusoids at the fundamental frequency f_0, they can be expressed in terms of complex phasors X and Y (we assume here normalization with respect to the peak value):

$$x(t) = \text{Re}\left[|X|\exp(j2\pi f_0 t + j\angle X)\right]$$
$$y(t) = \text{Re}\left[|Y|\exp(j2\pi f_0 t + j\angle Y)\right].$$

Since powers are proportional to the square of the signal magnitude, we have $P_{in} \propto |X|^2$, $P_{out} \propto |Y|^2$. Therefore, the $P_{in} - P_{out}$ characteristic is equivalent to the $|X| - |Y|$ curve or to the $P_{in} - |Y|$ curve. Similarly, the phase delay $\phi = \angle Y - \angle X$ can be represented, as a function of the input power, as the $P_{in} - \phi$ curve but also, equivalently, as the $|X| - \phi$ curve.

Furthermore, we can introduce a complex function of a real variable (the so-called *descriptive function* F_P) such as:

$$Y = F_P(|X|)X = F_P(P_{in})X.$$

In small signal conditions, F_P coincides with the amplification $A = Y/X$ (voltage or current according to the meaning of the input and output variables). In LS conditions, the magnitude of F_P yields the AM–AM distortion curve of the amplifier, providing information on the small-signal gain and on gain compression. The phase of F_P yields the AM–PM curve, describing the variation of phase delay as a function of the input signal magnitude or power.

The descriptive function model thus obtained under single-tone excitation can be approximately extended to the case where x is a narrowband signal around the carrier at f_0. In this case, y is the output signal component around the carrier (i.e., neglecting out-of-band harmonics and intermodulation products); this is also a narrowband signal even if we allow for some spectral regrowth. Both the input and the output signals can now be represented in terms of slowly varying phasors as:

$$x(t) = \text{Re}\left[|X(t)|\exp(j2\pi f_0 t + j\angle X(t))\right]$$
$$y(t) = \text{Re}\left[|Y(t)|\exp(j2\pi f_0 t + j\angle Y(t))\right],$$

8.2 Characteristics of Power Amplifiers

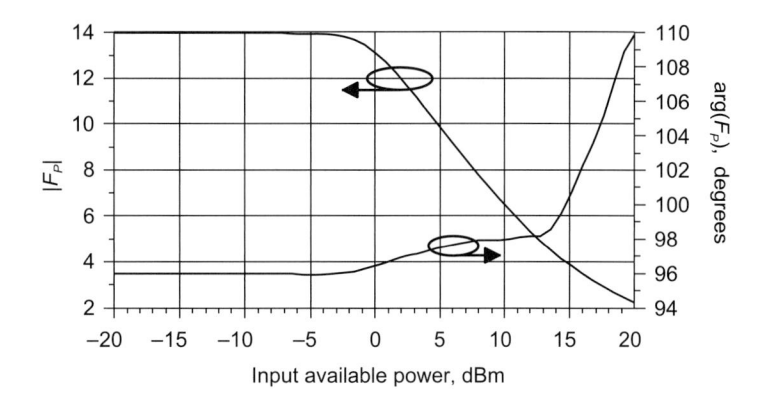

Figure 8.5 AM–AM and AM–PM characteristics of a class A power amplifier as a function of the input power.

where the output and input phasors (also denoted as complex envelopes) are related as:

$$Y(t) = F_P\left(|X(t)|\right) X(t) = F_P\left(P_{in}(t)\right) X(t).$$

Notice that now P_{in} is the instantaneous input power. For a narrowband system, the descriptive function representation is exact when memory effects within the signal bandwidth are negligible; this happens, for instance, when the system under consideration is nonlinear but without memory, see Example 8.7. An example of the behavior of $|F_P(P_{in})|$ (the AM–AM characteristic) and of $\angle F_P(P_{in})$ (the AM–PM characteristic) for a quasi-linear (class A) amplifier is shown in Fig. 8.5; for low input powers the AM–AM curve is constant and coincides with the small-signal gain, and so is the AM–PM curve; for large input powers we have gain compression and a phase modulation. Notice that the term AM–PM refers to the fact that the amplitude modulation of the input causes, due to the system nonlinearity, a spurious phase modulation of the output.

The descriptive function model can be conveniently exploited to analyze the behavior of any amplifier in the presence of input interferers; see Example 8.3.

Example 8.3 The input signal of a low-noise amplifier is a narrowband modulated signal x of complex envelope $X(t)$ centered around f_0. Suppose that at a frequency f_1, close enough to f_0, a strong interferer is received by the input antenna and appears at the LNA input as a large sinusoid x_1 of phasor X_1. Discuss the amplifier behavior and show that the interfering signal negatively affects the amplifier response to the wanted signal, leading to the so-called desensitization or blocking of the receiver. Show that this effect would not be present in an ideally linear amplifier.

Solution
In the presence of the wanted signal only the amplifier amplification is $A = \lim_{|X(t)| \to 0} F_P(|X(t)|)$, the small-signal amplification. Since the interferer is close in frequency to f_0, the amplifier response to it can be described by the same descriptive

function, with an amplification $A_1 = F_P(|X_1|) \ll A$. Considering both x and x_1 we have that the resulting signal $x_2 = x + x_1$ can be written as:

$$x_2(t) = \text{Re}\left[\left(X(t) + X_1 e^{j(\omega_1 - \omega_0)t}\right) e^{j\omega_0 t}\right] = \text{Re}\left[X_2 e^{j\omega_0 t}\right],$$

where:

$$X_2(t) = X(t) + X_1 e^{j(\omega_1 - \omega_0)t}.$$

However, since the interferer is much stronger than the wanted signal, we have $|X_2(t)| \approx |X_1|$ and therefore the system output will be a narrowband signal with complex envelope Y_2 such as:

$$Y_2(t) = F_P(|X_2(t)|) X_2(t) \approx F_P(|X_1|) X_2(t)$$
$$= A_1 X_1 e^{j(\omega_1 - \omega_0)t} + A_1 X(t) \approx A_1 X_1 e^{j(\omega_1 - \omega_0)t}.$$

In other words, the gain compression associated with the interferer also affects the wanted signal, and the amplifier output is dominated by the interferer. Notice that in an ideally linear amplifier $A_1 = A$ and the amplifier response to the interferer can be eliminated, at least in principle, by filtering, since the gain of the wanted signal is not affected.

8.2.3 Dynamic Range

The dynamic range defines the interval of the input powers where the amplifier operates with a satisfactory signal over noise ratio. It is often referred to as the Spurious Free Dynamic Range (SFDR). For low input power, the amplifier quality is limited by noise; for high power, by distortion. The minimum acceptable input power, called the amplifier sensitivity $P_{in,S}$, is the minimum input power at which a prescribed signal over noise ratio is obtained. For $P_{in} < P_{in,S}$ the amplifier signal over noise ratio is below the minimum acceptable value, and for an input power corresponding to the *noise floor* $P_{in} = P_{in,NF}$ the S/N ratio is unity. The definition of the maximum acceptable input power is somewhat arbitrary, and corresponds to an output IMP$_3$ power equal to the output power corresponding to $P_{in} = P_{in,NF}$.[4] Notice that the SFDR concept does not hold for power amplifiers only, but for any amplifier (low noise, high gain).

The SFDR can be obtained from the third-order intermodulation products intercepts as follows. First, let us define the amplifier sensitivity $P_{in,S}$; this is the input power yielding a desired signal over noise ratio on the load $(S/N)_L$, with a given amplifier noise figure F. In such conditions:

$$F = \frac{(S/N)_{in}}{(S/N)_L} = \frac{P_{in,S}}{k_B TB} \frac{1}{(S/N)_L}$$

---

[4] The SFDR is often defined not as an interval, but as the ratio of the upper and the lower limits, expressed in dB; given the amplifier sensitivity, the interval can be recovered.

and therefore:
$$P_{in,S} = F \cdot k_B TB \cdot (S/N)_L.$$

For $(S/N)_L = 1$ the input power is the noise floor $P_{in,NF}$; we then have $P_{in,NF} = F \cdot k_B TB$, while the sensitivity is related to the noise floor as $P_{in,S} = P_{in,NF} \cdot (S/N)_L$.

We have so far identified the lower limit of the SFDR. The upper limit of the SFDR corresponds to the condition:

$$P_{out}(\text{IMP}_3) = \frac{G_{op}^{ss} P_{in,max}^3}{\text{IIP}_3^2} = G_{op}^{ss} \cdot P_{in,NF},$$

i.e., solving:

$$P_{in,max} = \text{IIP}_3^{2/3} \cdot P_{in,NF}^{1/3}.$$

Finally, defining the SFDR as a ratio we find:

$$\text{SFDR} = \frac{P_{in,max}}{P_{in,S}} = \frac{\text{IIP}_3^{2/3} \cdot P_{in,NF}^{1/3}}{P_{in,NF} \cdot (S/N)_L} = \frac{\text{IIP}_3^{2/3} \cdot P_{in,NF}^{-2/3}}{(S/N)_L},$$

i.e., in log scale:

$$\text{SFDR}|_{dB} = \frac{2}{3}\left[\text{IIP}_3|_{dBm} - P_{in,NF}|_{dBm}\right] - (S/N)_L|_{dB}.$$

8.2.4 Efficiency and Power Added Efficiency

Let us define as P_{DC} the average power supplied by the DC bias to the amplifier. The amplifier load typically is DC decoupled (through a capacitive DC block integrated in the bias T) and tuned, meaning that all higher-order harmonics are shorted or open-circuited at the amplifier output. Thus, the DC supply power is partly converted into the RF output power at the fundamental ($P_{out} = P_{RF}$) and partly dissipated into heat, mainly within the active device, either in DC or after frequency conversion. The *power efficiency* η (also called drain or collector efficiency in common source or common emitter FETs and bipolars, respectively) is the ratio between the RF output power and the power from the DC bias:

$$\eta = \frac{P_{out}(f_0)}{P_{DC}}.$$

In the above definition, $P_{out}(f_0)$ refers to a single-tone test, but the definition can be easily extended in the presence of a multi-tone or modulated input signal. Another useful parameter is the *power added efficiency*, PAE, defined as:

$$\text{PAE} = \frac{P_{out}(f_0) - P_{in}(f_0)}{P_{DC}} = \eta\left(1 - \frac{1}{G_{op}}\right). \tag{8.2}$$

We have that PAE tends to the efficiency in the presence of large gain. Notice that the gain in (8.2) is the operational gain, since from the energy standpoint the RF power added by the device is $P_{out} - P_{in}$.

Efficiency impacts on the amplifier operation in several ways. On the one side, low efficiency obviously means a larger DC power consumption to obtain the same RF output power. In many systems, the PA stage contributed to a significant fraction of the overall power budget. On the other side, low efficiency also means that a large power is dissipated within PA stage, thus requiring a more complex and expensive thermal management strategy in order to keep the circuit temperature increase within tolerable limits. The development of high-efficiency PAs is, however, fraught with a number of limits; as discussed in Sec. 8.4, Class A amplifiers have a maximum theoretical efficiency of 50 percent only, that becomes 78 percent in Class B; the theoretical efficiency of Class C can approach 100 percent but at the expense of an unacceptable gain penalty. Non-traditional amplifier classes can achieve theoretical efficiencies of 100 percent by generating drain (collector) voltage and current waveforms that are nonzero in complementary intervals; in this way, the instantaneous power dissipated by the device vanishes and all DC power is converted into RF. Strategies to achieve this result are the harmonic loading (Class F) and the so-called switching amplifiers (e.g., Class E), see Sec. 8.7.1 and Sec. 8.7.2, respectively.

8.2.5 Input Matching

For the sake of definiteness, let us concentrate to the single-tone test; extension to other cases is trivial. Given the input incident power $P^+(f_0)$ and the power $P^-(nf_0)$ reflected at nf_0 (the device does generate harmonics, that also appear at the device input) we can define the input power reflection coefficient R_n as:

$$R_n = \frac{P^-(nf_0)}{P^+(f_0)} = \frac{|b_1(nf_0)|^2}{|a_1(f_0)|^2},$$

where port 1 is the amplifier input and a_1 is the incident power wave at the fundamental, $b_1(nf_0)$ the reflected power wave at frequency nf_0. In a similar way we define an input reflection coefficient:

$$\Gamma_{in,n} = \frac{b_1(nf_0)}{a_1(f_0)}.$$

The input power at fundamental is related to the incident input power as:

$$P_{in}(f_0) = (1 - R_1) P^+(f_0).$$

In a multi-tone test the reflected power can include both harmonics and IMPs. Notice that the harmonic power generated at the input of the device can be reflected back by the input termination; this can be exploited in the PA optimization, see, e.g., [3].

8.3 Power Amplifier Classes

Consider a simple compound semiconductor FET power stage (Fig. 8.6) undergoing a single-tone test. The input (v_{GS}) is driven by a sinusoidal voltage centered around the gate bias V_{GG}; the input voltage range may vary from a maximum value equal to

8.3 Power Amplifier Classes

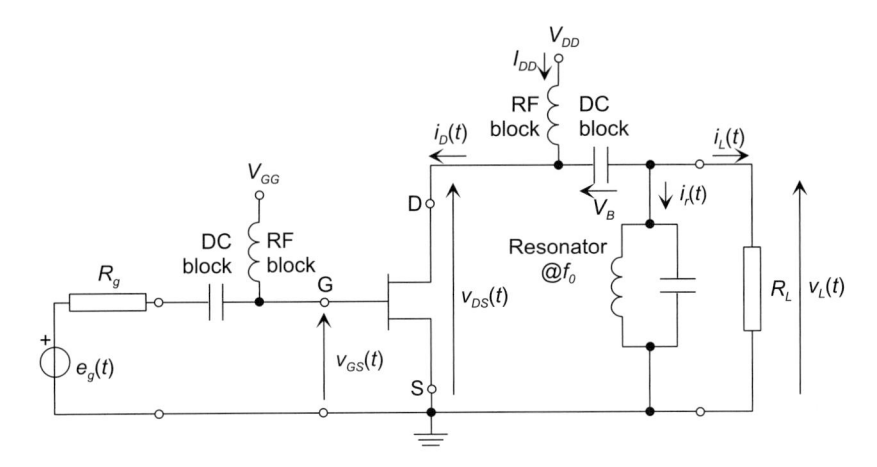

Figure 8.6 Power amplifier in *single-ended* configuration with tuned load. According to the input bias voltage and to the input voltage swing the amplifier can be class A, AB, B or C. The structure of the input and output bias Ts is made explicit (DC capacitive and RF inductive blocks; the RF block is often referred to as *RF choke*).

(approximately) 0 and a minimum value above, at or below the device threshold voltage. For simplicity, let us neglect the output device capacitance and device parasitics. As already remarked, the power stage generates harmonics of the fundamental; to avoid useless power dissipation, the output is connected to a *tuned load*, i.e., only a neighborhood of the fundamental can reach the load resistor, while all the harmonics are shorted to ground (a convenient solution taking into account that the device output is a current generator). The DC bias is also decoupled from the load through a DC blocking capacitor integrated in the bias T. According to the input gate bias and the input voltage swing, the output current waveform can be either a full or a clipped sinusoid, yielding the so-called *traditional* amplifier classes A, B and C (with the intermediate case AB). In detail:

- In Class A amplifiers, the minimum total input voltage (including both the input signal and the input voltage DC bias) is always larger than the device threshold voltage.[5] As a consequence, the output current is never zero, and the current waveform is almost sinusoidal. In an *n*-type compound semiconductor FET the maximum peak-to-peak input voltage swing is from the (negative) threshold voltage to approximately 0, and the gate bias voltage allowing for the maximum swing is midway between the two extremes. Since the output current is almost sinusoidal, Class A amplifiers operate with minimal distortion but also with relatively low maximum efficiency. Notice that in a single-tone test the load voltage is purely sinusoidal but in a two-tone test in-band IMPs also are present in the load current.
- In Class B amplifiers, the output voltage is below threshold for exactly half of the period, and the gate bias is at threshold. As a consequence, the output current is

---

[5] The name "class A" interpreted as "best class" may derive from the fact that in this class distortion is minimum.

(approximately) a half-wave rectified sinusoid. Despite the high nonlinearity of the amplifier, thanks to the tuned load the load current is almost sinusoidal. Class B amplifiers exhibit larger distortion when compared to class A and lower gain, but a better efficiency.
- Class AB is an intermediate case between class A and class B, meaning that the input voltage is below threshold for less than half a period. The load current is a clipped sinusoid and efficiency, gain and distortion are intermediate between class A and B.
- In Class C amplifiers, the input bias is below threshold and the output current is different from zero for more than half a period. Class C exhibit stronger distortion and lower gain vs. class B, but also a larger efficiency. However, 100 percent efficiency is achieved, in the limit, with zero gain; this limits the practical utility of class C amplifiers.

According to the traditional classification, the amplifier class clearly depends on the drain current *circulation angle* α, such as:

$$\frac{\alpha}{2\pi} = \frac{t_{on}}{T},$$

where T is the period of the waveform and t_{on} is the time interval where the current is different from zero. In class A, $t_{on} = T$ and $\alpha = 2\pi$; in class B, $t_{on} = T/2$ and $\alpha = \pi$; in class C, $t_{on} < T/2$ and $\alpha < \pi$. Notice that class A and B correspond to a well defined conduction angle, while class C (and also class AB) to an interval where the conduction angle may vary with the applied RF excitation.

Practical microwave amplifiers can range from class A to class AB, while class B and even worse C are uncommon, apart from particular amplifier designs. In recent years, novel amplifier classes have been however proposed, exploiting almost all of the available letters of the alphabet. For the sake of brevity, we will only discuss class F and class E amplifiers. Class F amplifiers use the so-called harmonic loading principle, where the load at each harmonic is controlled so as to yield the maximum efficiency. Class E amplifiers are a significant example of the so-called switching amplifiers, where the input voltage is a digitally modulated signal, that drives the output from the ON to the OFF state. In such amplifiers, the active device is operating more like a controlled switch than as an analog device.

Finally, all the configurations discussed so far are examples of single-ended amplifiers. Push-pull configurations are also possible, either exploiting complementary devices (when available), or directional couplers imposing the correct phase shift to the output stage. For the sake of brevity such configurations will not be discussed in detail.

8.4 The Quasi-Linear Power Amplifier (Class A)

The class A amplifier can be considered as an extension of the small-signal amplifier, the main difference being the proper choice of the DC working point and load termination to achieve maximum output power rather than maximum gain. Moreover, distortion

8.4 The Quasi-Linear Power Amplifier (Class A)

(e.g., third-order intermodulation products) has to be controlled in the design of a class A power amplifier; since the IMP$_3$ power grows like P_{in}^3 while the output power as P_{in}, we have that CIMR$_3 \propto P_{in}^{-2}$. This means that the carrier to intermodulation ratio can be improved by reducing the input power level, i.e., by operating the device in *backoff* with respect to the saturation or 1 dB compression input power. We denote the input backoff as IBO and the resulting output backoff as OBO. Since the prescribed output power is a design goal, operation in backoff implies to exploit a device with a larger gate periphery (or collector area) and therefore with a larger DC dissipation with respect to the case where a device is used operating in compression or power saturation. Ultimately, backoff is a way to improve the class A amplifier linearity at the expense of efficiency.

To introduce the class A amplifier basics, let us evaluate first the maximum RF power and the efficiency of a class A amplifier with tuned load. The amplifier is based on a compound semiconductor FET (e.g., a HEMT) with (off state) breakdown voltage $V_{DS,br}$ and knee voltage (corresponding to the onset of drain current saturation) $V_{DS,k}$. The threshold voltage is $V_T < 0$ and the maximum gate voltage is $V_{GS,M} \approx 0$ V. Finally, the maximum current is $I_{D,M} \approx I_{DSS}$ (with reference to Schottky gate FETs in which the maximum current is obtained for $V_{GS} \approx 0$). The waveforms of a class A amplifier working, in a single-tone test, at the limit of the linear region (i.e., at the onset of power saturation) are shown in Fig. 8.8. The gate voltage is approximately sinusoidal and varies between $V_{GS} = V_T < 0$ and $V_{GS} = 0$, with gate bias voltage $V_{GG} \approx V_T/2 < 0$. The drain bias $V_{DD} \approx (V_{DS,k} + V_{DS,br})/2$ is located midway between the knee and the breakdown voltage, to allow for the maximum voltage swing, see Fig. 8.7. If the load resistance (and therefore the load line) is properly chosen, this bias choice also implies that the DC drain current is $I_{DD} \approx I_{DSS}/2$. Let us define now the peak value of the sinusoidal component of the drain voltage:

$$v_{DS,M} = \frac{V_{DS,br} - V_{DS,k}}{2} \approx \frac{V_{DS,br}}{2}$$

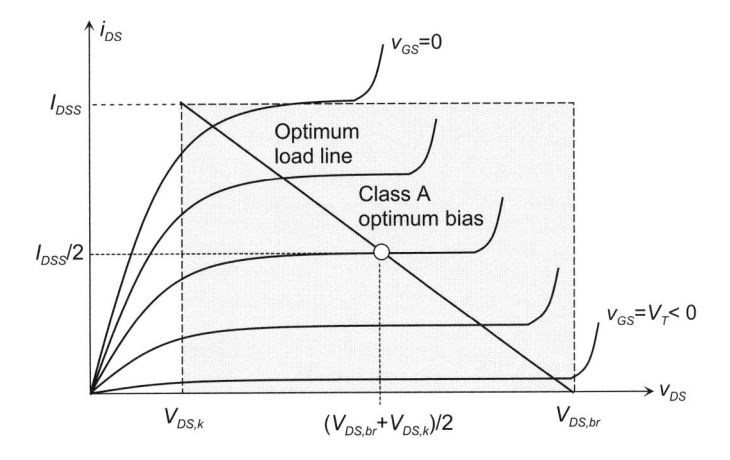

Figure 8.7 Load line of class A amplifier with optimum load.

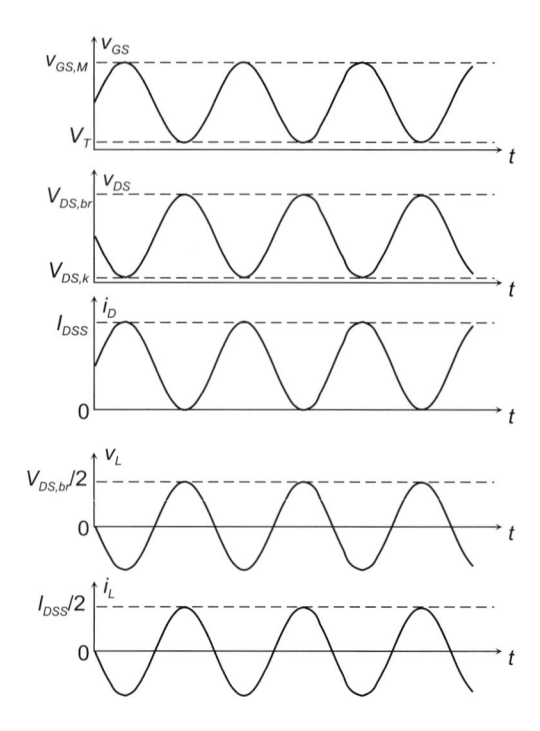

Figure 8.8 Waveforms of a class A amplifier with optimum tuned load and maximum output power.

and the peak value of the sinusoidal component of the drain current:

$$i_{D,M} = \frac{I_{DSS}}{2}.$$

The load resistance R_L at the fundamental (remember that, at the amplifier load, the DC component is blocked and the harmonics are shorted by the resonator) is designed so as to implement, at the same time, the maximum voltage and current swings. Taking into account that the DC working point is in the center of the useful rectangle of the output DC characteristics ($V_{DD} \approx V_{DS,br}/2$, $I_{DD} \approx I_{DSS}/2$) the optimum load resistance allowing for the voltage and current simultaneous swings $v_{DS,M}$ and $i_{D,M}$, respectively, corresponds to the value:

$$R_{Lo} = \frac{v_{DS,M}}{i_{D,M}} \approx \frac{V_{DS,br}}{I_{DSS}}.$$

Taking into account that i_L (v_L) only has the sinusoidal component of i_D (v_{DS}), we obtain that the average load power is given, in the optimum conditions (bias and load resistance) as:

$$P_{RF,M} = \frac{1}{2} v_{DS,M} i_{D,M} \cos\theta = \frac{1}{2} v_{DS,M} i_{D,M} \approx \frac{V_{DS,br} I_{DSS}}{8}. \quad (8.3)$$

Since at the fundamental the load is a resistor, the phase angle θ between the load voltage and the load current is zero. This also confirms that the total optimum load seen by the

8.4 The Quasi-Linear Power Amplifier (Class A)

output current device generator is resistive, with no reactive component. A more detailed analysis of the currents and voltages appearing in the output of the tuned load amplifier is discussed in Example 8.5.

The DC power supplied by the bias is the product of the DC components of the drain current and voltage (we assume that negligible DC power is dissipated by the input, approximated as an open circuit):

$$P_{DC} = \frac{V_{DS,br} I_{DSS}}{4}.$$

Notice that P_{DC} is independent from the signal level. With no input signal, such a power is entirely dissipated by the device. In the presence of an input signal, an increasing fraction of the DC power is converted into RF power at the output. In optimum conditions the maximum class A efficiency is given by:

$$\eta_{A,M} = \frac{P_{RF,M}}{P_{DC}} = \frac{V_{DS,br} I_{DSS}}{8} \cdot \left(\frac{V_{DS,br} I_{DSS}}{4}\right)^{-1} = 0.5.$$

Thus, the maximum efficiency of a class A amplifier is 50 percent. If the input power is backed off, the efficiency decreases linearly. In fact:

$$\eta_A = \frac{P_{RF}}{P_{DC}} = \frac{P_{RF,M}}{P_{DC}} \frac{P_{RF}}{P_{RF,M}} \approx \eta_{A,M} \frac{P_{in}}{P_{in,M}} = \eta_{A,M} \cdot \text{IBO}.$$

Notice that backoff can derive from several scenarios. In the presence of a constant envelope input (e.g., a digital frequency modulated signal) backoff is a strategy to improve linearity at the expense of efficiency. However, if the input signal dynamically varies it power, the average efficiency becomes worse, since low instantaneous input power leads to low efficiency.

Example 8.4 Discuss the design of the resonator exploited in the tuned load of a class A amplifier.

Solution
In class A power amplifiers, a tuned load is used where a series or (more commonly) a parallel resonator blocks higher-order harmonics from dissipating power in the load resistance R_L. The same occurs in class A to C amplifiers (see Sec. 8.6) while in class F and E amplifiers (see Sec. 8.7.1 and 8.7.2, respectively) resonators are used to suitably shape the output waveform. While distributed solutions based on transmission lines can be in principle adopted, a lumped element resonator is the most compact solution in integrated circuits.

The series (parallel) resonators considered in the discussion of power amplifiers are ideal, i.e., they behave as short (open) circuits at the resonant frequency f_0, and as open (short) circuits for $f \neq f_0$. Some criteria are provided here for the practical design of such elements. For the sake of simplicity, we initially neglect the effect of losses.

Resonance at $\omega_0 = 2\pi f_0$ can be achieved by any inductance and capacitance value satisfying the condition:

$$\omega_0^2 LC = 1.$$

Let us focus first on series resonator; a suitable choice of L_S and C_S can be derived by considering the resonator impedance far from the resonance, where the resonator should ideally behave as an open circuit; in practice, we require that the series resonator impedance $Z_S(\omega)$ be suitably larger (by a factor $K > 1$) than the load resistance R_L at the fundamental harmonic $n_h \omega_0$:

$$|Z_S(n_h\omega_0)| = \frac{n_h^2 - 1}{n_h \omega_0 C_S} > K \cdot R_L. \tag{8.4}$$

Typically, $K = 10$ can be sufficient. It follows:

$$L_S > \frac{n_h}{(n_h^2 - 1)\omega_0} K \cdot R_L, \quad C_S = \frac{1}{\omega_0^2 L_S}.$$

In most cases, condition (8.4) has to be imposed at the second harmonic ($n_h = 2$). A dual behavior holds for parallel resonators, where a condition on the parallel resonator admittance Y_P being suitably larger than the load admittance $Y_L = 1/R_L$ yields:

$$C_P > \frac{n_h}{(n_h^2 - 1)\omega_0} \frac{K}{R_L}, \quad L_P = \frac{1}{\omega_0^2 C_P}.$$

Resonator parasitic losses, neglected so far, imply that at resonance the series resonator resistance be $R_S > 0$ and the parallel resonator conductance be $G_P > 0$. While no impact follows on the out of resonance behavior discussed above, the design should take into account for the constraints $R_S \ll R_L$, $G_P \ll 1/R_L$.

Example 8.5 Discuss the frequency components (DC, fundamental, harmonics) of voltages and currents (i_L, i_D, i_r, I_{DD} and v_L, v_{DS}, V_B, respectively) in the output loop of a tuned load class A device amplifier stage (see Fig. 8.6) under a single-tone test, and derive the expression of the device load line.

Solution
In time domain, the following voltage and current Kirchoff equations hold:

$$v_L(t) + V_B = v_{DS}(t)$$
$$i_L(t) + i_D(t) + i_r(t) = I_{DD},$$

where V_B is the voltage across the DC block (ideally DC only), I_{DD} is the current from the bias (ideally DC only because of the bias T RF blocking inductor), i_r is the resonator current. Balancing the harmonics we have at DC:

$$V_{DS}(DC) = V_B = V_{DD}$$
$$V_L(DC) = 0$$
$$I_L(DC) = I_r(DC) = 0$$
$$I_D(DC) = I_{DD}.$$

8.4 The Quasi-Linear Power Amplifier (Class A)

In fact, at DC the load is shorted by the resonator, so that the load DC current is certainly zero; but since the sum of the load and resonator DC currents is also zero owing to the DC block, also the resonator DC current vanishes. The DC drain current coincides with the bias current. Concerning the fundamental f_0 we have:

$$V_{DS}(f_0) = V_L(f_0)$$
$$I_L(f_0) = -I_D(f_0),$$

i.e., the fundamental component of the load and output voltages coincide, while the load current is the drain current with a minus sign. At the harmonics we finally have:

$$V_{DS}(nf_0) = V_L(nf_0) = 0$$
$$I_L(nf_0) = 0$$
$$I_r(nf_0) = -I_D(nf_0);$$

in fact, the device output is shorted at nf_0 by the DC block and by the resonator, and so is the load; as a consequence, the load current is zero and the harmonics of the drain current flow in the resonator. Taking into account that:

$$V_L(f_0) = V_{DS}(f_0) = R_L I_L(f_0) = -R_L I_D(f_0),$$

we can write the time-domain v_{DS} as:

$$v_{DS}(t) = V_{DD} - R_L I_D(f_0) \cos \omega_0 t$$

with components at DC and at the fundamental, while the harmonic distortion is in the drain current:

$$i_D(t) = I_{DD} + I_D(f_0) \cos \omega_0 t + I_D(2f_0) \cos 2\omega_0 t + \ldots$$

The load voltage and current are of course purely sinusoidal at the fundamental. Notice that if the drain current harmonics are negligible we can write:

$$v_{DS}(t) - V_{DD} = -R_L I_D(f_0) \cos \omega_0 t$$
$$i_D(t) - I_{DD} = I_D(f_0) \cos \omega_0 t,$$

i.e., combining:

$$(i_D - I_{DD}) = -\frac{(v_{DS} - V_{DD})}{R_L}. \tag{8.5}$$

Eq. (8.5) describes the device load line; the instantaneous working point is the intersection of the load line with the device characteristic:

$$i_D(t) = I_D(v_{DS}(t), v_{GS}(t)).$$

Example 8.6 A class A amplifier stage has a thermal resistance $R_\theta = 2\,°C/mW$. Taking into account that $I_{DSS} = 10$ mA, $V_{DS,br} = 14$ V and that the bias point is the optimum class A bias, estimate the temperature variation between two conditions: (a) zero input signal; (b) maximum input signal. Assume 20 dB operational gain.

Solution

The maximum class A power is:

$$P_{RF,M} = \frac{I_{DSS}V_{DS,br}}{8} = \frac{10 \cdot 14}{8} = 17.5 \text{ mW};$$

we have $P_{in} = 0$ in case (a) and $P_{in} = 17.5/100 = 0.175$ mW in case (b), since $G_{op} = 100$. The DC power does not depend on the input signal level and is:

$$P_{DC} = \frac{I_{DSS}V_{DS,br}}{4} = \frac{10 \cdot 14}{4} = 35 \text{ mW},$$

while the device dissipated power is:

$$P_{\text{diss}} = P_{DC} - P_{RF} = 35 - 100 \cdot P_{in,M} \cdot \frac{P_{in}}{P_{in,M}} = \left(35 - 17.5\frac{P_{in}}{P_{in,M}}\right) \text{ mW}.$$

Correspondingly, the device temperature increase is:

$$\Delta T = R_\theta P_{\text{diss}} = 2 \cdot \left(35 - 17.5\frac{P_{in}}{P_{in,M}}\right) \text{°C} = \left(70 - 35\frac{P_{in}}{P_{in,M}}\right) \text{°C}.$$

Thus the device cools down with increasing input signal power, going from 70 °C (a) to 35 °C (b).

8.4.1 Class A Design and the Load-Pull Approach

We have shown that the maximum class A power can be obtained when the device output drain current generator is connected to a resistive load of optimum value $R_L = R_{Lo} \approx V_{DS,br}/I_{DSS}$. An approximate strategy for class A design starts from the simplified, unilateral equivalent circuit in Fig. 8.9. As a first step, the optimum bias point has to be chosen at half of the maximum current and half of the breakdown voltage (approximately). Notice that the bias point is imposed by the bias Ts (not shown in the circuit in Fig. 8.9) and is not influenced by the choice of the RF load. As a second step, the output matching network is implemented so as to obtain, at the terminals of the $i_D(v_{DS}, v_{GS})$ output current generator, the optimum admittance:

$$Y'_L = 1/R_{Lo}.$$

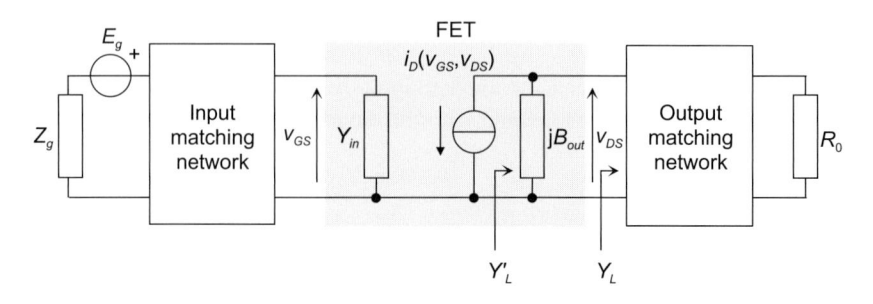

Figure 8.9 Class A amplifier scheme with output matching. For brevity, bias Ts have been omitted.

Thus, if jB_{out} is the device capacitive output susceptance at the fundamental, the output matching network should transform the external load R_0 into the admittance:

$$Y_L = 1/R_{Lo} - jB_{out}$$

thus compensating for the internal output device susceptance. As a final step, the input matching network can be designed so as to achieve input power matching, as in maximum gain small-signal design. Stability issues can be addressed as in the small-signal case.

If the load condition deviates vs. the optimum class A condition, the maximum output power decreases with respect to the optimum. In fact, if the imaginary part of the output susceptance is not compensated for, the $\cos\theta$ power factor imposes a power penalty and the dynamic load line becomes an ellipse. Supposing however that the reactive compensation is performed and that only the real part of the load varies, for load resistances larger or smaller than the class A optimum, the small-signal MAG will be achieved; for such a load, the small-signal gain will be larger than in the optimum class A load, but power saturation shall occur earlier and with lower saturation power. If the load resistance deviates from both the large-signal (LS) and small-signal (SS) optimum condition, a lower SS gain and LS saturation power is obtained, see Fig. 8.10.

The analysis of the output power when varying the load impedance can be performed according to a simplified approach first proposed by Cripps [4]. The method closely follows an experimental procedure, whereby the output of the device is connected to a tunable (mechanically or electronically) load, and the Smith chart is swept while monitoring the output power for a given compression level. The procedure is termed *load-pull measurement* and the constant output power curves on the load reflectance Smith chart are called *load-pull curves*. From the load-pull curves the maximum power and related optimum load reflectance can be experimentally identified. In small-signal conditions

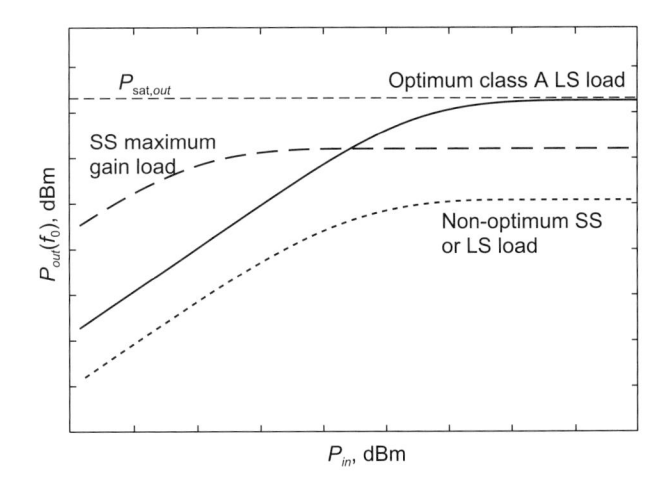

Figure 8.10 Class A single-tone $P_{in} - P_{out}$ curve with optimum large-signal (LS) load, optimum small-signal (SS, maximum gain) load, and non-optimum LS or SS load. $P_{out}(f_0)$ is the output power at the fundamental.

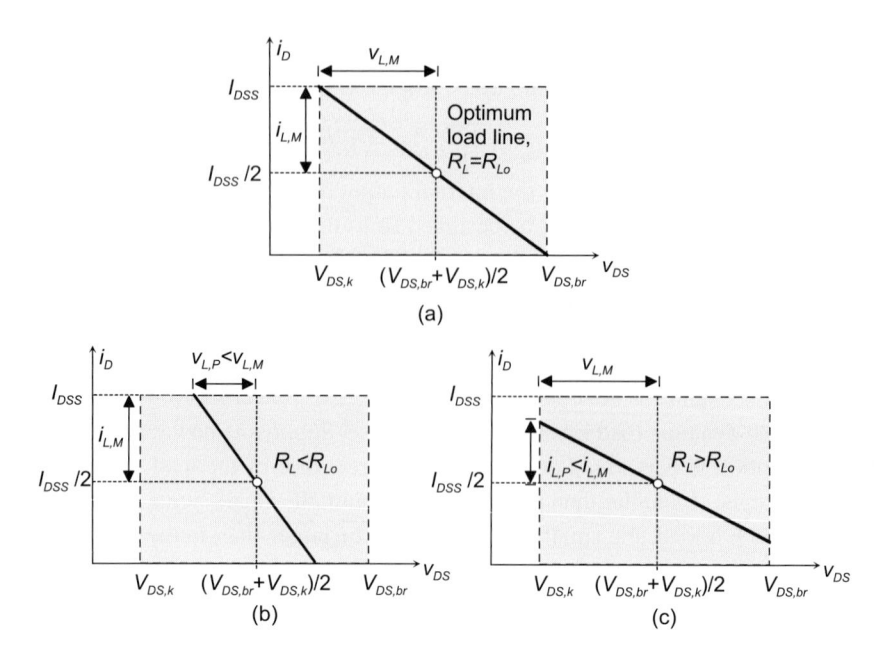

Figure 8.11 Class A amplifier with optimum (maximum power) load (a) and non-optimum load, (b) and (c).

the load-pull curves are constant gain circles; Cripps' theory allows the shape of the load-pull curves at the limit of power saturation to be estimated as a set of circle arcs. Experimental load-pull curves can be approximated as ellipses.

We now summarize Cripps' theory. Consider Fig. 8.11; in optimum load conditions (a) the maximum power is obtained:

$$P_{RF,M} = \frac{1}{2} v_{L,M} i_{L,M},$$

where $v_{L,M}$ and $i_{L,M}$ are the maximum peak voltage and current, respectively, corresponding to the limits of the device linearity. Suppose now, see Fig. 8.11 (b), that $R_L < R_{Lo}$; the output current swing is maximum ($i_{L,M}$ peak value), but the output voltage is limited (e.g., with a non-maximum swing, peak value $< v_{L,M}$); therefore:

$$i_{L,P} = i_{L,M}$$
$$v_{L,P} = |Z'_L| i_{L,M}$$

where $i_{L,P}$ is the peak value of the fundamental load current, $v_{L,P}$ the corresponding voltage, and $Z'_L = 1/Y'_L$, where Y'_L is the admittance seen by the output current device generator. In such conditions, the RF load power is:

$$P_{RF} = \frac{1}{2} i_{L,M} v_{L,P} \cos\theta = \frac{1}{2} i^2_{L,M} \text{Re}\left[Z'_L\right] = P_{RF,M} \frac{1}{R_{Lo}} \text{Re}\left[\frac{1}{Y_L + jB_{out}}\right].$$

Since $v_{L,P} < v_{L,M}$ we also have $|Z'_L| i_{L,M} < R_{Lo} i_{L,M}$; thus, the case of voltage limitation occurs if $|Z'_L| < R_{Lo}$, i.e., if:

8.4 The Quasi-Linear Power Amplifier (Class A)

$$\frac{1}{|Y_L + jB_{out}|} < R_{Lo}. \tag{8.6}$$

Suppose now, see Fig. 8.11 (c), that $R_L > R_{Lo}$; the peak output current swing will be smaller than the maximum value $i_{L,M}$, while the peak voltage swing is maximum, i.e.:

$$v_{L,P} = v_{L,M}$$
$$i_{L,P} = |Y'_L| v_{L,M}.$$

In this case, therefore, the current is limited. In such conditions, the RF load power is:

$$P_{RF} = \frac{1}{2} i_{L,P} v_{L,M} \cos\theta = \frac{1}{2} v_{L,M}^2 \mathrm{Re}\left[Y'_L\right] = P_{RF,M} R_{Lo} \mathrm{Re}\left[Y_L + jB_{out}\right].$$

Since $i_{L,P} < i_{L,M}$ we also have $|Y'_L| v_{L,M} < R_{Lo}^{-1} v_{L,M}$ and therefore the current limitation occurs if $|Y'_L| < R_{Lo}^{-1}$, i.e., if:

$$\frac{1}{|Y_L + jB_{out}|} > R_{Lo}, \tag{8.7}$$

a condition complementary to (8.6). To discuss the behavior of the load power vs. the load resistance, consider a simple case where the load is resistive with resistance $R'_L = 1/G'_L$. We then have, for $R'_L < R_{Lo}$:

$$P_{RF} = \frac{1}{2} i_{L,M}^2 R'_L = P_{RF,M} \frac{R'_L}{R_{Lo}},$$

while, for $R'_L > R_{Lo}$,

$$P_{RF} = \frac{1}{2} v_{L,M}^2 \frac{1}{R'_L} = P_{RF,M} \frac{R_{Lo}}{R'_L}.$$

Defining as Γ_L the reflection coefficient of R'_L vs. R_{Lo}, the output power is shown in Fig. 8.12 vs. Γ_L and compared to the variation occurring in the small-signal case (and only due to load mismatch vs. the optimum). In the LS case the power penalty vs. the optimum is much heavier than in the SS case already for small deviations.

To analyze the load-pull curves, let us start from the case where $B_{out} = 0$ and R_{Lo} is the normalization impedance. The optimum condition now corresponds to the center of the Smith chart ($\Gamma_L = 0$), while conditions (8.6) and (8.7) are expressed, having defined the normalized impedance $z_L = Z_L/R_{Lo}$, as $|z_L| > 1$ or $|z_L| < 1$, respectively. The limiting condition $|z_L| = 1$ can be translated, in terms of reflection coefficients, into:

$$|\Gamma_L + 1| = |\Gamma_L - 1|,$$

i.e., the corresponding Γ_L is purely imaginary. Therefore the Smith chart is divided in two regions, the left one ($\mathrm{Re}\left[\Gamma_L\right] < 0$) where (8.6) holds, and the right one ($\mathrm{Re}\left[\Gamma_L\right] > 0$) where (8.7) holds. For $\mathrm{Re}\left[\Gamma_L\right] < 0$ the output power is:

$$P_{RF} = P_{RF,M} \frac{R_L}{R_{Lo}} = P_{RF,M} r_L,$$

while, for $\mathrm{Re}\left[\Gamma_L\right] > 0$:

$$P_{RF} = P_{RF,M} R_{Lo} G_L = P_{RF,M} g_L,$$

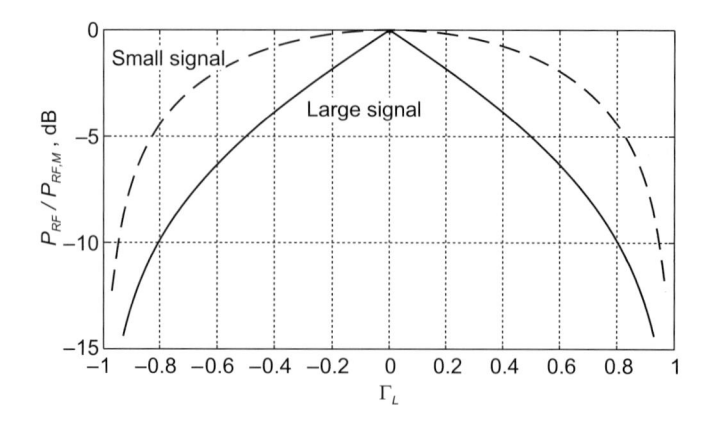

Figure 8.12 LS and SS behavior of the output power vs. the variation of the load resistance vs. the optimum value.

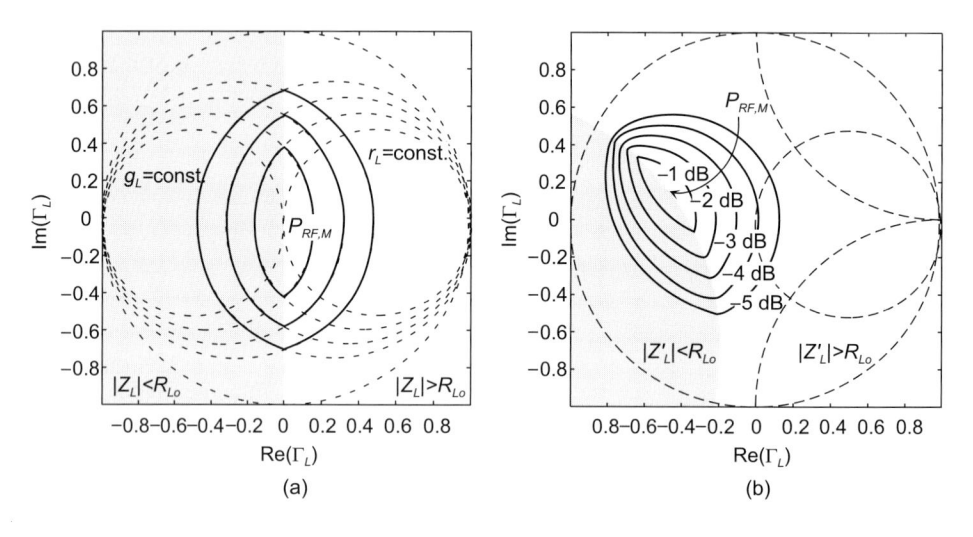

Figure 8.13 Constant load power curves in the Smith chart with a resistive load and $B_{out} = 0$, Z_L normalized vs. R_{Lo} (a); the same, but with $B_{out} \neq 0$ (b).

where R_L and G_L are the real parts of the load impedance and admittance, respectively, while r_L and g_L are their normalized values. It follows that for $\text{Re}\,[\Gamma_L] < 0$ the constant power curves coincide with the constant resistance circles, while for $\text{Re}\,[\Gamma_L] > 0$ those coincide with the constant conductance circles, see Fig. 8.13 (a). Notice that the constant output power curves are markedly different from the constant gain circles. In the general case where $B_{out} \neq 0$, conditions (8.6) and (8.7), corresponding in the Z_L plane to two regions separated by a circle, lead to the same kind of division in the Γ_L plane, thanks to the property of the $Z_L \rightarrow \Gamma_L$ transformation to map circles into circles. In each region the constant power curves are proportional to the normalized load resistance and conductance as seen from the output device current generator, according to the scheme in Fig. 8.13 (b). From the experimental standpoint, since power saturation and waveform clipping is not abrupt, as postulated in the simplified model, the angular points are not observed and there is a continuous transition between the small-signal behavior (where the curves are circles) to the large-signal regime (where experimental curves can be approximated as ellipses).

8.5 Analysis of Distortion and Power Saturation in a Class A Amplifier

Active electronic devices exhibit several causes for nonlinearity, that in turn yield the nonlinear behavior of amplifiers. We can approximately group them in two classes, focusing, for the sake of definiteness, on FETs. The FET transcharacteristics is not linear, and can be approximated (see the discussion in Sec. 5.8.3) by a second or higher-order polynomial of the driving voltage v^*. This means that, already in small-signal conditions, some soft nonlinear behavior is present, causing the generation of harmonics and of intermodulation products. This is occurring far before power saturation takes place; a convenient analytical model that can be exploited for the analysis is the so-called *power series model*. With increasing driving voltage, however, an almost abrupt limitation to the FET drain current occurs, caused by the upper limit to the drain current (remember that the gate-channel junction cannot be brought in direct bias) or to the drain voltage (caused by the insurgence of breakdown for large v_{DS} and the onset of the triode or linear region for low v_{DS}) or, if the output load is the optimum class A load, to both. This strongly nonlinear behavior is the main cause of power saturation, and ultimately dictates the maximum output power that the device can provide. We will therefore divide the discussion in two parts, referring for the sake of definiteness to a class A amplifier, one devoted to harmonic and IMP generation through a power series model (Sec. 8.5.1), the second devoted to an approximate analysis of power saturation (Sec. 8.5.2).

8.5.1 Power-Series Analysis of Compression and Intermodulation

Consider a FET device in quasi-static condition (i.e., neglecting reactive effects). Suppose that the device is excited by a generator with open-circuit signal voltage $e_G(t)$ and generator resistance R_G, and that its output is connected to a resistive load R_L. Both the generator and the load include a DC bias generator. Assume that the signal output load current i_L ($i_L = -i_D$) can be expressed, in quasi-static conditions, as a power series of the generator signal voltage:

$$i_D(t) = -i_L(t) = a_1 e_G + a_2 e_G^2 + a_3 e_G^3 \ldots$$

The coefficient a_1 coincides with the device transconductance. In the following analysis, we will truncate the series to the third term, since the first three terms include the main information on harmonic and IMP generation. Notice that if the input device capacitance is an open circuit and parasitics are neglected the driving voltage v^* coincides with e_G; in such conditions $P_{in} = 0$ and the analysis can be conveniently carried out using as the input power the input available power.

Let us consider first the single-tone test case, where:

$$e_G(t) = E_G \cos\theta$$

with $\theta = \omega_0 t + \phi$. Evaluating the power series up to the third term we obtain:

$$\begin{aligned}
i_D(t) &= a_1 E_G \cos\theta + a_2 (E_G \cos\theta)^2 + a_3 (E_G \cos\theta)^3 \\
&= \frac{1}{2}a_2 E_G^2 + \left(a_1 E_G + \frac{3}{4}a_3 E_G^3\right)\cos\theta + \frac{1}{2}a_2 E_G^2 \cos 2\theta + \frac{1}{4}a_3 E_G^3 \cos 3\theta \\
&= I_{D0} + I_{D1}\cos\theta + I_{D2}\cos 2\theta + I_{D3}\cos 3\theta.
\end{aligned}$$

The load (drain) current includes a DC term (from the second-order term of the power series), a fundamental related both to the first-order term and to the third-order one, and a second and third harmonic related to the second-order and third-order terms, respectively. Neglecting the DC term (that is however affecting the output DC bias of the device) we can evaluate the signal power on the load associated with the fundamental and harmonics:

$$\begin{aligned}
P_L(f_0) &= \frac{1}{2}R_L I_{D1}^2 = \frac{1}{2}R_L \left(a_1 E_G + \frac{3}{4}a_3 E_G^3\right)^2 \\
&= \frac{1}{2}R_L a_1^2 E_G^2 + \frac{3}{4}R_L a_1 a_3 E_G^4 + \frac{9}{32}R_L a_3^2 E_G^6 \\
P_L(2f_0) &= \frac{1}{2}R_L I_{D2}^2 = \frac{1}{2}R_L a_2^2 E_G^4 \\
P_L(3f_0) &= \frac{1}{2}R_L I_{D3}^2 = \frac{1}{2}R_L a_3^2 E_G^6.
\end{aligned}$$

Taking into account that the available generator power at the fundamental is (remember that E_G is a peak value):

$$P_{av,in} = \frac{E_G^2}{8R_G} \rightarrow E_G^2 = 8R_G P_{av,in},$$

we can also write:

$$P_L(f_0) = 4R_L R_G \left(a_1 + 6a_3 R_G P_{av,in}\right)^2 P_{av,in} \qquad (8.8)$$

$$P_L(2f_0) = 32 a_2^2 R_G^2 R_L P_{av,in}^2 = K_2 P_{av,in}^2 \qquad (8.9)$$

$$P_L(3f_0) = 256 a_3^2 R_G^3 R_L P_{av,in}^3 = K_3 P_{av,in}^3. \qquad (8.10)$$

The second and third harmonic load power follow the previously introduced power law, at least for input available power low enough to neglect contributions coming from higher-order terms. Taking into account that if $a_1 > 0$ typically $a_3 < 0$, i.e., the third-order term is compressive rather than expansive, we immediately obtain from (8.8) that the $P_{in} - P_{out}$ characteristics has a compressive nature, where the first term is positive and linearly depends on the input available power, while the second term, depending on the cube of the input available power, is negative. From (8.8) we can derive an estimate (based on the coefficients a_1 and a_3) of the 1 dB compression point. Taking into account that -1 dB corresponds to a factor $\alpha = 10^{-1/10} \approx 0.79433$ we have:

$$4R_L R_G \left(a_1 - 6|a_3| R_G P_{av,in}\right)^2 P_{av,in} = \alpha \cdot 4a_1^2 R_G R_L P_{av,in},$$

8.5 Analysis of Distortion and Power Saturation in a Class A Amplifier

from which:
$$P_{av,in}\big|_{1\text{dB}} = \frac{1-\sqrt{\alpha}}{6R_G}\frac{a_1}{|a_3|} = \frac{1.8125\times 10^{-2}}{R_G}\frac{a_1}{|a_3|}. \tag{8.11}$$

Thus, the 1 dB compression point input power increases (as obvious) with decreasing magnitude of the third-order term of the power series. If we consider reactive effects imposing a low-pass filter action between the generator voltage and the driving voltage v^*, we obtain that the 1 dB compression point corresponds to larger input powers for increasing frequency f_0, simply because a larger power is needed to achieve the same value of the driving voltage v_{GS} with increasing f_0.

We now come to the two-tone test case, where we assume for simplicity that the two tones have the same amplitude:

$$e_G(t) = E_G \cos\theta_1 + E_G \cos\theta_2$$

with $\theta_i = \omega_i t + \phi_i$, $i = 1, 2$. Evaluating the power series up to the third term we obtain:

$$\begin{aligned}i_D(t) &= a_1(E_G\cos\theta_1 + E_G\cos\theta_2) + a_2(E_G\cos\theta_1 + E_G\cos\theta_2)^2 \\ &\quad + a_3(E_G\cos\theta_1 + E_G\cos\theta_2)^3 \\ &= E_G^2 a_2 + \left(E_G a_1 + \frac{9}{4}E_G^3 a_3\right)\cos\theta_1 + \left(E_G a_1 + \frac{9}{4}E_G^3 a_3\right)\cos\theta_2 \\ &\quad + E_G^2 a_2 \cos(\theta_1-\theta_2) + E_G^2 a_2\cos(\theta_1+\theta_2) \\ &\quad + \frac{3}{4}E_G^3 a_3 \cos(2\theta_2-\theta_1) + \frac{3}{4}E_G^3 a_3\cos(2\theta_1-\theta_2) \\ &\quad + \frac{3}{4}E_G^3 a_3 \cos(\theta_1+2\theta_2) + \frac{3}{4}E_G^3 a_3\cos(2\theta_1+\theta_2) \\ &\quad + \frac{1}{2}E_G^2 a_2\cos 2\theta_1 + \frac{1}{2}E_G^2 a_2\cos 2\theta_2 + \frac{1}{4}E_G^3 a_3\cos 3\theta_1 + \frac{1}{4}E_G^3 a_3\cos 3\theta_2,\end{aligned}$$

where we identify the DC conversion term, the components at the fundamental frequencies, the second-order intermodulation products (both out of band), the third-order IMPs with a minus sign (in band), the third-order IMPs with a plus sign (both out of band) and finally the second and third harmonics of the fundamental tones. If we restrict the analysis to the in-band output tones we obtain:

$$\begin{aligned}i_D(t) &\approx \left(E_G a_1 + \frac{9}{4}E_G^3 a_3\right)\cos\theta_1 + \left(E_G a_1 + \frac{9}{4}E_G^3 a_3\right)\cos\theta_2 \\ &\quad + \frac{3}{4}E_G^3 a_3 \cos(2\theta_2-\theta_1) + \frac{3}{4}E_G^3 a_3\cos(2\theta_1-\theta_2),\end{aligned}$$

i.e., denoting as $i_D(t)$ the total current at the fundamentals and in-band third-order intermodulation products:

$$i_D(t) = I_D^{(1)}\cos\theta_1 + I_D^{(2)}\cos\theta_2 + I_D^{(+)}\cos(2\theta_2-\theta_1) + I_D^{(-)}\cos(2\theta_1-\theta_2) \tag{8.12}$$

that is:

$$I_D^{(1)} = \left(a_1 + \frac{9}{4}a_3 E_G^2\right)E_G \tag{8.13a}$$

$$I_D^{(2)} = \left(a_1 + \frac{9}{4}a_3 E_G^2\right) E_G \tag{8.13b}$$

$$I_D^{(+)} = \frac{3}{4}a_3 E_G^3 \tag{8.13c}$$

$$I_D^{(-)} = \frac{3}{4}a_3 E_G^3, \tag{8.13d}$$

we have:

$$P_L(f_1) = \frac{1}{2}R_L \left(I_D^{(1)}\right)^2 = \frac{1}{2}R_L \left(a_1 + \frac{9}{4}a_3 E_G^2\right)^2 E_G^2$$

$$P_L(f_2) = \frac{1}{2}R_L \left(I_D^{(2)}\right)^2 = \frac{1}{2}R_L \left(a_1 + \frac{9}{4}a_3 E_G^2\right)^2 E_G^2$$

$$P_L(2f_2 - f_1) = \frac{1}{2}R_L \left(I_D^{(+)}\right)^2 = \frac{1}{2}R_L \left(\frac{3}{4}a_3 E_G^3\right)^2 = \frac{9}{32}a_3^2 R_L E_G^6$$

$$P_L(2f_1 - f_2) = \frac{1}{2}R_L \left(I_D^{(-)}\right)^2 = \frac{1}{2}R_L \left(\frac{3}{4}a_3 E_G^3\right)^2 = \frac{9}{32}a_3^2 R_L E_G^6.$$

We assume as the input available power the power of one of the two tones:

$$P'_{av,in} = \frac{E_G^2}{8R_G} \rightarrow E_G^2 = 8R_G P'_{av,in},$$

from which:

$$P_L(f_1) = P_L(f_2) = 4R_L R_G \left(a_1 + 18 a_3 R_G P'_{av,in}\right)^2 P'_{av,in} \tag{8.14}$$

$$P_L(2f_2 - f_1) = P_L(2f_1 - f_2) \equiv P_{out}\,(\text{IMP}_3)$$
$$= 144 a_3^2 R_G^3 R_L P'^3_{av,in} = H_3 P'^3_{av,in}, \tag{8.15}$$

where H_3 is a *transducer* conversion gain and the transducer gain in linearity is:

$$G_t^{ss} = 4R_L R_G a_1. \tag{8.16}$$

Let us compare the single-tone (8.8) and two-tone characteristics (8.14) at the fundamental(s); assuming for the two-tone case the input power as the total power $P_{av,in} = 2P'_{av,in}$ and as the output power $P_L(f_1) + P_L(f_2)$ we obtain, respectively:

$$P_L(f_0) = 4R_L R_G \left(a_1 - 6\,|a_3|\,R_G P_{av,in}\right)^2 P_{av,in}$$

$$P_L(f_1) + P_L(f_2) = 4R_L R_G \left(a_1 - 9\,|a_3|\,R_G P_{av,in}\right)^2 P_{av,in}.$$

It is therefore evident that, in the two-tone case, compression occurs (slightly) earlier than in the single-tone case. This is basically due to the different PAPR of the two signals. As a consequence, also the two-tone 1 dB compression point will be slightly different from the one detected in the single-tone test.

From (8.15) we see that the IMP$_3$ power is indeed proportional to the third power of the input available power through the third-order coefficient a_3 of the power series. An approximate relation can be derived between the third-order intermodulation product intercept IIP$_3$ and the input power corresponding to the 1 dB compression point in the

8.5 Analysis of Distortion and Power Saturation in a Class A Amplifier

single-tone test. We use as the input variable the power corresponding to one of the tones; from the definition of IIP_3 in (8.1) we obtain, using (8.16) and (8.15):

$$IIP_3 = \sqrt{\frac{G_t^{ss} P_{av,in}'^3}{P_{out}(IMP_3)}} = \sqrt{\frac{4R_L R_G a_1^2 P_{av,in}'^3}{144 a_3^2 R_G^3 R_L P_{av,in}'^3}} = \frac{a_1}{6|a_3|R_G},$$

where we have used the transducer gain instead of the operational gain as in (8.1). However, from (8.11) we have

$$P_{av,in}\big|_{1dB} = 1.8125 \times 10^{-2} \frac{a_1}{|a_3|R_G} = 1.8125 \times 10^{-2} \cdot 6 \cdot IIP_3 = 0.10875 \cdot IIP_3,$$

or, in log units, the IIP_3 is about 9.6 dB larger than the 1 dB compression point. Notice that in this relation we compare the input power of a single-tone $P_{in} - P_{out}$ corresponding to 1 dB compression with the IIP_3 of a two-tone test expressed in terms of the input available power of one tone only. (Of course we should be consistent in using, in both cases, as the input variable the input available power or the input power.) A more consistent definition would result from deriving the compression point from the two-tone $P_{in} - P_{out}$; from (8.14) we obtain:

$$P'_{av,in}\big|_{1dB} = \frac{1 - \sqrt{\alpha}}{9 R_G} \frac{a_1}{|a_3|} = \frac{1.2083 \times 10^{-2}}{R_G} \frac{a_1}{|a_3|}, \qquad (8.17)$$

from which:

$$P'_{av,in}\big|_{1dB} = 1.2083 \times 10^{-2} \cdot 6 \cdot IIP_3 = 7.2498 \times 10^{-2} \cdot IIP_3$$

corresponding to a difference of 11.4 dB between the third-order intermodulation intercept and the 1 dB compression point.

Example 8.7 A FET transcharacteristic can be approximated as a third-order polynomial as:

$$i_D(t) = a_1 e_G + a_2 e_G^2 + a_3 e_G^3.$$

Derive the corresponding descriptive function model around a carrier frequency f_0 and show that the model correctly predicts the in-band third-order intermodulation products and also the two-tone output current at the fundamental.

Solution
Since the starting model is memoriless, the carrier frequency is immaterial. Starting from a single-tone test where $e_G = E_G \cos\theta$, we have already derived the expression of the fundamental output current as:

$$i_{D,ib}(t) = I_{D,ib} \cos\theta = \left(a_1 + \frac{3}{4} a_3 E_G^2\right) E_G \cos\theta,$$

where $i_{D,ib}$ denotes the frequency components of the output current close to the carrier, i.e.:

$$I_{D,ib} = F_P(E_G) E_G, \quad F_P = a_1 + \frac{3}{4} a_3 E_G^2.$$

Notice that, due again to the memoryless nature of the input–output relationship, the phasor is real and there is no phase delay between the input and the output. Moreover, the descriptive function does not include the second-order term of the power series, but only the odd-order terms. Let us now introduce a two-tone input signal, $e_G(t) = E_G \cos\theta_1 + E_G \cos\theta_2$. We can write:

$$e_G(t) = \operatorname{Re}\left[E_G e^{j\theta_1} + E_G e^{j\theta_2}\right] = \operatorname{Re}\left[E_G\left(1 + e^{j\Delta\theta}\right)e^{j\theta_1}\right],$$

with $\Delta\theta = \theta_2 - \theta_1$, so that:

$$\begin{aligned}I_{D,\text{ib}}(t) &= F_P\left(E_G\left|1 + e^{j\Delta\theta}\right|\right) E_G\left(1 + e^{j\Delta\theta}\right) \\ &= a_1 E_G\left(1 + e^{j\Delta\theta}\right) + \frac{3}{4}a_3\left|1 + e^{j\Delta\theta}\right|^2 E_G^3\left(1 + e^{j\Delta\theta}\right).\end{aligned}$$

Taking into account that:

$$\left|1 + e^{j\Delta\theta}\right|^2 = \left(1 + e^{j\Delta\theta}\right)\left(1 + e^{-j\Delta\theta}\right) = 2 + e^{j\Delta\theta} + e^{-j\Delta\theta},$$

we obtain:

$$\begin{aligned}I_{D,\text{ib}}(t) &= a_1 E_G\left(1 + e^{j\Delta\theta}\right) + \frac{3}{4}a_3\left|1 + e^{j\Delta\theta}\right|^2 E_G^3\left(1 + e^{j\Delta\theta}\right) \\ &= a_1 E_G\left(1 + e^{j\Delta\theta}\right) + \frac{3}{4}a_3\left(2 + e^{j\Delta\theta} + e^{-j\Delta\theta}\right)\left(1 + e^{j\Delta\theta}\right) E_G^3 \\ &= \left(a_1 + \frac{9}{4}a_3 E_G^2\right) E_G + \left(a_1 + \frac{9}{4}a_3 E_G^2\right) E_G e^{j\Delta\theta} \\ &\quad + \frac{3}{4}a_3 E_G^3 e^{-j\Delta\theta} + \frac{3}{4}a_3 E_G^3 e^{j2\Delta\theta}.\end{aligned}$$

Recovering the time-domain output current (remember that $\theta = \omega t$):

$$i_{D,\text{ib}}(t) = \operatorname{Re}\left[I_{D,\text{ib}}(t) e^{j\theta_1}\right],$$

we have:

$$\begin{aligned}I_{D,\text{ib}}(t) e^{j\theta_1} &= \left(a_1 + \frac{9}{4}a_3 E_G^2\right) E_G e^{j\theta_1} + \left(a_1 + \frac{9}{4}a_3 E_G^2\right) E_G e^{j\theta_2} \\ &\quad + \frac{3}{4}a_3 E_G^3 e^{j(\theta_1 - \Delta\theta)} + \frac{3}{4}a_3 E_G^3 e^{j(\theta_2 + \Delta\theta)}.\end{aligned}$$

Thus:

$$\begin{aligned}i_{D,\text{ib}}(t) &= \left(a_1 + \frac{9}{4}a_3 E_G^2\right) E_G \cos\theta_1 + \left(a_1 + \frac{9}{4}a_3 E_G^2\right) E_G \cos\theta_2 \\ &\quad + \frac{3}{4}a_3 E_G^3 \cos(2\theta_1 - \theta_2) + \frac{3}{4}a_3 E_G^3 \cos(2\theta_2 - \theta_1),\end{aligned}$$

i.e., comparing with (8.12), we have the same result as in (8.13).

8.5.2 Power Saturation in Class A

The power series technique yields a simplified picture of power compression and distortion that is valid from small-signal conditions to the onset of power compression. Power saturation can be investigated by invoking a strong nonlinear mechanism limiting the maximum swing of the output current, of the output voltage or both, depending on the slope of the device load line. For the sake of definiteness, suppose that the limiting mechanism acts on the output (drain) current, that, for very large values of the input voltage swing, becomes a clipped sinusoid, see Fig. 8.14. The output current waveform admits the Fourier expansion:

$$i_D(t) = \frac{I_{DSS}}{2} + \sum_{n=1,3,5...}^{n=\infty} I_n \cos(n\omega t),$$

where the current bias is at 50 percent of the maximum value, as in the optimum class A bias condition. We are interested, in particular, to the first harmonic coefficient, since the load is tuned and all other harmonics do not dissipate power. We have:

$$I_1 = I_M \left[1 - \frac{2}{\pi} \left(\cos^{-1} \xi - \xi \sqrt{1-\xi^2} \right) \right], \tag{8.18}$$

(see [5, Example 15.14-1]) with:

$$\xi = \frac{I_{DSS}}{2I_M},$$

where I_{DSS} is the drain saturation current while I_M is the peak value of the unclipped waveform, in turn proportional to the v_{GS} swing and therefore to $\sqrt{P_{av,in}}$. Let us define as $\hat{P}_{av,in}$ the input available power such as $I_M = I_{DSS}/2$, leading the device to the limit of nonlinear operation. We clearly have:

$$\xi = \sqrt{\frac{\hat{P}_{av,in}}{P_{av,in}}}. \tag{8.19}$$

Taking into account that in linearity we have $P_{out} = G_{av} P_{av,in}$ while, for $P_{av,in} > \hat{P}_{av,in}$,

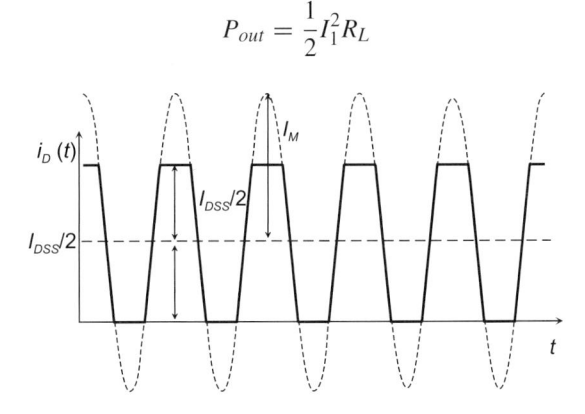

$$P_{out} = \frac{1}{2} I_1^2 R_L$$

Figure 8.14 Clipping model for drain current in class A.

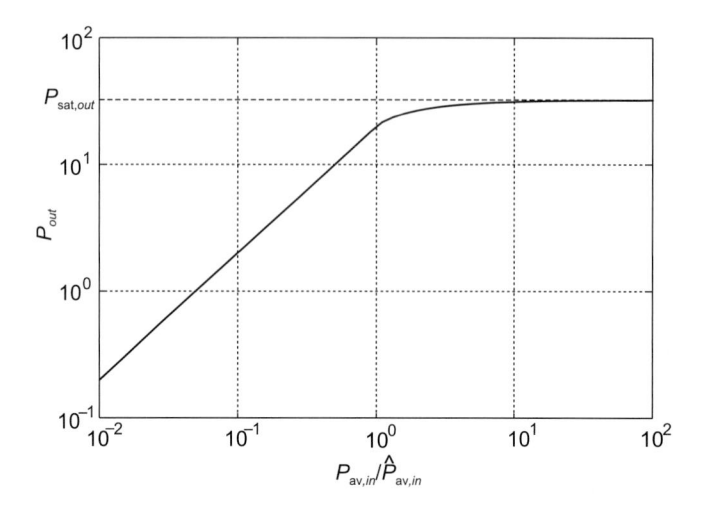

Figure 8.15 Model for power saturation of a class A amplifier: $P_{in} - P_{out}$ curve.

and that:
$$P_{out}\left(\hat{P}_{av,in}\right) = \frac{1}{2}I_M^2 R_L = G_{av}\hat{P}_{av,in},$$

we obtain the following expressions for the output power:
$$P_{out} = G_{av}P_{av,in}, \quad P_{av,in} \leq \hat{P}_{av,in}$$
$$P_{out} = G_{av}P_{av,in} \cdot \left[1 - \frac{2}{\pi}\left(\cos^{-1}\xi - \xi\sqrt{1-\xi^2}\right)\right]^2, \quad P_{av,in} > \hat{P}_{av,in},$$

where ξ is defined in (8.19). The resulting behavior of the $P_{in} - P_{out}$ curve is shown in Fig. 8.15; the saturation power corresponding to $\hat{P}_{av,in}/P_{av,in} \to 0$ and $\xi \to 0$ is, taking into account that $I_1/I_M \to 4\xi/\pi = (4/\pi)\sqrt{\hat{P}_{av,in}/P_{av,in}}$ for $\xi \to 0$:
$$P_{sat,out} = \frac{16}{\pi^2}G_{av}\hat{P}_{av,in},$$

i.e., the saturation power is 1.62 times the output power at the limit of linearity (or 2.1 dB above it).

8.6 Nonlinear Amplifier Classes: AB, B, C

For the sake of definiteness, let us consider a single-tone test. Contrarily to class A amplifiers, in class AB, B and C the input waveform drives the device below threshold for an increasing fraction of the period (exactly half a period in class B, more than half of a period in class C). As a consequence, the output current waveform is an asymmetric clipped sinusoid, with a strong harmonic content. The amplifier behavior is therefore, at a device level, strongly nonlinear; however, due to the tuned load, only the fundamental component of the current will flow in the load resistance. In some cases (e.g., in class B) the output power is, at least in low-power conditions, proportional to the input power; in other cases (class AB and C) the output power increases with the input power, but in a sublinear way. The main reason of introducing class AB to C amplifiers is the increase in efficiency with respect to class A, under two respects: not only the peak efficiency is larger than 50 percent, but also the efficiency decrease in the presence of signal backoff is weaker than in class A amplifiers. Nonlinear amplifiers, however, are inferior to class A amplifiers both in linearity and in gain; moreover, contrarily to class A behavior, backoff is not always an effective way to significantly improve the amplifier distortion. As already stressed, the traditional power amplifier nonlinear classes described in the present section are not able to achieve in practice a 100 percent ideal efficiency, due to the large class C gain penalty; this is possible instead with more innovative concepts like the so-called harmonic loading (class F) and switching (class E) approaches. For convenience, we start with the analysis of the class B amplifier and then discuss the general case (class A to C) in which the amplifier circulation angle is varied from 2π to 0.

8.6.1 Single-Ended Class B Amplifier

For simplicity, let us consider a single-tone test. In the class B amplifier the device input bias is at threshold ($V_{GG} = V_T < 0$), see Fig. 8.17. Thus, the output current is zero in the absence of an input signal. However, the input signal generates a DC drain current component, leading to the amplifier output *self-bias* corresponding to a DC drain current that is proportional to the signal level. The voltage drain bias V_{DD} is instead imposed by the bias T; its value is determined by maximizing the swing of the drain voltage from $V_{DS,k}$ to $V_{DS,br}$, and is in fact the same as in class A. If the load resistance is designed according to its optimum value (corresponding to the maximum drain current sweep) we obtain the drain current and drain voltage waveforms in Fig. 8.17. Notice that, since the drain current is a half-wave rectified sinusoid while v_{DS} is a sinusoid with a DC offset (because of the presence of the tuned load: see the discussion in Example 8.5, that holds for both class A and class B or C), the load line becomes a load curve ideally made of two straight lines, as shown in Fig. 8.16. Let us begin the analysis by developing the drain current into a Fourier series up to the fundamental (higher-order harmonics in fact will flow through the output resonator); we have, taking into account that in maximum power conditions the peak drain current amplitude is I_{DSS}, that the drain current is, in the same conditions:

$$i_{D,M}(t) \approx \frac{I_{DSS}}{\pi} + \frac{I_{DSS}}{2}\sin\omega t + \ldots.$$

By properly choosing the drain bias midway between the knee and breakdown voltages, so as to allow for the maximum drain voltage swing, we obtain the following waveform, that includes a DC and a fundamental component:

$$\begin{aligned}v_{DS}(t) &= \frac{V_{DS,k} + V_{DS,br}}{2} - \frac{V_{DS,br} - V_{DS,k}}{2}\sin\omega t \\ &\approx \frac{V_{DS,br}}{2} - \frac{V_{DS,br}}{2}\sin\omega t,\end{aligned}$$

Power Amplifiers

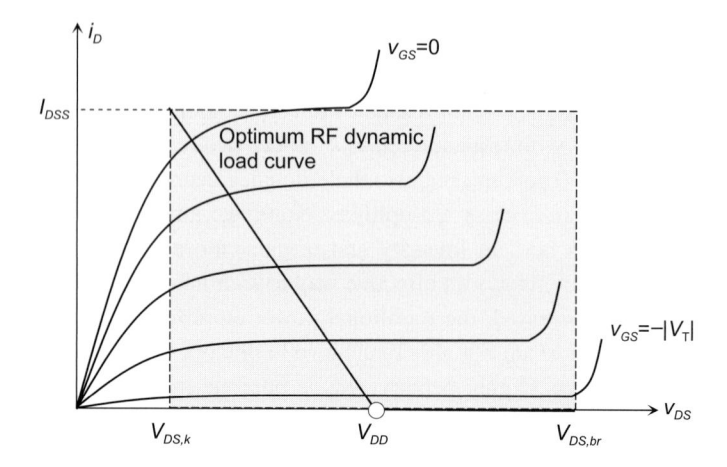

Figure 8.16 Load curve of a class B amplifier with tuned load in maximum output power conditions and optimum load.

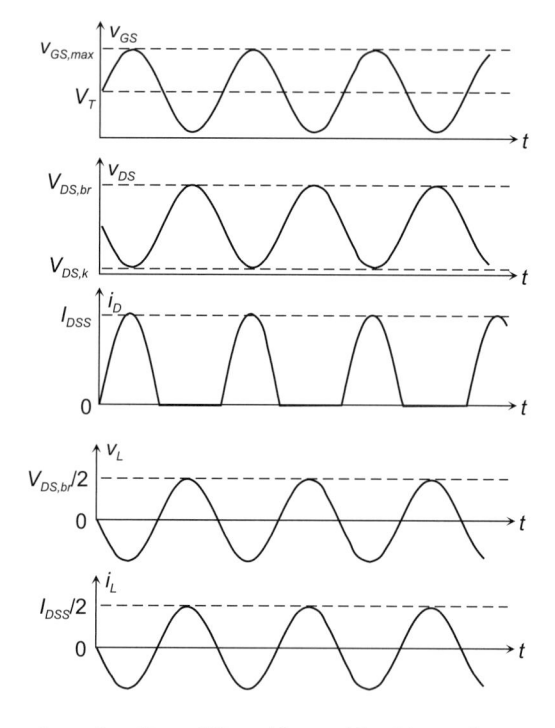

Figure 8.17 Waveforms for a class B amplifier with tuned load in maximum output power conditions and optimum load.

with drain bias:

$$V_{DD} \approx \frac{V_{DS,br}}{2}.$$

The power from the DC power supply (P_{DC}) can be derived as the product of the DC components of i_D and v_{DS}; in maximum output power conditions $P_{DC} = P_{DC,M}$ where:

$$P_{DC,M} = \frac{V_{DS,br} I_{DSS}}{2\pi}.$$

The optimum class B load resistance can be evaluated by considering the ratio between the load voltage and the load current in maximum power conditions:

$$i_L(t) = -i_{D1}(t) = -\frac{I_{DSS}}{2} \sin \omega t$$

$$v_L(t) = v_{DS1}(t) = -\frac{V_{DS,br}}{2} \sin \omega t,$$

where i_{D1} and v_{DS1} are the fundamental components of i_D and v_{DS}, respectively. Thus, we have:

$$R_{L,o} = \frac{v_L(t)}{i_L(t)} \approx \frac{V_{DS,br}}{I_{DSS}},$$

corresponding to the same result as in class A (as discussed later, the optimum resistance generally varies from class A to class C). With the optimum load, the maximum RF load power will be:

$$P_{RF,M} = \frac{1}{2}\left(\frac{I_{DSS}}{2}\right)^2 R_{L,o} = \frac{1}{2}\left(\frac{I_{DSS}}{2}\right)^2 R_{L,o} = \frac{I_{DSS} V_{DS,br}}{8},$$

yielding the maximum efficiency:

$$\eta_{B,M} = \frac{P_{RF,M}}{P_{DC,M}} = \frac{\pi}{4} \approx 78\%.$$

The class B amplifier has therefore a larger maximum efficiency than the class A amplifier. Moreover, the efficiency drop in the presence of backoff (i.e., for output powers less than the maximum) is less significant than is class A. In fact, in class B the DC dissipated power depends on the output power level; its maximum (corresponding to maximum output power) was denoted as $P_{DC,M}$. If the drain current is smaller than the maximum value, we will have:

$$i_D(t) \approx \frac{I_{D,p}}{\pi} + \frac{I_{D,p}}{2} \sin \omega t + \ldots = -\frac{2I_L}{\pi} - I_L \sin \omega t + \ldots,$$

where $I_{D,p} < I_{DSS}$ is the peak value of the drain current waveform and I_L is the peak value of the fundamental component of the load current. The efficiency can be generally written as:

$$\eta_B = \frac{P_{RF}}{P_{DC}} = \frac{P_{RF}}{P_{DC}} \frac{P_{RF,M}}{P_{RF,M}} \frac{P_{DC,M}}{P_{DC,M}} = \eta_{B,M} \frac{P_{RF}}{P_{RF,M}} \frac{P_{DC,M}}{P_{DC}}. \quad (8.20)$$

However, denoting as I_L the peak load current, with $I_{L,M} = I_{DSS}/2$, we also have:

$$\frac{P_{RF}}{P_{RF,M}} = \frac{\frac{1}{2} I_L^2 R_{L,o}}{\frac{1}{2} I_{L,M}^2 R_{L,o}} = \frac{I_L^2}{I_{L,M}^2} \rightarrow \frac{I_L}{I_{L,M}} = \sqrt{\frac{P_{RF}}{P_{RF,M}}}$$

$$\frac{P_{DC,M}}{P_{DC}} = \frac{\frac{1}{\pi} V_{DS,br} I_{L,M}}{\frac{1}{\pi} V_{DS,br} I_L} = \frac{I_{L,M}}{I_L} = \sqrt{\frac{P_{RF,M}}{P_{RF}}}.$$

Therefore, we obtain from (8.20):

$$\eta_B = \eta_{B,M} \frac{P_{RF}}{P_{RF,M}} \sqrt{\frac{P_{RF,M}}{P_{RF}}} = \eta_{B,M} \sqrt{\frac{P_{RF}}{P_{RF,M}}} = \eta_{B,M} \cdot \sqrt{\text{OBO}}. \qquad (8.21)$$

Thus, if the input power is backed off with respect to the maximum value, the efficiency decreases as the square root of the backoff rather than linearly with the OBO as in class A.

While the class B amplifier has a better maximum efficiency with respect to the class A amplifier, and also a more favorable behavior in backoff, the class B gain will be lower than in class A. In fact, in maximum output power conditions the input voltage peak to peak swing is $2|V_T|$ (class B) rather than $|V_T|$ (class A). Thus, the needed generator available power is four times larger in class B than in class A. However, the maximum output power is the same in both classes; it follows that the class B transducer gain is 6 dB below that of class A:

$$G_{t,B} \approx G_{t,A} - 6 \text{ dB}.$$

This gain penalty is made even worse in typical FETs, where the transconductance drops close to the threshold voltage. This suggests, in practical amplifiers, to avoid strict class B and rather use the intermediate class AB, where the device is biased at the input above threshold.

8.6.2 From Class A to Class C Amplifier

Class A and B amplifiers can be considered as particular cases of a tuned-load amplifier whose input voltage has maximum value $v_{GS} \approx 0$ and whose minimum value becomes increasingly negative, from $v_{GS} = -|V_T|$ (class A in maximum power conditions) to $v_{GS} \approx -2|V_T|$ (class B in maximum power conditions) to $v_{GS} < -2|V_T|$ (class C). While in class A the device always operates above or at threshold, in class B the drain current is zero for half of the period; in class C $i_D = 0$ for more than half of the period. We can therefore perform a general analysis of the traditional amplifier classes (from A to C) by varying the *circulation angle* α from 2π (class A) to 0 (the limit of class C). The drain current can be written, as a function of time and the circulation angle:

$$i_D(\theta(t)) = \begin{cases} I_q + I_p \cos\theta, & 0 \leq \theta \leq \alpha/2 \\ 0, & \alpha/2 \leq \theta \leq \pi, \end{cases} \qquad (8.22)$$

where $\theta = 2\pi t/T$, T period. The circulation angle $\alpha = 2\pi t_{on}/T$ denotes the time interval t_{on} where $i_D > 0$. Thus $\alpha = 2\pi$ in class A, $2\pi < \alpha < \pi$ in class AB, $\alpha = \pi$ in class B, $\alpha < \pi$ in class C. The behavior of the drain current for several values of the circulation angle is shown in Fig. 8.18, where the parameters I_q and I_p are defined. The condition corresponding to zero current is, from (8.22):

$$\cos(\alpha/2) = -I_q/I_p,$$

8.6 Nonlinear Amplifier Classes: AB, B, C

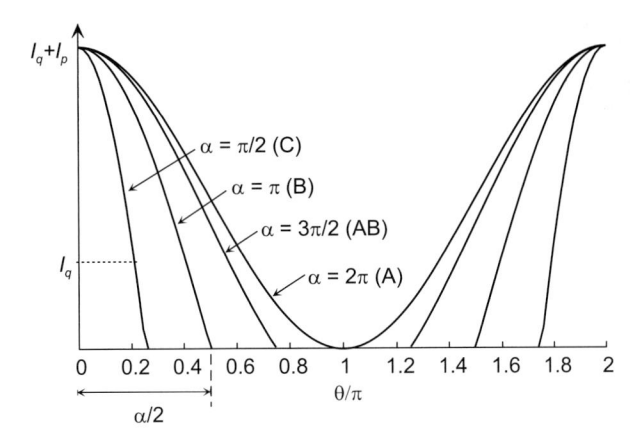

Figure 8.18 Behavior of the drain current for different values of the circulation angle α, going from class A to C.

while the maximum current is:

$$I_P = I_p + I_q.$$

Notice that in maximum power conditions we have $I_P = I_{DSS}$. Eq. (8.22) can be therefore written in the form:

$$i_D(\theta(t)) = \begin{cases} I_P \dfrac{\cos\theta - \cos(\alpha/2)}{1 - \cos(\alpha/2)}, & 0 \leq \theta \leq \alpha/2 \\ 0, & \alpha/2 \leq \theta \leq \pi. \end{cases}$$

Since the load is tuned, the output device voltage will only include the fundamental and DC components, according to the analysis already performed in class A and B tuned load amplifiers. If the load resistance is optimum (i.e., allowing for the maximum output voltage swing), the load voltage (equal to the fundamental component of v_{DS}) will have peak amplitude $V_{DS,P} \approx V_{DS,br}/2$. Concerning the input voltage v_{GS}, this is a sinusoidal waveform of peak value $V_{GS,P}$ and DC value $V_{GS,DC}$; it can be generally written as:

$$v_{GS}(\theta(t)) = V_{GS,P} \cos\theta + V_{GS,DC}. \tag{8.23}$$

Notice that, in maximum power conditions, $v_{GS} = -|V_T|$ for $\theta = \alpha/2$ and its maximum value is, in a FET, approximately 0. This leads to the conditions:

$$v_{GS}(\alpha/2) = V_{GS,P} \cos(\alpha/2) + V_{GS,DC} = -|V_T| \tag{8.24}$$

$$v_{GS}(0) = V_{GS,P} + V_{GS,DC} \approx 0, \tag{8.25}$$

from which we obtain:

$$V_{GS,DC} = -\frac{|V_T|}{1 - \cos(\alpha/2)} \tag{8.26}$$

$$V_{GS,P} = \frac{|V_T|}{1 - \cos(\alpha/2)}. \tag{8.27}$$

To analyze the behavior of the drain current it is convenient to expand it in Fourier series, with coefficients:

$$I_n = \frac{1}{\pi} \frac{I_P}{1 - \cos(\alpha/2)} \int_{-\alpha/2}^{\alpha/2} [\cos\theta - \cos(\alpha/2)] \cos(n\theta)\, d\theta$$

$$= \frac{2I_P}{\pi} \frac{n \sin(\alpha/2) \cos(\alpha n/2) - \cos(\alpha/2) \sin(\alpha n/2)}{n(1 - n^2)[1 - \cos(\alpha/2)]}.$$

The average (DC) value is therefore:

$$I_0 = \frac{1}{2\pi} \frac{I_P}{1 - \cos(\alpha/2)} \int_{-\alpha/2}^{\alpha/2} [\cos\theta - \cos(\alpha/2)]\, d\theta$$

$$= \frac{I_P}{2\pi} \frac{2\sin(\alpha/2) - \alpha \cos(\alpha/2)}{1 - \cos(\alpha/2)},$$

while the harmonic coefficients are:

$$I_1 = \lim_{n \to 1} \frac{2 I_P}{\pi} \frac{n \sin(\alpha/2) \cos(\alpha n/2) - \cos(\alpha/2) \sin(\alpha n/2)}{n(1 - n^2)[1 - \cos(\alpha/2)]} = \frac{I_P}{2\pi} \frac{\alpha - \sin(\alpha)}{1 - \cos(\alpha/2)}$$

$$I_2 = -\frac{I_P}{\pi} \frac{2 \sin(\alpha/2) \cos(\alpha) - \cos(\alpha/2) \sin(\alpha)}{3[1 - \cos(\alpha/2)]}$$

$$I_3 = -\frac{I_P}{\pi} \frac{3 \sin(\alpha/2) \cos(3\alpha/2) - \cos(\alpha/2) \sin(3\alpha/2)}{12[1 - \cos(\alpha/2)]}$$

$$I_4 = -\frac{I_P}{2\pi} \frac{4 \sin(\alpha/2) \cos(2\alpha) - \cos(\alpha/2) \sin(2\alpha)}{15[1 - \cos(\alpha/2)]}.$$

The behavior of the harmonic components of i_D (normalized with respect to I_P) is shown in Fig. 8.19. The DC component decreases from class A to AB to B to C, while the fundamental component increases from class A to AB to decrease in class B to the same value as in class A. The higher harmonics generally increase from class A to class C.

Let us consider now the optimum load resistance (conductance) from class A to class C. Take into account that in all classes $V_{DD} = V_{DS,DC} \approx V_{DS,br}/2$ to allow for maximum output voltage swing $V_{DS,P} \approx V_{DS,br}/2$. The load conductance can be obtained as the ratio between the peak fundamental component of the load current $-I_1$ and the peak component of the load voltage $V_{DS,br}/2$:

$$G_{Lo}(\alpha) = \left[\frac{I_P}{2\pi} \frac{\alpha - \sin\alpha}{1 - \cos(\alpha/2)}\right]\left[\frac{V_{DS,br}}{2}\right]^{-1}$$

$$= \frac{I_{DSS}}{V_{DS,br}} \left[\frac{1}{\pi} \frac{\alpha - \sin\alpha}{1 - \cos(\alpha/2)}\right] = G_{Lo,A} \left[\frac{1}{\pi} \frac{\alpha - \sin\alpha}{1 - \cos(\alpha/2)}\right].$$

The behavior of the optimum conductance is shown in Fig. 8.20 (right-hand scale) with respect to the circulation angle. Notice that the optimum load is the same in class A and B, while in deep class C it becomes an open circuit – meaning that in such conditions no power reaches the load.

8.6 Nonlinear Amplifier Classes: AB, B, C

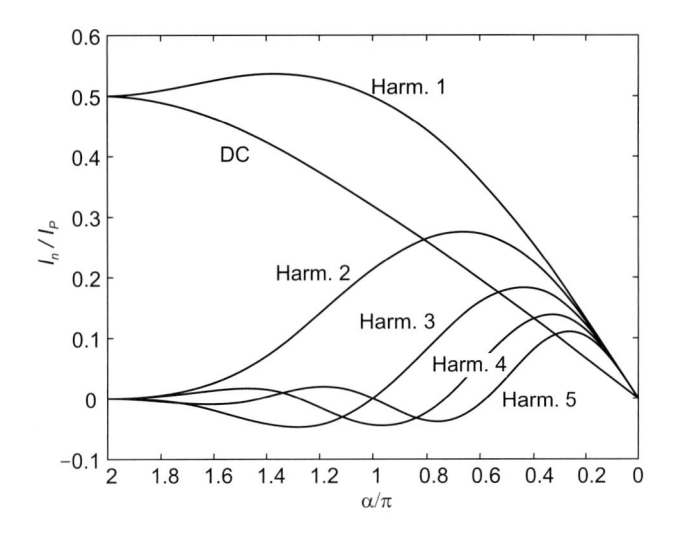

Figure 8.19 Behavior of the harmonic components of i_D as a function of the circulation angle α.

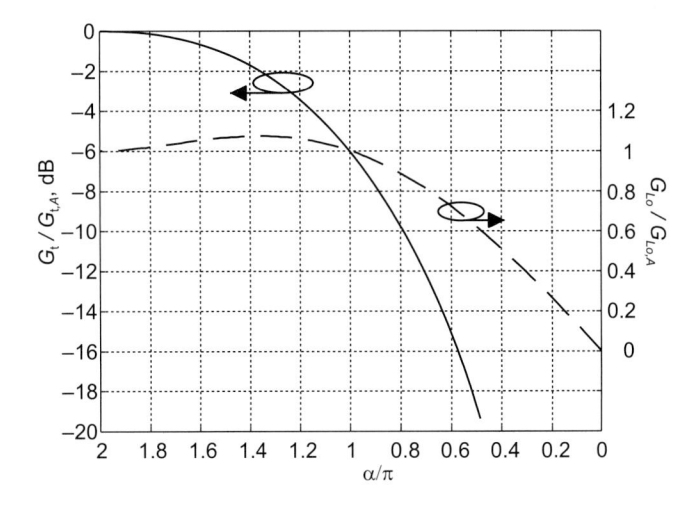

Figure 8.20 Behavior of the transducer gain from class A to C as a function of the circulation angle (left-hand scale); behavior of the optimum load conductance as a function of the circulation angle (right-hand scale) normalized vs. the optimum class A conductance.

Let us now estimate the efficiency and power added efficiency from class A to class C. The DC power from the power supply is:

$$P_{DC} = V_{DS,DC} I_0,$$

where $V_{DS,DC} \approx V_{DS,br}/2$ to allow for the maximum output voltage swing. The RF power on the load in maximum power conditions is, on the other hand:

$$P_{RF} = \frac{1}{2} I_1 V_{DS,P} = \frac{1}{2} I_1 \cdot \frac{V_{DS,br}}{2} = \frac{I_1 V_{DS,DC}}{2};$$

therefore, the maximum efficiency (i.e., the efficiency corresponding to the maximum output power conditions) will be:

$$\eta(\alpha) = \frac{P_{RF}}{P_{DC}} = \frac{I_1}{2I_0} = \frac{1}{2}\frac{\alpha - \sin(\alpha)}{2\sin(\alpha/2) - \alpha\cos(\alpha/2)}.$$

The PAE depends on the amplifier operational gain. However, evaluating the operational gain with the simplified device model exploited here (with zero input power) is impossible; we will therefore derive a relation between the class A transducer gain and the transducer gain for arbitrary α, and assume that it approximately holds also for the operational gain. Let us denote as $G_t(\alpha)$ the transducer gain for arbitrary circulation angle; we have (R_g is the generator internal resistance):

$$G_t(\alpha) = \frac{\frac{I_1 V_{DC}}{2}}{\frac{V_{GS,P}^2}{4R_g}} = \frac{I_P V_{DC} R_g}{\pi V_T^2}[\alpha - \sin(\alpha)][1 - \cos(\alpha/2)],$$

but in class A, $G_t(2\pi) \equiv G_{t,A}$, where:

$$G_{t,A} = \frac{4 I_P V_{DC} R_g}{V_T^2}.$$

Therefore:

$$G_t(\alpha) = G_{t,A}\frac{[\alpha - \sin(\alpha)][1 - \cos(\alpha/2)]}{4\pi}.$$

The transducer gain decreases from class A to C, to finally vanish in deep class C, as shown in Fig. 8.20, left hand scale. In particular, in class B:

$$G_{t,B} = G_t(\pi) = \frac{1}{4}G_{t,A},$$

leading, as already discussed, to a 6 dB gain penalty of class B vs. class A.

With the approximation:

$$G_{op}(\alpha) \approx G_{op,A}\frac{[\alpha - \sin(\alpha)][1 - \cos(\alpha/2)]}{4\pi},$$

we can now evaluate the power added efficiency as:

$$\text{PAE}(\alpha) = \eta(\alpha)\left[1 - \frac{1}{G_{op}(\alpha)}\right] \approx \frac{1}{2}\frac{\alpha - \sin(\alpha)}{2\sin(\alpha/2) - \alpha\cos(\alpha/2)}$$
$$\times \left\{1 - \frac{1}{G_{op,A}}\frac{4\pi}{[\alpha - \sin(\alpha)][1 - \cos(\alpha/2)]}\right\}.$$

The efficiency and the PAE are shown in Fig. 8.21 as a function of the circulation angle; for the PAE, two values (10 dB and 100 dB) have been assumed for the class A gain. The efficiency increases monotonically from class A to deep class C, but the 100 percent efficiency limit achieved in class C is practically useless, since it corresponds to zero gain. The PAE behavior reveals that the maximum efficiency is achieved for large circulation angles only for very high class A gain.

8.6 Nonlinear Amplifier Classes: AB, B, C

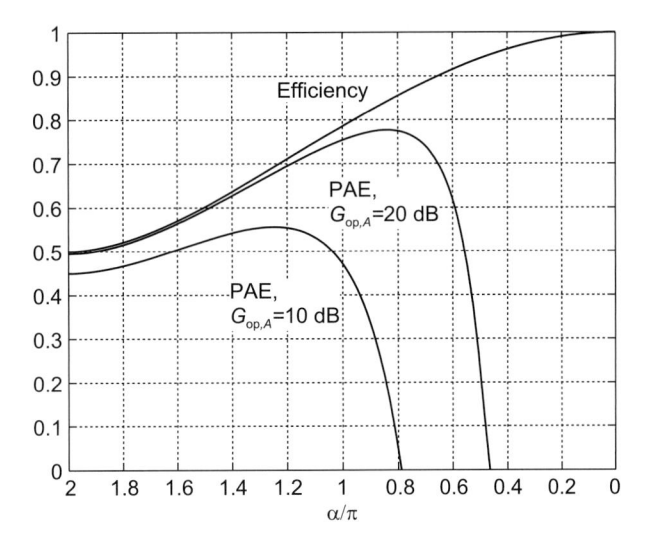

Figure 8.21 Efficiency and power added efficiency as a function of the circulation angle for several values of the class A operational gain.

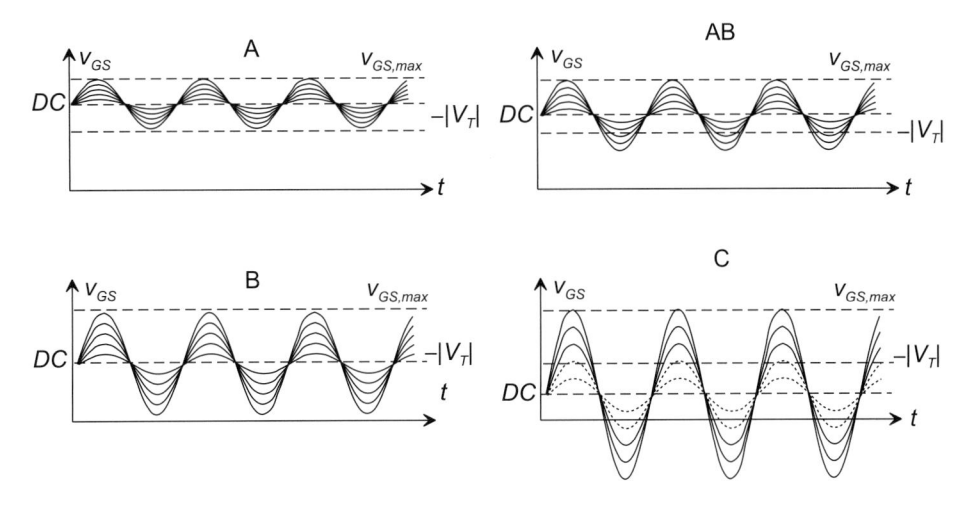

Figure 8.22 Increasing the input available power in different operation classes: class A, α is constant, independent of the v_{GS} peak amplitude; class AB, for low input power the device operates in class A, then α decreases with the v_{GS} peak amplitude; class B, α is independent of the v_{GS} peak amplitude; class C, for low input power the device is turned off, after turn-on α depends on the v_{GS} peak amplitude.

While in class A and B the circulation angle is independent from the input power, leading to a linear $P_{in} - P_{out}$ relationship before compression, this is not any more true for class AB or class C amplifiers, where the circulation angle, and therefore the amplifier gain, depends on the input power, see Fig. 8.22. Notice that in class C the amplifier is turned off for low input power, since $v_{GS} < -|V_T|$; after turn-on the circulation angle will depend on the input power.[6] Therefore, the $P_{in} - P_{out}$ characteristic of a class AB or class C amplifier is sublinear, see Example 8.8 for the class AB case.

Example 8.8 Consider a class AB amplifier. Analyze the effect of power backoff due to the input power dependence of the circulation angle and derive the $P_{in} - P_{out}$ relation.

Solution
Suppose that the input bias of the class AB amplifier is set so as to maximize the input voltage swing. In such conditions, the circulation angle is α_M. We have then from (8.23), the input waveform, repeated here for convenience:

$$v_{GS}(\theta(t)) = V_{GS,P} \cos\theta + V_{GS,DC}$$

with $V_{GS,DC}$ and $V_{GS,P}$ given, respectively, as a function of the circulation angle by (8.26) and (8.27) evaluated for $\alpha = \alpha_M$:

$$V_{GS,DC} = -\frac{|V_T|}{1 - \cos(\alpha_M/2)}$$

$$V_{GS,P} = \frac{|V_T|}{1 - \cos(\alpha_M/2)}.$$

We now back the amplifier off the maximum power conditions by decreasing the peak input voltage amplitude to $V'_{GS,P} < V_{GS,P}$; since the input bias voltage is kept constant we have:

$$v_{GS}(\theta(t)) = V'_{GS,P} \cos\theta + V_{GS,DC},$$

while the new circulation angle α is given by the condition:

$$v_{GS}(\alpha/2) = V'_{GS,P} \cos(\alpha/2) + V_{GS,DC} = -|V_T|,$$

or:

$$\cos(\alpha/2) = -\frac{V_{GS,DC} + |V_T|}{V'_{GS,P}} = \frac{\cos(\alpha_M/2)}{1 - \cos(\alpha_M/2)} \frac{|V_T|}{V'_{GS,P}}.$$

The input available power now reads:

$$P_{av,in} = \frac{\left(V'_{GS,P}\right)^2}{4R_g} = \frac{(V_{GS,P})^2}{4R_g} \left(\frac{V'_{GS,P}}{V_{GS,P}}\right)^2 = P_{sat,in} \left[\frac{\cos(\alpha_M/2)}{\cos(\alpha/2)}\right]^2,$$

where $P_{sat,in}$ is the input power corresponding to the maximum output power condition; we thus have that the circulation angle depends on the input available power as:

$$\cos(\alpha/2) = \sqrt{\frac{P_{sat,in}}{P_{av,in}}} \cos(\alpha_M/2). \tag{8.28}$$

---

[6] This turn-on feature is exploited in the Doherty amplifier, see Sec. 8.7.3.

8.6 Nonlinear Amplifier Classes: AB, B, C

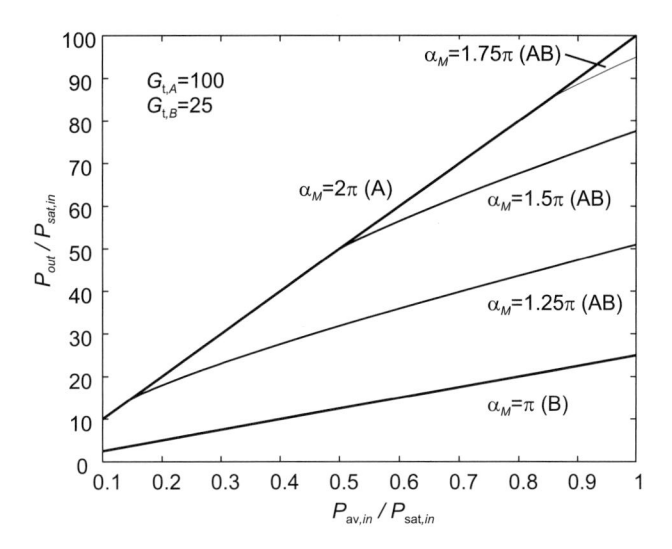

Figure 8.23 Nonlinear behavior of class AB amplifiers as a function of the input power. We assumed $G_{t,A} = 100$ and $P_{sat,in} = 1$ mW. Notice that the plot is in natural, not log, units. See Example 8.8 for the discussion.

Therefore, the output power will be:

$$P_{out} = G_t(\alpha) P_{av,in} = G_{t,A} \frac{[\alpha - \sin(\alpha)][1 - \cos(\alpha/2)]}{4\pi} P_{av,in}.$$

In other words, the output power generally is a nonlinear function of the input power, since $\alpha = \alpha(P_{av,in})$, see (8.28). Notice that, when decreasing the input power, the amplifier is back in class A operation; this happens when $\alpha = 2\pi$, i.e., when:

$$V'_{GS,P} \leq \frac{\cos(\alpha_M/2)}{1 - \cos(\alpha_M/2)} |V_T|.$$

When class A operation is reached, the amplifier ultimately remains in class A also for lower power. Thus, the class AB $P_{in} - P_{out}$ is a straight line with slope 1 (in log scale) for low power, where the device operates in class A; for large enough input power the device enters class AB and the circulation angle increases with increasing power, leading to a sublinear characteristic, as shown in Fig. 8.23. Note that this behavior is typical of class AB (and also of class C), while both in class A and B the circulation angle does not depend on the input power.

The intermodulation product behavior of class AB, B and C amplifiers cannot be described in a simple way as for the class A amplifiers. As shown in Fig. 8.24, the typical behavior of intermodulation products as a function of the input power does not follow a third-power law as in class A, and, more important, shows the so-called *sweet spots* where the IMP_3 undergo cancellation. In the example shown just one sweet spot is present, but multiple sweet spots are also detected in practical amplifiers.

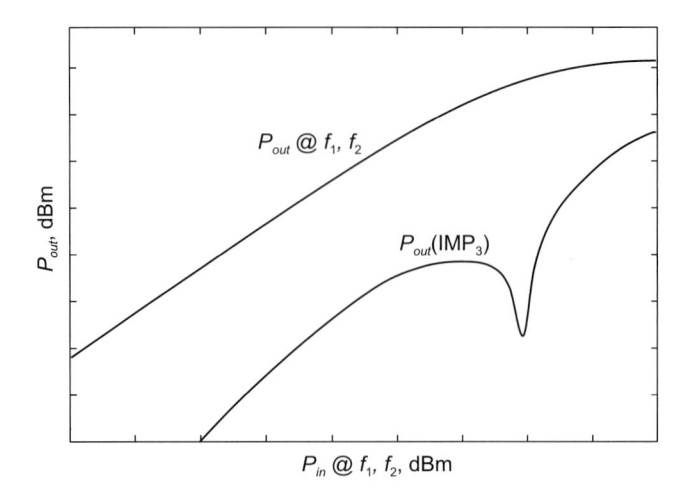

Figure 8.24 Example of two-tone test of a class AB amplifier. The intermodulation product behavior vs. the input power does not follow a third-power law and *sweet spots* are present.

8.7 High-Efficiency Amplifiers

In the conventional amplifier classes (class A to C) the amplifier maximum efficiency can be only increased at the expense of a larger gain penalty. As a limiting case, class C efficiency can reach 100 percent but with zero gain. Moreover, the efficiency deteriorates in the presence of output backoff, leading to a low average efficiency when signals with large PAPR are considered. During the last few years, strategies have been devised to design amplifiers with a theoretical 100 percent efficiency but with a nonzero gain; examples are the Class F amplifier (Sec. 8.7.1) and the Class E switching amplifier (Sec. 8.7.2). Along a different line, amplifier architectures have been developed able to maintain a large efficiency in a certain range of output backoff; a typical example is the Doherty amplifier, see Sec. 8.7.3.

8.7.1 Harmonic Loading: The Class F Amplifier

The efficiency of the class B amplifier can be significantly increased by providing a proper load to the signal harmonics at the device output. The technique is called *harmonic loading*, leading to the so-called *class F* amplifier [6]. To clarify the harmonic load concept, consider a single-ended class B amplifier, in which the load impedance $Z_L(nf_0)$ at each harmonic can be controlled. Suppose that the input voltage is the same as in class B; the drain current will have a half-wave rectified sinusoid waveform, see Fig. 8.25. Let us postulate an output v_{DS} waveform having complementary support with respect to i_D. This is possible if the drain voltage waveform is a square wave, see Fig. 8.25.[7] In such conditions, the instantaneous power dissipated by the device is zero,

---

[7] Another possible solution is to square the output current and make the output voltage sinusoidal; this is the so-called *inverse class F* amplifier, see [1].

8.7 High-Efficiency Amplifiers

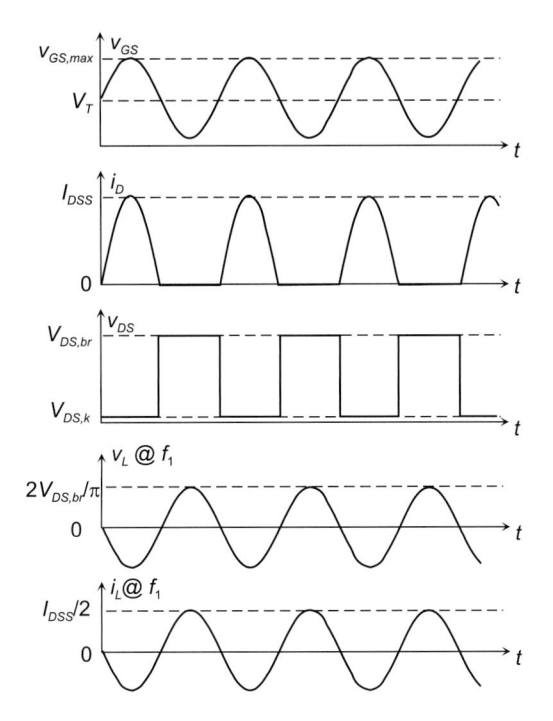

Figure 8.25 Waveforms of class F amplifier.

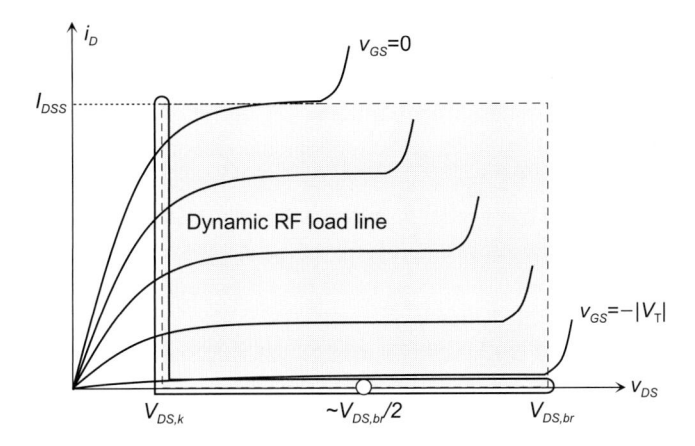

Figure 8.26 Ideal load curve for a class F amplifier.

and therefore the power from the DC source is completely converted into RF, with 100 percent efficiency. The resulting behavior of the dynamic load line is shown in Fig. 8.26.

Expanding into a Fourier series the drain voltage:

$$v_{DS}(t) = V_{DS,br} \left[\frac{1}{2} - \frac{2}{\pi} \sum_{n=1,3,5...} \frac{1}{n} \sin(n\omega t) \right],$$

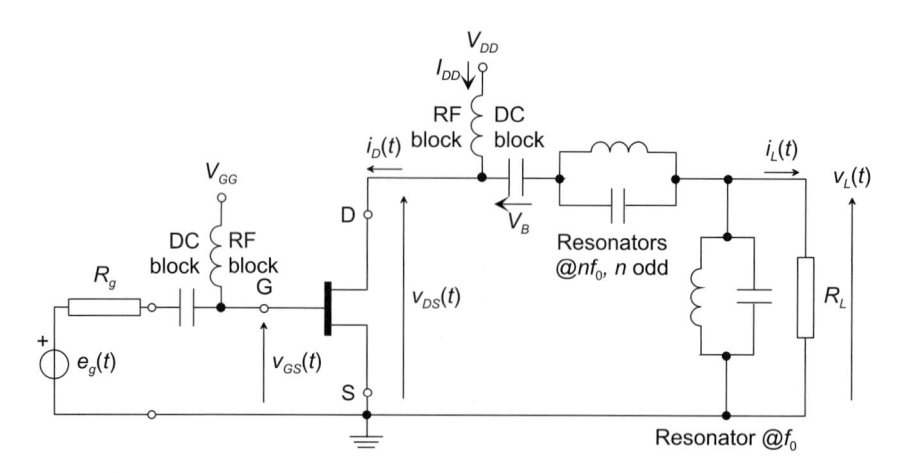

Figure 8.27 Class F amplifier with harmonic control through lumped resonators.

we see that the drain voltage is zero at all even-order harmonics, apart from the DC that is, as usual, $V_{DS,br}/2$ to allow for the maximum output voltage swing. For the drain current we have instead:

$$i_D(t) = I_{DSS}\left[\frac{1}{\pi} + \frac{1}{2}\sin(\omega t) - \frac{2}{\pi}\sum_{n=2,4,6...}\frac{1}{n^2-1}\cos(n\omega t)\right],$$

so that all odd-order harmonics, apart from the first one, are zero. The device should therefore connected to a load that is a short circuit at all even-order harmonics, while it is an open at all odd-order harmonics. As a matter of principle, such a load can be implemented as in Fig. 8.27. At the fundamental, all odd-order harmonics resonators are short circuits, while the parallel resonator at the fundamental is an open, and a resistive load is seen. At all even-order harmonics the odd-order harmonic resonator are shorts, and so is the fundamental resonator; thus the load at even-order harmonics is a short. Finally, at all odd-order harmonics one of the resonators is an open in series, and therefore the load seen by the device is an open.

Let us now evaluate the optimum load resistance and RF power for an ideal class F amplifier. Taking into account that the fundamental component of the load current is equal to the fundamental component of the drain current, with a minus sign, we have that the RF load power is, in maximum power conditions:

$$P_{RF} = \frac{1}{2}\cdot\frac{2V_{DS,br}}{\pi}\cdot\frac{I_{DSS}}{2} = \frac{V_{DS,br}I_{DSS}}{2\pi},$$

larger than the class A or B power; the optimum load will be:

$$R_{Lo} = \frac{4V_{DS,br}}{\pi I_{DSS}}.$$

The DC power from the bias supply is the product of the DC component of the drain current and voltage, i.e.:

$$P_{DC} = \frac{V_{DS,br}}{2} \cdot \frac{I_{DSS}}{\pi} = P_{RF},$$

i.e.:

$$\eta_F = 100\%.$$

Thus, the theoretical efficiency of a class F amplifier with control over all harmonics is 100 percent. The practical implementation of the harmonic loading is however difficult for two reasons: first, high-frequency resonators having an almost ideal behavior cannot be easily implemented due to losses; second, the output device capacitance cannot be effectively compensated at high frequency, ultimately shorting the load. The harmonic control strategy can be therefore optimized taking into account that only a limited number of harmonics (typically up to the third) can be actually controlled. For a complete analysis on the effect of controlling the second and third harmonic see [3, Ch. 5, 7, 9]; a discussion on the achievable class F efficiency with harmonic control up to the fifth harmonic is reported in [6].

If the fundamental frequency f_0 is low enough to make the effect of the output device capacitance negligible also at higher harmonics, and if the large size of a transmission element designed at a low f_0 is not an issue, a distributed class F implementation making use of a quarter-wave transformer is possible, see Fig. 8.28. At the fundamental, the quarter-wave line of impedance Z_0 transforms the load Z_0^2/R_{Lo} into R_{Lo}; at $2f_0$ the quarter-wave line has length $\lambda/2$ and therefore the load seen by the device is a short; at $3f_0$ the line works again as an impedance transformer and the device load is an open. The same behavior follows at all even and odd harmonics, respectively. The approach is of course limited by the frequency dispersion of the line effective permittivity and by the line losses, that increase with increasing frequency.

The class F amplifier as described so far is intrinsically narrowband, due to the presence of the resonant load. This limitation is overcome by means of the so-called *continuous class F* amplifier, see [7, 8], where the load is controlled at the fundamental

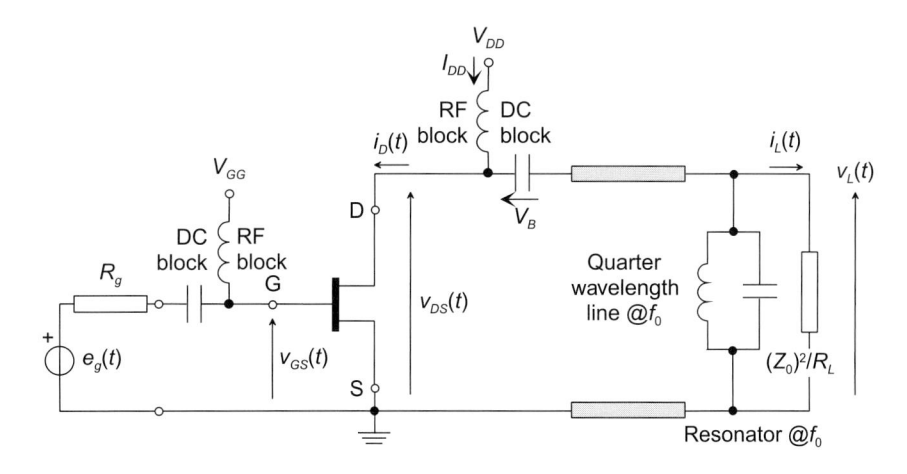

Figure 8.28 Class F amplifier with quarter wave transformer.

and at the second and third harmonic. The continuous class F design is based on the fact that complementary (or almost complementary) output voltage and current waveforms can be obtained for a comparatively wide range of first- and second-harmonic terminations (the so-called *design space*) while keeping the third-order termination reactive. This allows a set of terminations to be synthesized that, one wide frequency range, enable the output waveforms to satisfy the complementarity condition. As an example, [7] presents a 0.5-1.2 GHz amplifier with flat 40 dBm output power, 10 dB gain, and efficiency larger than 65 percent from 0.55 to 1.10 GHz. (Because of the low-order approach the theoretical maximum efficiency will be anyway below 100 percent.)

8.7.2 The Switching Class E Power Amplifier

Power amplifiers are classified as *switching amplifiers* when active devices operate as controlled switches, as in digital circuits, rather than as output current generators controlled by the input voltage or current. For the sake of definiteness, let us focus on FET-based implementations. In switching amplifiers, the gate voltage only controls the FET output impedance. If the input voltage drives the device below threshold (OFF state), the drain current is approximately zero, and the FET almost behaves an open circuit, with large output impedance between drain and source. In the ON state, the drain-source voltage is low, the device operates in the linear or triode region, and the drain current is decided by the load connected to the device output. When compared to class B or C amplifiers, switching amplifiers have the same behavior in the OFF state, but a markedly different one when the device operates above threshold. In order to achieve digital operation, the gate voltage could ideally be a square wave with a low level below threshold, and a high level large enough to overdrive the FET allowing it to operate in the linear or triode region, i.e., with a low drain-source voltage. In practice, an input square-wave voltage is difficult to obtain due to the large number of harmonics involved, and the input voltage is therefore sinusoidal, swinging from below threshold to a value providing the required overdrive; if the FET supports a large input overdrive (as in MOS devices), this can well approximate the switching effects of a square wave.

Switching amplifiers clearly introduce a new amplification paradigm, since the output power is not proportional to the input power, as for conventional amplifiers. However, such systems will be anyway able to transfer, with an increased power, the information content from the input to the output: in fact, phase and frequency input modulations are maintained at the output. Moreover, switching amplifiers are able to minimize the power dissipated by the active devices, thus achieving, at least ideally, 100 percent efficiency. Since at any time instant either the device v_{DS} (ON state) or the drain current i_D (OFF state) are zero, the instantaneous power dissipated by the device $v_{DS}i_D$ is always, ideally, null. We can conclude that switching amplifiers can provide high-efficiency power amplification of purely phase or frequency (analog or digital) modulated signals, where information does not depend on the signal amplitude.

After this overview, let us now focus now on the Class E amplifier, a quite popular high-efficiency switching amplifier introduced and patented by Sokal in 1975 [9, 10]; for a complete treatment see [3, 11]. Class E amplifiers can ideally achieve 100 percent

8.7 High-Efficiency Amplifiers

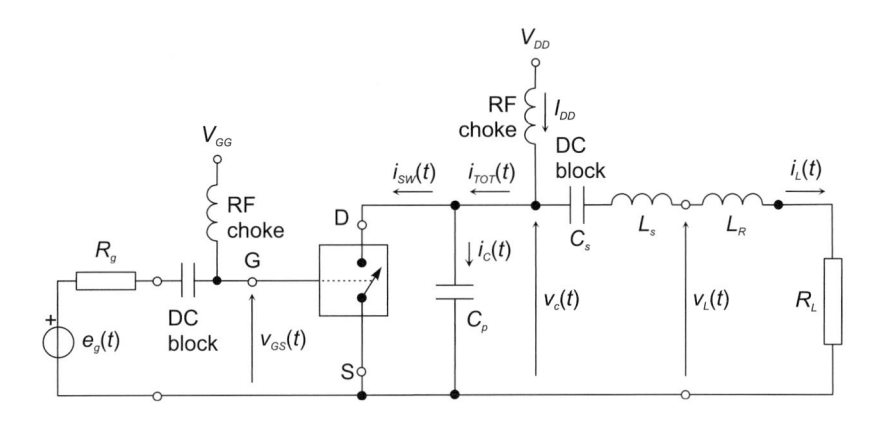

Figure 8.29 Electrical scheme of class E switching amplifier.

efficiency; however, to this aim the zero active device instantaneous dissipation already mentioned is not enough. To maximize efficiency, we also have to deliver to the load no power at the harmonics (remember that only the load fundamental power is relevant for the input-output information transfer). A classical scheme of Class E amplifier is reported in Fig. 8.29; a resonator in series with the load is present, that is transparent to the fundamental, but isolates the load from the higher harmonics. As in other amplifier classes, a bias T (the RF choke, together with a DC blocking capacitor that coincides here for simplicity with C_s) is needed to block the RF signal from entering the DC supply. Finally, the circuit schematic does not include any dissipative elements (e.g., resistors) that would decrease the amplifier efficiency.

A closed-form analysis of this circuit can be performed with the following simplifying assumptions:

1. the series resonator centered at the fundamental is lossless, with a large enough quality factor Q to block all load current harmonics, besides the DC component;
2. the RF choke is ideal, with DC bias current I_{DD};
3. the active device is either an open (OFF state) or a short (ON state) circuit, according to whether the applied gate voltage is below or above threshold, respectively.

Although these assumptions are not exactly verified in real-world implementations, they enable to gain insight into the amplifier operation and to derive first-approximation design expressions for the output circuit elements, that can be used as a starting point for CAD optimization.

From assumptions 1 and 2, we obtain that (1) the load current i_L is purely sinusoidal, i.e., $i_L(t) = -I_{RF} \sin \omega_0 t$ where ω_0 is the fundamental angular frequency, and (2) the total current $i_{TOT}(t) = i_{SW}(t) + i_C(t)$ (where i_{SW} is the drain current of the active device, modeled as a switch in Fig. 8.29, and $i_C(t)$ is the current in the capacitor C_P) is $i_{TOT}(t) = -i_L(t) + I_{DD}$, i.e., it has the same waveform as $i_L(t)$, apart from a DC offset.

The input voltage forces the device in the ON state during the T_{ON} interval and in the OFF state during the T_{OFF} interval, respectively, see Fig. 8.30; the waveform duty cycle δ_c is defined as the ratio between the ON interval and the total waveform period:

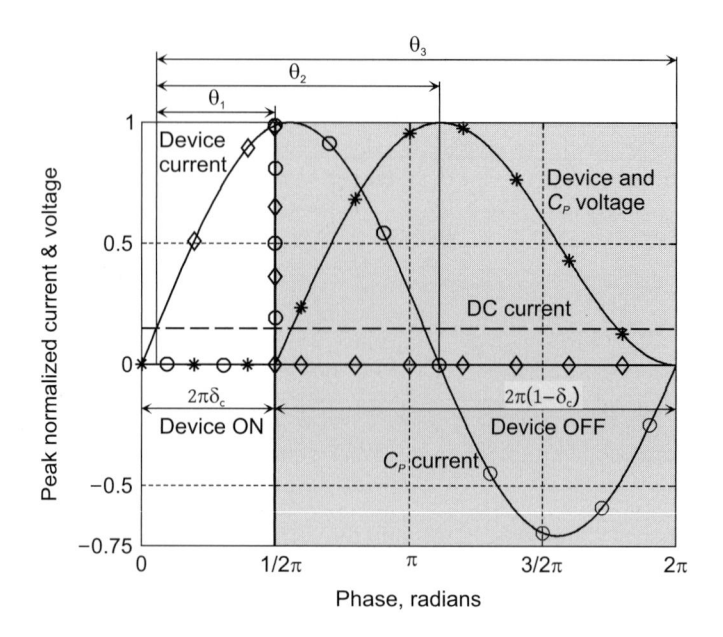

Figure 8.30 Waveforms of class E amplifier: device (diamonds) and capacitor (circles) current, normalized to the peak value of i_{TOT}; output device voltage (asterisks), normalized to the peak value.

$$\delta_c = \frac{T_{ON}}{T_{ON} + T_{OFF}}.$$

The analysis of the components of i_{TOT} immediately yields:

$$i_C(t) = 0, \quad t \in T_{ON}$$
$$i_C(t) = i_{TOT}(t) = I_{DD} + I_{RF} \sin \omega_0 t, \quad t \in T_{OFF}$$
$$i_{SW}(t) = i_{TOT}(t) = I_{DD} + I_{RF} \sin \omega_0 t, \quad t \in T_{ON}$$
$$i_{SW}(t) = 0, \quad t \in T_{OFF}.$$

Fig. 8.30 shows the device current $i_{SW}(t)$ (diamonds) and the capacitor current $i_C(t)$ (circles) for $\delta_c = 0.25$, normalized with respect to the current peak value; the two currents have complementary support and their sum is i_{TOT}. Notice that the ON interval is shaded in light gray, the OFF interval in dark gray. Let us consider now the voltage across the device output, v_{SW}. During the ON interval $v_{SW} = 0$ according to assumption 3; during the OFF interval it coincides with the voltage across C_P, and can hence be obtained by integrating the capacitor current $i_C(t)$:

$$v_{SW}(t) = \frac{1}{C_P} \int_{\theta_1/\omega_0}^{t} i_C(t) \cdot dt = \frac{I_{DD}(\omega_0 t - \theta_1) - I_{RF}(\cos \omega_0 t - \cos \theta_1)}{\omega_0 C_P}, \quad (8.29)$$

where $\theta_1 = \omega_0 t_1$ is the angle corresponding to the ON interval. The device voltage v_{SW}, normalized to its peak value, is shown in Fig. 8.30 (asterisks) for $\delta_c = 0.25$.

8.7 High-Efficiency Amplifiers

A few remarks are now in order:

1. The capacitor current $i_C(t)$ is zero for $\omega_0 t_2 = \theta_2$ and $\omega_0 t_3 = \theta_3$ (see Fig. 8.30). This imposes a relationship between the sinusoidal component of i_{TOT} (equal, apart from the sign, to the load current) and I_{DD}. In fact, since $I_{DD} + I_{RF} \sin\theta_2 = I_{DD} + I_{RF} \sin\theta_3 = 0$, we have:

$$\frac{I_{DD}}{I_{RF}} = -\sin\theta_2 = -\sin\theta_3, \qquad (8.30)$$

where, since $\theta_3 > \theta_2 > \pi$, $\sin\theta_2 = \sin\theta_3 < 0$ and thus $I_{DD}/I_{RF} > 0$.

2. In steady-state conditions, the capacitor charges and discharges completely during the OFF interval, so that its charge (and therefore its voltage) is zero during the ON interval. Therefore, the total area under the $i_C(t)$ curve during the OFF interval is zero; this also implies that at the end of the OFF interval $v_{SW}(t_3) = 0$ (point $\omega_0 t_3 = \theta_3$ in Fig. 8.30). More precisely, for $t \to t_3$, $v_{SW} \to 0$ with zero time derivative. Taking into account (8.29) this implies:

$$\frac{I_{DD}}{I_{RF}} = \frac{\cos\theta_3 - \cos\theta_1}{\theta_3 - \theta_1}. \qquad (8.31)$$

3. The switch peak current I_{PK} is given by:

$$I_{PK} = I_{DD} + I_{RF} = I_{DD}\left(1 - \frac{1}{\sin\theta_3}\right) = I_{RF}(1 - \sin\theta_3). \qquad (8.32)$$

Clearly, one must ensure that $I_{PK} < I_M$ where I_M is the maximum current the device can sustain ($I_M \approx I_{DSS}$ for a Schottky gate FET).

4. The DC supply drain voltage V_{DD} is given by the DC value of $v_C(t)$, and therefore it can be derived from its time average. Taking into account (8.29), (8.30) and (8.31), we have:

$$V_{DD} = \frac{\omega_0}{2\pi}\int_{\theta_1/\omega_0}^{\theta_3/\omega_0} v_C(t)\mathrm{d}t = -I_{RF}\frac{(\sin\theta_3 - \sin\theta_1)^2}{4\pi\omega_0 C_p \sin\theta_3} = I_{DD}\frac{(\sin\theta_3 - \sin\theta_1)^2}{4\pi\omega_0 C_p (\sin\theta_3)^2}. \qquad (8.33)$$

5. The switch peak voltage $V_{PK} < V_M$, where $V_M \approx V_{DS,br}$, the device breakdown voltage, occurs at $\omega_0 t_2 = \theta_2$. In fact, for $t = t_2 = \theta_2/\omega_0$ the capacitor current is zero and therefore the capacitor and device voltages have a maximum. Considering (8.29) and taking into account (8.30) we obtain:

$$V_{PK} = v_{SW}(t_2) = \frac{I_{DD}}{\omega_0 C_P}\frac{\sin\theta_2(\theta_2 - \theta_1) + \cos\theta_2 - \cos\theta_1}{\sin\theta_2}. \qquad (8.34)$$

6. The load impedance Z_L imposes the ratio between the load voltage and current at the fundamental frequency. Since the fundamental component of v_{SW} (that coincides with the load voltage) has both in-phase and quadrature components with respect to the load current i_L, the load impedance must be complex, a rather unique feature of class E amplifiers. From the analysis, we obtain that the imaginary part of Z_L is positive, and can therefore synthesized by the inductance L_R, see Fig. 8.29. Let us define as $V_{I,SW}$ and $V_{Q,SW}$ the in-phase and quadrature components of v_{SW} with

respect the load current; taking into account that i_L is assumed as sinusoidal, we immediately have:

$$V_{I,SW} = \frac{\omega_0}{\pi} \int_0^T v_{SW}(t) \sin \omega_0 t \cdot dt$$

$$V_{Q,SW} = \frac{\omega_0}{\pi} \int_0^T v_{SW}(t) \cos \omega_0 t \cdot dt,$$

where T is the period of the fundamental. Using (8.29), accounting for (8.30), (8.31), and considering that $v_{SW} \neq 0$ only during the OFF interval, i.e., for $\theta_1/(2\pi) < t < \theta_3/(2\pi)$, we obtain after straightforward but lengthy manipulations the following expressions:

$$V_{I,SW} = -I_{RF} \frac{1}{2\pi\omega_0 C_P} (\sin\theta_3 - \sin\theta_1)^2 \tag{8.35}$$

$$V_{Q,SW} = -I_{RF} \frac{1}{2\pi\omega_0 C_P} \left(\theta_3 - \theta_1 + \frac{\sin 2\theta_3}{2} + \frac{\sin 2\theta_1}{2} - 2\sin\theta_3\cos\theta_1 \right). \tag{8.36}$$

Assuming $Z_L = R_L + j\omega_0 L_R$ we have, taking into account that the fundamental component of i_L is $-I_{RF}$:

$$V_{I,SW} = -I_{RF} R_L$$

$$V_{Q,SW} = -I_{RF} \omega_0 L_R,$$

we finally obtain from (8.35) and (8.36) the relationships:

$$R_L = \frac{1}{2\pi\omega_0 C_P} (\sin\theta_3 - \sin\theta_1)^2 \tag{8.37}$$

$$\omega_0 L_R = \frac{1}{2\pi\omega_0 C_P} \left(\theta_3 - \theta_1 + \frac{\sin 2\theta_3}{2} + \frac{\sin 2\theta_1}{2} - 2\sin\theta_3\cos\theta_1 \right). \tag{8.38}$$

Let us now verify that the amplifier efficiency is, ideally, 100 percent, as expected. To this aim, using (8.37) and (8.31) and taking into account (8.33), we have that the RF output power P_{RF} equals the power from the DC supply, P_{DC}:

$$P_{RF} = \frac{1}{2} I_{RF}^2 R_L = \frac{1}{4\pi\omega_0 C_P} (\sin\theta_3 - \sin\theta_1)^2 \frac{I_{DD}^2}{\sin\theta_3^2} = V_{DD} I_{DD} = P_{DC}. \tag{8.39}$$

As the last step, we need to relate the phases θ_1, θ_2 and θ_3 to the phase of the gate control voltage phase. Since $\theta_3 - \theta_1 = 2\pi(1 - \delta_c)$ (see Fig. 8.30), we can rewrite (8.31) as:

$$\cos(\theta_1 - 2\pi\delta_c) - \cos\theta_1 = \sin(\theta_1 - 2\pi\delta_c)(\theta_3 - \theta_1) = \sin(\theta_1 - 2\pi\delta_c)2\pi(1 - \delta_c).$$

Applying well known trigonometric relationships, after isolating terms $\cos\theta_1$ and $\sin\theta_1$, we finally find the expression of θ_1 as a function of δ_c:

$$\theta_1 = \pi \cdot K_\pi - \arctan\left[\frac{\cos(2\pi\delta_c) - 1 - 2\pi(1-\delta_c)\sin(2\pi\delta_c)}{2\pi(1-\delta_c)\cos(2\pi\delta_c) + \sin(2\pi\delta_c)}\right], \tag{8.40}$$

8.7 High-Efficiency Amplifiers

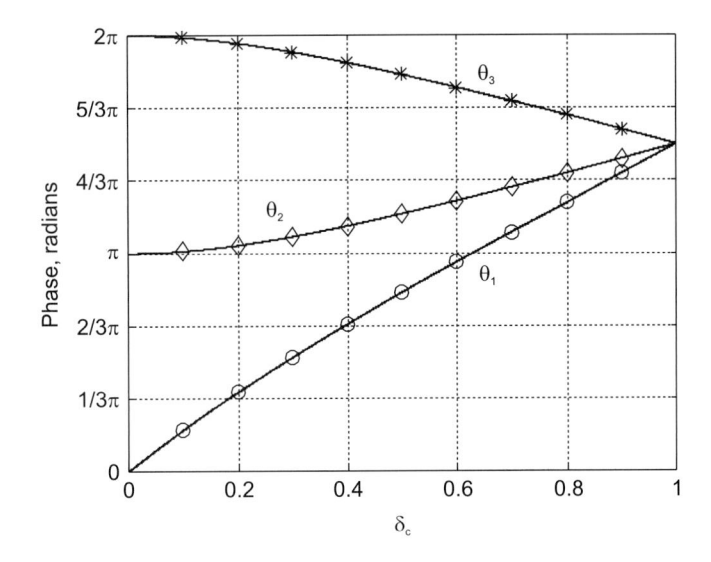

Figure 8.31 Behavior of the angles θ_1 (circles), θ_2 (diamonds), and θ_3 (asterisks), defined in Fig. 8.31, vs. δ_c.

where $K_\pi = 0$ when $\delta_c \leq \delta_{c0} \approx 0.285$, and $K_\pi = 1$ when $\delta_c > \delta_{c0}$.[8] From θ_1, the angles θ_2 and θ_3 can be identified by inspection of Fig. 8.30 as:

$$\theta_3 = \theta_1 + 2\pi(1 - \delta_c) \tag{8.41}$$
$$\theta_2 = 3\pi - \theta_3 = \pi(1 + 2\delta_c) - \theta_1. \tag{8.42}$$

The above expressions of θ_1, θ_2, and θ_3 as a function of δ_c are the basis for the amplifier design; the behavior of θ_1, θ_2, and θ_3 when sweeping δ_c from 0 to 1 is reported in Fig. 8.31, that can be used for the amplifier graphical design instead of the analytical expressions (8.40), (8.42) and (8.41).

We now come to the amplifier design, taking into account the constraints dictated by the device limits, i.e., $I_{PK} \leq I_M$, and $V_{PK} \leq V_M$, where I_{PK} and V_{PK} are defined in (8.32) and (8.34), respectively, I_M and V_M are the maximum drain current and maximum drain (breakdown) voltage V_M. In the following design procedure we will assume for the previous inequalities the upper limit, i.e., $I_{PK} = I_M$ and $V_{PK} = V_M$. Having assigned δ_c and ω_0, the design steps are as follows:

1. Derive θ_1, θ_2 and θ_3 from (8.40), (8.42) and (8.41), or graphically from Fig. 8.31;
2. Derive the DC current I_{DD} (graphically from Fig. 8.33, trace with asterisks), and I_{RF} using (8.32):

[8] The parameter δ_{c0} corresponds to a zero of the arctan argument denominator. The value $\delta_{c0} \simeq 0.28485$ is the first numerical solution of the transcendental equation $\tan(2\pi \delta_c) = 2\pi(\delta_c - 1)$. The two-valued parameter K_π ensures that $\sin \theta_2 = \sin \theta_3 < 0$ for any δ_c.

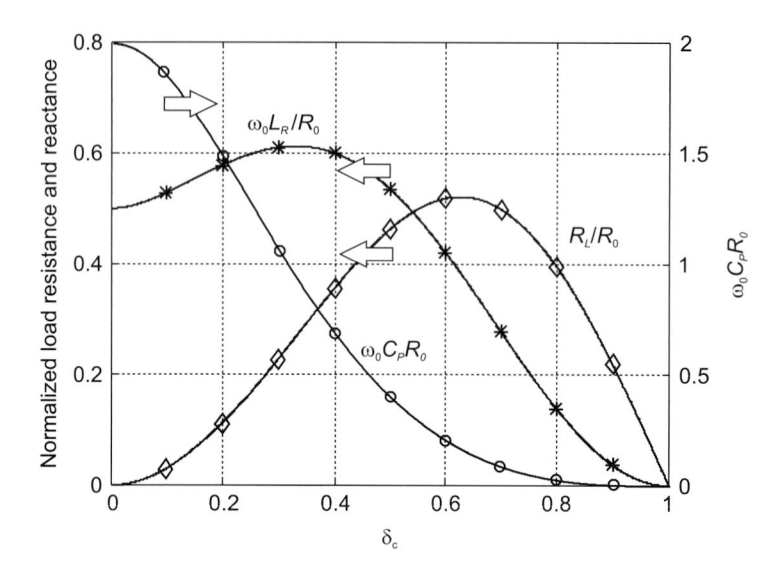

Figure 8.32 Left axis: real (diamonds) and imaginary (asterisks) part of load impedance, normalized to class A resistance for maximum power R_0, vs. δ_c. Right axis: C_P susceptance (circles) normalized to $1/R_0$ vs. δ_c.

$$I_{DD} = I_{PK}\frac{\sin\theta_3}{\sin\theta_3 - 1} = I_M\frac{\sin\theta_3}{\sin\theta_3 - 1}$$

$$I_{RF} = I_{PK}\frac{1}{1 - \sin\theta_3} = I_M\frac{1}{1 - \sin\theta_3}.$$

3. Obtain the capacitor susceptance $\omega_0 C_P$ from (8.34) or, graphically, from Fig. 8.32, trace with circles:

$$\omega_0 C_P = \frac{1}{R_0}\frac{\sin\theta_2(\theta_2 - \theta_1) + \cos\theta_2 - \cos\theta_1}{\sin\theta_2 - 1}, \qquad (8.43)$$

where $R_0 = V_M/I_M$ is the optimum load resistance for class A and B power amplifiers. From $\omega_0 C_P$, the capacitance C_P can be derived, since ω_0 is known.

4. Find the DC bias voltage using (8.33) and previous expressions, or graphically from Fig. 8.33, trace with diamonds:

$$V_{DD} = V_M \frac{(\sin\theta_2 - \sin\theta_1)^2}{4\pi \sin\theta_2 [\sin\theta_2(\theta_2 - \theta_1) + \cos\theta_2 - \cos\theta_1]}.$$

5. Find the load resistance R_L and reactance $\omega_0 L_R$ from (8.37) and (8.38), or graphically from Fig. 8.32, trace with diamonds and asterisks, respectively:

$$R_L = R_0 \frac{(\sin\theta_3 - \sin\theta_1)^2(\sin\theta_2 - 1)}{2\pi[\sin\theta_2(\theta_2 - \theta_1) + \cos\theta_2 - \cos\theta_1]}$$

$$\omega_0 L_R = R_L \frac{\theta_3 - \theta_1 + \sin\theta_3\cos\theta_3 + \sin\theta_1\cos\theta_1 - 2\sin\theta_3\cos\theta_1}{(\sin\theta_3 - \sin\theta_1)^2},$$

8.7 High-Efficiency Amplifiers

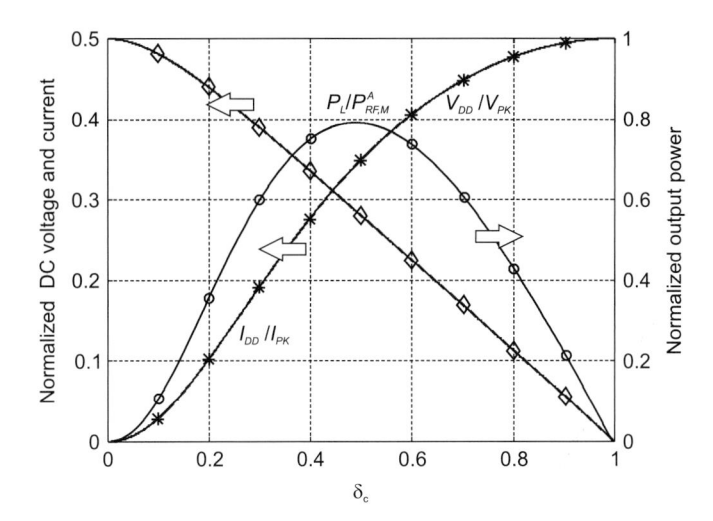

Figure 8.33 Left axis: DC voltage normalized to V_{PK} (diamonds), DC current normalized to I_{PK} (asterisks), vs. δ_c. Right axis: Output power normalized to class A maximum power (circles), vs. δ_c.

where R_0 is again the optimum class A load. From the load reactance the inductance can be derived, since ω_0 is known.

6. Evaluate the RF output power as:

$$P_L = P_{RF,M}^A \frac{2(\sin\theta_3 - \sin\theta_1)^2}{\pi(\sin\theta_3 - 1)[\sin\theta_2(\theta_2 - \theta_1) + \cos\theta_2 - \cos\theta_1]}. \tag{8.44}$$

The expression yields the class E maximum output power as a function of δ_c (through θ_1, θ_2 and θ_3), and of the maximum class A power $P_{RF,M}^A = V_M I_M/8$, see (8.3). The RF power is shown in Fig. 8.33, trace with circles, as a function of δ_c.

From Fig. 8.33 we see that a value of the duty cycle δ_c exists, that maximizes the output power P_L. The optimum value can be derived numerically and is $\delta_c \approx 50\%$, and, in the optimum condition, $P_L \approx 0.8 \cdot P_{RF,M}^A = 0.8 \cdot P_{RF,M}^B$, where $P_{RF,M}^B$ is the maximum class B power. The maximum power is therefore 80 percent of the maximum power that can be obtained, with the same device, in class A or B; this power penalty corresponds however to a much better (albeit less than 100 percent in practice) efficiency. We conclude that, when no other constraints forcing the choice of a specific duty cycle are specified, class E amplifiers are typically designed to operate with $\delta_c \approx 50\%$.

Some comments are now in order concerning the capacitance value derived in (8.43), that completely neglects the device internal output capacitance C_{DS}. This can be approximately taken into account introducing a modified external capacitance C_P' such as:

$$C_{DS} + C_P' = C_P \rightarrow C_P' = C_P - C_{DS}.$$

Since C_{DS} is fixed, this expression, together with (8.43) defines, for a specific device, the maximum class E operating frequency $f_{E,\max}$, corresponding to $C'_P = 0$ or $C_P = C_{DS}$; working at higher frequency would in fact require a negative nonphysical value of C'_P. We obtain from (8.43):

$$f_{E,\max} = \frac{1}{2\pi C_{DS} R_0} \frac{\sin\theta_2(\theta_2 - \theta_1) + \cos\theta_2 - \cos\theta_1}{\sin\theta_2 - 1}. \tag{8.45}$$

Considering that (see Fig. 8.32) the capacitor susceptance monotonically decreases with δ_c, no optimum value for δ_c exists maximizing $f_{E,\max}$. For the typical value $\delta_c = 0.5$ we obtain from (8.45):

$$f_{E,\max} \simeq 0.06/(C_{DS} R_0) = 0.06 I_M/(C_{DS} V_M).$$

For the same value $\delta_c = 0.5$, Fig. 8.32 also yields $R_L \approx 0.46 R_0$ and $\omega_0 L_R \approx 0.53 R_0$.

8.7.3 The Doherty Amplifier

During the last few years, the request for increasing data rate in wireless systems has forced the adoption of increasingly complex signal modulation schemes. An example is the QAM modulation, see Sec. 1.2.1; QAM uses baseband signals with simultaneous phase and amplitude modulation, with a constellation including up to 1024 points deployed on a 32 × 32 regular grid. The RF signal derived from the upconversion of the QAM modulation, and ultimately handled in the transmission chain by the power amplifier, clearly exhibits a variable envelope. In such cases, the average efficiency of the power amplifier, related to the average output power, can be completely unsatisfactory even when using a power amplifier with large maximum efficiency. In fact, defining the Peak to Average Power Ratio (PAPR) as the ratio between the maximum and the average signal power, we have for the NQAM scheme (see Example 8.2):

$$\text{PAPR}|_{\text{dB}} = 4.77 + 20\log_{10}\left(\sqrt{N} - 1\right) - 10\log_{10}(N - 1).$$

For instance, in a QAM constellation of 1024 equiprobable symbols we have PAPR ≈ 5 dB. In this case, a class A amplifier with 50 percent theoretical maximum efficiency (scaling down like the backoff), the average efficiency will be around 16 percent only. In a variable envelope signal context, the power amplifier maximum efficiency is therefore important, but the behavior of the efficiency as a function of the of the backoff is even more important. Since the efficiency of conventional amplifiers (e.g., class A and B) drops with increasing backoff, proper amplifier architectures have to be devised to obtain constant efficiency vs. backoff, at least in a certain backoff range, thus allowing for the efficient amplification of high-PAPR signals.

In the QAM example discussed above, a large PAPR occurs in a single transmitted channel. Another reason for a large difference between the maximum and the average power arises in multi-channel amplifiers, where different channels are activated in a random way according to the existing traffic. Notice, however, that while in the last case the dynamics of the variation of the power level is comparatively slow, in the QAM case such a variation has the same bandwidth of the transmitted signal.

Several techniques have been proposed in the past to maintain the amplifier efficiency close to the maximum, also in the presence of several dB of output power backoff. One of these is *envelope tracking*, where the power amplifier bias is dynamically adapted to the instantaneous envelope, so that the power amplifier always works as it were under maximum power conditions. However, envelope tracking requires a complex control system to be implemented around the amplifier, and has limited bandwidth.

A simpler solution, entirely based on the architecture of the power amplifier, is the Doherty power amplifier (DPA). This scheme was initially conceived by William Doherty in 1936 [12] within the framework of FM radio broadcasting, and implemented with the vacuum tube technology available at that time. The DPA has gained a renewed interest during the last few years, in a completely different technological context, as a high-efficiency amplifier solution in the presence of high-PAPR digitally modulated signals.

The DPA is composed of two power stages (here, FET based) working in tandem, and interacting is such a way as to keep the overall DPA efficiency (almost) independent of the applied power level, at least in a certain backoff range. The two combined power stages are referred to as the Main and the Auxiliary (or Peak) amplifier. For low input power, the output power is only provided by the Main amplifier. The Auxiliary amplifier is switched on after the input power has reached a certain level; after switch-on, it modulates the Main load, and adds its power to the output power generated by the Main. The Main amplifier typically is class B (or AB, or possibly F), while the Auxiliary can be conveniently implemented as a class C amplifier; this allows the Auxiliary to be turned on automatically at a certain input power level and input voltage swing.

The DPA principle is based on the fact that the output power and efficiency of power amplifiers generally depend on the load termination, that determines the slope of the output dynamic load line in the $I_D - V_{DS}$ plane. In classical amplifier schemes, the load is constant, and the maximum output power is obtained when the instantaneous working point, swinging along the load line, reaches the two limiting conditions of the amplifier dynamics (the off-state breakdown and the linear region). In maximum output power conditions, also the efficiency is maximized, while a further increase in the input power drives the device into compression, thus reducing the power added efficiency.

In the Doherty scheme, the Main load resistance is initially constant and set to a value larger than the optimum one, that is proportional to the ratio between the maximum output voltage and current swings. This suboptimal load forces the Main to reach its maximum voltage swing for a certain input power leading to the output power $P_{out} = \gamma^2 P_{M,M}$, $0 \leq \gamma \leq 1$, where $P_{M,M}$ is the Main maximum output power. In such conditions, the output current is lower than the maximum device current. When $P_{out} > \gamma^2 P_{M,M}$, the Main load resistance is progressively reduced, thus allowing the output current to increase, with the same maximum voltage swing, up to the maximum current condition. This allows the output power to further increase while keeping the efficiency constant and independent from the output power.

The modulation of the load resistance of the Main amplifier is obtained by exploiting, for input powers above a certain threshold, the (initially off) Auxiliary stage. This is driven by a fraction of the RF input signal, obtained, e.g., through a divider or directional

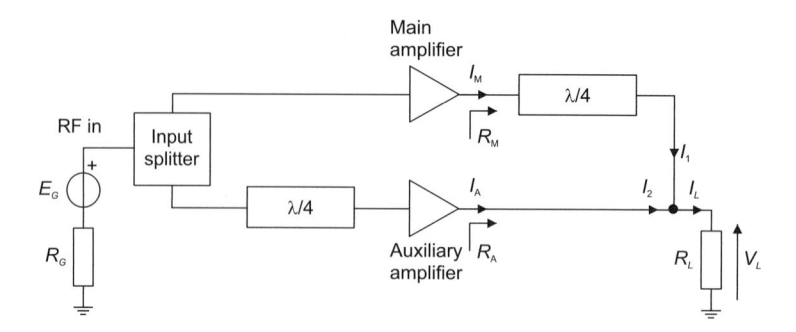

Figure 8.34 Scheme of the Doherty amplifier: the input splitter divides the input power in two parts, one driving the Main amplifier, the other the Auxiliary (Peak) amplifier. The Auxiliary stage only turns on above a certain value of the input power while the power combining at the output is performed through an impedance inverter.

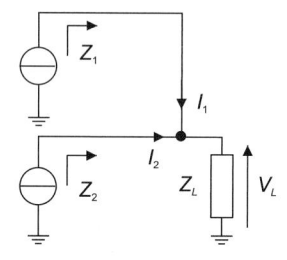

Figure 8.35 Active load modulation principle.

coupler. Alternative Doherty-like schemes make use of an auxiliary amplifier independently driven by a RF signal upconverted from the baseband signal, rather than by a fraction of the RF amplifier input power.

Fig. 8.34 shows the basic scheme of the DPA. To discuss the DPA main features, we will first model the devices outputs as ideal current sources. Let us consider first the circuit of Fig. 8.35, where the output of the Main (above) and Auxiliary (below) amplifiers are shown as ideal current generators synchronously controlled by the respective input voltages. The two generators are directly connected in parallel to the amplifier load impedance Z_L. We initially suppose that the two generator currents are in phase.

Since the output voltage is:

$$V_L = Z_L \cdot (I_1 + I_2),$$

the equivalent input impedances Z_1 and Z_2 seen by the two generators are:

$$Z_1 = \frac{V_L}{I_1} = Z_L \cdot \left(1 + \frac{I_2}{I_1}\right) \quad (8.46)$$

$$Z_2 = \frac{V_L}{I_2} = Z_L \cdot \left(1 + \frac{I_1}{I_2}\right).$$

Therefore, the effective load impedance seen by the Main amplifier generator of current I_1 depends on the current ratio I_2/I_1 between the Auxiliary and Main amplifier output generators. However, a number of issues have to be solved in order to practically exploit the active load modulation scheme involving the Auxiliary amplifier. First, the current ratio I_2/I_1 should be modulated (as a function of I_1) only when I_1 has reached a certain threshold level $\gamma I_{M,M}$ (where $I_{M,M}$ is the maximum Main current) corresponding to the maximum voltage swing. Secondly, $|Z_1|$ should *decrease* with increasing I_1 to keep the voltage swing of the Main amplifier constant and equal to its maximum value. This condition implies a decrease of $|Z_1|$ and therefore of I_2/I_1, for increasing I_1.

The first issue can be solved by applying an offset between the Auxiliary and Main current generators, in such a way that the Auxiliary generator is initially inactive, and is turned on when the Main amplifier current I_M becomes larger than the threshold level $\gamma I_{M,M}$. This allows the ratio I_2/I_1 to increase as a function of the input power even if $dI_1/dP_{in} = dI_2/dP_{in}$. Considering the second point, since $I_2/I_1 > 0$ (as the two currents are in phase), the load modulation in (8.46) always provides $|Z_1| > |Z_L|$, contrarily to the need to decrease the main amplifier load with increasing Auxiliary and Main currents. The problem can be solved by inserting, between the Main amplifier output and the load Z_L, an impedance inverter. The inverter transforms the increase of $|Z_1|$ in a decrease of the impedance seen at the main amplifier output; the corresponding modified output scheme is shown in Fig. 8.36, where the impedance inverter is implemented through a quarter-wave transformer. Fig. 8.36 (a) refers to the condition where the Main generator only is active, while the Auxiliary is off. In Fig. 8.36 (b) the Main output current is above threshold, and both the Main and the Auxiliary amplifiers are active.

The currents I_M and I_A of the Main and Auxiliary amplifiers can be expressed as:

$$I_M = G I_{M,M} \text{ for } 0 \leq G \leq 1 \tag{8.47}$$

$$I_A = \begin{cases} 0 & \text{for } I_M \leq \gamma I_{M,M}, \, 0 \leq \gamma \leq 1 \\ I_{M,M} \cdot T \cdot \dfrac{G-\gamma}{1-\gamma}, & \text{for } I_M \geq \gamma I_{M,M}, \, 1 \geq G \geq \gamma, \end{cases} \tag{8.48}$$

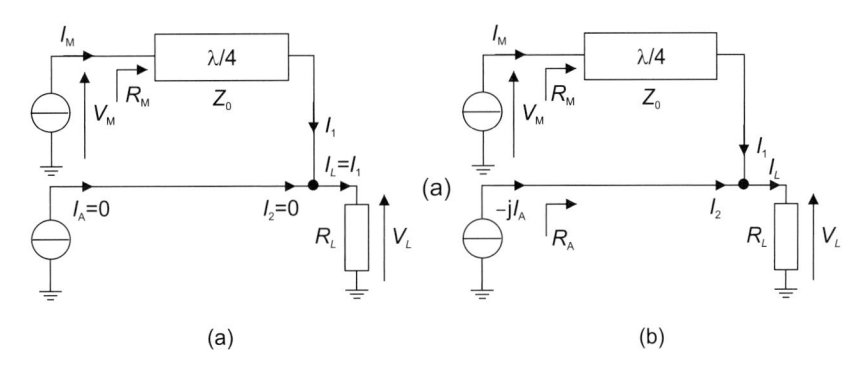

Figure 8.36 Output circuit of Doherty amplifier: (a) the main amplifier is active while the peak amplifier is off; (b) both the main and the peak amplifiers are active.

where G is a normalized dimensionless parameter ideally defined as:

$$G = \sqrt{\frac{P_{in}}{P_{in,M}}} \approx \sqrt{\frac{P_{out}}{P_{out,M}}},$$

where $P_{in,M}$ and $P_{out,M}$ are the maximum input and output DPA power, respectively, while:

$$T = \frac{I_{A,M}}{I_{M,M}} \tag{8.49}$$

accounts for the fact that the maximum current of the Main ($I_{M,M}$) and Auxiliary ($I_{A,M}$) amplifiers may be different. Finally, γ corresponds to the threshold value of G where the Auxiliary amplifier is turned on. The range $\gamma^2 < G^2 < 1$ also refers to the output power interval, normalized with respect to the maximum output power, where the efficiency should remain constant; the corresponding output backoff (OBO) range is $-20\log_{10}\gamma$ dB.

To continue the discussion, let us evaluate the admittance matrix of a quarter wave transformer with characteristic impedance $Z_0 = 1/Y_0$:

$$\mathbf{Y} = jY_0 \begin{pmatrix} 0 & 1 \\ 1 & 0 \end{pmatrix}.$$

Considering the scheme in Fig. 8.36 and adding the load resistance in parallel to the output of the transformer we have:

$$\begin{pmatrix} 0 & jY_0 \\ jY_0 & 1/R_L \end{pmatrix} \begin{pmatrix} V_M \\ V_L \end{pmatrix} = \begin{pmatrix} I_M \\ -jI_A \end{pmatrix},$$

where V_L is the load voltage, coinciding with the output voltage of the Auxiliary amplifier, while V_M is the output voltage of the Main amplifier. Solving we obtain:

$$V_M = \left(I_M \frac{Z_0}{R_L} - I_A\right) Z_0 \tag{8.50}$$

$$V_L = -jI_M Z_0, \tag{8.51}$$

where the additional $\pi/2$ phase delay ($-jI_A$ instead of I_A), introduced by the delay line in Fig. 8.34 at the input of the Auxiliary amplifier, compensates for the same delay caused by the impedance transformer at the output of the Main amplifier. From (8.47), (8.48), (8.50) and (8.51) we obtain the load resistance seen by the Main (R_M) and Auxiliary (R_A) amplifier outputs, respectively:

$$R_M = \frac{V_M}{I_M} = \begin{cases} \dfrac{Z_0^2}{R_L} & \text{for } 0 \leq G \leq \gamma,\ \gamma \leq 1 \\[6pt] \left(\dfrac{Z_0}{R_L} - T\dfrac{1-\gamma/G}{1-\gamma}\right) Z_0 = \left(\dfrac{1}{R_L} - \dfrac{1}{R_A}\right) Z_0^2 & \text{for } \gamma \leq G \leq 1 \end{cases} \tag{8.52}$$

$$R_A = \frac{V_L}{-jI_A} = \begin{cases} \infty & \text{for } 0 \leq G \leq \gamma,\ \gamma \leq 1 \\[6pt] \dfrac{Z_0}{T} \dfrac{1-\gamma}{1-\gamma/G} & \text{for } \gamma \leq G \leq 1. \end{cases} \tag{8.53}$$

From (8.52) and (8.53) the resistances seen from the Main and Auxiliary modules can be derived before the Auxiliary turn on and in maximum current conditions:

1. resistance seen from the Main before the Auxiliary turn-on:

$$R'_M \equiv R_M|_{G<\gamma} = \frac{Z_0^2}{R_L}; \qquad (8.54)$$

2. resistance seen from the Main at maximum current:

$$R''_M \equiv R_M|_{G=1} = \left(\frac{Z_0}{R_L} - T\right) Z_0; \qquad (8.55)$$

3. resistance seen from the Auxiliary amplifier before turn-on:

$$R'_A \equiv R_A|_{G<\gamma} = \infty; \qquad (8.56)$$

4. resistance seen from the Auxiliary at maximum current:

$$R''_A \equiv R_A|_{G=1} = \frac{Z_0}{T}. \qquad (8.57)$$

Before the Auxiliary amplifier is switched on, the efficiency of the Main increases with increasing P_{in} according to the behavior of its class of operation. To obtain maximum efficiency for $I_M = \gamma I_{M,M}$, the Main output resistance R'_M must satisfy the condition:

$$R'_M \approx \frac{1}{\gamma} \frac{V_{DS,br}}{I_{D,M}} = \frac{R^M_{Lo}}{\gamma},$$

where R^M_{Lo} is optimum load resistance in class B (also class A) operation. From (8.54) we obtain a design equation for R_L:

$$\frac{Z_0^2}{R_L} = \frac{R^M_{Lo}}{\gamma} \rightarrow R_L = \frac{\gamma Z_0^2}{R^M_{Lo}}. \qquad (8.58)$$

When the Main yields the maximum current, its load should be $R''_M = R^M_{Lo}$; we thus have from (8.54):

$$\left(\frac{Z_0}{R_L} - T\right) Z_0 = \frac{R^M_{Lo}}{\gamma} - TZ_0 = R^M_{Lo}. \qquad (8.59)$$

A last relationship can be finally obtained by considering the load seen by the Auxiliary amplifier R''_A when both the Auxiliary and the Main yield the maximum current. In such conditions, the DPA can be considered as a combined amplifier, with a possible unbalance between branches, given by factor T. In the maximum power condition, the Auxiliary stage must be loaded with its optimum termination R^A_{Lo}. From (8.57), (8.58) and (8.59) we obtain:

$$Z_0 = T R^A_{Lo} \qquad (8.60)$$

$$T = \sqrt{\frac{R^M_{Lo}}{R^A_{Lo}}\left(\frac{1}{\gamma} - 1\right)} \qquad (8.61)$$

$$R_L = R^A_{Lo}(1 - \gamma). \qquad (8.62)$$

Power Amplifiers

As already mentioned, the Main and Auxiliary amplifiers typically are implemented with a class B and C amplifier, respectively. To allow for a simpler analysis, however, we will from now on assume that the Auxiliary simply is a scaled version of the Main amplifier with a maximum current multiplied by a factor T and the same breakdown voltage, while the Auxiliary turn-on is properly taken care by a separate drive control. In such case, equations (8.60), (8.61) and (8.62) simplify, since:

$$R_{Lo}^A = R_{Lo}^M/T, \tag{8.63}$$

yielding:

$$Z_0 = R_{Lo}^M \tag{8.64}$$

$$T = \frac{1}{\gamma} - 1 \tag{8.65}$$

$$R_L = \gamma R_{Lo}^M. \tag{8.66}$$

Accounting for these relationships, (8.52) and (8.53), become:

$$R_M = \begin{cases} \dfrac{R_{Lo}^M}{\gamma} & \text{for } 0 \leq G \leq \gamma,\, \gamma \leq 1 \\ \dfrac{R_{Lo}^M}{G} & \text{for } \gamma \leq G \leq 1 \end{cases} \tag{8.67}$$

$$R_A = \begin{cases} \infty & \text{for } 0 \leq G \leq \gamma,\, \gamma \leq 1 \\ R_{Lo}^A \dfrac{1-\gamma}{1-\gamma/G} & \text{for } \gamma \leq G \leq 1. \end{cases} \tag{8.68}$$

The behavior of the conductances R_M^{-1} and R_A^{-1} seen by the Main and Auxiliary amplifier, normalized with respect to the respective conductance for maximum power, are plotted in Fig. 8.37 as a function of the square root of the applied backoff, for $\gamma = 0.5$ and $\gamma = 0.25$. In the former case $T = 1$ and both the Main and the Auxiliary stage have the same maximum current; this is the most commonly adopted solution, providing a high-efficiency region down to 6 dB backoff vs. the maximum output power. For $\gamma < 0.5$, $T > 1$, leading to a wider high-efficiency region but also to the need of an Auxiliary device with a larger periphery of the final stage device than the Main.

After the Auxiliary turn-on, the conductance seen by the Main linearly increases with the drain current. Thus, the Main output voltage swing (from the FET knee voltage to the breakdown voltage) is pinned to the value corresponding to the Auxiliary turn-on, independent of the backoff level. As the currents of the Main and Auxiliary modules add up in phase on the load, the Auxiliary amplifier not only modulates the Main load, but also contributes to the output power. The power delivered by the Main and Auxiliary stages to the load can be derived as a function of the square root of the backoff, represented by G, by combining (8.67) and (8.68), that provide the Main and Auxiliary load resistances, with (8.47) and (8.48), yielding the related currents. The load power can be now evaluated as the square of the peak load current, multiplied by the load resistance; normalizing to the maximum Main and Auxiliary output powers:

8.7 High-Efficiency Amplifiers

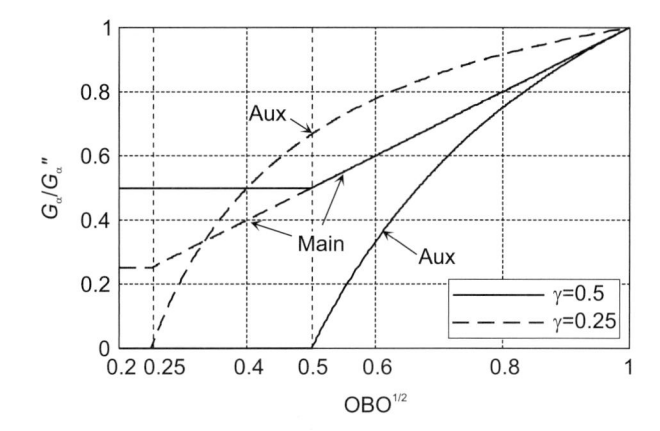

Figure 8.37 Load conductance G_α (α = Main or Aux) seen by the Main and Auxiliary amplifier normalized to the conductance G_k'' for maximum output power. The values $\gamma = 0.25$ (dashed line) and $\gamma = 0.5$ (continuous line) correspond to a 6 and 12 dB backoff high-efficiency region, respectively.

$$P_{M,M} = \frac{R_{Lo}^M I_{M,M}^2}{2} \tag{8.69}$$

$$P_{A,M} = \frac{R_{Lo}^A I_{A,M}^2}{2} = \frac{R_{Lo}^M (TI_{M,M})^2}{2T} = TP_{M,M} = \frac{1-\gamma}{\gamma} P_{M,M}, \tag{8.70}$$

where we have taken into account (8.49) and (8.63), we obtain:

$$\frac{P_M}{P_{M,M}} = \begin{cases} \dfrac{G^2}{\gamma} & \text{for } 0 \leq G \leq \gamma, \, \gamma \leq 1 \\ G & \text{for } \gamma \leq G \leq 1 \end{cases} \tag{8.71}$$

$$\frac{P_A}{P_{A,M}} = \begin{cases} 0 & \text{for } 0 \leq G \leq \gamma, \, \gamma \leq 1 \\ \dfrac{G(G-\gamma)}{1-\gamma} & \text{for } \gamma \leq G \leq 1. \end{cases} \tag{8.72}$$

The maximum DPA output power $P_{D,M}$ results, using (8.70):

$$P_{D,M} = P_{M,M} + P_{A,M} = P_{M,M}(1+T) = \frac{P_{M,M}}{\gamma}. \tag{8.73}$$

Finally, taking into account (8.65), (8.73), (8.71), (8.72), the PDA output power P_D can be readily derived as:

$$P_D = P_M + P_A = P_{D,M} G^2, \quad 0 \leq G \leq 1. \tag{8.74}$$

The normalized Main, Auxiliary and DPA output powers are shown in Fig. 8.38.

If the Main and Auxiliary stages adopt an active device with the same periphery ($T = 1$, $\gamma = 1/2$), we have $P_{D,M} = 2P_{M,M}$, see Fig. 8.38. Under maximum power conditions, the DPA can therefore be regarded as a two-way combined stage. Since $I_{A,M} = TI_{M,M}$ while the Auxiliary is active in a normalized input voltage range

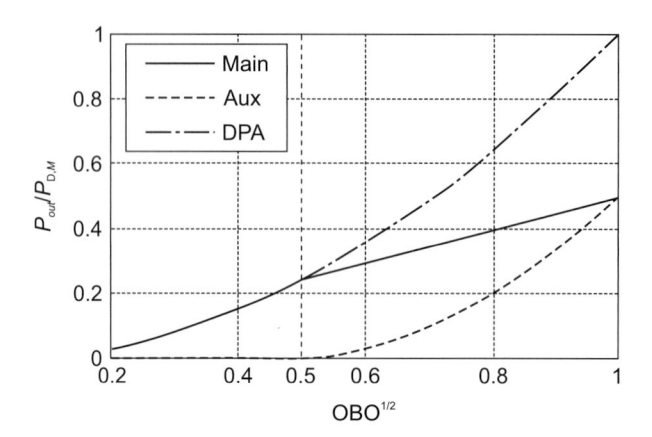

Figure 8.38 Main, Auxiliary and DPA output power, normalized to the maximum DPA output power, as a function of the square root of the output backoff, for $\gamma = 0.5$.

$\gamma \leq G \leq 1$, the Main and Auxiliary device transconductances are approximately (assuming a linear transcharacteristic) related as:

$$g_{m,A} = \frac{T g_{m,M}}{1-\gamma} = \frac{g_{m,M}}{\gamma}.$$

For $\gamma = 0.5$ and $T = 1$, we obtain $g_{m,A} = 2 g_{m,M}$. Notice that the condition on transconductances roughly translates into a similar condition on the Auxiliary and Main stage gains, whose implementation requires particular attention since stages having the same maximum current would typically have the same transconductance. The Auxiliary stage gain can be properly controlled, e.g., by adding a driver stage, or by properly unbalancing the input splitter.

Before analyzing the DPA efficiency, a discussion is in order on the DPA practical implementations. While field-effect or bipolar transistors can be exploited for the Main and the Auxiliary stages, the different turn-on behavior of the Auxiliary and the Main can be controlled in several ways. Confining the discussion to *n*-channel FET stages, the Auxiliary can be biased at a more negative gate potential than the class-B Main, thus obtaining a class C Auxiliary stage. As an alternative strategy, that is however hardly compatible with monolithic implementation, the Main and Auxiliary stages can adopt active devices with different threshold voltage. Taking into account the efficiency–linearity compromise, practical implementations of the DPA often adopt a class AB (rather than B) Main and a class C Auxiliary module. Since both the AB and the C stages (see Sec. 8.6.2) have variable circulation angle and variable optimum load as a function of the input power, the approximate design equations (8.64), (8.65) and (8.66), based on the assumption of the Auxiliary stage being a scaled version of the Main, should be properly revisited, see [3] for a complete analysis. In general, owing to its complexity, the DPA design can be only effectively carried out by means of CAD tools exploiting realistic device nonlinear models.

8.7 High-Efficiency Amplifiers

A simple, approximate model for the amplifier efficiency will be discussed here, where a DPA with class B Main and Auxiliary stages is considered. The practical implementation of this configuration is rather difficult, since it requires the Main and Auxiliary blocks to be separately driven. However, the results can be readily expressed in closed form and are consistent with the behavior of more realistic class AB and C DPAs. Taking into account that in class B the circulation angle is constant, and therefore the ratio between the peak RF current I_{RF} and its average value (the DC current I_{DC}) is independent from the input power level ($I_{DC}/I_{RF} = 2/\pi$), we have for the DC powers in the Main and Auxiliary modules, respectively:

$$P_{\alpha,DC} = V_{\alpha,DC} I_{\alpha,DC} = V_{\alpha,DC} I_\alpha \frac{2}{\pi}, \quad \alpha = \text{M, A}$$

$$P_\alpha = R_\alpha \frac{I_\alpha^2}{2}, \quad \alpha = \text{M, A};$$

thus, the Main (η_M) and Auxiliary (η_A) stage efficiencies are:

$$\eta_\alpha = \frac{P_\alpha}{P_{\alpha,DC}} = \frac{\pi}{4} \frac{R_\alpha I_\alpha}{V_{\alpha,DC}}, \quad \alpha = \text{M, A},$$

i.e., using (8.47), (8.48), (8.67) and (8.68) we obtain:[9]

$$\eta_M = \begin{cases} \dfrac{4}{\pi} \cdot \dfrac{R_{Lo}^M G I_{M,M}}{\gamma V_{M,DC}} & \text{for } 0 \le G \le \gamma, \gamma \le 1 \\ \dfrac{4}{\pi} \cdot \dfrac{R_{Lo}^M I_{M,M}}{V_{M,DC}} & \text{for } \gamma \le G \le 1 \end{cases} \quad (8.75)$$

$$\eta_A = \begin{cases} \text{ind. for } 0 \le G \le \gamma, \gamma \le 1 \\ \dfrac{4}{\pi} \cdot \dfrac{R_{Lo}^A G I_{M,M} T}{V_{A,DC}} & \text{for } \gamma \le G \le 1. \end{cases} \quad (8.76)$$

The maximum efficiency is achieved for $G = \gamma$ (and remains constant for $G > \gamma$) in the Main module (8.75) and for $G = 1$ in the Auxiliary module (8.76):

$$\eta_{M,M} = \frac{4}{\pi} \cdot \frac{R_{Lo}^M I_{M,M}}{V_{M,DC}}$$

$$\eta_{A,M} = \frac{4}{\pi} \cdot \frac{R_{Lo}^A (T I_{M,M})}{V_{A,DC}}.$$

The normalized efficiencies can be now expressed as:

$$\frac{\eta_M}{\eta_{M,M}} = \begin{cases} \dfrac{G}{\gamma} & \text{for } 0 \le G \le \gamma, \gamma \le 1 \\ 1 & \text{for } \gamma \le G \le 1 \end{cases}$$

$$\frac{\eta_A}{\eta_{A,M}} = \begin{cases} 0 & \text{for } 0 \le G \le \gamma, \gamma \le 1 \\ G & \text{for } \gamma \le G \le 1. \end{cases}$$

---

[9] Notice that the efficiency of the Auxiliary stage in the off state is indeterminate, since both the RF and the DC power are zero.

We now analyze the DPA overall efficiency η_D. For $G < \gamma$ we have of course $\eta_D = \eta_M$. To evaluate the DPA efficiency for $G \geq \gamma$ we take into account that if the Main and Auxiliary stages are class B, then their maximum efficiency is $\eta_{M,M} = \eta_{A,M} \equiv \pi/4$. Therefore, for $G \geq \gamma$:

$$\eta_M = \frac{P_M}{P_{M,DC}} = \eta_{M,M}, \quad \eta_A = \frac{P_A}{P_{A,DC}} = G\eta_{A,M} = G\eta_{M,M},$$

i.e., using (8.71), (8.72), (8.73), we have:

$$P_{M,DC} = \frac{P_M}{\eta_{M,M}} = \frac{GP_{M,M}}{\eta_{M,M}}$$

$$P_{A,DC} = \frac{P_A}{G\eta_{M,M}} = \frac{(G-\gamma)P_{A,M}}{(1-\gamma)\eta_{M,M}} = \frac{(G-\gamma)P_{M,M}}{\gamma\eta_{M,M}},$$

and the Doherty DC power $P_{D,DC}$ results as:

$$P_{D,DC} = P_{M,DC} + P_{A,DC} = \frac{P_{M,M}}{\eta_{M,M}}\left(\frac{G\gamma + G - \gamma}{\gamma}\right)$$

$$= \frac{P_{D,M}}{\eta_{M,M}}(G\gamma + G - \gamma). \qquad (8.77)$$

Taking into account (8.74) and (8.77), the efficiency η_D of the DPA for $G \geq \gamma$ results as:

$$\eta_D = \frac{P_D}{P_D^{DC}} = \frac{P_{D,M}G^2}{\frac{P_{D,M}}{\eta_{M,M}}(G\gamma + G - \gamma)} = \eta_{M,M}\frac{G^2}{\gamma(G-1) + G}.$$

The maximum DPA efficiency ($G = 1$) is $\eta_{D,M} = \eta_{M,M} \equiv \pi/4$, that corresponds to the maximum class B efficiency. Notice that the same maximum efficiency is also obtained for $G = \gamma$. The normalized DPA efficiency can be finally written for $G \geq 0$ as:

$$\frac{\eta_D}{\eta_{D,M}} = \begin{cases} \dfrac{G}{\gamma} & \text{for } 0 \leq G \leq \gamma,\ \gamma \leq 1 \\ \dfrac{G^2}{\gamma(G-1) + G} & \text{for } \gamma \leq G \leq 1. \end{cases}$$

The Main, Auxiliary and DPA efficiencies are shown in Fig. 8.39 as a function of the square root of the backoff for $\gamma = 0.5$ and $\gamma = 0.25$. Above the Auxiliary turn-on, the DPA efficiency has a minimum for:

$$G_{\eta,m} = \frac{2\gamma}{1+\gamma},$$

where:

$$\frac{\eta_{D,m}}{\eta_{D,M}} = \frac{4\gamma}{(1+\gamma)^2} = \frac{G_{\eta,m}^2}{\gamma}.$$

The minimum of the normalized drain efficiency and the corresponding backoff value are reported in Fig. 8.40 as a function of γ expressed in dB. For the typical case $\gamma = 0.5$ (corresponding to -6 dB on the horizontal axis of Fig. 8.40) the normalized minimum

8.7 High-Efficiency Amplifiers

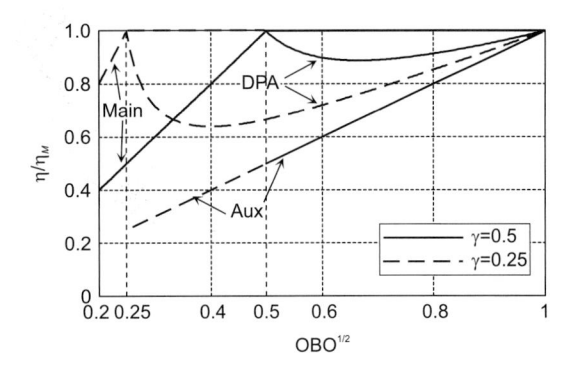

Figure 8.39 Efficiency of the Main, Auxiliary and Doherty amplifier (DPA) for $\gamma = 0.5$ (continuous curves) and $\gamma = 0.25$ (dashed curves) as a function of the square root of the output backoff. For $\gamma = 0.25$ the high-efficiency region extends to 12 dB OBO while for $\gamma = 0.5$ it extends to 6 dB OBO.

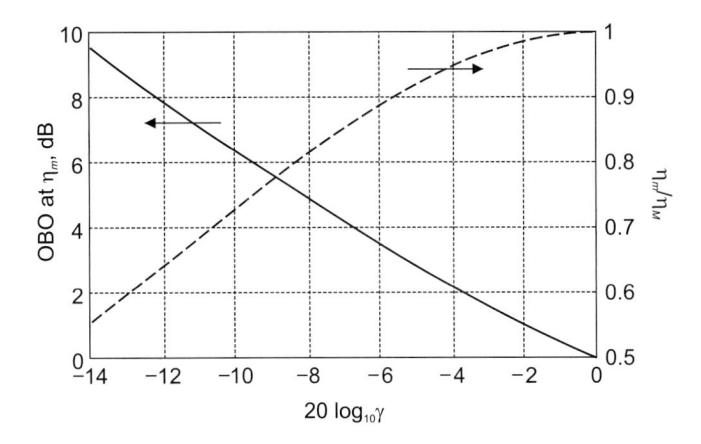

Figure 8.40 Normalized minimum DPA efficiency after the Auxiliary turn-on (right axis), and corresponding backoff (left axis), as a function of γ in dB.

efficiency at a 3.5 dB backoff is $\eta_{D,m}/\eta_{D,M} = 0.89$ corresponding to an absolute 70 percent efficiency for class B implementation.

A comprehensive review of the most relevant Doherty realizations can be found in [13, 14], where several variants of the original Doherty configuration are reported, together with a discussion of the DPA features in terms of output power, operating frequencies, linearity and technological implementations. Fig. 8.41 and Fig. 8.42, respectively, report two examples of a hybrid [15] and a MMIC [16] Doherty implementation. The former adopts two CREE devices (CGH40010 GaN HEMTs on SiC) operating in Class AB (Main stage) and C (Auxiliary stage). The amplifier works on the 3–3.6 GHz It band and it is designed taking into account the specifications of WiMAX 3.5 GHz applications; the output power is above 2 W, with a 10 dB gain and an efficiency better than 55 percent on the typical 6 dB backoff region. The Auxiliary stage gain is made larger than that of the Main stage one through uneven input power splitting. This

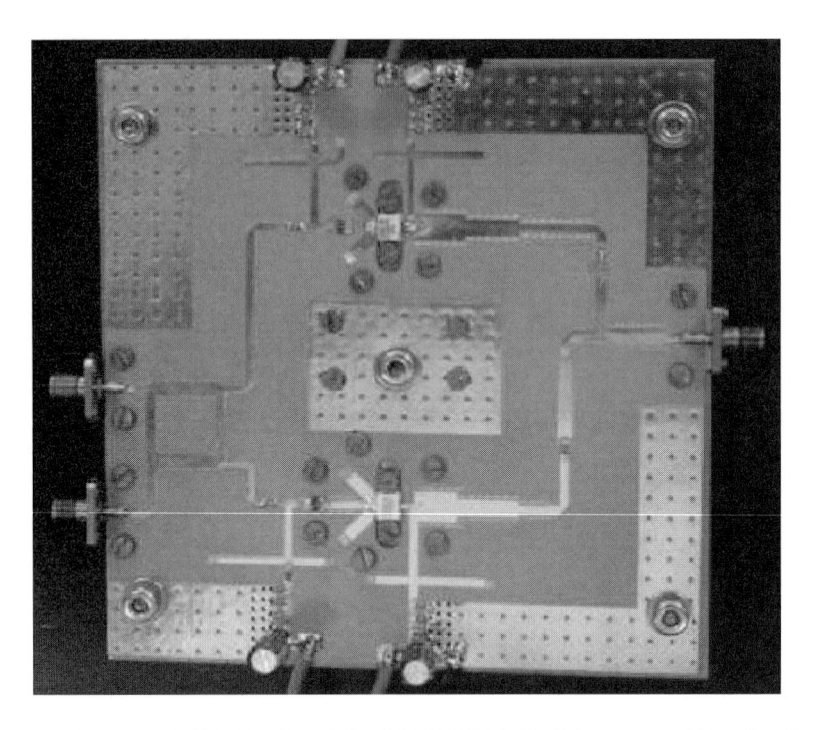

Figure 8.41 WiMAX Hybrid DPA. Copyright ©2012 IEEE. All rights reserved. Reprinted, with permission, from [15].

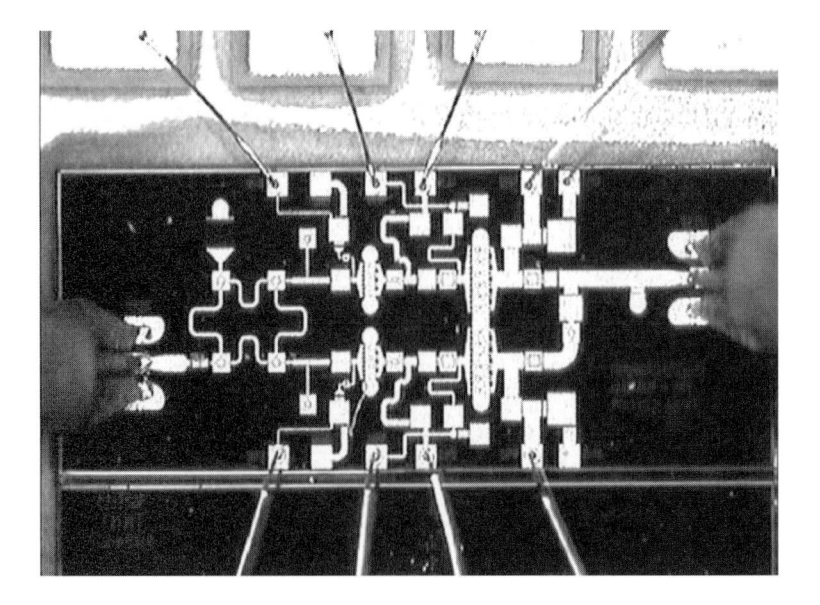

Figure 8.42 K-band monolithic DPA. Copyright ©2014 IEEE. All rights reserved. Reprinted, with permission, from [16].

technique is commonly adopted in practical DPA implementations; an alternative is to add to the Auxiliary stage a driver stage intended to boost its gain [17]. The monolithic DPA [14] again features a Main class B and an Auxiliary class C stage; it is conceived for K-band applications, and adopts the Qorvo 0.15 μm PHEMT GaAs foundry process. The output power is around 1 W at 24 GHz, with 38 percent and 20 percent efficiency at saturation and at 6 dB backoff, respectively. A driver module in both the Main and the Auxiliary stages avoids the typical low-gain DPA weakness, allowing around 12 dB gain to be achieved.

8.8 Layout and Power Combining Techniques

Power combining of active devices can be performed according to several approaches. At a device level, power device layouts often exploit parallel combination of stages, that is equivalent, in a FET, to having a total gate periphery $W = NW'$, where N is the number of paralleled stages. Notice that the width of the single gate finger typically decreases with the operating frequency. In the parallel combination the device current increases with the same output voltage. Another technique is the series combination of stages, also referred to as *device stacking*; in this cases two or more devices are connected in series, allowing for an increase of the output voltage at constant output current. While parallel or series combination of stages is better done at a device level, combination of multi-gate devices matched on a reference impedance (e.g., 50 Ω) is typically carried out through power dividers and combiners, such as the Wilkinson divider discussed in Sec. 4.6.1. An example of combination of four device through a two-stage Wilkinson divider/combiner is shown in Fig. 8.43 (a) while Fig. 8.43 (b) introduces a possible four-way Wilkinson combiner and divider.

Power combining can be also performed in a less conventional way through other strategies. Distributed and matrix amplifiers, typically introduced within a small-signal context, can be interpreted as non-conventional ways to combine the power of several devices. Notice that in a power distributed amplifier operating in class A design rules will be different with respect to the small-signal case, due to the need of maximizing the output power for each stage of the DAMP, see e.g., [18, Ch. 8]. Finally, both the balanced amplifier and the Doherty amplifier can be seen, in a way, as power combining strategies.

8.9 Linearization Techniques

Many recent communication standard require very high levels of linearity to the power stage. Owing to the trade-off between efficiency and linearity in a PA, the required low distortion levels cannon be conveniently achieved by exploiting class A amplifiers in backoff conditions. Several strategies to improve the linearity of the power stage, called *linearization techniques* [19, 20], have been proposed in the past; here we will only refer

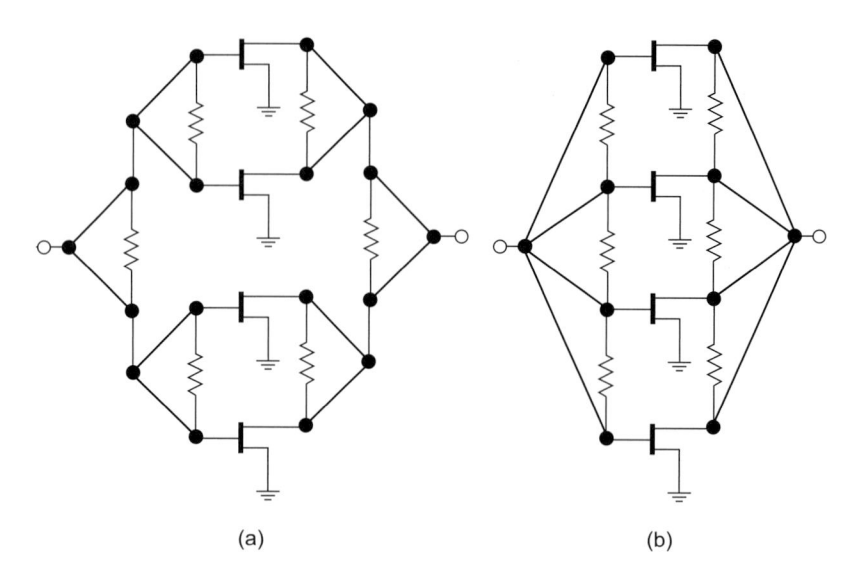

Figure 8.43 Wilkinson power combining: (a) through cascading of a two-way divider and combiner; (b) through a four-way divider and combiner.

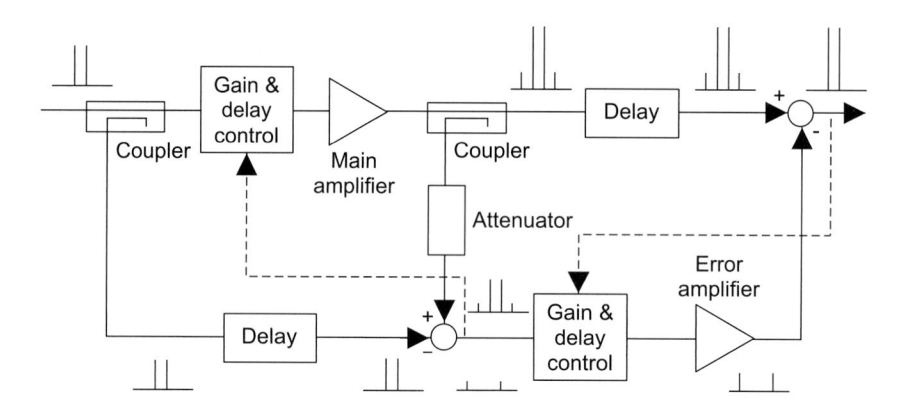

Figure 8.44 Scheme of a feedforward linearizer.

to two approaches, the *feedforward* [21] and the *predistortion* [22, 23]. A third approach, based on feedback [24, 25], will not be considered here.

The feedforward linearizer is based on a cancellation strategy. As shown in Fig. 8.44, where the system undergoes, for simplicity, a two-tone test, a fraction of the input signal is sampled and brought on a parallel branch. At the same time, the input signal is amplified by the power amplifier, that will add intermodulation distortion. Part of the output signal is again sampled, and an attenuator and a phase shifter on the parallel branch are adjusted in such a way that a difference signal is created in the parallel branch, only including the signal distortion. The distortion contribution is then amplified by the error amplifier is such a way that the amplitude of the distortion becomes, at the output, equal

8.9 Linearization Techniques

to the one in the amplified signal. In the last step, the distortion contribution is subtracted to the distorted signal from the power amplifier, and an undistorted output signal is recovered. A number of adaptive gain and phase control blocks are present in the main and parallel branch (see Fig. 8.44, dashed lines) so as to have a fine control on the cancellation. While the feedforward system is comparatively wideband and can achieve spectacular improvements in the $CIMR_3$ (of the order of 30 dB), its performance in terms of overall efficiency is not overly good, also because the error amplifier should be extremely linear; a class A amplifier is therefore needed, whose design is however made less critical by the fact that its output power level is of the order of the intermodulation products of the main amplifier only.

Predistortion is based on an entirely different open-loop approach, in which the input signal is preconditioned in order to compensate for the distortion introduced by the amplifier stage. For a comprehensive treatment, see [23]. This strategy was already exploited in the seventies for linearizing power traveling wave tubes for satellite applications [26].

For simplicity, let us keep the discussion to the system level and consider a PA with input x, in cascade to a predistorter generating and output y that is fed into the PA generating the final output z, see Fig. 8.45 (a). Assume that all signals are narrowband and that a descriptive function approach can be used. Introducing the complex slowly varying envelopes $X(t), Y(t), Z(t)$ the predistorter and the PA are thus described by the functions:

$$Y(t) = F_{pr}\left(|X(t)|\right) X(t) \equiv G_{pr}\left(X(t)\right)$$
$$Z(t) = F_{PA}\left(|Y(t)|\right) Y(t) \equiv G_{PA}\left(Y(t)\right),$$

therefore we have, imposing that the cascade be equivalent to a linear amplifier with constant complex amplification A:

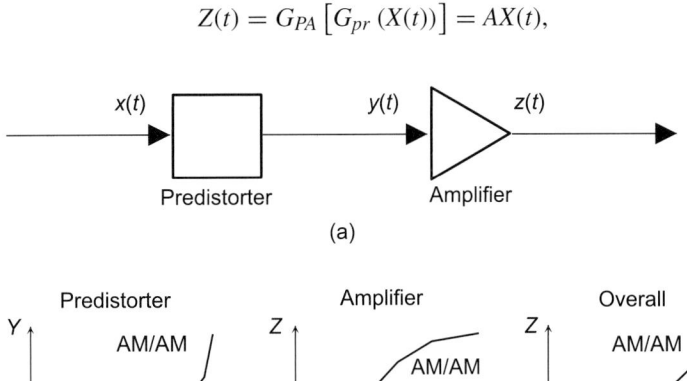

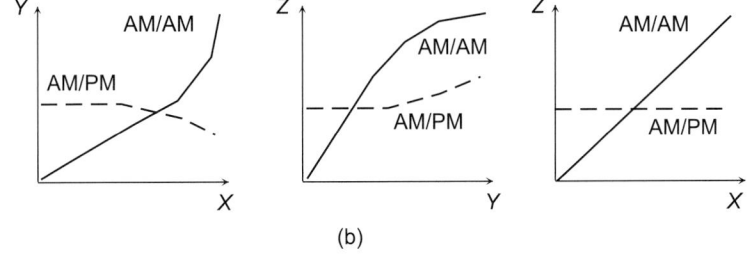

$$Z(t) = G_{PA}\left[G_{pr}\left(X(t)\right)\right] = AX(t),$$

Figure 8.45 Scheme of a predistorter – PA cascade (a). Input–output characteristics of the predistorter, PA and their cascade (B).

from which:

$$G_{pr}(X(t)) = G_{PA}^{-1}(AX(t)),$$

where G_{PA}^{-1} denotes the inverse of G_{PA}. In other words, the predistortion function is a generalized inverse of the amplifier descriptive function, meaning that applying G_{PA} to G_{pr} leads to $Z = AX$ rather that to $Z = X$. The AM/AM characteristic of the predistorter should be complementary to the one of the power amplifier, an the same should happen for the AM/PM characteristic, see Fig. 8.45, (b).

The ideal predistorter descriptive function can be easily implemented as the inverse of the PA characteristic only if the PA has no memory effects within its bandwidth. Moreover, the variation in the PA characteristics with aging could lead to a mismatch between the predistorter and the PA, that can be corrected by setting up a feedback loop leading to the so-called adaptive predistortion [22]. Finally, predistortion can be implemented at several levels of the input.

In RF predistortion the predistorter operates directly on the RF input signal of the amplifier [27]. RF predistorters are traditionally implemented with diode circuits, with the aim of providing, at least in a certain range of input powers, an expansive $P_{in} - P_{out}$ characteristic. They are simple and potentially wideband, but cannot be controlled in an adaptive way. Predistortion can be, on the other hand, carried out in a digital way on the baseband $I(t)$ and $Q(t)$ components of the input signal before digital to analog conversion [28, 29]. Working with digital signals allows a feedback loop to be implemented where the output signal of the amplifier is downconverted and sampled, the PA model is extracted in real time into a look-up table and the digital predistortion function is derived in an adaptive way. Despite its complexity, digital adaptive predistortion is able to obtain results comparable with the feedforward technique, but with better overall efficiency. Finally, an intermediate solution for predistortion is to perform it at IF [30], either in an analog way, or, if the system implements a digital IF, again digitally.

8.10 Questions and Problems

8.10.1 Questions

1. To obtain the maximum output power in a Class A power amplifier, the load line must:
   1. have a slope related to the maximum voltage and current sweep
   2. the slope should match the device output resistance
   3. be as vertical as possible in the $i_D - v_{DS}$ plane.
2. Draw the schematic of a class A power amplifier with tuned load.
3. Draw the schematic of a class B power amplifier with tuned load.
4. Discuss the efficiency of a tuned load power amplifier in class A and class B.
5. What is a class B amplifier self bias? Does a class B device dissipate power in the absence of an input signal?
6. Discuss the efficiency, gain and optimum load resistance from class A to class C amplifiers.
7. Discuss the bias of Class A and class B amplifiers with tuned load.

8. Motivate the choice of the tuned load in a power amplifier and discuss its impact on the amplifier bandwidth.
9. In a class A power amplifier the optimum output working point is:
   1. at half of the maximum current, approximately at half of the maximum voltage
   2. at half of the maximum current and at the maximum voltage
   3. at the maximum current and maximum voltage.
10. Discuss the input bias point of a power amplifier from class A to class C.
11. In a class B power amplifier, the optimum output bias point of the device is:
    1. at zero drain current, approximately half of the maximum voltage
    2. at maximum drain current, zero voltage
    3. at maximum voltage and half of the maximum current.
12. Discuss the load-pull design technique for power amplifiers.
13. The third-order intermodulation products $2f_1 - f_2$ and $2f_2 - f_1$ are particularly dangerous because they:
    1. fall close to the fundamentals
    2. have larger intensity than the corresponding products with positive sign
    3. are more easily generated by the device nonlinearity.
14. Explain why a descriptive function model derived from a power series model only includes odd-order powers.
15. Discuss the two-tone test for the intermodulation products of a power amplifier. How does the $CIMR_3$ behave with increasing input power?
16. A transistor with small-signal maximum output gain matching:
    1. has higher small-signal gain of an amplifier matched for maximum power, but lower saturation power.
    2. has better linearity than the amplifier matched for maximum power
    3. has lower small-signal gain of an amplifier matched for maximum power, and lower saturation power.
17. Explain why the single-tone and the two-tone $P_{in} - P_{out}$ at the fundamental are not exactly the same in compression.
18. What is the approximate relationship between the third-order IMP intercept and the 1 dB compression point in a class A amplifier?
19. Explain why the gain of a class AB amplifier may depend on the input power.
20. Compare the class C and the class F maximum efficiency.
21. Explain the operation of an ideal class F amplifier.
22. Explain the operation of an ideal class E amplifier.
23. What is the amplifier backoff? How does it influence the amplifier linearity?
24. The 1 dB compression point corresponds to an input power that:
    1. is constant with frequency
    2. decreases with frequency
    3. increases with frequency.
25. Compare the class A and class B amplifier efficiency in backoff conditions.
26. Define the Power Added Efficiency of a power amplifier. How does the PAE vary from class A to class C?

27. Consider an amplifier in which the load is DC decoupled. The DC power dissipated by the device (and converted into heat) in the presence of an input signal:
   1. always coincides with the DC dissipated power in the absence of a signal, regardless of the class of the amplifier
   2. never coincides with the DC dissipated power in the absence of a signal, apart from in a class A amplifier in small-signal operation
   3. coincides with the DC dissipated power in the absence of signal only in a perfectly linear amplifier.
28. The temperature of the active device in a class A power amplifier:
   1. is maximum at zero input signal
   2. is minimum at zero input signal
   3. does not depend on the input signal.
29. The temperature of the active device in a class B amplifier:
   1. is maximum at zero input signal
   2. is minimum at zero input signal
   3. does not depend on the input signal.
30. Describe the main power amplifier linearization approaches.
31. Discuss the trade-off between efficiency and linearity in the backoff of a class A amplifier.
32. Explain the motivation behind the Doherty scheme.
33. Why does a Doherty amplifier make use of a class AB or B main and a class C auxiliary stage?
34. Explain why the cutoff frequency of a device exploited in a switching amplifier should be much larger than the operating frequency.

8.10.2 Problems

1. In a class A power amplifier the gain in linearity is 20 dB and the 1 dB compression corresponds to an input power of 0 dBm. What is the output power at the 1 dB compression point?
2. A class A power amplifier, with a single-tone input of 100 μW, has a second harmonic output power 10 nW. The 1 dB compression point is at 1 mW input power. What is the second harmonic output power for an input of 200 μW?
3. A class A power amplifier, with a two-tone input of 100 μW, has a CIMR$_3$ of 60 dB. The 1 dB compression point is at 1 mW input power. What is the CIMR$_3$ for an input of 200 μW?
4. A HEMT has 16 V breakdown voltage, 0.5 V knee voltage (onset of current saturation) and maximum current of 500 mA. What is the maximum class A output power? What is the optimum load and optimum class A working point?
5. A class A power amplifier working at 1 dB compression point has CIMR$_3$ = 15 dB. The output power is 20 W and the gain in linearity is 20 dB; the efficiency is 35 percent in the working point. Evaluate the input backoff needed to increase the CIMR$_3$ up to 30 dB. Evaluate the efficiency in the initial condition and in backoff and the corresponding power-added efficiency.

6. A receiver stage has bandwidth $B = 2$ GHz, noise figure of 4 dB, output signal over noise ratio of 20 dB. Assuming input thermal noise at 300 K, evaluate the sensitivity and noise floor of the receiver. Suppose than that the third-order intermodulation product intercept be $\text{IIP}_3 = 10$ dBm; evaluate the Spurious Free Dynamic Range of the receiver.
7. An amplifier is designed on a 50 Ω load to provide, at low frequency, small signal available gain $G_1 = 25$ dB. The 1 dB compression point is for an input power $P_{in,1} = 10$ dBm while the saturation power is $P_{sat,1} = 36$ dBm. Evaluate the corresponding performances for a balanced amplifier closed on 50 Ω in which two identical amplifiers are connected in tandem.

References

[1] S. Cripps, *Advanced techniques in RF power amplifier design*, ser. Artech House Microwave Library. Artech House, 2002.

[2] S. Cripps, *RF power amplifiers for wireless communications*, ser. Artech House Microwave Library. Artech House, 2006.

[3] P. Colantonio, F. Giannini, and E. Limiti, *High efficiency RF and microwave solid state power amplifiers*, ser. Microwave and Optical Engineering. John Wiley & Sons, 2009.

[4] S. C. Cripps, "A theory for the prediction of GaAs FET load-pull power contours," in *Microwave Symposium Digest, 1983 IEEE MTT-S International*, May 1983, pp. 221–223.

[5] R. C. Dorf and J. A. Svoboda, *Introduction to electric circuits*. John Wiley & Sons, 2010.

[6] F. H. Raab, "Maximum efficiency and output of class-f power amplifiers," *IEEE Transactions on Microwave Theory and Techniques*, vol. 49, no. 6, pp. 1162–1166, Jun. 2001.

[7] V. Carrubba, A. L. Clarke, M. Akmal, J. Lees, J. Benedikt, P. J. Tasker, and S. C. Cripps, "The continuous class-f mode power amplifier," in *Microwave Integrated Circuits Conference (EuMIC), 2010 European*, Sep. 2010, pp. 432–435.

[8] V. Carrubba, A. L. Clarke, M. Akmal, J. Lees, J. Benedikt, P. J. Tasker, and S. C. Cripps, "On the extension of the continuous class-f mode power amplifier," *IEEE Transactions on Microwave Theory and Techniques*, vol. 59, no. 5, pp. 1294–1303, May 2011.

[9] N. O. Sokal and A. D. Sokal, "High-efficiency tuned switching power amplifier," Patent US 3 919 656, 1975.

[10] N. O. Sokal and A. D. Sokal, "Class E – A new class of high-efficiency tuned single-ended switching power amplifiers," *IEEE Journal of Solid-State Circuits*, vol. 10, no. 3, pp. 168–176, Jun. 1975.

[11] A. Grebennikov and N. Sokal, *Switchmode RF power amplifiers*. Elsevier, 2009.

[12] W. H. Doherty, "A new high efficiency power amplifier for modulated waves," *Proceedings of the Institute of Radio Engineers*, vol. 24, no. 9, pp. 1163–1182, Sep. 1936.

[13] A. Grebennikov and S. Bulja, "High-efficiency Doherty power amplifiers: historical aspect and modern trends," *Proceedings of the IEEE*, vol. 100, no. 12, pp. 3190–3219, Dec. 2012.

[14] V. Camarchia, M. Pirola, R. Quaglia, S. Jee, Y. Cho, and B. Kim, "The Doherty Power Amplifier: Review of Recent Solutions and Trends," *IEEE Transactions on Microwave Theory and Techniques*, vol. 63, no. 2, pp. 559–571, Feb. 2015.

[15] J. M. Rubio, J. Fang, V. Camarchia, R. Quaglia, M. Pirola, and G. Ghione, "3.6-GHz wideband GaN Doherty power amplifier exploiting output compensation stages," *IEEE Transactions on Microwave Theory and Techniques*, vol. 60, no. 8, pp. 2543–2548, Aug. 2012.

[16] R. Quaglia, V. Camarchia, T. Jiang, M. Pirola, S. D. Guerrieri, and B. Loran, "K-band GaAs MMIC Doherty power amplifier for microwave radio with optimized driver," *IEEE Transactions on Microwave Theory and Techniques*, vol. 62, no. 11, pp. 2518–2525, Nov. 2014.

[17] L. Piazzon, P. Colantonio, F. Giannini, and R. Gioffré, "Asymmetrical Doherty power architecture with an integrated driver stage in the auxiliary branch," *International Journal of RF and Microwave Computer-Aided Engineering*, vol. 24, pp. 498–507, 2014.

[18] N. Kumar and A. Grebennikov, *Distributed power amplifiers for RF and microwave communications*. Artech House, 2015.

[19] A. Katz, "Linearization: Reducing distortion in power amplifiers," *IEEE Microwave Magazine*, vol. 2, no. 4, pp. 37–49, 2001.

[20] J. Wood, *Behavioral modeling and linearization of RF power amplifiers*. Artech House, 2014.

[21] N. Pothecary, *Feedforward linear power amplifiers*. Artech House, 1999.

[22] A. Saleh and J. Salz, "Adaptive linearization of power amplifiers in digital radio systems," *Bell System Technical Journal*, vol. 62, no. 4, pp. 1019–1033, 1983.

[23] F. M. Ghannouchi, O. Hammi, and M. Helaoui, *Behavioral modeling and predistortion of wideband wireless transmitters*. John Wiley & Sons, 2015.

[24] J. L. Dawson and T. H. Lee, *Feedback linearization of RF power amplifiers*. Springer Science & Business Media, 2007.

[25] S. W. Chung, J. W. Holloway, and J. L. Dawson, "Open-loop digital predistortion using Cartesian feedback for adaptive RF power amplifier linearization," in *2007 IEEE/MTT-S International Microwave Symposium*. IEEE, 2007, pp. 1449–1452.

[26] C. Bremenson and D. Lombard, "Linearization of a satellite transmission channel in a time division multiple access system," in *3rd International Conference on Digital Satellite Communications*, vol. 1, 1975, pp. 144–151.

[27] M. S. Hashmi, Z. S. Rogojan, and F. M. Ghannouchi, "A flexible dual-inflection point RF predistortion linearizer for microwave power amplifiers," *Progress In Electromagnetics Research C*, vol. 13, pp. 1–18, 2010.

[28] F. Ghannouchi and O. Hammi, "Behavioral modeling and predistortion," *IEEE Transactions on Microwave Theory and Techniques*, vol. 10, no. 7, pp. 52–64, Dec. 2010.

[29] L. Sundstrom, M. Faulkner, and M. Johansson, "Quantization analysis and design of a digital predistortion linearizer for RF power amplifiers," *IEEE Transactions on Vehicular Technology*, vol. 45, no. 4, pp. 707–719, 1996.

[30] A. Bernardini and S. De Fina, "Analysis of different optimization criteria for IF predistortion in digital radio links with nonlinear amplifiers," *IEEE Transactions on Communications*, vol. 45, no. 4, pp. 421–428, 1997.

9 Microwave Measurements

9.1 Introduction

This chapter aims at providing an overview of the experimental characterization techniques of linear and nonlinear one- and two-ports.[1] After a short review of some basic instrumentation tools (the power meter, the vectorial voltmeter, the spectrum analyzer) in Sec. 9.2, the characterization of linear one- and two-ports, such as passive circuits and linear amplifiers, based on the measurement of power waves (see Sec. 3.2.2) and the related circuit-oriented parameters are discussed. As a first example, one-port measurements are introduced, whose fundamental tool is the so-called reflectometer (Sec. 9.3). The analysis of the reflectometer already allows the main issues involved in microwave measurement set-ups to be discussed, such as: the use of directional couplers to separate incident and reflected waves at the port where the Device Under Test (DUT) is connected; the need to perform measurements at a suitably low Intermediate Frequency (IF) rather than at RF to improve accuracy; the need to *calibrate* the measurement system to correct for systematic errors in the instrumentation.

In Sec. 9.4 the discussion is extended to the characterization of two-port linear devices adopting the two-port Vector Network Analyzer (VNA), the key instrument for high-frequency characterization first introduced by Hewlett Packard, around 1970 [2, 3]. Downconversion approaches are discussed in Sec. 9.4.1, while a number of calibration strategies are reported in Sec. 9.4.2.

Large-signal rather than linear-only characterizations are fundamental in the experimental assessment of power amplifiers. In Sec. 9.5 the most important aspects related to load-pull measurements are discussed. Sec. 9.6 is devoted to system-level characterization of RF components under modulated signal excitation, with particular attention to power amplifiers. Finally, Sec. 9.7 presents an overview of noise measurement techniques.

9.2 Basic Microwave Instrumentation Tools

Microwave measurements exploit a number of tools that are also commonly found in low-frequency electronic or electrical characterizations. Their function and principle of operation will be briefly recalled here.

---

[1] Several reference books deal with the issue of microwave linear and nonlinear measurements; the reader may refer to a more in-depth treatment to [1] and references therein.

Power meters are instruments able to measure the power dissipated by a resistive load being part of the meter. For microwave measurements a typical load may have a 50 Ω impedance. Several approaches can be applied to implement the power meter, see, e.g., [4]; a traditional set-up includes a bolometer (i.e., a resistor whose resistance changes with temperature and therefore with the dissipated power, also called a thermistor) connected to a resistance bridge that is balanced for zero dissipated power and zero heating, but provides a nonzero voltage output when the thermistor heats up. To obtain temperature compensation with respect to variations in the ambient temperature T_a, two equal bridges can be used, of which only one is connected to the RF signal, while the other is not excited and therefore is kept at T_a. The resulting signal is then digitized in order to display, or further process it. Power meters directly operate at RF; moreover, due to the problems related to the implementation of narrowband filters at high frequency, their input bandwidth can be narrow enough to exclude harmonics, but not in-band distortion.

From the dissipated power, the load resistance being known, the voltage across the load can be determined. This is the principle of the *scalar voltmeter*, whose purpose is, for a time-varying periodic waveform, to provide the effective value of the applied voltage, proportional to power. The bandwidth limitations of the scalar voltmeter are the same as for the power meter. A different approach has to be pursued in a *vector* or *vectorial voltmeter*, where not only the magnitude but also the phase of the measured voltage has to be derived with respect to some reference phase. Digital vector voltmeters [5] act by digitally sampling the voltage waveform, that is then interpolated so as to provide the relative phase vs. a reference signal, e.g., by considering the waveform zero crossings. Any information concerning the magnitude of the waveform (such as the effective value or the minimum and maximum value) can be easily recovered from the digital interpolation.

From the above discussion, it is clear that, while power meters can directly characterize a microwave signal, the phase measurement implicit in the vector voltmeter does require the digital sampling of the waveform and can therefore be hardly performed at microwaves. The need therefore arises to downconvert, as in a radio receiver, the microwave signal to an intermediate frequency low enough to allow for efficient A/D conversion. Besides, IF downconversion allows the system to be flexible in the frequency range, since the frequency agility is achieved by sweeping a local oscillator that also provides a phase reference to the whole system, while the IF frequency is kept constant.

Spectrum analyzers are instruments able to measure the power spectrum of the input signal, see, e.g., [6]. Also in this case the instrument has a receiver with a swept local oscillator, that scans a prescribed RF frequency range and converts it to IF, where it is detected and displayed or digitized via A/D conversion. The resolution of the spectrum analyzer is related to the IF bandwidth and is critical when very close spectral lines have to be separately detected, as in the analysis of in-band intermodulation products. In more recent approaches A/D conversion of the signal allows Fast Fourier transform techniques to be exploited to evaluate the signal spectrum. More complex spectrum analyzers, equipped with a quadrature receiver, can be used to measure modulated signals. As a last point, this approach can also be applied to characterize the noise power on a

9.3 The Reflectometer

prescribed bandwidth, again provided by the IF passband filter. Notice that there is an obvious compromise between the sweep speed and the IF bandwidth; a very narrow IF bandwidth improves the spectrum analyzer resolution but requires a very slow speed.

Power waves can be ultimately characterized through voltage measurements under well defined port terminations. However, to this aim we need a device able to separate the forward (a) and backward (b) power waves at a port, making them available for individual measurement by means of two voltmeters, see Fig. 9.1. This is the coupled-line directional coupler discussed in Sec. 4.2. Using the port numbering in Fig. 9.1, the scattering matrix of the coupler is, at centerband:

$$\underline{S} = \begin{pmatrix} 0 & -j\sqrt{1-C^2} & C & 0 \\ -j\sqrt{1-C^2} & 0 & 0 & C \\ C & 0 & 0 & -j\sqrt{1-C^2} \\ 0 & C & -j\sqrt{1-C^2} & 0 \end{pmatrix},$$

where $C = (Z_{0e} - Z_{0o})/(Z_{0e} + Z_{0o})$ is the coupling, and the load and generator impedance ensuring matching at all ports is $\sqrt{Z_{0e}Z_{0o}}$, see (4.6), where Z_{0e} and Z_{0o} are the characteristic impedances of the even and odd modes of the coupled-line coupler, respectively. From the scheme in Fig. 9.1, where two voltmeters V_{m1} and V_{m2} are connected at port 3 and 4, respectively, and port 2 is connected to the DUT (Device Under Test) while port 1 is connected to a generator, we have:

$$b_3 = Ca_1$$
$$b_4 = Ca_2$$
$$b_2 = -j\sqrt{1-C^2}a_1,$$

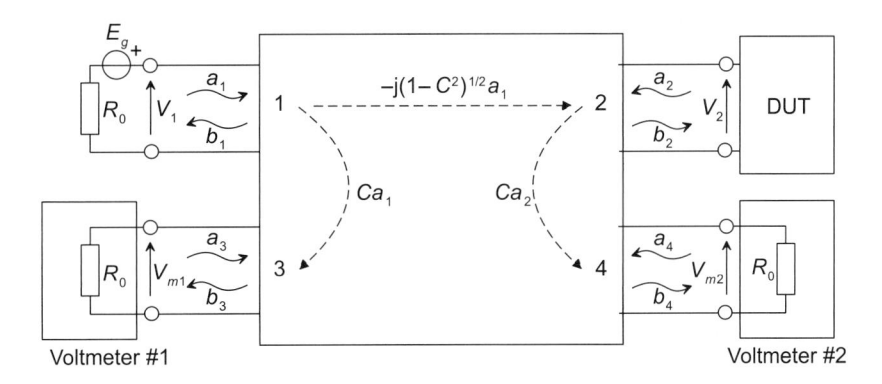

Figure 9.1 Separating and measuring the power waves at the DUT: the directional coupler setup. Two vectorial voltmeters are connected at ports 3 and 4.

i.e., the power waves incident into (b_2) and reflected by (a_2) the DUT are proportional to the power waves b_3 and b_4, respectively:

$$b_3 = j\frac{C}{\sqrt{1-C^2}} b_2$$

$$b_4 = C a_2.$$

Assuming that the two voltmeters have input impedance $Z_i = R_0$, we have:

$$V_{m1} = \sqrt{R_0}\, b_3 \;\rightarrow\; V_{m1} = \frac{jC\sqrt{R_0}}{\sqrt{1-C^2}} b_2 \;\rightarrow\; b_2 = -j\frac{\sqrt{1-C^2}}{C\sqrt{R_0}} V_{m1} \quad (9.1)$$

$$V_{m2} = \sqrt{R_0}\, b_4 \;\rightarrow\; V_{m2} = C\sqrt{R_0}\, a_2 \;\rightarrow\; a_2 = \frac{1}{C\sqrt{R_0}} V_{m2}. \quad (9.2)$$

The directional coupler therefore allows the forward and reflected waves to be separated at the DUT port and measured at ports 3 and 4.

The directional coupler setup described so far can be exploited to implement the reflectometer, i.e., a measurement system able to characterize the scattering parameter (the reflection coefficient) of a one-port. The reflectometer also is a key element of a two-port scattering parameter measurement system, the two-port VNA. A scheme of the reflectometer is shown in Fig. 9.2; port 0 is connected to a source, port 1 to the DUT, while ports 3 and 4 are connected to the voltmeters 1 and 2, respectively.[2] While in the initial discussion the directional coupler was assumed ideal and the voltmeters

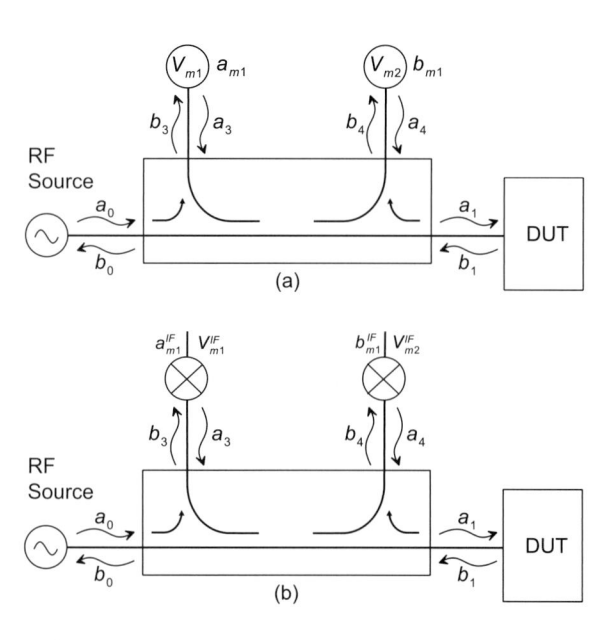

Figure 9.2 One-port reflectometer: (a) voltmeters directly connected to the RF outputs; (b) voltmeters measuring the IF downconverted signals.

---

[2] This unconventional numbering with a missing port 2 is justified in the discussion of the two-port VNA where port 1 and 2 will be connected to the DUT.

9.3 The Reflectometer

matched, we will here consider a more general coupler scattering matrix, with possibly mismatched voltmeters. Let us denote the coupler S-matrix as:

$$\mathbf{S} = \begin{pmatrix} \rho & \alpha & \beta & \gamma \\ \alpha & \rho & \gamma & \beta \\ \beta & \gamma & \rho & \alpha \\ \gamma & \beta & \alpha & \rho \end{pmatrix}. \tag{9.3}$$

The coupler is assumed reciprocal and symmetrical, but with some mismatch of the coupler ports, since $S_{ii} = \rho \neq 0$. Moreover, the coupler may be lossy and with finite directivity, since $\gamma \neq 0$. We also allow for some mismatch in the voltmeters, i.e.:

$$a_3 = \Gamma_V b_3 \tag{9.4}$$
$$a_4 = \Gamma_V b_4, \tag{9.5}$$

where Γ_V is the voltmeter reflection coefficient (assumed to be equal for simplicity for both voltmeters, a condition that can be easily relaxed). The voltmeter readings are now given by:

$$V_{m1} \equiv a_{m1} = \sqrt{R_0}(1 + \Gamma_V) b_3 \tag{9.6}$$
$$V_{m2} \equiv b_{m1} = \sqrt{R_0}(1 + \Gamma_V) b_4. \tag{9.7}$$

Notice that a_{m1} and b_{m1} are the voltmeter readings related, in an ideal system, to the power waves a_1 and b_1. Combining the coupler and the voltmeter constitutive relationships obtained from (9.3) and (9.4), (9.5), we obtain a linear system of six equations in the eight power waves a_i, b_i ($i = 0, 1, 3, 4$), see Fig. 9.2. In fact, however, b_3 and b_4 can be derived from the voltmeter readings according to (9.6) and (9.7) and are therefore known. All other power waves can be now expressed as a linear combination of b_3 and b_4 and hence of the voltmeter readings a_{m1} and b_{m1}. Focusing only on the power waves a_1 and b_1, at the DUT input, we obtain:

$$a_1 = E_{11} a_{m1} + E_{12} b_{m1} \tag{9.8}$$
$$b_1 = E_{21} a_{m1} + E_{22} b_{m1}, \tag{9.9}$$

where the system coefficients are:

$$E_{11} = \frac{\Gamma_V(\alpha\beta - \gamma\rho) + \gamma}{\sqrt{R_0}(1 + \Gamma_V)(\gamma^2 - \beta^2)}$$

$$E_{12} = \frac{\Gamma_V(\alpha\gamma - \beta\rho) - \beta}{\sqrt{R_0}(1 + \Gamma_V)(\gamma^2 - \beta^2)}$$

$$E_{21} = \frac{\Gamma_V(2\alpha\beta\rho - \alpha^2\gamma - \beta^2\gamma + \gamma^3 - \gamma\rho^2) - \alpha\beta + \gamma\rho}{\sqrt{R_0}(1 + \Gamma_V)(\gamma^2 - \beta^2)}$$

$$E_{22} = \frac{\Gamma_V(\alpha^2\beta - 2\alpha\gamma\rho - \beta^3 + \beta\gamma^2 + \beta\rho^2) + \alpha\gamma - \beta\rho}{\sqrt{R_0}(1 + \Gamma_V)(\gamma^2 - \beta^2)}.$$

Notice that (9.8) and (9.9) reduce (9.2) and (9.1), respectively (with $1 \to 2$ in the port numbering) if the coupler is ideal and the voltmeters are matched, i.e., for $\rho = \gamma = \Gamma_V = 0$, $\alpha = -j\sqrt{1 - C^2}$, and $\beta = C$.

From (9.8) and (9.9) we see that a_1 and b_1 can be derived from the voltmeter readings a_{m1} and b_{m1} once the coefficients E_{11}, E_{12}, E_{21}, and E_{22} are given. In practice, however, the coupler parameters and voltmeter reflection coefficient are not exactly known and therefore a *calibration* procedure has to be set up to derive E_{11}, E_{12}, E_{21}, and E_{22} from measurements carried out on a one-port of known reflection coefficient (the *calibration standard*). The calibration standard can be, e.g., a short circuit (for which $b_1 = -a_1$), an open circuit (for which $b_1 = a_1$), or a resistance whose value has been measured independently. Notice in particular that if the standard is the normalization resistance R_0, $b_1 = 0$ and a_1 is undetermined. We call such standards *ratiometric* because they set the ratio b_1/a_1 only, i.e., the reflection coefficient at port 1, but not the individual values of a_1 and b_1; these aspects will be discussed in further detail in Sec. 9.4.2 devoted to VNA calibration. Ratiometric standards are however enough to calibrate the one-port system in order to measure the DUT reflection coefficient, Γ_DUT. In fact, taking the ratio of (9.9) and (9.8) we obtain:

$$\Gamma_\text{DUT} = \frac{b_1}{a_1} = \frac{a_{m1} + \dfrac{E_{12}}{E_{11}} b_{m1}}{\dfrac{E_{21}}{E_{11}} a_{m1} + \dfrac{E_{22}}{E_{11}} b_{m1}} = \frac{a_{m1} + e_{12} b_{m1}}{e_{21} a_{m1} + e_{22} b_{m1}}, \qquad (9.10)$$

where $e_{ij} = E_{ij}/E_{11}$. From (9.10) we see that the calibration procedure only needs to identify 3 parameters e_{12}, e_{21}, and e_{22}, thus requiring the measurement of 3 independent standards of known reflection coefficients $\Gamma_\text{STD}^{(i)}$, $i = 1\ldots 3$. Rewriting (9.10) we obtain the linear system:

$$a_{m1}^{(i)} + e_{12} b_{m1}^{(i)} - e_{21} \cdot a_{m1}^{(i)} \Gamma_\text{STD}^{(i)} - e_{22} \cdot b_{m1}^{(i)} \Gamma_\text{STD}^{(i)} = 0, \quad i = 1, 2, 3,$$

where $a_{m1}^{(i)}$ and $b_{m1}^{(i)}$, $i = 1, 2, 3$ are the readings of voltmeter 1 and 2, respectively, when the standard i is connected as the DUT. Using, e.g., as standards a short circuit, an open circuit and the normalization resistance we have $\Gamma_\text{STD}^{(1)} \equiv \Gamma_s = -1$, $\Gamma_\text{STD}^{(2)} \equiv \Gamma_o = 1$, $\Gamma_\text{STD}^{(3)} \equiv \Gamma_{R_0} = 0$ and the system becomes:

$$\begin{aligned} a_{m1}^s + e_{12} b_{m1}^s + e_{21} \cdot a_{m1}^s + e_{22} \cdot b_{m1}^s &= 0 \\ a_{m1}^o + e_{12} b_{m1}^o - e_{21} \cdot a_{m1}^o - e_{22} \cdot b_{m1}^o &= 0 \\ a_{m1}^{R_0} + e_{12} b_{m1}^{R_0} &= 0, \end{aligned}$$

where the voltmeter readings for the three standards have been denoted with the superscripts o, s and R_0, respectively. The linear system can be easily solved to find e_{12}, e_{21}, and e_{22}; once these coefficients have been identified, the reflectometer is calibrated and the reflection coefficient of a generic DUT can be derived from the voltmeter readings using (9.10). Note that in principle the calibration procedure must be carried out at each of the frequencies of interest, leading to frequency-dependent coefficients e_{12}, e_{21}, and e_{22}.

Some further comments are needed on the practical implementation of the reflectometer. As already discussed, while scalar measurements can be performed at RF, phasors cannot be measured in amplitude and phase at high frequency (RF) with the accuracy required. To circumvent the problem, the power waves at ports 3 and 4 are downconverted to a suitable intermediate frequency (IF) according to the scheme already discussed in the heterodyne receiver (Sec. 1.2.2). Of course downconversion should be coherent for both channels involved, i.e., the same local oscillator should be used. Downconversion introduces, in a systematic way, a linear amplitude scaling and a phase delay; therefore, accurate IF voltmeters can be exploited to obtain the readings a_{m1}^{IF} and b_{m1}^{IF} where $a_{m1}^{\text{IF}} = k_a a_{m1}$, $b_{m1}^{\text{IF}} = k_b b_{m1}$ and k_a, k_b are complex constants. Since the IF and RF readings are proportional we have from (9.10):

$$\Gamma_{\text{DUT}} = \frac{a_{m1} + e_{12}b_{m1}}{e_{21}a_{m1} + e_{22}b_{m1}} = \frac{a_{m1}^{\text{IF}} + \frac{e_{12}k_a}{k_b}b_{m1}^{\text{IF}}}{e_{21}a_{m1}^{\text{IF}} + \frac{e_{22}k_a}{k_b}b_{m1}^{\text{IF}}} \equiv \frac{a_{m1}^{\text{IF}} + e'_{12}b_{m1}^{\text{IF}}}{e_{21}a_{m1}^{\text{IF}} + e'_{22}b_{m1}^{\text{IF}}}, \qquad (9.11)$$

i.e., also the calibration procedure required to identify e'_{12}, e_{21}, and e'_{22} can now be carried out at IF using RF standards. Moreover, IF processing can easily include analog to digital (A/D) conversion and further digital data handling. For the sake of brevity, in the next sections the IF measured waves will be denoted omitting the IF superscript.

In practical wideband systems a critical point is given by the coupler bandwidth and directivity. Broadband couplers (e.g., multi-stage) are exploited with even-odd mode velocity matching (compensation) techniques, like serrated lines and capacitor loading; alternatively, purely TEM structures like striplines can be introduced as coupled line couplers, see the discussion in Sec. 4.3.2. An alternative broadband solution, effective however only in the lower microwave range (i.e., below a few GHz) is provided by the so-called *directional bridge* [7], a resistance bridge shown in simplified form in Fig. 9.3. As shown in Example 9.1, the bridge allows a signal proportional to the DUT reflection coefficient to be measured by the matched voltmeter closed on port 3. If the DUT and RF source ports are exchanged, the voltmeter instead measures a power wave proportional to the power wave entering the DUT. The configuration shown has to be completed by proper balancing-unbalancing units and the bridge resistances can be dimensioned in order to provide variable coupling, see, e.g., [8].

Example 9.1 Evaluate the scattering matrix of the directional bridge in Fig. 9.3 and show that the signal on the matched meter is proportional to the DUT reflectance.

Solution
The scattering matrix of the bridge coupler can be conveniently derived from the resistance matrix, that can be evaluated by inspection as:

$$\mathbf{R} = \begin{pmatrix} 2R_0 & 2R_0 & R_0 \\ 2R_0 & 3R_0 & 2R_0 \\ R_0 & 2R_0 & 2R_0 \end{pmatrix}.$$

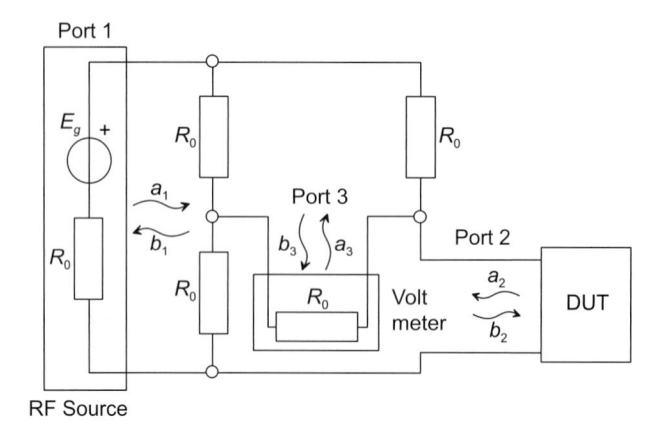

Figure 9.3 Simplified directional bridge connected to a DUT, to a signal generator and to a matched voltmeter.

Using:

$$\mathbf{S} = (\mathbf{R} - R_0\mathbf{I})(\mathbf{R} + R_0\mathbf{I})^{-1},$$

we obtain:

$$\mathbf{S} = \begin{pmatrix} 0 & \frac{1}{2} & 0 \\ \frac{1}{2} & 0 & \frac{1}{2} \\ 0 & \frac{1}{2} & 0 \end{pmatrix}.$$

The coupling bridge therefore operates like a 3dB coupler between ports 1 and 2, while port 3 is isolated, and a 3dB coupler between ports 2 and 3, port 1 being isolated; the bridge is moreover matched at all ports. In scalar form, we have:

$$b_1 = \frac{1}{2}a_2$$
$$b_2 = \frac{1}{2}a_1 + \frac{1}{2}a_3$$
$$b_3 = \frac{1}{2}a_2.$$

However, the voltmeter is matched and therefore $a_3 = 0$ and $b_2 = a_1/2$; moreover, $a_2 = \Gamma_{\text{DUT}} b_2$. We therefore have:

$$b_3 = \frac{1}{2}a_2 = \frac{1}{2}\Gamma_{\text{DUT}} b_2 = \frac{1}{4}\Gamma_{\text{DUT}} a_1$$
$$\frac{b_3}{a_1} = \frac{1}{4}\Gamma_{\text{DUT}}.$$

Thus a measurement of the ratio b_3/a_1 is proportional to the load reflectance. If the DUT is connected at port 1 and the RF source at port 2, $b_1 = a_2/2 = b_3$; thus the power wave measured by the voltmeter coincides with the power wave entering the DUT. In conclusion, two cascaded coupling bridges allow the incident and reflected waves to be measured at the DUT port.

9.4 The Vector Network Analyzer

The solutions implemented in the one-port reflectometer can be extended to a two-port or N-port measuring tool, the Vector Network Analyzer (VNA). In the present section only the two-port version will be discussed, since N-port characterization can be carried out by repeated two-port measurements. The VNA is based on the heterodyne downconversion of the measured signals to IF where the relevant signals are digitally converted and processed. Moreover, since the IF downconverted signals on several channels must be vectorially measured, coherent downconversion has to be ensured by a unique LO. To allow for accurate measurements, the calibration procedures already introduced in the one-port reflectometer case have to be extended to the two-port VNA. The scheme of a two-port VNA with four independent acquisition channels is reported in Fig. 9.4. The scheme is completed with the switches required to commute the stimulus injection from port 1 to port 2, and to terminate the other on a load resistance that may coincide with the scattering parameter normalization resistance R_0 resulting from calibration.

9.4.1 Downconversion Module Solutions

Downconversion of the signals to be measured in the VNA can be carried out through two techniques, mixing or subsampling. The popularity of either method appears to have varied throughout the VNA evolution, subsampling having been initially preferred, while today mixing is the most common choice. We will briefly review here the two approaches.

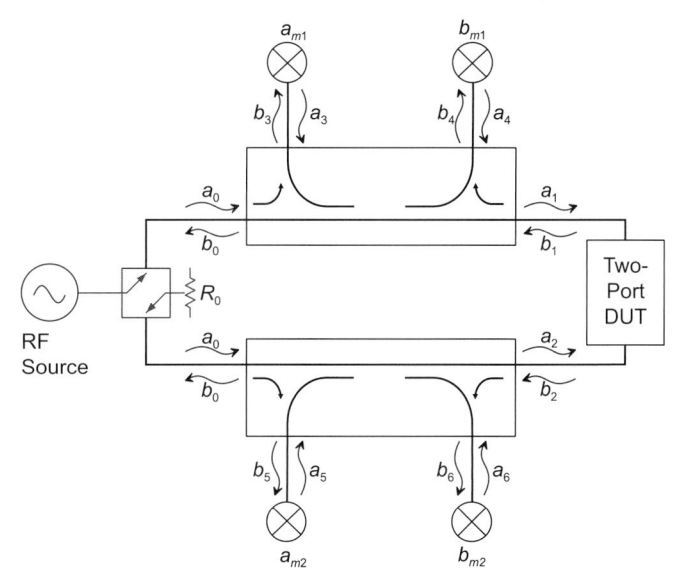

Figure 9.4 Two-port VNA scheme. The input source and matched load are switched between port 1 and 2 while the directional couplers allow the power waves to be separated and measured after downconversion to IF.

Subsampling downconversion is based on the RF signal being switched on and off at a certain sampling rate, thus downconverting the RF spectrum to IF. In the subsampling downconverter, a voltage-controlled oscillator (VCO) generates a train of pulses with width t_W and repetition frequency f_{IF} ($\omega_{IF} = 2\pi f_{IF}$) in the kHz/MHz range, that drives the sampling switch. For $t_W \to 0$ the switch driving signal is a series of Dirac pulses; in that case the input (at ω_0) and output (downconverted and after low-pass filtering) spectra are, assuming the RF and LO initial phase to be zero:

$$x_{RF}(t) = A \sin \omega_0 t \to X_{RF}(\omega) = jA\pi \left[\delta(\omega - \omega_0) - \delta(\omega + \omega_0)\right]$$

$$Y(\omega) = jA\pi \sum_{n=-\infty}^{\infty} \left[\delta(\omega + \omega_0 - n\omega_{IF}) - \delta(\omega - \omega_0 - n\omega_{IF})\right] \to$$

$$\to Y_{IF}(\omega) = jA\pi \left[\delta(\omega + \omega_0 - \omega_{IF}) - \delta(\omega - \omega_0 - \omega_{IF})\right].$$

For the sake of simplicity, the downconversion loss is assumed to be unity. The Phase Locked Loop (PLL) corrects the VCO frequency locking it to one of the RF waves (the reference wave), so as to isolate, after low-pass filtering, the desired replicas centered at IF. In this way the VCO also synchronously drives all switches ensuring that the four IF signals preserve the relative phase and amplitude relationships of the RF signals. Thanks to coherent and symmetrical downconversion, the phasor ratios defining the scattering parameters are unaffected by the process and can therefore be measured at IF. In practice, the switch driving signal exhibit finite pulse width $t_W \neq 0$; this means that the signal replicas are modulated by the Fourier transform $S(\omega)$ of the square pulse:

$$S(\omega) = \frac{\sin\left(\frac{\omega t_W}{2}\right)}{\frac{\omega t_W}{2}} \exp\left(j\frac{\omega t_W}{2}\right).$$

Considering, for example, a VNA with maximum frequency 20 GHz, with the IF receiver at $f_{IF} = 1$ MHz, this means that the maximum allowed t_W is 50 ps, the value for which $\omega t_W/2 = \pi$ and therefore the corresponding IF signal has its first zero. It is therefore clear that the major issue for the subsampling downconversion scheme is not related to the VCO frequency but rather concerns the capability to generate extremely short pulses in order to avoid an unacceptable attenuation of the IF translated replica that would compromise the VNA dynamic range.

Downconversion mixing was already introduced in Sec. 1.2.2; from the RF, LO and IF signals we have:

$$x_{RF} = A_{RF} \cos(\omega_{RF} t + \psi_{RF})$$
$$x_{LO} = A_{LO} \cos(\omega_{LO} t + \psi_{LO})$$
$$y_{IF} = G_c A_{RF} \cos\left[(\omega_{RF} - \omega_{LO} t) + \psi_{RF} - \psi_{LO}\right],$$

where G_c is the mixer conversion gain, depending on the amplitude A_{LO} of the local oscillator signal. In four-channel VNAs exploiting downconversion mixing, four identical mixers are used, driven by the same LO (phase locked to the reference channel) to ensure coherent downconversion. Mixer unbalance can be anyway taken care of in the calibration stage. Notice that, since the IF is unique, the mixer downconversion scheme requires wideband mixers; in fact, while the mixer frequency response can be compensated for through the calibration, if the mixer conversion loss is too large this will compromise the VNA dynamic range.

9.4.2 VNA Calibration

It was clear from their first appearance on the market that the main limitation to the accuracy of microwave VNAs is related to the systematic errors [9]. They are due to the non-ideality of their components, such as the couplers finite directivity and mismatch, mixer asymmetries, phase shifts and losses of cables. Such effects make the magnitude and phase relationships between waves at the receiver section different from those at the DUT port. Moreover, such discrepancies cannot be compensated for once and for all, because of component aging and thermal effects, but also considering the customization of the individual DUT mountings that the users may apply to their components, e.g., to carry out on-chip rather than in-package characterizations. A calibration process is therefore required, as already stressed in the discussion of the reflectometer, in order to de-embed from the raw, uncalibrated, measured data at the receiver section an accurate estimate of the DUT parameters.

Calibration standards are simple (one- or two-port) passive devices whose behavior can be accurately assessed from their physical and geometrical parameters through independent measurements or electromagnetic simulations. Already mentioned one-port standards are short and open circuits and matched terminations. Examples of two-port standards are transmission lines with known electrical length and characteristic impedance (the so-called *line* standard) but also lines with zero length (called the *thru* standard). The set of standard devices required to complete a two-port VNA calibration procedure represents the *calibration standard sequence*, often denoted from the initials of the standards exploited: e.g., the TRL calibration [10, 11] requires a Thru and a Line as two-port standards plus a Reflection (i.e., a load providing some reflection coefficient) as the one-port standard.

Not necessarily all standards have to be perfectly known. The so-called *self-calibration algorithms* adopt partially unknown devices whose undefined parameters are obtained as by-products of the calibration procedure. For instance in TRL calibration, as shown in Sec. 9.4.4, the Line must have known characteristic impedance but its length may be unknown, while the Reflection Γ is arbitrary. In the SOLR (Short Open Load Reciprocal) calibration algorithm [12], three completely specified one-port standards are needed (Short, Open and Load) together with a two-port whose only requirement is reciprocity.

From a conceptual standpoint, the systematic error removal procedure requires the following steps:

1. Definition of the error model: the relationship between the instrument readings and the DUT quantities of interest are evaluated. In general, the two sets are connected by a linear relationship whose parameters are called the *error coefficients*.

2. The calibration procedure: the error coefficients are identified by measuring a set of reference circuits, i.e., the calibration standards.
3. De-embedding: the error model with known coefficients is exploited to extract or de-embed from the raw data the correct DUT characteristics.

9.4.3 Defining the Error Model

Extending the discussion of the reflectometer calibration, see Sec. 9.3, we can conclude that also in the VNA case a linear relationship exists between the waves at the DUT section (a_1, b_1, a_2, b_2) and the IF voltmeter readings (a_{m1}, b_{m1}, a_{m2}, b_{m2}; the subscript refers to the reflectometers connected at port 1 and 2, respectively), see the two-port VNA scheme in Fig. 9.4. We thus have:

$$a_1 = E^1_{11} a_{m1} + E^1_{12} b_{m1}$$
$$b_1 = E^1_{21} a_{m1} + E^1_{22} b_{m1}$$
$$a_2 = E^2_{11} a_{m2} + E^2_{12} b_{m2}$$
$$b_2 = E^2_{21} a_{m2} + E^2_{22} b_{m2},$$

where coefficients E^i_{jk} are the error coefficients. In the linear model adopted we only exclude that the voltmeter readings at port i may be influenced by the waves at port j with $j \neq i$, an assumption that holds true for frequencies up to 40–50 GHz; above that the electromagnetic coupling between the internal VNA parts cannot be disregarded, and the error model has to account for leakage between ports [13, 14]. In matrix form we obtain:

$$\begin{pmatrix} a_1 \\ b_1 \end{pmatrix} = \begin{pmatrix} E^1_{11} & E^1_{12} \\ E^1_{21} & E^1_{22} \end{pmatrix} \begin{pmatrix} a_{m1} \\ b_{m1} \end{pmatrix} = \mathbf{X}_1 \begin{pmatrix} a_{m1} \\ b_{m1} \end{pmatrix} \quad (9.12)$$

$$\begin{pmatrix} a_2 \\ b_2 \end{pmatrix} = \begin{pmatrix} E^2_{11} & E^2_{12} \\ E^2_{21} & E^2_{22} \end{pmatrix} \begin{pmatrix} a_{m2} \\ b_{m2} \end{pmatrix} = \mathbf{X}_2 \begin{pmatrix} a_{m2} \\ b_{m2} \end{pmatrix}, \quad (9.13)$$

where $\mathbf{X}_1$ and $\mathbf{X}_2$ are the error coefficient matrices of the reflectometers at port 1 and 2, respectively. The four IF reading can be interpreted introducing the pseudo transmission matrix $\mathbf{T}^M$ between the instrumentation readings at first and second reflectometer:

$$\begin{pmatrix} a_{m1} \\ b_{m1} \end{pmatrix} = \mathbf{T}^m \begin{pmatrix} a_{m2} \\ b_{m2} \end{pmatrix}. \quad (9.14)$$

Notice that $\mathbf{T}^m$ satisfies the definition of the transmission matrix discussed in Sec. 3.2.8; we call it as a pseudo matrix since it does not account for any actual transit of signal. However, $\mathbf{T}^m$ can be readily identified by imposing the DUT two independent excitations (e.g., one at port 1 and another at port 2); the IF voltmeter readings provide 4 linearly independent equations in the elements of $\mathbf{T}^m$ that can therefore be obtained by solving a linear system. Using (9.12), (9.13) and (9.14) we obtain:

$$\begin{pmatrix} a_1 \\ b_1 \end{pmatrix} = \mathbf{X}_1 \mathbf{T}^m \mathbf{X}_2^{-1} \begin{pmatrix} a_2 \\ b_2 \end{pmatrix} = \mathbf{T}_{\text{DUT}} \begin{pmatrix} a_2 \\ b_2 \end{pmatrix},$$

i.e., the DUT transmission matrix is:

$$\mathbf{T}_{\text{DUT}} = \mathbf{X}_1 \mathbf{T}^m \mathbf{X}_2^{-1}. \tag{9.15}$$

From $\mathbf{T}_{\text{DUT}}$ the scattering matrix $\mathbf{S}_{\text{DUT}}$ (see Sec. 3.2.8) can be finally derived using the transformation:

$$\mathbf{S} = \frac{1}{T_{12}} \begin{pmatrix} T_{22} & -\Delta_T \\ 1 & -T_{11} \end{pmatrix}, \tag{9.16}$$

where $\Delta_T = T_{11}T_{22} - T_{12}T_{21}$. Eq. (9.15) includes the *raw measurement* $\mathbf{T}^m$ and the matrices $\mathbf{X}_1$ and $\mathbf{X}_2$ representing the *error boxes* of the VNA. The error model that can be interpreted as a pair of two-ports, $\mathbf{X}_1$ and $\mathbf{X}_2$, sandwiching the DUT and interposing between it and the errorless VNA ports where the measurements are collected.

9.4.4 Calibration Algorithms

The goal of this section is define the strategy for the identification of the error box transmission matrices $\mathbf{X}_1$ and $\mathbf{X}_2$. Linear characterization is generally based on the measurement of ratios of voltages, currents or power waves; from this standpoint, high-frequency calibration standards impose the ratios between power waves rather than their absolute values. This suggests that only seven elements of the matrices $\mathbf{X}_1$ and $\mathbf{X}_2$ are actually needed. In fact, if we normalize $\mathbf{X}_1$ ($\mathbf{X}_2$) with respect to the element E_{11}^1 (E_{11}^2), we obtain the normalized matrices $\mathbf{x}_1$ and $\mathbf{x}_2$ as:

$$\mathbf{x}_1 = \frac{1}{E_{11}^1} \mathbf{X}_1 = \begin{pmatrix} 1 & \frac{E_{12}^1}{E_{11}^1} \\ \frac{E_{21}^1}{E_{11}^1} & \frac{E_{22}^1}{E_{11}^1} \end{pmatrix} \equiv \begin{pmatrix} 1 & e_{12}^1 \\ e_{21}^1 & e_{22}^1 \end{pmatrix} \tag{9.17}$$

$$\mathbf{x}_2 = \frac{1}{E_{11}^2} \mathbf{X}_2 = \begin{pmatrix} 1 & \frac{E_{12}^2}{E_{11}^2} \\ \frac{E_{21}^2}{E_{11}^2} & \frac{E_{22}^2}{E_{11}^2} \end{pmatrix} \equiv \begin{pmatrix} 1 & e_{12}^2 \\ e_{21}^2 & e_{22}^2 \end{pmatrix} \tag{9.18}$$

and (9.15) can be rewritten as:

$$\mathbf{T}_{\text{DUT}} = \alpha \mathbf{x}_1 \mathbf{T}^m_{\text{DUT}} \mathbf{x}_2^{-1}, \tag{9.19}$$

where $\alpha = E_{11}^1 / E_{11}^2$. This clearly shows that only seven unknowns have to be identified in the calibration (the six elements of $\mathbf{x}_1$ and $\mathbf{x}_2$ and the parameter α). Seven is therefore also the number of independent measurements needed to identify the error box parameters. A general unique mathematical formulation which can be applied to almost all two-port VNA calibration procedures can be found in [15]. However, for a better comprehension of the mathematical algorithms and their physical implications we separately present in the following subsections three examples of well-known calibration procedures.

QSOLT Calibration

The QSOLT calibration technique [16] was introduced as a "quick" variety of the SOLT procedure (from Short, Open, Load and Thru) [12]. It requires three one-port standards (short, open and load), plus a two-port standard, the thru. The Load is often referred to as the *match*, since it imposes the normalization impedance R_0.

The reflection coefficients of the one port standards vs. R_0 are known and will be denoted as Γ_1, Γ_2 and Γ_3 for the short, open and load, respectively (notice that the open and short reflection coefficients do not actually depend on the choice of R_0). The corresponding measured quantities at the reflectometer connected to port 1 will be denoted as Γ_1^{m1}, Γ_2^{m1} Γ_3^{m1}, where $\Gamma_i^{m1} = b_{m1}/a_{m1}$ when the ith standard is connected to port 1. From (9.12), accounting for (9.17) we obtain:

$$a_1 = E_{11}^1 \left(a_{m1} + e_{12}^1 b_{m1} \right)$$
$$b_1 = E_{11}^1 \left(e_{21}^1 a_{m1} + e_{22}^1 b_{m1} \right)$$

and therefore, closing port 1 on standard i, $i = 1, 2, 3$, the standard reflection coefficient reads:

$$\Gamma_i = \frac{b_1}{a_1} = \frac{e_{21}^1 + e_{22}^1 \Gamma_i^{m1}}{1 + e_{12}^1 \Gamma_i^{m1}}.$$

We therefore obtain 3 linear equations in the unknowns e_{21}^1, e_{12}^1 and e_{22}^1:

$$e_{21}^1 - \Gamma_1 \Gamma_1^{m1} e_{12}^1 + \Gamma_1^{m1} e_{22}^1 = \Gamma_1 \quad (9.20)$$
$$e_{21}^1 - \Gamma_2 \Gamma_2^{m1} e_{12}^1 + \Gamma_2^{m1} e_{22}^1 = \Gamma_2 \quad (9.21)$$
$$e_{21}^1 - \Gamma_3 \Gamma_3^{m1} e_{12}^1 + \Gamma_3^{m1} e_{22}^1 = \Gamma_3, \quad (9.22)$$

whose inversion yields, in matrix form:

$$\begin{pmatrix} e_{21}^1 \\ e_{12}^1 \\ e_{22}^1 \end{pmatrix} = \begin{pmatrix} 1 & -\Gamma_1 \Gamma_1^{m1} & \Gamma_1^{m1} \\ 1 & -\Gamma_2 \Gamma_2^{m1} & \Gamma_2^{m1} \\ 1 & -\Gamma_3 \Gamma_3^{m1} & \Gamma_3^{m1} \end{pmatrix}^{-1} \begin{pmatrix} \Gamma_1 \\ \Gamma_2 \\ \Gamma_3 \end{pmatrix}. \quad (9.23)$$

The three error box parameters of the first reflectometer are thus derived. We now exploit the thru standard, whose scattering matrix is the unity matrix; the transmission matrix of the thru, $\mathbf{T}_4$, is therefore:

$$\mathbf{T}_4 = \begin{pmatrix} 0 & 1 \\ 1 & 0 \end{pmatrix},$$

while the measured thru transmission matrix will be denoted as $\mathbf{T}_4^m$. The elements of such a matrix can be readily identified from the ratios of voltmeter readings at port 1 and 2 taking into account that the linear system holds:

$$T_{4,11}^m + T_{4,12}^m \frac{b_{m2}}{a_{m2}} = \frac{a_{m1}}{a_{m2}}$$
$$T_{4,21}^m + T_{4,22}^m \frac{b_{m2}}{a_{m2}} = \frac{b_{m1}}{a_{m2}}.$$

Finally, since:

$$\mathbf{T}_4 = \alpha \mathbf{x}_1 \mathbf{T}_4^m \mathbf{x}_2^{-1},$$

the three error box parameters of the second reflectometer and α are derived from:

$$\frac{1}{\alpha}\mathbf{x}_2 = \begin{pmatrix} 1/\alpha & e_{12}^2/\alpha \\ e_{21}^2/\alpha & e_{22}^2/\alpha \end{pmatrix} = \mathbf{T}_4^{-1} \mathbf{x}_1 \mathbf{T}_4^m,$$

thus completing the calibration procedure. Once $\mathbf{x}_1$ and $\alpha\mathbf{x}_2^{-1}$ are known, the DUT transmission matrix $\mathbf{T}_{\text{DUT}}$ can be de-embedded from the raw VNA measurement $\mathbf{T}_{\text{DUT}}^m$ using (9.19). Notice that the calibration procedure and error box evaluation must be carried out at all measurement frequencies.

TRL Calibration

The TRL calibration (Thru Reflect Line) [10, 11] probably is the most used today especially in on-chip measurements, where the DUT reference planes are set on the tips of the probes used to contact the DUT. It requires a Thru and a Line as two-port standards, and only one one-port standard (the Reflect). The Thru, a direct connection between the DUT ports, can be referred to also as the *null device*. The Line is a transmission line of arbitrary length, its only requirement being to have a known and almost frequency-independent characteristic impedance, that will become the normalization resistance R_0. When compared to other calibration procedures, the TRL calibration has the advantage of exploiting a wideband component (the transmission line) rather than a concentrated load whose behavior as a function of frequency is often difficult to accurately predict and characterize.

To exploit TRL in broadband characterization, we have to consider that transmission lines have a periodic behavior vs. frequency. In particular, for frequencies at which the line length is a multiple of a half wavelength, the Line becomes a Thru, causing failure or poor accuracy of the calibration. To solve this problem the multiple line TRL approach is used, where lines with several lengths are measured during calibration and the Line-Thru pair is chosen so as to avoid lines with length too close to an integer number of half wavelengths at the measurement frequency [17]. Notice that the Thru is, in practice, a line with length $L_T \ll L$, where L is the length of the Line; in such cases the reference planes of the calibration are places at the center of the Thru. In the following formulae one should set $L \rightarrow L - L_T$ for a finite length Thru. Fig. 9.5 reports an example of TRL standards in coplanar waveguide with coplanar-coaxial transitions (whose section is shown). The horizontal dashed line denotes the position of the reference planes.

An additional advantage of TRL derives from the need of a single one-port reflection standard, to be measured at both ports. Moreover, a variant of TRL exists, as will be described at the end of the subsection, where the one port standard can be completely unknown, provided that the same device is measured at both ports. In this case, a 180 degree phase uncertainty in the reflection parameters has to be resolved; a relatively easy task to be accomplished for example through the measure of a DUT with roughly appreciable phase.

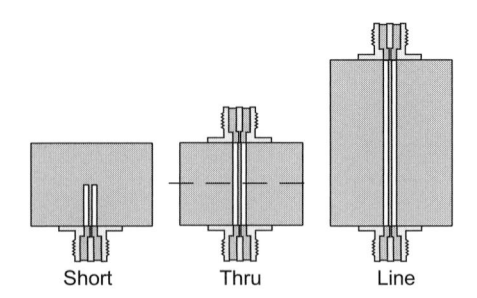

Figure 9.5 Examples of coplanar waveguide TRL standards with coaxial connectors (shown as a section). The Reflection standard here is a short.

The scattering and transmission matrix of the Thru and the Line with respect to the Line characteristic impedance $Z_0 \equiv R_0$ read, respectively:

$$\mathbf{S}_T = \begin{pmatrix} 0 & 1 \\ 1 & 0 \end{pmatrix} \rightarrow \mathbf{T}_T = \begin{pmatrix} 1 & 0 \\ 0 & 1 \end{pmatrix}$$

$$\mathbf{S}_L = \begin{pmatrix} 0 & e^{-\gamma L} \\ e^{-\gamma L} & 0 \end{pmatrix} \rightarrow \mathbf{T}_L = \begin{pmatrix} e^{-\gamma L} & 0 \\ 0 & e^{-\gamma L} \end{pmatrix},$$

where $\gamma = \alpha + j\beta$ and L are the complex propagation and length of the line. As in previous cases, the measured transmission matrices of the Thru and Line, $\mathbf{T}_T^m$ and $\mathbf{T}_L^m$, can be readily identified from IF voltmeter readings. From (9.19) we obtain:

$$\mathbf{T}_T^m = \frac{1}{\alpha} \mathbf{x}_1^{-1} \mathbf{T}_T \mathbf{x}_2$$

$$\mathbf{T}_L^m = \frac{1}{\alpha} \mathbf{x}_1^{-1} \mathbf{T}_L \mathbf{x}_2$$

and therefore, introducing the matrices $\mathbf{R}_M$ and $\mathbf{R}_N$, we have:

$$\mathbf{R}_M = \mathbf{T}_L^m \left(\mathbf{T}_T^m\right)^{-1} = \mathbf{x}_1^{-1} \mathbf{T}_L \mathbf{T}_T^{-1} \mathbf{x}_1 = \mathbf{x}_1^{-1} \Lambda \, \mathbf{x}_1 \quad (9.24)$$

$$\mathbf{R}_N = \left(\mathbf{T}_T^m\right)^{-1} \mathbf{T}_L^m = \mathbf{x}_2^{-1} \mathbf{T}_T^{-1} \mathbf{T}_L \mathbf{x}_2 = \mathbf{x}_2^{-1} \Lambda \, \mathbf{x}_2, \quad (9.25)$$

where Λ is a diagonal matrix:

$$\Lambda = \mathbf{T}_L \left(\mathbf{T}_T\right)^{-1} = \left(\mathbf{T}_T\right)^{-1} \mathbf{T}_L = \begin{pmatrix} e^{-\gamma L} & 0 \\ 0 & e^{-\gamma L} \end{pmatrix}.$$

Since Λ is diagonal, (9.24) and (9.25) are similarity transformation pairs and $\mathbf{x}_1^{-1}$ and $\mathbf{x}_2^{-1}$ must be the eigenvector matrices of $\mathbf{R}_M$ and $\mathbf{R}_N$, respectively. Notice that evaluating the eigenvalue and eigenvectors of matrices $\mathbf{R}_M$ and $\mathbf{R}_N$ the complex propagation constant of a line with known length L is obtained as a by product of the calibration procedure. However, since the two eigenvectors of $\mathbf{R}_M$ and $\mathbf{R}_N$ are defined to a constant, the step does not allow the error box matrices to be evaluated completely.

9.4 The Vector Network Analyzer

For the sake of definiteness, let us focus on the matrix $\mathbf{x}_1^{-1}$, whose columns are (proportional to) the eigenvectors of $\mathbf{R}_M$. We can express this matrix as:

$$\mathbf{x}_1^{-1} = \begin{pmatrix} av_1^M & bv_2^M \\ a & b \end{pmatrix} = b \begin{pmatrix} \frac{a}{b}v_1^M & v_2^M \\ \frac{a}{b} & 1 \end{pmatrix},$$

where v_1^M and v_2^M are known and the constants a and b are arbitrary. Inverting we obtain:

$$\mathbf{x}_1 = \frac{1}{a(v_1^M - v_2^M)} \begin{pmatrix} 1 & -v_2^M \\ -\frac{a}{b} & \frac{a}{b}v_1^M \end{pmatrix}.$$

Notice that the prefactor can be considered as a normalization constant to be merged with α. Neglecting the prefactor and comparing with (9.17) we can readily identify $-a/b = e_{21}^1$; we therefore obtain for $\mathbf{x}_1$ (and similarly for $\mathbf{x}_2$) the expressions:

$$\mathbf{x}_1 = \begin{pmatrix} 1 & -v_2^M \\ e_{21}^1 & -e_{21}^1 v_1^M \end{pmatrix}$$

$$\mathbf{x}_2 = \begin{pmatrix} 1 & -v_2^N \\ e_{21}^2 & -e_{21}^2 v_1^N \end{pmatrix},$$

where v_1^N and v_2^N are the eigenvector normalization constants of $\mathbf{x}_2^{-1}$ and the relevant prefactors have been merged into coefficient α. The two coefficients e_{21}^1 and e_{21}^2 can be now derived from the measurement at port 1 and 2 of a one-port standard (that can be different for the two-ports). Let us define as Γ_i and Γ_i^m the actual and measured reflection coefficients of the standard at port i, $i = 1, 2$. Using (9.20) we have, for port 1 and 2, respectively:

$$e_{21}^1 - \Gamma_1 \Gamma_1^m e_{12}^1 + \Gamma_1^m e_{22}^1 - \Gamma_1 = e_{21}^1 + \Gamma_1 \Gamma_1^m v_2^M - \Gamma_1^m v_1^M e_{21}^1 - \Gamma_1 = 0$$

$$e_{21}^2 - \Gamma_2 \Gamma_2^m e_{12}^2 + \Gamma_2^m e_{22}^2 - \Gamma_2 = e_{21}^2 + \Gamma_2 \Gamma_2^m v_2^N - \Gamma_2^m v_1^N e_{21}^2 - \Gamma_2 = 0,$$

i.e., solving:

$$e_{21}^1 = \frac{\Gamma_1 \left(1 - \Gamma_1^m v_2^M\right)}{1 - \Gamma_1^m v_1^M} \tag{9.26}$$

$$e_{21}^2 = \frac{\Gamma_2 \left(1 - \Gamma_2^m v_2^N\right)}{1 - \Gamma_2^m v_1^N}. \tag{9.27}$$

To complete the calibration the seventh coefficient α appearing in (9.19) has to be determined. To this purpose, the measured matrix of the Thru standard $\mathbf{T}_T^m$ is derived inverting (9.19):

$$\mathbf{T}_T^m = \frac{1}{\alpha} \mathbf{x}_1^{-1} \mathbf{x}_2,$$

i.e.:

$$\mathbf{T}_T^m = \frac{1}{\alpha e_{21}^1 (v_1^M - v_2^M)} \begin{pmatrix} v_1^M e_{21}^1 e_{21}^2 - v_2^M & v_2^M v_2^N - v_1^M v_1^N e_{21}^1 e_{21}^2 \\ e_{21}^1 e_{21}^2 - 1 & v_2^N - v_1^N e_{21}^1 e_{21}^2 \end{pmatrix}$$

and, deriving the measured scattering matrix of the thru:

$$S_T^m = \frac{1}{v_2^M v_2^N - v_1^M v_1^N e_{21}^1 e_{21}^2} \begin{pmatrix} v_2^N - v_1^N e_{21}^1 e_{21}^2 & \alpha^{-1} e_{21}^1 (v_1^N - v_2^N) \\ \alpha e_{21}^1 (v_1^M - v_2^M) & v_2^M - v_1^M e_{21}^1 e_{21}^2 \end{pmatrix}. \quad (9.28)$$

The parameter α can be therefore derived from the measured thru S_{21} (or S_{12}) as:

$$\alpha = S_{T,21}^m \frac{v_1^M v_1^N e_{21}^1 e_{21}^2 - v_2^N v_2^M}{e_{21}^1 (v_2^M - v_1^M)}. \quad (9.29)$$

This completes the identification of the seven coefficients of the TRL calibration.

A possible modification of TRL consists in replacing the one-port standard with a load of unknown reflection coefficient, provided that the same standard is used at both ports. From (9.28) we obtain:

$$S_{T,11}^m = \frac{v_2^N - v_1^N e_{21}^1 e_{21}^2}{v_2^M v_2^N - v_1^M v_1^N e_{21}^1 e_{21}^2},$$

from which we can evaluate the product $e_{21}^1 e_{21}^2$ as:

$$e_{21}^1 e_{21}^2 = \frac{v_2^N \left(v_2^M S_{T,11}^m - 1 \right)}{v_1^N \left(v_1^M S_{T,11}^m - 1 \right)}. \quad (9.30)$$

Measuring the same one-port unknown device with reflection coefficient Γ at both ports, we have from (9.26) and (9.27) imposing $\Gamma_1 = \Gamma_2 = \Gamma$:

$$e_{21}^1 e_{21}^2 = \Gamma^2 \frac{\left(1 - \Gamma_1^m v_2^M\right)\left(1 - \Gamma_2^m v_2^N\right)}{\left(1 - \Gamma_1^m v_1^M\right)\left(1 - \Gamma_2^m v_1^N\right)}, \quad (9.31)$$

where Γ_1^m and Γ_2^m are the measured reflectivities corresponding to the unknown load being connected to port 1 and 2, respectively. Equating the right-hand sides of (9.30) and (9.31) we can derive Γ as:

$$\Gamma = \pm \sqrt{\frac{\left(S_{T,11}^m - 1/v_2^M\right)\left(\Gamma_1^m - 1/v_1^M\right)\left(\Gamma_2^m - 1/v_1^N\right)}{\left(S_{T,11}^m - 1/v_1^M\right)\left(\Gamma_1^m - 1/v_2^M\right)\left(\Gamma_2^m - 1/v_2^N\right)}}.$$

Apart from the sign uncertainty, Γ can be used in (9.26) and (9.27) instead of Γ_1^m and Γ_2^m to identify e_{21}^1, e_{21}^2; the coefficient α can be finally extracted from (9.29), thus completing the calibration procedure. The uncertainty on the sign of Γ affects the sign of e_{21}^1, e_{21}^2 and α; this changes the sign of the scattering matrix reflectance elements S_{11} and S_{22} but not those of the direct and reverse transmission coefficients S_{21} and S_{12}. The correct sign can be readily identified at the end of the procedure, measuring a one-port device whose reflectance has an approximately known phase (at least with an error less than 180 degrees), and choosing, at each frequency, the more coherent sign determination. As a final remark, the modified TRL only requires the precise definition of the characteristic impedance of the Line standard. Owing to these features, the TRL calibration is today the preferred choice in all applications where the two-port is provided with the same waveguide termination (e.g., coaxial or microstrip) at all ports.

SOLR Calibration

The SOLR calibration (from Short – Open – Load – Reciprocal) [18] is often referred to as the *Unknown Thru* calibration. In fact, the device used as a two-port standard can be arbitrary, provided it is reciprocal, a rather mild constraint that is typically satisfied by all passive two-ports. The SOLR technique is essential whenever a Thru is not available, e.g., because the DUT has different connectors at the two-ports, or when the two-ports are realized in different waveguiding structures (e.g., in a transition between a coaxial cable and a rectangular waveguide), or when the connectors are anyway incompatible (like a male–male or female–female pair). In addition to the Reciprocal two-port device, the SOLR calibration requires three one-port standards (Short, Open and Load) to be measured at both ports; the two Loads will be the normalization impedances to which the measured parameters are referred. The SOLR calibration may also be critical in the position of the reference planes, since the three one-port standards require identical access structures to ensure a well-defined location of the DUT planes.

The first part of the calibration is performed as in the QSOLT case and consists in the measurement of the three standards connected at the first and then at the second port. This enables $\mathbf{x}_1$ and $\mathbf{x}_2$ to be identified using (9.23) for both ports. The α coefficient is derived from the measurement of the reciprocal two-port. From (9.16) we immediately see that, since in a reciprocal device $S_{12} = S_{21}$, this implies that the determinant of the transmission matrix of the reciprocal device $\mathbf{T}_R$ is unity. From (9.19) we obtain, taking into account the properties of the product of matrices and having identified the elements of $\mathbf{T}_R^m$:

$$\Delta(\mathbf{T}_R) = 1 = \alpha^2 \frac{\Delta(\mathbf{x}_1)\Delta(\mathbf{T}_R^m)}{\Delta(\mathbf{x}_2)},$$

where $\Delta(\mathbf{A})$ is the determinant of matrix $\mathbf{A}$. We therefore obtain:

$$\alpha = \pm\sqrt{\frac{\Delta(\mathbf{x}_2)}{\Delta(\mathbf{x}_1)\Delta(\mathbf{T}_R^m)}}.$$

As in the modified TRL, the correct sign of α has to resolved; notice that the sign of α only affects the transmission parameters S_{12} and S_{21} (S_{11} and S_{22} are unaffected). The sign of α can be selected, e.g., by measuring the transmission delay introduced by a line having known length and whose phase rotation can therefore be easily estimated.

9.5 Load and Source Pull Characterization

The measurement techniques described so far aim at characterizing N-ports operating in linear or small-signal conditions. An exhaustive linear model can be identified by exciting the DUT with a single tone swept over a large bandwidth and by measuring, as a function of frequency, the DUT response in terms of the signal complex phasors (voltages, currents or power waves). However, many important microwave subsystems are quasi-linear or nonlinear, thus requiring proper characterization techniques able to monitor the DUT under a large-signal excitation. We will focus here on the two-port characterization of power amplifiers, whose aim is to identify the main PA features, such as power compression and saturation, identification of the optimum load and source reflectances, evaluation of the harmonic or intermodulation distortion and of the AM/AM or AM/PM conversion. In large-signal characterization the DUT typically is excited by a narrowband signal of increasing complexity (single tone, multi-tone, modulated signal); due to the DUT nonlinearity the measurement system should allow for the power wave characterization of each of the generated frequencies. Moreover, when compared to the linear VNA, the system should permit to estimate the absolute value of the signals generated at the fundamentals, harmonic or intermodulation products, not only of signal ratios; to this purpose, the measurement setup has to able to measure absolute power levels.

In load-pull (source-pull) measurement systems the two-port load (source) is varied, at the fundamental frequency and possibly at some of the harmonics, in order to optimize two-port performances, like the output power at some desired level of compression, the efficiency, the intermodulation distortion. Notice that source-pull techniques can also be exploited to identify the optimum low-noise source impedance. For the sake of brevity we will in what follows generally refer to such systems as *load-pull* systems.

Load-pull systems can be classified according to the complexity of the measurement they are able to perform. In scalar systems only the power is measured and therefore only the output power and the power gain can be optimized. VNA-based vectorial systems also have the possibility to characterize the input–output phase relationship and therefore to analyze AM/AM and AM/PM distortion. Harmonic load-pull systems can vary the load at a set of harmonics and/or characterize the device output at the harmonics [19, 20]. Another classification concerns the way the load is tuned, either mechanically (passive load-pull systems) or electronically (active load-pull systems). In the following sections some relevant examples are provided.

9.5.1 Scalar Systems

In scalar systems the test signal source is single-tone and two tuners are used to vary the source and load impedances, while a power meter is exploited to measure the power on the load, see Fig. 9.6 (a). Typically, the input power is not measured, only the output power, and only the transducer gain can be identified on the basis of the generator known input available power. The output power meter has to be coupled through a very selective bandpass filter to exclude from the measurements the in-band distortion (like the third-order intermodulation products).

This simple solution can be enhanced by adding at the output a spectrum analyzer able to characterize harmonics and intermodulation products. A two-tone test can also be performed by modulating the signal generator with a low-frequency sinusoid. Due to the scalar nature of the system, features like the AM/PM conversion cannot be characterized.

Moreover, the source and load impedances are not measured in real time by the system, and therefore an accurate tuner pre-calibration is needed, together with the de-embedding of the power meter readings, to identify the output power at the DUT output

9.5 Load and Source Pull Characterization

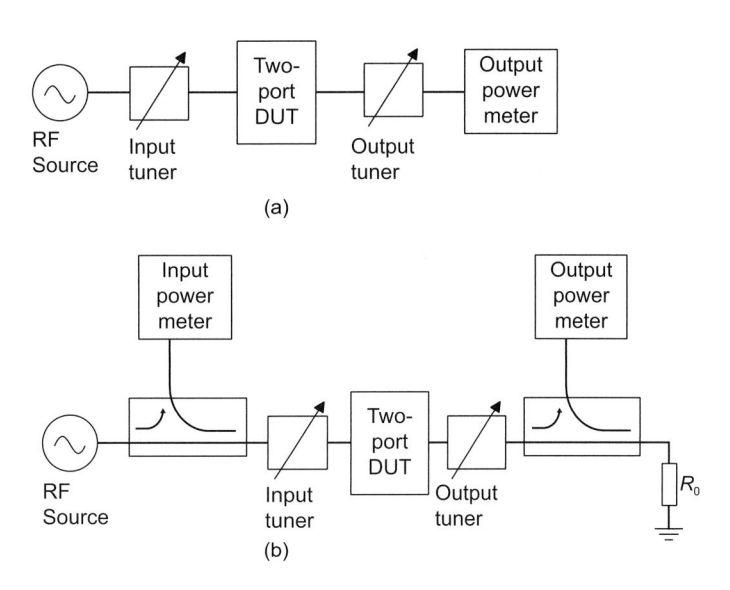

Figure 9.6 Scalar source/load-pull systems based on power meters: (a) simple version with load power measurement only; (b) version with directional couplers and input power measurement.

reference planes. Mechanical tuners, typically used in this low-budget solution, must ensure repeatability to allow for accurate measurements.

A modified version of the scalar system includes two power meters that sample the input and output DUT powers by means of two directional couplers, see Fig. 9.6 (b). Due to the indirect sampling, the power meter dynamic range is reduced.

9.5.2 Vectorial Systems

In vectorial load-pull systems the DUT is connected to a VNA able to measure in real time the source and load reflection coefficients and the power waves at the DUT reference planes. Due to the coupler power losses, systems exploiting passive tuners cannot implement purely reactive loads and the load and source tuning range is limited; this limitation can be overcome by active load tuning, as discussed further on. The scheme of the vectorial system is shown in Fig. 9.7, that clearly is an extension of the linear VNA. However, while a ratiometric calibration is sufficient to the linear VNA, the vectorial system needs an additional step enabling the absolute power level of the power waves to be measured. To this aim, port 3 in Fig. 9.7 is closed on a power meter providing the needed power reference during the calibration phase, done after the small-signal calibration procedures already described. The vectorial load-pull system is in fact an integrated scattering parameter and large-signal measurement setup; thanks to the small-signal calibration approach the tuner pre-calibration is not needed any more and no constraints exist on the tuner repeatability. Finally, the downconversion to IF allows the system to be readily tuned to measure at IF the harmonic and intermodulation components (without the need of selective RF bandpass filters) of the output signals through

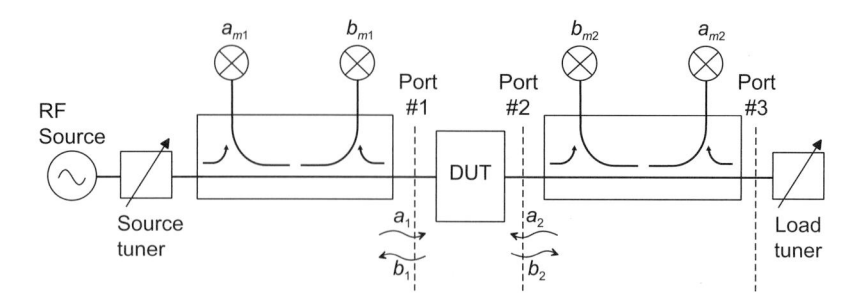

Figure 9.7 Vectorial source/load-pull system based on VNA.

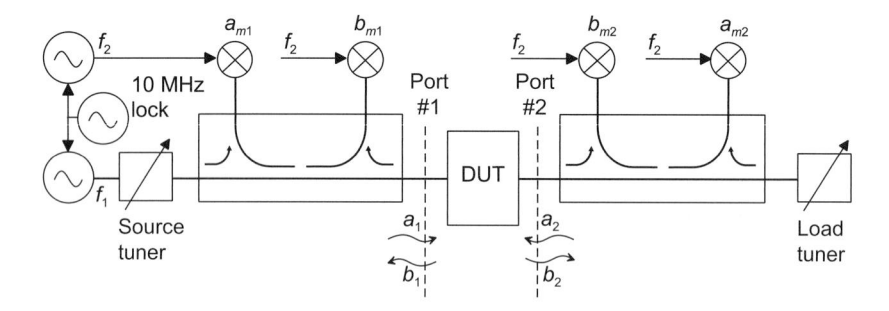

Figure 9.8 Vectorial source/load-pull system based on VNA with excitation at f_1 and downconversion at IF through a source at f_2, harmonic-locked with f_1 by means of a 10 MHz locking oscillator. If $f_2 = 2f_1$ the system measures, e.g., the second harmonic.

the setup in Fig. 9.8, where the two generators are phase locked at low frequency (the 10 MHz frequency exploited is a *de facto* standard), and the additional synthesizer, working at a frequency different from the excitation, is sent at the VNA reference channel. Since an absolute measurement of the two-port power waves is now available, all relevant quantities at the fundamental and at the harmonics or intermodulation products (input and output operational and available powers, input and output reflection coefficients, source and load reflectances, efficiency and power-added efficiency) can be characterized [21, 22, 23, 24].

A shortcoming of VNA-based load-pull systems is related to their frequency selectivity. While they are extremely effective in measuring signals resulting from multi-tone excitation with a low number of tones, modulated signals, having a continuous and comparatively wideband spectrum, cannot be accurately characterized.

9.5.3 Load Tuning Techniques

In load-pull systems two approaches can be implemented to tune the load or source impedances. Passive tuners exploit passive structures with mobile elements, driven manually or electrically; this is the preferred choice up to a few GHz thanks to the simplicity and low cost of the system. Pre-matching networks [25, 26] can be added to the tuner to achieve large magnitudes of the tuner reflectance; this issue becomes more critical with

9.5 Load and Source Pull Characterization

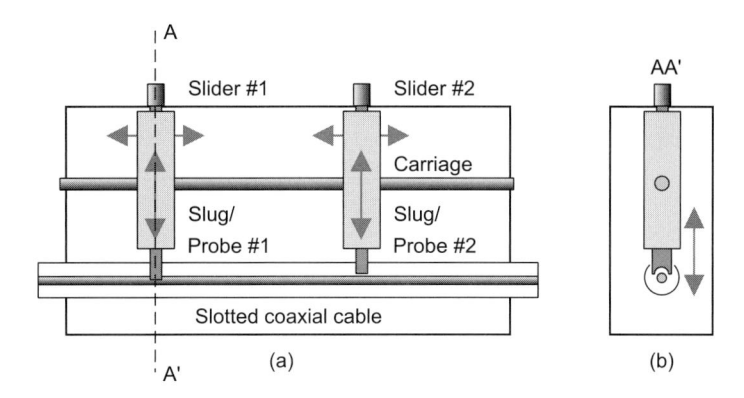

Figure 9.9 Example of passive manual tuner with two slugs/probes: (a) lateral section; (b) front section. The probes can be moved along the slotted cable and vertically through the adjusting screws.

increasing frequency and reflectometer losses, not to mention the harmonic tuning setup where the optimum harmonic load typically are reactive, i.e., at or close to the unit circle in the Smith chart [27]. The above issues, that become crucial at millimeter frequencies, are alleviated by the use of *active* tuners.

Examples of **passive tuners** are discussed in [28, 29, 30, 31, 32, 33]. A single-frequency passive tuner can be obtained in principle through a 50 Ω load with a transmission line stub in parallel of length l_s, followed by a line of length l, both l_s and l being adjustable. Practical tuners often exploit a slotted coaxial line with one or more slugs that can be moved along the line axis (to change the reflectance phase) or along the line radius (to change the reflectance magnitude). To increase accuracy, the position of the slug is varied through micromanipulators driven by stepper motors. A scheme of principle is shown Fig. 9.9. Passive tuners equipped with up to three slugs operating in different bands and therefore able to tune the load at the fundamental an harmonics are available today; the driving stepper motors are controlled through GPIB or LAN connections and ensure good repeatability of the load setting.

Owing to the tuner and set-up losses, passive tuners are limited in the magnitude of the achievable reflection coefficient; to alleviate the issue several solutions are available [25, 34, 26], such as changing the reference impedance of the measurement system with a tuner pre-matching or shifting the DUT load by pre-matching the device. In the first case the pre-matching network is interposed between the tuner and the reflectometer, in the second case between the reflectometer and the DUT. Passive tuners may integrate programmable pre-matching networks [26].

As already recalled, the main shortcomings of passive load-pull techniques are the difficult synthesization of reflectances close to unity, but also the complications in the tuners arising when harmonics have to be simultaneously controlled together with the fundamental. In **active load-pull systems** such problems are overcome by implementing an active, electronically controllable load. The active load is based on the re-injection, at the DUT ports, of the power waves a_1 and a_2 proportional to b_1 and b_2, respectively. The idea can be implemented according to two main approaches, which differ on how the

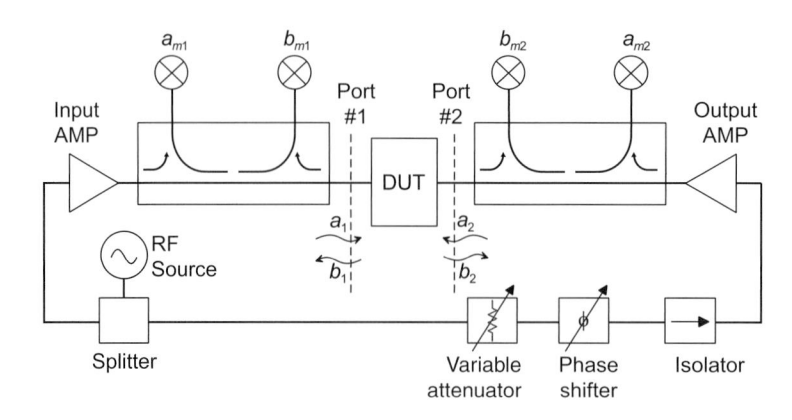

Figure 9.10 Active load-pull system based on the two-signal technique.

re-injected power is derived: the *two signal path* [35] and the *active loop* techniques [36]. In the active load approach, losses can be completely compensated for, and proper signal combination prior to re-injection allows for virtually independent control of the load at the relevant harmonics. Due to its active nature, however, the approach has stability issues that have to be kept under control; finally, the set up complexity and cost is much higher than in the passive case.

In the two-signal method, [35, 37] the scheme of Fig. 9.10, originally proposed by Takayama, can currently exploited to implement the active load. First, the source signal is split by a power divider; one of the resulting signals is sent to the input of the DUT, while the other signal is amplified, phase shifted, and finally sent to the output port of the DUT as the b_2 wave. Load tuning is obtained by adjusting the gain and phase shift. Thanks to the high isolation of the DUT this technique has little oscillation issues, and is the preferred one in millimeter wave systems [38]. However, a shortcoming of the two signal method is the difficulty of keeping the load constant for increasing input power, as needed in an amplifier power sweep characterization. In fact, when the DUT enters the compression region, the output power is not any more proportional to the input power, on which the re-injected power linearly depends. Thus, for any input power step the amplification and phase shift have to be adjusted in order to keep the load constant. Finally, the two-signal method can be extended to harmonic load control. Suppose for simplicity to control the fundamental and second harmonic loads only; this can be obtained by dividing the source signal at f_0 into three paths. The first is injected in the input port of the DUT, the second is amplified and phase shifted, the third is multiplied by two through a frequency multiplier, thus obtaining the second harmonic $2f_0$, amplified and phase shifted. Finally, the second and third signal are combined and injected into the DUT output, allowing for independent control of the load at f_0 and $2f_0$.

The second strategy is denoted the *active loop technique* [36], see Fig. 9.11. In the active loop technique a part of the signal reflected by the output DUT port is extracted through the output loop coupler, amplified, phase shifted and filtered at the fundamental before being re-injected into the DUT output port. The magnitude of the load reflection coefficient will depend on the total gain of the loop (including the variable attenuator and

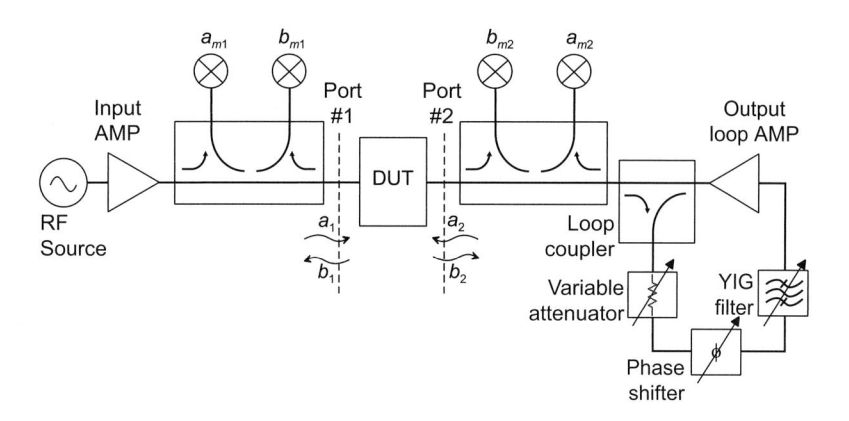

Figure 9.11 Active load-pull system based on the active-loop technique.

the output loop amplifier), while the phase can be adjusted by playing on the loop total phase shift. The YIG (Yittrium Iron Garnet) high Q, narrowband tunable filter [39] is introduced to prevent out-of-band oscillations in the loop. The method can be extended to harmonic control by splitting the loop coupler signal in several branches, each with a YIG filter centered around one of the harmonics [40].

9.6 System-Level Characterization

System-level characterization techniques aim at measuring components and subsystems in the TX and RX parts of the transceiver when excited by realistic modulated signals emulating specific (digital or analog) modulation formats. Focusing on the TX branch, the PA characterization under realistic modulated inputs is fundamental due to the PA nonlinearity, and cannot be completely replaced by simpler single- or two-tone tests. System-level PA characterization is an excellent verification tool and can be effectively exploited to optimize and design subsystems cascaded with the PA (e.g., baseband predistortion stages); however, it is too complex to be performed in the iterative optimization of the PA.

During the last few years, system-level measurement setups have been proposed as an evolution of the conventional, narrowband instrumentation [41]. Those are the Vector Signal Generator (VSG) and the Vector Signal Analyzer. The VSG includes a baseband waveform generator yielding a random signal compliant with some prescribed modulation format, a tunable upconversion stage to the RF carrier frequency, and an amplifier compliant with the power levels required by the standard under consideration. The VSA basically is a reconfigurable receiver, including a quadrature demodulation stage that extracts the in-phase and quadrature baseband components of the signal, converts them to digital and makes statistical analyses, e.g., of the average error vs. the ideal received signal. With proper synchronization and phase locking to a common 10 MHz oscillator the two instruments can be connected at the input and output stages of a VNA, see Fig. 9.12, thus allowing for the time- and frequency-domain vector measurement of the

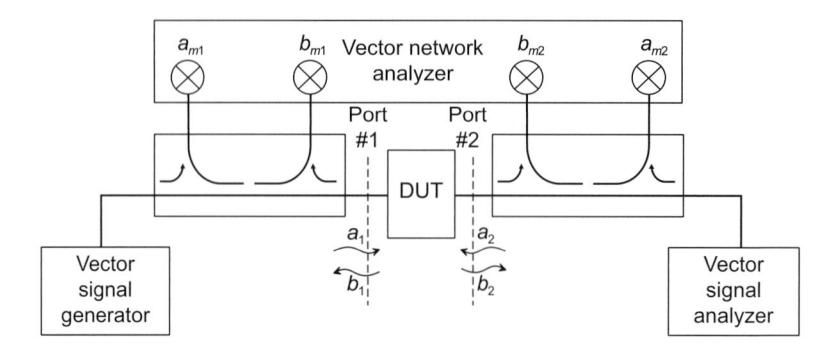

Figure 9.12 Scheme of system-level characterization setup.

DUT output. When applied to on-wafer measurements, special care should be applied to adjust the power levels to the DUT reference planes through ad-hoc calibration techniques, see [42].

A common average figure of merit for linearity that can be extracted from the output signal resulting from a complex modulated input excitation is the Error Vector Magnitude (EVM), already introduced in Sec. 8.2.1, which quantifies the difference between the ideal and the measured signal constellation. The VSA also has display facilities able to graphically show the output signal constellation in the I and Q axis, thus providing a generalized eye diagram of the output signal.

9.7 Noise Measurements

Noise measurements are performed through the *noise figure meter*. This instrument determines the noise added by the two-port DUT by comparing it to the noise from the DUT input noise source only. The noise figure meter (from now on NFM) includes a power meter or spectrum analyzer able to provide noise power measurements on a prescribed bandwidth B, some electronics to control the state of the noise source to be connected to the DUT, and a digital part for the elaboration and display of the result. To measure the noise figure of the DUT, a calibrated noise source is connected to the DUT input. Such a source may be, e.g., a diode switched on and off; when the diode is in the off state the noise temperature of the diode (source) is T_s^{OFF}, while when it is in the on state the noise temperature is T_s^{ON}. The off temperature T_s^{OFF} typically coincides with the ambient temperature (slightly different from the standard reference temperature $T_0 = 290$ K) while typical values for T_s^{ON} are in the range of 10^4 K. We define the parameter ENR (Excess Noise Ratio) as:

$$\mathrm{ENR} = \frac{T_s^{ON} - T_s^{OFF}}{T_0}.$$

The source ENR is assumed to be known; since T_s^{OFF} is the (known) ambient temperature, this allows the ON temperature to be derived as:

$$T_s^{ON} = T_s^{OFF} + \mathrm{ENR} \cdot T_0.$$

The ratio of the on/off state noise powers and therefore noise temperatures is often called the Y factor:

$$Y = \frac{P_{n,\text{avs}}^{\text{ON}}}{P_{n,\text{avs}}^{\text{OFF}}} = \frac{P_{n,s}^{\text{ON}}}{P_{n,s}^{\text{OFF}}} = \frac{k_B B T_s^{\text{ON}}}{k_B B T_s^{\text{OFF}}} = \frac{T_s^{\text{ON}}}{T_s^{\text{OFF}}},$$

where B is the measurement system bandwidth. Notice that the ratio of available powers coincides with the ratio of powers on a prescribed load, since the load is the same in the ON and the OFF states.

A popular technique for the measurement of the noise figure is called the *Y-factor method*, see [43, 44]. The method takes into account the fact that also the NFM adds noise to the signal and therefore has a noise figure F_{NFM} and a noise temperature T_{NFM} that should be estimated. Since the NFM is cascaded to the DUT, we can apply the Friis formula where the total noise figure is:

$$F = F_{\text{DUT}} + \frac{F_{\text{NFM}} - 1}{G_{\text{DUT}}},$$

and since:

$$F = 1 + \frac{T_2}{T_0},$$

where T_2 is the noise temperature of the cascade, we also have:

$$\frac{T_2}{T_0} = \frac{T_{\text{DUT}}}{T_0} + \frac{T_{\text{NFM}}}{T_0 G_{\text{DUT}}},$$

i.e.:

$$T_2 = T_{\text{DUT}} + \frac{T_{\text{NFM}}}{G_{\text{DUT}}}. \tag{9.32}$$

We suppose for simplicity that all elements are 50 Ω matched and neglect mismatch issues that may be corrected in a calibration phase.

In a *first step* the noise source is directly connected to the NFM, see Fig. 9.13 (a). The NFM performs two noise power readings, on a prescribed bandwidth B, with the source in the ON and OFF states, respectively, and evaluates the ratio:

$$Y_1 = \frac{P_{n1}^{\text{ON}}}{P_{n1}^{\text{OFF}}} = \frac{T_s^{\text{ON}} + T_{\text{NFM}}}{T_s^{\text{OFF}} + T_{\text{NFM}}},$$

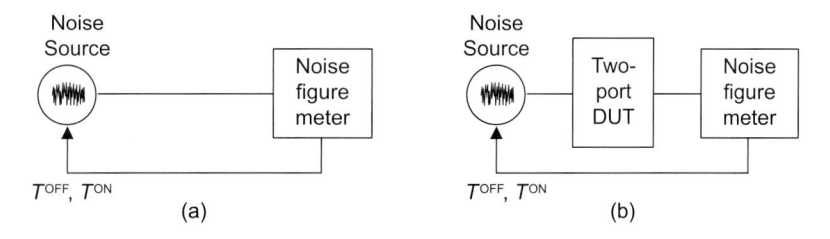

Figure 9.13 Noise parameter measurements setup: (a) Y-factor technique, first step, measurement of the noise source only; (b) Y-factor technique, second step, measurement of the DUT with the noise source at the input.

from which the NFM noise temperature can be estimated as:

$$T_{\text{NFM}} = \frac{T_s^{\text{ON}} - Y_1 T_s^{\text{OFF}}}{Y_1 - 1}. \tag{9.33}$$

In a *second step* the DUT is connected between the noise source, see Fig. 9.13 (b), and the NFM and the ratio is evaluated:

$$Y_2 = \frac{P_{n2}^{\text{ON}}}{P_{n2}^{\text{OFF}}} = \frac{T_s^{\text{ON}} + T_2}{T_s^{\text{OFF}} + T_2},$$

where T_2 is the total noise temperature of the cascade of the DUT and of the NFM. T_2 can be estimated applying again (9.33) as:

$$T_2 = \frac{T_s^{\text{ON}} - Y_2 T_s^{\text{OFF}}}{Y_2 - 1}.$$

Taking into account that uncorrelated random signals sum their power, we have:

$$P_{n1}^{\text{ON}} = P_{n,s}^{\text{ON}} + P_{n,\text{NFM}}$$
$$P_{n1}^{\text{OFF}} = P_{n,s}^{\text{OFF}} + P_{n,\text{NFM}}$$
$$P_{n2}^{\text{ON}} = G_{\text{DUT}} P_{n,s}^{\text{ON}} + P_{n,\text{DUT}} + P_{n,\text{NFM}}$$
$$P_{n2}^{\text{OFF}} = G_{\text{DUT}} P_{n,s}^{\text{OFF}} + P_{n,\text{DUT}} + P_{n,\text{NFM}},$$

where $P_{n,\text{DUT}}$ is the output noise power of the DUT and $P_{n,\text{NFM}}$ is the noise power due to the NFM. Subtracting the first two and second two equations, we have:

$$P_{n1}^{\text{ON}} - P_{n1}^{\text{OFF}} = P_{n,s}^{\text{ON}} - P_{n,s}^{\text{OFF}}$$
$$P_{n2}^{\text{ON}} - P_{n2}^{\text{OFF}} = G_{\text{DUT}} \left(P_{n,s}^{\text{ON}} - P_{n,s}^{\text{OFF}} \right),$$

from which the (transducer) gain of the DUT can be evaluated as:

$$G_{\text{DUT}} = \frac{P_{n2}^{\text{ON}} - P_{n2}^{\text{OFF}}}{P_{n1}^{\text{ON}} - P_{n1}^{\text{OFF}}}.$$

We can now from (9.32) compute the DUT noise temperature as:

$$T_{\text{DUT}} = T_2 - \frac{T_{\text{NFM}}}{G_{\text{DUT}}}$$

and the DUT noise figure as:

$$F_{\text{DUT}} = 1 + \frac{T_{\text{DUT}}}{T_0}.$$

The characterization of the minimum noise figure, optimum source impedance and parallel noise resistance or series noise conductance (expressing the sensitivity of the noise figure with respect to a variation of the load vs. the optimum value) can be carried out from (7.49):

$$F = F_{\min} + 4 g_n R_0 \frac{|\Gamma_G - \Gamma_{Go}|^2}{\left(1 - |\Gamma_G|^2\right) |1 - \Gamma_{Go}|^2} \equiv \frac{4 R_n}{R_0} \frac{|\Gamma_G - \Gamma_{Go}|^2}{\left(1 - |\Gamma_G|^2\right) |1 - \Gamma_{Go}|^2},$$

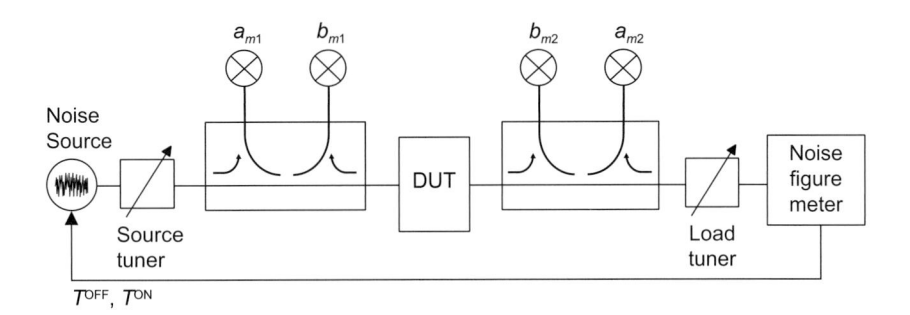

Figure 9.14 Setup for the measurement of the minimum noise figure, optimum source impedance, noise resistance.

by measuring the noise figure with a source-pull setup [21], see Fig. 9.14, able to vary Γ_G on the Smith chart. The output tuner can be anyway be exploited to obtain output matching to a NFM with 50 Ω impedance. A set of noise figure values for different loads are recorded and interpolated on the Γ plane by means of suitable polynomial least-square approximation; this way the minimum ($F_{\min}$) and the corresponding optimum generator reflectance Γ_{Go} are identified; g_n (R_n) is then derived from the curvature of the F surface around the minimum. Within this framework, the Y-factor technique can be applied but a more complex calibration step is needed since the setup now includes one or two tuners in cascade with the DUT, besides the VNA couplers, whose noise has to be de-embedded from the DUT. The VNA included in Fig. 9.14 is used to provide real-time measurements of the source reflectance as seen from the DUT input reference plane. For additional details see [45] and references therein.

9.8 Questions and Problems

9.8.1 Questions

1. Qualitatively explain the structure of a power meter and the role played by the bolometer.
2. Describe the temperature compensation method used in power meters.
3. Explain how power measurements can be carried out by a spectrum analyzer.
4. Describe two devices able to provide a separate measurement of incident and reflected power waves at a device port.
5. Explain the operation of the reflectometer.
6. Justify the need to perform vectorial voltage measurements at the intermediate frequency rather than at RF.
7. Describe the structure of a Vector Network Analyzer.
8. Justify the need to perform a calibration of the Vector Network Analyzer.
9. How many error coefficients have to be identified in a two-port VNA calibration?
10. Describe the principle of the QSOLT calibration technique.
11. Describe the principle of the TRL calibration technique.

12. Describe the principle of the SOLR calibration technique.
13. How are reference planes defined in the TRL calibration?
14. Explain the concept of the load-pull or source-pull measurements.
15. Comment on the advantages and disadvantages of passive tuners when compared to active load-pull systems.
16. Explain how the load-pull system can perform absolute power wave measurements rather than relative ones.
17. Describe two techniques to implement an active load.
18. Explain why active load-pull systems may have stability problems.
19. Describe the structure of a system able to characterize a two-port under modulated signal excitation.
20. Why are active load-pull techniques not viable for a two-port characterization under modulated signal excitation?
21. Describe the measurement set-up needed to perform noise figure measurements.
22. What is the Y-factor noise measurement technique? Explain.
23. How can we measure the optimum source impedance and the noise resistance of a device?
24. Explain the relationship between source-pull and noise measurements.

9.8.2 Problems

1. A reflectometer has the following readings when loaded by a short, an open and a matched standard:

$$a_{m1}^s = 0.98, \quad b_{m1}^s = -0.96,$$
$$a_{m1}^o = 0.98, \quad b_{m1}^o = 0.94,$$
$$a_{m1}^{R_0} = 0.99, \quad b_{m1}^{R_0} = 0.01.$$

Evaluate the error coefficients e_{12}, e_{21} and e_{22} and the reflectivity of a DUT with readings $a_{m1} = 0.98$, $b_{m1} = 0.94j$. What is the apparent reflectivity measured without the correction?

2. The Line of a TRL calibration set has a length $L = 1$ mm. The measured transmission matrices of the Line and of the Thru are:

$$\mathbf{T}_T^m = \begin{pmatrix} 0.89794 & 3.0829 \times 10^{-2} \\ 2.1044 \times 10^{-3} & 0.97875 \end{pmatrix}$$

$$\mathbf{T}_L^m = \begin{pmatrix} 0.72572 - 0.52727j & 2.4916 \times 10^{-2} - 1.8103 \times 10^{-2}j \\ 1.7008 \times 10^{-3} - 1.2357 \times 10^{-3}j & 0.79103 - 0.57472j \end{pmatrix}.$$

Find the complex propagation constant of the line.

3. An active load-pull system exploiting the active load technique needs to implement a load of impedance $Z_L = 50+j50$ Ω. Evaluate the loop gain and phase delay needed, supposing to use a loop coupler with 10 dB power coupling. The reference impedance is $R_0 = 50$ Ω.

4. An active load-pull system (active load technique) needs to implement a short circuit. Evaluate the loop gain and phase delay needed, supposing to use a loop coupler with 20 dB power coupling. The reference impedance is $R_0 = 50\ \Omega$.
5. A noise source has ENR $= 30$ and the OFF temperature is $T_s^{\text{OFF}} = 300$ K. We apply the Y-factor technique, obtaining in the first step (noise source directly connected to the noise figure meter) $P_{n1}^{\text{ON}} = 300\ \mu\text{W}$, $P_{n1}^{\text{OFF}} = 15\ \mu\text{W}$. In the second step (DUT connected) we obtain $P_{n1}^{\text{ON}} = 3500\ \mu\text{W}$, $P_{n1}^{\text{OFF}} = 250\ \mu\text{W}$. Find the T^{ON} of the noise source, and the DUT gain, noise figure and noise temperature.

References

[1] V. Teppati, A. Ferrero, and M. Sayed, *Modern RF and microwave measurement techniques*. Cambridge University Press, 2013.

[2] D. Rytting and S. Sanders, "A system for automatic network analysis," *Hewlett Packard J.*, vol. 21, pp. 2–10, Feb. 1970.

[3] R. Anderson and O. Dennison, "An advanced new network analyzer for sweep-measuring amplitude and phase from 0.1 to 12.4 GHz," *Hewlett Packard J.*, vol. 18, pp. 2–9, Feb. 1967.

[4] A. Fantom, *Radio frequency & microwave power measurement*, ser. IEE Electrical Measurement. Peter Peregrinus, 1990, vol. 7.

[5] F. Nagy, "A new digital vector voltmeter," *Measurement*, vol. 9, no. 1, pp. 44–48, 1991.

[6] A. D. Helfrick, *Electrical spectrum & network analyzers: a practical approach*. Academic Press, 2012.

[7] N. Drobotun and P. Mikheev, "A 300kHz–13.5 GHz directional bridge," in *European Microwave Conference (EuMC)*. Paris, FR: IEEE, Oct. 2015, pp. 287–290.

[8] J. P. Dunsmore, "Dual directional bridge and balun used as reflectometer test set," Oct. 1990, US Patent 4,962,359.

[9] B. Hand, "Developing accuracy specifications for automatic network analyzer systems," *Hewlett Packard J.*, vol. 21, pp. 16–19, Feb. 1970.

[10] N. Franzen and R. Speciale, "A new procedure for system calibration and error removal in automated s-parameter measurements," in *European Microwave Conference (EuMC)*, Hamburg, Germany, Sep. 1975, pp. 69–73.

[11] G. F. Engen and C. A. Hoer, "Thru-reflect-line: an improved technique for calibrating the dual six-port automatic network analyzer," *IEEE Transactions on Microwave Theory and Techniques*, vol. 27, no. 12, pp. 987–993, Dec. 1979.

[12] H. Eul and B. Schieck, "Reducing the number of calibration standards for network analyzer calibration," *IEEE Transactions on Instrumentation and Measurement*, vol. IM-40, pp. 732–735, Aug. 1991.

[13] A. Ferrero and U. Pisani, "A simplied algorithm for leaky network analyzer calibration," *IEEE Microwave Guided Wave Letters*, vol. MGWL-5, pp. 119–121, Apr. 1995.

[14] R. Speciale and N. Franzen, "Super – TSD: a generalization of the TSD network analyzer calibration procedure, covering n-port measurements with leakage," in *IEEE MTT-S Intenational Microwave Symposium Digest*, San Diego, CA, Jun. 1977, pp. 114–117.

[15] K. Silvonen, "A general approach to network analyzer calibration," *IEEE Transactions on Microwave Theory and Techniques*, vol. 40, no. 4, pp. 754–759, Apr. 1992.

[16] A. Ferrero and U. Pisani, "QSOLT: a new fast calibration algorithm for two-port S-parameter measurements," in *38th ARFTG Conference Digest*, San Diego, CA, Dec. 1991, pp. 15–24.

[17] R. Marks, "A multiline method of network analyzer calibration," *IEEE Transactions on Microwave Theory and Techniques*, vol. MTT-39, pp. 1205–1215, Jul. 1991.

[18] A. Ferrero and U. Pisani, "Two-port network analyzer calibration using an unknown 'thru'," *IEEE Microwave Guided Waves Letters*, vol. MGWL-2, pp. 505–507, Dec. 1992.

[19] F. Microwaves, "An affordable harmonic load pull setup," *Microwave Journal*, pp. 180–182, Oct. 1998.

[20] A. Ferrero and M. Pirola, "Harmonic load-pull techniques: an overview of modern systems," *IEEE Microwave Magazine*, vol. 14, no. 4, pp. 116–123, Jun. 2013.

[21] B. Hughes and T. P., "Improvements to on-wafer noise parameter measurements," in *36th ARFTG Conference Digest*, Monterrey, CA, Nov. 1990, pp. 16–25.

[22] D. Le and F. Ghannouchi, "Source-pull measurements using reverse six-port reflectometers with application to MESFET mixer design," *IEEE Transactions on Microwave Theory and Techniques*, vol. MTT-42, pp. 1589–1595, Sep. 1994.

[23] G. Madonna, M. Pirola, A. Ferrero, and U. Pisani, "Testing microwave devices under different source impedances: a novel technique for on-line measurement of source and device reflection coefficents," in *IMTC/99 Conf. Proc.*, Venezia, Italy, May 1999, pp. 130–133.

[24] G. Madonna and A. Ferrero, "Simple technique for source reflection coefficient measurement while characterizing active devices," in *53rd ARFTG Conference Digest*, Anaheim, CA, Jun. 1999, pp. 104–106.

[25] J. Sevic, "A sub 1 Ω load-pull quarter wave pre-matching network based on a two-tier TRL calibration," in *52nd ARFTG Conference Digest*, Santa Rosa, CA, Dec. 1998, pp. 73–81.

[26] C. Tsironis, "Prematched programmable tuners for very high VSWR testing," in *IEEE μAPS Microwave Application & Product Seminars*, Anaheim, CA, Jun. 1999.

[27] P. Colantonio, F. Giannini, and E. Limiti, "Nonlinear approaches to the design of microwave power amplifiers," *International Journal of RF and Microwave Computer-Aided Engineering*, vol. 14, no. 6, pp. 493–506, Nov. 2004.

[28] M. M. Co., *Automated tuner system user's manual, v.1.9*. Maury Microwave Corporation, 1998.

[29] F. M. Co., *Computer controlled tuner system user's manual, v. 6.0*. Focus Microwave Corporation, 1998.

[30] A. M. Co., *LP2 automated load-pull system user's manual*. ATN Microwave Corporation, 1997.

[31] F. Sechi, R. Paglione, B. Perlman, and J. Brown, "A computer controlled microwave tuner for automated load pull," *RCA Review*, vol. 44, pp. 566–572, Dec. 1983.

[32] V. Adamian, "2–26.5 GHz on-wafer noise and S-parameter measurements using a solid state tuner," in *34th ARFTG Conference Digest*, Dec. 1989, pp. 33–40.

[33] C. McIntosh, R. Pollard, and R. Miles, "Novel MMIC source-impedance tuners for on-wafer microwave noise-parameter measurements," *IEEE Transactions on Microwave Theory and Techniques*, vol. MTT-47, pp. 125–131, Feb. 1999.

[34] J. Sevic, "A sub 1 Ω load-pull quarter wave prematching network baser on a two-tier TRL calibration," *IEEE Transactions on Antennas and Propagation*, vol. 47, no. 2, pp. 389–391, Feb. 1999.

[35] Y. Takayama, "A new load-pull characterization method for microwave power transistor," in *IEEE MTT-S Intenational Microwave Symposium Digest*, Cherry Hill, NJ, Jun. 1976, pp. 218–220.

[36] G. Bava, U. Pisani, and V. Pozzolo, "Active load technique for load-pull characterization at microwave frequencies," *Electronics Letters*, vol. 18, n.4, pp. 178–179, Feb. 1982.

[37] S. Mazumder and P. Van der Puije, "'Two-signal' method of measuring the large-signal S-parameters of transistors," *IEEE Transactions on Microwave Theory and Techniques*, vol. MTT-26, pp. 417–420, Jun. 1978.

[38] B. Bonte, C. Gaquiere, E. Bourcier, C. Lemeur, and Y. Crosnier, "An automated system for measuring power devices in Ka-band," *IEEE Transactions on Microwave Theory and Techniques*, vol. MTT-46, pp. 70–75, Jan. 1998.

[39] P. Carter, "Magnetically-tunable microwave filters using single-crystal yttrium-iron-garnet resonators," *IRE Transactions on Microwave Theory and Techniques*, vol. 9, no. 3, pp. 252–260, May 1961.

[40] B. Hughes and P. Tasker, "Accurate on-wafer power and harmonic measurements of mm-wave amplifiers and devices," in *IEEE MTT-S Intenational Microwave Symposium Digest*, Albuquerque, NM, Jun. 1992, pp. 1019–1022.

[41] V. Teppati, A. Ferrero, V. Camarchia, A. Neri, and M. Pirola, "Microwave measurements – Part III: Advanced non-linear measurements," *IEEE Instrumentation Measurement Magazine*, vol. 11, no. 6, pp. 17–22, Dec. 2008.

[42] R. Quaglia, T. Jiang, and V. Camarchia, "Frequency extension of system level characterization and predistortion setup for on-wafer microwave power amplifiers," in *European Microwave Integrated Circuit Conference (EuMIC), 2014 9th*, Oct. 2014, pp. 488–491.

[43] Application Note 57-1, "Fundamentals of RF and microwave noise figure measurements," *Agilent Technologies, Palo Alto, California, USA*, 2010.

[44] Application Note 57-2, "Noise figure measurement accuracy: the Y factor method," *Agilent Technologies, Palo Alto, California, USA*, 2014.

[45] A. Cappy, "Noise modeling and measurement techniques," *IEEE Transactions on Microwave Theory and Techniques*, vol. 36, no. 1, pp. 1–10, Jan. 1988.

10 CAD Projects

10.1 Introduction

The present chapter describes a set of simulation or design exercises meant to apply, within a microwave Computer-Aided Design CAD environment, the concepts developed in the rest of the book. Each example is, potentially, a CAD laboratory trace. To allow the reader to replicate the examples, additional material is made available online in the form of a project in its specific CAD environment. We chose to develop the project on two well-known and widely used platforms, Microwave Office (MWO) [1] and ADS (Advanced Design Systems) [2].[1] To keep the presentation within reasonable limits (and also to avoid fast obsolescence due to the evolution of the CAD tool interfaces), for each example only a synthesis is provided with the main results in graphical form.

The examples closely follow the evolution of the text and are intended as short and tutorial in nature. Rather than reporting on full real-word designs, leading to a ready-to-go layout, they concentrate on specific aspects that are dealt with in the text on a more theoretical equation-based level. All figures are directly copied (sometimes with minor changes and a conversion to black and white) from the CAD tool user interface, their graphical format being in a way a part of the design experience of the user. Some of the CAD projects actually provide a CAD implementation of numerical examples presented throughout the text.

10.2 Microstrip Line and Stub Matching of a Complex Load

The project proposes a CAD implementation of Example 2.5. A 50 Ω line plus stub section (the stub in short circuit) is set up to match to 50 Ω the impedance $Z_L = 25 + j10$ Ω. The design frequency is chosen as 10 GHz. The electrical line length at 10 GHz is initially set equal to the approximate values found in Example 2.5, and a further optimization is carried out, obtaining slightly different values. Fig. 10.1 shows the resulting input impedance compared to the goal, $Z_i = 50 + j0$ Ω.

At the end of the optimization step, the electrical lengths are exploited in the LINECALC[2] tool to derive the geometrical features of a microstrip line on a 300

---

[1] The examples from Sec. 10.2 to Sec. 10.11 are in the MWO environment while those in Sec. 10.12 and Sec. 10.13 are for the ADS environment.
[2] Transmission line calculators are provided within most CAD tools; versions are also available online.

10.2 Microstrip Line and Stub Matching of a Complex Load

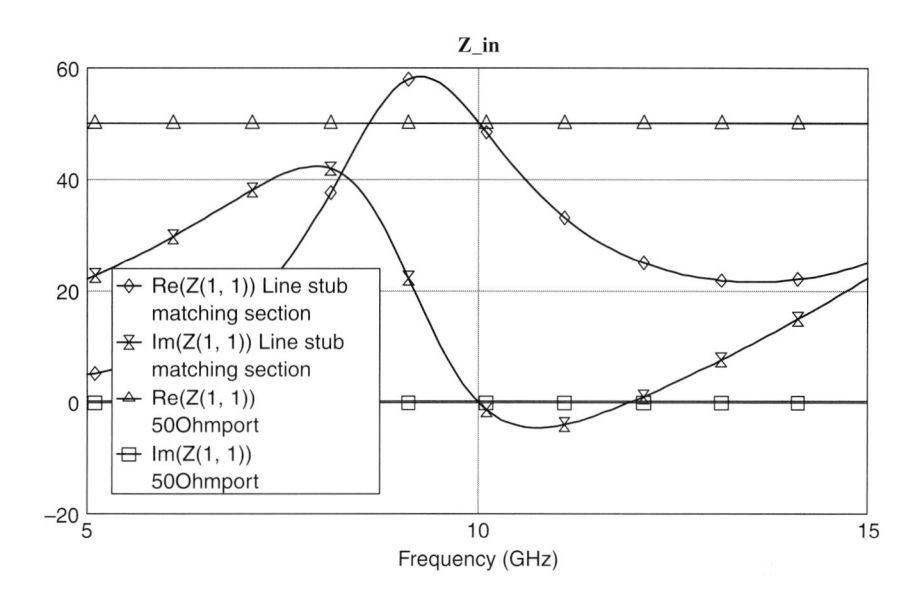

Figure 10.1 Input impedance of the ideal line and stub matching section (Sec. 10.2).

µm GaAs substrate, with gold metallization (2 µm thick) and substrate loss tangent $\delta = 0.001$. 50 Ω lines are again the goal. LINECALC provides the line and stub geometrical widths and lengths corresponding to the desired electrical lengths at 10 GHz.

The initial values of the lengths are inserted in a new schematic, shown together with the ideal line and stub matching section in Fig. 10.2 (b). This schematic also includes a microstrip T-junction model. Further, the microstrip line lengths are newly optimized to account for the parasitics introduced by the microstrip T-junction. The final result after optimization is compared with the initial approximation based on the ideal line lengths ("Microstrip Matching Section Starting"), neglecting also the presence of the T-junction, in Fig. 10.3. The initial approximation is reasonable, but further tuning is needed to achieve better matching.

The example can be repeated by starting the optimization of the ideal line matching section with electrical lengths very different from those obtained in Example 2.5. Line lengths have anyway to be constrained in the optimization to be shorter than 180 degrees in electrical angle, unless particular layout constraints suggest the need to use longer lines. Also in this case, however, since more than one solution is possible (with different line lengths and an inductive or capacitive input impedance short circuit stub), as suggested in Example 2.5, the optimization can lead to different solutions according to the initial point selected. It is therefore important to suitably choose the starting point if one particular solution is intended, e.g., on the basis of a *tuning* step. As additional exercise, a via hole can be added at the end of the shorted microstrip stub to further tune the stub length, or a solution with an open stub can be implemented accounting for the open-end microstrip extra capacitance.

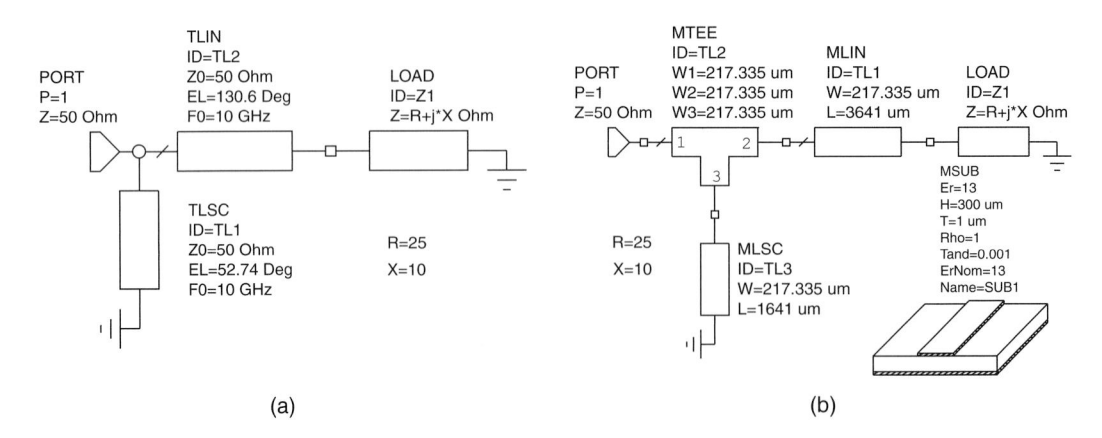

Figure 10.2 Ideal line and stub matching section (a) and microstrip version (b) (Sec. 10.2).

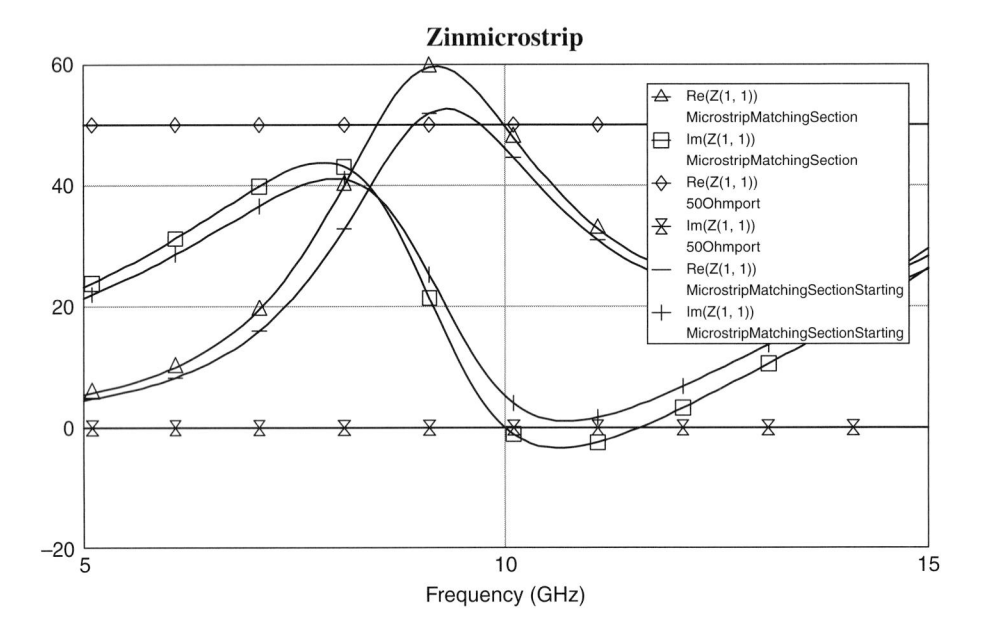

Figure 10.3 Input impedance of the microstrip line and stub matching section before and after length optimization accounting for the microstrip T-junction parasitics (Sec. 10.2).

10.3 Design of a 3 dB Directional Coupler at 20 GHz

The project compares a number of possible designs for a 3dB coupler on 50 Ω loads with a centerband frequency of 20 GHz. The design values are obtained through optimization and/or tuning. The designs considered are as follows:

1. a four-conductor Lange coupler (Fig. 10.4 (a));
2. an ideal branch-line coupler with microstrip lines but no T-junctions;

10.3 Design of a 3 dB Directional Coupler at 20 GHz   525

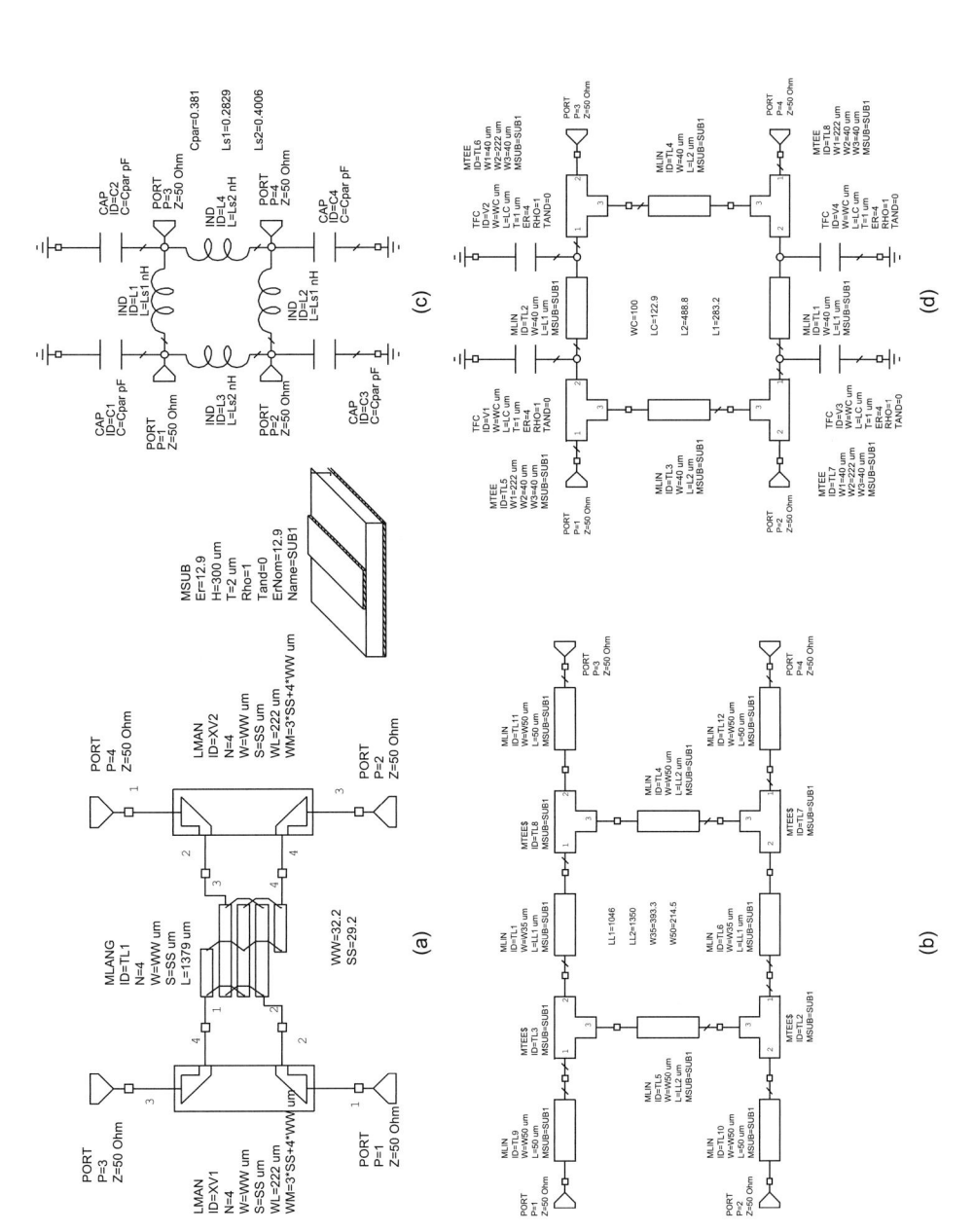

Figure 10.4 Four-conductor Lange coupler with input sections (a); branch-line coupler with microstrip lines and T-junctions (b); lumped branch-line coupler (c); lumped branch-line with "real" component implementation (d) (Sec. 10.3).

3. a more realistic ("real") branch-line coupler with microstrip lines and T-junctions (Fig. 10.4 (b));
4. an ideal lumped implementation of the branch-line coupler, made with ideal capacitors and inductors (Fig. 10.4 (c));
5. a more realistic ("real") implementation of the lumped branch line, exploiting thick film capacitors and short microstrip lines as the inductances (Fig. 10.4 (d)).

For each design, the magnitude (in dB) and the phase of the scattering parameters S_{11} (reflection at port 1), S_{21} (transmission to isolated port), S_{31} (transmission to coupled port), S_{41} (transmission to thru port) are evaluated. For brevity, only the magnitude of the scattering parameters for the Lange coupler (Fig. 10.5), the "real" distributed branch-line coupler (Fig. 10.6) and the "real" lumped branch line coupler (Fig. 10.7) are shown. Finally the phase difference between the *coupled* and *thru* ports is shown (graph "DeltaPhi34" in Fig. 10.8). The designs considered are always on a 300 μm GaAs substrate and in microstrip technology.

From the S parameter magnitudes, it is clear that the bandwidth of the Lange coupler is far wider than any of the branch-line implementations. Moreover, the Lange coupler is able to maintain the correct 90 degrees shift between the coupled and thru ports for a large frequency range, thanks to the wideband low values of the reflection coefficients $S_{11} = S_{22} = S_{33} = S_{44}$ and to the good isolation between port 1 and 2. On the other hand, the Lange coupler does not have spectacularly low centerband reflection coefficient and isolation, due to the mismatch between the even and odd mode velocities. Concerning the distributed branch-line coupler, its bandwidth is much narrower and the 90 degrees phase difference is observed only close to centerband, due to the deterioration of matching out of centerband; on the other hand, centerband isolation and matching

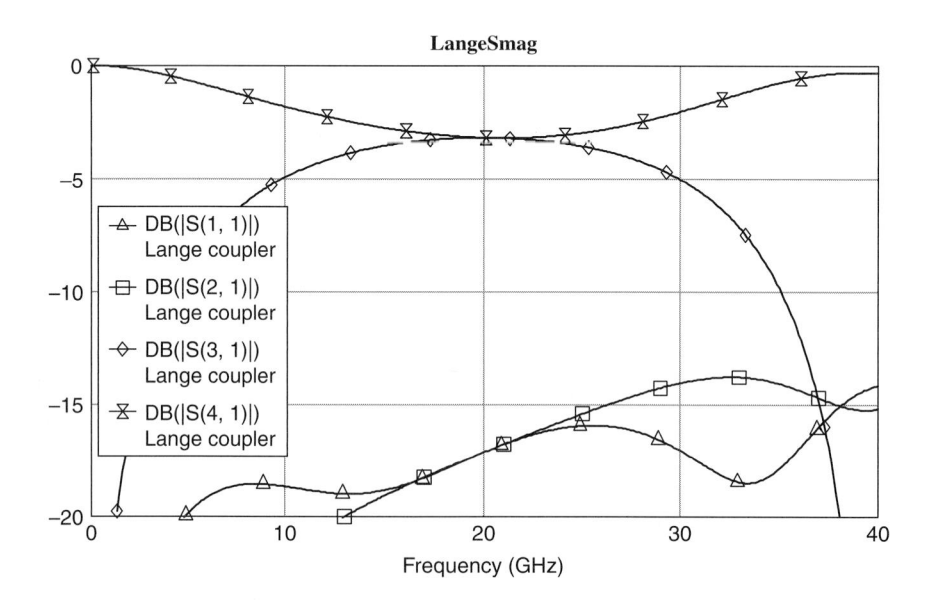

Figure 10.5 Magnitude of the scattering parameters of the Lange coupler (Sec. 10.3).

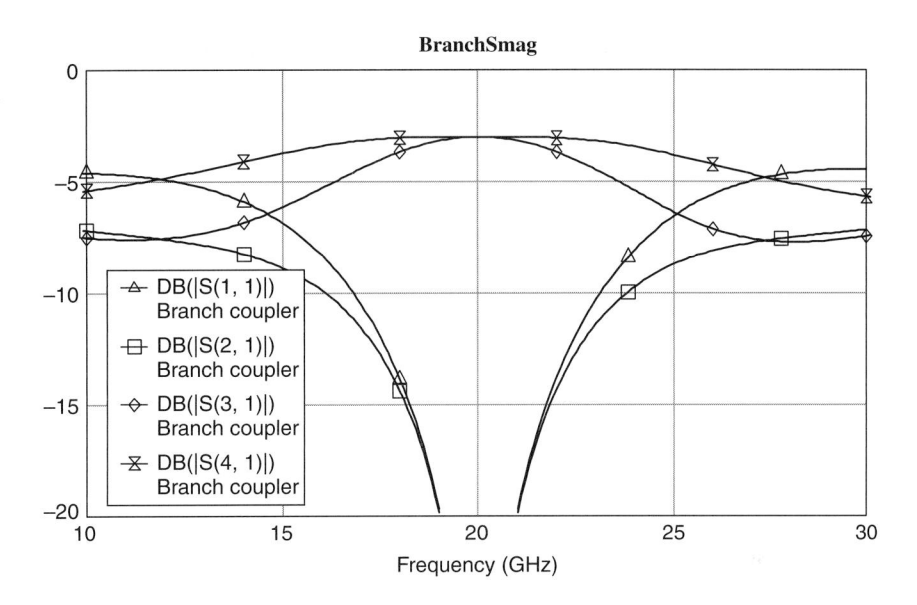

10.3 Design of a 3 dB Directional Coupler at 20 GHz

Figure 10.6 Magnitude of the scattering parameters of the branch-line coupler with microstrip lines and T-junctions (Sec. 10.3).

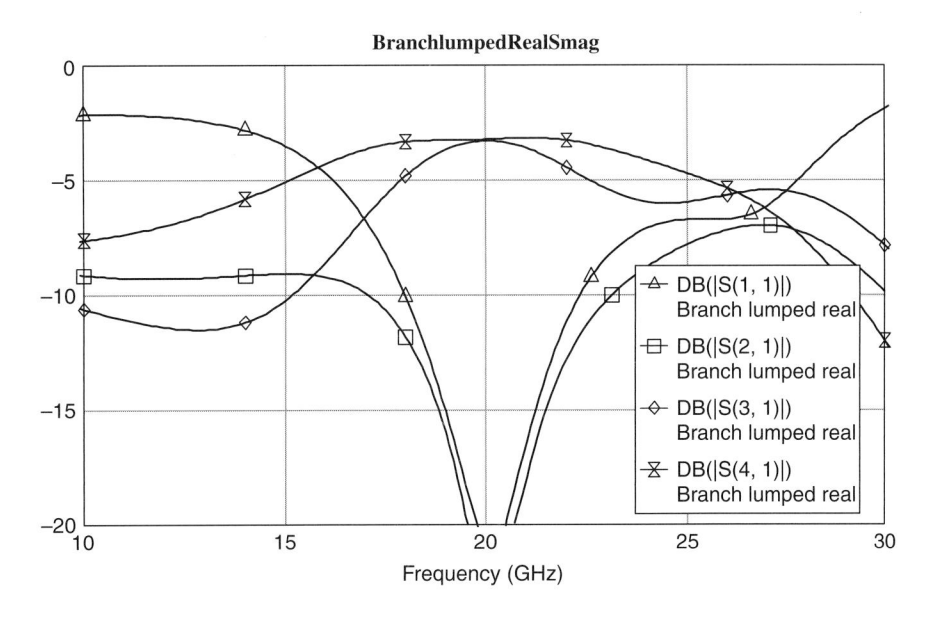

Figure 10.7 Magnitude of the scattering parameters of the lumped branch-line coupler with "real" component implementation (Sec. 10.3).

are much better than for the Lange design. The concentrated branch-line coupler has the same features as the distributed one, but an even narrower bandwidth.

Concerning the coupler footprint, the Lange coupler is approximately a rectangle of size 250 μm × 1400 μm. The branch-line in distributed form is almost a square with

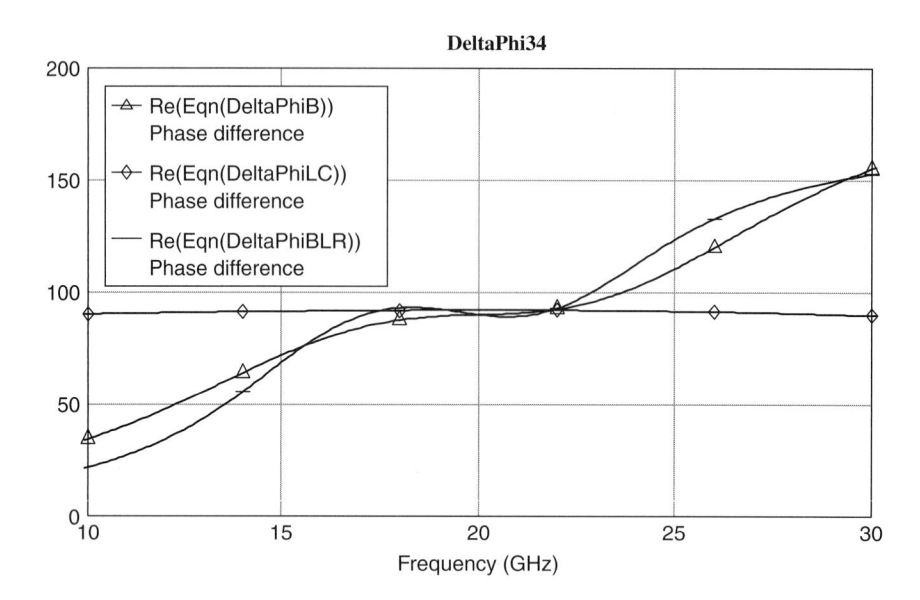

Figure 10.8 Phase difference between the coupled and thru ports for three coupler solutions: Lange coupler ("DeltaPhiLC"), branch line coupler with microstrips and T-junctions ("DeltaPhiB"), lumped branch coupler with "real" component implementation ("DeltaPhiBLR") (Sec. 10.3).

area 1000 μm × 1400 μm. The lumped version is more compact, 500 μm × 300 μm plus the four 120 μm × 100 μm capacitors. A more realistic design of the lumped version would of course require reference to a capacitor and inductor foundry library.

10.4 Fitting and Optimization of a Field-Effect Small-Signal Equivalent Circuit

The purpose of the example is to simulate the fitting and optimization of a small-signal FET equivalent circuit starting from the (pseudo) measured small-signal parameters. To this purpose, the example is divided into two parts. A first project generates the admittance parameters from a microwave FET equivalent circuit and the impedance parameters of the "cold FET" over the frequency range 1–50 GHz. The parameters of the equivalent circuit are from the literature [3] and have been rounded off so as to make them more readable. Some elements of the small-signal model are not included so as to make the FET equivalent circuit, see Fig. 10.9, correspond to that described in the text, see Fig. 5.22. The extrinsic elements are added externally. After generating the total admittance and cold FET impedance parameters, such parameters are saved in the Touchstone *s2p* format and MATLAB format.

A MATLAB script reads the *Y* and *Z* parameters and adds numerical noise to them in order to emulate real measurements. The *Y* parameters are saved in a text file that, after having been manually merged into a *s2p* file, is used by a second project as a "measurement" file. Both multiplicative and additive numerical noise (with uniform

10.4 Fitting and Optimization of a Field-Effect Small-Signal Equivalent Circuit

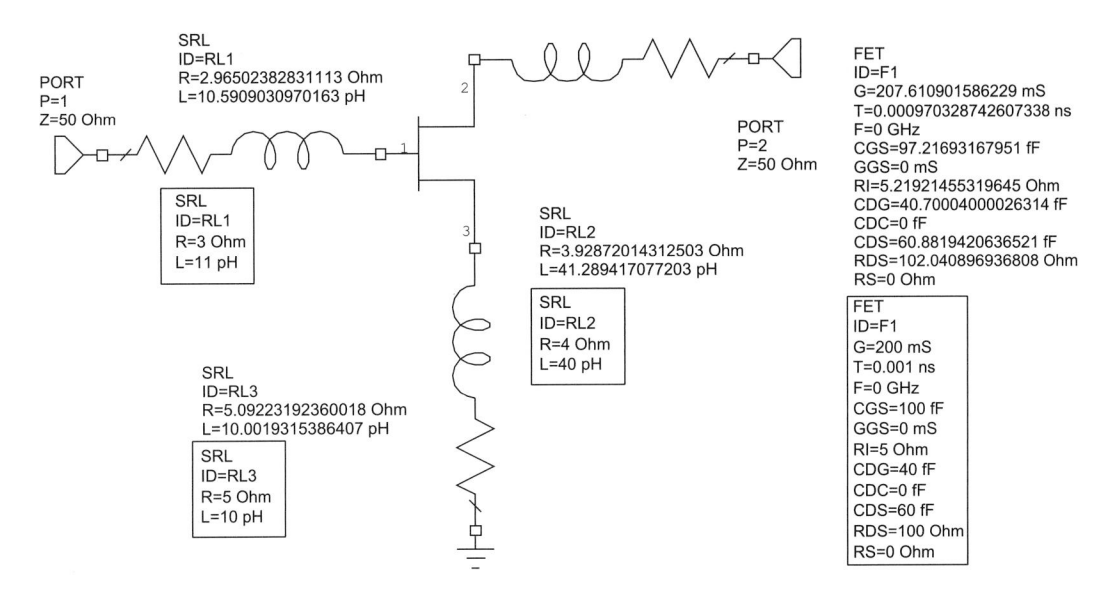

PORT
P=1
Z=50 Ohm

SRL
ID=RL1
R=2.96502382831113 Ohm
L=10.5909030970163 pH

SRL
ID=RL1
R=3 Ohm
L=11 pH

SRL
ID=RL2
R=3.92872014312503 Ohm
L=41.289417077203 pH

SRL
ID=RL2
R=4 Ohm
L=40 pH

PORT
P=2
Z=50 Ohm

SRL
ID=RL3
R=5.09223192360018 Ohm
L=10.0019315386407 pH

SRL
ID=RL3
R=5 Ohm
L=10 pH

FET
ID=F1
G=207.610901586229 mS
T=0.000970328742607338 ns
F=0 GHz
CGS=97.21693167951 fF
GGS=0 mS
RI=5.21921455319645 Ohm
CDG=40.70004000026314 fF
CDC=0 fF
CDS=60.8819420636521 fF
RDS=102.040896936808 Ohm
RS=0 Ohm

FET
ID=F1
G=200 mS
T=0.001 ns
F=0 GHz
CGS=100 fF
GGS=0 mS
RI=5 Ohm
CDG=40 fF
CDC=0 fF
CDS=60 fF
RDS=100 Ohm
RS=0 Ohm

Figure 10.9 FET small-signal model with external parasitics. The element values shown are the final fitted values. Initial values are framed (Sec. 10.4).

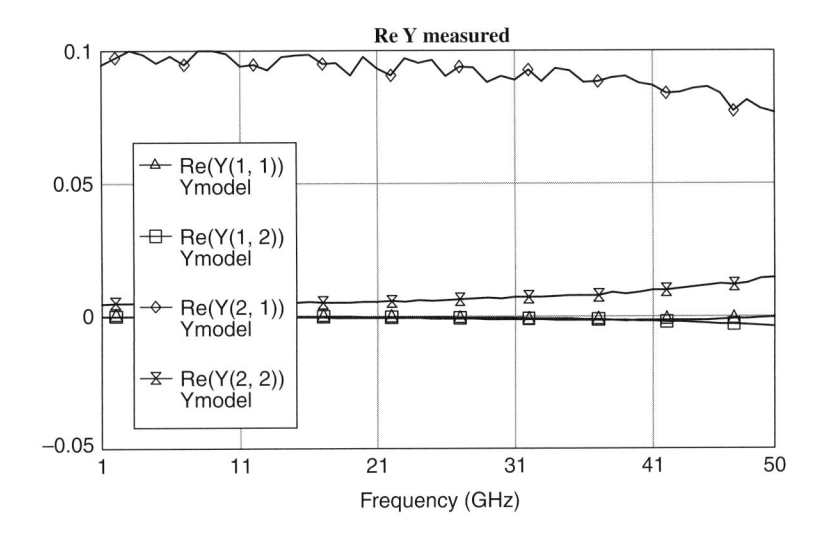

Figure 10.10 Real part of the "measured" admittance parameters (Sec. 10.4).

statistical distribution) are applied; the resulting admittance parameters (real and imaginary part) are shown in Fig. 10.10 and Fig. 10.11, respectively. The script extracts the average values of the parasitic resistances from the cold FET impedance matrix (real part); from the slope of the imaginary part. the parasitic inductances are estimate according to the procedure described in Sec. 5.8.2. As the last step, the extraction procedure in Sec. 5.8.2 is applied: from the "measured" Y matrix the Z matrix is derived frequency by frequency by inversion; then, the parasitic Z matrix is de-embedded, and finally the elements of the intrinsic equivalent circuit are explicitly derived frequency

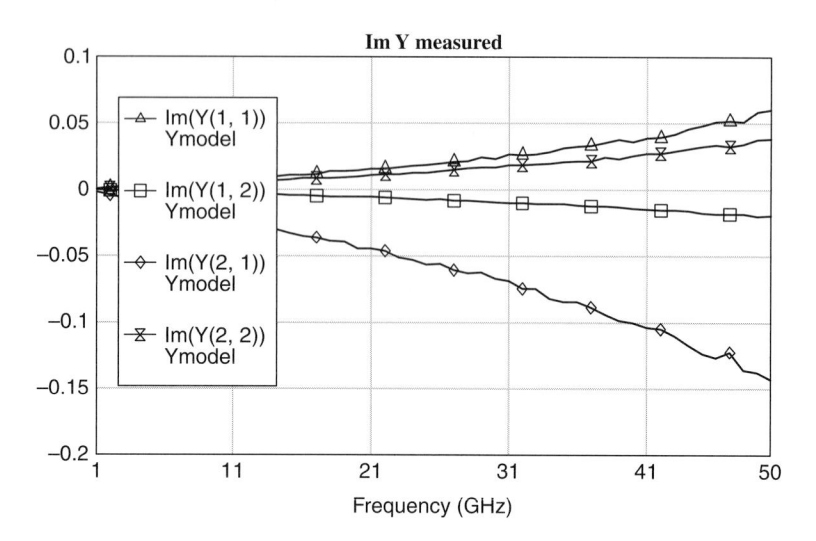

Figure 10.11 Imaginary part of the "measured" admittance parameters (Sec. 10.4).

by frequency (note that the MWO equivalent circuit uses the delay parameter $-\tau$ rather than τ as in Sec. 5.4.3). At the end of the script, the parameters of the equivalent circuit are averaged over the whole frequency band to obtain a starting value for the second project.

The second project performs the optimization of the equivalent circuit from the initial values generated by the MATLAB script. The starting values for optimization in the SS FET schematic have to be inserted by hand from the output of the MATLAB script. Then, optimization is launched to fit the model to the "measured" S parameters, affected by numerical noise. The optimization goal is to minimize the difference between the model and the "measured" scattering parameters. The comparison between the "measured" and fitted scattering parameters is shown in Fig. 10.12. Notice that the MATLAB script generates a set of random "measurements" that is varying every time, although the average "measured" values are the same. The present optimization exercise has some peculiarities with respect to "real life" model optimization: in fact, we know that the model topology is accurate, since the pseudo-measured small-signal parameters were generated by adding numerical noise to the simulated parameters from the same topology; moreover, we already know an approximate solution to the problem, i.e., the element values of the FET equivalent circuit.

By increasing the amount of numerical noise the initial estimate of the FET SS parameters can be made as poor as desired (also, arbitrary but controlled variations can be applied to the initial estimation to evaluate the effect). As a result, it is seen that if the initial guess is "reasonably close" to the goal, the optimization is very effective and fast; however, in the presence of a poor estimate the optimization algorithm (the one selected is conjugate gradient, others can be more effective if the initial guess is poor) fails to converge to a proper solution, and also produces unphysical element values. The optimization of the FET small-signal equivalent circuit is therefore no trivial exercise, unless a good initial estimate is available.

10.5 Small-Signal FET Stabilization

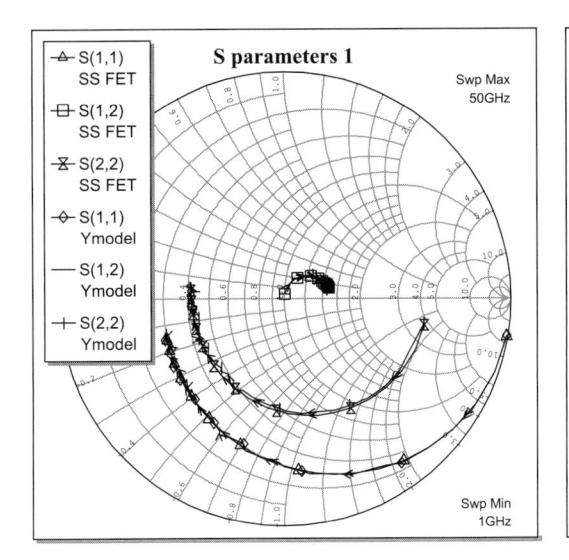

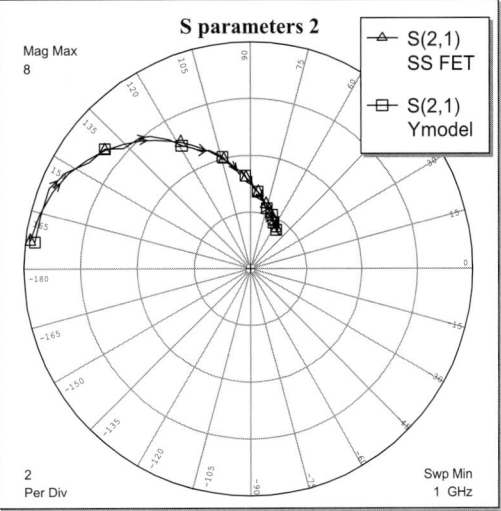

Figure 10.12 Scattering parameters from the "measured" ("Y model") and fitted ("SS FET") devices (Sec. 10.4).

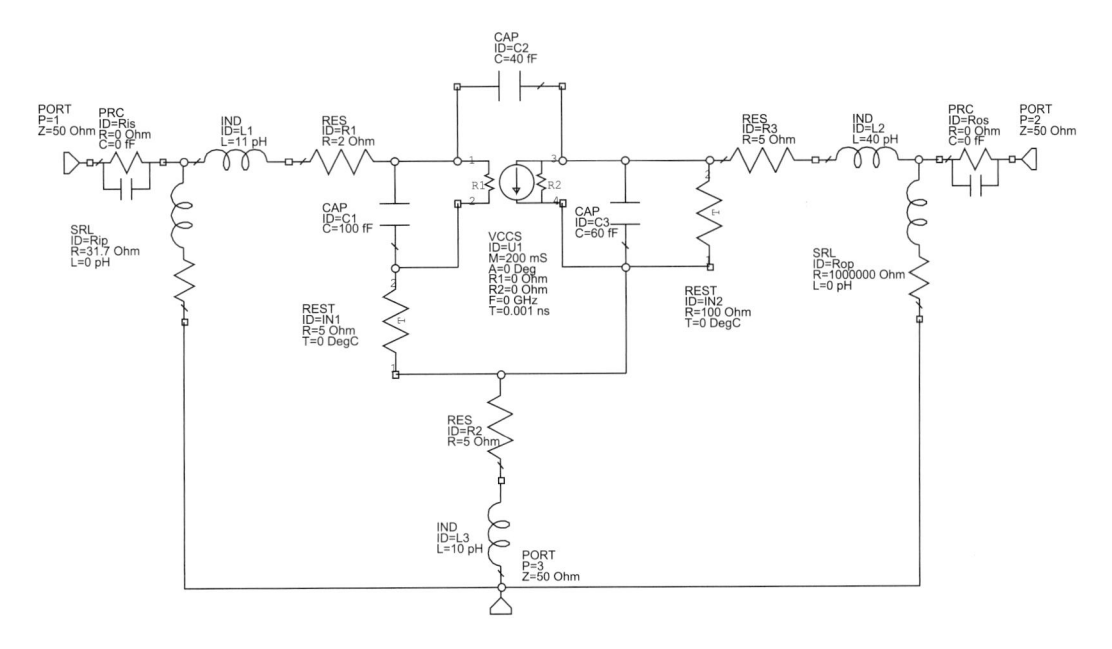

Figure 10.13 FET small-signal model with input/output, series/parallel stabilization blocks (Sec. 10.5).

10.5 Small-Signal FET Stabilization

In this project, a FET is analyzed in small-signal conditions and some stabilization schemes are added to the equivalent circuit. For simplicity, the reactive part of the stabilization network is disabled, so that the scheme is frequency independent. Fig. 10.14

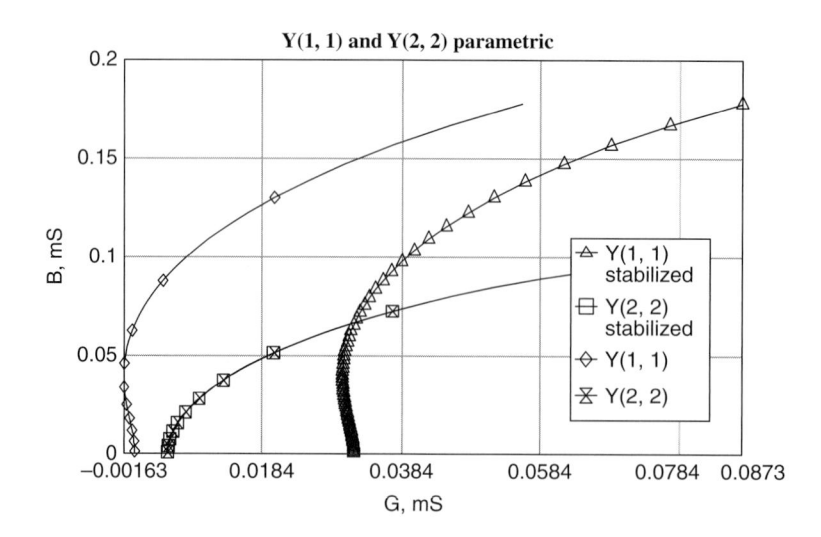

Figure 10.14 Input and output short-circuit admittances as a parametric function of frequency of the unstabilized and stabilized devices (Sec. 10.5). Notice that the two Y_{22} traces are super imposed.

shows in the admittance plane Y_{11} and Y_{22} parametrically as a function of the operating frequency from 1 to 70 GHz. The Y_{11} behavior of the unstabilized device shows a negative real part in a certain frequency range, suggesting to use as a preferred scheme a parallel input conductance. By playing with the *tune* tool it is readily realized that the other stabilization schemes have little effect or even (as the one based on the output parallel conductance) make stability worse. By properly tuning the input parallel conductance a value is obtained that makes the device unconditionally stable at all frequencies, as shown by the stability factors μ_1 and μ_2, see Fig. 10.15. The gains of the stabilized device are finally shown in Fig. 10.16 together with the MSG and the S_{21} of the unstabilized device.

10.6 Design of a Maximum Gain Amplifier at 15 GHz

The aim of the project is to describe the design of a simple, single stage, open loop small-signal amplifier. The design goal is to maximize the gain and simultaneously provide input and output matching for a frequency close to 15 GHz, with a bandwidth around 2 GHz. The design strategy corresponds to the simultaneous input and output conjugate (power) matching approach described in Section 6.9; prior to this, the amplifier in-band stability must be assessed.

The design starting point is the small-signal model of the active device. We adopt here a simple look-up table model based on the interpolation of the measured scattering parameters of a GaAs FET (available in the so-called Touchstone *s2p* format) between 6 and 26 GHz. While this model is well suited to the design of a small signal amplifier, it introduces the problem of interpolation and extrapolation of the experimental data,

10.6 Design of a Maximum Gain Amplifier at 15 GHz

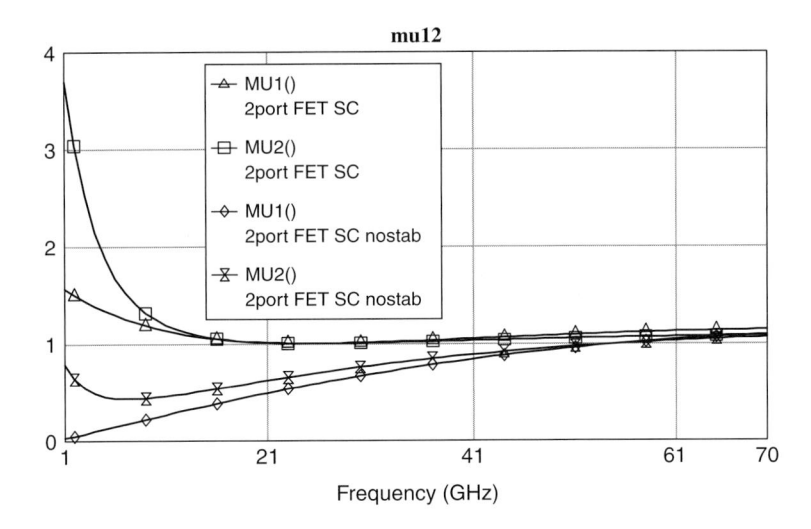

Figure 10.15 Single-parameter stability factors of the device before and after stabilization (Sec. 10.5).

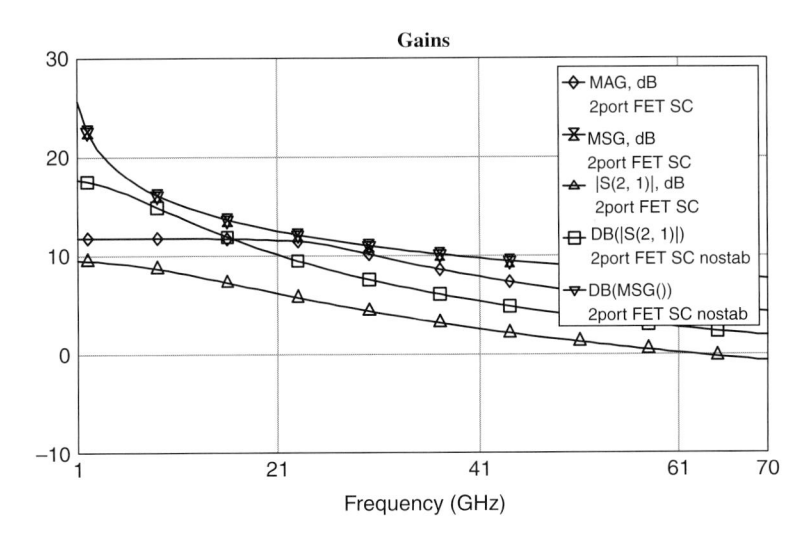

Figure 10.16 Gain parameters of the device before and after stabilization (Sec. 10.5). Notice that the MSG traces of the unstabilized and stabilized devices coincide.

pointing out that care must be taken to avoid simulation at frequencies far from the measured range, where the results can be completely unreliable. In particular, such a model does not allow any stability check to be carried out outside the measurement bandwidth. The stability check (Fig. 10.17), performed through the single stability parameter μ_1, reveals that the device is indeed unconditionally stable in the design band; however, no information is provided on stability outside of the measurement range, since the interpolated results are not reliable outside of that range. In other words, a low-frequency

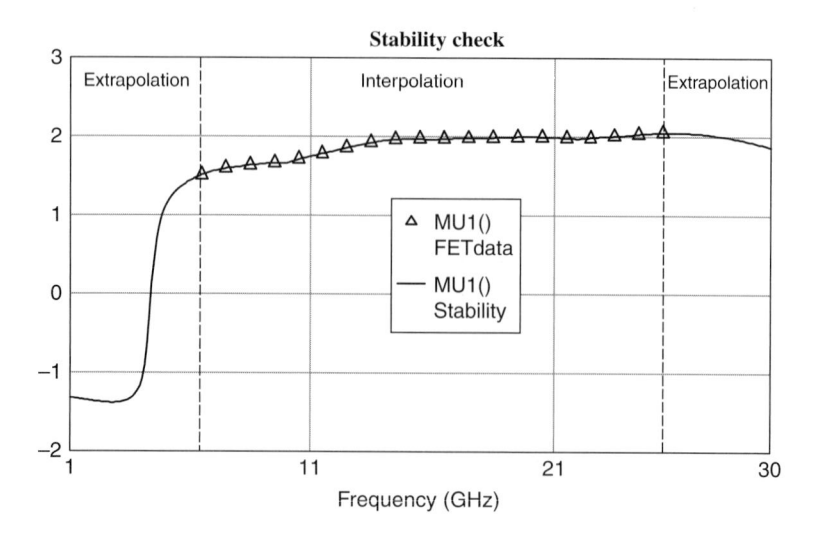

Figure 10.17 Stability parameters of the active device as a function of frequency. Triangles are the measured data, the curve is the interpolated model (Sec. 10.6).

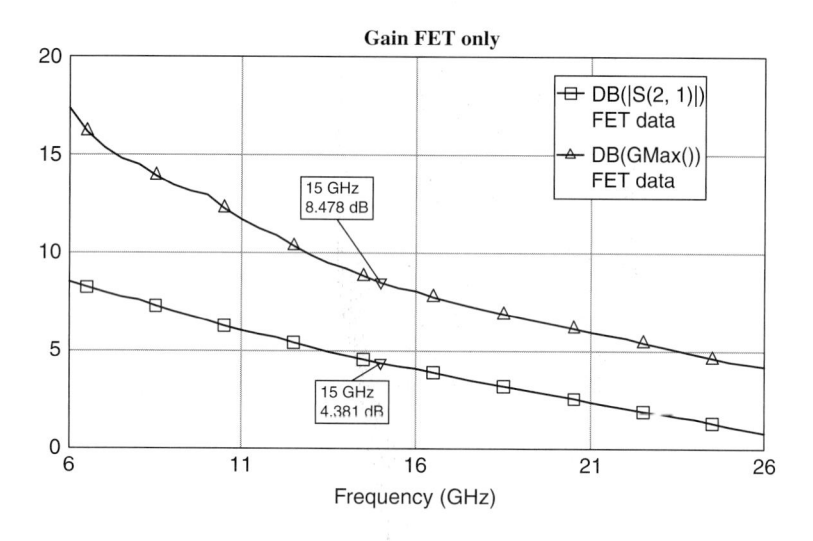

Figure 10.18 Maximum available gain (MAG) and S_{21} of the active device as a function of frequency (Sec. 10.6).

stability check cannot be performed on the basis of the available data, but would require the extraction of a wideband equivalent circuit model.

Since the device is unconditionally stable for all the frequencies of the measured range, the maximum gain the FET can provide is well defined and corresponds to the MAG. Fig. 10.18 reports the MAG (GMax, triangles) together with the magnitude of S_{21} on 50 Ω (squares) as a function of frequency. Extrapolating the MAG to higher frequencies, it can be seen that the maximum oscillation frequency of the device is

10.6 Design of a Maximum Gain Amplifier at 15 GHz

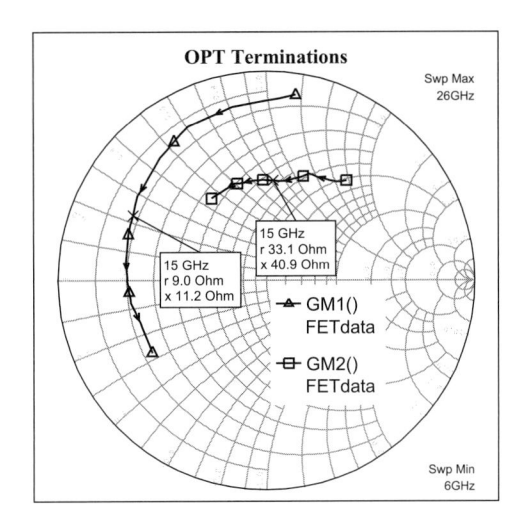

Figure 10.19 Optimum terminations for maximum gain as a function of frequency (Sec. 10.6).

around 35 GHz. The gain improvement vs. the 50 Ω termination that can be obtained through matching exceeds 4 dB at the design frequency of 15 GHz.

The next step corresponds to the identification of the unique termination pair (of reflection coefficient $\Gamma_{m1} = G_{m1}$ for the gate, and $\Gamma_{m2} = G_{m2}$ for the drain) for which the FET exhibits its maximum available gain. The CAD tool provides a direct evaluation of those two quantities as a function of frequency in the frequency range where the measured S parameters are available and where the device is unconditionally stable. Fig. 10.19 reports the reflection coefficient of such optimum terminations on the Smith chart. The two markers provide the values of the corresponding optimum impedances at the design frequency of 15 GHz. Such optimum impedances are the value that must be synthesized, through proper matching sections, starting from the external 50 Ω generator and load impedances.

The network chosen for the implementation of the two matching sections is a simple line plus stub topology. The characteristic impedances of the distributed elements are fixed to 50 Ω, while the line and stub lengths are optimized. Notice that in this simple case the same goal could be obtained by optimizing the cascade of the input matching section, FET model and output matching section. We prefer instead to separately address the design of the input and output matching sections. The structure of the matching network is shown in Fig. 10.20. The implementation uses a modular *subcircuit* approach, so that these subcircuits could be reused for both the input and the output matching networks, and for the simulation of the whole amplifier. The schematic within the subcircuit "Ideal Stub Line Filter" is reported in Fig. 10.20 (b).

The program optimizer goal is set to minimize, at 15 GHz, the absolute value of the difference between G_{m1} and the reflection coefficient seen at port P_1, see Fig. 10.20 (a), i.e., at the input of "Ideal Stub Line Filter". A similar procedure is applied for the output matching network. Fig. 10.21 reports the behavior of the input and output matching sections in terms of the input reflection coefficients with respect to G_{m1} and

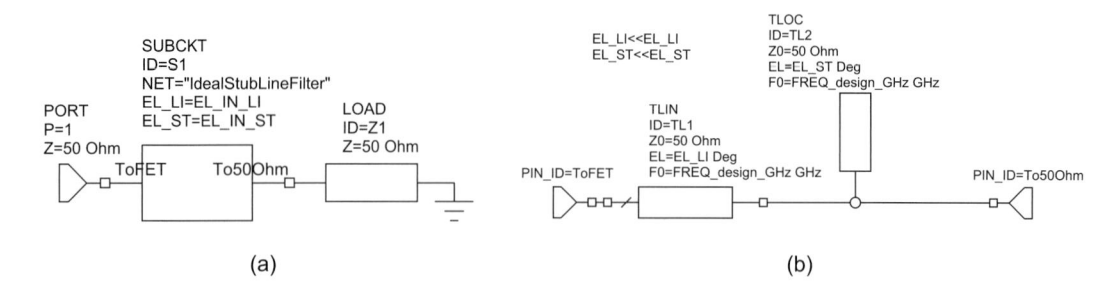

Figure 10.20 Schematic of the matching sections: (a) block diagram; (b) details on the stub plus transmission line matching section (Sec. 10.6).

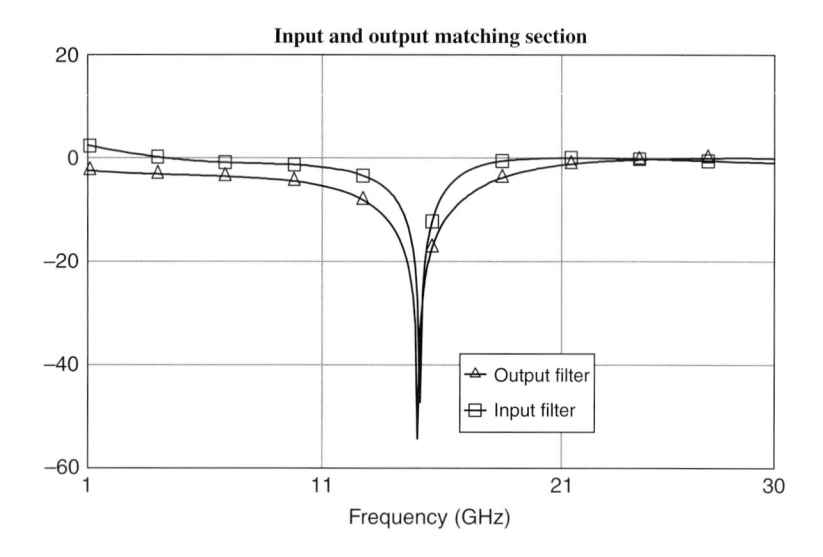

Figure 10.21 Reflection coefficients of the matching sections as a function of frequency (Sec. 10.6).

G_{m2}, respectively. The plot indicates that at 15 GHz the two optimum impedances are correctly synthesized.

Finally, the whole amplifier is simulated using the circuit shown in Fig. 10.22, and the amplifier gain is compared with the maximum gain of the FET, see Fig. 10.23. At 15 GHz, as expected, the matched amplifier $|S_{21}|^2$ equals the MAG of the FET at the same frequency.

In the next phase of design, the amplifier is designed in a microstrip environment; a duroid substrate available in the CAD library is exploited. Each ideal elements is converted into its microstrip counterpart using the transmission line calculator provided by the CAD environment, and the input and output matching section in the microstrip realization are assembled element by element. Moreover, microstrip layout-induced non-idealities are accounted for by inserting T-junction element and open-circuit microstrip stub models, accounting for the end fringing capacitance. After considering

10.6 Design of a Maximum Gain Amplifier at 15 GHz

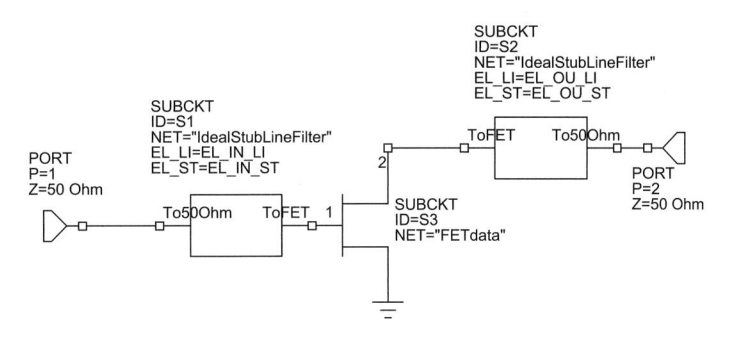

Figure 10.22 Schematic of the amplifier; the matching sections are realized with ideal lines (Sec. 10.6).

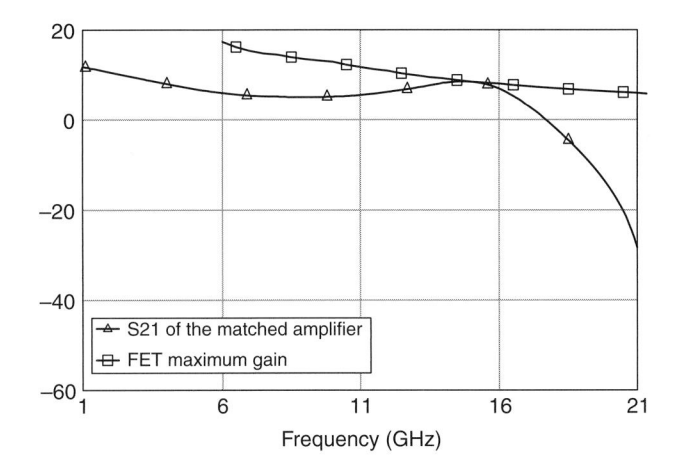

Figure 10.23 Maximum available gain and S_{21} of the matched amplifier (Sec. 10.6).

the effect of such non-idealities, the circuit will need a re-optimization. The microstrip implementation of the input and output matching sections is shown in Fig. 10.24 (b), while Fig. 10.24 (a) reports the schematic of the amplifier with microstrip matching sections. Also in this case, a modular approach through subcircuits has been adopted.

Fig. 10.25 finally compares the performances of the ideal line and microstrip implementation of the matched amplifier close to the design frequency of 15 GHz. Notice that the microstrip parasitics yield a (small) frequency shift that can be recovered through a final optimization carried out on the whole circuit (i.e., an optimization of the line and stub lengths of the matching sections).

Additional design steps that may be carried out in the design could consist in: (1) the extraction of the amplifier layout from standard or library cells available in the CAD tool; (b) a post-layout simulation check on the basis of the electromagnetic solver included in the tool. The final layout can be then generated and exported in a suitable format (e.g., *gds*) for processing through a thin-film hybrid technology. This stage would also need to design the device DC matching networks, properly connected to the FET through (in this case) lumped bias Ts.

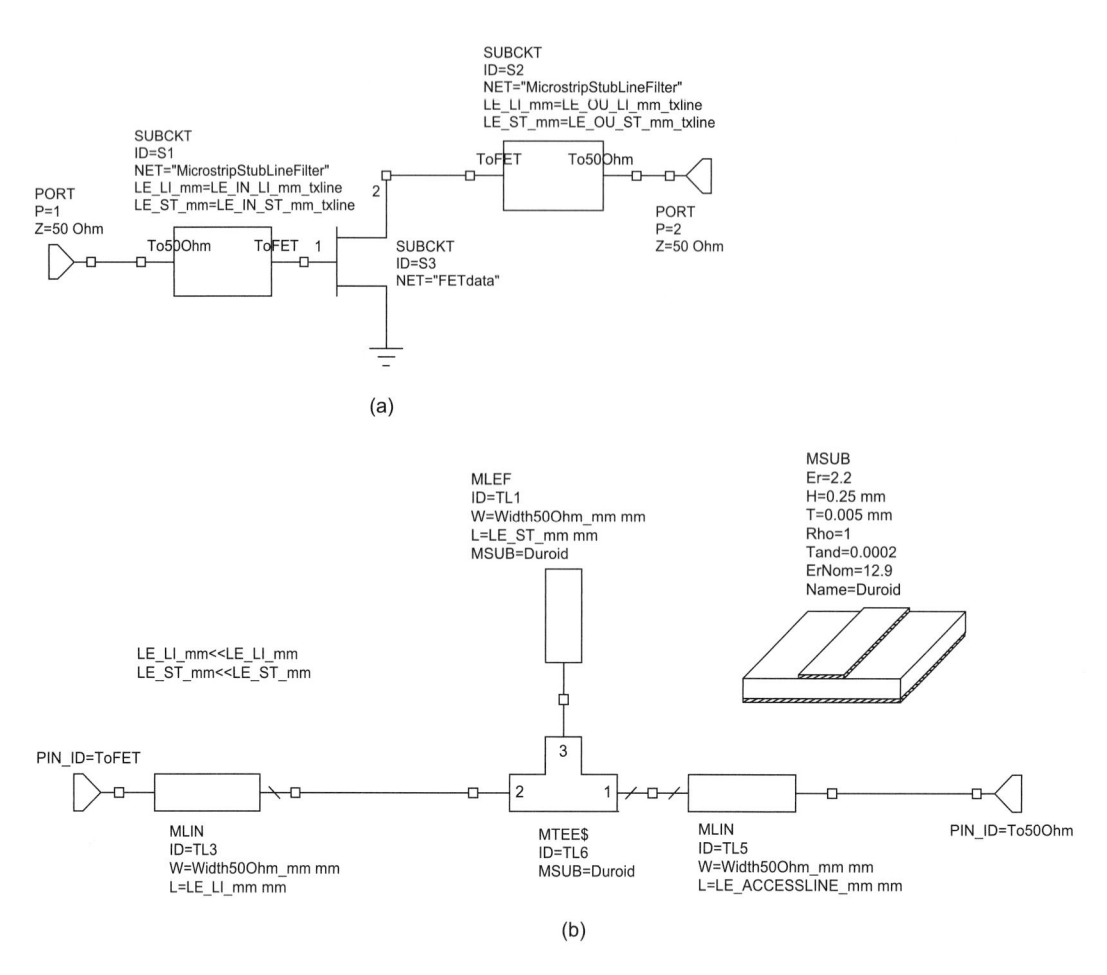

Figure 10.24 (a) Schematic of the amplifier with microstrip matching sections; (b) microstrip implementation of the matching sections: internal subcircuit schematic (Sec. 10.6).

10.7 Design of a Two-Stage Balanced Amplifier

The project concerns the design of a two-stage balanced amplifier with center frequency at 8 GHz and bandwidth in excess of 1 GHz. As a first step, a FET model with an OFF breakdown voltage around 12 V and a knee voltage around 2 V is biased close to the optimum class A point ($V_{GS} = -1$ V, $V_{DS} = 8$ V), also corresponding roughly to the bias where the device transconductance is maximum. Even if the device technology is not specified, consider that the here adopted Angelov (Chalmers) model, typically refers to a HEMT device, see Sec. 5.8.5. As it is, the FET is not unconditionally stable; stabilization is carried out by a set of input and output resistive networks shown in Fig. 10.26 (a). For the second stage, a scaled-up version of the same FET is used with a ×3 DC current. The same bias point is used and also in this case a stabilization network has to be introduced to make the second stage device unconditionally stable.

10.7 Design of a Two-Stage Balanced Amplifier

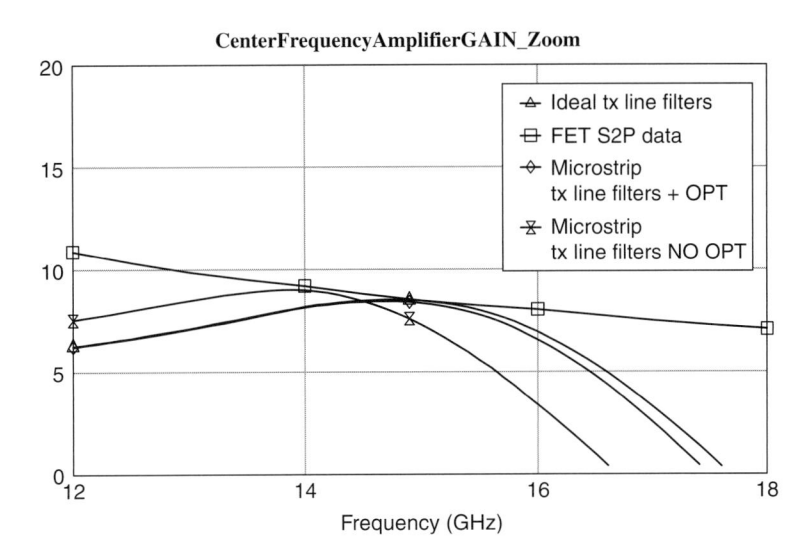

Figure 10.25 Maximum available gain (squares) and S_{21} of the matched amplifier with ideal lines (triangles), microstrip lines, non-optimized (hourglasses) and optimized (diamonds) (Sec. 10.6).

Although the circuit is intended as a small-signal amplifier, we also investigate the power behavior in compression conditions. It should be noticed that the amplifier exploits a resistive rather than tuned load; while no difference exists in small-signal conditions, this choice obviously is a penalty for the amplifier efficiency under compression.

A single-tone test on the stabilized device (without matching networks) is performed on the first and second stage, respectively, showing that the saturation power is around 21 dBm for the first stage and around 24 dBm for the second stage. Fig. 10.27 reports the DC curves of the first stage; to these DC curves the dynamic load line is superimposed for an input power at the onset of saturation. It can be noticed that the dynamic load line shows some reactive effects and that the 50 Ω load is actually close to the optimum load for a class A power amplifier. This is not the case for the larger second stage device.

As a starting point for the design of the two-stage amplifier that will be used in tandem in the balanced implementation, the behavior of the optimum input and output terminations is monitored for the two stages and a first design of the input and output matching networks, see Fig. 10.26 (b) and (c), is performed on a 50 Ω load using a simple lumped-parameter narrowband approach. Notice that analytical expressions (equivalent to those discussed in Sec. 6.9.1) are used to identify the ideal component values at a frequency slightly larger than the centerband design frequency (i.e., 10 GHz vs. 8 GHz); component value variations were added in the schematic to allow tuning and optimization in a further phase of the project.

For the sake of simplicity, the interstage matching network was implemented by cascading the output matching network of the first stage and the input matching network of the second stage, both designed to 50 Ω. This simple approach also allows, with some further redesign, to merge the two matching networks in a lumped or distributed

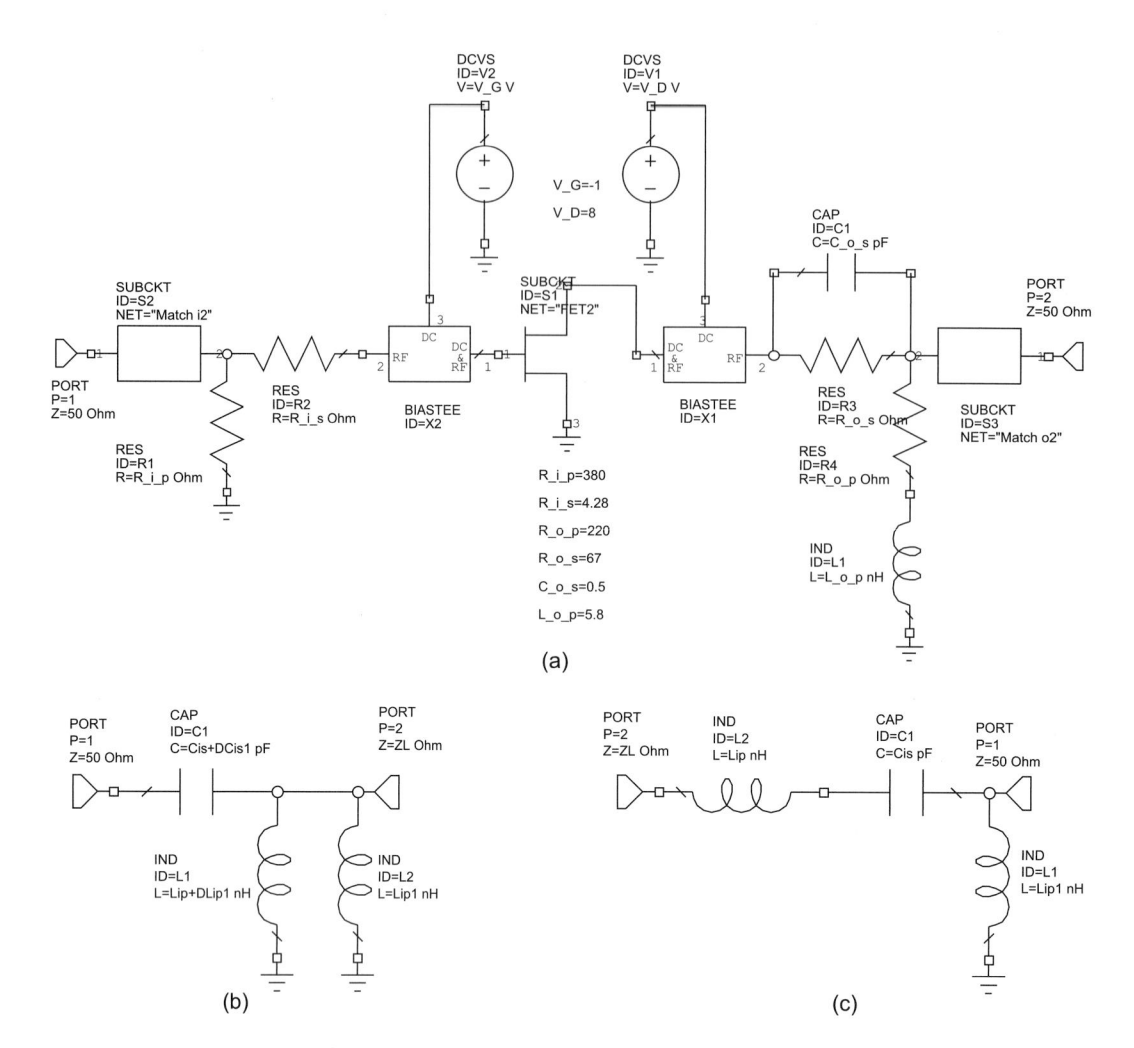

Figure 10.26 Schematic of the first amplifier stage (a) and of the input (b) and output (c) matching network for the tandem two-stage amplifiers exploited in the balanced configuration. The topology of the second stage is similar, only a larger device is used. The input and output stabilization networks are also shown (Sec. 10.7).

interstage matching network with simpler topology and reduced element count. In our case, perfect matching was anyway not a goal since a certain amount of mismatch will have to be introduced to allow for broadbanding gain equalization.

The two-stage amplifier is simply obtained by cascading the first and second amplifier stages. After cascading, the input matching networks of the first and second stages are tuned to flatten the gain and obtain a center frequency of 8 GHz with a bandwidth in excess of 1 GHz. Fig. 10.28 reports the gain and the input and output reflectances of the two-stage amplifier; we clearly see that the gain flattening (around 28 dB) was obtained at the expense of an unsatisfactory input and output matching, with reflectances well above -10 dB also at centerband (8 GHz).

10.7 Design of a Two-Stage Balanced Amplifier

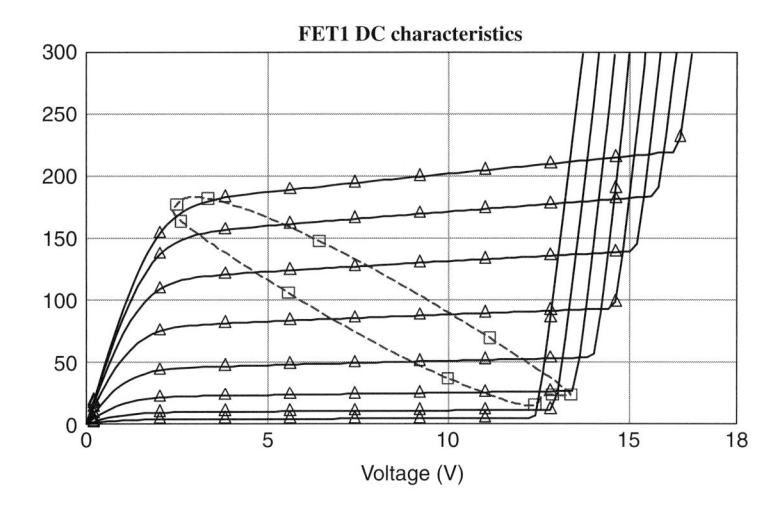

Figure 10.27 DC characteristics of the first stage of the tandem balanced amplifiers; the dynamic load line is shown at the onset of compression (Sec. 10.7).

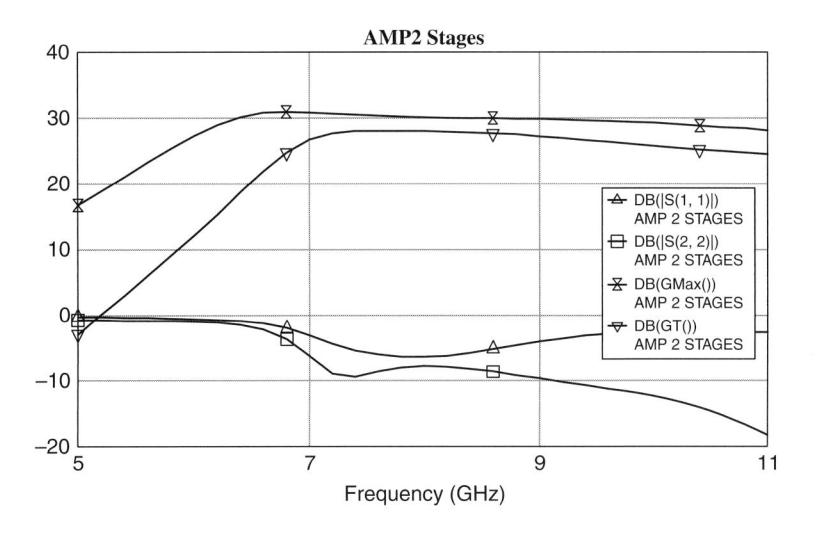

Figure 10.28 Maximum gain (hourglasses) and transducer gain (nablas) of one of the two identical two-stage tandem amplifiers to be used in the balanced configuration. Triangles and squares are the input and output reflectivities on a 50 Ω load, respectively (Sec. 10.7).

To implement the balanced amplifier, a Lange coupler is designed with a centerband frequency of 8 GHz on an alumina substrate, and two two-stage identical amplifiers are connected to the input and output coupler according to the classical configuration described in Sec. 6.11.1. Fig. 10.29 shows the balanced amplifier MAG and $|S_{21}|$ (almost coinciding), and the input and output 50 Ω reflectances. The balanced amplifier gain is very similar to the one of the two-stage version (around 28 dB) but this time the input

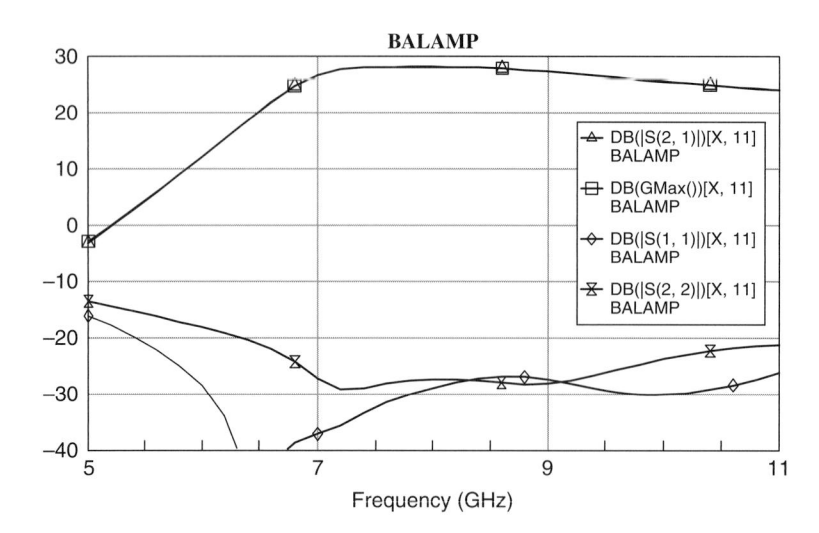

Figure 10.29 Maximum gain (squares), S_{21} (triangles) of the balanced amplifier in the linear region. Diamonds and hourglasses are the input and output reflectivities on a 50 Ω load, respectively (Sec. 10.7).

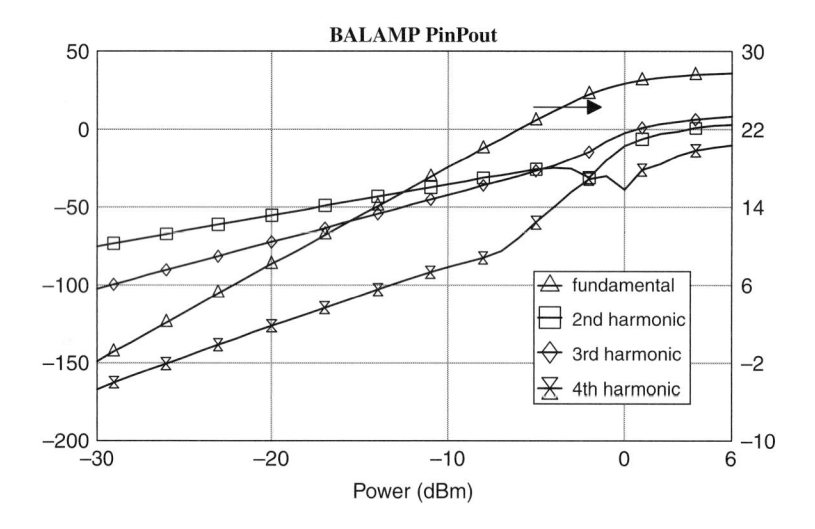

Figure 10.30 Harmonic single-tone $P_{in} - P_{out}$ of the balanced amplifier at 8 GHz. The output at the fundamental is on the right scale, those of the harmonics on the left scale (Sec. 10.7).

and output reflectances are well below −20 dB. Additional gain flattening was achieved by a slight final tuning of the matching sections.

Fig. 10.30 reports the balanced amplifier single-tone centerband (8 GHz) power sweep at the first four harmonics; according to theory, the saturation power of the balanced version should be 3 dB larger than the one of the second stage, and in fact it is more or less so (almost 28 dB against around 25 dBm). The power sweep can be repeated at several fundamental frequencies ranging from 7.2 GHz to 8.8 GHz; the output power

10.8 Design of Parallel Resistive Feedback Wideband Amplifiers

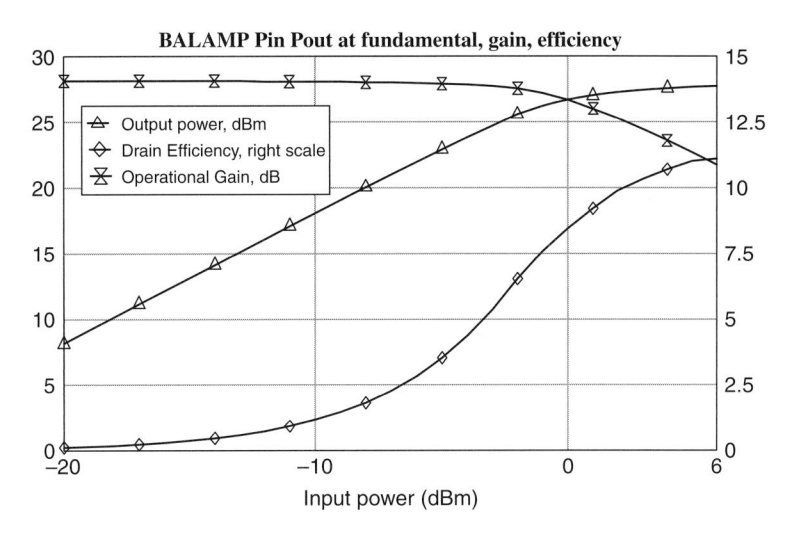

Figure 10.31 Output power, operational gain and drain efficiency of the balanced amplifier at 8 GHz (Sec. 10.7).

at the 1 dB compression point (around 0 dBm input power) varies from 26.3 to 26.7 dBm and is therefore constant enough on the design bandwidth. Fig. 10.31 finally reports the output power at the fundamental for the single-tone power sweep at 8 GHz, the operational gain and the drain efficiency as a function of the input power. The drain efficiency is reasonable since both amplifier stage work in class A (the first one also in backoff conditions) and the final stage is optimized for small signal operation, rather than with respect to the output power. The drain efficiency correctly scales as the output backoff.

10.8 Design of Parallel Resistive Feedback Wideband Amplifiers

Ideally, a combination of parallel and series resistive feedback should result in wideband closed-loop gain with input and output matching. This result is based in fact on an approximate theory neglecting parasitics and the output drain-source parallel resistance, see Sec. 6.12. The present example is carried out with a FET described by a two-temperature noise model; this allows the amplifier noise figure to be checked; that is not however a design goal. The FET cutoff frequency is intentionally chosen large (well above 100 GHz) to ensure that the open-loop FET parameters are good enough to allow for an easy closed-loop design.

The design was carried out mainly by tuning and shows that, even with the limited number of parameters at hand, it has to be based on a quite complex compromise between different performance. Let us consider first a few introductory remarks on the effect of the single components:

- Stability is a first concern, since the device, as it is, is not unconditionally stable, see Fig. 10.32. Inserting a feedback resistor stabilizes the device; however, a large

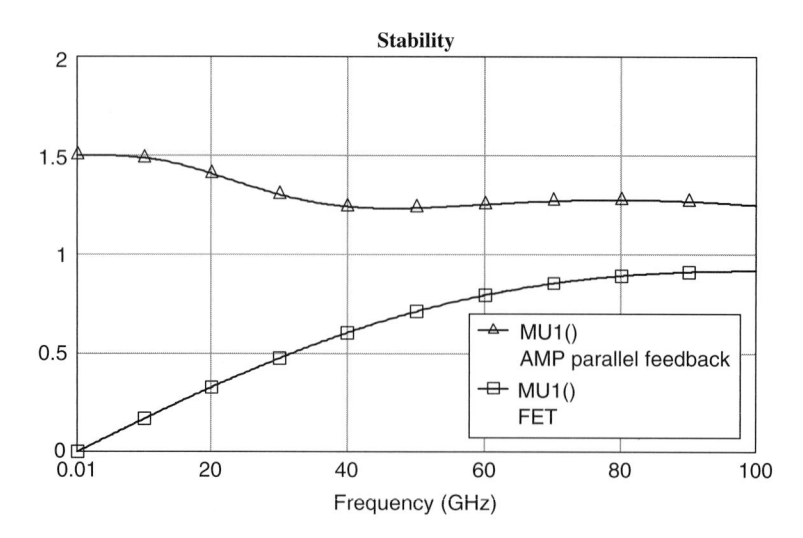

Figure 10.32 Stability parameter of the device exploited for the parallel feedback amplifier and stability of the amplifier (Sec. 10.8).

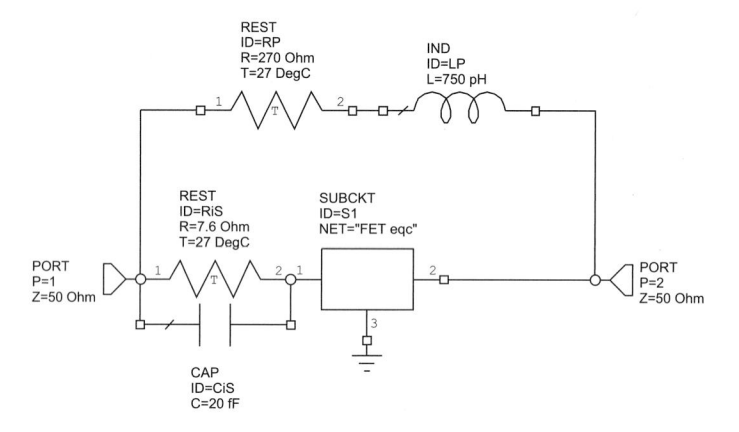

Figure 10.33 Schematic of the parallel feedback amplifier with input stabilization block (Sec. 10.8).

feedback resistance makes the bandwidth narrower, in such a way that roughly the gain-bandwidth product remains constant. Some additional high-frequency stabilization scheme has therefore to be exploited.

- The parallel feedback inductance (see Fig. 10.33; and additional DC block in the feedback path is not introduced for simplicity) produces the expected "inductive peaking" and may be exploited to flatten the gain roll-off at low frequency, but finally if the peaking is too large also the input reflection coefficient does not improve, and neither does stability.
- Input and output matching values are not extremely good but can be easily kept below 10 dB on a wide frequency band.

10.8 Design of Parallel Resistive Feedback Wideband Amplifiers

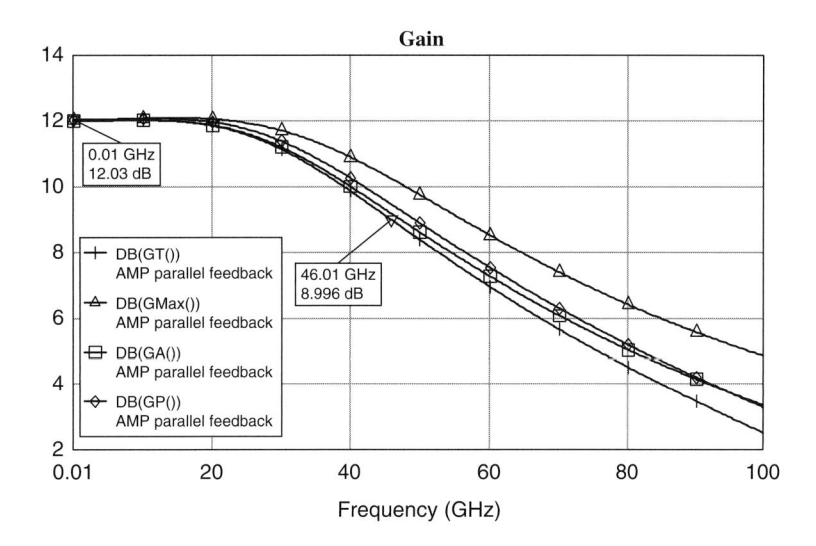

Figure 10.34 Gain of the parallel feedback amplifier (Sec. 10.8).

In conclusion, a parallel feedback resistance of 270 Ω was selected; this allows for a 12 dB low-frequency gain, while the 3 dB bandwidth is around 45 GHz, see Fig. 10.34; notice that the traces correspond to the available power, transducer and operational gains, and that these are slightly different owing to the imperfect matching on 50 Ω. Nevertheless, without the use of further device stabilization the feedback amplifier, that has been stabilized at low frequency by the effect of the feedback resistor, would still be potentially unstable above 50 GHz; to avoid this problem an additional *RC* stabilizing block was inserted at the device input, see the schematics in Fig. 10.33. The parallel *RC* block at the input appears to allow for stabilization up to 100 GHz (above that frequency any consideration on the basis of the simplified equivalent circuit neglecting parasitics appears to be critical). The *RC* values were manually tuned and some adjustment on the feedback parameters was additionally made. Fig. 10.35 points out that the input and output reflectances are indeed less than -10 dB in magnitude, on the whole amplifier 3dB bandwidth.

Some remarks are finally needed on the validity and approximation of the simplified design criteria for this structure, with reference to the discussion in Sec. 6.12. The design leads to $|S_{21f}| \approx 4$ while $|S_{21}| \approx 80$, of course at low frequency. According to the ideal design guidelines we should have:

$$R_p \approx R_0(1 + |S_{21f}|) = 250 \, \Omega;$$

the design value is in fact 270 Ω. We should also have:

$$R_s = R_0^2/R_p - 1/g_m = 50^2/270 - 1/0.2 = 4.5 \, \Omega.$$

This resistance was apparently neglected; however, the parasitic source resistance in the equivalent circuit already has a value of 5 Ω.

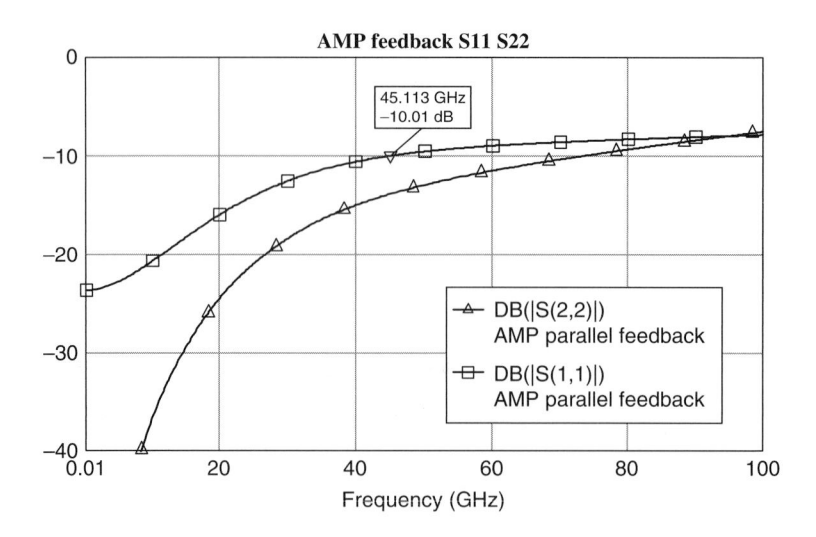

Figure 10.35 Input and output reflectances of the parallel feedback amplifier (Sec. 10.8).

A second project aims at exploring parallel feedback as a tool to implement a wideband (but not extremely wideband) amplifier, in this case centered at 20 GHz with a 5 GHz bandwidth. The 270 Ω parallel feedback resistance is again adopted and the optimization goal, implemented with lumped parameter input and output matching networks, is to have on the whole bandwidth an input and output reflectance ≤ -20 dB with a gain ripple within 0.2 dB. The resulting network is shown in Fig. 10.36. The peaking inductor was introduced to allow for some additional gain flattening.

After optimization, the network is further exploited to perform a sensitivity (yield) Monte Carlo analysis of the circuit. For simplicity, only the model transconductance is assumed to have a uniform distribution with a ±10 percent variation vs. the nominal value. The results for the affected parameters are shown in Fig. 10.37 (gain) and Fig. 10.38 (input and output reflectance). The use of resistive feedback should, in principle, alleviate the statistical fluctuations of the device elements on the amplifier response. The example also points out the possibility to realize wideband amplifiers around a center frequency through resistive feedback and IO matching, provided that the design frequency is much lower than the cutoff frequency of the device adopted.

10.9 Design of a Uniform DAMP with 40 GHz Bandwidth

The project aims at designing a discrete-cell distributed amplifier (DAMP) exploiting a somewhat idealized FET model shown in Fig. 10.39 (c) neglecting external parasitics and internal capacitive feedback. The device is exploited to implement a uniform distributed amplifier with 40 GHz bandwidth. Artificial lines whose basic cells are in Fig. 10.39 (b) for the drain and gate line, respectively, are exploited to connect the

10.9 Design of a Uniform DAMP with 40 GHz Bandwidth

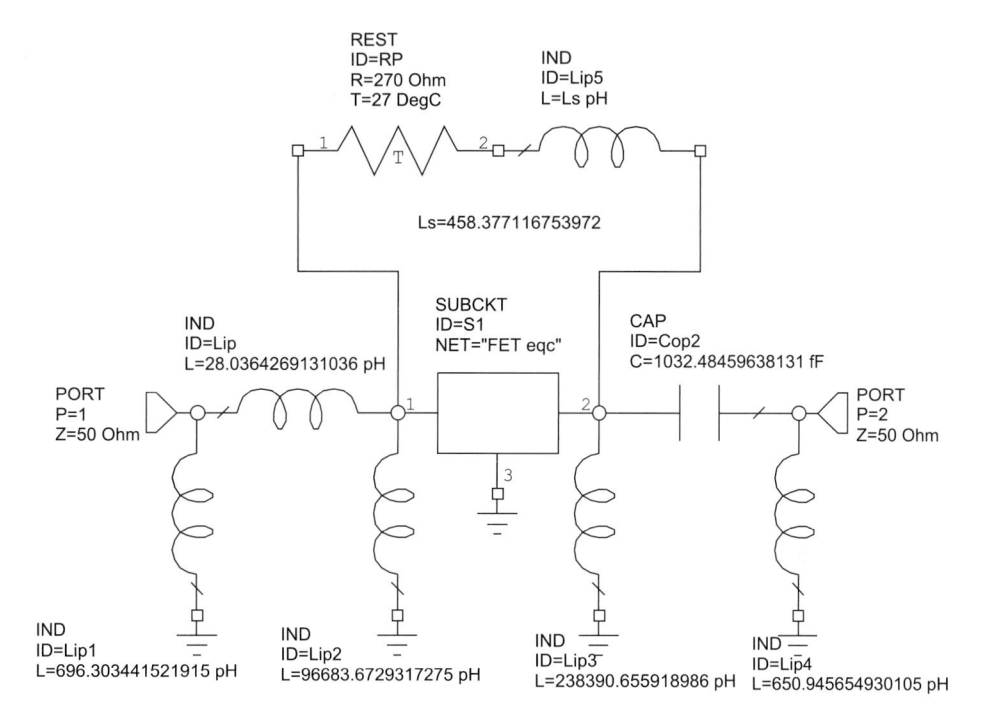

Figure 10.36 Medium-wideband parallel feedback amplifier with input and output matching (Sec. 10.8).

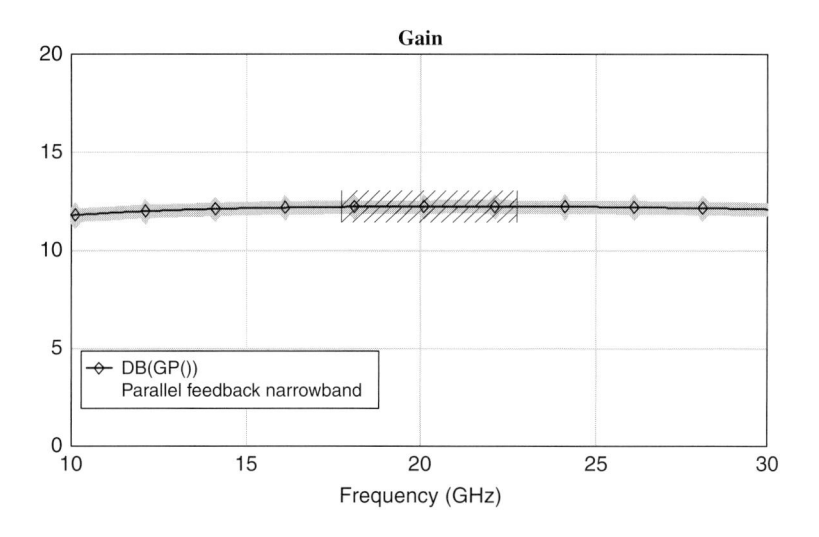

Figure 10.37 Gain of the parallel feedback amplifier with yield simulation (Sec. 10.8).

active elements. The drain cell also includes an additional capacitor needed to equalize the drain and gate line capacitances. The initial values of the inductances are chosen so as to achieve 50 Ω impedance on the gate and drain lines. With this choice, the cutoff frequency of the artificial lines is slightly above 60 GHz.

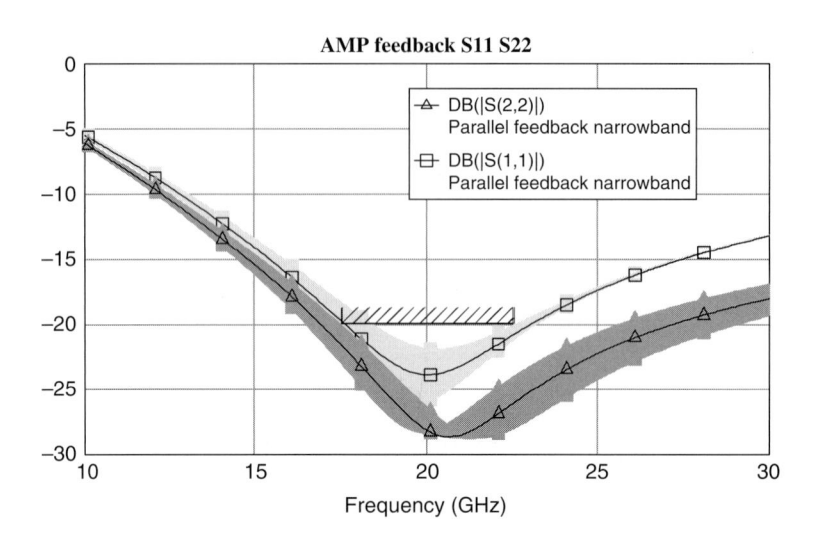

Figure 10.38 Input and output reflectances of the parallel feedback amplifier with yield simulation (Sec. 10.8).

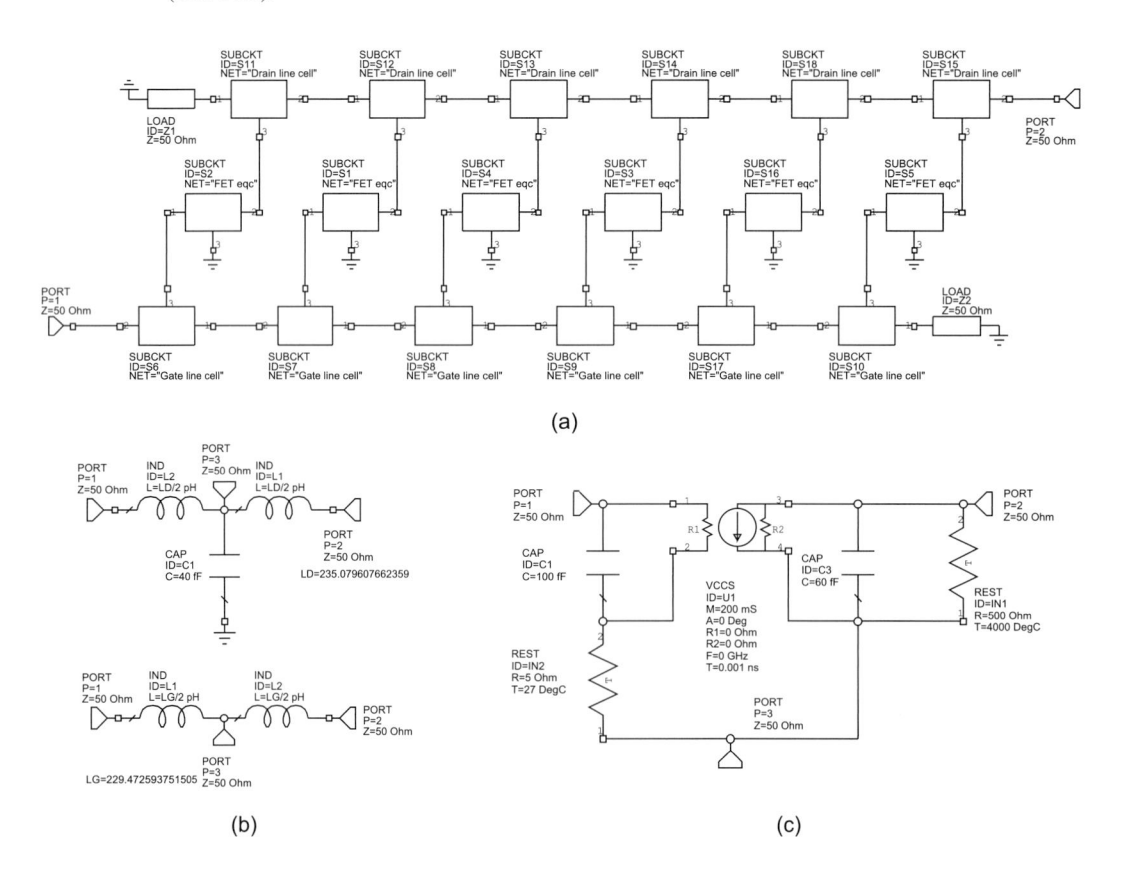

Figure 10.39 Schematic of the six-cell lumped DAMP (a); drain and gate cell ((b), above and below); ideal active device (c) (Sec. 10.9).

10.9 Design of a Uniform DAMP with 40 GHz Bandwidth

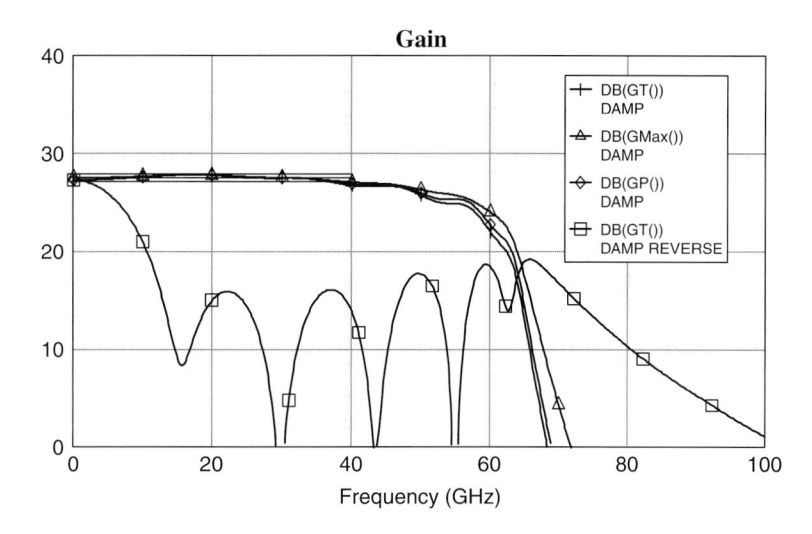

Figure 10.40 Six-cell lumped DAMP gain (Sec. 10.9).

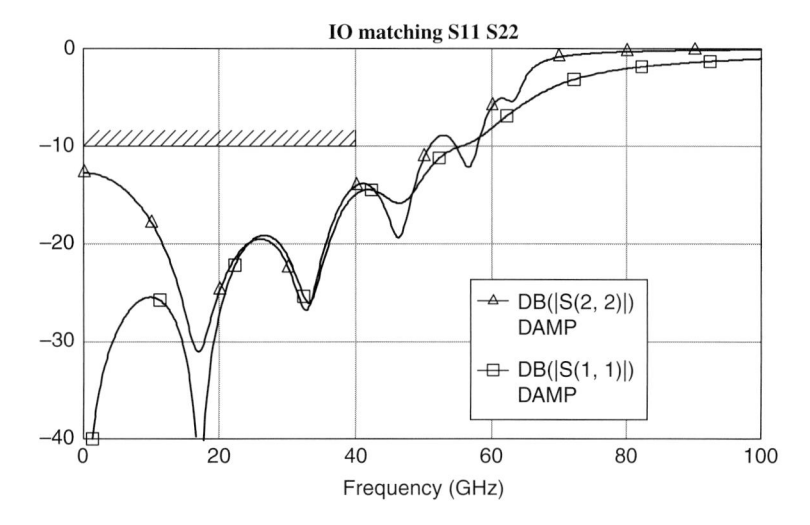

Figure 10.41 Six-cell lumped DAMP input and output reflectivities (Sec. 10.9).

Somewhat arbitrarily, $N = 6$ cells have been chosen for the implementation. Increasing the number of cells should yield a larger gain, but beyond a certain limit losses will lead to a decrease of the gain. The FET model also includes noise, although the noise figure is not a design goal.

The optimization process aims at obtaining a maximally flat power gain, together with input and output reflection coefficients less than -10 dB. The low-frequency gain is 27.5 dB and the obtained ripple up to 40 GHz is lower than 1 dB, while the input and output reflectances are well below -10 dB, see Fig. 10.40 and Fig. 10.41. The noise figure, shown in Fig. 10.42, left axis, is approximately below 2 in the amplifier bandwidth.

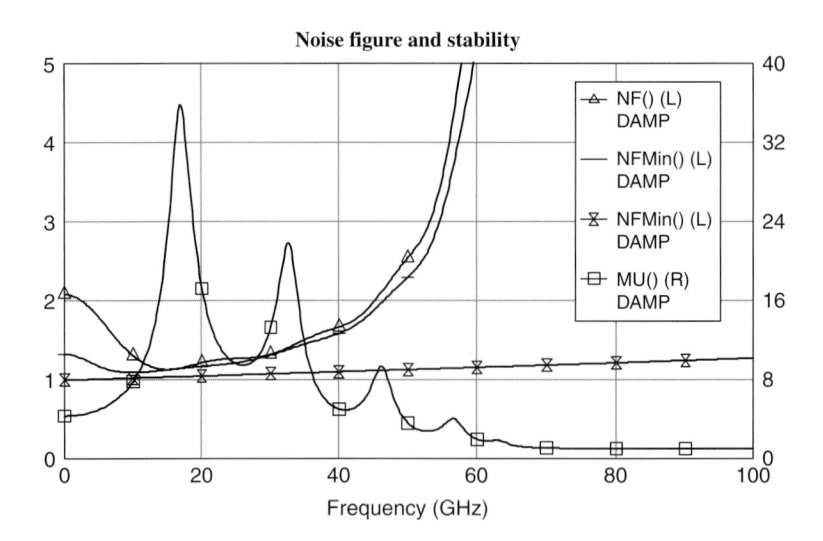

Figure 10.42 Noise figure (left axis) and stability factor μ_1 (right axis) of the six-cell lumped DAMP (Sec. 10.9).

The amplifier response is in agreement with the analytical model introduced in Section 6.13; Fig. 10.40 also shows the *reverse gain* of the DAMP between the input gate port and the counter-coupled port of the drain line. As expected from theory, the forward and reverse gains are equal at low frequency, but reverse gain drops comparatively quickly at high frequency. The amplifier is unconditionally stable on the whole bandwidth (see Fig. 10.42, right axis).

While the CAD simulation confirms the validity of the DAMP theory for an idealized device, it should be noticed that a real DAMP design including a full set of device parasitics and internal feedback elements can be challenging. Stability can be a concern, although for well-designed gate and drain lines the load of each stage should in theory correspond to 50 Ω. However, above the cutoff frequency of the artificial lines the FET stage loads can become reactive, thus compromising the amplifier stability. Notice that the design adopts a uniform structure, a solution that is not necessarily the optimum one when designing a DAMP with additional constraints on the power behavior.

10.10 FET Noise, Single-Ended and Balanced LNA Design

The project addresses the noise simulation of a FET represented by its two-temperature equivalent circuit (Fig. 7.21) and the design of a simple low-noise stage following the classical approach, based on the identification of the optimum source impedance, at a frequency of 20 GHz. The schematic of the two-temperature model of the FET is shown in Fig. 10.43, while Fig. 10.44 shows the noise figure on 50 Ω source resistance, the minimum noise figure, and the stability factor μ_1. The device is potentially unstable at the design frequency, but, to avoid increasing too much the noise figure, it is

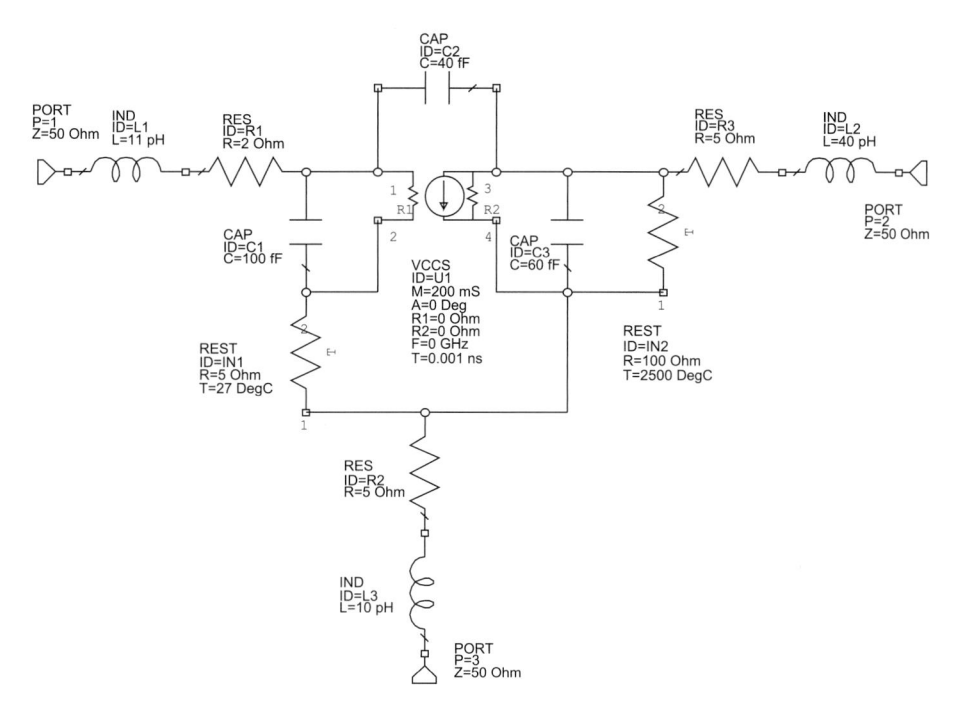

Figure 10.43 Two-temperature noisy FET equivalent circuit (Sec. 10.10).

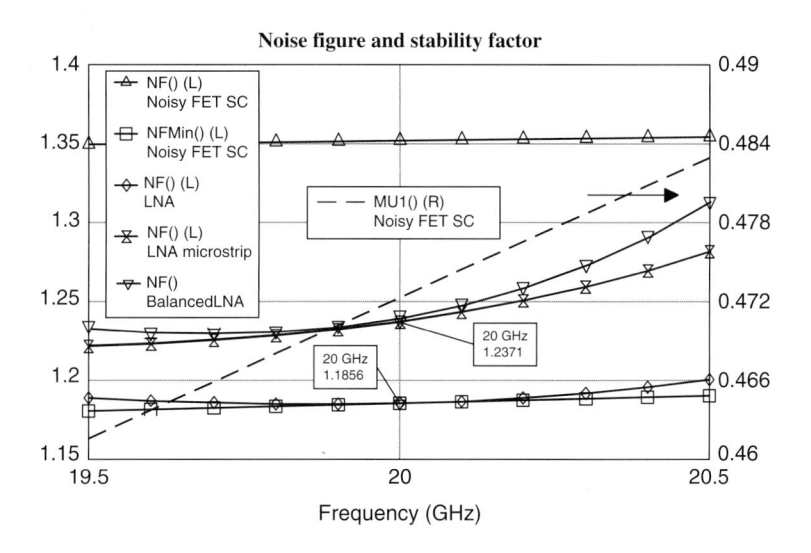

Figure 10.44 50 Ω noise figure, minimum noise figure and stability factor μ_1 (right axis) (Sec. 10.10).

decided to stabilize it out-of-band, leaving it potentially unstable at the design frequency. The out-of-band stabilization can be easily achieved by inserting an input parallel conductance that is disconnected through a resonator at the design frequency. However, for simplicity the out-of-band stabilization is not implemented in the schematics. The

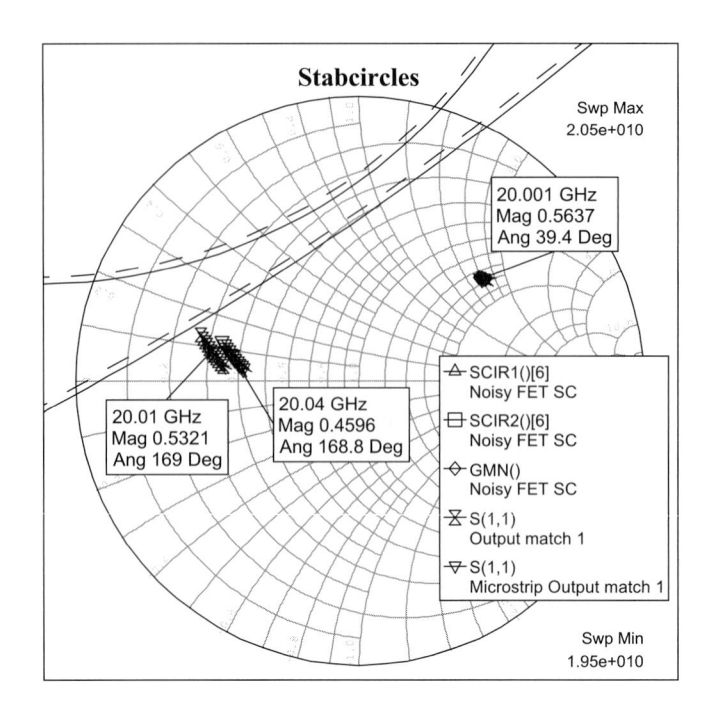

Figure 10.45 Input and output stability circles at 20 GHz; the optimum source load and the output load are also indicated in the Smith chart. The unstable region is marked by the dashed line. The optimum source load is the optimum noise source impedance, while the output load corresponds to output conjugate matching, according to the ideal line and microstrip matching section approach (Sec. 10.10).

minimum noise figure at 20 GHz is 1.18, in natural units, while the 50 Ω noise figure is around 1.35.

The LNA matching networks are implemented in ideal transmission line and then re-optimized in microstrip. The input matching section should transform the 50 Ω generator impedance into the optimum source impedance (minimizing the noise figure), while the output matching section should implement conjugate matching at the output port. Fig. 10.45 shows the input and output stability circles at 20 GHz, confirming that the input and output loads fall in the stable region of the Smith chart. The complete microstrip implementation of the LNA is shown in Fig. 10.43; the resulting performances are again plotted in Fig. 10.44 and Fig. 10.47.

The total gain at 20 GHz is around 17.8 dB for the ideal line implementation and 15.5 for the microstrip implementation. With the ideal implementation, the noise figure coincides with the minimum device noise figure, while the losses introduced by the microstrip implementation lead to a slightly larger noise figure (1.237 against 1.186 at 20 GHz). However, while the device has a good output matching with $|S_{22}| \approx 0$ at 20 GHz, a large input mismatch results, with $|S_{11}| \approx 0.87$.

The example discusses two issues in LNA design: the design with a potentially unstable device, if the noise figure obtained after in-band stabilization is unacceptably low;

10.10 FET Noise, Single-Ended and Balanced LNA Design

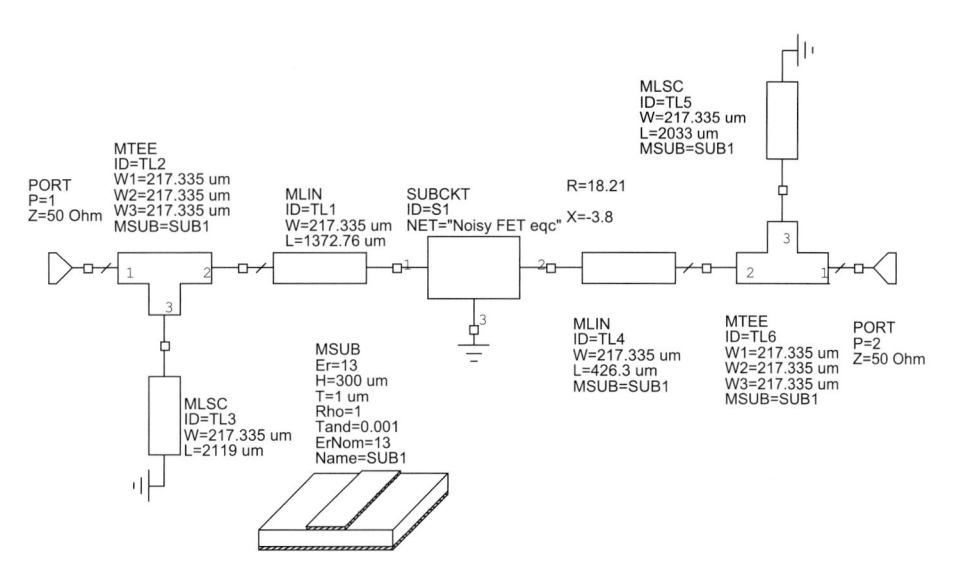

Figure 10.46 Microstrip implementation of the single-stage LNA (Sec. 10.10).

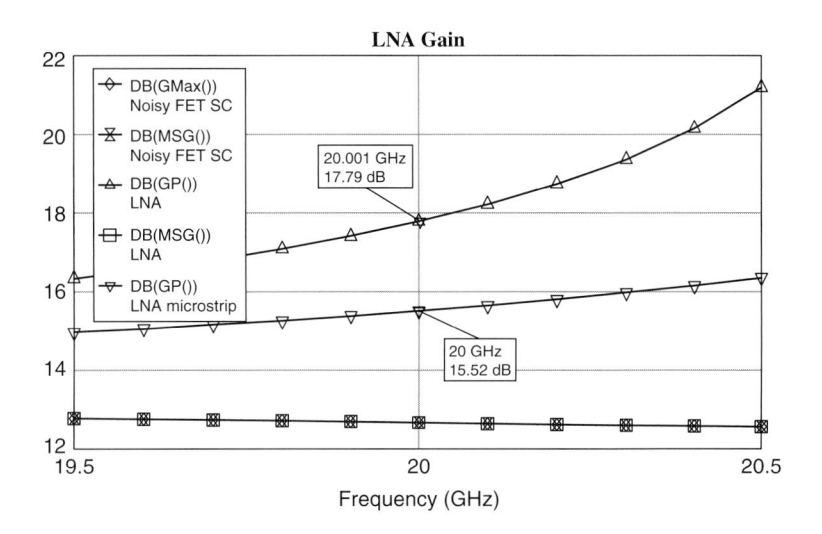

Figure 10.47 Gains of the single-stage LNA (Sec. 10.10). Notice that since the FET is not unconditionally stable the two traces "GMax" (MAG) and "MSG" of the "Noisy FET SC" coincide, and so does the trace "MSG" of the amplifier.

the severe input mismatch that may result from the noise-matched rather than power-matched input. In practice, a compromise can be sought between noise and power matching, accepting a larger noise figure but obtaining in change a better input match. Alternatively, techniques can be adopted to reduce the input mismatch through the balanced configuration.

A separate project is devoted to the implementation of a balanced LNA exploiting for each branch the noise-matched amplifier, and an input and output branch-line coupler,

554   CAD Projects

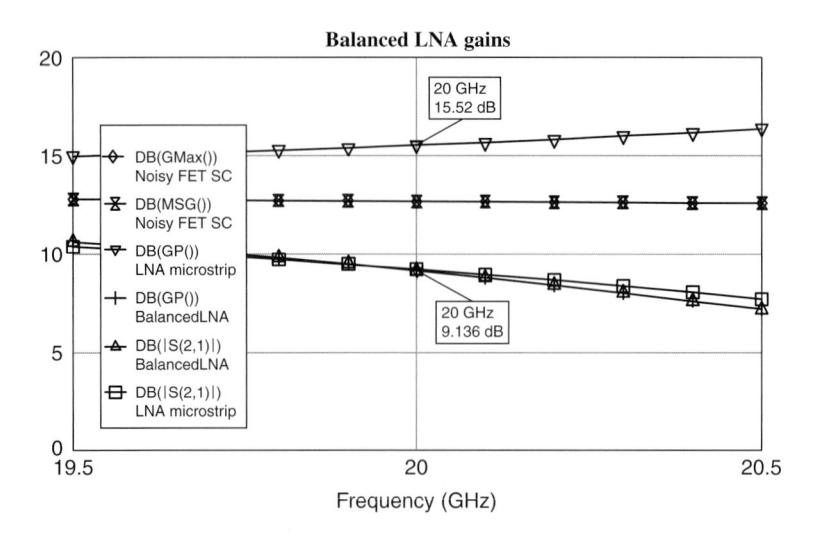

Figure 10.48 Gains of the single-ended and balanced LNA (Sec. 10.10).

see Sec. 10.3. The resulting LNA is now unconditionally stable at all frequencies; the minimum noise figure, see Fig. 10.44, is only marginally larger than the minimum noise figure of the single stage, owing to the input resistive loading implied by the couplers and input loads. As the equal values of the operational gain and of $|S_{21}|$ show, the amplifier is now very well matched to 50 Ω both at the input and output ports. The decrease in the operational gain from the single-stage to the balanced configuration, see Fig. 10.48, is due to the fact that the operational gain, being referred to the input power rather than to the input available power, increases in the presence of input mismatch; a comparison between available power gains (coinciding, at least at centerband, with $|S_{21}|^2$) would reveal that these do not vary from the single-stage to the balanced configuration.

10.11   Design of 5 GHz LNA with Source Inductive Series Feedback

The project aims at designing a 5 GHz low-noise amplifier with series inductive feedback. A complete equivalent circuit is used with a two-temperature noise model; noise in parasitic elements is also considered, see Fig. 10.43. To quickly estimate the optimum input loading conditions and noise figure of the "conventional" open-loop LNA, the optimum source impedance is evaluated and applied to the LNA input as the port impedance. Since the comparison is made in terms of available power gains, the output port is left unmatched.

Concerning the LNA with inductive feedback, a first approximation of the source and gate inductances is derived through tuning, thus achieving input matching. However, in such conditions the output S_{22} has magnitude greater than 1, meaning that the device is operating in the unstable region. In order to stabilize it in band, a parallel resistance is added at the device output, whose value is tuned so as to bring the corresponding S_{22}

10.11 Design of 5 GHz LNA with Source Inductive Series Feedback

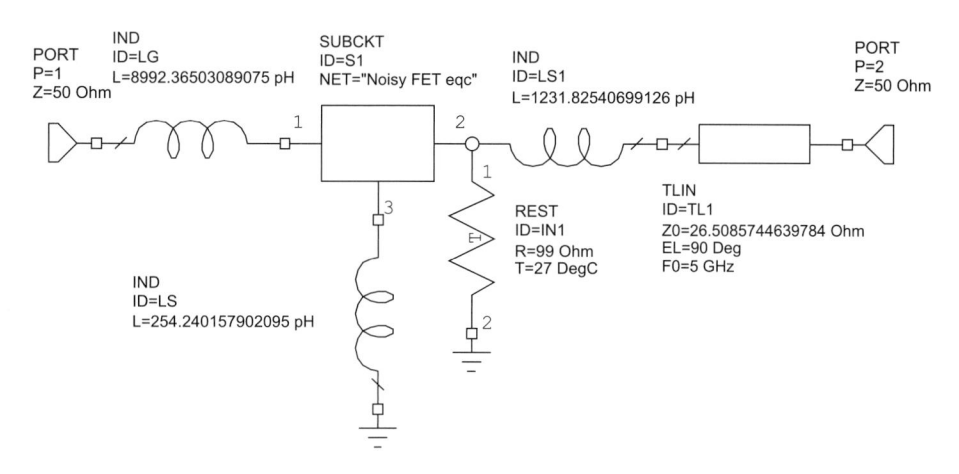

Figure 10.49 Schematic of the LNA with source inductive series feedback, output stabilizing resistance and quarter-wave output matching section (Sec. 10.11).

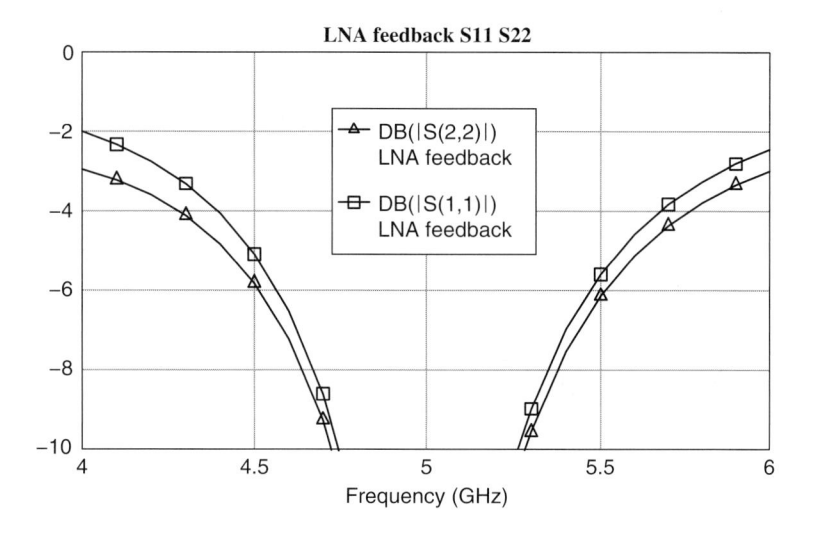

Figure 10.50 LNA with source inductive series feedback: input and output reflectances (Sec. 10.11).

within the Smith chart. (Out-of-band stabilization through a frequency-selective network can be added, but this step was not carried out here for brevity.) Then, a matching network made of a series inductor and of a quarter-wave transformer is added and a reasonable initial guess is obtained acting on the source and gate inductance, on the output inductance and on the line impedance. The schematic of the feedback LNA is shown in Fig. 10.49. On the basis of the initial guess, optimization is performed to minimize the input and output reflectances at 5 GHz, see Fig. 10.50.

The resulting available power gain is shown in Fig. 10.51 for the two solutions, the feedback and open loop amplifier with noise source matching. In this case, the feedback amplifier has a larger gain (by about 1 dB); the noise figure is a little worse (1.20 against

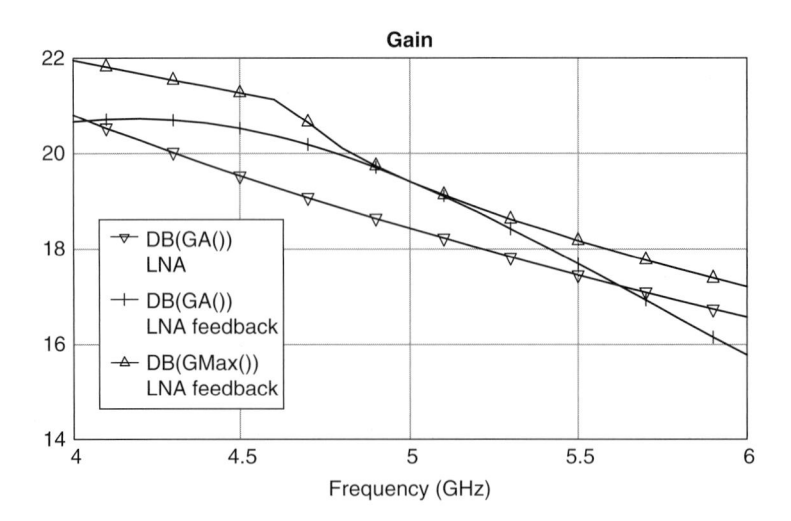

Figure 10.51 Available power gains of the conventional LNA and of the inductive feedback LNA (Sec. 10.11).

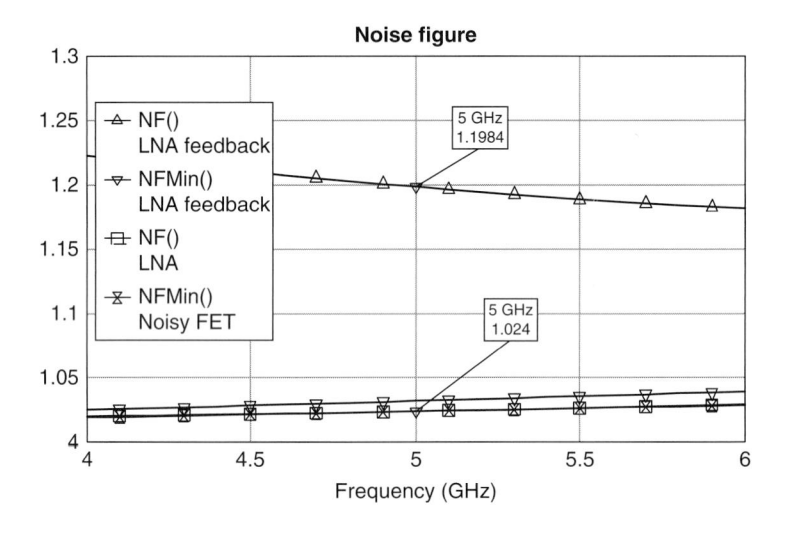

Figure 10.52 Noise figure of the inductive feedback LNA (Sec. 10.11).

1.02) for the LNA feedback solution, see Fig. 10.52, but the input matching is good on a 10 percent bandwidth, while it is extremely poor, as already discussed, for the noise matched solution. Some additional flattening of the gain around the centerband frequency could be achieved by tuning the output matching networks.

10.12 Design of a 3.5 GHz Narrowband Hybrid Class A Power Amplifier

This section presents the design of a class A narrowband power amplifier operating at 3.5 GHz, i.e., in the band of WiMax wireless systems, conceived as a hybrid module.

10.12 Design of a 3.5 GHz Narrowband Hybrid Class A Power Amplifier

The amplifier uses a tuned load (short at all harmonics). The device chosen is a GaN HEMT today commonly used in applications with a required output power around 10 W, i.e., the CREE GaN HEMT CGH 40010 [4] (now under Wolfspeed brand). The design steps discussed are of general validity and can be adapted to other active devices. From the data sheet [4], we see the device exhibits a maximum drain current around 3 A and a maximum DC drain voltage of around 30 V, yielding a theoretical class A output power in excess of 10 W. Concerning the frequency behavior, the suggested band, DC to 6 GHz, is well suited for a power stage at a target frequency of 3.5 GHz. For the sake of simplicity, a narrowband amplifier is considered, so that the design can be carried out at a single frequency. We will develop our CAD environment within the Keysight ADS suite [2], for which the factory built-in model is available. The model is in black-box form and therefore no direct access is available to the extrinsic/parasitic part of the device.

The suggested design strategy aims to find a trade-off among different figures of merit: output power, efficiency and output mismatch. The design procedure includes the following steps:

1. Choice of the device bias point.
2. Assessment of the device small signal behavior and stability at the chosen bias point. As already discussed, if the device is not unconditionally stable, stabilization networks have to be introduced both at the design frequency and out of band. If the in-band stabilization network unacceptably compromises the gain, the in-band potential instability has to be checked adopting a load termination well away from the output stability circle.
3. Identification of an output termination able to provide a reasonable compromise between output power, efficiency and output mismatch.
4. Design of the output matching network according to the technology selected.
5. Design and implementation of the input network to ensure input matching.
6. Test of the behavior of the complete chain made by the amplifier and the input and output matching networks, and final tuning of the matching networks.

All of these steps, with the exception of 3, are typical of any amplifier design; we will therefore focus on step 3, which is peculiar of the design strategy of power amplifiers. A last remark concerns the monolithic vs. hybrid realization; while the same steps hold in the former case, the better active and passive component repeatability, combined with the lower parasitics of active devices (which ensure in turn a better accuracy of large signal models), make MMIC design even more straightforward.

As seen in Sec. 8.4.1, the design of a class A power amplifier relies on the proper choice of the load termination, that ensures the correct behavior of the dynamic load line at the intrinsic device drain terminals. Parasitic identification through DC and small signal simulations could be carried out, to account for the transformation they introduce on the extrinsic load and hence derive the optimum load. This approach is however rather simplistic, since it assumes that the device output, at least at the intrinsic level, is actually behaving as a current generator. The approach followed here is based on the load-pull concept, albeit at a simulation stage.

CAD Projects

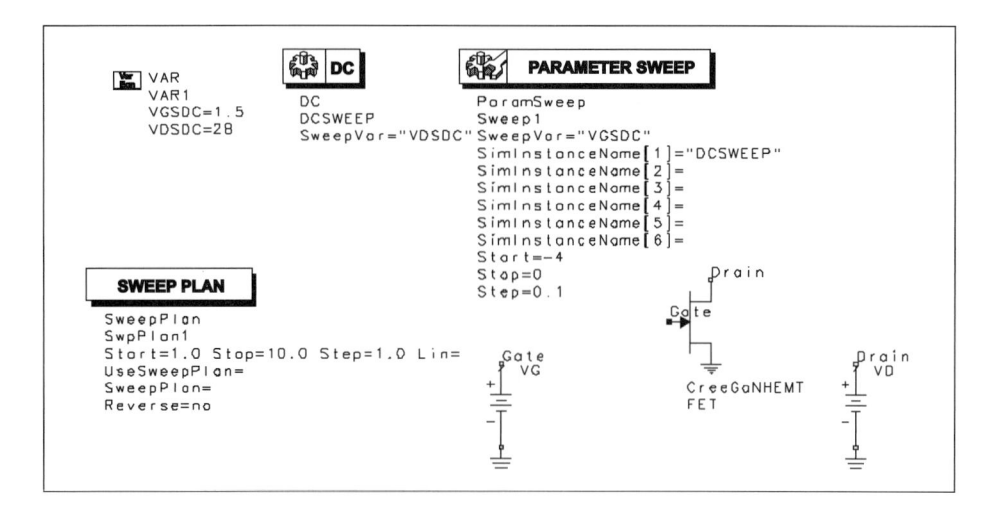

Figure 10.53 ADS schematic exploited to obtain the device DC characteristics (Sec. 10.12).

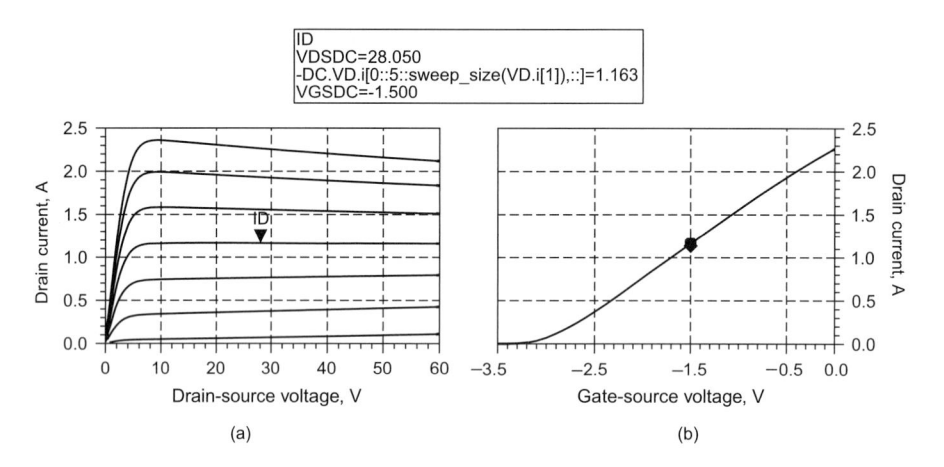

Figure 10.54 Drain current as a function of the drain (a) and gate (b) voltage (Sec. 10.12).

Before starting the load-pull simulation, the device bias must be chosen on the basis of the device drain and transfer characteristics. A library, embedding all the schematics used for the PA design, has been implemented in ADS and a schematic specifically dedicated to DC simulations has been created within this library, a sketch of which is reported in Fig. 10.53. The result obtained by stepping the gate voltage with a 0.5 V step and sweeping V_{DS} between 0 and 60 V is reported in Fig. 10.54. Moving of marker *ID* in Fig. 10.54 (a), the transcharacteristic can be visualized in Fig. 10.54 (b) for the drain voltage point that *ID* is *marking*. In this way, the bias point can be chosen according to, for example, to the bias drain current level and to the linearity of the transcharacteristic. The position of the marker *ID* in Fig. 10.54 shows the bias point adopted for the class A PA design. We select $V_{GS} = -1.5$ V, roughly corresponding $I_{DSS}/2$, while for V_{DS} the value of 28 V has been adopted, as suggested by the data sheet.

10.12 Design of a 3.5 GHz Narrowband Hybrid Class A Power Amplifier

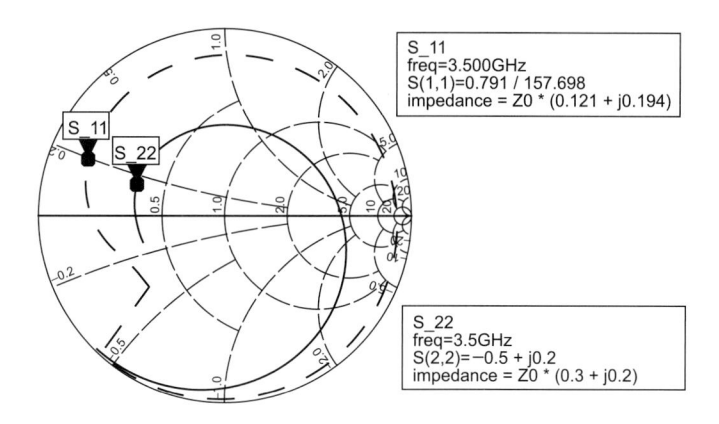

Figure 10.55 Active device input and output reflectances S_{11} and S_{22} between 1 and 20 GHz (Sec. 10.12).

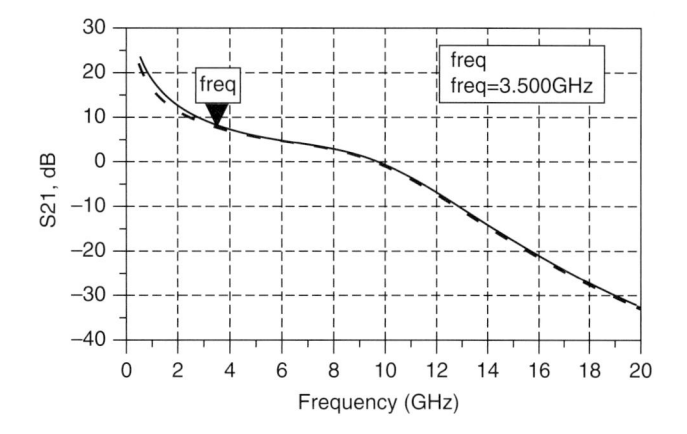

Figure 10.56 Active device $|S_{21}|$ between 1 and 20 GHz, with (dashed trace), and without (continuous trace) stabilization network (Sec. 10.12).

The next design step consists in evaluating the small signal behavior and stability at the bias point. Also in this case, a schematic controlling the design phase has been created. The scattering parameters S_{11} and S_{22} at $V_{GS} = -1.5$ V, $V_{DS} = 28$ V, simulated between 1 and 20 GHz, are reported in Fig. 10.55, while Fig. 10.56 shows $|S_{21}|$ in dB. Concerning stability, the Linville K factor, reported in Fig. 10.57, shows potential instability issues in the frequency range 3–11 GHz. A parallel RC network has been added in series to the gate to make the device unconditionally stable. With a tuning process, the values $R = 15\ \Omega$ and $C = 6$ pF have been identified, yielding an almost negligible loss in terms of $|S_{21}|$. Fig. 10.56 and Fig. 10.57 report $|S_{21}|$ and the Linville factor, respectively, with and without the input stabilization network.

As the next step of design, the large signal analysis of the active element is carried out under the form of load-pull map of the device output power and drain efficiency

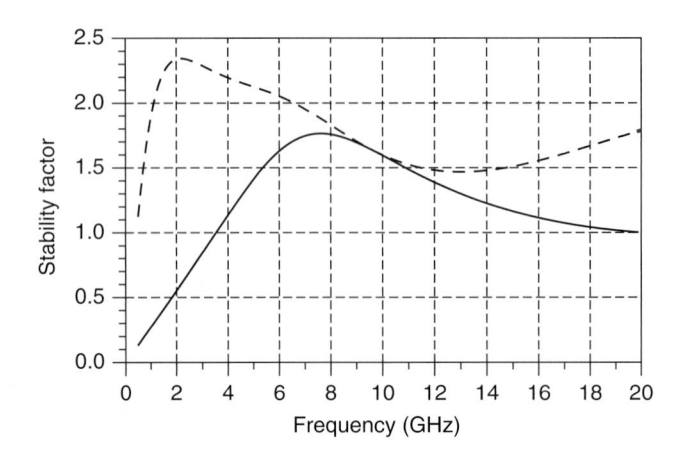

Figure 10.57 Linville stability factor between 1 and 20 GHz, with (dashed trace) and without (continuous trace) stabilization network (Sec. 10.12).

under specified power compression conditions, assumed as the design figures of merit. To this aim, another ad hoc schematic in the same library is created, where the input power sweep is carried out on a grid of output loads covering the whole Smith chart, with radial and phase steps $\Delta|\Gamma| = 0.05$ and $\Delta\angle\Gamma = 5$ degrees, respectively. Notice that experimental load-pull techniques typically cover only a subset of the Smith chart, because real measurements are time consuming, and some loads may lead the device to unstable and potentially self-destructive behavior; such issues of course do not exist in simulation.

The contour plots of the output power and drain efficiency at a prescribed gain compression (e.g., 1 dB), are then plotted on the load Smith chart exploiting scripts in the ADS Application Extension Language (AEL). The results are reported in Fig. 10.58 (a) for the output power, and in Fig. 10.58 (b) for the drain efficiency.

The choice of an optimum load can now be carried out by inspection, taking advantage of the ADS presentation page of the load-pull data that has been designed as a console where the designer can directly tune the load through the *marker* tool, thus simultaneously controlling all the optimization parameters: the output power at 1 dB gain compression, the drain efficiency and the resulting small-signal output reflectance. The ADS marker tool operates like a generalized tuner, where output loads are selected on the Smith chart corresponding to the output power or efficiency load-pull contours, by simply clicking and dragging the marker.[3] The console then automatically shows, for the selected load, the output power sweep and the corresponding efficiency and operational gain, and the dynamic load line superimposed to the DC characteristics, see Fig. 10.59, for a fixed value of the input and output power, again selected by means of the click and draw marker tool. The output mismatch is controlled by locating the

---

[3] In fact, two separate markers are associated with loads in the output power and efficiency Smith chart for greater flexibility.

10.13 Design of a 3.5 GHz Narrowband Hybrid Class B and AB Power Amplifier

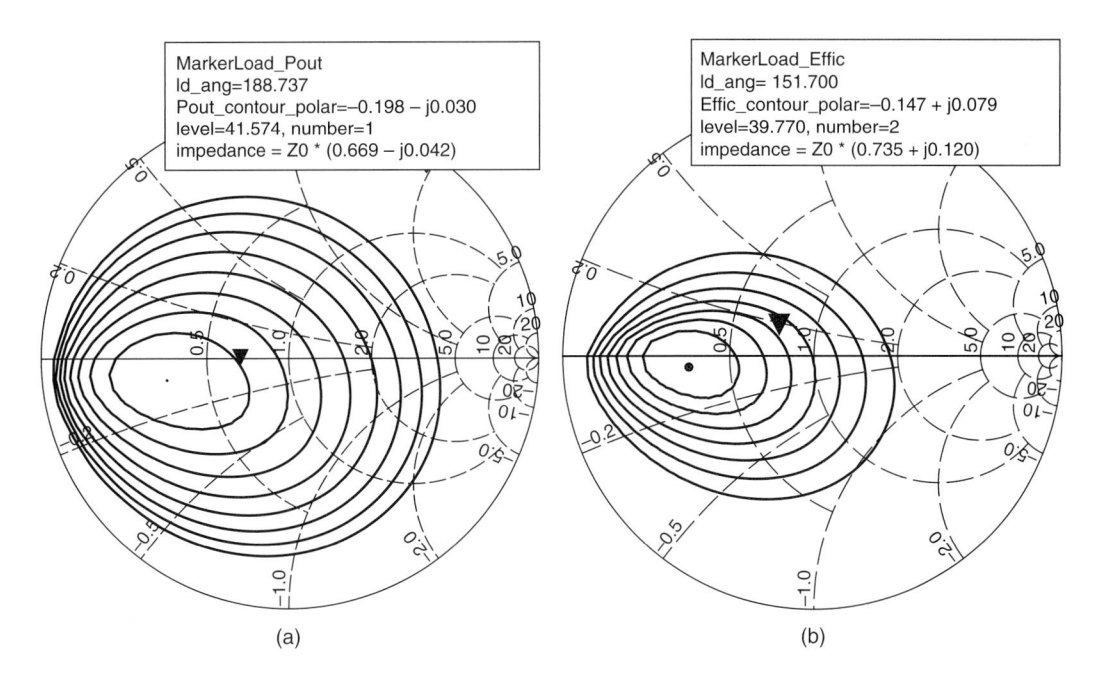

Figure 10.58 Output power (a) and efficiency (b) contour plot at 1 dB gain compression (Sec. 10.12).

marked impedance on a Smith chart were constant mismatch contours (in fact, circles) $|\Gamma_{out}|$ = const. are plotted, Fig. 10.59, bottom rightmost plot; the accumulation point in the load Smith chart corresponds to small-signal conjugate matching. On the basis of the data available, a satisfactory termination can be identified (the *selected load*).

In our case, to trade-off with the output power, efficiency and output mismatch constraint, the output load $\Gamma_L = -0.2 + j\,0.02$, corresponding to an impedance of $33 + j\,1.4 \ \Omega$ has been selected. The output power, efficiency and gain for this termination are reported in Fig. 10.60. Notice that theoretical efficiency in excess of 50 percent is obtained in strong overdrive conditions. Concerning the output mismatch, a value around 13 dB is expected, as can be read from marker *OutM* in Fig. 10.59.

10.13 Design of a 3.5 GHz Narrowband Hybrid Class B and AB Power Amplifier

The same strategy discussed in Sec. 10.12 has been applied to the design of two narrowband amplifiers exploiting the same CREE active device, but operating in class B and AB, respectively. The two amplifiers use a tuned load (short at all harmonics). The class B amplifier is biased at $V_{GS} = -3$ V (i.e., at pinchoff) and $V_{DS} = 28$ V as suggested by the manufacturer. The class AB amplifier is biased at 10 percent of the maximum current and at the same $V_{DS} = 28$ V. Using the simulated load-pull maps the load yielding good compromise between efficiency and output power is identified. For both cases, the selected load is found to be close to 17 Ω. The dynamic load line of the class B amplifier,

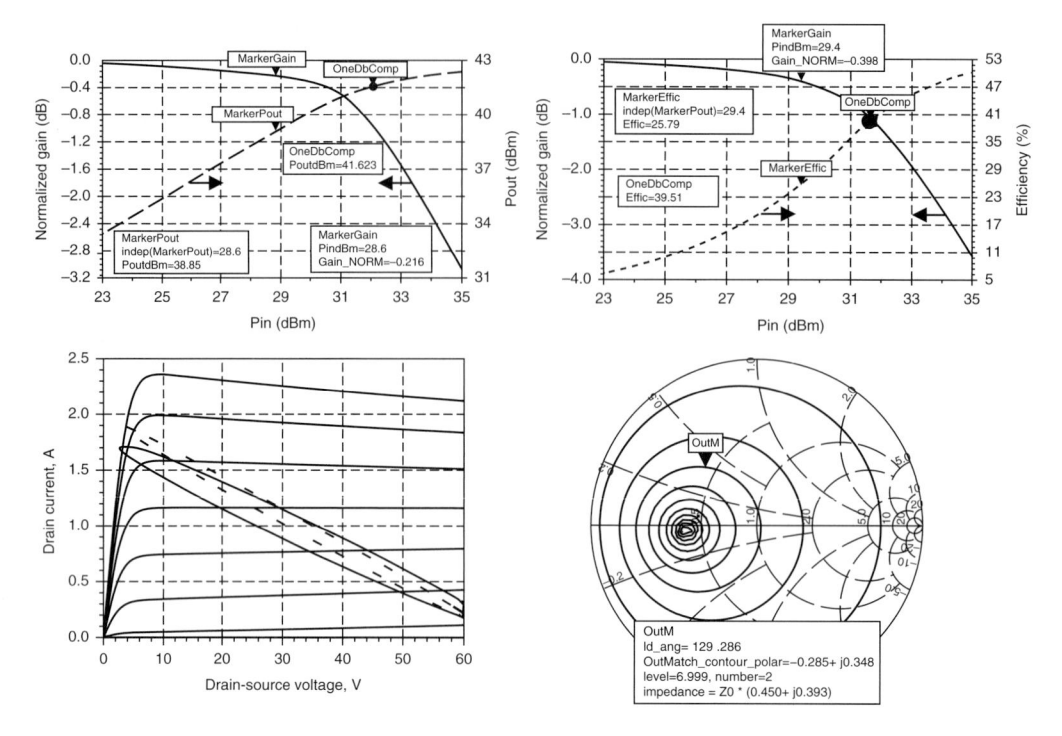

Figure 10.59 ADS control console for the choice of the optimum load (from top left, to bottom right): output power and normalized gain vs. input available power for the load corresponding to the marker of Fig. 10.58 (a); efficiency and normalized gain vs. input available power for the load corresponding to the marker of Fig. 10.58 (b); dynamic load lines superimposed to the DC device characteristics for the above two loads (continuous line, marker in Fig. 10.58 (a); dashed line, marker in of Fig. 10.58 (b)) and the marked input power levels; output mismatch contour. (Sec. 10.12).

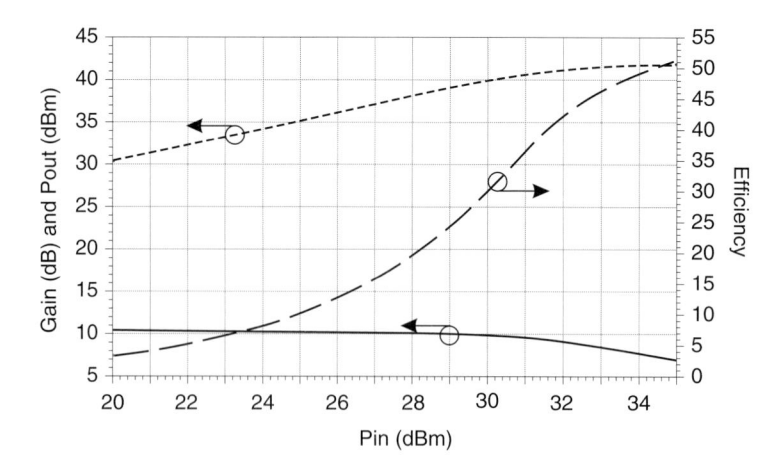

Figure 10.60 Class A amplifier output power (short dashed line), efficiency (long dashed line) and gain (continuous line) vs. available input power, for the selected load $\Gamma_L = -0.2 + j0.02$ (Sec. 10.12).

10.13 Design of a 3.5 GHz Narrowband Hybrid Class B and AB Power Amplifier

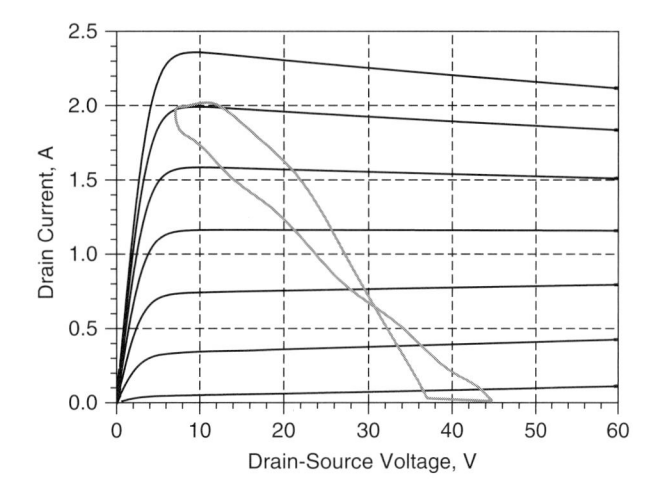

Figure 10.61 Dynamic load line superimposed to the device drain characteristics: class B amplifier with a load close to 17 Ω. The output power is 41.5 dBm, close to 1 dB compression point (Sec. 10.13).

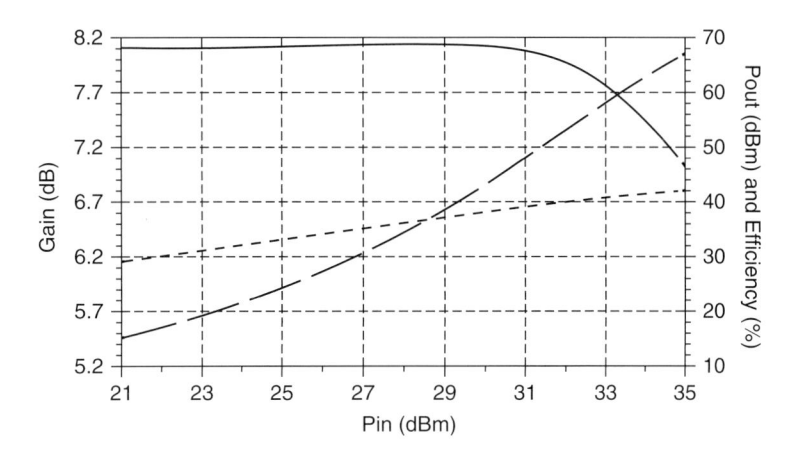

Figure 10.62 Output power (right axis, short dashed line), efficiency (right axis, long dashed line) and gain (left axis, continuous line) vs. available input power, for a load close to 17 Ω, for the class B amplifier (Sec. 10.13).

superimposed to the DC characteristics, is reported in Fig. 10.61 for an output power of 41.5 dBm, close to 1 dB compression point; the efficiency, gain and output power as a function of the available input power are shown in Fig. 10.62. A gain penalty around 2.5 dB with respect to the class A implementation can be seen comparing with Fig. 10.60. Fig. 10.63 and Fig. 10.64 show the corresponding results for the class AB amplifier. As expected, the maximum efficiency of the class B amplifier is larger than for the class A amplifier, and the behavior in the presence of backoff is much more favorable. The class AB amplifier exhibits, in low-power condition, a gain is similar to that of class A.

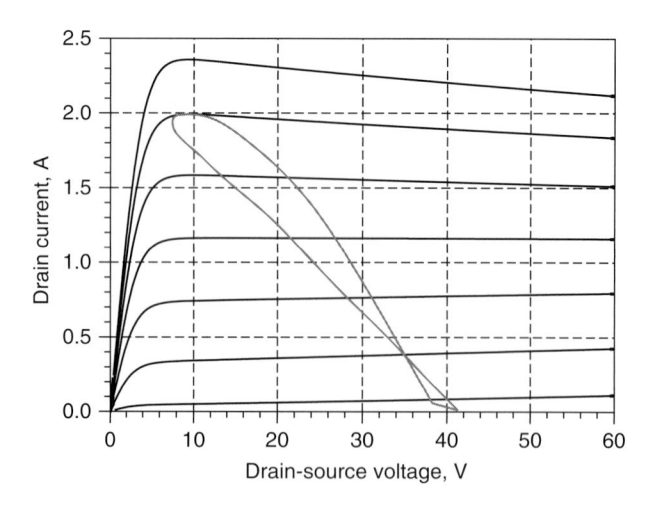

Figure 10.63 Dynamic load line superimposed to the device drain characteristics: class AB amplifier, with a load close to 17 Ω. The output power is 41.4 dBm, for a compression of 2 dB (Sec. 10.13).

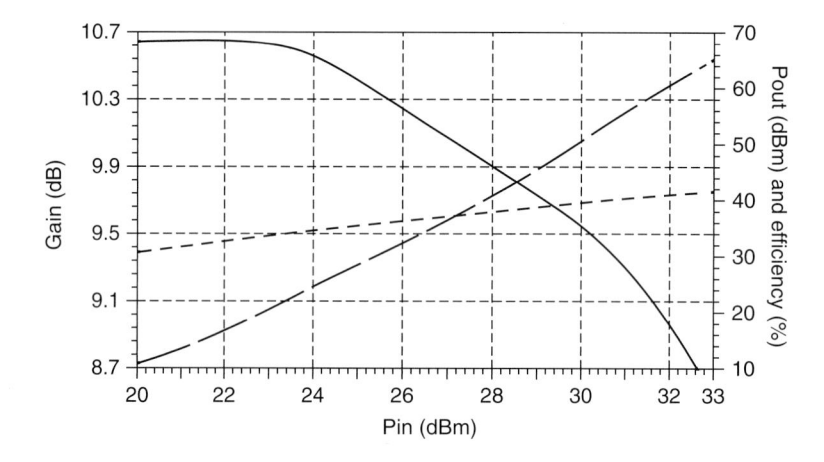

Figure 10.64 Output power (right axis, short dashed line), efficiency (right axis, long dashed line) and gain (left axis, continuous line) vs. available input power, for a load close to 17 Ω, for the class AB amplifier. (Sec. 10.13).

Since in class AB the circulation angle depends on the power level, the gain decreases towards that of class B, at high power level. As a consequence, an output power similar than those of class A and B is reached, but at an higher compression level.

References

[1] "Microwave Office," www.awrcorp.com/products/ni-awr-design-environment/microwave-office.

[2] "Advanced Design Systems," www.keysight.com/en/pc-1297113/advanced-design-system-ads.
[3] A. Miras and E. Legros, "Very high-frequency small-signal equivalent circuit for short gate-length inp hemts," *IEEE Transactions on Microwave Theory and Techniques*, vol. 45, no. 7, pp. 1018–1026, Jul. 1997.
[4] "Wolfspeed," www.wolfspeed.com/media/downloads/317/CGH40010-Rev4_0.pdf.

Index

Active loads
   active loop method, 512
   two-signal method, 512
Adjacent Channel Power Ratio (ACPR), 418
Amplifiers
   noise floor, 422
   sensitivity, 422
   Spurious Free Dynamic Range (SFDR), 422
Analysis of linear and nonlinear circuits, 108
   frequency-domain approaches, 108
   time-domain approaches, 108
Associated gain (LNA), 383
Available power gain, 268

Balanced amplifier
   centerband scattering parameters, 324
   examples, 326
   matching, 324
   operation, 321
   out-of-centerband behavior, 324
Bias T, 318
Bipolar small- and large-signal equivalent circuits, 254
Bipolar transistors (BJT)
   base transport factor, 235
   common emitter current gain, 235
   cutoff and maximum oscillation frequency, 236
   design rules, 236
   emitter bandgap narrowing, 236
   emitter efficiency, 235
Bonding wires
   inductance, 71
   inductance with ground plane correction, 71
Branch line coupler
   analysis, 162
   evaluation of the S-matrix, 165
   lumped-parameter, 168
   matching impedance, 164
   S-matrix, 164

CAD projects
   design of 3.5 GHz class A power amplifier, 556
   design of 3.5 GHz class AB and B power amplifier, 561
   design of 5 GHz LNA with source inductive series feedback, 554
   design of a maximum gain amplifier at 15 GHz, 532
   design of a two-stage balanced amplifier at 8 GHz, 538
   design of a uniform DAMP with 40 GHz bandwidth, 546
   design of parallel resistive feedback wideband amplifiers, 543
   FET noise, single-ended and balanced LNA design, 550
   FET stabilization, 531
   fitting of FET small-signal equivalent circuit, 528
   microstrip line and stub matching of a complex load, 522, 524
Carrier to intermodulation ratio (CIMR), 417
Cascode transistor, 346
Circuit optimization techniques, 123
Circulation angle, 426
Coaxial cable
   losses, 53
   parameters, 53
Coplanar waveguide, 59
   conductor and dielectric attenuation, 61
   finite-thickness substrate, parameters, 60
   frequency dispersion, 61
   infinitely thick substrate, parameters, 60
   structures, 59
Coupled lines
   capacitance matrix, 131
   coplanar, 129
   even and odd impedances, 134
   even and odd mode, 132
   even and odd mode capacitances, 133
   examples, 134
   inductance matrix, 131
   microstrip, 129

Index

symmetrical, 129
telegraphers' equation, 131
Coupled microstrips, 135
   analysis formulae, 135
   even and odd mode impedances, 137
   even and odd mode permittivities, 137
Crystal structure, 193
   cubic and hexagonal, 193
   wurtzite cell, 194

Dielectric loss angle, 34
Directional bridges, 495
Directional couplers, 139
   applications, 139
   branch-line, 140
   coupled line, 140
   coupling, 140
   directivity, 140
   interdigitated, 140
   isolation, 140
   parameters, 139
   rat race, 140
   summary, 181
   transmission, 140
   used in reflectometers, 491
Distributed amplifier, 336
   cascode-cell configuration, 346
   continuous, 336
      amplification, 340
      analysis, 338
      loss-limited bandwidth, 340
      reverse amplification, 340
      synchronous coupling, 338
   discrete-cell, 342
      artificial lines, 343
      cutoff frequency, 345
      effect of losses, 345
      synchronous coupling condition, 344
   examples, 347
   structure, 336

Electromagnetic models, 89
Electron affinity, 193
Enabling technologies for microwave circuits
   comparison between monolithic and hybrid implementation, 19
   hybrid integrated circuits, 18
   hybrid substrates, 19
   monolithic microwave integrated circuits (MMICs), 19
   planar hybrid or integrated circuits, 18
   waveguide circuits, 18
Error Vector Magnitude (EVM), 418

Fano's limit, 301
Feedback amplifier
   analysis, 327
   constraints on the open-loop gain, 330
   limits of the ideal design, 331
   narrowband, 327
   peaking inductor, 331
   scattering matrix, 329
   wideband, 327
FET bias networks, 320
FET measurement oriented noise models
   two temperature model, 387
Field-Effect Transistors (FET), 205
   comparison between cutoff and maximum oscillation frequencies, 232
   cutoff frequency, 213
   DC characteristics, 207
   DC model, channel pinch-off saturation, 209
   DC model, velocity saturation, 229
   heterojunction-based (HEMTs), 206
   maximum oscillation frequency, 213
   MESFETs, 205
   small-signal model, 211
Frequency bands, 3
   IEEE classification, 3
Frequency-domain analysis of nonlinear circuits, 108
Friis formula, 390

Generator-load power transfer, 262
   generator directly connected to the load, 264
   input reflection coefficient, 266
   power gains, 268
   power transfer in loaded two-ports, 265

Harmonic Balance, 108
   envelope simulation, 119
   multi-tone excitation, 112
      artificial frequency mapping, 115
      multidimensional Fourier Transform, 114
      oversampling method, 114
   single-tone excitation, 108
Heterojunction Bipolar Transistors (HBT), 234
   cutoff frequency, 240
   GaAs based, 241
   Gummel plot, 240
   InP based, 241
   model, 237
   SiGe, 241
   small-signal equivalent circuit, 240
Heterostructures, 199
   affinity rule, 203
   hetero- and homotype, 204
   lattice-matched, 199
   strained or pseudomorphic, 199
High Electron Mobility Transistor (HEMT), 222
   Chalmers model, 253
   conventional, GaAs-based, 223

Index

DC model (velocity saturation), 229
GaN-based, 226
InP, GaAs, GaN-based, 222
large-signal equivalent circuits, 252
metamorphic (MHEMT), 225
modulation-doped heterojunction, 222
PHEMT charge control, 226
pseudomorphic, GaAs-based, 224
pseudomorphic, InP-based, 224
supply layer, widegap, 223
Hybrid and monolithic microwave integrated circuits
coplanar circuits, 20
distributed implementation, 20
examples, 22
lumped implementation, 19
microstrip circuits, 20
Hybrid ring
matching impedance, 172
S-matrix, 172
use to sum/subtract analog signals, 173

Input–output power curve ($P_{in} - P_{out}$)
1 dB compression point, 416
harmonic intercepts, 416
third-order intermodulation intercept, 417
Interference couplers, 162
branch line, 162
hybrid ring, 172
Intermodulation distortion, 416
Adjacent Channel Power Ratio (ACPR), 418
carrier to intermodulation ratio (CIMR), 417
Error Vector Magnitude (EVM), 418
modulated signal test, 418
odd-order intermodulation products, 416
third-order intermodulation intercept, 417
third-order intermodulation intercept measurement, 418
Intermodulation products, 16
odd order, 16
Ionization, 193

Lange coupler
design formulae, 160
design procedure, 161
folded, 160
unfolded, 160
Layout generation techniques, 123
Linear n-ports
equivalent circuit models, 87
frequency-domain modeling, 87
power waves and scattering parameters, 89
review of two-port representations, 89
Linear two-ports
hybrid representations, 90
parallel representation, 90
series representation, 90

transmission representations, 90
Linear amplifier classes, 300
according to bandwidth, 300
according to matching networks, 300
according to topology, 300
distributed amplifiers, 301
high gain, 300
low noise, 300
Linear amplifiers
balanced amplifier, 321
DC bias, 318
design steps, 303
distributed amplifier, 334
feedback, 326
frequency-dependent stabilization, 316
gain-bandwidth tradeoff, 334
low noise, 392
matching section design, 306
open loop, wideband, 320
open-loop, narrowband, 304
multi-stage, 305
single-stage, 305
stabilization through series or parallel resistances, 314
Linear and nonlinear blocks, 15
bilinear components, 17
effects of nonlinearity, 15
harmonic generation, 15
intermodulation distortion, 15
linear with memory, 15
modeling, 87
nonlinear with memory and memoryless, 15
small-signal operation, 17
Linearization, 481
Linville stability coefficient, 283
Load and source pull measurements, 507
active load tuning, 510
passive tuners, 508, 511
vectorial, 509
Load- and source-pull measurements
scalar, 508
Low-noise amplifiers
alternative designs, 394
associated gain, 383
classical design, 392
classical design issues, 393
common gate or base, 394
available power gain, 396
input matching, 395
noise figure, 397
stability, 398
noise–gain tradeoff, 383
series inductive feedback, 398
analysis, 399
available power gain, 401
input matching, 400

noise figure, 401
noise figure optimization, 402
optimum design procedure, 404
power optimization, 403
transformer implementation, 399

Matching sections
cascaded line adapter, 312
distributed, 310
generalized quarterwave transformer, 313
line and stub adapter, 311
lumped, 307
quarterwave transformer, 310
stub and quarterwave transformer, 311
Maximum Available Gain (MAG), 291
Maximum Stable Gain (MSG), 292
Maximum Unilateral Gain (MUG), 295
MESFET large-signal equivalent circuits, 249
cubic Curtice, 251
quadratic Curtice, 249
Microstrip line, 54
characteristic impedance, 54
conductor attenuation, 55
design formulae, 56
dielectric attenuation, 55
effective permittivity, 55
frequency dispersion, 55
model, 54
parameter examples, 56
Microwave and millimeter wave transistors, 20
Gallium Antimonide, 20
Gallium Arsenide, 20
Gallium Nitride, 20
Indium Phosphide, 20
Si-based, 20
Silicon-Germanium, 20
Microwave and millimeter wave vacuum tubes, 20
Microwave circuit enabling technologies, 18
Microwave FETs
extracting small-signal equivalent circuits from measurements, 242
large-signal equivalent circuits, 246
measurement-based equivalent circuits, 242
small- and large-signal circuit models, 242
small-signal equivalent circuits
extraction of intrinsic elements, 243
parasitic de-embedding, 242
Microwave Integrated Circuits – (M)MICs
layout, 74
automatic layout tools, 80
coplanar circuits, 75
from netlist to layout, 80
hybrid layout examples, 78
integrated layout examples, 79
line discontinuities, 75
microstrip circuits, 75

lumped parameters components, 65
bonding wire inductance, 71
capacitor figures of merit, 72
capacitors, 71
external (chip) components, 73
inductors, 66
quality factor and resonant frequency of inductors, 71
resistors, 72
resistors, capacitors, inductors, 66
spiral inductors, 68
transmission line implementations, 65
packaging, 81
connectors and transitions, 82
planar lines
coupling and radiation losses, 62
transmission lines, 51
coaxial cable, 53
coplanar line, 52
examples, 51
microstrip line, 52
Microwave linear amplifiers, 261
low-noise amplifiers, 261
Microwave measurements
load-pull, 507
modulated signals, 513
noise, 514
power measurements, 507
power meter, 489
reflectometer, 491
S parameter measurement, 491
spectrum analyzer, 489
system-level, 513
Vector Network Analyzer, 497
voltmeter, 489
Microwaves, 1
Millimeter waves, 1
Minimum noise figure, 374, 381
bipolar transistors, 389
evaluation, 381
parallel model, 381
series model, 381
field-effect transistors, from *PRC* model, 386
field-effect transistors, Fukui formula, 387
noise resistance, 374
optimum source impedance, 374
relation with associated gain, 388
sensitivity vs. generator impedance, 374
MOSFETs, 222
Multiconductor buses, 129
Multiconductor line couplers, 155
analysis, 157
centerband length, 158
closing impedance, 158
coupling, 158
Lange coupler, 160

Index

Networks with random generators, 359
  evaluating the exchanged power, 360
  symbolic phasor approach, 359
Noise
  1/f or flicker noise, 368
  definition, 352
  diffusion noise, 367
  excess noise, 367
  generation-recombination noise, 367
  intrinsic, 352
  physical origin, 366
  thermal noise, 367
  generator, 366
Noise circles, 383
Noise figure, 372
  cascaded two-ports, 390
  definition, 373
  evaluation from the parallel model, 375
  evaluation from the series model, 374
  Friis formula, 390
  measurement, 514
  sensitivity vs. minimum, 383
  system definition, 373
Noise Lange invariant N, 374
Noise measure, 392
Noise measurements
  excess noise ratio, 514
  noise figure meter, 514
  noise resistance, 516
  optimum source impedance, 516
  Y-factor technique, 515
Noise models
  active devices, 385
    bipolar transistors, 389
    field-effect transistors, 385
      PRC model, 385
      two-temperature model, 387
  junction devices, 371
  passive devices, 368
    generalized Nyquist law, 369
    Nyquist law, 369
Noise parallel resistance, 374
Noise parameters
  circuit, 366
  system, 372
Noise series conductance, 374
Noise temperature, 373
Noisy n-port, 366
Noisy linear n-ports, 362
  equivalent circuits, 363
  one-ports, 362
  one-port equivalent circuit, 362
Nonlinear components
  behavioral models, 88
  equivalent circuit models, 88
  system-level models, 88

Volterra series approach, 88
Normalized impedance and admittance, 46

Operational power gain, 268

Phonons, 198
Power amplifiers, 411
  $P_{in} - P_{out}$ curve, 416
  Adjacent Channel Power Ratio (ACPR), 418
  AM–AM and AM–PM characteristics, 420
  blocking by an interferer, 421
  characteristics, 412
  class A, 426
    1 dB compression point, 439
    analysis of power saturation, 443
    analysis of tuned load operation, 429
    approximate design, 432
    efficiency in backoff, 429
    load-pull design, 433
    maximum efficiency, 429
    maximum power, 428
    optimum load, 428
    power series analysis, 437
    relation between 1 dB compression point and IMP$_3$ intercept, 441
    single- and two-tone compression, 440
  class AB gain, 452
  class AB, B, C, 444
  class B, 445
    efficiency with backoff, 448
    gain, 448
    maximum efficiency, 447
    optimum load, 447
  class E, 460
    analysis, 461
    load impedance, 463, 466
    maximum operating frequency, 467
    output power, 467
  class F, 456
    broadbanding, 459
    continuous, 459
    implementations, 459
    optimum load, 458
  classes, 411, 424
    circulation angle, 426
    traditional (A, AB, B, C), 425
  continuous class F, 459
  Crest Factor, 414
  Descriptive function model, 420
  Doherty amplifier, 468
  dynamic range (SFDR), 422
  efficiency, 423
  Error Vector Magnitude (EVM), 418
  from class A to C, 448
    efficiency and PAE, 450
    gain, 452

    intermodulation behavior, 455
    optimum load, 450
    sweet spots, 455
  gain, 412
  harmonic conversion gain, 415
  input matching, 424
  intermodulation conversion gain, 416
  layout and power combining, 481
  linearization, 481
    feedforward, 482
    predistortion, 483
  modulated signal test, 414
  Narrowband (baseband) model, 420
  operational gain, 414
  Peak to Average Power Ratio (PAPR), 414
  power-added efficiency, 423
  purpose, 411
  single-ended vs. push–pull, 426
  single-tone test, 412
  switching amplifiers, 460
  third-order intermodulation intercept, 417
  transducer gain, 414
  tuned load, 415
  Tuned load (resonator) design, 429
  two-tone test, 412
  upper limit to dynamics, 423
Power combiners and dividers, 173
  Wilkinson divider, 174
Power gains
  available power gain, 268
  maximizing the available power gain, 274
  maximizing the operational gain, 270
  maximizing the transducer gain, 275
  maximum power transfer, 268
  operational power gain, 268
  transducer power gain, 268
Power meter, 489
Power series model
  1 dB compression point in class A amplifier, 439
  analysis of compression and intermodulation in class A amplifier, 437
Power waves
  continuity conditions between two connected ports, 106
  definition, 92
  input power expression, 93
  relation with voltages and currents, 92
  solution of a network in terms of, 106
  two-port transmission matrix, 105
Power waves and S parameters
  rationale and motivation, 91
Power waves and scattering matrix
  equivalent circuit, 100
  forward and backward wave generators, 100

QSOLT calibration, 502

Quantum dot, 204
Quantum well, 203, 204
  carrier confinement, 203
Quantum wire, 204
Quasi-TEM lines
  dielectric losses, 43
  effective permittivity, 40
  examples, 38
  frequency behavior of effective permittivity, 42
  metal losses, 43
  parameters, 38
  relationship between the p.u.l. inductance and air capacitance, 40

Radio frequencies, 1
Random processes
  correlation and power spectrum, 354
  correlation and self-correlation function, 354
  correlation coefficient, 356
  correlation function and spectrum, 356
  ensemble average, 353
  quadratic mean, 355
  r.m.s. or effective value, 356
  review, 353
  stationary and ergodic, 353
  symbolic phasor notation, 358
  through linear systems, 357
  time average, 353
  white and colored processes, 355
Receiver
  building blocks, 14
  digital IF, 14
  direct conversion, 8, 13
  heterodyne, 8
  quadrature demodulator, 13
Reflection coefficient, 43
  behavior along a transmission line, 46
  properties of mapping between Γ and Z planes, 47
  relation with impedance and admittance, 46
Reflectometer, 491
  calibration, 493
  directional bridges, 495
  directional coupler issues, 495
  IF voltmeters, 494
RF transceiver, 7
  building blocks, 8
    amplifiers, 9
    filters, 9
    mixers, 9
  image frequency, 10
  Implementation examples, 20
  receiver, 8
  transmitter, 8
Rollet stability coefficient, *see* Linville stability coefficient

Scattering matrix
  conversion formulae from Z, Y, h parameters (two-port), 97
  definition, 93
  derivation from series or parallel representation, 94
  direct evaluation of the elements, 102
  normalization resistance, 95
  properties, 97
  reference plane shift, 103
  representation of autonomous n-port, 93
Scattering parameters
  measurement, 491
Semiconductor alloys, see Alloys, 199
  AlGaAs, 201
  AlGaN, 202
  III-N alloys, 202
  InAlAs, 201
  InGaAs, 202
  InGaAsP, 201
  InGaAsSb, 202
  quaternary, 200
  SiGe, 202
  ternary, 200
  Vegard and Abeles law, 200
Semiconductors, 193
  bandstructure, 193
  compound families: III-V, IV-IV, II-VI, 192
  substrates, 202
    metamorphic, 202
Signal modulation, 4
  I and Q baseband signals, 6
  analog and digital examples, 6
  review, 5
Skin penetration depth, 32
Smith chart, 47
  constant reactance circles, 48
  constant resistance circles, 48
  unit circle, 48
SOLR calibration, 507
Spectrum analyzer, 490
Stability and power matching, 275
System-level measurements, 513

Third-order intermodulation intercept, 417
Transducer power gain, 268
Transmission lines, 28
  RG, RC and LC (high frequency) regimes, 36
  attenuation, 31
  characteristics of substrates, 36
  complex characteristic impedance, 31
  complex propagation constant, 31
  dispersive behavior, 32
  effective permittivity, 30
  frequency dependence of conductance, 34
  frequency dependence of resistance, 32
  frequency-domain analysis, 30
  input impedance of loaded line, 43
  input impedance of quarter-wave and half-wave line, 45
  input impedance of short line, 45
  internal inductance, 32
  lossless, 29
  lossy, 30
  per-unit-length parameters, 29
  phase velocity, 29
  propagation constant, 30
  surface impedance, 33
  telegraphers' equations, 28
Transmitter, 8
Transport properties
  drift, 197
  mobility, 198
  saturation velocity, 198
  velocity overshoot, 198
TRL calibration, 503
Two-conductor symmetrical coupler, 144
  analysis, 147
  compensation, 155
  coupling, 151
  frequency behavior, 149
  matching impedance, 151
  scattering parameters, 149
  velocity mismatch, 152
Two-port stability, 276
  behavior vs. frequency, 297
  examples, 295
  managing conditional stability, 292
  one-parameter criteria, 289
  one-parameter criteria, proof, 289
  stability circles, 279
  stability circles and constant gain contours, 292
  stability conditions, 279
  stability criteria, 283
  two-parameter criteria, 283
  two-parameter criteria, proof, 284
  unilateral two-port, 293
Two-port stability and power matching, 290

Unilateral two-port
  definition, 293
  Maximum Unilateral Gain (MUG), 295
  unilateral transducer gain, 295
  unilaterality index, 293

Vector Network Analyzer, 497
  calibration, 499
  calibration algorithms, 501
  downconversion mixing, 498
  downconversion solutions, 497
  error model, 500
  structure, 497

Index 573

subsampling downconversion, 497
Vector Network Analyzer calibration
   QSOLT, 502
   SOLR, 507
   TRL, 503
Voltmeter
   scalar, 490

vectorial, 490

Wilkinson divider
   design and S-matrix, 175
   direct analysis, 178
   lumped-element, 179